18th International School on Condensed Matter Physics (ISCMP): Challenges of Nanoscale Science: Theory, Materials, Applications 2014

Journal of Physics: Conference Series Volume 558

Varna, Bulgaria
1-6 September 2014

Editors:

Doriana Dimova-Malinovska
Diana Nesheva
Marina T. Primatarowa
Julia Genova
Alexander G. Petrov

ISBN: 978-1-63439-676-9
ISSN: 1742-6588

Printed from e-media with permission by:

Curran Associates, Inc.
57 Morehouse Lane
Red Hook, NY 12571

Some format issues inherent in the e-media version may also appear in this print version.

Printed by Curran Associates, Inc. (2015)

For permission requests, please contact the Institute of Physics
at the address below.

Institute of Physics
Dirac House, Temple Back
Bristol BS1 6BE UK

Phone: 44 1 17 929 7481
Fax: 44 1 17 920 0979

techtracking@iop.org

Additional copies of this publication are available from:

Curran Associates, Inc.
57 Morehouse Lane
Red Hook, NY 12571 USA
Phone: 845-758-0400
Fax: 845-758-2634
Email: curran@proceedings.com
Web: www.proceedings.com

Table of contents

Volume 558

2014

18th International School on Condensed Matter Physics: "Challenges of Nanoscale Science: Theory, Materials, Applications"
1–6 September 2014, Varna, Bulgaria

Accepted papers received: 3 November 2014
Published online: 3 December 2014

Preface

Papers

Invited Lectures

Posters

Journal of Physics: Conference Series **558** (2014) 011001 doi:10.1088/1742-6596/558/1/011001

We are delighted to present the Proceedings of the 18th International School on Condensed Matter Physics: *Challenges of Nanoscale Science: Theory, Materials, Applications*, organized by the Institute of Solid State Physics of the Bulgarian Academy of Sciences and chaired by Professor Alexander G Petrov. On this occasion the School was held in memory of Professor Nikolay Kirov (1943-2013), former Director of the Institute and Chairman between 1991 and 1998. The 18ISCMP was one of several events dedicated to the 145th anniversary of the Bulgarian Academy of Sciences in 2014, and was held in the welcoming Black Sea resort of St. Constantine and Helena near Varna, at the Hotel and Congress Centre Frederic Joliot-Curie. Participants from 16 countries delivered 32 invited lectures, and 71 contributed posters were presented over three lively and well-attended evening sessions. Manuscripts submitted to the Proceedings were refereed in accordance with the guidelines of the *Journal of Physics: Conference Series*, and we believe the papers published herein testify to the high technical quality and diversity of contributions.

A satellite meeting, *Transition Metal Oxide Thin Films - Functional Layers in Smart Windows and Water Splitting Devices: Technology and Optoelectronic Properties* was held in parallel with the School (http://www.inera.org, 3-6 Sept 2014). This activity, which took place under the FP7-funded project *INERA*, offered opportunities for cross-disciplinary discussions and exchange of ideas between both sets of participants. As always, a major factor in the success of the 18ISCMP was the social programme, headed by the organized events (Welcome and Farewell Parties) and enhanced in no small measure by a variety of pleasant local restaurants, bars and beaches.

We are most grateful to staff of the *Journal of Physics: Conference Series* for their continued support for the School, this being the third occasion on which the Proceedings have been published under its auspices.

Doriana Dimova-Malinovska, Julia Genova, Diana Nesheva, Alexander G Petrov and Marina T Primatarowa (Editors).

Journal of Physics: Conference Series **558** (2014) 011002 doi:10.1088/1742-6596/558/1/011002

Bulgarian Academy of Sciences

GEORGI NADJAKOV INSTITUTE OF SOLID STATE PHYSICS

SOFIA, BULGARIA

PROCEEDINGS

18 ISCMP
Varna 2014

18th INTERNATIONAL SCHOOL ON CONDENSED MATTER PHYSICS

"Challenges of Nanoscale Science: Theory, Materials, Applications"

(in memory of Nikolay Kirov)

Dedicated to the 145th anniversary of the Bulgarian Academy of Sciences

September 1st - September 6th, 2014
Varna, Bulgaria

Journal of Physics: Conference Series **558** (2014) 011002 doi:10.1088/1742-6596/558/1/011002

Organising Committee

Chairman

Alexander G. Petrov, Director, Georgi Nadjakov Institute of Solid State Physics, Bulgarian Academy of Sciences, Sofia, Bulgaria (director@issp.bas.bg).

Local Organising Committee

Marina Primatarowa (scientific secretary, prima@issp.bas.bg), Diana Nesheva, Doriana Dimova-Malinovska, Nicholay Tonchev, Hassan Chamati, Julia Genova, Radostina Kamburova, Emilia Vlaikova, Eleonora Popova, Lyubomila Dedinska, Christo Popov, Irina Velkova

International Programme Committee

Chairman: A.G. Petrov
Members: D. Alexandrov (Thunder Bay), V. Čelebonović (Belgrade), D. Dimova-Malinovska (Sofia), B. Dulmet (Besancon), A. Grechnikov (Moscow), M. Güneş (Mugla), A. Iglič (Ljubljana), C. Main (Dundee), D. Nesheva (Sofia), A. Nesrullajev (Mugla), M.C. Petty (Durham), M. Popescu (Bucharest), S. Reynolds (Dundee), K. Shimakawa (Gifu), M. Tomilin (St. Petersburg), Ph. Vanderbemden (Liege), A. Vaseashta (Washington DC).

Local Scientific Council

Chairman: A.G. Petrov
Members:
Bulgarian Academy of Sciences: S. Vodenicharov (President), O. Stoilova (Scientific Secretary, Nanosciences, Novel materials and Technologies)
Georgi Nadjakov Institute of Solid State Physics: D. Nesheva, M. Primatarowa, H. Chamati, N. Tonchev
Central Laboratory of Solar Energy and New Energy Sources: D. Dimova-Malinovska
Central Laboratory of Applied Physics: L. Kolaklieva (Director)
Sofia University - Faculty of Physics: V. Donchev, S. Russev (Heads of Departments).
Plovdiv University Paisii Hilendarski - Faculty of Physics and Engineering Technologies: T. Yovcheva (Head of Department).

Journal of Physics: Conference Series **558** (2014) 011003 doi:10.1088/1742-6596/558/1/011003

Nikolay Kirov Nikolov

1943 – 2013

Professor Nikolay Kirov Nikolov graduated in Chemistry from Sofia University 'St. Kliment Ohridski', Bulgaria, in 1968. He received his Ph.D. degree in 1973, D.Sc. in 1988 and was appointed to the position of Professor of Physics in the Institute of Solid State Physics (ISSP), Bulgarian Academy of Sciences (BAS), in 1989. Prof. Kirov rose to become an international authority in the area of molecular spectroscopy, and in particular in the vibrational spectroscopy of thermotropic liquid crystals.

Prof. Kirov was a scientist of high reputation both in Bulgaria and globally. His scientific papers, numbering some 150, focus on the physics of liquid crystals. His systematic investigations of the vibrational spectroscopy of a wide class of thermotropic liquid crystals were collected in the monograph *Vibrational spectroscopy of liquid crystals* (1984), in which the molecular structure and properties of various liquid crystals, vibrational assignments, orientational order parameters and their calculation by molecular dynamics are presented. A range of liquid crystal materials was categorised and reported in his second book *Atlas of vibrational spectra of liquid crystals* (1988). Both publications remain a valuable source of information for specialists in the field of molecular spectroscopy. The outstanding contributions and growing reputation of Prof. Kirov led to periods as Visiting Scientist in the Universities of Parma, Calabria and Wroclaw. His abilities as a scientific organizer and leader were also notable, and led to a long and distinguished period of service in the Institute of Solid State Physics, being first appointed in 1980 as Scientific Secretary, and subsequently as Head of the Laboratory of Optics and Spectroscopy (1990-1998) and Director of the Institute (1991-1999). He was Chairman of the Scientific Council of ISSP over this period, and a member of the Specialized Scientific Council on Condensed Matter Physics. Prof. Kirov was a member of the High Testimonial Committee of the Republic of Bulgaria during 1996.

Nikolay Kirov was a key figure in the organization, promotion and success of the International School on Condensed Matter Physics (ISCMP), where he was able to use his network of international contacts to excellent effect. He served as Scientific Secretary (1988-1990), Chairman and Editor of the ISCMP Proceedings (1992-1998), and as Chairman Emeritus (2002-2012).

Prof. Kirov is an Honorary Member of the Institute of Solid State Physics. He received the Honorary Diploma 'Nicola Obreshkov' in 1986 and the "Marin Drinov" Sign of Honour BAS (on a ribbon) in 1998, for achievements in the physical and mathematical sciences. He was also honoured outside his native Bulgaria, receiving a Gold Medal from Wroclaw University, Poland, in recognition of successful scientific collaborations in the area of molecular structure.

Minko Petrov and Steve Reynolds

Peer review statement

All papers published in this volume of *Journal of Physics: Conference Series* have been peer reviewed through processes administered by the proceedings Editors. Reviews were conducted by expert referees to the professional and scientific standards expected of a proceedings journal published by IOP Publishing.

Study of electron-irradiated silicon thin films using transient photocurrent spectroscopy

S Reynolds[1], O Astakhov[2] and V Smirnov[2]
[1] School of Engineering, Physics and Mathematics, University of Dundee, Dundee DD1 4HN, Scotland.
[2] IEK-5 Photovoltaik, Forschungszentrum Jülich, D-52425 Jülich, Germany.

E-mail: s.z.reynolds@dundee.ac.uk

Abstract. Electron irradiation of silicon thin films creates localised states, which degrade their opto-electronic properties. We present a series of transient photocurrent spectroscopy (TPC) measurements on electron-irradiated amorphous and microcrystalline silicon films, annealed at progressively increasing temperatures. This has enabled localised states associated with both dangling bonds and conduction band tails to be examined over a wide energy range. Trends in the evolution of the DOS following electron irradiation followed by isochronal annealing steps indicate reductions in the deep defect density, which correlate with spin density. We also find a steepening of the conduction band tail slope in amorphous silicon on annealing. Both defect density and tail slope may be restored close to as-prepared material values. Earlier CPM data are re-examined, and a similar trend in the valence band tail slope is indicated. Computer simulations predict that following e-irradiation changes in deep defect density primarily control solar cell performance, and will tend to obscure effects related to band tails.

1. Introduction

The effects of energetic charged particles on films of hydrogenated amorphous silicon (a-Si:H) [1-4] and microcrystalline silicon (μc-Si:H) [5-10] solar cells and the transport properties of single films [11-16] have been studied extensively, due mainly to potential applications in space PV. It is generally agreed that the predominant effect of radiation damage is to create silicon dangling bonds (DBs), detected as unpaired spins in ESR measurements, which are metastable and can largely be removed by thermal annealing. While there is debate over the detailed process of DB creation [4], the outcome is evidently similar to that of light-induced degradation (Staebler-Wronski effect, SWE) [17]. However higher spin densities, approaching 10^{19} cm^{-3}, can be generated by low-temperature e-irradiation, a figure only reached by light-soaking using pulsed lasers [18] or within small volumes [19].

The effect of increased defect densities on the performance of thin-film silicon solar cells is predominantly a consequence of increased carrier recombination in the absorber layer, which reduces the short-circuit current density J_{SC}, coupled with a reduction in electric field strength in certain regions due to increased trapped space-charge, which degrades the fill-factor *FF*. Reduction in open-circuit voltage V_{OC} is comparatively minor. Remarkably, cells heavily degraded by e-irradiation (J_{SC} < 1 mA/cm^2, FF < 30%) may be restored to performance approaching that of a pristine cell by annealing for an hour at 180 °C [7].

As well as improving our ability to predict performance of cells in space applications, defects created by laboratory e-irradiation may be used to gain a more general understanding of solar cell physics and material limitations. By using annealing steps to control the defect density in the absorber layer, laboratory measurements of solar cell properties, coupled with computer simulations, enable a detailed optoelectronic model and material parameter sets to be developed [8, 9].

Here we investigate the effect of 2 MeV electron fluences of up to 10^{17} cm^{-2} on thin film silicon using transient photocurrent (TPC) density of states (DOS) spectroscopy and ESR spin density measurements following successive 30 min. annealing periods between room temperature and 180 °C. Some earlier constant photocurrent (CPM) measurements are also re-visited. Results are linked to the performance of solar cells with varying defect distributions by computer simulations.

2. Experimental

2.1. Film deposition

Intrinsic thin film silicon layers were prepared by PECVD using optimized conditions [20], and deposited on glass substrates. The chamber pressure, RF power and silane concentration respectively were: a-Si:H samples: 3 mbar, 3 W, 10%; μc-Si:H samples: 1.5 mbar, 20 W, 4.9%. A substrate temperature of 180°C was used throughout. Crystallinity of μc-Si:H absorber layers was determined from Raman measurements.

2.2. e-irradiation and annealing protocols

Films on glass and ESR powder samples underwent identical 2 MeV electron bombardments and subsequent stepwise isochronal thermal annealing procedures. Irradiation was performed at approximately 100K in a liquid nitrogen flow cryostat. Samples were exposed to a beam of 2 MeV electrons with current density 5 $\mu A/cm^2$ for 50 min., giving a fluence of 9.4×10^{16} cm^{-2}. Further details may be found elsewhere [11].

Between the electron bombardment and commencement of the annealing and measurement cycles, samples were transported and stored in liquid nitrogen to prevent uncontrolled annealing of defects. Defect densities in the layers were determined using Electron Spin Resonance (ESR) measurements on a-Si:H and μc-Si:H powder samples [21] deposited under identical deposition conditions to the samples on glass substrates.

Spin densities following successive isochronal annealings are shown in figure 1. Identical annealing regimes were followed, so that ESR spin densities were aligned as closely as possible with the films measured using TPC.

2.3 Transient Photocurrent Spectroscopy

A VSL337 nitrogen laser and dye attachment generates a 3 ns width 620 nm wavelength pulse at the sample, defined by two Ag contacts deposited on the Si film separated by a 1 mm gap, across which a 300 volt dc bias was maintained. An initial carrier density of some 10^{16} cm^{-3} was established by adjusting the pulse intensity using neutral density filters. Following pre-amplification, the photocurrent decay was recorded by a Tektronix TDS3052 oscilloscope. Repetitive averaging was used to reduce the noise to an acceptable level, and the data transferred to a PC for analysis, display and storage [22]. Each TPC current vs. time data file used subsequently to calculate the DOS is obtained from several sets of overlapping measurements (typically five), over progressively increasing time-scales from ns to seconds. Measurements were conducted at several experimental temperatures, typically 130, 160, 220, 250 and 298 K, giving plots of the DOS extending typically from 0.1 eV to 0.7 eV. Successive annealing steps (between 50 °C and 180 °C, in order of lowest temperature to highest temperature) were carried out on the films, followed by TPC measurements in the same cryostat as described above.

The TPC data files were then processed to extract the DOS present after each annealing step. A spectroscopic method was used in which the current-time data firstly undergo discrete Fourier transformation [23, 24] and are then analysed using techniques developed for modulated photocurrent spectroscopy (MPC). Within a multiple-trapping model, the DOS N_{ti} at the ith level, corresponding to

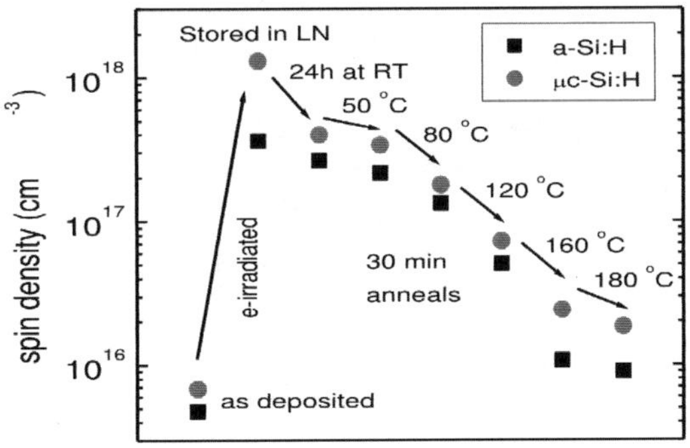

Figure 1. Spin density of a-Si:H and μc-Si:H samples measured after deposition, e-irradiation, transit at liquid nitrogen, 24 h at room temperature, and successive 30 min. annealings.

an energy depth E_ω is

$$N_{ti} \propto \frac{\sin(\phi(\omega))}{C_n|I(\omega)|}, \text{ where } E_\omega = kT\ln(\nu/\omega). \quad (1)$$

Here k is Boltzmann's constant, T the absolute temperature, ν the attempt-to-escape frequency and C_n the carrier capture coefficient. In this derivation C_n and ν are assumed constant for all localized states, and related by detailed balance to the band-edge DOS: $N_C = \nu / C_n$. An attempt-to-escape frequency of 10^{12} s^{-1} combined with an effective band-edge DOS of 10^{20} cm^{-3} was found to offer a suitable overlap between DOS sections measured at different temperatures, and was used throughout this work.

As the spectroscopic method is susceptible to noise, especially at short times, an alternative quantitative analysis is to force-fit the current-time data to the form appropriate for an exponential 'band-tail' of states [25]. Assuming multiple-trapping transport at an experimental temperature T below the conduction band tail temperature T_C, in the 'pre-transit' quasi-equilibrium regime,

$$I(t) \propto t^{-(1-\alpha)}, \quad (2)$$

where the dispersion parameter $\alpha = kT / kT_C$. One may thus obtain α and hence an assumed T_C from fitting a straight line to a log-log plot of the current-time data at short times and low temperatures, where the decay is controlled by band-tail thermalisation prior to the onset of deep trapping or recombination.

2.4 Solar cell simulations

Computer simulation of solar cells was carried out using the *SC-Simul* program developed at the University of Oldenburg. This is a one-dimensional numerical simulation which solves the Poisson equation, the continuity equations for electrons and holes, and the current transport equations including drift, diffusion, and thermionic emission over barriers, if present, in the valence and conduction bands. The thermal equilibrium and steady-state solutions are obtained by Newton iteration. Further details are provided elsewhere [26, 27].

3. Results and discussion

3.1. TPC Spectroscopy – general features

An example of current-time decays for a-Si:H and μc-Si:H at 220 K is shown in figure 2. From similar data recorded at experimental temperatures between 130 K and 360 K, we obtain the spectroscopic DOS plots for a-Si:H and μc-Si:H, shown in figures 3 and 4 respectively, proceeding through the 30 min. annealing steps identified in the legend. In both cases a general trend of decreasing 'integrated'

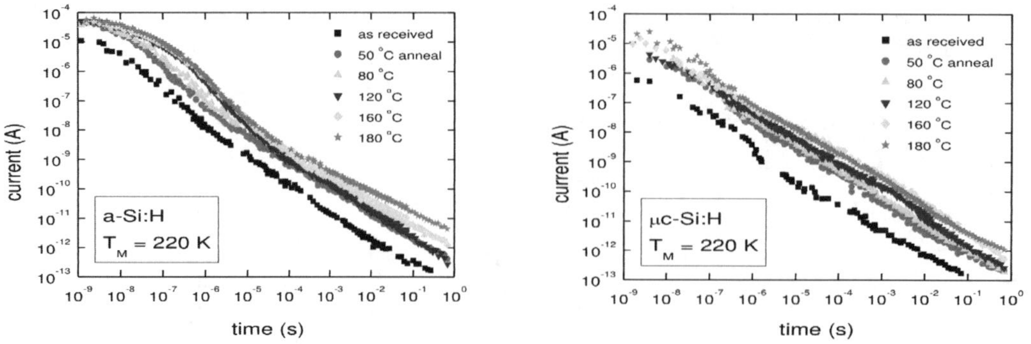

Figure 2. TPC current decay series for a-Si:H and μc-Si:H films measured at 220 K, following annealing of as-received samples for 30 min. at the temperatures indicated.

DOS with increasing anneal temperature is evident. In figure 5 we compare the deep defect density (at 0.5 eV) with the ESR spin density (N_S) data from figure 1. While the data are quite noisy, which may reflect the uncertainties introduced when matching the DOS sections obtained at different temperatures, the relationship is reasonably linear. A similar correlation has been reported previously for the CPM 'defect shoulder' vs. spin density for a-Si:H films, although the situation is less clear for μc-Si:H, with a correlation evident for low crystalline volume fraction films but no general trend for higher volume fractions [11]. Bronner et al [16] have found a systematic reduction in defect density on annealing μc-Si:H films when measured using MPC, although samples could not be returned to their as-deposited condition. Our TPC results therefore largely support previous findings in the case of a-Si:H. Given the rather arbitrary scaling procedure for the TPC DOS, little significance should be attached to the absolute TPC DOS magnitudes.

3.2. Conduction band tails

Switching attention to the shallower states, steepening of the band tail slopes with progressive annealing steps is evident, most clearly for a-Si:H as shown in figure 3. Trends in the μc-Si:H shallow states are less clear; the DOS sections obtained are noisier and overlap with temperature is poorer. An

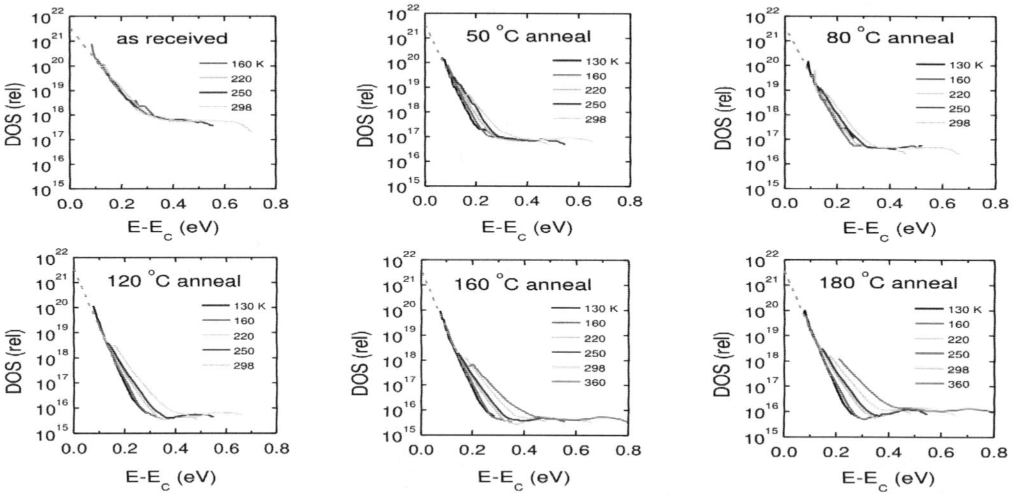

Figure 3. DOS plots for e-irradiated a-Si:H sample, as received and following successive 30 min. anneals. Dashed line is extrapolation to the conduction band edge (assumed 3×10^{21} units).

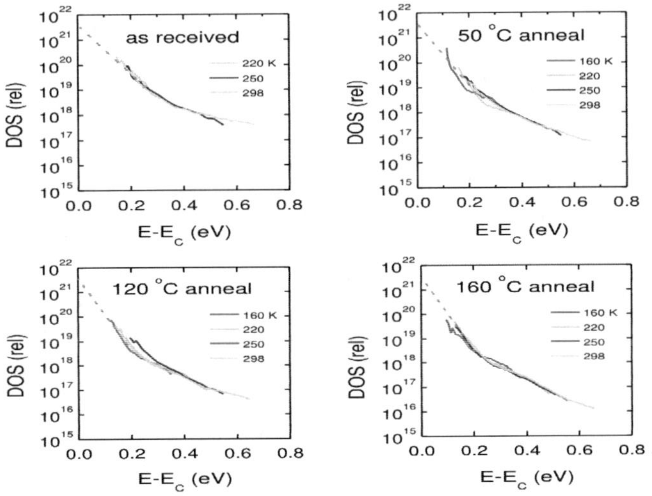

Figure 4. DOS plots for e-irradiated μc-Si:H sample, as received and following successive 30 min. anneals. Dashed line is extrapolation to the band edge (assumed 3×10^{21} units).

alternative measure of the band-tail DOS may be obtained using the force-fit approach defined by equation 2. Figure 6(a) indicates that for a-Si:H the short-time data at low temperatures (160 K data shown) does indeed follow a power law, and the band-tail slopes extracted by this method are plotted in figure 6(b). This confirms the trend evident in the spectroscopic data, and further reveals that the reduction in band-tail slope occurs predominantly after annealing at 120°C with a more gradual reduction between 120 °C and 180 °C.

Apparent inconsistencies in overlap between sections of the DOS in the region of the band tails evident in figure 3 are known to be due to resolution limitations in the MPC DOS method [24]. This becomes more apparent as annealing proceeds and the band tail becomes steeper, where it is only properly resolved below 220 K. The 'envelope' curve formed by the overlapping sections is however, reliable. A general comparison between a-Si:H and μc-Si:H in figures 3 and 4 shows that the band tail slope in the former case is somewhat steeper, in keeping with previous reports [28, 29].

Quantitative changes in CB tail states in a-Si:H following e-irradiation do not appear to have previously been reported explicitly in the literature. There have however been reports of changes in band tail state densities or slopes in a-Si:H following light soaking. Longeaud et al [30] and Roy et al

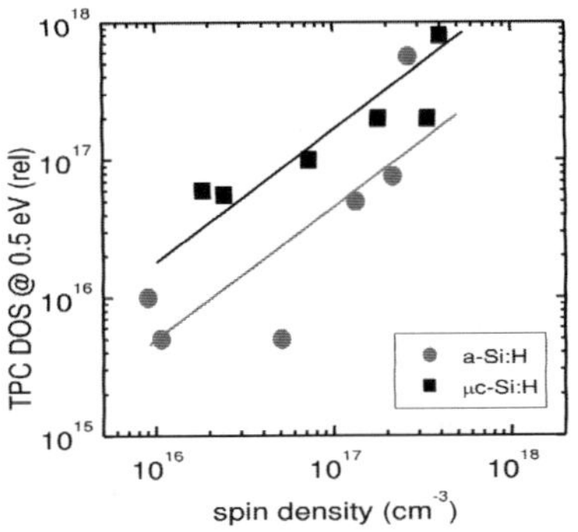

Figure 5. Correlation between relative TPC DOS at 0.5 eV and spin density N_S for a-Si:H and μc-Si:H samples. Lines correspond to an assumed linear relationship between N_S and DOS, with each positioned to provide a best fit.

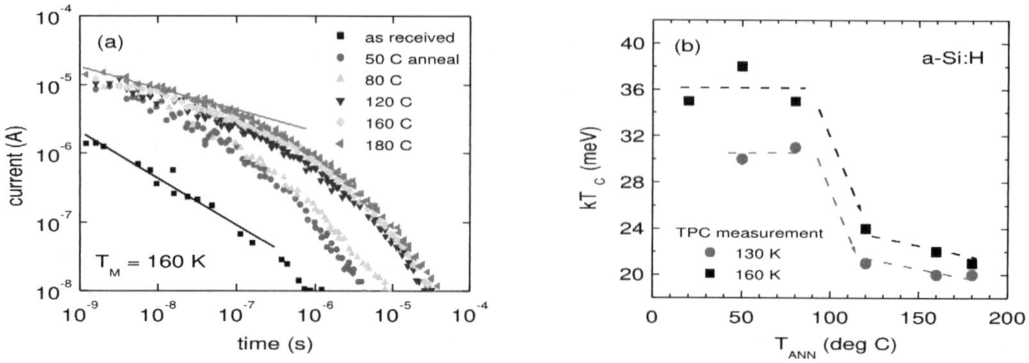

Figure 6. (a) Short-time TPC current decays for a-Si:H film measured at 160 K, following progressive annealing of as-received samples for 30 mins. as indicated. (b) Conduction band tail slopes calculated from best-fit lines as described in the text, *vs.* annealing temperature.

[31] carried out MPC measurements following increasing periods of light illumination, and found that above a certain exposure threshold, increased densities of CB tail states were created. However, in contrast to our results, the changes they report in CB tail states created above the exposure threshold could *not* be annealed out.

Band tails arise from intrinsic disorder in the amorphous structure, with distortions of the bond lengths and angles of tetrahedral coordination frozen-in as a consequence of the deposition process. The implication is therefore that in addition to creating a high density of DBs, electron irradiation introduces greater disorder, or modifies the existing disorder, in a manner which broadens the CB tail.

3.3. Constant Photocurrent Method and Valence Band Tail

TPC spectroscopy enables the DOS of unoccupied artifact states, above the Fermi level, to be examined. In order to probe *occupied* states in undoped a-Si:H, other techniques such as the constant-photocurrent method (CPM), are required. The CPM optical absorption spectrum $\alpha(E_P)$ integrates transitions in which electrons in filled localised states absorb the photon energy E_P and enter transport states above the CB mobility edge E_C. The term $d\alpha/dE_P$ is thus proportional to the DOS at $(E_C - E_P)$ [32]. Other transitions are also possible, but probably contribute little here. We have used CPM data from a previous publication [12] to construct and examine the DOS below E_F in a-Si:H, as shown in figure 7(a). While the data are somewhat noisy due to numerical differentiation, the trends exhibited are sufficiently clear. As reported earlier, the density of deep states is increased by e- irradiation, and reduced by isochronal annealing. However these data also reveal systematic changes in the VB tail slope, plotted vs. annealing temperature in figure 7(b). The trend is quite similar to that observed in the TPC data plotted in figure 6(b), with a clear reduction in band tail slope occurring after anneals at 80°C and 120°C. The peak apparent in figure 7(a) above 1.6 eV is an artifact of strong optical absorption in the film ($\alpha t > 1$ invalidates the CPM assumptions).

3.4. Band tails, e-irradiation and Disorder

While an association between exponential band tails and disorder in bond lengths and angles in silicon thin films has long been accepted, understanding of the detailed linkage on an atomic scale is still developing. Computer modelling studies of amorphous silicon systems [33, 34] suggest that while tail states arise from a large number of atoms in the ensemble, there is a statistical (but not necessarily causal) association between short bonds and CB tail states, and long bonds and VB tail states.

For example Koch et al [35] demonstrate a clear correlation between increasing deposition temperature between 50°C and 150°C, and decreasing E_U. It is also known that film stress [36] and short and medium-range order [37], both linked to atomic structure, are dependent on growth

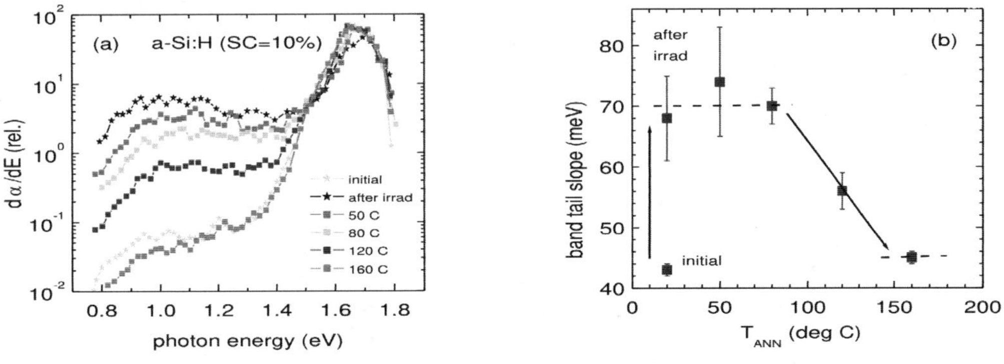

Figure 7. (a) DOS derived from CPM spectrum, (b) Characteristic VB tail energy calculated from best-fit lines to band tail regions, vs. annealing temperature.

conditions. Danesh et al [38] observed a correlation between E_U and the TO phonon frequency ω_{TO} obtained from Raman spectroscopy, and found a systematic reduction in E_U with increasing ω_{TO} which supports a linkage between steepening band tails and increased structural order.

There are reports of changes in structural order [39] and increases in volume [40] following light-soaking of a-Si:H samples, associated with structural changes in the lattice. In common with the SWE, and with e-irradiation induced changes in optical and electronic properties, these mechanical changes are also metastable, and samples may be thermally annealed back to, or close to, the as-deposited state.

The implication is therefore that e-irradiation results in reduced structural order, and further, that subsequent thermal annealing may improve or even restore the original degree of order to a-Si:H films. We have not yet carried out structure-sensitive measurements such as Raman spectroscopy or x-ray diffraction on e-irradiated samples. However, Köhler et al [41] have done so in the case of strongly light-soaked samples, which suggest changes in ω_{TO} tantalisingly close to the resolution limit. One may surmise that the e-irradiation regime used here is more aggressive than extensive light-soaking, given that up to 2 orders of magnitude more dangling-bond defects are created. These key experiments linking structural order and transport properties remain to be carried out.

3.5. Computer modelling of a-Si:H solar cells

Computer modelling enables material parameters to be incorporated into device structures and simulated to see if they might account for observed device behaviour. The 'inner workings' of the device are exposed, and may be used for more detailed analysis and to make informed suggestions for improvement. Modelling is thus a powerful tool in 'closing the loop' on structure-property relationships and performance optimisation. We modelled the a-Si:H device structure using *SC-Simul* [26, 27], and incorporated a range of defect densities and band tail slopes suggested by the above experiments to see if test results on e-irradiated cells can be reproduced. Material and device parameters are the default values used by *SC-Simul* except where otherwise stated. Summaries are given in figures 8 and 9. The experimental solar cell data is that presented in Astakhov et al [9].

The simulations in figure 8 were performed with band tail slopes of 25 meV (CB) and 50 meV (VB) to represent a 'typical' device, in which only the deep defect density is varied. It can be seen that the experimental data are reasonably well reproduced over a wide range of defect densities for both 0.3 μm and 1 μm absorber layers. The main discrepancy is with the value of V_{OC}, which is too low in the simulation. This may be due to a need to adjust the p- and n-layer simulation parameters, as the default settings were used here. The abrupt fall in V_{OC} for $N_S > 10^{17}$ cm^{-3} is also not reproduced, and remains a topic for future investigation.

The simulations in figure 9 represent our first attempt at modelling the effects of degradation of the

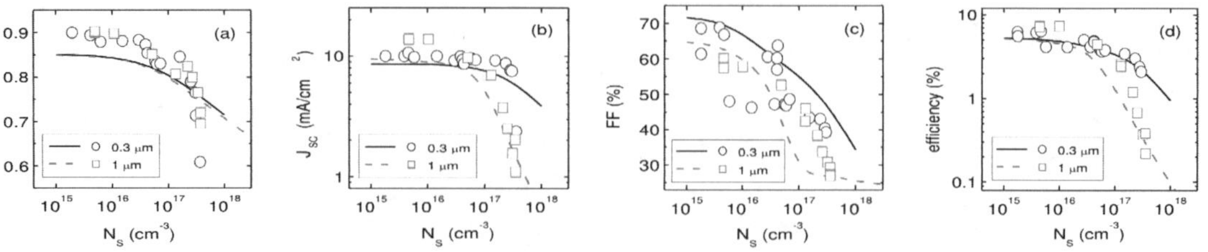

Figure 8. Effect of e-irradiation and subsequent annealing on PV parameters: (a) VOC, (b) JSC, (c) FF, (d) efficiency, measured for a-Si:H solar cells, with i-layer thicknesses of 0.3 μm and 1 μm. Lines are computer simulations obtained by changing defect density in line with experimental spin density.

band tails due to e-irradiation. The defect density is varied across a series of curves for band tail slopes encompassing those measured experimentally. Device performance is found from these simulations to be quite insensitive to the CB tail slope; and not much happens until it approaches the value of the VB tail slope. Thus what is shown in figure 9 is largely due to changes in VB tail slope. For an i-layer thickness of 0.3 μm, efficiency for $N_D < 3\times10^{16}$ cm^{-3} varies gradually with kT_V and roughly halves between 40 meV and 70 meV. For $N_D > 10^{17}$ cm^{-3} the curves coalesce, becoming independent of tail slope. For an i-layer thickness of 1 μm, behaviour is more complex for $N_D > 10^{17}$ cm^{-3}, with broader band tails predicted to actually increase efficiency. This may be due to a redistribution of charge between defects and tail states, which improves the electric field profile at some critical point in the device. This rather counter-intuitive result requires more detailed analysis of the model predictions.

Experimentally, we have no means of independently changing the band tail slopes by e-irradiation without simultaneously affecting the deep defect density, and indeed the two may be fundamentally inseparable [34]. According to the simulations, N_D needs to be below 10^{16} cm^{-3} to reliably distinguish the influence of the band tails on cell performance. This may be the case in modern device-quality material, as discussed by Liang et al [42] from a different perspective.

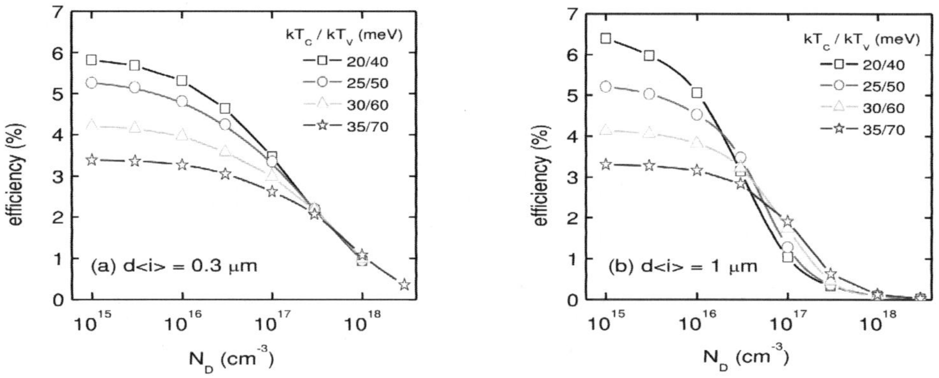

Figure 9. Simulated PV efficiency *vs.* defect density N_D for a range of band tail slopes, for i-layer thicknesses of (a) 0.3 μm, (b) 1 μm.

4. Conclusions

Hydrogenated amorphous and microcrystalline silicon thin films irradiated with 10^{17} electrons cm^{-2} contain a density of dangling-bond defects approaching 10^{18} cm^{-3}. These films have been investigated using the transient photocurrent method to measure the density of localised states above the Fermi level vs. energy depth. It is found that the deep DOS detected by TPC increases in keeping with the spin density following e-irradiation, and decreases in a similar way after isochronal thermal annealing at progressively increasing temperatures. We have also observed changes in the conduction band-tail slope, which reaches a maximum value after e-irradiation of some 36 meV, compared with typically 20 to 22 meV in as-deposited samples. e-irradiated samples may be returned close to the as-deposited state following annealing at 180°C for 30 mins. The behaviour of the Urbach tail following e-irradiation has been reviewed, and evidence for proportionate changes in valence band tail slope is present. This reaches a value of some 70 meV after e-irradiation, and returns to the as-prepared value of 45 meV when fully annealed. The greatest reduction in slope occurs in both cases after annealing at around 120°C. Since band tail slopes are thought to be associated with the degree of disorder, we suggest that the doses of e-irradiation used here may be sufficient to bring about metastable changes in structural order in the amorphous 'lattice'. As the band-tail effects anneal in a way which mirrors that of deep defects, it is suggested that both may be linked by the action of mobile hydrogen in relieving strain in the structure, as a by-product of passivating dangling bonds. Since such high defect densities (and by inference, degree of disorder) cannot readily be generated by light-soaking these effects appear not to have been observed, or at least reported, previously. Natural variations in 'intrinsic' disorder, typically introduced by varying the material preparation conditions, can be correlated with the TO phonon peak in the Raman spectrum. However, detecting any additional disorder caused by light-soaking appears to be at the limit of such measurements. Because e-irradiation gives rise to one to two orders of magnitude more dangling bonds than standard light-soaking regimes, it seems possible that the hydrogen passivation of this larger density of metastable defects may be sufficient to enable a more definite connection between structural order and transport properties to be made in future.

Acknowledgments
The authors thank Charlie Main, Rudi Brüggemann and co-workers for their contributions over many years to the development of DOS spectroscopy and transport simulations. We are grateful to technical staff at IEK-5, Forschungszentrum Jülich who assisted in preparation and characterisation of samples, and to Florian Köhler and Christophe Longeaud for a number of most useful recent discussions.

References

[1] Srour J R, Palko J W, Lo D H, Liu S H, Mueller R L and Nocerino J C 2009 *IEEE Trans. Nucl. Sci.* **56** 3300
[2] Srour J R, Vendura G J, Lo D H, Toporow C M C, Dooley M, Nakano, R P and King E E 1998 *IEEE Trans. Nucl. Sci.* **45** 2624
[3] Wang K, Lord K, Woodyard J R 2000 *IEEE PV Specialists Conference* p 1057
[4] van Swaaij R A C M M and Klaver A 2008 *J. Non-Cryst. Solids* **354** 2464
[5] Kuendig J, Goetz M, Shah A, Gerlach L and Fernandez E 2003 *Sol. Energ. Mat. Sol. Cells* **79** 425
[6] Meillaud F, Vallat-Sauvain E, Niquille X, Dubey M, Shah A and Ballif C 2006 *J. Non-Cryst. Solids* **352** 1851
[7] Astakhov O, Smirnov V, Carius R, Petrusenko Yu, Borysenko V, Böttler W and Finger F 2012 *J. Non-Cryst. Solids* **358** 2198
[8] Smirnov V, Astakhov O, Carius R, Pieters B E, Petrusenko Yu, Borysenko V and Finger F 2012 *Appl. Phys. Lett.* **101** 143903
[9] Astakhov O, Smirnov V, Carius R, Pieters B E, Petrusenko Yu, Borysenko V and Finger F 2014 *Sol. Energ. Mat. Sol. Cells* **129** 17
[10] Smirnov V, Astakhov O, Carius R, Petrusenko Yu, Borysenko V and Finger F 2012 *Jpn. J. Appl. Phys.* **51** 022301
[11] Astakhov O, Carius R, Petrusenko Yu, Borysenko V, Barankov D, Finger F 2007 *Mat. Res. Soc. Symp. Proc.* **989** 3
[12] Astakhov O, Carius R, Finger F, Petrusenko Yu, Borysenko V and Barankov D 2009 *Phys. Rev. B* **79** 104205

[13] Liao N M, Li W, Liu Z, Jiang Y D, Ma N, Li Y, Wu Z M and Li S B 2008 *Phil. Mag. Lett.* **88** 871
[14] Brüggemann R, Bronner W and Mehring M 2001 *Sol. St. Commun.* **119** 23
[15] Bronner W, Mehring M and Brüggemann R 2002 *Phys. Rev.* B **65** 165212
[16] Bronner W, Kleider J P, Brüggemann R and Mehring M 2003 *Thin Solid Films* **427** 51
[17] Stutzmann M, Jackson W B and Tsai C C 1985 *Phys. Rev.* B **32** 23
[18] Stutzmann M, Rossi M C and Brandt M S 1994 *Phys. Rev.* B **50** 11592
[19] Reynolds S, Anand S, Meftah A and Smirnov V 2012 *J. Non-Cryst. Solids* **358** 2202
[20] Smirnov V, Das C, Melle T, Lambertz A, Hülsbeck M, Carius R, Finger F 2009 *Mater. Sci Eng.* B **159-160** 44
[21] Xiao J, Astakhov O and Finger F 2011 *Jpn. J. Appl. Phys.* **50** 071301
[22] Reynolds S, Main C, Webb D P and Rose M J 2000 *Philos. Mag.* **80** 547
[23] Main C 1997 *Mat. Res. Soc. Symp. Proc.* **143** 167
[24] Main C 2002 *J. Non-Cryst Solids* **299-302** 525
[25] Tiedje T, Cebulka J M, Morel D L and Abeles B 1981 *Phys. Rev. Lett.* **46** 1425
[26] M Rösch, PhD thesis (2003), Universität Oldenburg http://oops.uni-oldenburg.de/247/
[27] Brüggemann R and Rösch M 2005 *J. Optoel. Adv. Mater.* **7** 65
[28] Reynolds S, Smirnov V, Finger F, Main C and Carius R 2005 *J. Optoel. Adv. Mater.* **7** 91
[29] Kleider J P, Longeaud C, Brüggemann R and Houzé F 2001 *Thin Solid Films* **383** 57
[30] Longeaud C, Roy D and Teukam Hangouan Z 2000 *Appl. Phys. Lett.* **77** 3604
[31] Roy D, Longeaud C and Saadane O 2002 *J. Non-Cryst. Solids* **299-302** 511
[32] Pierz K, Mell H and Terukov J 1985 *J. Non-Cryst. Solids* **77–78** 547
[33] Fedders P A, Drabold D A and Nakhmanson S 1998 *Phys. Rev* B **58** 15264
[34] Khomyakov P A, Andreoni W, Afify N D and Curioni A 2011 *Phys. Rev. Lett.* **197** 255502
[35] Koch C, Ito M and Schubert M 2001 *Sol. Energ. Mat. Sol. Cells* **68** 227
[36] Tzanetakis P 2003 *Sol. Energ. Mat. Sol. Cells* **78** 369
[37] Köhler F, Zimmermann T, Muthmann S, Gordijn A and Carius R 2014 *IEEE J. Photovoltaics* **4** 4
[38] Danesh P, Toneva A, Savatinova I and Liarkapis E 1996 *J. Non-Cryst. Solids* **204** 265
[39] Gupta S, Katiyar R S, Weisz S Z and Balberg I 2000 *J. Non-Cryst Solids* **266-269** 496
[40] Nonomura S, Yoshida N, Gotoh T, Sakamoto T, Kondo M, Matsuda A and Nitta S 2000 *J. Non-Cryst. Solids* **266-269** 474
[41] Köhler F, private communication
[42] Liang J, Schiff E A, Guha S, Yan B and Yang J 2006 *Appl. Phys. Lett.* **88** 063512

Information Loss from Technological Progress

P D Townsend
Physics building, University of Sussex, Brighton BN1 9QH, UK

E-mail: pdtownsend@gmail.com

Abstract. Progress in electronics and optics offers faster computers, and rapid communication via the internet that is matched by ever larger and evolving storage systems. Instinctively one assumes that this must be totally beneficial. However advances in software and storage media are progressing in ways which are frequently incompatible with earlier systems and the economics and commercial pressures rarely guarantee total compatibility with earlier systems. Instead, the industries actively choose to force the users to purchase new systems and software. Thus we are moving forward with new technological variants that may have access to only the most recent systems and we will have lost earlier alternatives. The reality is that increased processing speed and storage capacity are matched by an equally rapid decline in the access and survival lifetime of older information.

This pattern is not limited to modern electronic systems but is evident throughout history from writing on stone and clay tablets to papyrus and paper. It is equally evident in image systems from painting, through film, to magnetic tapes and digital cameras. In sound recording we have variously progressed from wax discs to vinyl, magnetic tape and CD formats. In each case the need for better definition and greater capacity has forced the earlier systems into oblivion. Indeed proposed interactive music systems could similarly relegate music CDs to specialist collections.

The article will track some of the examples and discuss the consequences as well as noting that this information loss is further compounded by developments in language and changes in cultural views of different societies.

1. Survival of technology

Throughout history there has been a steady increase in understanding and use of technology, from manufacture of stone and metal tools to glass, and the plastics, metals and semiconductors of the present day. It is therefore very tempting to believe that we are acquiring not only more data and information, but also a deeper level of knowledge and perhaps some wisdom. Unfortunately this is not true as the new technologies exist because they have displaced, or made obsolete, the earlier alternatives. In general that means that many of the former skills and understanding of the relevant science and production methods have been lost. It is also disconcerting to realise that the expertise to generate and innovate technological progress resides in a very tiny fraction of the population. The reality is that a smaller percentage of the population now understand our key technologies. Trivial examples are that 50 years ago most families had members who could undertake many of the repairs on a car engine or equipment in the home. Now it is rare to even recognise the components of the engine and most home devices have electronics and electronic chips which are only amenable to experts with complex monitoring systems.

Knowledge has never been permanently retained so loss is not a new phenomenon. Our brain certainly has limited retention and reliability but we assume that we can always have secure tangible

Journal of Physics: Conference Series **558** (2014) 012002 doi:10.1088/1742-6596/558/1/012002

records available to deal with important data such as bank accounts, photographs, letters and personal or family matters. Such items, when committed to paper or photographs can certainly survive our life span and many archives have historic documents and family records that have survived many hundreds of years, plus family photos even from the 19th century. Technological progress has modified this as instead of pen and ink, or typewriters, we have advanced to sophisticated word processors on our computers and electronic data storage in a range of formats and on numerous types of storage system. Similarly black and white photographs moved into colour film and then to digital cameras (and mobile phones). Acquisition has therefore speeded up very rapidly, image processing to edit images, and corrections and mass production of text for storage or electronic communication have become the norm. Indeed many people probably keep no hardcopy records and correspondence is solely by email etc. Writing, correction, mass production and transmission have therefore greatly speeded up. This is 21st century progress.

Associated problems are now emerging. For example the improvements in computer speed, software and storage have taken place, not by improving earlier designs but by leaping into new structures and software which almost invariably are incompatible with earlier versions. Therefore if we had our correspondence and private information on early systems and stored it on floppy discs as used on older computers then we will now have a problem. Once we replaced the computers and bought new software and word processing packages much of the prior information is inaccessible. Further, we need upgrades to receive documents in the new formats. As I will emphasise later most software becomes obsolete or unsupported after around 10 years, and similarly the storage improvements and new formats are rarely accessible after 20 years. These are time scales that are shorter than our human memories and definitely give a situation where we will never pass on information to future generations. The other key factor is that storage on commercial systems (or even home computers) is far from secure in terms of device failure, deliberate malicious access to destroy, change or use the data, and certainly in terms of any network communications of the data it is likely to be monitored both by government agencies and an ever increasing band of criminals. Cyber crimes are a boom industry with profits now measured in the range approaching a billion dollars a year.

2. Reasons for information loss

Skills, data, information and knowledge are often acquired laboriously over long periods of time therefore it may seem unfortunate that they are lost. The reasons are diverse and broadly fall into the following categories of (A) natural disasters; (B) human generated disasters such as wars; (C) religious and political doctrine; (D) failure of the storage media by loss or decay and (E) loss of the language and culture in which the information was written. In addition we may not accept knowledge when it has been acquired or offered because of (F) instinctive rejection or blinkered precondition viewpoints; (G) cognitive dissonance (that means we do not accept ideas which are far from our experience); or (H) religious and ideological dogma.

2.1 Natural disasters

These are inevitable and although we may have little control over avoiding and predicting them the majority of events such as earthquakes, volcanoes, tsunamis or small meteorite impacts tend to be localised. Although they may destroy a local region they rarely have long term global effects. Historically, a major volcanic event at Santorini not only destroyed the Minoan civilization but the consequent disruption in global weather patterns also caused collapse of other regions, including the dynasty in China at that time. The large scale examples of major meteor impacts and ice ages are clearly different and if they are repeated we will be content with survival, even if most of our technology vanishes. Nevertheless some natural events have caused unexpected problems in manufacture of technological items, for example parts of laser printers were made at a single factory that was destroyed in an earthquake. Other equipment examples may emerge and certainly key materials, such as the minerals providing rare earth ions, tend to originate from rather few sources so political restrictions could have major world wide impact.

A more critical natural problem exists as the result of particle emission from sunspots. The clouds of ions and disturbance of our magnetic field would previously have offered beautiful examples of the Aurora Borealis but, prior to any electronics; the only effects would have been psychological or interpreted in terms of an act of the gods (i.e. in the same way as solar or lunar eclipses). Nevertheless geological ice records show that major sun spot activity happens relatively frequently with patterns of major events every few hundred years. A modest event in 1859 disrupted the telegraph system in America. If we realise that this was the simplest of electrical pulse technologies using Morse code then it is clear that electrically it was a significant electrical storm. In recent years there have been several occasions when electrical grids have failed because of solar activity. Of far greater concern is that when the next major sunspot emission takes place we are extremely likely to be catastrophically affected, not just because of our total reliance on electronics, but more importantly that the electrical grid network could be destroyed. This means total loss of all facilities such as electricity, water, gas, petrol pumps and food supplies. Even for smaller events there is the potential danger of the loss of satellites plus substantial interference with ground based electronics. Destroying satellites would have impact on many areas from communication, satellite navigation, radio and TV as well as all the military defence systems, and not least, communication via mobile (cell) phones. Aircraft would be grounded and many computers and electronic networks could potentially be destroyed. The damage would probably include access to internet and media storage systems.

Electronic interference is inevitable but the consequences have probably rarely been considered in the depth that is essential for survival of society as it currently operates, however a recent international group considered a scenario comparable with the modest 1859 event and suggested low cost safeguards that should be implemented now [1, 2]. Without such precautionary measures power failure could exist for several years, with insurance predictions in trillions of dollars, even assuming that governments could operate.

A further example of disasters that are outside of our control is the spread of disease and plagues. These are an inevitable part of evolution with an average of about 3 pandemics per century. In former times, when many people were illiterate and information or music were passed by personal contact, then a plague could curtail this route and all the prior knowledge of a region could be lost. Previously we were cushioned from such situations as transport speeds were low and diseases spread relatively slowly. For example in the Black Death of the Middle Ages land transport was typically spreading the disease at less than 5 km per day. By contrast, at the end of the 1914-1918 war the influenza epidemic travelled far more rapidly and killed more than 20 million people (i.e. more than in the fighting) and this was the result of better transport. By contrast every major city and country is now linked by air travel so a new virus can be disseminated across the world within a 24 hour period (i.e. often faster than symptoms can develop). We need therefore to be aware that widely held knowledge and information may survive the next major epidemic, but in many areas of technology the skills are highly localised and one could imagine a scenario where all the world experts in a topic were struck down at a conference in an exotic location either by disease, tsunami or a plane crash.

2.2 Human generated disasters

Here one could write volumes but in terms of knowledge loss, or suppression, the major villains are wars, religious fanaticism and fear of knowledge which would undermine particular beliefs. Persecutions motivated by racism and religion have, and continue to destroy countries and kill or exile specific groups of people so the knowledge and diversity of culture of the countries involved then suffer. There are many political examples of invaders destroying all literature of the country that has been invaded, and also killing those with knowledge. From the viewpoint of the invaders this is a sensible strategy as it removes dissent and makes it easier to impose the new regime. Religious persecutions and enforced conversions of an invaded nation are similarly viewed as standard practice. Scientific knowledge can easily be seen as undesirable by those who base their ideas on faith, not quantifiable repeatable observations, and a familiar historic example came with the difficulty in accepting that we are just a planet circulating the Sun, not the centre of the Universe. The reality downgraded the faith that we are the epitome of all creatures. Further, the current astronomical observations that many other stars have planetary partners are likely to cause religious problems if there is evidence of other life forms on those distant planets.

Accidental fires have also destroyed documents and information that were stored in major collections. One such was the loss of the Ancient Library of Alexandria. This site, of library plus research institution, accumulated many thousands of key documents from all over the Mediterranean region from its founding in the 3rd century BCE. It was the greatest centre for study in the region. One, or more, fires destroyed it and the irreplaceable contents. Several of these fires were related to warfare or religious intolerance.

2.3 Intuitive rejection of new ideas

Somewhat surprisingly information is often unconsciously rejected or distorted as part of our normal signal processing. Our instincts, honed for rapid response and survival, mean we make instant decisions as to a danger or situation. All subsequent evidence is then processed to support our initial view, even if it is wrong. There are classic examples of this over zealous decision making. One such is found in the courtroom where juries tend to make an instant initial assessment of the person on trial, and then ignore or re-interpret evidence that disagrees with this first view. Such a subconscious error is equally a common feature of research where we have a scientific model, and then bend our interpretations to make data agree with it. This is not dishonesty, but an inherent human response. A third well documented example is seen in analysis of X-ray images to detect cancer. If the analyst sees a definite cancer site in one region then this will become a primary focus and, despite efforts to look at the rest of the image, the eyes and brain will return to the first site and not recognise secondary ones elsewhere.

We are also more likely to accept new ideas that are close to facts or ideas that we already understand or believe, and reject anything that is too radical (independent of the evidence). The degree of acceptance will increase if the presenter is particularly famous. This effect, called Cognitive Dissonance, is very unfortunate for outsiders to a field who arrive with no preconceived prejudice and clearly see facts that the members of the topic area cannot recognise because it is not in their training or would conflict with senior figures in the field. A classic example was the case of Wegener who, although a meteorologist, noted that maps of South America and Africa had close similarity as though they might have been joined. He then made studies of fossils and rock strata in equivalent matching regions and found a close match. Consequently he proposed the continents had drifted apart. His work was totally rejected because he was of the wrong discipline and the idea was too radical, and it was nearly 50 years before his model of plate tectonic movements became widely accepted.

2.4 Information loss from changing language and material decay

Writing has existed for several thousand years with examples from the Middle East of simple Cuneiform lines made in soft mud that was then hardened, to various styles of picture coding such as Egyptian hieroglyphics or the forerunners of Chinese calligraphy. Both these Western civilizations eventually faded from prominence, their languages changed and new styles of writing evolved. This resulted in the existence of many examples of the writing but without any memory of the language or how to interpret the signs. Decoding such scripts has therefore been challenging. Hieroglyphic analysis was helped by discovery a stone in 1799 with the same information from 196 BCE written in Egyptian Hieroglyphics, Greek and Demotic with a decree from Ptolemy V. This was called the Rosetta Stone. Other lost languages such as the Minoan Linear A and B were decoded by inspired imaginative analysis. For factual writing and records we therefore have a gained a moderate understanding when such examples exist. Records of victories or justification for invasions may be clearly written but they are invariably produced by the victors, so here the knowledge lost may be the truth. Interpretation of more subtle texts of opinion, philosophy or comments are far more difficult because not only do we not know the nuances of the phrases used, but just as in modern language, these can change dramatically with time. We are also totally guessing at the cultural and religious

Journal of Physics: Conference Series **558** (2014) 012002 doi:10.1088/1742-6596/558/1/012002

implications of text written in earlier periods. The situation has not changed as languages are now evolving extremely rapidly because of easy modern communication. For example viewing films or TV programmes from 50 years ago (or less) offers insights into a different world, and the content often seem so dated that it almost inconceivable that they were the norm of that period. This changing situation is obvious for many people still alive today. Listening to teenage slang and the way words can alter, or reverse in meaning within a year, underlines that language and context are incredibly transient. The speed of the changes has been exacerbated by faster communications.

Survival of text and pictures is equally determined by the long term stability of the materials used and here we move from philology into more clearly defined chemistry and materials science. Writing on clay tablets or stone carvings may have been slow and tedious to produce but the materials have survived several thousand years. Indeed the numbers that exist are remarkably numerous once we realise that the early populations were extremely small, and most of the people were illiterate. There is no way that our generation will leave an equivalent fraction of the words that we have written. Progress to papyrus, vellum and later writing surfaces certainly speeded up document production but the materials have had a progressively shorter life expectancy. Materials such as papyrus and vellum could mechanically crumble, be burned or reused for other purposes, and in many cases depended strongly on the climate. For example the Dead Sea Scrolls are some two millennia old but have survived because they existed in dry desert conditions. The inks available at the time have also faded although in some cases infrared reading can reveal early writing, and sometimes later work written on the same parchment. Nationally important archival material has often been preserved but this is invariably because the parchments etc have neither been in a war zone, fire or poor climatic conditions.

Survival of paintings shows identical problems and one can contrast the pictures of the Last Supper painted by Leonardo da Vinci around 1494-8, with the Raphael School of Athens painted in ~1500. The Last Supper was a fresco on a damp North facing wall next to kitchens, in a room that was later used as an army barracks. The original may have been magnificent but by 1642 it had virtually vanished and all the current reproductions are based on copies and the very different attempts at restoration. Hence colours have changed or people face in different directions according to the restorer. By contrast the School of Athens is in excellent condition. We have not lost the painting but almost totally lost the hidden meaning. Art experts will offer opinions as to the symbolism of the statues and central figures, and identify the faces as images (real or imagined) of the various key faces displayed. This is lost knowledge for most viewers and the art literature suggests that there is far from universal agreement among the experts. The key theme here is that even if we have an ancient painting or text in good condition we invariably fail to recognise all the encoded information of the mythology, references to events at the time, or symbolisms of dress, gems, flowers and hand gestures etc. Basically we can never understand the work of earlier generations with the same mind set as those for whom the work was created. Similarly many people enjoy music written several centuries ago and play it with reproduction instruments at the pitch used at the time of composition. Undoubtedly pleasurable, but it is from our perspective, not that of the earlier time. Indeed this is a universal situation over all musical genres from all periods whether Baroque, early jazz or the hits of 20th century pop.

2.5 Recording of music and images

Rather than move directly to modern electronic technology it is informative to trace the way music and pictures have been recorded and broadcast. In the mid 19th century sound was used to mechanically imprint patterns of wax drums. It needed an intensely loud volume, the playback sound quality was poor, and playback damaged the recording. Nevertheless it showed potential. Improvements moved the method consecutively to wax discs and then, with the advent of triode amplifiers and microphones, on to shellac records offering some 4 minutes of sound. This was long enough for dance music with a wide public appeal and it triggered developments on to longer playing vinyl discs which eventually ran for about 30 minutes per side. Attempts to make more compact and/or portable systems used magnetic tape; this was followed in the mid 1980s by the introduction of the CD which could record up to 80 minutes per side. Sound quality steadily improved as did all the associated electronics of amplifiers, speakers and headphones. More important than the actual progress is to recognise that each new technology had a lifetime of around 20 or so years before it was overtaken. Further, the progress was enabled by totally new systems which made the old equipment obsolete. Commercially this was also desirable as it opened an expanding sales market for equipment and the need to rerecord music with the modern methods so even the older records etc were outdated. It also appealed each time to new younger audiences and was not limited to any specific style of music. Later recording may exist on versions of MP3 or computer files as these have enormous storage capacity. However, the format is not necessarily of the same sound quality as the CD but for the mass market this no longer seems critical either because of the style of music that they are used for, or as they merely provide background sound.

Image capture followed a similar pattern in that the earliest photographic emulsions emerged in the 1820s. Their speed and fidelity improved and moved from glass plates to more flexible cellulose supports etc and expanded into the silent cinema images. By the 20th century resolution and film speed increased (plus movie sound tracks). The cinema further drove a move into colour (and stereo sound). Video tapes competed with film and several alternatives battled for 20 years until they were in turn superseded by DVD formats. On-line viewing of films has even displaced these for many people.

A related set of developments has occurred in TV broadcasting from the tiny low resolution screens of the 1930s to a variety of line formats and signal encoding which have increased the size, colour and picture quality. A comparison of the original TV with modern high definition images is remarkable. Further, the response speed for action events and signals encoded for stereo viewing are still lively topics. The new systems are of course unable to play older recordings. Not surprisingly when we analyse the survival of each format the lifetime is still decreasing and modern TV systems have lifetimes of probably 10 years before they are replaced with new models.

3. Progress, obsolescence and information loss with electronics and communication

In the 21st century we may be justifiably proud of the technological achievements, especially from the last 150 years. In long range communication in the 1850s this was by digital electrical pulses with Morse code at a few letters per second. The move to optical pulse coding and glass fibres required a major technological effort in the production of transparent materials, electronics to encode pulses, laser sources compatible with fibres, switching and wavelength multiplexing to send many channels down the same fibre at different wavelengths. Receivers, decoders and in-line optical fibre amplifiers were all developed to maintain a steady improvement in capacity times distance (i.e. a measure of the fibre performance). As the user we did not appreciate that each new advance was made with new technology that replaced earlier ones that had reached their limits. However, from our viewpoint the performance is now some 10^{12} times better than the ancient Morse code (i.e. a phenomenal advance in 150 years).

More obvious to the public have been the improvements in electronics for faster computer chips, CCD cameras with more pixels at lower cost, and smaller and ever larger electronic storage systems. Initially computer chip density was doubling in under ~2 years (popularly known as Moore's law); computer speeds were therefore increasing; CCD "pixels per dollar" increased logarithmically, for example by 100 times during the period between1994 to 2004; and hard disc capacity has been rising at a similar rate. Marketing of computers, cameras and stores strongly advertised such progress as it encouraged purchase of new state of the art equipment with totally new software. Unfortunately more transistors per unit area imply they are smaller and so eventually the advances have hit minimum size limits, and the logarithmic improvements with time have currently ended. Instead the commercial focus is on mobile phones with applications and new models released annually (invariably incompatible with some features of earlier ones).

4. The emerging pattern

The pattern is always the same. Once a technology has a mass market it evolves by innovations that are rarely compatible with earlier versions. The time taken to obsolescence depends on the sophistication. Simpler items, such as equipment for playing music probably lasted 20 years, in part because there is a new generation that will purchase it, even if it is a large investment. More high technology examples of computers, software, CCD cameras or mobile telephones have progressively shorter half lives before they need to be replaced with times ranging from say 5 years to one (in order to have the fashionable mobile phone). Almost always the innovations come from young engineers who are also among the prime market for the products. So to them the associated obsolescence is irrelevant as they want it and can afford it, plus they understand how to use it. They fail to recognise that not only does this induce obsolescence of equipment and older stored information but it progressively isolates and undermines the sections of society who cannot afford the changes, or perhaps because of ageing, cannot comprehend how to use the new items. Therefore many retired people still prefer hand written correspondence and "real" people on a telephone help line. The computer driven society is both alienating them and undermining their self esteem. Indeed trying to find a state of the art modern mobile with a keypad readable and workable by a pensioner is almost impossible (except for very basic telephone functions).

Increasing wealth and new electronics generally speed up both the progress and the rate at which earlier systems become obsolete. The consumer who is content with the original system often has no choice to keep it as for example the costs of replacements and repairs favour the new purchase. In some cases, such as TV, no signals are broadcast which match older receivers. Information loss from obsolescence of earlier storage media is inevitable.

The pattern equally predicts that current items, such as CDs, are definitely approaching the end of their favoured position and for music the opportunities for interactive CDs merely require some new electronic processors so that we can adjust the sound to suit our home audio systems and room acoustics, or alter the balance between the various performers on the disc. These are highly desirable features and within the scope of current technology.

5. Estimating information survival lifetimes

Future historians will have major problems finding access to correspondence and ideas or journals that have become solely electronic as most will either be in obsolete formats, encrypted, corrupted or erased. Major data archives and libraries away from the security of access to the internet will be fewer in number. Whereas those items connected to the outside world via the computer, or stored on "cloud" systems are remarkably vulnerable to failure of the provider, malicious attack or, in the event of a political strategy have reduced access. Indeed it is predicted that in the next major world conflict the first attacks will be corruption and destruction of the electronic communications of the opposing side. (NB One major power recently invested 50,000 euros in electronic typewriters so as to minimise computer espionage). Note also that "cloud" service providers have not yet been in business for long periods of time and storage may require payment which will cease when a person dies and bank accounts are frozen. This will cause deletion of material. Home items on CD (or later systems) will need frequent backup to state of the art access and this is precisely the type of task that will not (or cannot) be done by the elderly, so family records and photos will vanish very rapidly.

One also assumes that access will continue whereas in reality fibre optic overloads will occur as links are used even more extensively for social networking, advertising, spam etc. For long distances most fibre optic links are under the seas and there are many regions where systems are channelled in clusters. This makes them vulnerable to natural, accidental or terrorist interference (e.g. near the Suez canal there have been two recent events where some 60% of the East-West capacity was temporarily lost).

In the early development of computer chips the speed of competitive progress resulted in an identifiable logarithmic pattern of improvement (Moore's 'Law') so one can ask if there is an equivalent description which predicts how fast information will be lost. Not least because of the technological advances. To avoid Cognitive Dissonance I will offer a model in terms of familiar science of Thermodynamics (Zeroth plus three laws), Exponential decay (as for radioactivity) and the Heisenberg uncertainty principle (i.e. a trade off between different factors).

There is an equivalent of a half life in the survival of storage media. The constant will differ between say stone tablets and CD storage but the pattern is that technology is the driving force producing a shorter survival time. By contrast the speed of writing and communicating is expanding phenomenally (e.g. social networks of Facebook, Twitter, blogs etc commit facts and trivia to the internet at ever increasing rates and Facebook rose from zero to a billion users in 10 years). So with my model of a Thermodynamics pattern of laws I have:

A zeroth assumption as:

- The speed at which information was written or prepared is the inverse of the survival time.

This offers a Heisenberg equivalent first law as:

- The (writing speed) x (the quantity stored) x (the survival time) is a constant (S).

A second law is in the style of radioactive decay:

- Information and storage media decay, or become lost, exponentially with time. The half life depends on the media.

A third law matches the non-attainability of absolute zero:

- All information is eventually lost.

This is good thermodynamics.

One can guess at numbers to add for the decay constant but in principle my first law has a sensible pattern relating writing speed, information content and half life survival time. As a simplistic attempt one might compare copyist from different epochs, such as those writing on clay tablets, the high sped scribe of the Doomsday Book and a good 20th century copy typist using an electric typewriter. My examples are adjusted for a year output as though there were a 24 hour working day and half lives are defined as from production to present.

Clay tablets at 365 pages per year have survived 5000 years gives S of ~1.8 million.

Doomsday scribe 2000 pages with books that exist 900 years later gives S of ~ 1.8 million.

20th century electric typewriter copyist at ~9 pages per hour with paper and print fading after 25 years offers an estimate of S around 2 million.

The pattern is encouragingly consistent. As a final test we might consider the writing of a doctoral thesis but here it is necessary to account for all the preparation time, experiments, calculations and graphics production. Since the work will involve a computer many aspects will be speeded up by a factor of say 10^4. With this major assumption we can then predict the half life for a thesis which took a year to prepare and had 150 pages. The first law offers 150 x 10^4 x (half life) $\approx 2\ 10^6$. Hence the thesis survival time is a little over a year. This is probably true since key items will be published and after the viva the thesis will be consigned to a library and mostly forgotten. Whilst such numbers are unimportant in detail they nevertheless support the pattern of advancing technologies causing an ever shorter survival time for the output.

6. Social isolation and loss of creativity

The preceding examples are tangible and often quantifiable but of at least equal significance are the negative social effects of new technologies. These appear in many guises. For example one often sees small children in push chairs or playgrounds being totally ignored by the parent who is concentrating on chatting on a mobile phone. Instead of stimulating creativity and socialising most children now play ready made computer games in isolation from the family (often with games that have a highly violent content). Similarly there is an essential need to conform to peer group pressure and so they speak incessantly with other children by phone, Facebook, Twitter etc. This means this generation has lost the time to daydream and have reflective self assessment of what they want from life, and it undermines their future creativity. For many, family meals, or even those in cafes with friends, are not focussed on the immediate people but on banal phone usage. One can only guess at the consequences of this social pressure and subjugation of private and original thinking. Further, lack of exercise and

Journal of Physics: Conference Series **558** (2014) 012002 doi:10.1088/1742-6596/558/1/012002

overeating is causing obesity and subsequent medical problems, so this is the first generation where the children are less fit than their parents were at the same age.

At the other end of the age spectrum both the elderly (and poor) are severely disadvantaged if they do not have ready access and capability to operate on-line communication for searching for bargains, finding out vital information from local governments and/or understand how to use on-line banking and simultaneously recognise scams and viruses. These are daily problems that are more serious and immediate than losing old correspondence or photographs. Other difficulties now arise during travelling, for example in London the buses do not accept money but require pre-paid electronic cards or bank cards; one of the major bridges needs pre-booking by mobile and card instead of a toll booth. For those without mobile phones and computers (roughly 10% in the UK), or visitors and travellers, these are highly inconvenient.

Overall advancing technology is both undermining creativity and serious face-to-face social interactions and progressively isolating both the poor and the elderly.

7. Summary

The aim was to emphasise that improvements in technology offer some advantages for our lives but invariably former skills are lost and equipment becomes obsolete. This has the major consequence that storage media are progressively more transient and less secure. Our reliance solely on modern technology is equally hazardous as in the event of international conflicts, or even major sunspot emissions, many of the communication and storage systems will either fail and/or be disrupted. Corruption and unwarranted access to data is similarly an increasing hazard, not only from monitoring by Governments and industries, but also from malicious and criminal activities. Indeed computer crime is a boom industry.

Virtually all aspects of my comments are in the public domain with internet access. In the spirit of ignoring hard copy, and relying on modern technology for fast access, minimal references are provided (e.g. from tectonic plates to paintings of Raphael) however internet sites will offer considerable detail in every case. The development of recording is covered by [3, 4] and [5] includes future ideas for CD recording and obsolescence of earlier instruments.

References

[1] Dale A 2014 *Physics World* **27** 6

[2] http://ow.ly/yDOXp (a report on sunspot events and how they may relate to our survival)

[3] Lebrecht N 2007 *Maestros, Masterpieces and Madness* (London: Allen Lane/Penguin).

[4] Day T 2002 *A Century of Recorded Music: Listening to Musical History* (New Haven: Yale University Press) ISBN 0300094019

[5] Townsend PD 2014 *Sounds of Music – The impact of technology on musical appreciation, and composition* (Amazon) ISBN 978-1-291-80795-0.

Physical properties of cholesteric liquid crystals – carbon nanotube dispersions

N V Usol'tseva[1], M V Smirnova[2], V V Sotsky[1] and A I Smirnova[1]

[1] Nanomaterials Research Institute, Ivanovo State University, Ermak str., 39, 153025 Ivanovo, Russia
[2] Ivanovo Institute of the State Firefighting Service, Stroiteley pr., 33, 153040 Ivanovo, Russia

nv_usoltseva@mail.ru

Abstract. The mesomorphic and visco-elastic properties of cholesteric liquid crystal – multi-wall carbon nanotube dispersions were investigated with respect to temperature and nanotube concentration. The important rheology characteristics, such as dynamic viscosity, elastic modulus and loss modulus were studied. For the first time a 3D–model of cholesteric liquid crystal at the temperature of cholesteric phase existence was obtained by molecular dynamics simulation method. A weak effect of nanotube on the molecular orientation of the liquid crystal molecules in the N* phase was found.

1. Introduction

Anisotropic structures – liquid crystals (LC) have a number of unique properties that made it possible to use them as easily manageable materials, e.g. for electronics. Helical liquid crystal phases are the most promising but the least studied in this aspect. The perspective of their application is caused by the presence of spontaneous ordering of electric dipoles in the ferroelectric liquid crystal phase, which gives an opportunity to create systems for optical information processing [1]. One of the ways to optimize properties of mesogenic and other materials is creation of compositions based on them.

Setting a task of studying the compositions of helical LC phases with multi-walled carbon nanotubes (MWCNTs), we considered MWCNTs as sensitizers having unique physical characteristics in terms of electrical conductivity, developed surface area, stability under physical and chemical stresses. We suggested that the creation of MWCNT dispersions in these materials optimizes some of their parameters and leads to new spheres of application.

Previously, we investigated phase transition temperatures, electro-optic and dielectric properties of ferroelectric liquid crystal – multi-wall carbon nanotubes dispersions [2]. In this paper we summarize the results of the study of MWCNTs impact on the molecular orientation of the liquid crystal molecules, mesomorphic and visco-elastic properties of cholesteric mesophases, specifically in N*-phase, as the function of nanotube concentration and temperature.

2. Materials and methods

The representatives of cholesterol esters – cholesteryl oleate **I** (*Cr* – 5 °C *N** 35.2 °C *Iso*); cholesteryl tridecylate **II** (*Cr 63.3 °C SmA 78.8 N* 84.2 °C Iso*); and cholesteryl myristate **III** (*Cr 71.4 °C SmA 79.3 °C N* 84.0 °C Iso*) were chosen as liquid crystal host materials:

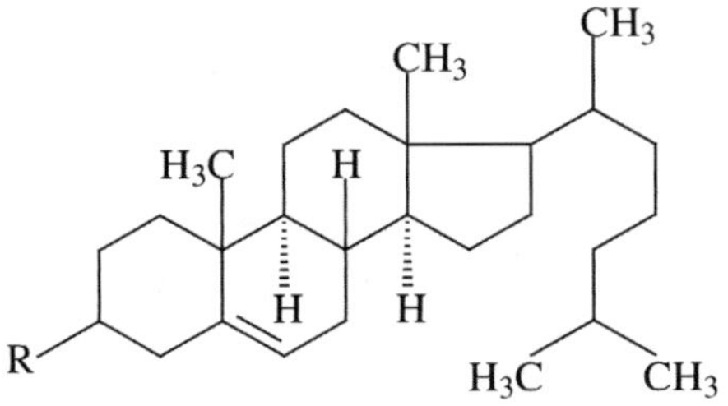

I: R = $CH_3(CH_2)_7CH{=}CH(CH_2)_7COO$–
II: R = $CH_3(CH_2)_{11}COO$–
III: R = $CH_3(CH_2)_{12}COO$–

The multi-wall carbon nanotubes (MWCNTs, Taunit-M) with an outer diameter of 8-15 nm, an inner diameter of 4-8 nm and a specific surface area of 300 m^2/g were provided by Nanotechcenter, Tambov, and were used as active dopants. They were produced by chemical vapour deposition using a Ni/Mg catalyst [3]. To identify the carbon nanotube, Raman spectroscopy before sonification (HoribaJobin Yvon LabRam HR800 spectrometer equipped with a He–Ne laser (λ=633 nm) operating at 20mW) and atomic force microscopy (Veeco Nanoscope Dimension 3100 AFM operating in the tapping mode with Si cantilever) were employed. The most probable length of the nanotubes was approximately 900 nm [4].

The ChLC–MWCNTs dispersions were prepared using an ultra sound sonication method [4]. The phase transition temperatures of all ChLC–MWCNTs dispersions were determined using a Leitz LaborLux 12 Pol polarizing microscope combined with a Mettler FP–82 hot stage.

The viscoelastic parameters of the respective materials depending on temperature and nanotube concentration were investigated using RheoStress RS 600 rheometer (Haake) with a cone-and-plate sensor system (cone angle – 1°, diameter – 20 mm, gap size between the cone and the plate – 0.055 mm) in the oscillation test mode. The measurements were carried out in the SmA and N* temperature range at 6.28 rad/s (1 Hz) angular velocity. The temperatures were controlled by TC501 Haake controller within 0.1 °C. During the harmonic oscillation mode measurements the samples were exposed to sinusoidally varying voltage [5]. This mode belongs to the class of non-destructive measurement of the material [6].

The initial experimental data to analyze the ChLC-compositions behaviour were the temperature dependence of elastic modulus G', loss modulus G'' and complex viscosity η^*.

Modulus G' (elastic or storage modulus) is a measure of stored and released energy for the oscillation period per volume unit of this material. Modulus G'' (the viscous or loss modulus) is a measure of the energy transformed to heat during one period of sinusoidal strain.

The value of the complex viscosity η^*, describing the total resistance to a dynamic shear, can also be obtained from oscillatory tests [6]:

$$\eta^* = \frac{G^*}{\omega}, \qquad (1)$$

where G^* – complex modulus, ω – angular velocity [rad/s]: $\omega = 2\pi f$, where f – frequency [Hz].

The molecular dynamics simulations of pure ChLC sample (cholesterol myristate) as well as its composition with a carbon nanotube were carried out with a MDsimGrid software [7] and software modules [8, 9] on the Intel Core i7-960 computer system, equipped with two GeForce GTH 580 graphics controllers. To accelerate the calculations of van der Waals forces we used an algorithm with Verlet lists (Verlet Neighbour Lists) [10] and transferred the computational tasks from the CPU to the GPU [11]. The system of classical equations of atoms motion was solved taking into account the AMBER force field (Assisted Model Building with Energy Refinement) [12].

3. Results and discussion

3.1. Effect of MWCNTs on mesomorphic properties of ChLC

Studies of mesomorphism and phase transition temperatures of the ChLC–MWCNTs mixtures (**I–III**) showed that the doping of MWCNTs into the ChLC matrix (0.005 ÷ 0.04 wt. %) leads to a shift of the phase transition temperatures into lower temperature region and significantly influences the SmA–Cr transition temperature. The temperature range of the SmA phase for compounds **II** and **III** essentially expands (approximately 13 °C) with increasing nanotube concentration [3]. It was confirmed by DSC technique, which determines phase transition more precisely compared to microscopy.

At the same time optical polarizing investigations showed that the LC phases textures of pure and doped materials were identical. It means that the MWCNTs did not disturb liquid crystal molecular orientation at the scale of optical wavelengths or larger. However, the polygonal texture of the crystalline phase of compounds **II** and **III** with MWCNTs, which is formed on cooling, obtained periodic striations resembling "fingerprint" texture that was not observed in the individual compounds (figure 1).

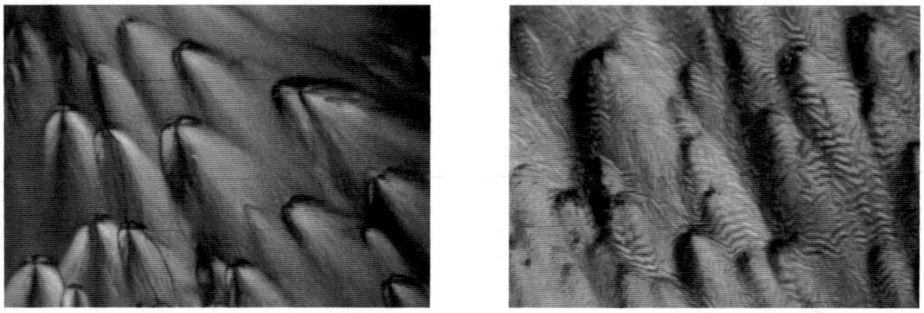

a *b*

Figure 1. Photographs of polygonal texture of crystalline phase: a) individual compound **II**; b) composition of compound **II** with nanotubes (MWCNTs concentration – 0.01 wt. %), on cooling, crossed polarizers, × 250 [3].

3.2. Visco-elastic properties of ChLC– MWCNTs dispersions

Focusing on the ChLC visco-elastic characteristics can be explained by practical importance of these properties, which determine LC material behavior in external (e.g., shear) fields.

A representative of cholesterol esters – cholesteryl myristate **III** was selected as the ChLC matrix. The measurements of the visco-elastic parameter of the given material and its composites with MWCNTs were carried out in the deformation mode within a wide temperature range of LC phases existence and at different frequencies.

Figure 2 shows temperature dependence of dynamic moduli and complex viscosity of compound **III** at angular velocity $\omega = 6.28$ rad/sec (1 Hz) and 0.01° amplitude [5].

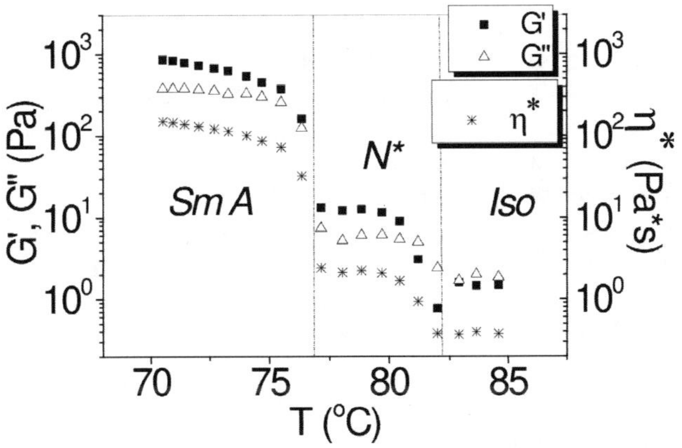

Figure 2. Temperature dependences of the dynamic moduli and complex viscosity of the cholesteryl myristate **III**.

The graph shows that the temperature dependences of η*, *G′* and *G″* of ChLC have qualitatively similar appearance, while the temperature curve corresponds to the general tendency to decreasing of the material order parameter with temperature increasing. Loss modulus *G″* prevails over storage modulus *G′* in the temperature range of isotropic phase existence, which indicates the behaviour of a viscous material. However, with temperature decrease in the transition region into N* phase, the reaction of the sample on the sinusoidal voltage exposure changes from viscous to elastic ($G' > G''$), later on, at the phase transition of the second order from N* to SmA the material becomes more viscoelastic. Most importantly, the increase of the dynamic moduli *G′* and *G″* in SmA phase as compared to N* phase is over an order of magnitude larger [5]. Such behaviour of ChLC materials under dynamic test mode was also mentioned in [13, 14].

The visco-elastic properties of ChLC compositions with respect to temperature and nanotube concentrations (0.01 and 0.02 wt. %) (figure 3) at different frequencies were investigated for the first time [5].

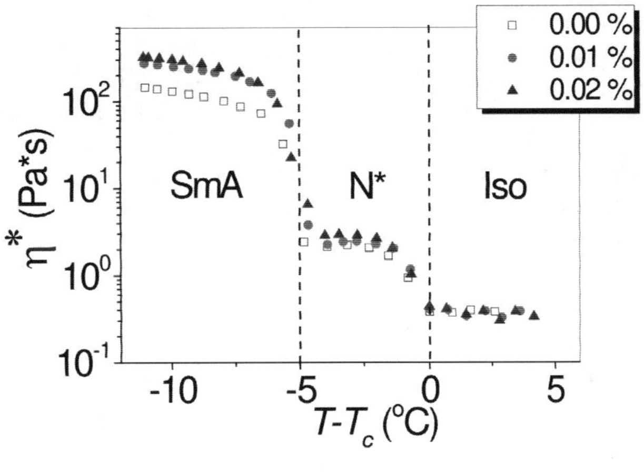

Figure 3. Temperature dependences of the complex viscosity of the ChLC – MWCNTs nano-composites with different MWCNTs concentrations (wt. %).

Doping of the nanotubes into the studied ChLC sample at concentration ≤ 0.02 wt. % does not affect the complex viscosity in the temperature range of the isotropic phase existence, however, it leads to a 25%-increase of η^* in N* phase and to an almost double increase in SmA phase [5, 15].

The experimental data on the temperature dependence of dynamic moduli of G' and G'' of ChLC – MWCNTs compositions are shown in figure 4. The peculiarity of the results obtained during the research of the samples was the increase of the elastic modulus G' value with a simultaneous decrease of the viscous modulus G'' value in the isotropic phase when adding MWCNTs (figure 4). Thus, in the Iso-phase at a certain concentration of carbon nanotubes, the LC material exhibits viscoelastic behaviour, which is probably caused by the formation of an additional spatial structure by the particles of the filler (percolation mesh). It is most likely that the rheological behaviour of the ChLC composites in the isotropic phase is determined by MWCNTs interaction [5].

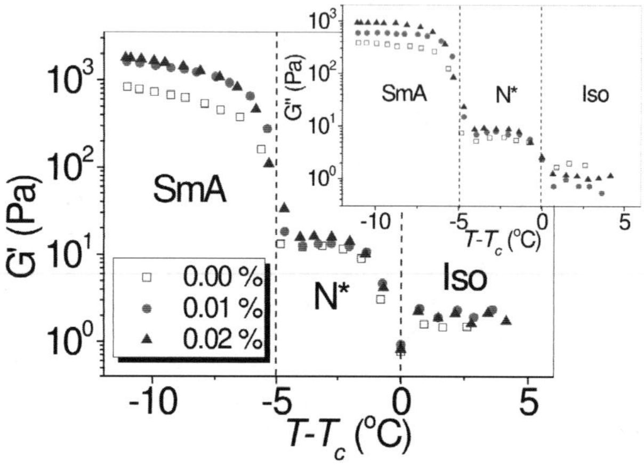

Figure 4. Temperature dependence of dynamic moduli of the ChLC – MWCNT dispersions at different MWCNTs concentrations (wt. %).

This assumption confirms the need of introducing CNTs in the LC material at the mesomorphic state temperatures to implement the process of disaggregation of the carbon dopant and its stabilization in the volume of the matrix [19]. At the same time, the introduction of MWCNTs into ChLC at a concentration of 0.02 wt. % leads to ~ 25% increase of G' and G'' in N* phase and to their 2.5-times increase in SmA phase. This probably indicates more efficient incorporation of MWCNTs into smectic phase [5, 15].

The frequency dependence of the studied viscoelastic characteristics, especially at low frequencies, is of a particular interest in the estimation of the dynamic investigation results [6]. Since MWCNTs introduction into the ChLC–matrix has no significant impact on the viscoelastic properties of N* phase, we studied the behaviour of the dynamic parameters η^*, G' and G'' of the investigated ChLC samples in the frequency range from 0.1 to 20 Hz in the temperature range of SmA phase existence $T-T_c = -8$ °C (figure 5) [5].

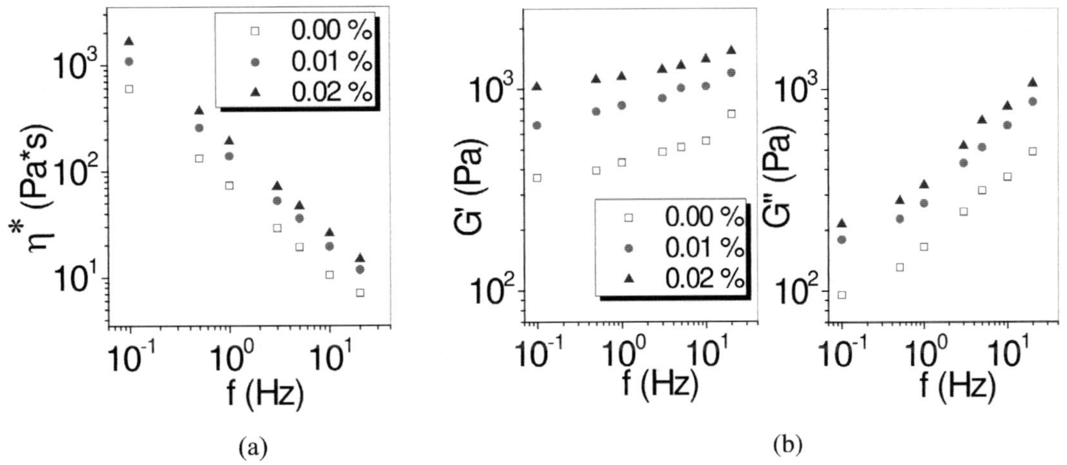

Figure 5. Frequency dependence of the complex viscosity (a) and dynamic moduli G' and G'' (b) of ChLC – MWCNTs dispersions.

The investigation of the dynamic characteristics of the compositions with MWCNTs in SmA phase showed that at the increase of the deformation frequency from 0.01 to 20 Hz the value of η* for the ChLC materials linearly decreases by more than two orders of magnitude (figure 5), while the storage and loss moduli get approximately twice as big. The G''(f) curve increases faster than the G'(f) curve, thus showing that at high frequencies the ChLC composite tends to exhibit viscous flow [5].

3.3. Molecular dynamics simulation of the ChLC – CNT system

The investigation of the supramolecular organization of compound **III** (cholesteryl myristate) and its composite with CNT was conducted using MDsimGrid software which makes calculations on the graphics processing units [7–9].

The molecular dynamic simulation of the pure material and its composition with CNT of *chair type* (a length of 40 Å and a diameter of 10 Å) was performed in the following way: preliminary optimized models of 524 molecules (60260 atoms) of the test compound (in case of the one-component system) or 524 molecules with a single-walled CNT (in case of the composition with the nanotube) were placed in the cubical box of side L about 75 Å. In order to prevent the influence of the initial arrangement of molecules on further calculations, the systems were "annealed": the simulation was performed at the temperature of 1000 K during 1500 ps under the periodic boundary conditions. Then the systems were driven to local thermodynamic equilibrium in the canonical NPT–ensemble under the external pressure of 1 atm. and the given temperature [16].

The molecular dynamics simulations were performed under cooling in the temperature range of isotropic, cholesteric, smectic A and crystalline phases during 1500 ps. The radial distribution of the centres of mass of the molecules was recorded in the course of the numerical experiments.

The simulations resulted in radial distribution functions (RDF) of the centres of mass of the molecules or pair correlation functions $g(r)$ [17]:

$$g(r) = \frac{\rho(r)}{\rho} = \frac{1}{4\pi r^2 \rho}\frac{dN(r)}{dr}, \qquad (2)$$

where $\rho(r)$ – local number density of points in a spherical shell of thickness dr, at a distance r from a fixed point, ρ – average density, $dN(r)$ – the number of points in a spherical shell of volume dV.

The RDF describes the correlation between arbitrarily chosen atomic positions (centres of mass of the molecules) at different moments of time, in other words, it characterizes the most probable distances between molecules. Exploring the given graphics of the correlation functions for height and width of the maxima it is possible to describe the peculiarities of the nearest neighbours in the simulated system, and identify the trends in the particle packing changes, depending on temperature.

Figure 6 shows the radial distribution functions of cholesterol myristate as an individual compound (figure 6a) and its composition with CNT (figure 6b).

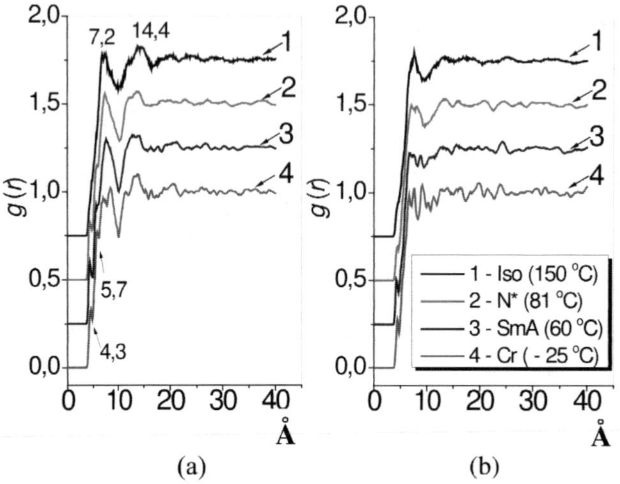

Figure 6. Radial distribution functions of cholesteryl myristate (a) and composite with CNT on its base (b) at a temperature of existence of isotropic (150 °C), cholesteric (81 °C), smectic A (60 °C) and a crystalline (– 25 °C) phases.

On curve *1*, corresponding to the RDF at the temperature of isotropic liquid in the range from 5 to 15 Å, there are two peaks of approximately equal amplitude: $r = 7.2$ and 14.4 Å, describing the short-range order of molecular arrangement. At large distances (over 20 Å) there is no correlation between the molecules, which describes the isotropic phase state. Besides the two initial peaks, there appear another two in the range from 4 to 6 Å on curves *2* and *3*, respectively corresponding to N* and SmA LC phases. The first peak with the position of the molecular mass centre $r = 4.3$ Å corresponds to molecular thickness, while the second one with $r = 5.7$ Å corresponds to molecular width. However, since the RDF value in these points is small (g(r) = 0.33 and 0.64, correspondingly), it is unlikely that the molecules will be located at such distances from each other. In the crystalline state, there is splitting of the first and second intensive peaks on curve *4*, which corresponds to a change in the short-range order and is characterized by molecular mobility decrease. It can be noticed that there is some periodicity in the distribution of the molecular mass centres compared to the RDF of certain mesophases, corresponding to the crystalline state.

Analyzing the results obtained, we can assume that the two distinct peaks: $r = 7.2$ and 14.4 Å (with the value of $g(r) > 1$) describing the short-range order of molecules arrangement express the two stable states corresponding to the arrangement of the molecules relatively to each other. Given that $r = 14.4$ Å is half the length of the molecule, the spatial arrangement of the molecules will look as in figure 7 and, finally, determines the anisotropy of the properties in the system [16].

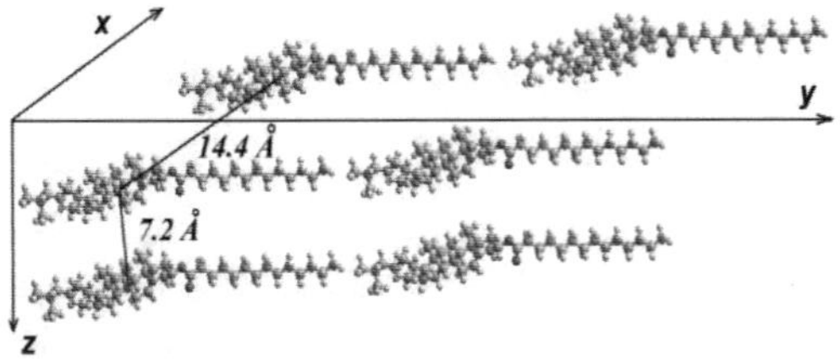

Figure 7. Spatial arrangement of the cholesteryl myristate molecules.

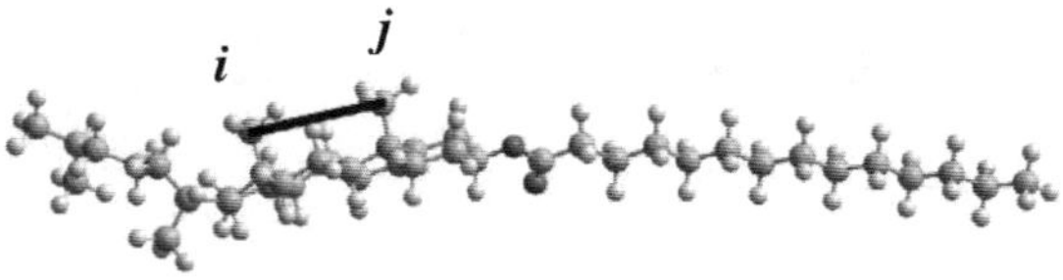

Figure 8. Atomistic model of cholesteryl myristate molecule depicting vector selected as conditional axis direction of the molecule.

A visual analysis of the simulated system made it possible to get additional data about the supramolecular organization of the studied compound in the cholesteric phase. In order to do it, a vector between two atoms ***i*** and ***j*** was selected as conditional direction of the molecule axis, which characterizes the molecule orientation in space (figure 8).

As a result of the molecular dynamic calculations, using the distribution vector, we got for the first time a 3D–model of a one-component system for compound **III** (figure 9) (at the temperature of the cholesteric liquid crystalline phase existence) [16].

Figure 9. Schematic representation of the simulated system at the cholesteric LC phase using a vector selected as a conditional image of the molecular axis.

As it can be seen in the schematic representation, a periodically ordered structure with an element of chirality (twisted nematic), corresponding to its mesomorphic state, is clearly observed in the volume model of the ChLC sample obtained during the experiment.

In order to determine the impact of the nanotube on the orientation order of the molecules in N* phase, we introduced the order parameter S' showing the orientation of the long molecular axis in its relation to the nanotube (4).

The order parameter of the neat ChLC material was calculated according to equation:

$$S = \frac{3}{2}\left(\langle cos^2\theta\rangle - \frac{1}{3}\right) \cong 0{,}44\,, \tag{3}$$

where θ is the angle between the long molecular axis direction and the director.

For the system with single wall CNT:

$$S' = \frac{3}{2}\left(\langle cos^2\beta\rangle - \frac{1}{3}\right) \cong 0{,}15\,, \tag{4}$$

where β is the angle between the long molecular axis direction and the axis directed along nanotube.

Based on the numerical calculations, we draw a conclusion that in the N* phase CNT has a weak impact on the orientation of the ChLC molecules and probably locates in the liquid crystal defects [16].

4. Conclusions

1. Doping of MWCNTs into the ChLC-matrix (0.005–0.04 wt. %) leads to significant expansion of the temperature range of SmA phase existence with preservation of LC-textures, typical for pure compounds. Polygonal texture of crystalline phase displays "fingerprint" striations which were not observed in pure ChLC material.
2. For the first time the visco-elastic properties of ChLC–MWCNT compositions have been investigated with respect to temperature and nanotube concentration and at different frequencies. At the addition of MWCNTs the increase of the elastic modulus *G'* value with simultaneous decrease of viscous modulus *G"* values were established in isotropic phase. In N* phase an increase of *G'* and *G"* values (up to 25%) and in SmA phase – by 2.5 times was found.
   At the increase of the deformation frequency from 0.01 to 20 Hz the value of η^* linearly decreases by more than two orders of magnitude, while *G'* and *G"* moduli increase ~by 2 times.
3. For the first time the 3D-model of pure ChLC compound **III** (at the temperature of cholesteric mesophase existence) was obtained. On the basis of numerical calculations slight impact of CNT on orientation of ChLC molecules was established.

Acknowledgements: The authors are thankful to the Ministry of Education and Science of the Russian Federation for financial support of this work (Grant № 4.106.2014K).

References

[1] Lagerwall S T 1999 *Ferroelectric and antiferroelectric liquid crystals* (Weinheim: Wiley–VCH) p 427
[2] Yakemseva (Smirnova) M, Dierking I, Kapernaum N, Usol'tseva N and Gisselmann F 2014 *Eur. Phys. J.* **37** (DOI 10.1140/epje/i2014-14007-4)
[3] Yakemseva (Smirnova) M V, Novikov I V and Usol'tseva N V 2010 *Liq. Cryst. & Appl. Russ. J.* **4** 98 (in Russian)
[4] Usol'tseva N V and Yakemseva (Smirnova) M V 2013 In *Organic and hybrid nanomaterials: tendencies and perspectives: monograph* (Ed. by Razumov V F, Kljuev M V; Ivanovo : Ivanovo State University, 2013 512 p.) Ch. **7** pp 228–280 (in Russian)
[5] Yakemseva (Smirnova) M V and Usol'tseva N V 2013 *Liq. Cryst. & Appl. Russ. J.* **2** 90 (in Russian)
[6] Schramm G 2003 *A practical approach to rheology and rheometry* (M.: Koloss) translated from English by I A Lavygin, Ed. by V G Kulichihin, 2003 p 312 (in Russian)
[7] Sotsky V V 2011 System for molecular dynamics simulation MDSimGrid *Certificate № 2011613855 registered in the Software registry of the Federal Institute of Industrial Property* 18.05.2011 (in Russian)
[8] Sotsky V V and Gabdulsadykova G F 2011 Software module for molecular dynamics simulation MDSimGrid-CPU *Certificate № 2011615296 registered in the Software registry of the Federal Institute of Industrial Property* 07.07.2011 (in Russian)
[9] Sotsky V V 2011 System for molecular dynamics simulation MDSimGrid-GPU *Certificate № 2011613854 registered in the Software registry of the Federal Institute of Industrial Property* 18.05.2011 (in Russian)
[10] Ivanov V A, Rabinovich A L and Khohlov A R 2009 *Computer simulation methods for*

researching polymers and biopolymers (Moscow, KD Librkom) p 696 (in Russian)

[11] Sotsky V V 2011 *Liq. Cryst. & Appl. Russ. J.* **3** 77 (in Russian)

[12] Cornell W D, Cieplak P, Bayly C I, Gould I R, Merz K M, Ferguson D M, Spellmeyer D C, Fox T, Caldwell J W and Kollman P A 1995 *J. Am. Chem. Soc.* **117** 5179

[13] Nakamura N, Maekawahara H, Hanasaki T and Yamaguchi T 2000 *Mol. Cryst. Liq. Cryst.* **352**(1) 125

[14] Hanasaki T, Fukui K, Kawamura A and Shimada O. 2009 Temperature dependence of dynamic viscoelastic properties for cholesteryl alkanoates and dicholesteryl alkanedioates *Proc.* 7th *Int. Conf. «Lyotropic Liquid Crystals and Nanomaterials» (Ivanovo, Russia, 2009)* p 15

[15] Yakemseva (Smirnova) M V and Usol'tseva N V 2011 Rheological investigations of cholesteric mesogen – carbon nanotube dispersions *Proc. 7th Annual European Rheology Conf. «AERC», (Suzdal, Russia, 2011)* p 134

[16] Smirnova M V and Sotsky V V 2014 Molecular dynamic simulations of cholesteric liquid crystal – carbon nanotube system *Conf. "Young science in a classical university" (Ivanovo, Russia, 2014)* p 22 (in Russian)

[17] Bushuev Ju G 2001 Structural properties of liquids with different types of intermolecular interactions on the data of computer simulation *Dissertation for the Doctor of Chemistry degree: speciality 02.00.04 Ivanovo, 2001* p 345 (in Russian)

Atomic diffusion on vicinal surfaces: step roughening impact on step permeability

B Ranguelov and M Michailov

Institute of Physical Chemistry, Bulgarian Academy of Sciences
Acad. G. Bonchev Str., block 11, 1113 Sofia, Bulgaria

E-mail: mike@ipc.bas.bg

Abstract. The problem of mass transport in material science for systems with reduced dimensionality holds special academic and technological attention since the fine diffusion control of adatoms could initiate exotic nanoscale patterning at epitaxial interfaces. The present study brings out important details of the atomic diffusion mechanisms on vicinal surfaces, accounting for the subtle competition between an external field imposed on the migrating adatoms and the roughening of the steps bordering the atomic terraces. The computational model reveals a temperature gap for breakdown of step permeability in the vicinity of the step roughening transition and sheds light on recently observed experimental results for atomic step dynamics on Si surfaces. The present study also demonstrates the extended capability of atomistic models in computer simulations to unravel simultaneous effects, to distinguish between them, and finally to assess their specific contribution to experimentally observed complex physical phenomena.

1. Introduction

The surface diffusion phenomena play substantial role in the theoretical models of crystal growth and related physical scenarios for mass transport on epitaxial interfaces [1-4]. In this important field for materials science, an extensively studied problem is the step dynamics of vicinal surfaces. In some specific cases the enhanced step mobility causes formation of regularly spaced narrow groups of steps followed by step-free large terraces on the interface [5]. This effect, known as step bunching, has a large impact on the processes of nanoscale patterning of epitaxial interfaces [5-7]. The surface atomic diffusion is the basic phenomenon that controls step bunching and related mechanisms of step mobility. In the presence of external electromigration force acting on adatoms on Si surfaces as a result of the current heating the crystal, the atomic diffusion does not follow a simple Einstein-walk behavior [4]. The migration of adatoms is preferentially oriented in direction of the electric current which changes dramatically the surface step dynamics. An important role in this process plays the phenomenon of step permeability (or transparency), i.e. the ability of adatoms arriving at the terrace edge to overcome the step diffusion barrier and to jump to the upper terrace level. The step permeability opens a way for still unraveled temperature dependence of the step distribution on the crystal surface to be theoretically explained. This distribution varies from regular equidistant monatomic steps to bunches of steps and vice versa as the temperature (respectively the heating current) increases [5-8]. Hence, the step permeability influences the local adatom concentration on the terraces and is therefore expected to have essential impact on the terrace nucleation processes and the step propagation [3, 7].

In the present study, based on atomistic simulation experiments, we reveal important details of step permeability diffusion kinetics of adatoms in the vicinity of the edges of atomic terraces. We discuss the role of external filed imposed on migrating atoms and the impact of fine atomic structure of steps on step permeability. We suggest a computational model and physical scenario for experimentally observed variation of step permeability as a function of temperature.

2. Physical and computational model

In the present study we consider homoepitaxial interface of atomically smooth Cu(111) crystal surface with a large single monatomic terrace and random adsorption of atoms and atomic clusters of the same material on top of the interface, figure 1. The system size is $L_x = 76$, $L_y = 44$ and $L_z = 5$ lattice units in the corresponding space dimensions. Depending on the specific diffusion problem, we also apply an external field to direct adatoms and clusters in specific crystallographic direction and to evaluate the impact of the external force on the atomic diffusion kinetics. The computational model implies classical canonical Monte Carlo (MC) sampling with many-body Tight Binding Second Moment Approximation of Electron Density of States (TB SMA) potential between interacting atoms [8, 9].

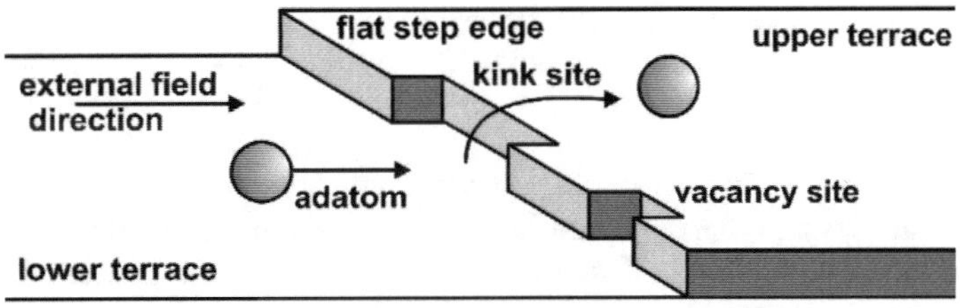

Figure 1. Interface atomic scale morphology of monatomic step on vicinal surface and adatoms on which an external field acts. The adatom displacement on the upper terrace presents the step permeability phenomenon.

This semi-empirical TB SMA potential has been extensively applied in contemporary atomistic simulations of transition metals and intermetallic alloys [2, 10-12]. Its basic advantage relates to the ability of reproducing real physical properties of surfaces and interfaces including point defects, dislocations, vacancies, grain boundaries [9]. In the present many-body TB SMA interactions model, the energy$E_i$of atom i in the simulated system is a sum of repulsive, E_i^r, and attractive, E_i^b, interaction terms: $E_i = E_i^r + E_i^b$. These terms are given by the expressions:

$$E_i^r = A\sum_j exp\left[-p\left(\frac{r_{ij}}{r_0}-1\right)\right] \tag{1}$$

$$E_i^b = -\xi\left\{\sum_j exp\left[-2q\left(\frac{r_{ij}}{r_0}-1\right)\right]\right\}^{\frac{1}{2}} \tag{2}$$

where r_{ij} is the distance between atoms i and j, and r_0 is the nearest-neighbour distance in the metal lattice. The parameters A, p, q and ξhave values obtained by fitting to experimental data for the cohesive energy, lattice parameters and independent elastic constants of the metal system studied. For our monatomic homoepitaxial system Cu/Cu(111) interface we used the A, p, q and ξ values for Cu given in table I of Ref. 9. The energy calculation is performed over all atoms within a sphere of radius $3r_0$.

Journal of Physics: Conference Series **558** (2014) 012004 doi:10.1088/1742-6596/558/1/012004

The statistical distribution is canonical, *i.e.* for an ensemble with fixed number of particles at constant temperature. In the state of equilibrium, the system energy is minimal and fluctuates around a constant value. Full lattice dynamics of the substrate, terrace and adatoms ensures complete thermodynamic relaxation of the system. The simulation experiments are done in the temperature range of 400 to 700 K. Hence, we can vary the adatom diffusion energy and the equilibrium atomic structure (density of kinks and vacancies) of the steps. The initial configuration at which all simulations start is a homoepitaxial interface with monatomic terrace on it and random distribution of adatoms on the lower terrace level. The external field having energy, E_e, is applied exclusively in direction normal to the stepson the terrace. The strength of E_e is calculated as a part of Cu-Cu interaction energy, E_{Cu-Cu}, of the TB SMA potential given by Eq. 1 and Eq. 2. The step permeability P_s is defined as the ratio of the number $N_p$of atoms on the terrace at a given moment, and the initial number N_{all} of adatoms migrating towards the terrace:

$$P_s = \frac{N_p}{N_{all}} \quad (3)$$

The model enables to gain insight into the diffusion kinetics of adsorbed atoms and the variety of diffusion energy barriers depending on the fine atomic morphology of stepped surfaces.

3. Results and discussion

The step permeability takes place when an adatom arriving at the terrace periphery has energy exceeding the step diffusion barrier, E_s^d, and therefore can jump with a certain probability on the upper terrace, figure 1. In the general case, the adatom diffusion energy is related tothe adatom thermal energy, kT, which makes the adatoms walk randomly in all space directions of the interface. Hence, the probability an adatom to jump on the upper terrace level is controlled by the system temperature. More complicated is the case of atomic diffusion in presence of external fieldimposed on the moving particles. Then the adatom energyE_a is enhanced and can be expressed by the sum of two terms:

$$E_a = E_T + E_e \quad (4)$$

where E_T and E_e are the thermal and external field energies, respectively. It is essential to note that in step bunching experiments the external field acts only in one direction which is perpendicular to the terrace steps. Typical example is the electromigration force, acting on Si adatoms on Si(111) surfaces and originating from the current, heating the sample. This is in contrast to the thermal energy which causes isotropic thermal fluctuations non-confined in space. Let us point out that an adatom jump to upper terrace level requires atomic shift in direction normal to the substrate. Therefore, the second term in Eq.4, contributes to the adatom move to the upper terrace level only in the case when the thermal energy already renders a fluctuation in direction normal to the interface. The energy contribution analysis based on Eq. 4 suggests an important conclusion related to the external field impact on step permeability phenomena. Being consisted of two terms with different physical origin, the adatom energy impact on adatom diffusion will depend on the dominant term in Eq. 4. Therefore, the step permeability could be suppressed or enhanced just by variation of the relative contribution of E_T or E_e. This result suggests that adatoms even with total energy E_a exceeding the step diffusion energy barrier could or could not lead to step permeability depending on the dominant term in the energy expression. Therefore, diffusion kinetics at low temperature under strong external field will differ substantially from the diffusion kinetics at high temperature under weak external field even when in both cases the absolute values of the adatom diffusion energies are identical. Hence, in a system satisfying the general requirement $E_a > E_s^d$, for adatom jump on the second terrace level, the step permeability could take or could not take place depending on the impact of thermal fluctuations resulting from non-zero system temperature.

3.1. Step permeability at constant temperature

Following the above analysis, we have performed simulation experiments to evaluate separately the influence of the system temperature and the magnitude of the external filed. Apparently, at constant temperature the thermal energy of migrating adatoms is fixed and therefore the diffusion energy is controlled solely by the external field imposed on the adatoms. The Monte Carlo simulation data show that step permeability diffusion kinetics can be realized only if a certain critical strength of the external field is reached. The diffusion kinetics at 400 K in weak, $E_e = 0.1E_{Cu-Cu}$, external field is shown in figure 2. It is seen that the adatom energy is not sufficient for the adatoms to overcome the step diffusion barrier and therefore the step is impermeable. Keeping the temperature constant, 400 K, we have gradually increased the external field strength up to $0.32E_{Cu-Cu}$. The snapshots in figure3 give clear evidence of complete step transparency for adatoms at overcritical external field. The value of $0.17E_{Cu-Cu}$was evaluated as the critical magnitude providing the adatom energy required to overpass the step diffusion barrier.

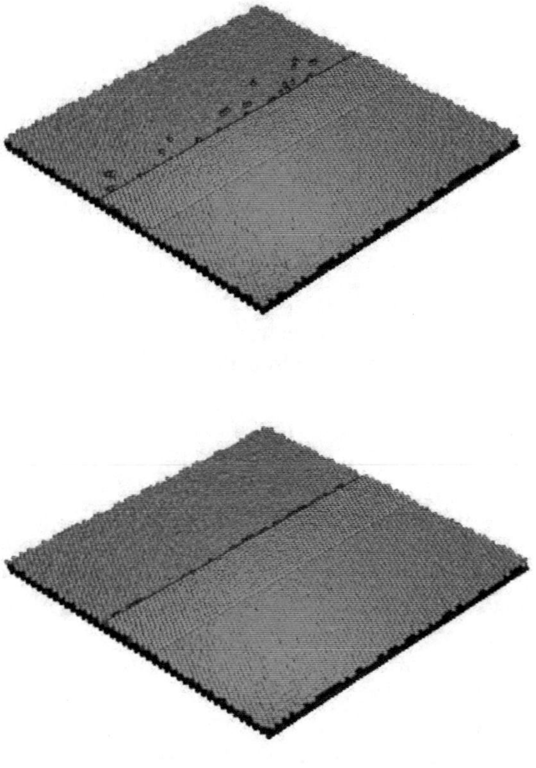

Figure 2. Atomic diffusion at 400 K in a weak external field. All adatoms “decorate” the step edge. The adatom diffusion energy is below the step energy diffusion barrier. Snapshots timescale is 1000 MCS (up) and 50000 MC steps (down). Green balls – substrate atoms, gray balls – terrace atoms, red balls – adatoms.

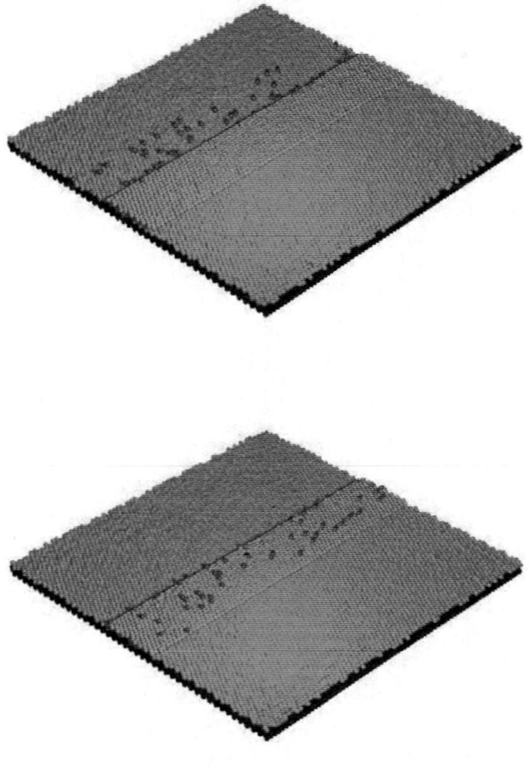

Figure 3. Atomic diffusion at 400 K and resulting step permeability at overcritical external field. The steps are transparent since the adatom energy exceeds the step diffusion barrier preventing their jump on the terrace. Snapshots timescale is 1000 MCS (up) and 50000 MC steps (down). Green balls – substrate atoms, gray balls – terrace atoms, red balls – adatoms.

The variation of P_s as a function of time is demonstrated in figure 4. The stronger external field causes also faster diffusion kinetic and enhanced step transparency for migrating adatoms. This is distinctly pronounced in the initial time between 1x10^3 and 3x10^3 MCS where the permeability P_s strongly depends on the external field intensity.

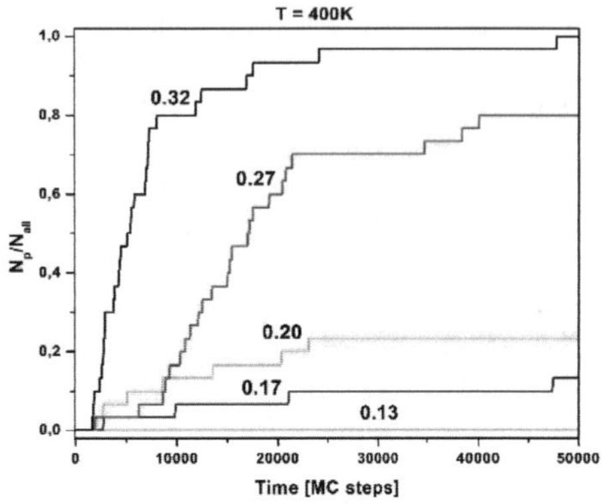

Figure 4. Step permeability at constant temperature, 400 K, and variable external field imposed on adatoms. The field strength value E_e is related to E_{Cu-Cu} interaction energy in the TB SMA computational model. Its critical value for step transparency is evaluated to $0.17E_{Cu-Cu}$.

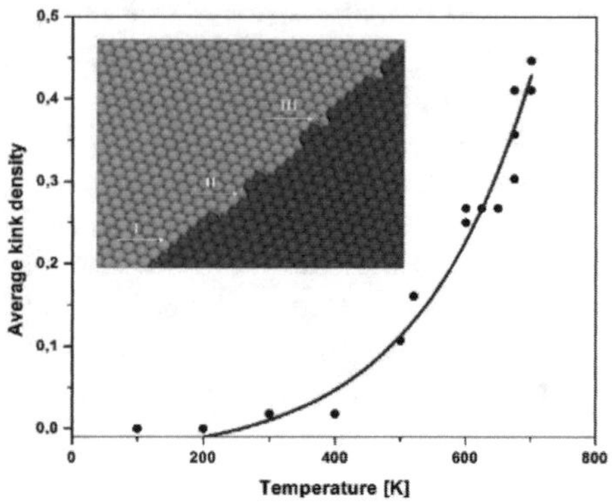

Figure 5. Average kink density of atomic B-step (see Ref. 1-3) of the terrace as a function of temperature. The step roughening transition takes place at about 700 K., see also figure 7. The inset arrows indicate I - flat step, II - kink site, III - atomic vacancy site.

It is essential to point out that step transparency was not observed at temperatures below 200 K despite the great strength of the external field which increases the adatom energy to $E_a > E_s^d$. This is due to the considerably weaker thermal fluctuations of adatoms and therefore lower probability for normal to the substrate displacements.

3.2. Step roughening impact on step permeability

The second problem of the present study deals with the dependence of step diffusion barriers on the fine atomic structure of steps. The adatoms can be attached to the terrace at energetically different places on the step and therefore the diffusion barrier on flat step could be significantly different from the diffusion barrier in a kink or step vacancy position, see figure 1 and figure 5 inset. This analysis requires detailed knowleadge of the step morphology at the atomic scale including kink density and step vacancy distribution [13]. Being a temperature function, the equilibrium kink density defines also the step roughening transition at the point of annulling the step free energy. The impact of atomic step roughness on step permeability requires evaluation of the average kink density at fixed temperature. The equilibrium kink distribution between 200 and 750 K is shown in figure 5. Let us note that the symmetry of fcc(111) surface orientation provides two types, A- and B-steps on the terrace with different step free energies and therefore different densities of kinks and different step roughening temperatures [1-3]. In the present study we analyse kink density exclusively for B-steps for which the step transparency is under investigation. The kink density temperature dependence defines two different ranges of step permeability. The first, 300 K $<T<$ 500 K, relates to atomically smooth steps and the step transparency is controlled by the external field acting on adatoms. This diffusion scenario is shown in figure 6. The terrace being in thermodynamic equilibrium at 400 K has atomically smooth steps which are completely transparent for adatoms in overcritical external field. In contrast at high temperature the step morphology is atomically rough with a large number of energetically new adsorbtion sites, figure 7. These sites, having higher diffusion barriers compared to the adsorbtion sites on a flat step, compleately block the adatoms and prevents their jumps on the upper terrace level. Hence, in the second temperature range, T

> 700 K, the step permeability and corresponding diffusion kinetic are entirely controled by the kinks density at the terrace edges.

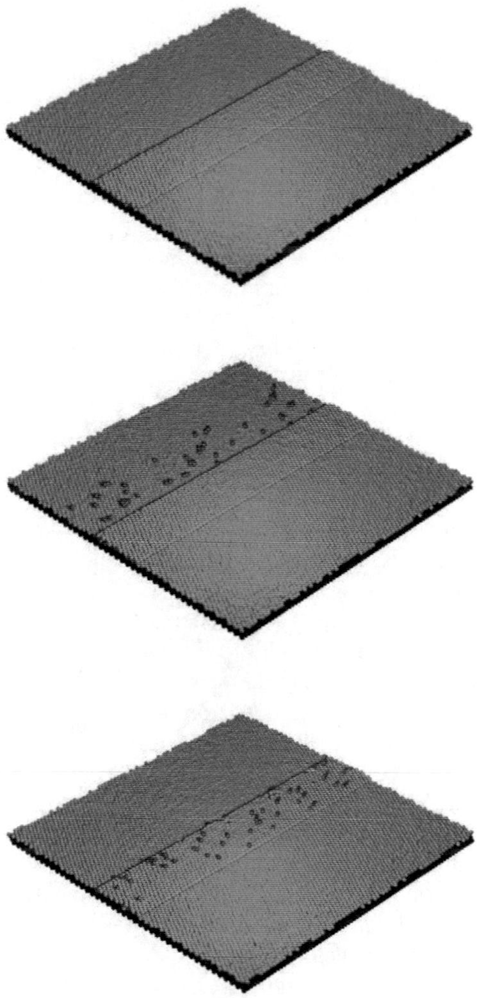

Figure 6. Diffusion kinetics at atomically smooth terrace steps in equilibrium, *T*=400 K. The step is entirely transparent for adatoms coming from the lower terrace level. The external filed imposed on adatoms $0.2E_{Cu-Cu}$ is overcritical for adsorbtion sites on flat step. Green balls – substrate atoms, gray balls – terrace atoms, red balls – adatoms.

Figure7. Diffusion kinetics at atomically rough terrace steps in equilibrium, *T*=700 K. The step permeability is tightly suppressed by the high density of kinks and vacancies on the step. The external filed imposed on adatoms is $0.2E_{Cu-Cu}$ same as those in figure 6. Green balls – substrate atoms, gray balls – terrace atoms, red balls – adatoms.

Increasing the system temperature above its critical value for step roughening transition, the adatom energy becomes again sufficiency high for the adatoms to overcome the step diffusion barriers at kink or step vacancy sites. As a result, the steps are again transparent for adatoms and step permeability diffusion kinetics becomes operative again. Let us point out that the enhanced atomic scale surface dynamics originating from elevated system temperature also supports the mass transport via atomic exchange between different levels of terraces.

The present computational model reveals the temperature impact on the surface dynamics at the atomic scale. The temperature contributes to the adatom diffusion energy required for step permeability in a physically different way compared to the impact of the external field. The system temperature moulds simultaneously the surface structure and the surface dynamics of both steps and substrate atoms. Therefore, it shapes in a specific way the thermodynamic fluctuations of all atoms on the interface. In contrast, the external field impact is restricted only to the adatoms diffusing towards the terrace. Hence, in the adatom critical energy required for step permeability the contribution of the external field is not equivalent to those of the temperature. For instance, at low T the atomic step is expected to be non-permeable even for adatoms having overcritical diffusion energy for flat step. The reduced permeability is a result of low surface dynamics of atoms and therefore low thermodynamic fluctuations. Restricted fluctuations of atoms could decrease significantly the probability for atomic exchange at the terrace edges. In contrast, at high T and the same value of adatom diffusion energy (compensated by weak external field) the step will be transparent for diffusing adatoms because of the enhanced atomic surface dynamics.

4. Conclusion

The present study provides atomistic insight into the step permeability diffusion kinetics of adatoms on crystalline vicinal surfaces. Based on detailed analysis of diffusion energy barriers at different sites of atomic steps (flat step area, kinks, and step vacancies) it reveals the step roughening impact on step permeability diffusion mechanism. On that physical background, the present model clearly defines three temperature-dependent diffusion scenarios of step permeability. The first scenario is realized at temperatures far below the critical temperature of step roughening transition, T_c. Here the atomic diffusion takes place on atomically smooth surface towards atomically smooth steps of the terrace. In this temperature region, depending on the strength of the external field imposed on adatoms, the step permeability could be entirely restricted or favoured. In this way, the resulting mass transfer between different levels of atomic terraces can be turned on or off. The second scenario operates at temperatures where the terrace steps are atomically rough. Because of the high density of kinks and step vacancies, the adatoms are blocked at the step edges and permeability collapses despite the enhanced value of the adatom thermal energy. The third scenario comes about at temperatures exceeding sufficiently T_c. In this case the adatom energy surpasses the diffusion energy barriers at kink sites and the step permeability diffusion kinetics comes again into play. Finally, the presented computational model opens a way to distinguish between the specific impacts of the thermal energy and the energy of the external field on the adatom diffusion on vicinal surfaces. While the variation of the system temperature influences the surface structure at the atomic scale by changing the energy barriers corresponding to smooth or rough interface, the external filed only contributes to the adatom diffusion energy. The simulation results and the observed in the present study diffusion kinetics are in line with recent experimental findings of step bunching phenomena on Si surfaces.

Acknowledgment:
The authors acknowledge the financial help of Project BG051PO001-3.3.06-0038 funded by OP Human Resources Development 2007-2013 of EU Structural Funds.

References

[1] Brune H 1998 *Surf. Sci. Rep.* **31** 121
[2] Michailov M 2011 in: *"Nanophenomena at Surfaces: Fundamentals of Exotic Condensed Matter, Properties*" Springer Series in Surface Science, Ed. M. Michailov, vol. 47, Chapter 6

Journal of Physics: Conference Series **558** (2014) 012004 doi:10.1088/1742-6596/558/1/012004

[3] Krug J and Michely T 2004 *Islands, Mounds and Atoms* (Berlin - Springer)
[4] Yongsunthon R, Tao C, Rous P and Williams E 2011in: *"Nanophenomena at Surfaces: Fundamentals of Exotic Condensed Matter Properties*" Springer Series in Surface Science, Ed. M. Michailov, vol. 47, Chapter 5
[5] Latyshev A, Krasilnikov A, Aseev A andStenin S 1990 *Surface Science* **227**, 24
[6] Stoyanov S, 1997 *Surface Science* **370** 345
[7] Ranguelov B andStoyanov S 2009 *Surface Science* **603** 2907
[8] Landau D P and Binder K 2000 *A Guide to Monte Carlo Simulations in Statistical Physics*, (Cambridge University Press)
[9] Cleri F and Rosato V 1993 *Phys. Rev. B* **48** 22
[10] Georgiev N and Michailov M 1998 *Phys. Rev. B 58 13895*
[11] Michailov M 2009 *Phys. Rev. B* **80** 035425
[12] Michailov M 2004 in: "Computer Simulations Studies in Condensed Matter Physics XVI" Springer Series Proc. in Physics, Vol. 95, p. 226-249, Eds. Landau D P, Lewis S and Schettler H, (Springer)
[13] Filimonov S and Hervieu Yu, 2004 *Surface Science* **553**, 133

Challenges in the areal measurement of surface roughness and shape at the micro and nanoscale

P C Montgomery[1*], M Guellil[1], P Pfeiffer[1], B Serio[2], F Anstotz[1], L Pramatarova[3] and S Roques[1]

[1] *Laboratoire des Sciences de l'Ingénieur, de l'Informatique et de l'Imagerie (ICube), Unistra-CNRS, UMR 7357, 23 rue du Loess, 67037 Strasbourg cedex, France.*
[2] *Laboratoire Energétique Mécanique Electromagnétisme (LEME) - EA 4416, 50 rue de Sèvres, 92410 Ville d'Avray, France.*
[3] *Institute of Solid State Physics, Bulgarian Academy of Sciences, 72 Tzarigradsko Chaussee Blvd., 1784 Sofia, Bulgaria.*
E-mail: paul.montgomery@unistra.fr

Abstract. Measuring surface roughness accurately at the micro and nano scale presents several challenges. While optical techniques can be used to rapidly measure large areas, significant variations can be found between results from different techniques on similar samples. In the present work, a comparison has been made between the results of two different systems using interference microscopy and AFM to make measurements at the same place on the same sample. Two samples were prepared on silicon wafers by marking them with a multi-scale pattern using a photoresist process of lithography from an optical mask, followed by reactive ion etching. One was left bare and the other was prepared with a rough layer of hydroxyapatite before measuring at the chosen positions. Comparison of the results showed that while the general shapes of the measured surface microstructures were similar, several differences were found. For example, there was a variation of up to 7 % between techniques in the measurement of the depths of the etched features and artefacts were also visible at square edges. These results show the need to pay careful attention to instrument calibration and probe/surface interactions in order to improve the accuracy of surface characterization of surface roughness and topography.

1. Introduction

Measuring surface roughness accurately at the micro and nano scale presents several challenges concerning the lateral resolution, the absolute height measurement, the accuracy and repeatability. Up until recently, the stylus profiler has been used to provide the most reliable measurements of surface roughness, due to its simplicity combined with the many advanced standards that now exist [1]. These ensure the uniformity of measurements and an independence of the measurements from the machine and the user. Over the past three decades, many new optical techniques as well as near field scanning techniques have been developed. Optical methods are of particular interest because of their rapidity of measurement over large areas [2]. One emerging difficulty though, is the variation in the results from stylus, near field and optical techniques when measuring similar samples due to technique dependent artefacts and probe/surface interactions. The recent ISO/CD 25178-600 standard on areal measurement has begun to address this problem [3]. The use of carefully chosen measurement standards that can be traced back to the metre is an important part of calibrating surface profilers [4].

Comparing the measurements made using different techniques on real samples is another useful way for benchmarking, to help in the understanding of the performance of each technique and to improve the overall measurement precision in real life applications. Such a study that was carried out within a European project to characterize microlens arrays revealed some significant disparities between results from different techniques [5]. This is all the more true when the surface structure deviates from a homogeneous and flat surface to one that is heterogeneous and rough. Comparative studies of silicon and rough poly-Si have been useful in revealing different artefacts and algorithm related variations in measurements between AFM and interference microscopy [6].

Coherence scanning interferometry (CSI) or white light scanning interferomety (WLSI) as it is also known, is a powerful technique for areal surface roughness measurement today because of its high axial resolution, its versatility on different types of samples and its ease of use [7, 8]. Being an optical technique, CSI also has its limits, making it necessary to compare measurements with those from other techniques to help in the understanding of the different sources of errors such as with SEM and AFM [9]. For example, it is well known that CSI can produce artefacts in the measurement of rough layers due to the steep surfaces present [10].

Where the results of the characterization of surface topography of materials are presented in the literature using several techniques on the same sample, comparative measurements are more often made in different places [6, 11]. Taking measurements from exactly the same place on the same sample with different techniques in practice is not easy. Nonetheless, such measurements are of great importance in order to be able to characterize and to understand important new materials today such as in nano- and biomaterials. It is also important for the characterization of etching processes in micro and nanotechnologies for which the homogeneity and the aspect ratio often depends on the position of the sample in the etching device [12]. While it is already difficult to make comparative measurements for μm sized structures, the challenges increase even more when the details required are sub-μm and nanometric in size.

For measuring the same place on the same sample with different microscopy techniques, there exist three main solutions:

1. Multi-mode measurement on the same microscope (the sample is not moved).
2. Position indexing using a calibrated *XY* table or sample holder [13] (the sample is moved between different microscopes).
3. Marking the sample (the sample is moved between different microscopes).

In the present work we are particularly interested in the third solution, in marking the sample. Many different types of techniques exist for marking, such as by using ink [14], mechanical or electrochemical etching, pulsed laser ablation [15], laser marking with a photonic jet [14], thin film deposition, and photolithographic etching. The choice of marking technique depends on several factors, such as the properties of the material (physical, chemical, topographic...), the resolution of marking required and attainable on a given sample and the size of the field of view necessary for the different characterization methods employed.

In the ICube laboratory we are interested in characterizing a wide variety of materials, ranging from semiconductor crystals and microelectronic components, various organic, optical and biomaterials through to cement pastes. Tests carried out using laser marking of cement paste materials and a thin film evaporated Al layer on a Si wafer, while allowing the identification of a particular zone, did not allow a high enough resolution to be attained to be able to make the same linear profile in exactly the same areas under the microscopes.

To attain a higher positional resolution, the photolithographic etching technique on silicon wafers was therefore chosen [16]. A multi-scale photolithographic mask was developed and used to mark two samples of silicon wafers with features etched to a depth of about 2.4 μm and 6.6 μm respectively. The first sample was left bare to provide a smooth surface with simple etched levels. The second was prepared with a partial layer of hydroxyapatite deposited from a solution of simulated body fluid (SBF) [17] to give a partially filled, rough layer on the smooth levels of the etched silicon surface. Using the numbered squares, the same area was then found and measured with two different systems using

Journal of Physics: Conference Series **558** (2014) 012005 doi:10.1088/1742-6596/558/1/012005

interference microscopy (a Leitz-Linnik research microscope and a Zygo NewView 7200 microscope) and an AFM microscope (Park XE70). The samples were also imaged with ESEM (environmental SEM).

The measurements of the two dimensional cross sectional profiles and 3D views of the surface topography of the same area using the different techniques were then compared using the MountainsMap® 6 analytical measurement tool. The results of the comparisons are presented in order to reveal the similarities and the differences in the measurements between the different techniques. Particular attention is paid to average depth measurements of the etched features, the edge shapes and the presence of artefacts. A discussion is given concerning the different origins of the differences in measurements in terms of calibration, probe/surface interaction and the precision of placing the cursor in the data analysis software.

2. Measuring the same sample in different places

As mentioned previously, in materials characterization, comparative measurements of surface topography, if at all made are done so either on different samples that are similar or on the same sample in different areas [6, 9]. An example is given in figure 1 of the use of SEM, CSI and AFM to measure the roughness of a thick layer of HA on silicon substrate.

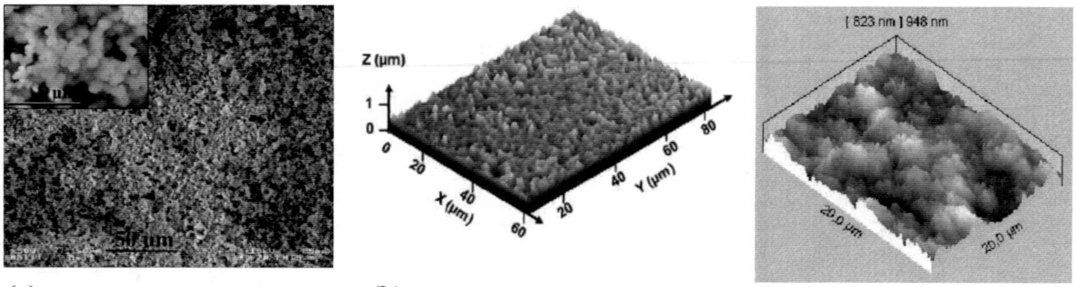

(a) (b) (c)

Figure 1. Comparative measurements of the surface topography of similarly prepared rough layers of hydroxyapatite on different Si substrates using (a) SEM, (b) the Leitz-Linnik microscope and (c) AFM.

Such comparative measurements are obviously useful, the different techniques often confirming the general topographic and structural characteristics of the sample surface. Nonetheless, it is clear that the results are slightly different. While the SEM image shows a higher lateral resolution and the detailed spongy nature of the HA layer together with the HA nanocrystals at the higher resolution, quantitative measurements can only be approximate. On the other hand, the CSI and AFM results provide quantitative data concerning the measured heights of the layer at different points and the roughness values [11]. Any quantitative data can only be representative of the layer. Problems arise when comparing measurements between different techniques since the values are often different due to differences between sample positions and instrument performance. Any artefacts present in the measurements would also be more difficult to detect.

By measuring the same place on the same sample, a more rigorous comparison can be carried out in the measurements between different techniques in terms of absolute values of roughness and height as well as the presence of different artefacts.

3. Measuring the same sample in the same place

In order to achieve measurement at the same place on the same sample using marking methods, various methods of marking were investigated. These included laser marking of cement paste samples, Al layer evaporation on a silicon wafer and the use of a photolithographic mask combined with etching on silicon wafers. The method used for growing the hydroxyapatite layers is also described. Details are given of the different topographic measurement techniques used, together with the analytical procedures employed to compare the different surface shape measurements.

3.1. Initial studies using laser and Al layer marking

Some first tests were carried out using laser marking on cement paste materials, consisting of a 1 mm numbered grid of 200 μm sided squares written with 30 μm wide lines (figure 2(a)). A thin film evaporated Al layer using a 1.5 mm square mask was also made on a Si wafer (figure 2(d)). While these two marking techniques allowed the successful identification of a particular zone under the optical and SEM microscopes, they could not be used to make comparisons between profiles. The resulting marks on each sample were too rough and inaccurate to be able to find the same place with sufficient precision to be able to compare the line profiles. This was more so the case for the AFM measurements on the Si sample with its limited field of view.

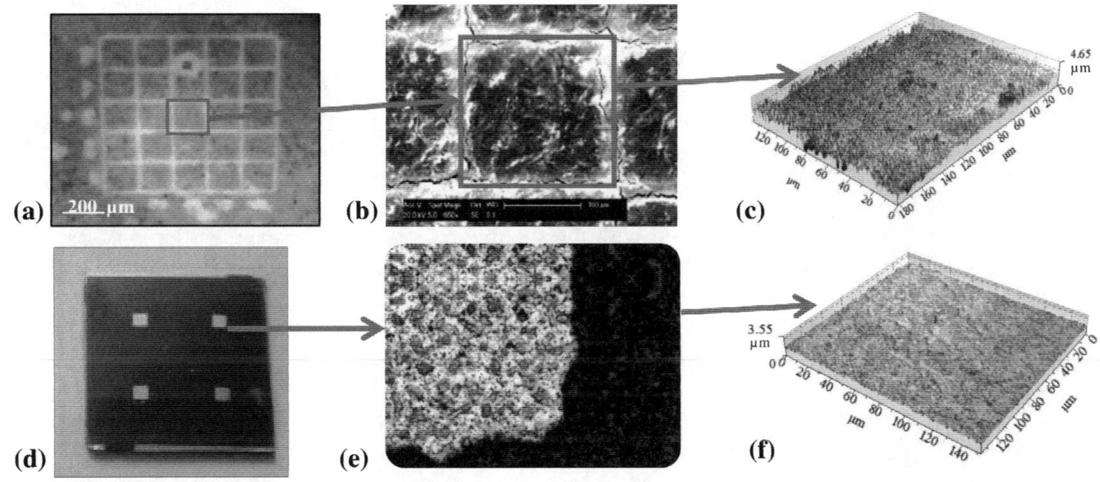

Figure 2. Comparison of results of marking using laser etching on cement (a) optical microscopy, (b) SEM, (c) the Leitz-Linnik microscope and Al layer on Si using (d) direct view, (e) optical microscopy and (f) the Leitz-Linnik microscope.

3.2. Preferred method chosen: multi-scale test pattern using photolithography

The preferred method chosen for marking was that using UV photolithography and RIE etching on smooth silicon wafers [6]. The challenge for positioning a sample under different microscopes is to have a test pattern with geometrical characteristics that allow the localization of different regions on a surface that may vary by several orders of magnitude in size, from 1 mm^2 to 1 μm^2 so that exactly the same zone can be identified.

To allow the rapid and efficient localization of a given position, a specific mask was developed based on a multi-scale geometry. The photomask was designed using LayoutEditor©, a computer-aided design software for editing MEMS designs and for use in integrated circuit fabrication. A special macro command in the C++ language was programmed to produce a multi-scale pattern of numbered squares with different geometries and sizes for a rapid and efficient localization of sample features from mm^2 to μm^2. The portion of interest is shown in figure 3(a). The test pattern was etched in two samples of silicon wafers to depths of 2.4 μm and 6.6 μm respectively. Standard photolithographic procedures were employed, using a positive photoresist, direct contact in a mask aligner and RIE etching. The resulting etched mask pattern on one of the silicon wafers with the zone of interest is shown in figure 3(b) using ESEM.

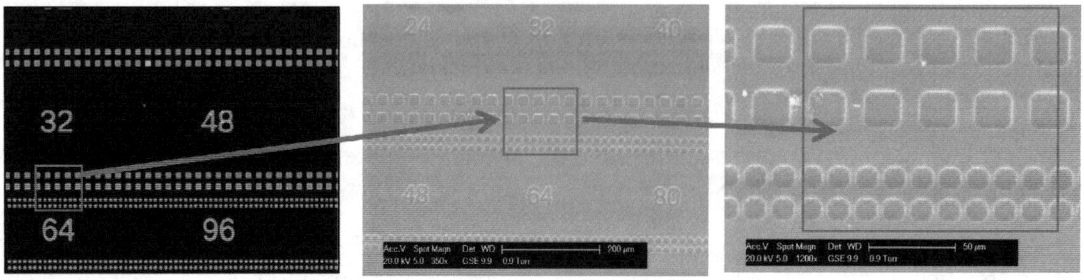

(a) **(b)** **(c)**

Figure 3. Details of the test pattern developed, showing (a) the photolithographic test pattern developed in CAD and (b) the numbered square pattern etched in a Si wafer observed by SEM and (c) details of the exact zone measured (red square) observed by SEM.

The rough layer used consisted of hydroxyapatite (HA), an inorganic material that is chemically similar to the mineral component of bones, teeth and hard tissues in mammals ($Ca_{10}(PO_4)^6(OH)^2$). The study of synthetically grown layers of HA and composite materials incorporated with nanodiamonds, polymers, etc. is important today in the fields of human implants and new biological scaffolds [17]. Amongst the many parameters used to characterize these materials, the quantified measurement of the topology of these rough layers is a key aspect [11]. So as to be able to measure the same zone with the different techniques, the second silicon wafer etched with the square test pattern to an average depth of 6.6 µm was used as a substrate on which to grow a partial layer of HA, with enough HA to be measured while leaving spaces to be able to recognize the pattern.

The HA layer was deposited on the etched silicon using a supersaturated simulated body fluid (SBF), an aqueous solution that resembles the inorganic composition and concentration of human blood plasma. The solution was prepared by dissolving reagent-grade chemicals in doubly distilled water according to the SBF method [18]. The samples were immersed for 3 hours and 25 min in the solution under natural conditions (37°C, pH 7.4) while being mixed with a magnetic stirrer.

3.3. Growth technique used for hydroxyapatite layers

The different microscope techniques used to characterize the samples are the following:

i) Leitz-Linnik microscope

The CSI interference microscopy system used was developed at ICube, based on the principle of white light scanning interferometry and described in [6, 11]. The system is based on a Leitz-Linnik microscope equipped with a Linnik type objective (x50, NA = 0.85), giving a lateral resolution of R_{lat} = 0.43 µm in visible light, an axial resolution of R_{ax} = 1 to 15 nm depending on the algorithm used and nature of the surface to be measured and a maximum field of view of 180 µm x 140 µm with a Prosilica CE1380 camera (1360x1024 pixels). The software used (CPM 2.2) for the control, acquisition and processing was developed in-house using LabView. The algorithm used for fringe processing and surface roughness characterization was an improved version of the visibility measurement technique [6] derived from Teager-Kaiser energy operators [19]. Rapid image acquisition was used (sequence of successive images) with axial steps of 90 nm, slight noise reduction (low pass 3x3 Gaussian filter) and envelope peak interpolation using second order spline fitting.

Journal of Physics: Conference Series **558** (2014) 012005 doi:10.1088/1742-6596/558/1/012005

ii) Zygo NewView 7200 microscope

The Zygo NewView 7200 interference microscope used was a system equipped with x50 Mirau interference objective, a 150 µm piezo-electric vertical stage, a digital B&W camera (640x480 pixels) and Zygo proprietary image analysis software. The field of view is 140 x 110 µm and the system has a quoted axial resolution of ~ 0.1 nm with a precision of < 0.75 % over 150 µm. The lateral resolution is $R_{lat} = 0.52$ µm.

iii) AFM (Park XE70) microscope

The AFM microscope used was a Park XE 70 model working in the non-contact mode. The tip used was a non-contact high frequency point probe. The field size used was 45x45 µm for 512x512 pixels, giving a lateral resolution of 90 nm. This value can be improved by scanning over a smaller area with a higher number of pixels. For example, scanning over 2x2 µm for 1000x1000 pixels gives a lateral resolution of 2 nm.

iv) ESEM environmental microscope

The electron microscope employed was an ESEM environmental system under a pressure of 5 Torr.

3.4. Analytical procedures for roughness analysis

For analyzing and comparing the measurements from the different microscopes, MountainsMap® 6 software (from Digital Surf) was used. This enables false color images of the height data, 3D views, 2D surface line profiles and height and roughness measurements to be made in a uniform way from the different measurement sources. For comparing the line profiles from the same place, a line profile 45 µm in length, corresponding to the width of the AFM measurements, was taken from the middle of the same etched square in the measurements from each of the microscopes. The square was identified using the numbering system.

4. Results of surface roughness measurements

The results of the surface topography measurements made with the different microscopes on the Si sample and HA on Si are now presented.

4.1. Results on patterned silicon wafer

A first comparison of the three sets of 3D measurements on the same area of bare patterned Si wafer (figure 4) shows certain broad similarities in the results.

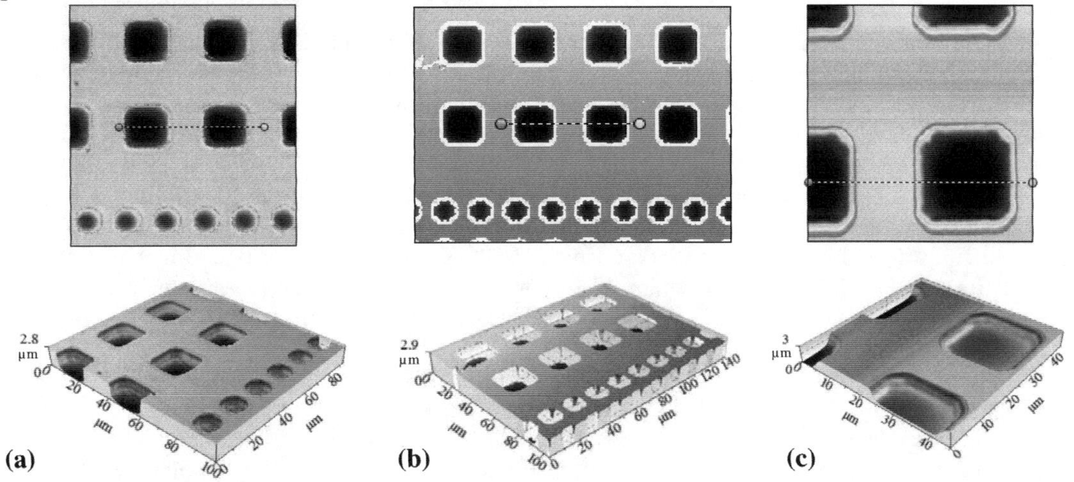

Figure 4. Comparison of measurements of the same etched squares in Si using (a) the Leitz-Linnik microscope, (b) the Zygo NewView 7200 microscope and (c) the Park XE70 AFM microscope.

As would be expected, square depressions of a similar size were found in each case. While the 2D surface profiles (figure 5) confirm the resemblance between the three measurements, on closer study, several differences can be observed. The most significant difference is the variation between the values of the average depths of the etched squares found by each measurement technique (Table 1).

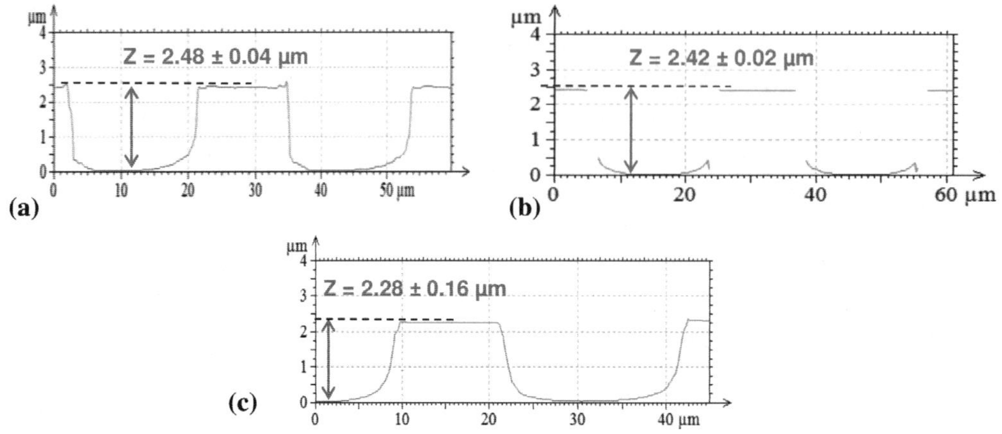

Figure 5. Comparison of 2D profiles of the same etched squares in Si measured using (a) the Leitz-Linnik microscope, (b) the Zygo NewView 7200 microscope and (c) the Park XE70 AFM microscope.

Table 1. Comparison of depth measurements of etched squares (figure 5) using each technique.

Technique	Average measured depth	Estimated depth uncertainty	Estimated percentage uncertainty
Leitz-Linnik	2.48 µm	± 0.04 µm	± 1.6 %
Zygo NewView 7200	2.42 µm	± 0.02 µm	± 0.75 %
Park XE70 AFM	2.28 µm	± 0.16 µm	± 7 %

The estimations of the uncertainties of measurements for each technique were made in the following way. For the Leitz-Linnik measurements, the main contributions to the uncertainty come from errors in the piezo step positioning, the envelope determination algorithm and the reference mirror flatness, resulting in an overall uncertainty of ± 0.04 µm (± 1.6 %). Concerning the Zygo NewView 7200, the uncertainty of ± 0.02 µm is calculated from the quoted accuracy of ± 0.75 % of the depth measurement.

For the AFM measurement the main contributions to the measurement uncertainty come from the non-linearity at depths greater than a few hundred nm together with noise from acoustic vibrations, leading to an estimated uncertainty of ± 0.16 µm (± 7 %). The uncertainty of the AFM measurements could be improved by optimizing the choice of the control parameters for the measurement. Since these measurements were made, the AFM microscope has been placed in an acoustic enclosure, which also improves the uncertainty values.

Another significant difference between the 2D profiles can be observed near to the measurements of the edges in all three cases. For both of the interferometric measurements, edge effects are visible. For the Leitz-Linnik this error is in the form of an over estimation of the top of the edge (figure 5(a)) and for the Zygo it is in the form of an under estimation at the bottom of the edge (figure 5(b)). These errors appear to be similar to the well-known "batwing" artefacts when measuring a step height near to the coherence length of the light used due to mixing of signals coming from the top and bottom of the edge [20]. In addition, for the Zygo results (figure 5(b)), the missing measurement points between the top and bottom of the edges are due to the lack of fringe signals on the steep slopes because of the limited numerical aperture of the objective [10]. Finally, for the AFM measurements, the edges can be seen to be rounded off, which is a result of the convolution between the tip shape and the edge of the etched square. The degree of rounding off was also found to vary for different tip scanning speeds and sample orientations.

4.2. Results on the HA layer on patterned silicon wafer

For these particular samples of HA layers on etched silicon patterns, measurements were not possible using AFM because of the high layer thickness and depth of the etched squares (6.6 μm) and the limited dynamic range of 7 μm of the instrument. It is well known that HA is very difficult to characterize with AFM [11].

A first comparison of the different 3D measurements on the same region (figure 6) again shows certain broad similarities in the results, indicating a partially rough layer on square depressions. The presence of individually recognizable large clumps of HA, such as the one indicated by the large arrow in figure 6, confirms that exactly the same area had indeed been measured by each technique and the success of the numbered multi-scale square pattern used.

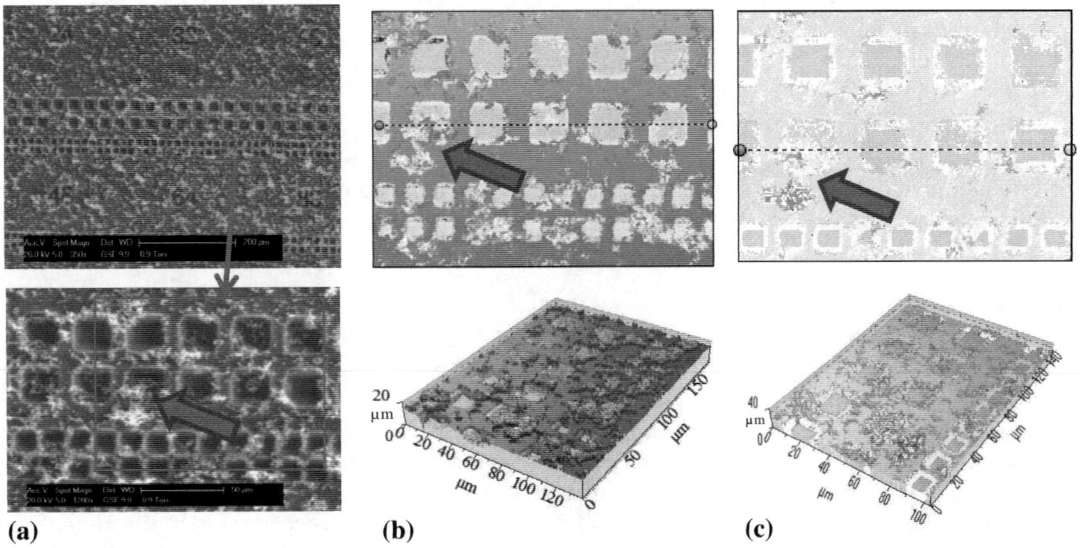

Figure 6. Comparison of the surface shape of the same place of hydroxyapatite layer deposited on etched squares in Si measured using (a) SEM, (b) the Leitz-Linnik microscope and (c) the Zygo NewView 7200 microscope. Blue arrows indicate same clump of HA.

A comparison of the 2D line profiles made at the same place (figure 7) between the measurements on the Leitz-Linnik and Zygo NewView 7200 also shows broad similarities as well as the presence of differences and artefacts. An exact comparison of the height measurements of the HA layer between the two systems is more difficult than just with the bare etched silicon squares due to the greater variation in surface height with axial position of the rough HA layer.

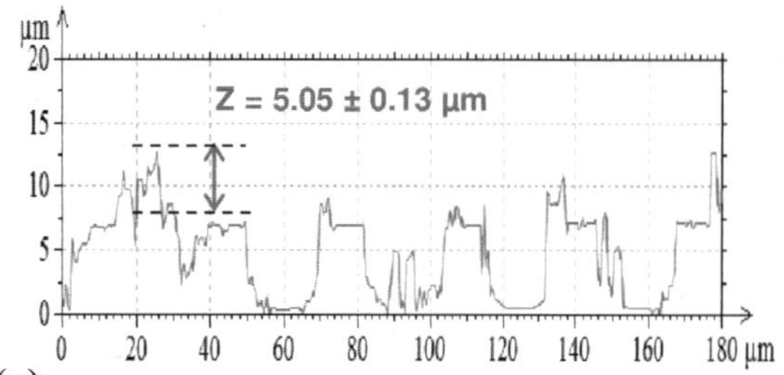

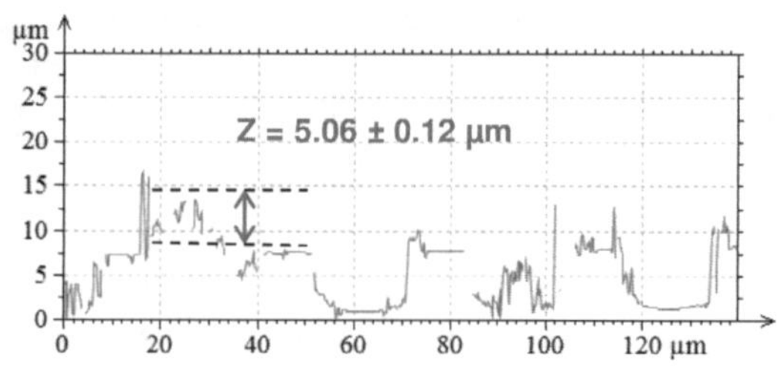

Figure 7. Comparison of 2D profiles of hydroxyapatite layer deposited on etched squares in Si measured using (a) the Leitz-Linnik microscope and (b) the Zygo NewView 7200 microscope.

Nonetheless, by careful positioning of the cursor lines in the two measurements, the two profiles obtained (figure 7) can be seen to be fairly similar. A comparison of the heights at the places indicated is shown in Table 2. The uncertainties in measurement were calculated as in section 4.1.

Table 2. Comparison of depth measurements of etched squares (figure 7) using each technique.

Technique	Average measured depth	Estimated depth uncertainty	Estimated percentage uncertainty
Leitz-Linnik	5.05 µm	± 0.20 µm	± 4 %
Zygo NewView 7200	5.06 µm	± 0.20 µm	± 4 %

The higher values of uncertainty in measurement on the HA layer compared with the results on the bare silicon are due to the inaccuracy from the placing of the measurement cursor in the image and the large lateral variation in roughness of the HA.

5. Conclusions

The aim of this work was to develop a technique of sample marking that enabled the topographic surface measurement of the same place on a given sample using different microscopes. Such a method allows the comparison of measurements at the same place so as to be able to reveal the specific artefacts and performance limitations of each microscopy method so as to be able to improve their measurement precision. After studying the laser marking of cement samples, the use of evaporated Al marks on silicon and photolithographic etching of a numbered, multi-scale square pattern on silicon, the latter technique was chosen for the purposes of the study. The etch pattern allowed the successful identification of the same zone of analysis under different microscopes for surface measurement. The microscopes used were interference microscopy (of different types), AFM and ESEM.

Two silicon samples were prepared with the etched patterns and one was covered with a partial layer of HA. Measurement of the same zone of 20 µm and 10 µm sided squares and comparison of the the 3D measurements and 2D line profiles showed a broad similarity in the results and confirmation that the same zone had been identified under each microscope. A more detailed study showed variations in the measurement of step heights between the different techniques, of up to 1.6 % between the interference techniques on the bare silicon sample, mainly due to errors from the piezo positioning, the envelope algorithm and the reference mirror flatness. This uncertainty increased to 7 % for the AFM measurements, due to the non-linearity over the depth range measured and noise from acoustic vibrations.

The difference between the results from the two interference techniques increased to 4 % for measurements on the rough HA layer, mainly due to the additional errors from the lateral positioning of the measurement cursor on a rough surface. Edge effects are also clearly visible due to probe/material interactions in both interferometry (optical effects) and AFM (convolution with the probe tip shape).

The overall conclusions of this work are that while topographic surface measurements using different techniques of microscopy give broadly similar results, variations exist between the exact details and roughness values. Differences can be observed between the absolute depth measurement values and between the details on step edges and rough surfaces. The use of different techniques on exactly the same zone in areal measurement is a useful solution for revealing the performance, artefacts and limits of each technique in real life applications. We have thus shown that great care needs to be taken in choosing the right technique for measuring different types of samples as well as in the correct setup of the specific measuring parameters and calibration of the z measurements. Moreover, this work is a starting point in the very broad research topic, requiring further work to reveal the different parameters that must be mastered and improved at the level of characterization, and to find the most appropriate choice of marking technique for a given material to lead to accurate areal roughness measurements.

Acknowledgements
Thanks are given to C. Heitz for the growth of the hydroxyapatite layers and to C. Munzer for providing the cement paste sample.

References

[1] Leach R K 2010 *Fundamental principles of engineering nanometrology* (Elsevier, Amsterdam)
[2] Leach R K 2011 *Optical measurement of surface topography* (Springer, Berlin)
[3] ISO/CD 25178-600 2013 Geometrical product specifications (GPS) – surface texture: areal – part 600. Nominal characteristics of areal surface topography measuring instruments (Geneva: International Organization for Standardization)
[4] Leach R K Giuscaa C L and Coupland J M 2012 *Proc. SPIE* **8430** 84300H 1
[5] Ottevaere H, Schwider J, Kacperski J, Steinbock L, Weible K, Kujawinska M and Thienpont H 2008 *Proc. SPIE* **6995** 69950J.1-69950J.9
[6] Montgomery P C, Salzenstein F, Montaner D, Serio B and Pfeiffer P 2013 *Proc. SPIE* **8788** id. 87883G 11
[7] Benatmane A Montgomery P C Fogarassy E and Zahorsky D 2003 *Applied Surface Science* **208-209** 189
[8] Schmit J Reed J Novak E and Gimzewski J K J 2008 *Opt. A: Pure Appl. Opt.* **10** 064001 1
[9] Montgomery P C and Gall-Borrut P 1996 *Institute of Physics Conference Series* **149** 147
[10] Coupland J M and Lobera J 2010 *Strain* **46** 69
[11] Pecheva E Montgomery P Montaner D Pramatarova L 2007 *Langmuir* **23** 3912
[12] Deng J W Dua K Wua B Sui M L 2013 *Intermetallics* **34** 75
[13] Galeano J A Sandoz P Gaiffe E Launay S Robert L Jacquot M Hirchaud F Pretet J L and Mougin C 2011 *Biomedical Optics Express* **2** 1307
[14] Abdurrochman A Lecler S Fontaine J Mermet F Meyrueis P Tumbelaka, B Y and Montgomery P C 2014 *Proc. SPIE* **9135** 913523 1
[15] Tsukamoto M Kayahara T Nakano H Hashida M Katto M Fujita M Tanaka M and Abe N 2007 *J. Phys.: Conf. Ser.* **59** 666
[16] Guellil M Montgomery P C Pfeiffer P & Serio B 2014 *Proc. SPIE* **9132** 913204 1
[17] Pecheva E Pramatarova L Maitz M F Pham M T and Kondyuirin A 2004 *Appl. Surf. Sci.* **235** 176
[18] Kokubo T Kushitani H Sakka S Kitsugi T and Yamamuro T 1990 *J. Biomed. Mater. Res.* **24** 721
[19] Salzenstein F Montgomery P C Montaner D and Boudraa A O 2005 *Eurasip J. Appl. Signal Process.* **17** 2804
[20] Harasaki A and Wyant J C 2000 *Appl. Opt.* **39** 2101

Journal of Physics: Conference Series **558** (2014) 012006 doi:10.1088/1742-6596/558/1/012006

The Hubbard model: useful for stretchable nano-materials?

V Celebonovic

Institute of Physics,Univ.of Belgrade,Pregrevica 118,11080 Zemun-Belgrade,Serbia

E-mail: `vladan@ipb.ac.rs`

Abstract. The aim of this lecture is to test the consequences of stretching and doping on the electrical conductivity of the Hubbard model. It is motivated by a recent experiment in which conductivity of stretchable nanomaterials was studied,and which showed that conductivity tends to zero but remains finite for considerable values of the strain. The Hubbard model reproduces such behavior for certain values of the parameters,and for the hopping,band filling and lattice "constant" dependent on the strain.

1. Introduction

According to standard condensed matter physics, some materials are electrical conductors because of the overlap of electronic orbitals on adjacent atoms. As a consequence, diminishing the overlap (for example by stretching a material) leads to a decreasing overlap of the orbitals and ultimately to a decrease of the conductivity.

On the other hand, certain technological issues, as varied as flexible conductors,t-shirts and cardiostimulating implants [1],[2] require materials which are conducting and stretchable, and whose conductivity remains as high as possible with increasing applied strain. The interest which the scientific community and the industry have in such materials is illustrated by a singular detail: a search on `http://www. sciencemag.org` gave 6749 papers published on the topic "stretchable electronics" from 2006 to the end of August 2014.

The aim of this lecture is to try to discuss the problem of "stretchable conductors" (SC) within a particular well known model in condensed matter physics - the so called Hubbard model (HM). No attempt will be made to fit any particular set experimental data,such as [1]. Practically speaking, it will be attempted to discuss the consequences of stretching and the dependence on stretching of various parameters of the HM on its electrical conductivity. The lecture is divided into several sections. As the calculations pertain to nanomaterials, the next section starts with a definition of nanomaterials and a brief description of the experiment which motivated this work [1]. The third part is devoted to a brief reminder of the HM, and the fourth part to a discussion of the influence of stretching and doping on the conductivity of the HM. The final part contains the conclusions.

2. Definition and experiments

2.1. Definition

According to [3] nanomaterials are in the simplest way defined as materials with at least one external dimension in the size range from approximately 1-100 nanometers. This is clear and simple, but it does not take into account materials which have only a size fraction in the nano scale, or which contain agregates of nano particles [4]. In the present paper the definition proposed in [2] will be considered as satisfactory.

2.2. A recent experiment

In the experiment analyzed in [1], the possibility of making a stretchable conductor consisting of polyurethane containing spherical gold nanoparticles deposited in two different ways was demonstrated. The nanoparticles had a mean diameter of 13.0 ± 0.3 nm, and they were citrate stabilized which made them negatively charged. Polyurethane, which is positively charged, was used as a polymeric partner for these particles, and thin films were made by two different techniques : layer-by layer deposition (LBL) and vacuum-assisted flocculation (VAF). The thickness of the films thus obtained was 2.0 ± 0.2 μm. Films obtained by the two methods had different properties, but lamination led to an increase of the conductivity, and improved their stretchability. Strain applied to an object of initial length l_0 so that it achieves length l is defined by $\epsilon = (l - l_0)/l_0$. For example, a strain of $\epsilon = 0.6$ applied uniaxialy on a $5LBL$ film led to a decrease of conductivity by two orders of magnitude. This is illustrated on figure 1, taken from [1].

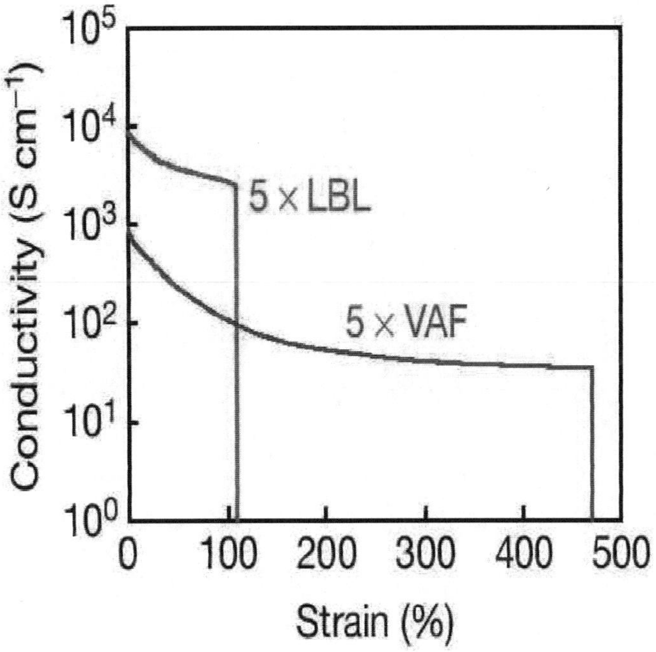

Figure 1. Conductivity as a function of strain for 5 LBL and 5 VAF films.

However,as noted in [1], the values of the conductivity obtained in this experiment for high strain were higher than the corresponding values obtained for carbon-nanotube-based materials. Full details of the experiment are avaliable in [1], but it clearly indicates the problem: how do these materials manage to stay in their conducting phase at such high values of strain? In the remainder of this paper it will be attempted to analyze this problem by using the (HM).

3. The Hubbard model

The HM has been treated in considerable detail in the lectures given by the present author at two earlier sessions of this school [7,8].Only the essential detail will be repeated here. The HM was proposed by the British physicist John Hubbard during his work on the microscopic theory of the metal to insulator (MI) transition [9]. The Hamiltonian by which the HM is defined contains only two terms: the "free" kinetic term H_0 and the "interaction" term H_I

$$H = H_0 + H_I \tag{1}$$

Complexity quickly becomes visible when this Hamiltonian is written completely. In two dimensions,using the second quantisation formalism, it has the form:

$$H = -\sum_{ij}\sum_{\sigma} t_{ij} c_{i\sigma}^{+} c_{j\sigma} + \frac{1}{2}\sum_{ijkl}\sum_{\sigma\sigma'} < ij|v|kl > c_{i\sigma}^{+} c_{j\sigma'}^{+} c_{l\sigma'} c_{k\sigma} \tag{2}$$

In spite of simplifying assumptions introduced by Hubbard, and the fact that it was proposed nearly 60 years ago, the HM has so far been solved analytically only in the $1D$ case [10]. Extensive work is going on aiming at solving the so called extended HM in $2D$ [11]. The interest of this model for the condensed matter community can be appreciated by the number of papers which use it in some way. A search at `http://prola.aps.org` performed at the time of this writing (middle of July 2014) with the keywords "Hubbard model" has given 4858 papers containing those words in the title or abstract published in Phys.Rev.Lett. and Phys.Rev.B in the period 2000.-2014.

Knowing the Hamiltonian of a system opens the possibility of calculating its electrical conductivity. When the Hamiltonian commutes with the current operator j, this calculation is not complicated. Problems can arise when $[j, H] \neq 0$, which is the usual case in realistic calculations. There is no general "prescription" for such situations.

One of the methods by which it can be done is the so called "memory function" method developed in the early 1970s as a consequence of previous work by Kubo. It was recently reviewed in [5]. The algorithm for the calculation of the conductivity within this method consists of the following equations [6]:

$$\chi_{AB}(\omega) =<< A; B >>= -i \int_0^{\infty} \exp i\omega t < [A(t), B(0)] > dt \tag{3}$$

where $A = B = [j, H]$, j denotes the current operator, and

$$\sigma(\omega) = i\frac{\omega_P^2}{4\pi\omega}[1 - \frac{\chi(\omega)}{\chi(0)}] \tag{4}$$

where σ denotes the electrical conductivity, ω is the complex frequency, ω_P is the plasma frequency, $\chi(\omega)$ is the frequency dependent susceptibility and $\chi(0)$ the static susceptibility. Some details of the derivation of the expression for the conductivity are given in [7-8] and references given there. The result, for the case of the $1D$ HM is

$$\sigma_R(\omega) = (1/2\chi(0))(\omega_P^2/\pi)[\omega^2 - (bt)^2]^{-1}(Ut/N^2)^2 \times S \tag{5}$$

and the symbol S denotes the following function

$$\begin{aligned} S = 42.49916 \times (1 + \exp(\beta(-\mu - 2t)))^{-2} + 78.2557 \times \\ (1 + \exp(\beta(-\mu + 2t\cos(1 + \pi))))^{-2} + (bt/(\omega + bt)) \times \\ (4.53316 \times (1 + \exp(\beta(-\mu - 2t)))^{-2} + \\ 24.6448(1 + \exp(\beta(-\mu + 2t\cos(1 + \pi)))))^{-2}) \end{aligned} \tag{6}$$

where U is the Hubbard U, $b = -4 \times [1 + \cos(1 - \pi)] = -1.83879$, N is the number of lattice sites, t is the hopping and μ denotes the chemical potential of the electron gas on a $1D$ lattice given by [7]

$$\mu = \frac{(\beta t)^6 (ns - 1)\,|t|}{1.1029 + .1694(\beta t)^2 + .0654(\beta t)^4} \tag{7}$$

In eq.(7) the symbol s denotes the lattice constant and n the band filling. The immaginary part of the conductivity is given by

$$\sigma_I = \frac{\omega_P^2}{4\pi\omega}\left(1 - \frac{\chi_R}{\chi(0)}\right) \tag{8}$$

where $\chi_R(\omega)$ denotes the real part of the susceptibility,given by

$$\chi_R(\omega) \cong [128U^2 t^3 cos^2((1 - 2\pi)/2)]/[1 + \exp(\beta(-\mu - 2t)))^2 N^4 (\omega + 2bt)^2)] \tag{9}$$

The plasma frequency ω_P is given by $\omega_P^2 = 4\pi e^2 n_e/m$, and e, n_e, m are the electron charge, number density of the background electron gas approximated as a jellium [6] and the electron mass respectively, while β is the inverse temperature . This calculation pertains to the $1D$ HM.

The simplest $2D$ case,a square lattice of side length a can be treated as an extension of these results. If the lattice sides are denoted by x and y,the electric current flowing through such a system can be expressed as

$$\vec{j} = j_x \vec{e_x} + j_y \vec{e_y} \tag{10}$$

where $\vec{e_x}$ and $\vec{e_y}$ are unitary vectors of the two lattice axes.The total current is given by

$$j^2 = j_x^2 + j_y^2 \tag{11}$$

As by definition $j = \sigma E$ where σ is the electrical conductivity and E the electric field,and assuming that $E_x = E_y = E$, it finally follows that

$$\sigma_{2D}^2 = \sigma_x^2 + \sigma_y^2 \tag{12}$$

It is assumed here that the conductivities along the two axes are mutually independent.

4. Applications

4.1. Constant hopping

Equations (3)-(7) give the possibility of calculating the electrical conductivity of a $1D$ HM. Some examples of the results of such calculations are given in [12] and later work. The problem is the choice of numerical values of various parameters. The following values of the parameters were used: $N = 20$, $U = 4t$, $\chi_0 = 1/3$, $\omega_P = 12t$, $\omega = 2.9t$, $b = -1.83879$, $\beta = 11600/T$, $s = s_0(1 + \epsilon)$ and ϵ is the strain.

Figure 2 shows the electrical conductivity as a function of strain for $T = 116K$ and two arbitrarily chosen values of the band filling: $n = 0.9$ and $n = 1.2$. It was assumed that $t = 0.01eV$ and $s_0 = 1$. The conductivity was normalized to 1 at $T = 116K$,$n = 0.8$,$s_0 = 1$,$\epsilon = 0$. Data presented on the figure refer to the $1D$ HM. The conductivity of a $2D$ HM, assuming that the lattice is square, can be calculated by eq.(10). In the simplest case of a square lattice and equal conductivities along the two lattice axes, the total lattice conductivity is

$$\sigma_{2D} = \sigma_x \sqrt{2} \tag{13}$$

where σ_x denotes the conductivity along one of the lattice axes.

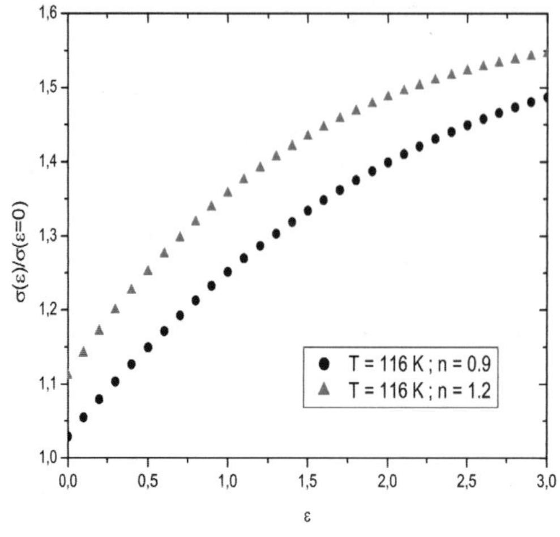

Figure 2. Conductivity as a function of strain for T = 116 K,n=0.9 and n = 1.2.

Figure 2 shows the electrical conductivity of a $1D$ HM as a function of the applied strain for two values of the band filling Clearly, the function $\sigma(\epsilon)$ is an increasing function of the strain, which could have highly interesting practical implications. However, it was assumed in the calculation that t is a constant, while in physical reality t depends on the strain. In the remainder of this paper the dependence of t on the strain will be taken into account.

4.2. Strain dependence of the hopping and the band filling

In order to make a physically realistic calculation of the electrical conductivity of a $1D$ HM as a function of the applied strain, one must take into account the strain dependence of the hopping energy t and of the band filling n. The physical origin of t is the overlap of the electronic wave functions in atoms on adjacent lattice sites,so it can be expected that it depends on the lattice constant. The band filling is defined as $n = N_e/L$, where N_e is the number of electrons and L is the length of the lattice. Due to strain, $L(\epsilon) = L_0(1+\epsilon)$,so it follows that $n(\epsilon) = n_0/(1+\epsilon)$. The mathematically simplest possible example of the dependence of t on the applied strain has the form

$$t = t_0(1-\epsilon) \tag{14}$$

where t_0 denotes the value of the hopping under strain zero. Performing the calculation of the conductivity for $n = 0.9$,$T = 116K$, and two values of the hopping t, normalized to 1 at the same point as for figure 2, one gets figure 3. Note that a larger value of t leads to a bigger value of the conductivity.

The conductivity is clearly a decreasing function of the applied strain,which is in agreement with the experimental results presented in figure 1. Calculating the overlap of electronic wave functions in arbitrarily complex molecules (or atoms) under arbitrary conditions would be a complicated task. However, a realistic approximate solution to the problem can be found by

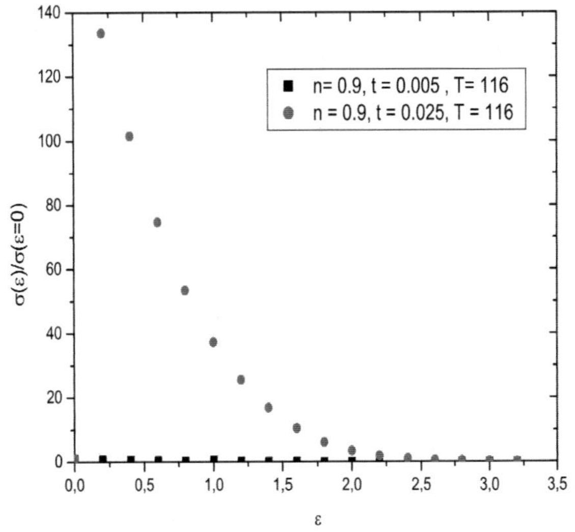

Figure 3. Conductivity as a function of strain for T = 116 K,n=0.9 and two values of t.

standard quantum mechanics. Using results on the overlap of atomic wave functions (for example [13]), it follows that the dependence of t on the interatomic distance r can be expressed as

$$t = t_0 \times [1 + r + (1/3)r^2] \times \exp(-r) \tag{15}$$

where the distance r is measured in the units of the Bohr radius a_0. This relation can be reformulated as a function of the form $t(\epsilon)$, where ϵ denotes the strain. Using the definition of strain it follows that

$$t = t_0 \exp[-(1+\epsilon)\frac{r_0}{a_0}] \times [1 + (\frac{r_0}{a_0})(1+\epsilon) + (1/3)(\frac{r_0}{a_0})^2(1+\epsilon)^2] \tag{16}$$

Developing into series, and retaining only the first order terms in ϵ one gets

$$t \cong 0.86 t_0 (1 - 0.286\epsilon) \tag{17}$$

which is approximately similar to the mathematical form of eq.(14). The conductivity of a $1D$ HM as a function of the applied strain, but in a low temperature case, is presented on figure 4

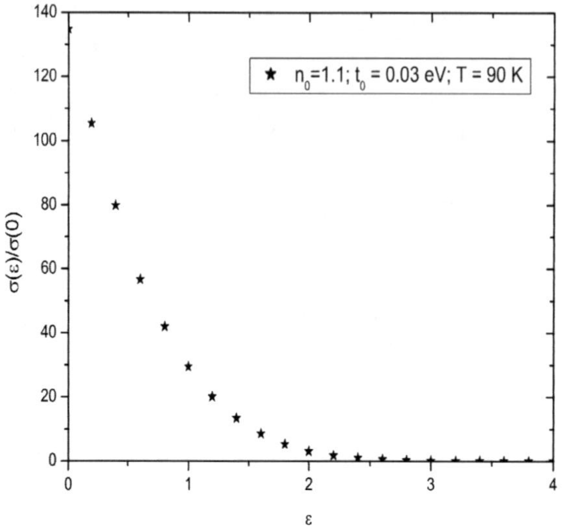

Figure 4. Conductivity as a function of strain for T = 90K,n=1.1.

Figure 5 shows a comparison of conductivities for two values of the temperature. Note the difference of nearly two orders of magnitude between the values of conductivity with no strain applied and at the highest values of strain for which the calculation was performed.

Figure 6 represents the conductivity as a function of applied strain for a square lattice.

4.3. The two dimensional case

All the calculations discussed so far are relevant to the $1D$ HM. This is a good illustration of the procedure,but the real "challenge" is the conductivity of a $2D$ model,and its dependence on the applied strain. For the case of a square lattice this can be determined using eq.(12). In the case of a material having a small value of the ratio σ_y/σ_x, it can easily be shown that the conductivity reduces just to σ_x.

However, in the case of material characteristics along the two axes being similar, eq.(12) can give interesting results. One such example is shown on figure 7, obtained with $n_{0x} = 1.1$,$n_{0y} = 1.3$, $T = 150K$ and $t_0 = 0.01eV$ along both lattice axes. Note that the ratio given on the vertical axis is $\sigma_{2D}(\epsilon)/\sigma_x$.

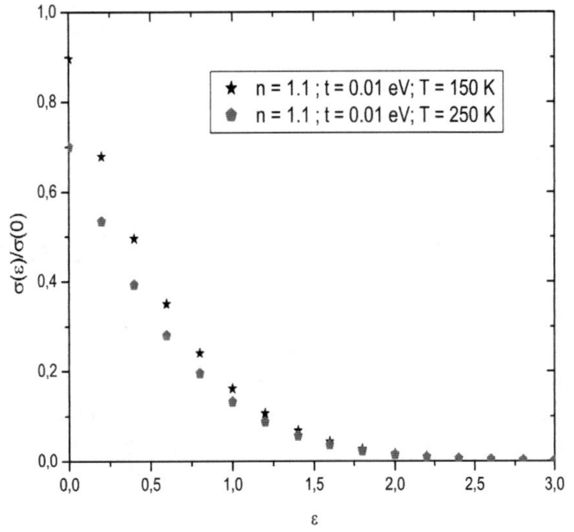

Figure 5. The function $\sigma(\epsilon)$ for two different values of T.

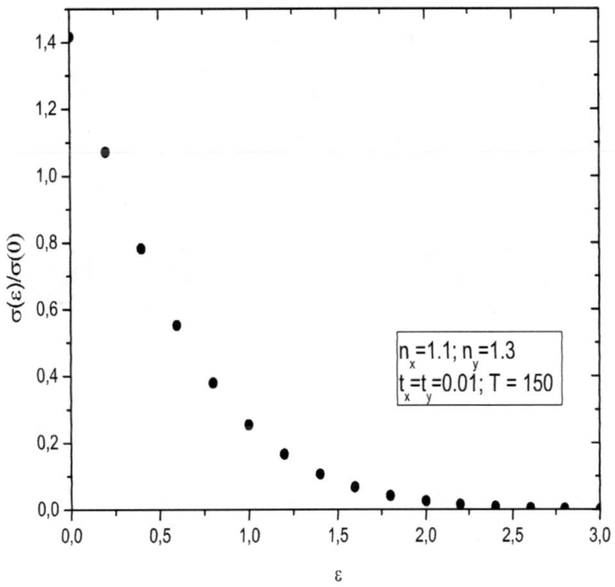

Figure 6. The function $\sigma(\epsilon)$ for a $2D$ square lattice.

Journal of Physics: Conference Series **558** (2014) 012006 doi:10.1088/1742-6596/558/1/012006

5. Discussion and conclusions

The basic aim of this paper was to test the applicability of the Hubbard model to the nanomaterials used for stretchable electronics. The question from the title does not have a simple answer.

In the simplest case, assuming that all material parameters are constant in spite of the fact that it is being stretched, one gets a strange result - the conductivity increases with the applied strain. This is obviously unphysical, and can be traced back to the assumption that material parameters are independent of strain.

Taking into account the strain dependence of t and n (the hopping and the band filling) gives results in agreement with the experimental data. This agreement means that the HM remains conductive for high values of the strain applied. However, more experimental data should be taken into account before stating that the HM is applicable to stretchable electronic materials. The HM in $2D$ behaves in the same way.

Although the general behavior of $\sigma(\epsilon)$ agrees with the experimental data, the shape this function for small values of ϵ does not. This will be investigated in the future. A preliminary possibility is that the strain dependence of some parameter has not been taken into account.

Acknowledgments
This paper was prepared within the research project 174031 of the Ministry of Education,Science and Technological Development of Serbia

References

[1] Kim Y, Zhu J, Yeom B, Di Prima M, Su X, Kim J-G, Yoo S J, Uher C and Kotov N 2013 *Nature* **500** 59
[2] Park M, Park J and Jeong U 2014 *Nano Today* **9** 244
[3] `http://web.stanford.edu/dept/EHS/prod/researchlab/IH/nano/what_are_nanomaterials.html`
[4] Kreyling W G, Semmler-Behnke M and Chaudhry Q 2010 *Nano Today* **5** 165
[5] Balucani U, Lee M H, Tognetti V 2003 *Phys Rep* **373** 409
[6] Götze W and Wölfle P 1971 *Phys Rev* **B6** 1226.
[7] Celebonovic V 2010 *J.Phys Conf Ser* **253** 012004
[8] Celebonovic V 2012 *J.Phys Conf Ser* **398** 012009
[9] Hubbard J 1963 *Proc R Soc London* **A 276** 238
[10] Lieb E H and Wu F Y 1968 *Phys Rev Lett* **20** 1445
[11] Harir S, Bennai M and Boughaleb Y 2008 *Physica Scripta* **78** 025701
[12] Celebonovic V 1997 *Phys Low-Dim Struct* **3/4** 65
[13] Gasiorowicz S 1974 *Quantum Physics*, (New York:John Wiley Inc.)

X-ray sensing materials stability: influence of ambient storage temperature on essential thermal properties of undoped vitreous selenium

D Tonchev[1,4], H Mani[2], G Belev[3], I Kostova[4] and S Kasap[1]

[1]Department of Electrical and Computer Engineering, University of Saskatchewan, Saskatoon, SK, S7N 5A9, Canada
[2]Analogic Canada Corporation, 4950 Levy Street, Saint-Laurent, Quebec H4R 2P1 Canada
[3]Canadian Light Source (CLS), University of Saskatchewan, Saskatoon, SK, S7N0X4 Canada
[4]Department of Chemical Technology, Plovdiv University "Paisii Hilendarski", Plovdiv 4000 Bulgaria

E-mail: dan.tonchev@usask.ca

Abstract. Amorphous selenium (a-Se) is currently used in x-ray image detectors as an x-ray photoconductor. Normally a-Se films used in device applications are fabricated by the evaporation of vitreous bulk material loaded into boats in a typical vacuum deposition system. The resistance against crystallization is an important factor in both film and bulk forms of a-Se. Previous work has indicted that the resistance to crystallization is surprisingly more pronounced around 35 °C [1]. In this work we have therefore examined the essential thermal properties of vitreous selenium (99.999%) samples that have been stored at different temperatures. The thermal characterization experiments involved a series of DSC (Differential Scanning Calorimetry) measurements in which have monitored the glass transition and melting endotherms, and the crystallization exotherm in heating–cooling–heating scans. In DSC experiments, a sample would be heated to a temperature above the melting temperature, equilibrated, then cooled at a fixed rate down to 20 °C, then equilibrated and finally scanned again under a heating schedule. The samples were isothermally stored at temperatures corresponding to 18, 35 and 55 °C. The thermal analysis results show that there are distinct differences in the thermal properties. We have examined the stability in terms of the difference in the crystallization onset temperature (T_c) and the onset of glass transition temperature (T_g). We also examined the Hruby coefficient (K_{gl}) of these samples, that is $K_{gl} = (T_c - T_g)/(T_m - T_c)$ where T_c is the crystallization onset temperature and T_m is the melting onset temperature. We have found K_{gl} to depend on the storage temperature. Surprisingly, we observed that the Hruby coefficient is actually larger at 35 °C compared to the values at 18 and 55 °C.

1. Introduction

Amorphous selenium (a-Se) is currently used in X-ray image detectors as an X-ray photoconductor [1]. Normally a-Se films used in device applications are fabricated by the evaporation of vitreous bulk material loaded into metal boats in a conventional vacuum deposition system. For reproducible a-Se based devises, it is highly desirable that the evaporated and deposited film material is fully vitreous; the

Journal of Physics: Conference Series **558** (2014) 012007 doi:10.1088/1742-6596/558/1/012007

bulk material is usually vitreous a-Se pellets or shots prepared by melt quenching techniques. We therefore need to monitor the amorphous state of the starting bulk material during storage. The resistance against crystallization is an important factor in both film and bulk forms of a-Se. Previous work on Se photodarkening has indicated that an increased resistance to crystallization appears at around 35 °C [2, 3]. In fact, according to Reznik and coworkers, high-field avalanche rushing photodetector (HARP) video tubes provided the best operation at 35 °C, which avoid defect related blemishes in the image; these blemishes occurred during operation only at room temperature (RT) or below. It was also been found that only when Se-based film devices are kept close but below the glass temperature at 35°C, the photodarkening is totally reversible, whereas at RT there is a reversible and an irreversible component. The reversible photodarkenig recovers after the cessation of photoexcitation whereas the irreversible part does not; the relaxation of these changes is characterized by different time constants. In this work we have examined the essential thermal properties of vitreous selenium (99.999%) bulk samples that have been stored at different temperatures without illumination, and discuss the results in term of possible structural and thermodynamic changes in these materials.

2. Experimental

We used pure a-Se (bulk granules of about 30-40 mg weight each) obtained from former Noranda Corporation, Canada, 20 years ago. The a-Se is rated as PR-grade (photoreceptor grade with purity 99,999%). The starting material was in the amorphous state with stable (aged) properties. We kept a large number of samples in a temperature controlled environment at 55 °C, 17.5 -18 °C and 35 °C for up to 5 months. The humidity was maintained constant at around 40% all the time, and measured by a Therminder device. The effect of water impurity on the crystallization of thin selenium film materials was studied by Legros et al.[4].

Differential scanning calorimetry (DSC) measurements were performed using TA Instruments DSC 2910 with an attached RCS (Refrigerating Cooling System) unit that allows heating and cooling. A piece of granule of the sample (about 20 mg in mass) was weighed and sealed in Al hermetic pan and then scanned under a schedule that has 3 scans: heating, cooling and heating again, i.e. an H-C-H schedule. The heating starts at 20 °C up to 250 °C, then cooling down to 20 °C and heating immediately starts again to 250 °C. The evaluation of thermal analysis results was performed by TA Instruments Universal Analysis (UA) software. Error bars are inside plots symbols.

From the DSC thermograms, we obtained T_g, T_c and T_m. The stability S of these glasses was assessed using the relationship $S = T_c - T_g$. The measured glass transition, crystallization and melting temperatures were then used in the Hruby equation [5]. The Hruby parameter (K_{gl}) is a numerical measure of the glass-forming tendency and is given by

$$K_{gl} = \frac{T_c - T_g}{T_m - T_c}, \quad (1)$$

where T_c is the crystallization onset temperature, T_g is the glass transition onset temperature and T_m is the melting onset temperature.

3. Results and discussion

Based on previously published papers regarding the importance of 35 °C for a-Se, we started the temperature controlled environment experiments with the above storage temperature. As shown in figure 1, that glass transition temperature (T_g) does not change much after the initial first drop. However, the glass stability $S = T_c - T_g$ increases because of the higher glass crystallization temperature as is shown in figure 2. These results affect the Hruby parameter, which shows an increase with the storage time (figure 3). This is an indication that the material is becoming more stable after about one month of storage time, following which the pure a-Se material stability does not change significantly (DSC scans were preformed periodically up to 5 months of storage time).

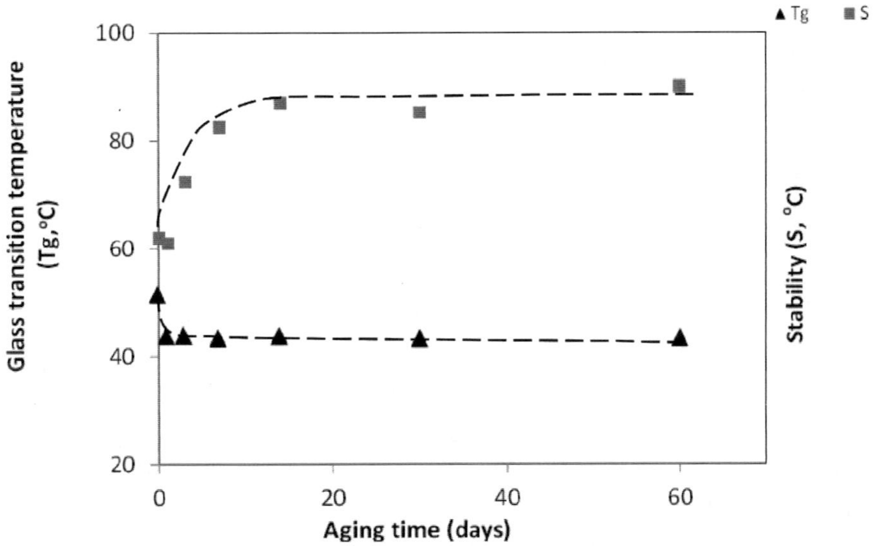

Figure 1. Glass transition temperature (T_g) and glass stability (S) of pure a-Se vs aging time when the storage temperature is kept constant at 35 °C in dark (dashed lines are guides to the eye only).

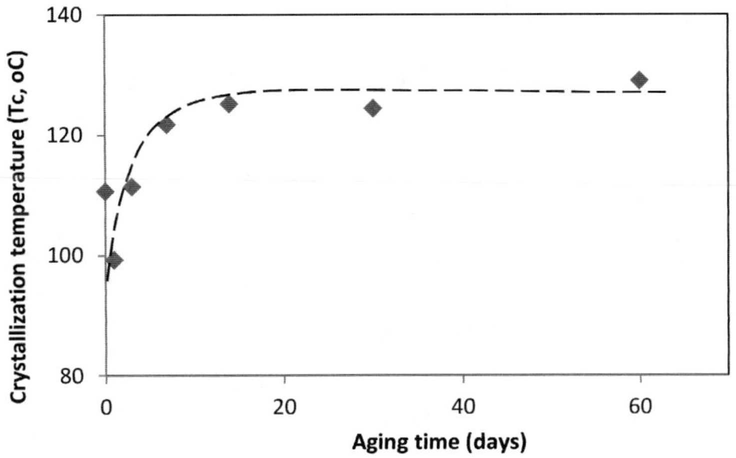

Figure 2. Crystallization onset temperature (T_c) of pure a-Se vs. aging time when the storage temperature is kept constant at 35 °C in the dark (the dashed line is a guide to eye).

We compared these results above at storage temperature 35°C with the same pure a-Se bulk samples but kept at 55 °C and 17.5 °C - 18 °C of storage temperature. The T_g remains unchanged as a function of anneal time in all cases corresponding to the storage temperature of 17.5 °C-18°C and also at RT. At storage temperatures higher than the RT, or 35 °C and 55 °C, the T_g initially drops, which is about 7-8 °C, following which T_g becomes practically unchanged as shown in figure 4. Relaxation enthalpies follow the same tendency as well when the crystallization enthalpies do not change i.e. these particular enthalpies are independent of the storage time.

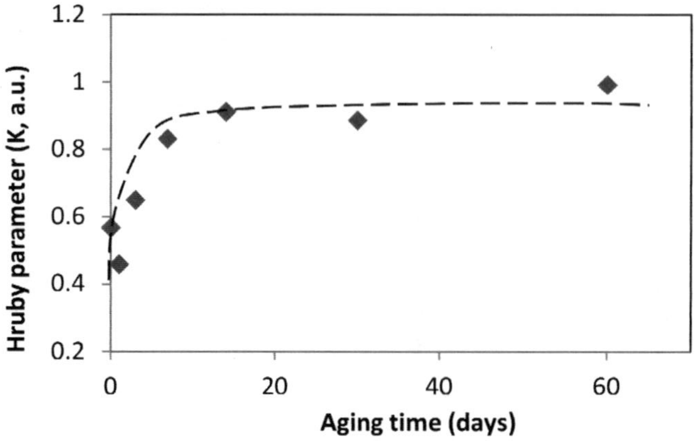

Figure 3. Hruby parameter (K_{gl}) of pure a-Se vs. aging time when the storage temperature is kept constant at 35 °C in the dark (the dashed line is a guide to the eye).

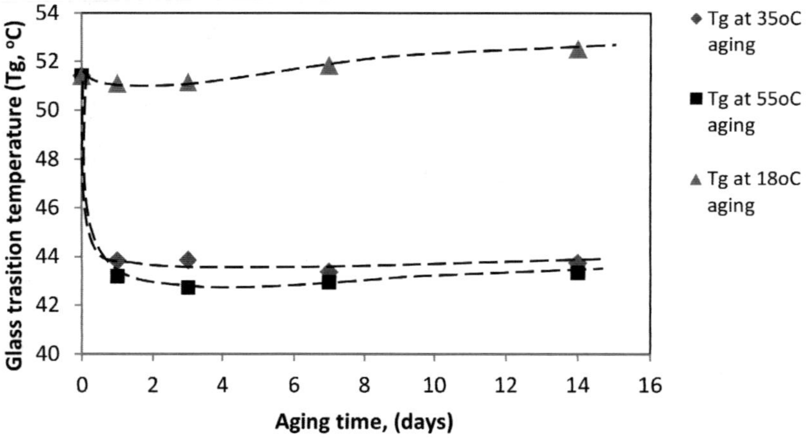

Figure 4. Glass transition temperature (T_g) of pure a-Se vs aging time when the storage temperature is kept at 35 °C, 55 °C and 17.5 °C in the dark (the dashed lines are guides to the eye).

Looking at the crystallization temperatures as shown in figures 4 it is clear that the only increase of crystallization onset temperatures is observed when the samples are kept at 35 °C; and are practically unchanged when the samples are stored at 17.5 °C – 18 °C and RT. The crystallization temperature shows a tendency to decrease at 55°C storage temperature.

Because the calculations of the glass stability and Hruby parameter (K_{gl}) are based on DSC results, it is expected that the only increase of these parameters will take place when the samples are stored at 35 °C as shown in figure 5 where it can be seen that at 17.5 °C — 18 °C, the crystallization temperature is almost unchanged but drops at 55 °C of storage temperature.

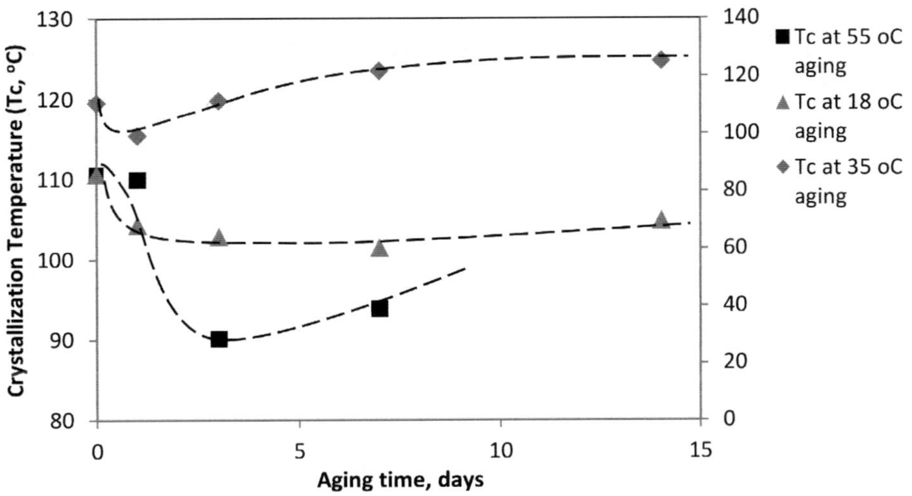

Figure 5. Crystallization onset temperatures (T_c) of pure a-Se vs. aging time when the storage temperature is kept at 35 °C, 55 °C and 17.5 °C – 18 °C in the dark (the dashed lines are guides to the eye).

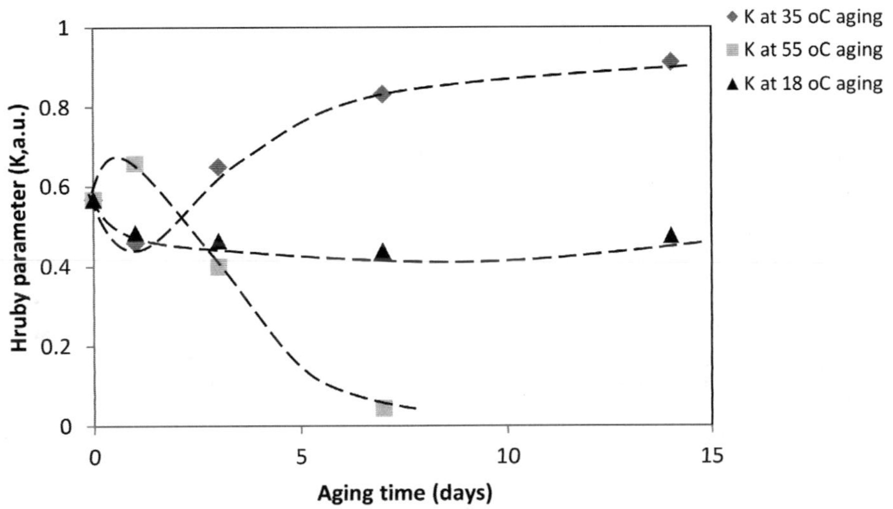

Figure 6. The Hruby parameter (K_{gl}) of pure a-Se vs. aging time when the storage temperature is kept at 35 °C, 55 °C and 17.5 °C — 18 °C in the dark (the dashed lines are guides to the eye).

It is well know the higher is the Hruby parameter, the better is the glass forming tendency. This is also a way to estimate a given material's resistivity against crystallization. The higher is the difference between the glass onset crystallization temperature T_c and the glass transition temperature T_g, and the lower the difference between the melting onset temperature T_m and the crystallization onset temperature, T_c the material is considered to be more resistive to crystallization according to Hruby theory. There is a correlation between the glass forming ability, glass stability and materials resistance to crystallization. For a comparison of glass stability parameters, the reader is referred to the discussions given by Nascimento et al [6].

The DSC overlay thermograms presented at figures 7 and 8 show the evolution of crystallization upon cooling and heating for a-Se samples kept at 35 °C and 55 °C after 1 and 3 days of storage time i.e. at a temperature below but higher than the RT (35 °C) and at a temperature above the glass transition temperature of a-Se (55 °C). It is immediately clear that the DSC scan signatures are totally different. The crystallization enthalpy at cooling for samples kept at 35 °C decreases, i.e. the material becomes more resistive to crystallization. For samples that are kept at 55 °C, the crystallization enthalpy at cooling increases with storage time i.e. material is less resistive to crystallization.

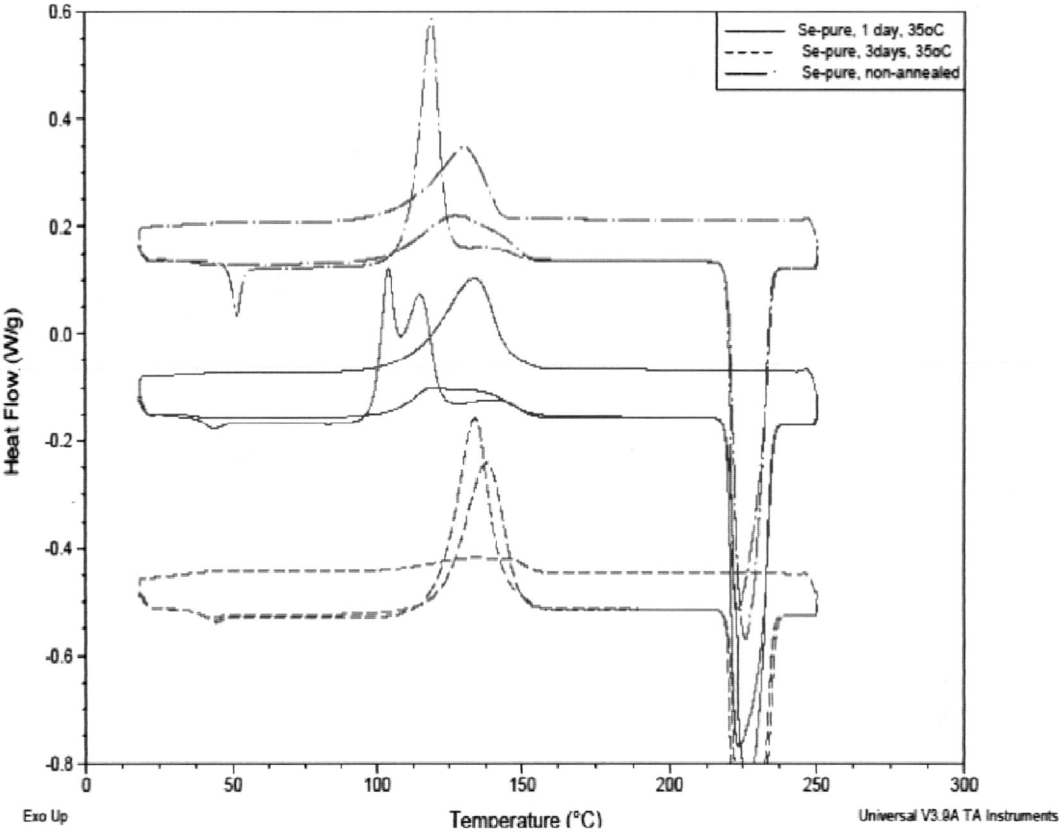

Figure 7. The evolution of DSC thermograms from H-C-H scans after the samples have been kept at 35 °C for different durations of storage (aging) time.

The type of behaviour of glass transformation properties of a-Se stored at different temperature can be explained by well-established models for enthalpy relaxation; and there are many previous works describing the phenomena of structural changes with storage time of amorphous materials including for a-Se. For example, in the earlier works of Stephens [7] reports on the viscosity and structural relaxation changes of a-Se vacuum evaporated films.

The thermal properties of pure a-Se such as the glass transition and crystallization depend on the synthesis and thermal history. Once quenched, the glass forms a network under stress, that is, the structure has a higher enthalpy than the extrapolated liquid or equilibrium enthalpy. The excess enthalpy is released through structural relaxation. The structural relaxation time that controls the rate of relaxation in the enthalpy depends on the temperature and also on the instantaneous enthalpy, (see for example reference [8]). For samples aged at 35 and 55 °C, there is an initial drop in T_g as can be seen in figure 4. This is due to the fact that at these storage temperaturesnthe equilibrium enthalpy is higher than that at the RT, and the structure also relaxes more quickly than at the RT. Thus, the observed T_g is lower for the sample aged at 35 °C and still lower for the sample aged at 55 °C.

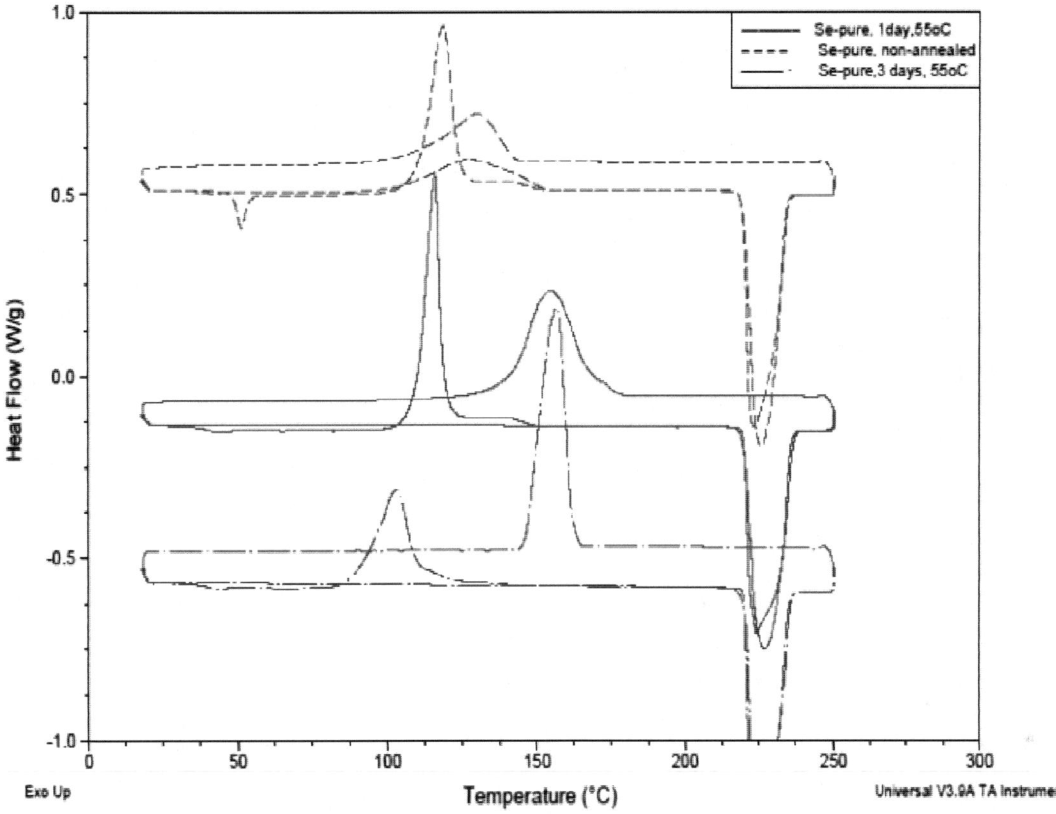

Figure 8. The evolution of DSC thermograms from H-C-H scans after the samples have been kept at 55 °C for different durations of storage (aging) time.

The most remarkable result in the present work is on the changes in the crystallization behavior.The crystallization process is a two-step process that has nucleation and growth. Either nucleation occurs first (heterogeneous nucleation) or continues during growth (homogeneous nucleation). Without an extensive detailed microscopic study, it is very difficult to separate the two processes. It would be reasonable to assume that in DSC experiments the crystallization onset temperature T_c probably corresponds to the onset of growth even though there may still be nucleation occurring at this time. Put differently, the onset marks a major nucleation and the start of growth. Further, the rate of crystal growth in a-Se is thermally activated without any unusual features (see, for example, ref [9]). Thus, based on the latter two assumptions, figure 5 indicates that the nucleation processes is supressed at 35 °C. This is a very unusual result. Perhaps because the sample is very close to the glass transition temperature, the structure is too floppy to allow any large interfacial strains to build-up, that is, the floppy structure reduces the surface tension between the floppy glassy structure and the crystalline state. At 55 °C, the crystallization onset temperature is lower than that at the RT, because these high temperatures simply encourage nucleation and growth. We believe the key factor for understanding the higher resistance to crystallization at 35 °C must rely on the surface energy factor between the crystalline embryos and the floppy glass matrix. This view is in line with a recent conclusion on the examination of photocrystallization of a-Se films, where the authors have concluded that "the observed growth rate between 260 and 302 K is driven by local strain, and that relaxation of this strain near T_g suppresses crystal growth until thermally assisted processes accelerate the photo-crystallization at higher temperatures [10].

4. Conclusions

Samples of pure aged amorphous Se were stored at temperatures below the RT at 17.5 °C – 18 °C, at temperature below T_g at 35 °C but higher than the RT and at 55 °C, which is above the glass transition of a-Se. Thermal analysis performed by DSC and the examination of the glass stability and Hruby parameters of these samples reveal the following results:

(a) Samples of aged pure a-Se stored at a temperature of 35 °C but below the glass transition temperature T_g are more resistive to crystallization. These samples possess higher glass stability (S) and Hruby parameter (K_{gl}) than the other samples, because the crystallization onset temperature T_c increases during the first two weeks of storage time and then stabilizes henceforth

(b) Samples of a-Se stored at temperatures below the RT at 17.5 °C – 18 °C, exhibit properties that do not change with the time.

(c) Samples of a-Se stored at temperatures above the T_g at 55 °C exhibit a decreased glass stability because the crystallization temperature T_c decreases, i.e. there is a reduction in the glass stability and the Hruby parameter of these samples.

(d) DSC glass transformation results, changes in glass stability and Hruby parameter of a-Se are discussed and explained in terms of structural relaxation of amorphous materials.

References

[1] Kasap S et al 2011 *Sensors* **11** 5112
[2] Reznik A et al 2006 *J. Appl. Physics* **100** 113506
[3] Reznik A et al 2009 *J.Mater.Sci:Matt. Electron.* **20** S111
[4] Legros A, Shi M , Mouton A, Selmani A 1995 J.Appl.Phys. **79** (5) 3048
[5] Hruby A 1972 *Czech. J. Phys.* **B** 1187
[6] Nascimento M et al 2005 *J.Non-Cryst.Solids* **351** 3296
[7] Stephens R 1978 *J. Appl. Phys.* **49** 5855
[8] Svoboda R, Pustková P, Málek J 2007 *J. Phys. Chem. Solids* **68** 850
[9] Hamada S, Sato T and Shirai T 1967 *Bull. Chern. Soc.* **40** (4) 864
[10] Tallman R, Weinstein B, Reznik A, Kubota M, Tanioka K and Rowlands J 2008 *J. Non-Cryst.Solids* **354** 4577

Journal of Physics: Conference Series **558** (2014) 012008 doi:10.1088/1742-6596/558/1/012008

Optical characterization of macro-, micro- and nanostructures using polarized light

P Petrik[1,2], N Kumar[4], G Juhasz[1], C Major[1], B Fodor[1,3], E Agocs[1], T Lohner[1], S F Pereira[4], H P Urbach[4] and M Fried[1,2]

[1] Institute for Technical Physics and Materials Science, Research Centre for Natural Sciences, Konkoly Thege Rd. 29-33, 1121 Budapest, Hungary

[2] Doctoral School of Molecular- and Nanotechnologies, Faculty of Information Technology, University of Pannonia, Egyetem u. 10, Veszprém, H-8200, Hungary

[3] Faculty of Science, University of Pécs, 7624 Pécs, Ifjúság útja 6, Hungary

[4] Department of Imaging Physics, Faculty of Applied Sciences, Delft University of Technology, P. O. Box 5046, 2600GA Delft, The Netherlands

E-mail: petrik@mfa.kfki.hu

Abstract. Reflection of light measured in a polarimetric, scatterometric and spectroscopic way allows the measurement of structures in a broad size range from large (meter) scales like photovoltaic panels down to small (nanometer) scales like nanocrystals. Optical metrology continues to be improved to measure those materials with increasing sensitivity and accuracy, typically in a form of thin films on high quality substrates. This review provides an overview of some recently developed or improved methods, e.g. divergent light source ellipsometry for the mapping of large surfaces for photovoltaic applications, Fourier scatterometry for the measurement of periodic structures with sizes comparable to the wavelength of illumination, as well as spectroscopy around the band gap photon energies to characterize nanostructures – without attempting completeness.

1. Introduction

There is a huge variety of optical metrologies for a large number of applications. One peculiarity of optical techniques is that the measurable feature sizes cover a range from sub-wavelength dimensions to large scales. While there is no upper size limit for imaging (e.g. astronomical telescopes), imaging of thin film properties with high accuracy on large surfaces is currently a hot topic, due to display and solar panel applications. For layer thicknesses larger than the wavelength of illumination, reflectometry is a simple and robust method. However, by measuring the phase between the reflections polarized parallel and perpendicular to the plane of incidence, the sensitivity can largely be increased and sub-wavelength layer thicknesses (even below one nanometer) can be measured [1–5].

These polarimetric techniques can be realized in a broad variety of instrumental configurations, of which one of the most popular and simple is rotating polarizer/analyzer ellipsometry [3]. Ellipsometric imaging has been realized in many configurations including microscopic imaging combined with spectroscopic ellipsometry for e.g. biosensors or graphene mapping [6–9], using translation stages or moving the complete ellipsometry heads over the sample to map it spot-by-spot [10, 11], or imaging large areas using extended illumination, as demonstrated in the case of divergent light source ellipsometry [12–18] in the next chapter.

If the characteristic feature size of the structures to be measured is close to the wavelength of illumination, scattering and diffraction can be utilized. Under these circumstances, the interaction between light and matter results in a characteristic pattern of the electric far field measured as a function of the reflection angle. This pattern can also be calculated using optical models of the structure and the structural parameters, which are much smaller than the resolution limit of geometrical optics, can be determined with high accuracy. The limitation is that this method can only be used for periodic structures. Optical diffraction and scattering measurements have been utilized extensively for a long time, using versatile configurations from goniometric scatterometry to Fourier scatterometry [19–27]. In Chapter 3, the latter method will be discussed briefly.

Finally, when the size of structures (e.g. semiconductor nanocrystals) is much smaller than the wavelength of the light used for the measurement, the characteristic size can be determined indirectly, by the influence of the crystal size on the electron band structure. When using spectroscopy and measuring dielectric functions in photon energy ranges close to the interband transition energies, the band structure can be scanned, since the imaginary part of the dielectric function is proportional to the joint density of electron states. Around the critical points of direct interband transition energies, the change of nanostructure has a large influence on the dielectric function, which allows a sensitive way of crystal size measurement (as discussed in Chapter 4) [4, 5, 28–38].

2. Mapping of large surfaces ($d > \lambda$)

From the viewpoint of optical characterization, a straightforward choice is to distinguish the regions of “macro”, “micro” and “nano” features by the relation between the feature size (d) and the wavelength of illumination (λ), as shown in the headings of this and the following chapters. In this chapter, we most specifically focus on the cases of $d \gg \lambda$ with the aim of extending the size, speed and precision of large area thin film measurement (most importantly for photovoltaic applications [11, 15, 17], but large displays and 450 mm wafer activities are significant targets as well).

Wafer mapping has been required in microelectronics for a long time (see e.g. Ref. [39] for ion implantation). The most sensitive optical mapping techniques involve polarimetry in a broad range of wavelengths and configurations [10–12, 15, 17, 40–58] (note that the sensitivity of polarimetric techniques allows not only a lateral but also a vertical depth profiling [59–62]). The most common and major approach, which emerged first and continues to be used, is to scan either the sample stage or the optical heads over the sample to measure point-by-point [10, 11, 47, 48, 50–53, 63] (even with microfocus-infrared [45]). A newer and nowadays rapidly improving method is imaging, which has been developed to measure areas from small [7, 9, 43, 54, 55, 57] to ever increasing sizes [12, 15, 17, 18, 49, 58]. It has to be pointed out that, parallel to instrumental development, optical modeling techniques are also undergoing a significant improvement [1–5, 64].

Film thickness imaging on large scales is challanging for several reasons. First of all, the flatness of the sample has to be minimized and/or taken into account. Second, the beam path and the optical configuration usually has to be adapted to a specific application. In some cases the sample to be investigated can be placed onto a large mapping tool [11], but for *in line* or even *real time* applications special constraints have to be taken into account, as demonstrated in figures 1 and 2. Both are utilizations of the spectroscopic line-scan version of divergent light source mapping ellipsometry [12, 15, 18] for *in line* applications. In these configurations a line-scan is realized, whereas the second index of the detector is used for spectroscopic information. In the arrangement of figure 1, 30 points of a 450 mm diameter wafer have to be measured

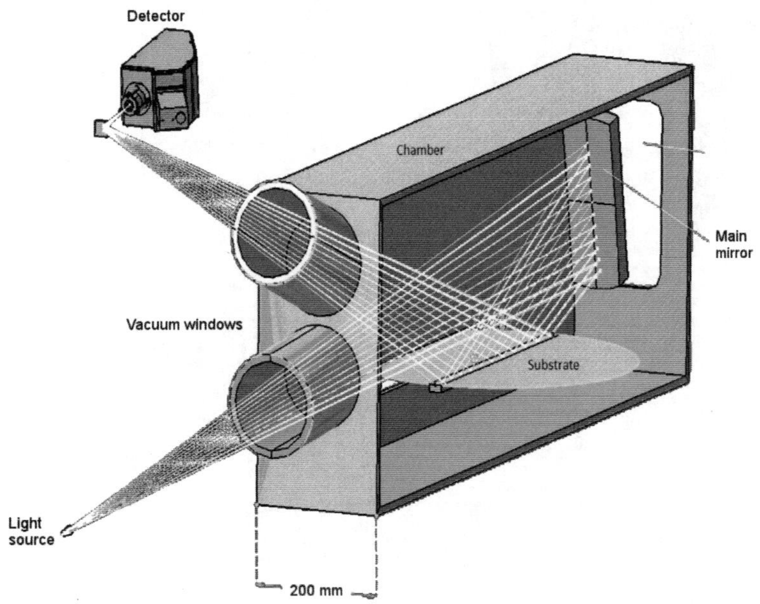

Figure 1. Integration of divergent light source mapping ellipsometry into a transport module of a cluster chamber. The substrate to be measured in the chamber has a diameter of 450 mm.

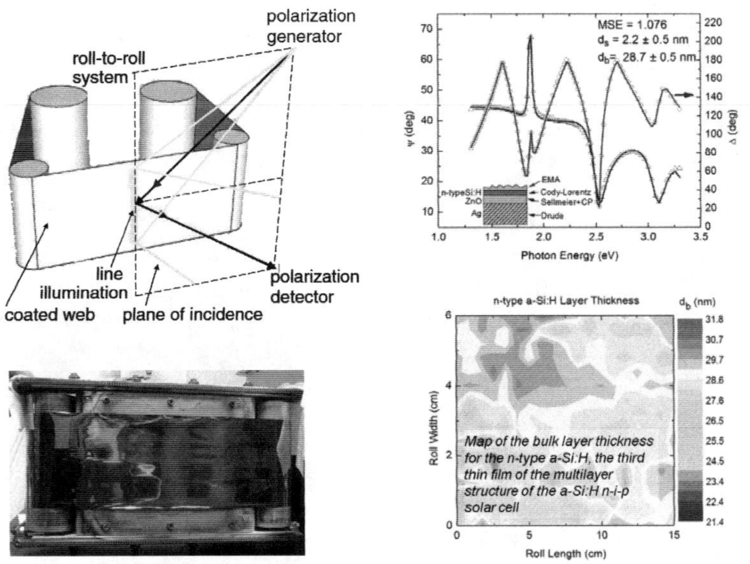

Figure 2. 2D-mapping by divergent light source ellipsometry in a roll-to-roll arrangement (top left). The equipment with the foil on the rolls is shown in the bottom left image. The measured and fitted spectra at one point of a line scan is shown in the top right graph, together with the optical model (see the inset). The determined thickness map is shown in the bottom right graph.

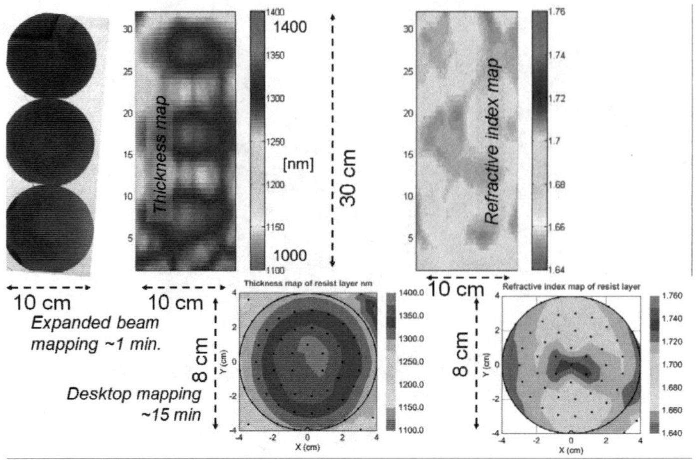

Figure 3. Thickness and refractive index maps using a line scan during translation (bottom graphs), verified by desktop mapping (bottom graphs).

along a line simultaneously, in the wavelength range of 350-1000 nm, within several seconds.[1] The 2D mapping can be realized by measuring along the line while moving the wafer in the transport module. This mapping is basically identical to that used for roll-to-roll mapping, as demonstrated in figure 2. One line is measured simultaneously in a broad wavelength range, and the mapping is performed by translating the roll, which can be part of the production procedure – this makes the tool integrable into an industrial environment in a straightforward way.

Currently, a line scan of 30 points during a translation of 10 cm can be measured within 1 minute, as shown in figure 3. The spectroscopic capability also allows the determination of the refractive index as a map. One of the near-future goals of the project is to scan a 60-cm line, and increase the speed so that, with a suitable traversal time, a 60 cm $\times$ 120 cm photovoltaic panel can be measured in 1800 points within $\approx$1 minute [18].

3. Metrology of photonic structures ($d \sim \lambda$)

Scatterometry has been applied for the measurement of microsctructure and periodic feature sizes for decades [27, 65–67]. As for large area thin film mapping presented in the previous chapter, polarimetry has an added value in this case as well, most importantly in terms of precision (e.g. critical dimension ellipsometry [68, 69] or infrared scattering ellipsometry [70]). The used angle and wavelength ranges as well as configurations cover a wide range, depending on the applications.

In this chapter we focus on Fourier scatterometry [19–27], which has the unique capability of measuring a broad range of scattering angles simultaneously, by illuminating and measuring using the same high numerical aperture objective (for the optical setup see figure 4). Using coherent illumination, the light is focused on the periodic structure with a spot size which is comparable with the period of the structure. When the sample is in focus, the intensity in each point of the back focal plane of the microscope objective uniquely corresponds to a given angle of reflection [19]. Therefore, by imaging the back focal plane of the objective, the spatial distribution of the intensity pattern on the CCD is directly linked to the angle distribution of the light reflected from the sample [26]. It has been shown previously that coherent illumination

[1] As part of the ENIAC E450EDL project.

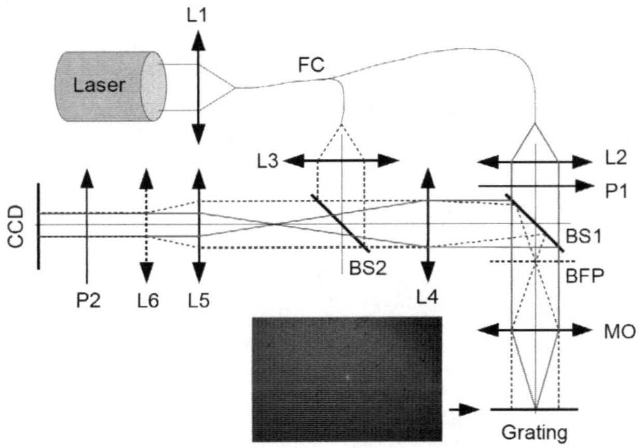

Figure 4. Setup of Fourier scatterometry. L1, L2, L3: collimating lences; L4, L5, L6: telescopic lenses; P1, P2: polarizers; BS1, BS2: beam splitters; MO: microscope objective; BFP: back focal plane; CCD: detector. The inset shows the case of imaging the grating to the CCD, together with the focused spot.

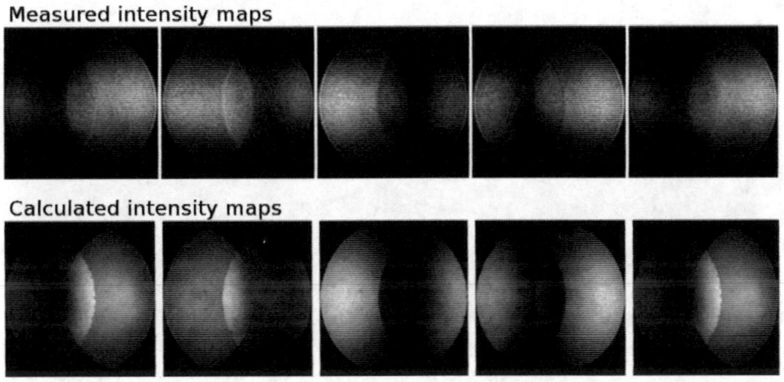

Figure 5. Measured (top row) and calculated (bottom row) far field intensity images for a 2-μm pitch silicon grating (duty cicle of 0.5) with a height of 148 nm, by scanning the focused spot evenly distributed over one period of the grating (represented by the different columns – the first and last images correspond to the same position with regard to the grating period). The wavelength of illumination was 633 nm.

provides a possibility of measuring the phase between the diffracted orders in the pupil, when properly scanning the focused spot along the investigated periodic structure. The capability of measuring the phase (as in the case of ellipsometry) increases the sensitivity significantly. The tool has recently been experimentally realized and the high sensitivity revealed [26].

Figure 5 shows a series of far field intensity images taken using different spot positions perpendicular to the grooves of a silicon grating. As emphasized for thin film mapping in the previous chapter, forward calculation of the far field based on a proper optical model is a crucial part of the procedure [71, 72]. Being a non-linear problem, *a priori* information on the structure is needed to have a suitable starting point of the evaluation procedure, from which the least squares method applied to the fit can converge to the global minimum. Comparison with reference methods like scanning electron microscopy (SEM) and atomic force microscopy verify the reliability and accuracy being within several nanometers for sub-micron grating periodicities. Furthermore, based on the large sensitivity of the far field interference pattern on the position of the focused spot, the method can also be used for spot positioning with nanometer precision [26].

4. Characterization of nanostructures ($d < \lambda$)

The optical characterization of nanostructures is based on the fact that the structural change of the materials changes its dispersion, which can be measured using spectroscopy. More specifically, the imaginary part of the dielectric function is proportional to the joint density of electron states. Since the dielectric function can be measured with a typical accuracy of down to 10^{-4} using sensitive optical techniques, minute structural changes can be followed even *in situ*, if proper optical models can be built concerning the structure [73–76].

As emphasized above, according to the indirect nature of most optical characterization methods, the accuracy of structural characterization largely depends on the application of proper optical models. These models are usually verified by direct reference methods (e.g. SEM), however, as soon as the model is verified, optical methods are capable of the high-sensitivity, fast and non-destructive measurement of the nanostructures, which is applied e.g. as a powerful *in line* or *real time* inspection method during thin film preparation and device manufacturing [77].

The most frequently applied method for the optical modeling of nanostructure is the effective medium approximation (EMA [28, 78]), which describes the material as a composite of distinct phases with sizes much smaller than the wavelength of characterization, but large enough to have a characteristic dielectric function. The two major problems with EMA are that the dielectric function of components are frequently different from those that can be found in databases (because of different preparations or due to size effects) and the way the screening effect and the anisotropy due to the non-uniformity of phase boundaries are taken into account. Nevertheless, there is a huge number of publications using this theory, usually with results that are in good agreement with reference methods.

A nice example of EMA is the characterization of porous materials, as demonstrated in figure 6. The ellipsometric measurement of electrochemically etched thin films in silicon can be modeled using EMA with components of single-crystalline silicon and void, which can be extended by a fine-grained polycrystalline silicon reference if the typical crystallite sizes of the remaining porous Si skeleton are very small [37, 79, 80]. By fitting the volume fraction of void, the optical density and in turn the porosity can be determined, as demonstrated for void gradients caused by ion implantation [81, 82]. The sensitivity also allows the determination of density gradients within the thin films.

By measuring absorbing materials with large dispersion spectroscopically, the fact that the penetration depth depends on the absorption can be utilized, as demonstrated in figure 7. Semiconductors have absorption peaks at the so called critical point photon energies corresponding to direct interband transitions. Figure 7 shows this fact at the example of polycrystalline silicon, which has a large absorption peak in the ultraviolet spectral range at wavelengths of around 300 nm. In this range, the imaginary part of the refractive index (extinction coefficient, k) almost reaches 5, and the corresponding optical penetration depth (OPD) is well below 10 nm (right-hand side axis and dashed line in figure 7). However, when the wavelengh is increased (from the peak value of $\approx$300 nm), k drops at wavelengths of around 400 nm and the penetration depth rapidly increases, reaching a value of 500 nm (the thickness of the polycrystalline silicon layer) at a wavelength of $\approx$550 nm.

When fitting the spectra measured by ellipsometry on the polycrystalline silicon layer (transmission electron microscopy image shown on the right-hand side of the graph), the

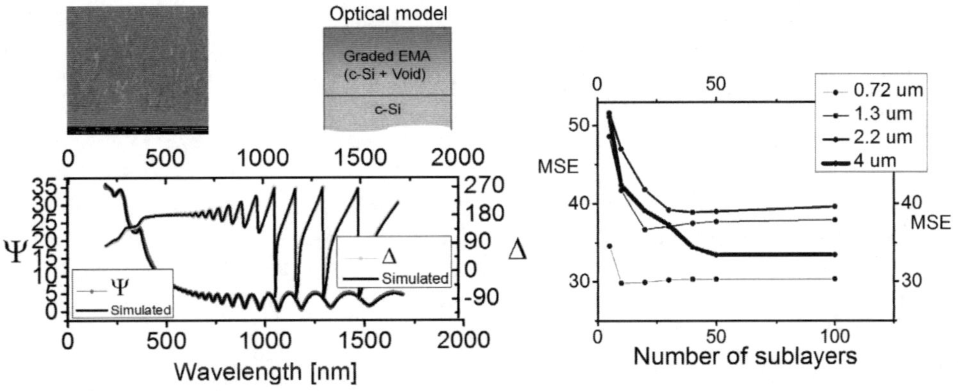

Figure 6. Graded effective medium model (insert at the top) for the characterization of porous silicon (top left image) with a thickness of $\approx 1.3\ \mu$m. The graph at the right-hand side shows that dividing the investigated layer into sub-layers, in order to determine void gradients, decreases the mean square error (MSE), i.e. increases the fit quality. However, using numerous sublayers may increase parameter correlations.

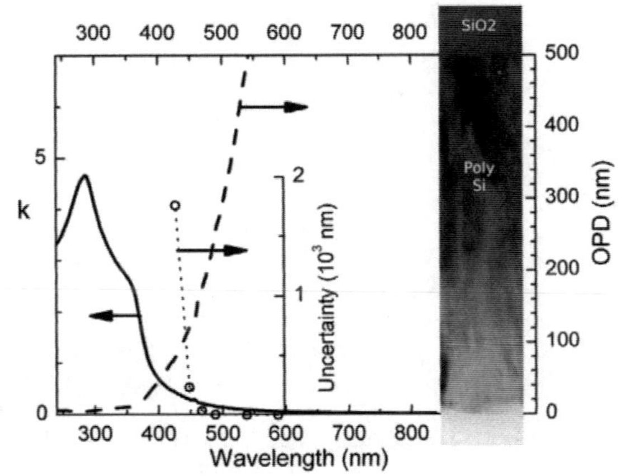

Figure 7. Extinction coefficient (k) and optical penetration depth (OPD) measured for an $\approx$500-nm polycrystalline silicon layer as a function of the wavelength, as well as the uncertainty of the fitted thickness of the polycrystalline silicon layer as a function of the upper limit wavelength of the range used for the fit [38].

uncertainty of the fitted layer thickness was determined for different wavelength ranges, in which the lowest wavelength was fixed at the value corresponding to the largest absorption ($\approx$300 nm) and the highest wavelength was changed from 600 to 430 nm. The open circles connected with a dotted line show that the uncertainty increases when the upper limit of the wavelength range decreases with increasing absorption and with decreasing penetration depth, due to the fact that the light does not sense the bottom interface of the layer, because of the increased absorption. It is also clear that the "information depth" is significantly larger than the "penetration depth" (calculated by $\lambda/(4\pi k)$), because the uncertainty is still low at a wavelength of 450 nm for which

OPD predicts ≈100 nm, although the layer thickness is ≈500 nm. We have shown recently that this effect can be used for a depth scan when plotting the fitted layer parameters as a funtion of the penetration depth, whereas the penetration depth can be varied by properly changing the wavelength range used for the fitting [38].

5. Conclusions

It has been that optical characterization methods are capable of measuring structural properties on a wide scale of sizes from macroscopic features that are larger than the wavelength of the illumination, down to nanoscale sizes which are much smaller than the wavelength. These goals are reached using versatile techniques including spectroscopy, diffraction, geometrical optics and polarimetry. Note that some of these techniques can be combined. For example, making imaging ellipsometry specroscopically, nanostructural information can be detected for each measured lateral position [18]. Mapping of non-uniform samples can also be combined with Fourier scatterometry, and Fourier scatterometry can be combined with spectroscopy.

Acknowledgments

Support from ENIAC E450EDL and KMR_12_1_2012_0225 projects, as well as the Hungarian-French Intergovernmental S&T Cooperation Programme grant Nr. TÉT_10-1-2011-0754 and by CampusFrance in the frame of a Balaton PHC project are greatly acknowledged.

References

[1] Fried M, Lohner T and Petrik P 2001 *Handbook of surfaces and interfaces of materials* vol 4 ed Nalwa H S (San Diego: Academic Press)
[2] Tompkins H G and Irene E G (eds) 2005 *Handbook of ellipsometry* (Norwich, NY: William Andrew)
[3] Fujiwara H 2007 *Spectroscopic Ellipsometry: Principles and Applications* (New York: Wiley)
[4] Petrik P 2012 *Nanocrystals - Synthesis, Characterization and Applications* ed Neralla S (http://dx.doi.org/10.5772/48732: InTech)
[5] Petrik P and Fried M 2012 *Ellipsometry at the Nanoscale* ed Losurdo M and Hingerl K (Heidelberg: Springer-Verlag)
[6] Jin G, Jansson R and Arwin H 1996 *Rev. Sci. Instrum.* **67** 2930
[7] Vaupel M, Yunfeng S and Zhimin Y 2008 *phys. stat. sol.* **205** 772
[8] Qi C, Zhu W, Niu Y, Zhu H G Z G Y, Meng Y H, Chen S and Jin G 2009 *Journal of Viral Hepatitis* **16** 822
[9] Wurstbauer U, Röling C, Wurstbauer U, Wegscheider W, Vaupel M, Thiesen P H and Weiss D 2010 *Applied Physics Letters* **97** 231901
[10] Huang Z, Chen J, Sestak M N, Attygalle D, Dahal L R and Mapes M R 2010 *IEEE Proceedings – Conference Record of the IEEE Photovoltaic Specialists Conference* **82805** 1678
[11] Dahal L R, Huang Z, Salupo C, Podraza N J, Marsillac S and Collins R W 2011 *IEEE Proceedings – 37th IEEE Photovoltaic Specialists Conference* **89752** 182
[12] Juhasz G, Horvath Z, Major C, Petrik P, Polgar O and Fried M 2008 *phys. stat. sol. (c)* **5** 1081
[13] Major C, Juhasz G, Petrik P, Horvath Z, Polgar O and Fried M 2009 *Vacuum* **84** 119
[14] Fried M, Juhasz G, Major C, Petrik P, Polgár O, Horváth Z and Nutsch A 2011 *MRS Proceedings* **1323** 157
[15] Fried M, Juhasz G, Major C, Nemeth A, Petrik P, Polgar O, Salupo C, Dahal L R and Collins R W 2011 *Thin Solid Films* **519** 2730
[16] Nemeth A, Attygalle D, Dahal L R, Aryal P, Huang Z, Salupo C, Petrik P, Juhasz G, Major C, Polgar O, Fried M, Pecz B and Collins R W 2011 *MRS Proceedings* **1321** 267
[17] Shan A, Fried M, Juhasz G, Major C, Polgar O, Nemeth A, Petrik P, Dahal L R, Chen J, Huang Z, Podraza N J and Collins R W 2014 *IEEE Journal of Photovoltaics* **4** 355
[18] Fried M 2014 *Thin Solid Films* **in press**
[19] Boher P, Petit J, Leroux T, Foucher J, Desieres Y, Hazart J and Chaton P 2005 *SPIE Proceedings* **5752** 192
[20] El Gawhary O, Kumar N, Pereira S F, Coene W M J and Urbach H P 2011 *Applied Physics B* **105** 775
[21] Kumar N, El Gawhary O, Roy S, Kutchoukov V G, Pereira S F, Coene W and Urbach H P 2012 *SPIE Proceedings* **8324** 83240Q
[22] Roy S, El Gawhary O, Kumar N, Pereira S F and Urbach H P 2012 *Journal of the European Optical Society - Rapid Publications* **7** 12031
[23] Kumar N, El Gawhary O, Roy S, Pereira S F and Urbach H P 2013 *SPIE Proceedings* **8788** 87881P

[24] Kumar N, El Gawhary O, Roy S, Pereira S F and Urbach H P 2013 *Journal of the European Optical Society - Rapid Publications* **8** 13048
[25] Roy S, Kumar N, Pereira S F and Urbach H P 2013 *Journal of Optics* **15** 075707
[26] Kumar N, Petrik P, Ramanandan G K P, El Gawhary O, Roy S, Pereira S F, Coene W M J and Urbach H P 2014 *Optics Express* **accepted**
[27] Endres J, Kumar N, Petrik P, Henn M A, Heidenreich S, Pereira S F, Urbach H P and Bodermann B 2014 *SPIE Proceedings* **9132** 913208–1
[28] Aspnes D E 1982 *Thin Solid Films* **89** 249
[29] Aspnes D E, Studna A A and Kinsbron E 1984 *Phys. Rev. B* **29** 768
[30] Logothetidis S, Polatoglou H M and Ves S 1988 *Solid State Commun.* **68** 1075
[31] Boultadakis S, Logothetidis S and Ves S 1992 *J. Appl. Phys.* **72** 364
[32] Asinovsky L M 1993 *Thin Solid Films* **233** 210
[33] Flueraru C, Gartner M, Dascalu D and Rotaru C 1996 *J. Phys. III France* **6** 225
[34] Borghesi A, Giardini M E, Marazzi M, Sassela A and Santi G D 1997 *Appl. Phys. Lett.* **70** 892
[35] Petrik P, Fried M, Lohner T, Berger R, Biró L P, Schneider C, Gyulai J and Ryssel H 1998 *Thin Solid Films* **313-314** 259
[36] Petrik P, Lohner T, Fried M, Biró L P, Khánh N Q, Gyulai J, Lehnert W, Schneider C and Ryssel H 2000 *J. Appl. Phys.* **87** 1734
[37] Petrik P, Fried M, Vazsonyi E, Basa P, Lohner T, Kozma P and Makkai Z 2009 *J. Appl. Phys.* **105** 024908
[38] Petrik P, Agocs E, Volk J, Lukacs I, Fodor B, Kozma P, Lohner T, Oh S, Wakayama Y, Nagata T and Fried M 2014 *Thin Solid Films* **in press**
[39] Keenan W A, Johnson W H and Yarling C B 1991 *Nucl. Instr. and Methods* **B55** 230
[40] Ohtsuka Y and Oka K 1992 *SPIE Proceedings* **1746** 42
[41] Chang Y S and Li S S 1995 *Solid-State Electronics* **38** 197
[42] Boher P, Defranoux C, Bourtault S and Stehle J L 1999 *SPIE Proceedings* **3677** 845
[43] Zhan Q and Leger J R 2002 *Applied Optics* **41** 4443
[44] Oka K and Kaneko T 2003 *Optics Express* **11** 1510
[45] Gensch M, Korte E H, Esser N, Schade U and Hinrichs K 2006 *Infrared Physics & Technology* **49** 74
[46] Hinrichs K, Gensch M, Esser N, Schade U, Rappich J, Kröning S, Portwich M and Volkmer R 2007 *Anal. Bioanal. Chem.* **387** 1823
[47] Roodenko K, Mikhaylova Y, Ionov L, Gensch M, Stamm M, Minko S, Schade U, Eichhorn K J, Esser N and Hinrichs K 2008 *Appl. Phys. Lett.* **92** 103102
[48] Krüger H, Kemnitz E, Hertwig A and Beck U 2008 *phys. stat. sol. (a)* **205** 821
[49] Major C, Juhasz G, Horvath Z, Polgar O and Fried M 2008 *phys. stat. sol. (c)* **5** 1077
[50] Weber J W, Hinrichs K, Gensch M, van de Sanden M C M and Oates T W H 2011 *Appl. Phys. Lett.* **505** 358
[51] Sago Y and Fujiwara H 2012 *Japanese J. Appl. Phys.* **51** 10NB01–1
[52] Rosu D, Petrik P, Rattmann G, Schellenberger M, Beck U and Hertwig A 2013 *Thin Solid Films* **in press**
[53] Aryal P, Attygalle D, Pradhan P, Podraza N J, Marsillac S and Collins R W 2013 *IEEE Journal of Photovoltaics* **3** 359
[54] Nielsen M M B and Simonsen A C 2013 *Langmuir* **29** 1525
[55] Röling C, Thiesen P, Meshalkin A, Achimova E, Abashkin V, Prisacar A and Triduh G 2008 *Appl. Phys. Lett.* **92** 103102
[56] Hinrichs K, Furchner A, Rappich J and Oates T W H 2013 *J. Phys. Chem.* **387** 1823
[57] Necas D, Ohlidal I, Franta D, Cudek V, Ohlidal M, Vodak J, Sladkova L, Zajickova L, Elias M and Vizd F 2013 *Thin Solid Films* **in press**
[58] Koops R, Sonin P, van Veghel M and Gawhary O E 2014 *J. Opt.* **16** 065701
[59] Vanhellemont J, Roussel P and Maes H E 1991 *Nucl. Instr. Meth.* **55** 183
[60] Fried M, Lohner T, Aarnink W A M, Hanekamp L J and van Silfhout A 1992 *J. Appl. Phys.* **71** 2835
[61] Lynch S, Murtagh M, Crean G M, Kelly P V and O'Connor M 1993 *Thin Solid Films* **233** 199
[62] Petrik P, Polgar O, Fried M, Lohner T, Khanh N Q and Gyulai J 2003 *J. Appl. Phys.* **93** 1987
[63] Berge C, Krasilnikova A and Masetti E 2000 *SPIE Proceedings* **4099** 319
[64] Fried M, Lohner T and Gyulai J 1997 *Effect of Disorder and Defects in Ion-Implanted Semiconductors: Optical and Photothermal Characterization* Semiconductors and Semimetals ed Cristofides C and Ghibaudo G (New York: Academic Press)
[65] Wurm M, Pilarski F and Bodermann B 2010 *Review of Scientific Instruments* **81** 023701
[66] Gross H, Richter J, Rathfeld A and Bär M 2010 *Journal of the European Optical Society Rapid Publications* **5** 10053
[67] Endres J, Bodermann B, Dai G, Wurm M, Henn M, Gross H, Scholze F and Diener A 2013 *Fringe Proceedings*

DOI: 10.1007/978-3-642-36359-7-128 695

[68] Hingst T, Marschner T, Moert M, Homilius J, Guevremont M, Hopking J and Elazami A 2003 *SPIE Proceedings* **5038** 274

[69] Huang H T and Terry F L 2004 *Thin Solid Films* **455-456** 828

[70] Guittet P Y, Mantz U and Weidner P 2004 *IEEE Proceedings - International Symposium on Semiconductor Manufacturing* **63503** 176

[71] Gross H, Model R, Bär M, Wurm M, Bodermann B and Rathsfeld A 2006 *Measurement: Journal of the International Measurement Confederation* **39** 782

[72] Model R, Rathsfeld A, Gross H, Wurm M and Bodermann B 2008 *J. Phys.: Conf. Series* **135** 012071

[73] Collins R W and Yang B Y 1989 *J. Vac. Sci. Technol.* **B7** 1155

[74] Collins R W, An I, Nguyen H V and Lu Y 1993 *Thin Solid Films* **233** 244

[75] Koh J, Fujiwara H, Koval R J, Wronski C R and Collins R W 1999 *J. Appl. Phys.* **85** 4141

[76] Collins R W and Ferlauto A S 2005 *Handbook of ellipsometry* ed Irene E G and Tomkins H G (Norwich, NY: William Andrew)

[77] Collins R W, An I, Lee J and Zapien J A 2005 *Handbook of ellipsometry* ed Irene E G and Tomkins H G (Norwich, NY: William Andrew)

[78] Bruggeman D A G 1935 *Ann. Phys. (Lepzig)* **24** 636

[79] Fried M, Lohner T, Polgar O, Petrik P, Vazsonyi E, Barsony I, Piel J P and Stehle J L 1996 *Thin Solid Films* **276** 223

[80] Lohner T, Fried M, Petrik P, Polgar O, Gyulai J and Lehnert W 2000 *Mat. Sci. and Engineering* **B69-70** 182

[81] Petrik P, Cayrel F, Fried M, Lohner T, Polgár O, Gyulai J and Alquier D 2005 *J. Appl. Phys.* **97** 123514

[82] Fodor B, Cayrel F, Agocs E, Alquier D, Fried M and Petrik P 2014 *Thin Solid Films* **in press** DOI: 10.1016/j.tsf.2014.02.014

Smectic A herringbone patterns

S Kralj[1,2], E Lacaze[3], G Cordoyiannis[1], Z Kutnjak[1,4]

[1] Jožef Stefan Institute, Jamova 39, 1000 Ljubljana, Slovenia
[2] Faculty of Natural Sciences and Mathematics, University of Maribor, Koroška 160, 2000 Maribor, Slovenia
[3] Institut des Nano-Sciences de Paris, 4 place Jussieu, 75005 Paris, France
[4] Jožef Stefan International Postgraduate School, Jamova 39, 1000 Ljubljana, Slovenia

E-mail: samo.kralj@ijs.si

Abstract. Two qualitatively different Sm*A* structures exhibiting herringbone-type layer patterns, to which we refer as the Defectless Smectic Herringbone (DSH) and the Dislocation Decorated Smectic Herringbone (DDSH) pattern are studied by a Landau-de Gennes-Ginzburg mesoscopic approach. Liquid crystal structures are described in terms of a nematic director field and a smectic complex order parameter. It is demonstrated that, in the proximity of the N-Sm*A* phase transition, a melting of smectic layers could be realised even for relatively weakly-tilted smectic layers in DSH patterns (i.e. $\theta_t \approx 10^0$) for type I Sm*A* phase. The width of melted region could be relatively large with respect to bulk values of the smectic characteristic lengths. In addition, a critical value of θ_t is determined at which a DDSH pattern is expected to appear.

1. Introduction

Recent years have brought increasing interest in soft nanocomposites [1]. Typically, such systems consist of a soft matrix containing nanoparticles, where each component contributes a specific desired system's property. In particular these materials have great potential to play important role in a wide variety of future tunable and multifunctional nanodevices [1,2].

Various liquid crystalline (LC) phases are often used as soft matrices [3,4]. Their main advantageous properties are softness, liquid character, optical transparency and anisotropy as well as an enormously large variety of different structures that they could exhibit. On the other hand, added nanoparticles (NPs) are used either to quantitatively anomalously enhance anisotropic LC properties [5] or to induce qualitatively new features [1].

Furthermore, LC structures could be exploited to trigger ordering [6,7], spatial localization and/or assembling [8,9,10,11,12,13] of immersed NPs within desired attracting regions. In those studies it was demonstrated that localized elastic distortions are very efficient attractors for appropriate (or adequately surface decorated) NPs. In view of potential applications it is of great interest to form different well controlled LC elastically distorted patterns as potential traps for NPs. In particular, well defined localisation of NPs forming linear nanowire-like structures could be achieved in elastically distorted LC structures exhibiting smectic translational ordering [11,12,13]. Recently, it was demonstrated [14] that the technique of scribing a substrate confining LC using the stylus of an atomic force microscope could be exploited to manipulate the smectic ordering in plan-parallel cells. Using this technique herringbone-type smectic layer structures were induced where the imposed bending of smectic layers efficiently enforced a localized melting of smectic ordering.

In this contribution we focus theoretically on structural properties of smectic herringbone-type patterns. We utilize a mesoscopic Landau-de Gennes-Ginzburg approach [3,15] in order to determine key parameters controlling the width of localized regions in which smectic ordering is melted. The paper is organized as follows. In Sec. 2 we introduce the model. The geometry of the problem and the parametrization used to describe smectic ordering are both introduced in Sec. 3. In Sec. 4 we numerically analyze the smectic patterns for cases without dislocations. The conditions at which dislocations could appear are estimated in Sec. 5. In the last section our results are summarized.

2. Model

The Landau-de Gennes-Ginzburg mesoscopic model [3,15] is used in standard terms of the nematic director field $\vec{n}$ and the complex smectic A (Sm*A*) order parameter $\psi = \eta e^{i\phi}$. The unit pseudo-vector $\vec{n}$ (i.e. orientations $\pm\vec{n}$ are physically equivalent) points along local uniaxial orientational ordering. The translational order parameter η measures the degree of the smectic layer ordering. The phase ϕ determines the position and orientation of smectic layers. The local smectic layer normal vector is given by $\vec{v} = \nabla\phi / |\nabla\phi|$. The smectic order parameter is related to the LC mass density spatial variation via the equation $\rho = \rho_0\left(1+\psi+\psi^*\right)$, where ρ_0 stands for the mass density in the nematic phase.

In terms of $\vec{n}$ and ψ we express the free energy density [3,15] as $f = f_0 + f_c^{(s)} + f_e^{(n)} + f_e^{(s)}$. The nematic elastic free energy is expressed as

$$f_e^{(n)} = \frac{K_1}{2}(\nabla\cdot\vec{n})^2 + \frac{K_2}{2}(\vec{n}\cdot\nabla\times\vec{n})^2 + \frac{K_3}{2}|\vec{n}\times\nabla\times\vec{n}|^2, \tag{1}$$

where the positive quantities K_1, K_2 and K_3 are the splay, twist and bend Frank nematic elastic constants, respectively. This contribution enforces a homogeneous alignment of $\vec{n}$ along a symmetry breaking direction.

The smectic condensation free energy density reads

$$f_c^{(s)} = \alpha_0(T-T_c)|\psi|^2 + \frac{\beta}{2}|\psi|^4. \tag{2}$$

Here α_0 and β are positive material constants and we consider the case of 2nd order bulk nematic-Sm*A* phase transition at $T=T_c$. Below T_c the condensation term imposes the bulk equilibrium value of the translational order parameter

$$\eta_{eq} = \sqrt{\frac{\alpha_0(T_c-T)}{\beta}}. \tag{3}$$

while $\eta_{eq}(T>T_c)=0$.

The smectic elastic term is given by

$$f_e^{(s)} = C_{\parallel}\left|(\vec{n}\cdot\nabla - iq_0)\psi\right|^2 + C_{\perp}\left|(\vec{n}\times\nabla)\psi\right|^2. \tag{4}$$

Here the wave vector $q_0 = 2\pi / d_0$ determines the bulk intrinsic smectic layer distance d_0 which is enforced by the positive smectic compressibility elastic constant $C_{\parallel}$ [15,16]. On the other hand, the positive smectic bend elastic constant $C_{\perp}$ enforces alignment of $\vec{n}$ along the smectic layer normal [15,16].

The quantity f_0 includes the free energy density contributions of the remaining degrees of freedom.

This relatively simple free energy density expression is sufficient to describe key phenomena of our interest. In following calculations we neglect the LC elastic anisotropy. We set $K_1 = K_2 = K_3 \equiv K$ and $C_{\parallel} = C_{\perp} \equiv C$. Note that on approaching the Sm*A* phase, by decreasing temperature in the nematic phase, the nematic twist and bend nematic elastic constant diverge as a result of the smectic pre-transitional fluctuations [3,15,16]. They enable the temporal formation of Sm*A*-like domains (their characteristic size is approximated by the smectic order parameter correlation length), which resist to the nematic twist and bend deformations. However, in the Sm*A* phase one can use "bare" values of K_2 and K_3 in our mesoscopic modelling, because the nematic twist and bend deformations are suppressed by the smectic constants, represented by *C*.

In addition to d_0, the important material dependent lengths of the systems are the smectic order parameter correlation length ξ and the nematic penetration length λ [3]. The quantity ξ estimates the distance over which the smectic degree of order recovers if it is locally perturbed. On the other hand, λ estimates the distance over which $\vec{n}$ roughly recovers its orientation along the smectic layer normal $\vec{v}$ if it is locally reoriented from this direction. Below T_c in bulk their values are estimated by [3]

$$\xi = \frac{\xi_0}{\sqrt{|r|}}, \tag{5}$$

$$\lambda = \frac{\lambda_0}{\sqrt{|r|}}, \tag{6}$$

$r = (T - T_c)/T_c$ is the reduced temperature, $\xi_0 = \sqrt{\frac{C}{\alpha_0 T_c}}$, and $\lambda_0 = \sqrt{\frac{\beta}{\alpha_0 T_c}}\sqrt{\frac{K}{Cq_0^2}}$. The temperature independent "smectic" Ginzburg parameter [3,14,15]

$$\kappa = \frac{\lambda}{\xi} = \frac{\lambda_0}{\xi_0}, \tag{7}$$

is also playing an important role in our study. Namely, it distinguishes between the type I and type II Sm*A* phase [15], corresponding roughly to materials characterized by $\kappa < 1/\sqrt{2}$ and $\kappa > 1/\sqrt{2}$, respectively. In the latter case dislocations are relatively more easily introduced into Sm*A* patterns if a smectic layer frustration is enforced.

3. Parameterisation and Euler-Lagrange equations

For reasons of computational convenience we introduce the scaled order parameter $\tilde{\psi} = \psi / \eta_{eq} = \tilde{\eta} e^{i\phi}$, and the dimensionless free energy density $\tilde{f} = f / (\alpha_0 (T_c - T)\eta_{eq}^2)$. We henceforth discard the tildes and express f in terms of the system's characteristic lengths (which are in principle known) as

$$f = -\eta^2 + \frac{\eta^4}{2} + \frac{q_0^2 \lambda^2 \xi^2}{2} |\nabla \vec{n}|^2 + \xi^2 \left(|(\vec{n} \cdot \nabla - iq_0)\psi|^2 + |(\vec{n} \times \nabla)\psi|^2 \right) \tag{8}$$

within the Sm*A* phase temperature regime. Note that the temperature dependence is hidden in the smectic material dependent lengths λ and ξ, see Eq.(5) and Eq.(6).

In our study we focus on LC ordering within the herringbone-type smectic structure [14] depicted in the figure 1 in the Cartesian coordinate system (x,y,z). The symmetry of the problem allows us to restrict to the (x,z) plane. Relatively far (i.e., $x >> \lambda$) from the herringbone's tip we set $\vec{n}$ to be tilted for a tilt angle θ_t with respect to the z–axis and the smectic layer distance d equals the equilibrium smectic layer distance d_0, i.e. $d = d_0$. On the other hand, LC molecules within the tip (at $x=0$) are aligned along the z-axis. Consequently, smectic layers tend to be locally dilated in this region. Namely, in absence of dislocations the smectic periodicity at x=0 is given by $q = 2\pi / d = q_0 / \cos\theta_t$. This smectic pattern roughly mimics the experimentally obtained herringbone-type patterns studied in [14].

In such LC configurations two qualitatively different equilibrium patterns could be formed, to which we refer as the Defectless Smectic Herringbone (DSH) pattern and the Dislocation Decorated Smectic Herringbone (DDSH) pattern. In the DSH pattern the smectic order parameter tends to be depressed at the herringbone tip and one dimensional $\eta = \eta(x)$ spatial variation is expected. In the DDSH pattern edge dislocations are introduced in order to partially relive the layer dilatation-imposed strain at the tip giving rise to two-dimensional spatial variation in degree of smectic ordering (i.e. $\eta = \eta(x,z)$). The former and latter patterns are expected in type I and type II SmA structures [15], respectively.

In this paper we focus on cases without dislocations. Consequently, it is convenient to parametrize the smectic phase factor as follows [16]

$$\phi = q(z+u), \tag{9}$$

where $q = 2\pi / d$ with $d = d_0 / \cos\theta_t$ describes smectic layer dilatation at x=0 due to bending of smectic layers and u represents the smectic layer displacement field. Furthermore, we express the nematic director field as

$$\vec{n} = (\sin\theta, 0, \cos\theta). \tag{10}$$

In DSH patterns the symmetry of the problem suggests $\theta = \theta(x)$ and u=$u(x)$, therefore

$$\begin{aligned} f = & -\eta^2 + \frac{\eta^4}{2} + \frac{q_0^2\lambda^2\xi^2}{2}\left(\frac{\partial\theta}{\partial x}\right)^2 + \\ & \xi^2\left(\left(\frac{\partial\eta}{\partial x}\right)^2 + \eta^2\left(\left(q\left(\sin\theta\frac{\partial u}{\partial x} + \cos\theta\right) - q_0\right)^2 + q^2\left(\cos\theta\frac{\partial u}{\partial x} - \sin\theta\right)^2\right)\right) \end{aligned} \tag{11}$$

One can minimize f with respect to $\frac{\partial u}{\partial x}$, yielding the relation

$$\frac{\partial u}{\partial x} = \frac{q_0}{q}\sin\theta. \tag{12}$$

With this in mind we finally obtain

$$f = -\eta^2 + \frac{\eta^4}{2} + \frac{q_0^2\lambda^2\xi^2}{2}\left(\frac{\partial\theta}{\partial x}\right)^2 + \xi^2\left(\left(\frac{\partial\eta}{\partial x}\right)^2 + \eta^2(q - q_0\cos\theta)^2\right). \tag{13}$$

The corresponding equilibrium Euler-Lagrange equations read

$$\xi^2\frac{\partial^2\eta}{\partial x^2} = -\eta + \eta^3 + \eta\left(\frac{q}{q_0} - \cos\theta\right)^2 q_0^2\xi^2, \tag{14}$$

$$\lambda^2\frac{\partial^2\theta}{\partial x^2} = 2\eta^2\left(\frac{q}{q_0} - \cos\theta\right)\sin\theta. \tag{15}$$

Due to the symmetry of the problem (see figure 1) the boundary conditions at *x*=0 read $\frac{\partial \eta}{\partial x}=0$ and $\theta=0$. At *x*=*R*, where we set *R*>max{ λ,ξ }, we impose $\frac{\partial \eta}{\partial x}=0$ and $\theta=\theta_t$. The latter conditions enforce an equilibrium smectic ordering at this distance. The equilibrium equations are solved numerically for a given set of parameters.

In our simulations in studying DSH patterns it is expected that $\eta(x)$ exhibits a minimal value at *x*=0 and recovers its bulk degree of ordering roughly at the distance $x \sim \xi$. On the other hand $\theta(x)$ monotonously increases from $\theta(x=0)=0$ to the value $\theta(x=R)=\theta_t$. One intuitively expects that on increasing x the condition $\vec{n} \sim (\sin\theta_t, 0, \cos\theta_t)$ would be realized at $x \sim \lambda$. However, in our simulations η and θ might recover the Sm*A* bulk-like structural properties at different distances. Namely, ξ and λ are estimated for "ideal" reference conditions. In deriving the expression for ξ the Sm*A* layer position and $\vec{n}$ are undistorted. Similarly, in the expression for λ, the Sm*A* layers are assumed to be undistorted, i.e., they form ~~an~~ uniform stack of layers with thickness $d=d_0$ and $\eta=\eta_{eq}$. In our cases there are deviations from these conditions. Furthermore, the smectic phase factor and director field are expected to respond on different length scales in ideal condition. Let us recall that ϕ plays the role of "gauge" field [4] in the Sm*A* phase which is known to respond to imposed frustrations on geometrically-imposed scales [4]. In contrast, $\vec{n}$ is expected to recover from locally induced distortions on the distance λ [15]. However, in our configurations both fields are distorted and coupled in a complex way as indicated in Eq.(12). For this reason we introduce the effective smectic order parameter correlation length ξ_{eff} and λ_{eff} as the distances from *x*=0, where the conditions $\eta_{eq}-\eta(\xi_{eff})=\frac{1}{2}\left(\eta_{eq}-\eta(0)\right)$ and $\theta(\lambda_{eff})=\frac{1}{2}\theta_t$ are realized, respectively. $(\xi/d_0=\lambda/d_0>3)$ the smectic melting becomes pronounced.

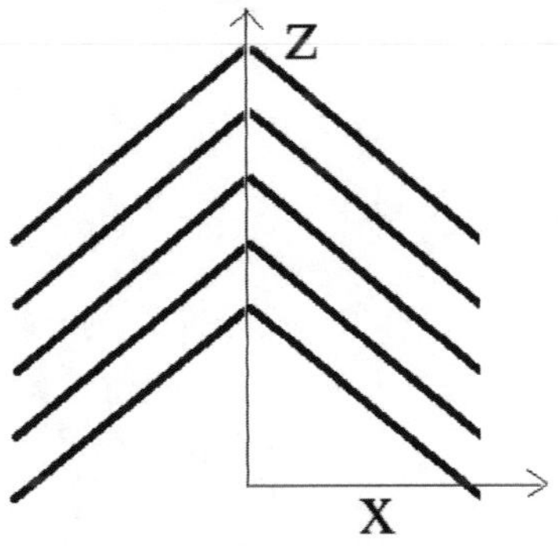

Figure 1. Schematic presentation of a herringbone-type Sm*A* pattern in the Cartezian (*x,z*) plane. Smectic layers are tilted for an angle θ_t with respect to the *z*-axis. LC molecules are forced to be aligned along the local normal of a smectic layer. The tip of the structure is placed at *x*=0. At the tip $\vec{n}$ is aligned parallel with the z-axis.

4. Defectless Smectic Herringbone pattern

We study first the DSH patterns. Of our special interest are conditions at which melting at the tip of a DSH pattern is essential. We consider both type I and type II Sm*A* LCs where we vary θ_t as well as the temperature. We probe values $\theta_t = 10^0$ and $\theta_t = 20^0$ corresponding to the minimal and maximal tilt angle experimentally realized in DSH-type structures in the type I Sm*A* studied in [14]. In our simulations a relatively low temperature deep in the Sm*A* phase roughly corresponds to $\min\{\xi,\lambda\} = d_0$. On approaching the 2nd order N-Sm*A* phase transition both λ and ξ tend to diverge.

In figure 2 we first plot the spatial dependences of θ / θ_t (dashed lines) and η / η_{eq} (full lines) for $\kappa = 1$, corresponding to the crossover regime between the Sm*A* I and II behaviour. A relatively large tilt angle ($\theta_t = 20^0$) and consequently the strong elastic strain at the tip of DSH patterns are imposed. At a relatively low temperature ($\xi / d_0 = \lambda / d_0 = 1$) smectic melting at the tip is relatively weak. Furthermore, distortions in η and θ recover on scales roughly given by ξ and θ as expected. However, at higher temperatures ($\xi / d_0 = \lambda / d_0 > 3$) the smectic melting becomes pronounced.

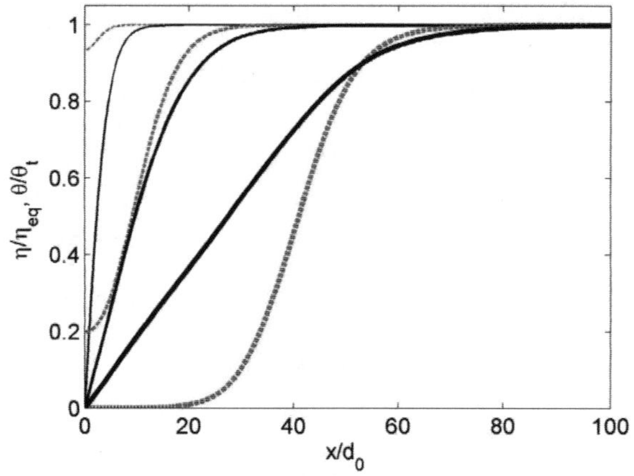

Figure 2. $\theta(x)$ (solid lines) and $\eta(x)$ (dashed lines) dependencies for different temperatures (thin lines: $\xi / d_0 = 1$; thick lines: $\xi / d_0 = 3$; very thick lines: $\xi / d_0 = 5$). $\theta_t = 20^0$, $\kappa = 1$.

In figures 3 we compare $\theta(x)$ and $\eta(x)$ responses in the type I and type II Sm*A* phases at a relatively low temperature ($\min\{\xi,\lambda\} = 1$). For $\theta_t = 20^0$ and type I Sm*A* (see figure 3a) the smectic order parameter is relatively reduced (for about 90% for $\kappa = 0.2$) at the tip. For the imposed lengths $\xi / d_0 = 5$ and $\lambda / d_0 = 1$ we obtain $\lambda_{eff} \approx 5.5 d_0$. On the other hand, only relatively weak departures from η_{eq} are observed for $\kappa = 5$ (the imposed lengths $\xi / d_0 = 1$ and $\lambda / d_0 = 5$) where we observed $\xi_{eff} \approx 12 d_0$.

Some further type I patterns (by imposing $\xi / d_0 = 10$ and $\lambda / d_0 = 1$) are shown in figure 3b for $\kappa = 0.1$. For $\theta_t = 20^0$ we obtain in this case a pronounced smectic melting at the tip and $\xi_{eff} \approx 20 d_0$. On the contrary, for $\theta_t = 10^0$ the imposed strain is too weak to melt the smectic ordering at relatively low temperatures. In figure 4 we demonstrate the changes in $\eta(x)$ and $\theta(x)$ behavior on increasing temperature towards the N-Sm*A* phase transition for $\theta_t = 10^0$ and $\kappa = 0.1$, where the initial configuration obtained for $\lambda / d_0 = 1$ is shown in figure 3b. One sees that, on increasing *T* (and consequently the value of λ), the degree of smectic ordering progressively decreases. A complete smectic melting at *x*=0 is reached for $\lambda \approx 5 d_0$. Above this value of λ, the width of melted region, represented by ξ_{eff}, begins to apparently increase on increasing *T*. Therefore, our study reveals that, for the type I Sm*A*, a total melting of smectic ordering can be realized at the tip of the herringbone-type structures even for relatively small values of the layer tilt (i.e., $\theta_t \approx 10^0$) that could be experimentally realized using AFM technique [14].

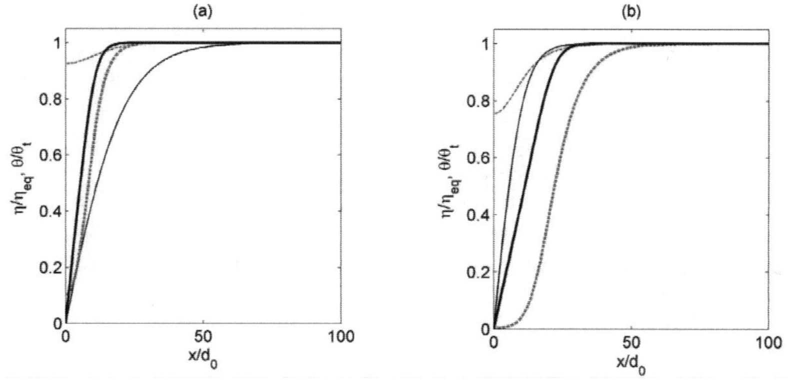

Figure 3. $\theta(x)$ (solid lines) and $\eta(x)$ (dashed lines) dependencies. (a) thin line: $\kappa = 5$, thick line $\kappa = 0.2$; $\theta_t = 20^0$. (b) thin line: $\theta_t = 10^0$, thick line $\theta_t = 20^0$; $\kappa = 0.1$.

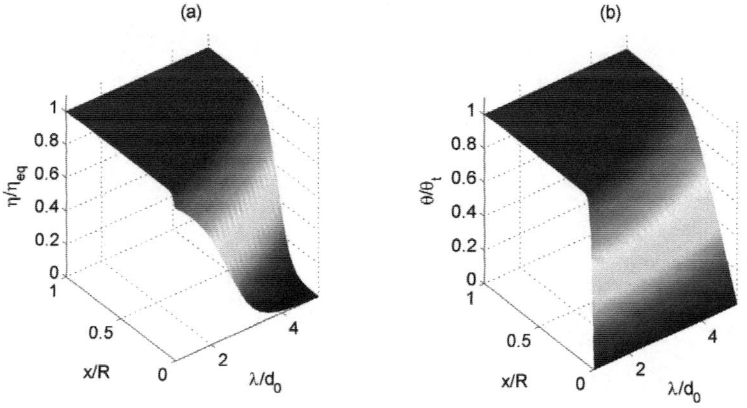

Figure 4. (a) $\eta(x)$ and $\theta(x)$ variations as a function of λ . $\kappa = 0.1$, $\theta_t = 20^0$.

5. Dislocation Decorated Smectic Herringbone patterns

In the following we derive the critical condition for which edge dislocations [3,17] are introduced into samples in order to partially relax the imposed elastic strain at the tip of herringbone-type patterns. For this purpose we estimate the free energy costs of the competing DDSM and DSM patterns. The critical condition is estimated by the energy balance of these patterns.

The basic steps of this approximate derivation are as follows. We take into account that at the herringbone tip the smectic layers tend to be dilated. Namely, geometrical reasons enforce the layer distance $d = d_0 / \cos\theta_t = 2\pi / q$ which is larger than the equilibrium spacing d_0. Consequently, the layer bending compressibility free energy density penalty at x=0 equals to [17]

$$f_e \square C\eta_{eq}^2 (q - q_0)^2 . \qquad (16)$$

This layer distortion roughly extends over the distance $\square\, \lambda$ from the tip along the z-axis. Assuming a homogeneous structure over a distance h along the z-axis the resulting total elastic free energy penalty $\Delta F^{(DSH)} = \iint f_e dxdz$ reads

$$\Delta F^{(DSH)} \square C\eta_{eq}^2 (q - q_0)^2 h\lambda . \qquad (17)$$

Next we assume that an edge dislocation appears within the distance h. We assume that the presence of the dislocation totally relaxes the layer strain at the tip. Therefore, for $d=d_0$ at x=0 the distance within which one dislocation is inserted equals to

$$h = \frac{d_0}{1 - \cos\theta_t} . \qquad (18)$$

The relationship follows from the condition $h(q_0 - q) = 2\pi$. Consequently, the compressibility penalty approximately vanishes at the tip area. However, the necessary topological presence of the edge dislocation introduces in the (x,z) plane a region of surface area $\square\, \pi\xi^2$ where the translational ordering is melted. The related total condensation penalty $\Delta F^{(DDSH)} = \iint f_c dxdz$ is approximated by

$$\Delta F^{(DDSH)} \square \frac{(\alpha(T - T_c))^2}{\beta} \pi\xi^2 . \qquad (19)$$

From the condition $\Delta F^{(DDSH)} = \Delta F^{(DSH)}$ an estimate is obtained for the critical tilt angle $\theta_t^{(c)}$ where the DDSH-type pattern could appear

$$\theta_t^{(c)} \square ArcCos\left(1 - \frac{1}{q_0\lambda}\right) . \qquad (19)$$

In the limit $\lambda \to \infty$ it follows $\theta_c^{(t)} = 0$ because melting does not cost any energy at the phase transition.

6. Conclusions

We have studied theoretically and numerically herringbone-type patterns in the liquid-crystalline SmA phase. This study has been inspired by experimental work [14] in which surprisingly broad region of melted smectic order at the tip of herringbone-type structures was reported in type I Sm*A*. By using the atomic force microscope, nanopatterned substrates that enforce such structures were prepared. In that study the smectic Ginzburg parameter was estimated in a range between 0.1 and 0.2. The tilt angles, characterising the herringbone patterns, were ranged between $\theta_t = 10^0$ and $\theta_t = 20^0$. The width of the melted region was substantially larger than the relevant bulk penetration or the smectic order parameter correlation length. Recent studies reveal that such localised melted regions could efficiently trap appropriate nanoparticles [8,9,10,11,12,13]. Therefore, it is of interest to understand the key parameters controlling the existence and the width of localised melted regions in these patterns.

In our study we have used Landau-de Gennes Ginzburg mesoscopic approach [3] in standard terms of the nematic director field and the complex smectic order parameter. We used minimal model in effectively two-dimensional system exhibiting the 2nd order N-Sm*A* phase transition to test the impact of material parameters and temperature on the width of the smectic melted region at the tip of herringbone patterns. We used single elastic constant approximations where the nematic and the smectic elasticities were approximated by the constants *K* and *C*, respectively.

In studying the degree of smectic melting in Defectless Smectic Herringbone (DSH) patterns we have assumed one-dimensional spatial variations $\eta(x)$ and $\theta(x)$. We analyzed the impact of the material, geometrical properties (represented by κ and θ_t, respectively) and the temperature on LC patterns. Our simulations reveal that relatively wide melted regions, with respect to the predicted bulk values of the relevant material characteristic lengths, could be realized even for relatively small tilt angles (θ_t □ 10^0) for high enough temperatures in the type I Sm*A* phase. Furthermore, we estimate conditions for which edge dislocations are expected to penetrate smectic patterns, giving rise to the Dislocation Decorated Smectic Herringbone (DDSH) pattern. In this derivation we assumed that main free energy penalties in DSH and DDSH patterns arise due to dilatation of smectic layers at the herringbone tip and the melting of smectic ordering within the cores of dislocations, respectively.

In near future we plan to study trapping of appropriate NPs to DSH and DDSH patterns both theoretically and experimentally. The former structures display planar trapping walls while the latter patterns consist of linear traps consisting of edge dislocations. We will calculate the mechanical forces on NPs immersed to different LC regions using the continuum approach presented in [17].

References

[1] Balazs A C, Emrick T and Russel T P 2006 *Science* **314** 1107
[2] Lin I H, Miller D S, Bertics P J, Murphy C J, de Pablo J J and Abbott N L 2011 *Science* **332** 1297
[3] De Gennes P G 1974 *The Physics of Liquid Crystals* (Oxford University, London)
[4] Palffy-Muhoray P 2007 *Phys. Today* **60** 54
[5] Li F H, Buchnev O, Cheon C, Clushchenko A, Reshetnyak V, Reznikov Y and Sluckin T J 2006 *Phys. Rev. Lett.* **97** 147801
[6] Lagerwall J, Scalia G, Haluska M, Dettlaff-Weglikowska U, Roth S and Giesselmann F, 2007 *Adv. Mater.* **19** 359.
[7] Schoot P, Popa-Nita V and Kralj S 2008 *J Phys. Chem.* B **112** 4512
[8] Kikuchi H, Yokota M, Hisakado M, Yang H and Kajiyama T 2002 *Nature Mater.* **1** 64
[9] Karatairi E, Rožič B, Kutnjak Z, Tzitzios V, Nounesis G, Cordoyiannis G, Thoen J, Glorieux C and Kralj S 2010 *Phys. Rev. E* **81** 041703
[10] Cordoyiannis G, Jampani V S R, Kralj S, Dhara S, Tzitzios V, Basina G, Nounesis G, Kutnjak Z, Tripathi C S P, Losada-Perez P, Jesenek D, Glorieux C, Muševič I, Zidanšek A, Amenitsch H and Thoen J 2013 *Soft matter* **9** 3956
[11] Zappone B, Lacaze E, Hayeb H, Goldmann M, Boudet N, Baroisdf P and Albae M 2011 *Soft Matter* **7** 1161
[12] Coursault D, Grand J, Zappone B, Ayeb H, Levi G, Felidj N and Lacaze E 2012 *Adv. Mater.* **24** 1461
[13] Yoon D K, Choi M C, Kim Y H, Kim M W, Lavrentovich O D, and Jung H 2007 *Nat. Mater.* **6** 866

[14] Wang R, Syed I M, Carbone G, Petschek R G and Rosenblatt C 2006 *Phys. Rev. Lett.* **97** 167802
[15] Lubensky T C and Renn S R 1990 *Phys. Rev. A* **41** 4392
[16] Kralj S and Sluckin TJ 1994 *Phys. Rev. E* **50** 2940
[17] Rosso R, Virga E G and Kralj S 2006 *Phys. Rev. E* **74** 061703

Journal of Physics: Conference Series **558** (2014) 012010 doi:10.1088/1742-6596/558/1/012010

Modeling of closed membrane shapes

S Penič[1*], L Mesarec[1*], M Fošnarič[1*], V Kralj Iglič[2], S Kralj[3,4], W Góźdź[5] and A Iglič[1]

[1]Laboratory of Biophysics and Laboratory of Bioelectromagnetics, Faculty of Electrical Engineering, University of Ljubljana, Traka 25, SI-1000 Ljubljana, Slovenia

[2]Laboratory of Clinical Biophysics, Faculty of Health Sciences, University of Ljubljana, Zdravstvena 5, SI-1000 Ljubljana, Slovenia

[3]Department of Physics, Faculty of Natural Sciences and Mathematics, University of Maribor, Koroka cesta 160, SI-2000 Maribor, Slovenia

[4]Joef Stefan Institute, P.O. Box 3000, SI-1000 Ljubljana, Slovenia

[5]Institute of Physical Chemistry, Polish Academy of Sciences, Kasprzaka 44/52, 01-224 Warsaw

*Authors contributed equally to this work.

E-mail: samo.penic@fe.uni-lj.si, mesarec.luka@gmail.com, miha.fosnaric@fe.uni-lj.si, veronika.iglic@fe.uni-lj.si, samo.kralj@um.si, wojciech.gozdz@gmail.com, ales.iglic@fe.uni-lj.si

Abstract. Closed biological membranes were considered within the spontaneous curvature model. Ground state membrane shapes were compared with Monte Carlo simulations in the thermal equilibrium, where membranes are subject to thermal fluctuations. The results of the two approaches correspond well with each other. The oblate discocyte membrane shapes are obtained in the ground state but can become metastable when thermal fluctuations are taken into account. The nematic ordering in oblate and stomatocyte vesicle membranes was also studied. It was confirmed that the net topological charge on the surfaces with the topology of a sphere was 2. On the oblate vesicle four topological defects, each with charge 1/2, assembled in the region exhibiting the highest Gaussian curvature. On the stomatocyte vesicle with six topological defects, each with charge 1/2, and two topological antidefects, each with charge −1/2, the latter assembled in the region with a negative Gaussian curvature. The position of topological defects is strongly curvature dependent.

1. Introduction

Biological cells are enclosed by a membrane that serves as a barrier separating the interior of the cell from its surroundings [1]. Biological membranes are actively or passively engaged in trans-membrane transport, encapsulation [2, 3], communication between cells and/or cell organelles and the cell's waste control. Biological membranes are complex aggregates, composed of lipid molecules, carbohydrates, proteins and other components [4].

The main components of biological membranes are lipid molecules that are organized in a molecular bilayer [5]. The thickness of the lipid bilayer is around 5 nm, which is several orders of magnitude smaller than the lateral dimensions of the lipid vesicles or cells they enclose. This enables us to use the continuum approach in the theoretical description of membrane surfaces. Lipid bilayers in normal conditions exhibit very low resistance to bending and are therefore subject to thermal fluctuations [6].

Journal of Physics: Conference Series **558** (2014) 012010 doi:10.1088/1742-6596/558/1/012010

Human erythrocytes have been intensively studied in the past. Much theoretical work has been devoted to understanding the origin of different shapes of red blood cells, such as discocytes or stomatocytes. One of the models that successfully predicted the stability these shapes is Helfrich's spontaneous curvature model [7]. Within the Helfrich's spontaneous curvature model stomatocytes and discocytes were calculated for the first time using the calculus of variation. The shapes were in good agreement with those observed in experiments.

Biological membranes can also exhibit in-plane ordering, for example lipid bilayers in regions of high curvature [8], or coated with proteins [9]. The shapes of viruses can also depend on the presence of disclinations in their protein shell [10]. A mimimal model capturing the main phenomena related to the orientational order in the membrane has been developed for colloids coated with a thin sheet of nematic liquid crystals, also referred to as nematic shells [11, 10]. Liquid crystal molecules are oriented within the tangent plane of the shell. In nematic shells with the topology of a sphere, topological defects are inevitable [12]. At the origin of topological defects, the orientational order is melted. The position of topological defects are curvature dependent [10, 13, 14, 15, 16]. On a spherical surface, the equilibrium configuration typically has four topological defects residing in vertices of a tetrahedron in order to maximize their mutual separation, which has been proved analytically [17] and confirmed experimentally [18].

In this work, we consider closed membrane surfaces with the topology of a sphere. We present different methods for studying membrane configurations we previously described in [19] and furthermore show some new, unpublished results. The paper is organized as follows. In Sec. 2, we introduce the theoretical models and briefly discuss the numerical procedures used to calculate equilibrium closed membrane shapes and nematic ordering. In Sec. 3, we present some of the results of our calculations and compare the closed membrane shapes in a ground state to those subject to thermal fluctuations. We also determine the nematic ordering on a membrane surface. In Sec. 4, we present our conclusions.

2. Models and methods

2.1. Bending energy of the membrane

Many theoretical approaches have been used in order to study the shape transformations of the closed shapes of biological membranes. In this work we used the model formulated by W. Helfrich [20] already in 1973, which treats the membrane as a homogeneous two-dimensional fluid surface characterized by the local bending energy density of the membrane in the form [7, 20, 21, 22, 23]:

$$f_b = \frac{\kappa}{2}(C_1 + C_2 - C_0)^2 + \overline{\kappa}\, C_1 C_2\ , \tag{1}$$

where κ and $\overline{\kappa}$ are the membrane bending stiffness and saddle-splay modulus, respectively, C_1 and C_2 are the principal curvatures, and C_0 is the spontaneous curvature of the surface. Similar models were proposed by Canham [24] and Evans [21].

The overall bending energy of the membrane, F_{tot}, can be obtained by integrating Eq. (1) over the whole membrane surface:

$$F_{tot} = \int_S f_b\, \mathrm{d}S, \tag{2}$$

where $\mathrm{d}S$ is an infinitesimal element of the vesicle area S. From the Gauss-Bonnet theorem we know that the last term on the right-hand side of Eq. (1), if integrated over the closed surfaces of fixed topology, is a constant.

In the present work, we consider closed membrane shapes with the topology of a sphere. The last term on the right-hand side of Eq. (1) does not influence our shape analysis and can therefore be neglected. The bending stiffness κ of various lipid bilayers was experimentally measured (see, for example, [25], [26], [27], [28], [29], [30] and references therein).

Journal of Physics: Conference Series **558** (2014) 012010 doi:10.1088/1742-6596/558/1/012010

2.2. Ground state calculations

We neglect membrane shape fluctuations due to thermal noise. We consider closed membrane shapes which are in the energy ground state and axisymmetric. To theoretically describe such shapes, we need to define the contour of the vesicle cross-section. Let the shapes be symmetric about the z-axis. The curve that defines the contour of the vesicle cross-section is rotated around the z-axis by 2π, which defines the surface of the vesicle. To describe the vesicle contour in the r-z plane, we introduce the arc length of contour s and angle, $\theta(s)$, between the tangent to the contour and the plane perpendicular to the rotation axis z. If $\theta(s)$ is known, the vesicle contour can be calculated by the following parametric equations:

$$z(s) = \int_0^s \sin\theta(s')\,\mathrm{d}s', \tag{3}$$

$$r(s) = \int_0^s \cos\theta(s')\,\mathrm{d}s', \tag{4}$$

where $r(s)$ and $z(s)$ are the coordinates of vesicle contour in r-z plane. We can express the function $\theta(s)$ in form of the Fourier series [31]:

$$\theta(s) = \frac{\theta_0}{L_s}s + \sum_{n=1}^{N} a_n \sin\left(\frac{n\pi}{L_s}s\right), \tag{5}$$

where L_s is the contour length, N is the number of Fourier modes and a_n are the Fourier amplitudes, which are calculated by minimizing the bending energy F_{tot}. If we want to restrict our calculations to the closed membrane shapes, we have to apply the following boundary conditions: $\theta(0) = 0$, $\theta(L_s) = \pi$, $r(0) = r(L_s) = 0$, which means that $\theta_0 = \pi$ in Eq. (5). The bending energy F_{tot} is basically the function of Fourier amplitudes a_n and the contour length L_s, therefore we need perform the numerical minimization of the function of many variables in order to calculate the equilibrium vesicle shapes [31, 32, 33]. In the minimization process, the constraints on the surface area and the volume of the vesicle have to be taken into account in order to ensure fixed value of the reduced volume v. The reduced volume v is defined as a ratio of the vesicle volume to the volume of the sphere with the same surface area.

2.3. Nematic shells

We have studied nematic ordering on the vesicles, which we calculated within the spontaneous curvature model as described above. We use the minimal model of in-plane ordering within vesicle membranes, which was developed to study nematic shells [11, 34]. Molecules are bound to lie in the local tangent plane of the surface. To describe the orientational ordering of molecules, we define the surface order tensor $\mathbf{Q}$. We introduce a local orthonormal basis $(\boldsymbol{e}_1, \boldsymbol{e}_2)$ in which $\mathbf{Q}$ is defined as [13, 35]:

$$\mathbf{Q} = q_0(\boldsymbol{e}_1 \otimes \boldsymbol{e}_1 - \boldsymbol{e}_2 \otimes \boldsymbol{e}_2) + q_m(\boldsymbol{e}_1 \otimes \boldsymbol{e}_2 + \boldsymbol{e}_2 \otimes \boldsymbol{e}_1),$$

where q_0 and q_m are scalar functions. In a diagonal form, tensor $\mathbf{Q}$ can be written as:

$$\mathbf{Q} = \lambda(\boldsymbol{n} \otimes \boldsymbol{n} - \boldsymbol{n}_\perp \otimes \boldsymbol{n}_\perp),$$

where λ and $-\lambda$ are eigenvalues of eigenvectors $\boldsymbol{n}$ and $\boldsymbol{n}_\perp$. Value λ is a measure for the orientational order with values in the interval $\lambda \in [0, 1/2]$. Value $\lambda = 0$ corresponds to the isotropic state with no orientational order at all, while the value $\lambda = 1/2$ corresponds to the maximal degree of the orientational order. Topological defects are points with no orientational order, therefore they usually occur where $\lambda = 0$. The basic characteristic property of the topological defect is its topological charge [36, 35], defined as the winding number of the vector field of molecules on the tangent plane at the defect [13]. Topological charge has positive value for topological defects and negative value for topological antidefects. The theorem of Poincaré [37] states that the net topological charge is determined by a surface topology and is equal to 2 for a sphere and all surfaces obtained by smoothly deforming a sphere.

The direction of molecules and the orientational order at any given point on a vesicle can be calculated from the values of q_0 and q_m. We minimize the free energy in order to calculate the nematic ordering. In the simplest form, the dimensionless free energy density is [35]:

$$\tilde{f} = \left(\frac{R}{\xi_0}\right)^2 \left(t\; Tr\tilde{\mathbf{Q}}^2 + \frac{1}{4}\left(Tr\tilde{\mathbf{Q}}^2\right)^2\right) + \frac{1}{2}\left|\tilde{\nabla}_s\tilde{\mathbf{Q}}\right|^2, \qquad (6)$$

where R is a characteristic length of the vesicle membrane and ξ_0 a nematic coherence length, which is the shortest length in the system, typically in the nanoscopic range. An ordered phase occurs below critical temperature, when t is negative. Operator $\tilde{\nabla}_s$ represents the surface gradient. In order to get the dimensionless expression for the energy density, we have scaled tensor $\mathbf{Q}$ and all the distances, which is why we used the tilde notation.

By integrating Eq. (6) over the whole surface of a vesicle membrane the total free energy is obtained and Monte Carlo method is used to minimize the total free energy. Values of $\tilde{q}_0$ and $\tilde{q}_m$ are randomly changed at random locations on the vesicle surface, until we reach the equilibrium configuration. The tilde notation denotes the scaled q_0 and q_m in the dimensionless expression for energy (Eq. (6)).

2.4. Thermal equilibrium calculations

To go beyond the ground state and model systems in the thermodynamic equilibrium, one needs to consider also thermal noise. Biological systems are often "soft enough" that thermal energy can induce important configuration changes of the system. As already mentioned in Sec. 1, lipid bilayers are, due to their low resistance to bending, subject to thermal shape fluctuations. In this section we describe the Monte Carlo simulations that take thermal fluctuations into account.

The fluid vesicle is represented by a set of N vertices that are linked by tethers (*i.e.* bonds) of variable length l in order to form a closed, randomly triangulated, self-avoiding network [38, 39]. The lengths of the tethers can vary between a minimal value, l_{min}, and a maximal value, l_{max}. The self-avoidance of the network is implemented by ensuring that no vertex can penetrate through the triangular network and that no bond can cut through another bond. In our simulations we assume $l_{max}/l_{min} = 1.7$.

The randomly triangulated network acquires its lateral fluidity from a bond flip mechanism [38, 39]. A single bond flip involves the four vertices of two neighboring triangles. The tether connecting the two vertices in a diagonal direction is cut and reestablished between the other two, previously unconnected, vertices.

The microstates of the vesicle are sampled according to the Metropolis algorithm, with the energy for a given microstate

$$E = F_{tot} - \Delta p\, V, \qquad (7)$$

where the first contribution is the elastic bending energy of the vesicle (Eq. (2)) and the second contribution accounts for the energy change with the change of the volume of the vesicle, V, due to the pressure difference, Δp, inside and outside of the vesicle. Our vesicle consists of a symmetric membrane (including the absence of a mismatch between the lateral areas of the two individual membrane leaflets), therefore we do not need to include the spontaneous curvature ($C_0 = 0$). Vesicle also does not change it's topology, which means that the Gaussian bending contribution (last term on the right-hand side of Eq. (1) integrated over the vesicle surface) is constant and therefore not taken into account in our calculations. The bending energy F_{tot} of

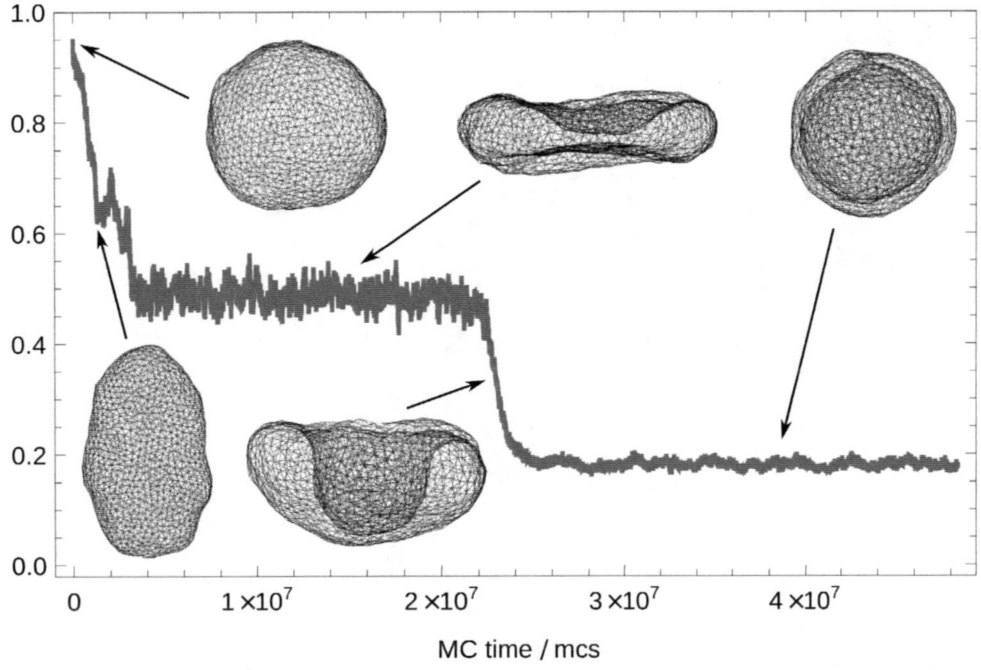

Figure 1. Monte Carlo evolution of the closed membrane vesicle from an initial quasi-spherical state towards an equilibrium stomatocyte state. The pressure difference is $\Delta p = -0.075$. The relative volume of the vesicle is presented as a function of the Monte-Carlo time (measured in mcs). The average relative volume in the equilibrium stomatocyte state is $\langle v \rangle = 0.181 \pm 0.007$, while in the metastable oblate discocyte state $\langle v \rangle = 0.50 \pm 0.02$. Above the curve, three snapshots of the vesicle are shown: the inital quazi-sphere, the discocyte and the stomatocyte. Below the curve, two snapshots of the vesicle are shown: the left configuration occurred during the transition from the quasi-sphere towards the metastable discocyte and the right configuration occurred during transition from the discocyte towards the equilibirum stomatocyte state.

the discretized vesicle (i.e., of the triangulated network) is calculated as described by Gompper and Kroll [39, 38]; for a recent review, see [40].

The evolution of the system is measured in *Monte Carlo sweeps* (mcs). One mcs consists of individual attempts to displace each of the N vertices by a random increment in the sphere with radius δ, centered at the vertex, followed by $R_{\mathrm{b}}N$ attempts to flip a randomly chosen bond. We denote R_{b} as the bond-flip ratio, which defines how many attempts to flip a bond are made per one attempt to move a vertex in one mcs. Note that the bond-flip ratio is connected to the lateral diffusion coefficient within the membrane, *i.e.* to the membrane viscosity [41, 42]. Diffusion also introduces a real time scale in the simulations and allows simulation of the dynamics of the modelled system (not considered in this work). In our simulations we have chosen $R_{\mathrm{b}} = 3$ and $\delta/l_{min} = 0.15$.

In our simulations the vesicle consists of $N = 1447$ vertices, which are connected with $3(N-2) = 4335$ bonds to form $N_{\mathrm{t}} = 2(N-2) = 2890$ triangles. The spherical vesicle has therefore a radius of approx. 13. During simulations the coordination number for each vertex (*i.e.* the number of nearest neighbors, Z) is allowed to vary between 4 and 8. For the bending stiffness of the vesicle we use $\kappa = 10\ k_BT$, where k_B is the Boltzmann constant and T the

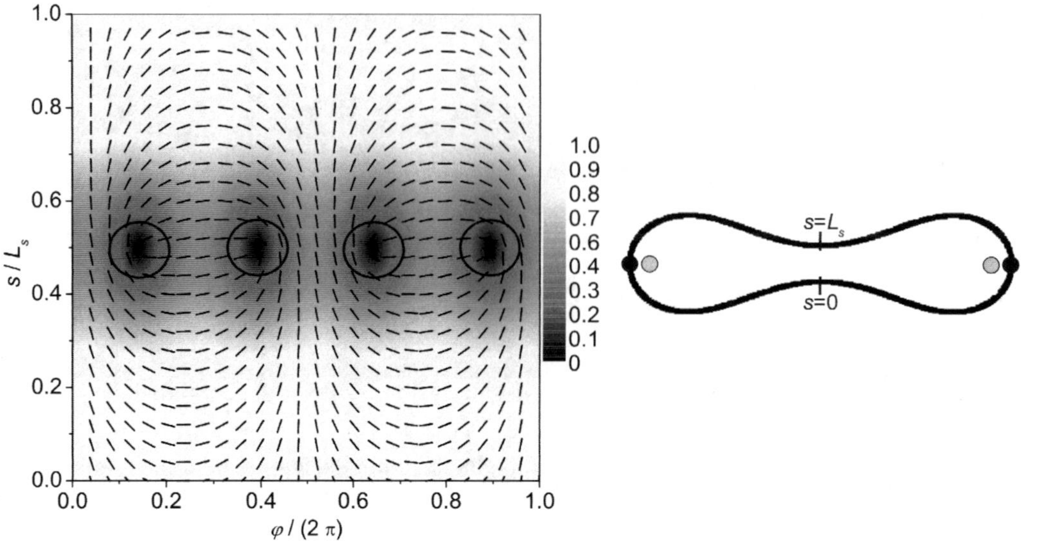

Figure 2. The calculated vector field in the membrane and contour plot of λ/λ_c ratio, where condensation value λ_c stands for the highest possible value of λ at given $t < 0$ (left panel). Configuration was calculated for $R/\xi_0 = 20$ and $t = -0.03$. The right panel shows the oblate vesicle shape with the reduced volume $v = 0.60$, calculated within the spontaneous curvature model for $C_0 = 0$. In the left panel topological defects with topological charge 1/2 are encircled with black lines. The right panel shows approximate positions of topological defects, which are represented by the black and grey dots.

absolute energy. In the following we use l_{min} as the unit length and $k_B T$ as the unit of energy.

3. Results and Discussion

The result of minimizing the bending energy F_{tot}, as defined in Section 2, yield ground state configurations of stomatocytic, oblate and prolate shapes, each of them being the equilibrium shapes in some range of the reduced volume v, as reported in [19]. Similar results were obtained in [43, 44]. In [19] we also compared minimal energy shapes (as described in Sec. 2.2) and thermal equilibrium shapes 2.4. The method described in Sec. 2.2 allows us to set the value for the reduced volume v, so that we are able to calculate the whole spectrum of shapes for different reduced volumes. In Monte Carlo simulations we have changed the parameter Δp (inside/outside pressure difference). Once we have obtained the equilibrium state for the given Δp, we were also able to calculate the average reduced volume $\langle v \rangle$ for that state, in order to compare thermal equilibrium and ground state calculations. Thermal equilibrium states obtained by Monte Carlo simulations correspond well to the ground state shapes calculated by the minimization of the bending energy.

Ground state oblate shapes are possible only in a small region of the values of the reduced volume v. By using Monte Carlo simulations we were not able to obtain the equilibrium but only "transient" oblate shapes. Figure 1 shows the Monte Carlo evolution of the vesicle from an initial quasi-spherical towards an equilibrium stomatocyte state. As can be seen, the vesicle spends a relatively long "time" in a metastable oblate discocyte state before it reaches the equilibrium state.

We calculated the nematic ordering of the oblate vesicle membrane with the reduced volume

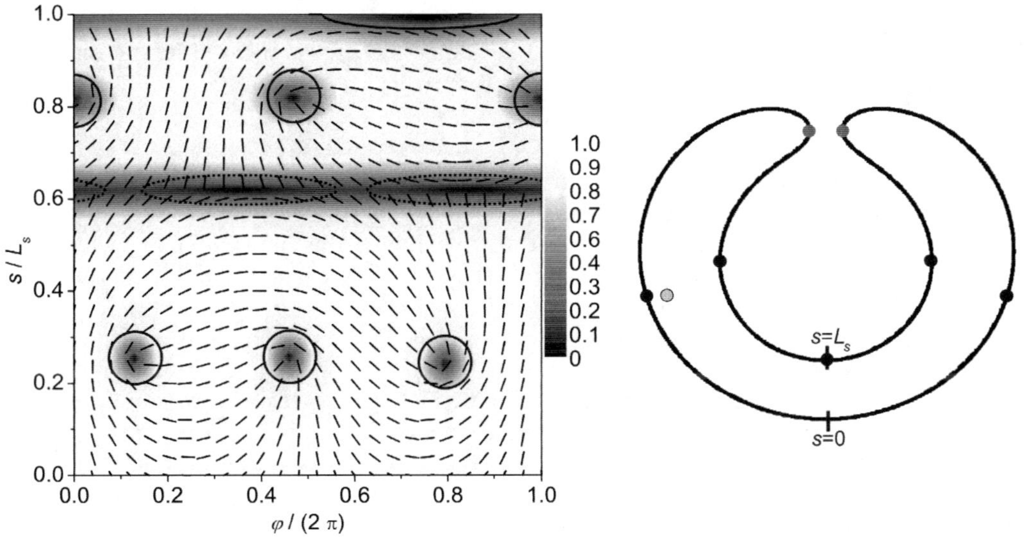

Figure 3. The calculated vector field in the membrane and contour plot of λ/λ_c ratio, where condensation value λ_c stands for the highest possible value of λ at given $t < 0$ (left panel). Configuration was calculated for $R/\xi_0 = 40$ and $t = -0.03$. The right panel shows stomatocyte vesicle shape with the reduced volume $v = 0.50$, calculated within the spontaneous curvature model for $C_0 = 0$. In the left panel topological defects with topological charge 1/2 are encircled with black lines, while antidefects with topological charge $-1/2$ are encircled with black dashed lines. The right panel shows approximate positions of topological defects and antidefects. The black and grey dots represent topological defects, while the red dots represent antidefects.

$v = 0.60$, which was calculated within the spontaneous curvature model. In figure 2 the topological defects are points where $\lambda = 0$ (dark red color). The equilibrium configuration has four topological defects, each with charge 1/2, which satisfies the Poincaré theorem [37]. Topological defects occur in the region of the vesicle with the highest positive Gaussian curvature. Regions of the vesicle exhibiting lower Gaussian curvature have a higher degree of orientational order (yellow color). The equilibrium configuration and the positions of topological defects are thus strongly curvature driven as previously described in [13, 16].

The nematic ordering in the membrane of the stomatocyte vesicle (figure 3) is more complex, because some of the membrane regions exhibit negative Gaussian curvature, which is favorable for topological antidefects. In the membrane regions with a negative Gaussian curvature two antidefects occur, each with topological charge $-1/2$. The stomatocyte is composed of two approximately spherical surfaces and the neck which connects them. Both spherical surfaces have a positive Gaussian curvature, while the neck region has a negative Gaussian curvature. The positive Gaussian curvature is favorable for topological defects. To this end, we can observe three topological defects on each spherical surface, each with charge 1/2. If we sum up all topological charges on the membrane of the stomatocyte vesicle, we get the net topological charge 2, which satisfies the theorem of Poincaré [37]. In figure 3, we observe that the "outer" surface of the stomatocyte has a higher degree of orientational order than the "inner" surface. That happens because the "inner" surface is more curved. Membrane regions with a higher absolute value of the Gaussian curvature always have a lower degree of orientational order. Regions of a vesicle membrane with small absolute values of Gaussian curvature have a higher degree of orientational order. Defects and antidefects are membrane points with a very low degree of order, so it is not energetically favorable for them to be located in these places. As we can observe in figure 2 and figure 3, defects and antidefects always appeared in the membrane regions with high absolute values of the Gaussian curvature.

4. Conclusions

Closed biological membranes (vesicles) were considered within the spontaneous curvature model. The membrane bending energy minimization was performed to obtain ground state (zero temperature) vesicle shapes, while the Monte Carlo simulations of randomly triangulated surfaces were used to obtain membrane configurations in the thermal equilibrium (non-zero temperature). The obtained configurations, using the two approaches, correspond well with each other. In Monte Carlo simulations we did not fix the volume of the vesicle, as in the ground state calculations. Instead of that we imposed pressure difference between both membrane sides to generate vesicles with different average volumes. The vesicle had spent a relatively long "time" in the metastable oblate/discocyte state, but it later reached the equilibrium stomatocyte state, since the thermal fluctuations overcame the energy barrier. The backward transition was not observed, indicating that the stomatocyte state, if obtained, has a lower mean energy than the metastable oblate state.

The nematic ordering in oblate and stomatocyte vesicle membranes was also studied. In both cases it was confirmed that the net topological charge on surfaces with the topology of a sphere is equal 2. On an oblate vesicle, we calculated the equilibrium configuration with four topological defects, each with charge $1/2$, assembled in a region exhibiting highest Gaussian curvature. On a stomatocyte vesicle we observed six topological defects (each with charge $1/2$) and two topological antidefects (each with charge $-1/2$). The latter assembled in the region with negative Gaussian curvature. The position of topological defects is strongly curvature dependent and can be controlled by changing the vesicle curvature.

References

[1] Alberts B, Johnson A, Lewis J, Raff M, Roberts K and Walter P 2008 *Molecular biology of the cell* (Garland Science)
[2] Gao H J, Shi W D and Freund L B 2005 *Proc. Nat. Acad. Sci. USA* **102** 9469–9474
[3] Doherty G J and McMahon H T 2009 *Annu. Rev. Biochem.* **78** 857–902
[4] Sackmann E 1990 *Can. J. Phys.* **68** 999–1012
[5] Lipowsky R and Sackmann E 1995 *Structure and Dynamics of Membranes: I. From Cells to Vesicles/II. Generic and Specific Interactions* (Elsevier)
[6] Schneider M, Jenkins J and Webb W 1984 *Biophys. J.* **45** 891–899
[7] Deuling H J and Helfrich W 1976 *Biophys J.* **16** 861–868
[8] Kralj-Iglič V, Babnik B, Gauger D R, May S and Iglič A 2006 *J. Stat. Phys.* **125** 727–752
[9] Kabaso D, Bobrovska N, Góźdź W, Gov N, Kralj-Iglič V, Veranič P and Iglič A 2012 *J. Biomech.* **45** 231–238
[10] Vitelli V and Turner A M 2004 *Phys. Rev. Lett.* **93** 215301
[11] Nelson D R 2002 *Nano Lett.* **2** 1125–1129
[12] MacKintosh F and Lubensky T 1991 *Phys. Rev. Lett.* **67** 1169
[13] Kralj S, Rosso R and Virga E G 2011 *Soft Matter* **7** 670–683
[14] Selinger R L B, Konya A, Travesset A and Selinger J V 2011 *J. Phys. Chem. B* **115** 13989–13993
[15] Nguyen T S, Geng J, Selinger R L and Selinger J V 2013 *Soft Matter* **9** 8314–8326
[16] Bates M A, Skačej G and Zannoni C 2010 *Soft Matter* **6** 655–663
[17] Vitelli V and Nelson D R 2006 *Phys. Rev. E* **74** 021711
[18] Fernández-Nieves A, Vitelli V, Utada A S, Link D R, Márquez M, Nelson D R and Weitz D A 2007 *Phys. Rev. Lett.* **99** 157801
[19] Mesarec L, Fošnarič M, Penič S, Kralj-Iglič V, Kralj S, Góźdź W and Iglič A 2014 *Adv. in Soft Matter* (in print)
[20] Helfrich W 1973 *Z. Naturforsch.* **28** 693–703
[21] Evans E A 1974 *Biophys. J.* **14** 923–931
[22] Shi Z and Baumgart T 2014 *Adv. Colloid Interface Sci.* **208** 76–88

[23] Boulbitch A, Simson R, Simson D, Merkel R, Häckl W, Bärmann M and Sackmann E 2000 *Phys. Rev. E* **62** 3974
[24] Canham P B 1970 *J. of Theor. Biol.* **26** 61–76
[25] Genova J, Zheliaskova A and Mitov M 2006 *Coll. Surf. A* **282-283** 420–422
[26] Pavlič J, Genova J, Zheliaskova A, Iglič A and Mitov M 2010 *J. Phys. Conf. Ser.* **253** 012064
[27] Genova J, Kralj-Iglič V, Iglič A, Marinov R and Bivas I 2012 *J. Phys. Conf. Ser.* **398** 012037
[28] Genova J 2013 *Advances in Planar Lipid Bilayers and Liposomes* vol 17 (Academc Press, London) pp 1–27
[29] Vitkova V, Mitkova D, Stoyanova-Ivanova A, Kozarev N and Bivas I 2012 *C.R. Acad. Bulg. Sci.* **65** 329–334
[30] Dimova R 2014 *Adv. Colloid Interface Sci.* **208** 225–234
[31] Góźdź W T 2005 *J. Phys. Chem. B* **109** 21145–21149
[32] Góźdź W T 2004 *Langmuir* **20** 7385–7391
[33] Góźdź W T 2006 *J. Phys. Chem. B* **110** 21981–21986
[34] Vitelli V and Nelson D R 2006 *Phys. Rev. E* **74** 021711
[35] Rosso R, Virga E G and Kralj S 2012 *Continuum Mech. Thermodyn.* **24** 643–664
[36] Kamien R D 2002 *Rev. Mod. Phys.* **74** 953–971
[37] Poincaré H 1886 *J. Math. Pures Et Appl.* **2** 151–217
[38] Gompper G and Kroll D M 2004 *Statistical Mechanics of Membranes and Surfaces* ed Nelson D, Piran T and Weinberg S (World Scientific, Singapore) pp 359–426 2nd edition
[39] Gompper G and Kroll D M 1996 *J. de Physique I* **6** 1305–1320
[40] Ramakrishnan N, Kumar P B S and Ipsen J H 2011 *Macromol. Theory Simul.* **20** 446–450 ISSN 1521-3919
[41] Noguchi H and Gompper G 2004 *Phys. Rev. Lett.* **93** 258102
[42] Noguchi H and Gompper G 2005 *Phys. Rev. E* **72** 011901
[43] Seifert U, Berndl K and Lipowsky R 1991 *Phys. Rev. A* **44**(2) 1182–1202
[44] Bahrami A H, Raatz M, Agudo-Canalejo J, Michel R, Curtis E M, Hall C K, Gradzielski M, Lipowsky R and RWeikl T 2014 *Adv. Colloid Interface Sci.* **208** 214–224

Journal of Physics: Conference Series **558** (2014) 012011 doi:10.1088/1742-6596/558/1/012011

An automated experiment for determination of thin film semiconductor transport parameters

A FathAllah, F Ventosinos and C Longeaud
Laboratoire de Génie Electrique de Paris, CNRS (UMR 8507), Supelec, Université Paris Sud, Université Pierre et Marie Curie, 11 rue Joliot Curie, 91190 Gif sur Yvette, France

E-mail : longeaud@lgep.supelec.fr

Abstract : Thin film semiconductors are more and more the major components of solar devices. Nevertheless, these materials are still the subjects of intense research to improve their properties and reach better cell efficiency. Many experiments have been dedicated to the study of their opto-electronic parameters, most of them using the photoconductivity property of these films. The different techniques use several flux spatial and temporal 'shapes', the light impinging on the sample being uniform or with interferences, steady, modulated or chopped. This is why these different techniques are usually performed with different benches and the state of the sample may change from one bench to the other. In this communication we show that the light 'shape' can be tailored for the needs of several techniques by means of an electro-optic modulator. Therefore, these techniques can be gathered on the same bench, the samples being subsequently studied under the same conditions of vacuum and temperature. We shall show how we have achieved an 'all-in-one' bench and present some experimental results to illustrate all the possibilities of this new bench.

1. Introduction

Thin film materials (Hydrogenated amorphous silicon (a-Si:H), CIGS, Sb_2S_3, organic polymers,...) are more and more integrated into solar energy conversion devices. For this reason, lots of research has been carried out for measuring their transport parameters so as to optimize the deposition conditions and their properties. Some of them (e.g. a-Si:H, CIGS) are currently integrated into solar modules whereas others are still under intensive research for future applications. Several techniques developed to determine thin films transport parameters are based on their photoconductivity property, and only the light excitation spatial and temporal 'shape' is varied from one experiment to the other.

Considering a subset of experiments in which the photon energy is such that both type of carriers are generated, we can quote the following methods. The simplest one is the steady-state photoconductivity (SSPC) in which one records the evolution of the film photoconductivity with temperature and/or generation rate. More sophisticated techniques were designed such as the transient photocurrent (TPC) technique [1, 2] in which, after shining a short light pulse onto the film, the transient current is recorded over decades of time. In the frequency domain the modulated photocurrent (MPC) technique was proposed by H. Oheda [3]. Other experiments are based on illuminating the film with a light grating created by two interfering laser beams. The grating can be fixed as in the steady state photocarrier grating (SSPG) [4], can move at a constant velocity along the film as in the moving grating technique (MGT) [5], can be modulated at a given frequency like in the modulated photocarrier grating (MPG) [6] technique, or is oscillating on top of the sample as in the Oscillating Photocarrier grating (OPG) technique [7].

Each of these techniques brings some insight into the transport parameters and/or density of states (DOS) of the studied thin film. One of the main issues is that they are usually performed on different benches. Therefore, the state of the studied sample may vary from one bench to the other. For instance some experiments are performed under vacuum at different temperatures whereas others are usually achieved at room temperature in the air, not to speak of the time that elapses in between two experiments for which the sample can evolve if left in the air as for organic thin films. The ideal bench would be an all-in-one bench on which most of these experiments can be carried out under the same conditions but the different intensities and temporal and spatial shapes of the light used for each of them make this task more difficult.

In this communication we show that at least some of them can be automated and gathered onto a single bench, the light being shaped by means of an electro-optic modulator (EOM). We first present shortly the main experiments that can be gathered on a single bench. Then we shall show how an EOM can be set to properly shape the light and, finally, we shall show some experimental results obtained on a-Si:H and bulk hetero-junction polymer illustrating the versatility of this new bench.

2. Experiments

All the samples studied by the experiments described in the following are built in coplanar geometry, the thin films being deposited on an insulating substrate and fitted on top by two parallel and ohmic electrodes between which a bias is applied.

The SSPC technique is one of the most widely used experiments. From the study of the evolution of the current in the dark, the activation energy of the conductivity and the energy position of the dark Fermi level can be estimated. With light shone onto the sample at different fluxes it is possible to deduce the evolution of the photoconductivity with temperature and flux as well as the mobility lifetime products of the majority carriers ($\mu\,\tau_{maj}$).

The SSPG technique was first proposed by D. Ritter *et al.* [4]. It is one of the few experiments from which an ambipolar diffusion length L can be measured that turns out to be that of the minority carriers if their mobility lifetime product ($\mu\,\tau_{min}$) is much smaller than that of majority carriers ($\mu\,\tau_{maj}$). For more details on this technique the reader may have a look at the review by R. Brüggemann [8]. The basic idea is to illuminate the sample with an interference array that generates two arrays of space charge in the sample. These two arrays might disappear by ambipolar diffusion, each array interacting with the other due to their opposite charge. If it is not the case, for instance if one of the carriers is particularly slow, an internal field array appears that alters the majority carrier current originating from the applied external bias. The interference array is achieved by splitting a laser beam into two beams that are subsequently directed onto the sample making an angle θ between them. One of the beams, the probe beam, is attenuated by a factor of say 30 after the beam splitter whereas the other one, the main beam, is not attenuated and fixes the equilibrium state of the sample. If the polarization of the light of both beams is parallel to the electrodes, interference fringes are created with a grating period Λ depending on θ, whereas if the polarization of the light of one beam is set perpendicular to the electrodes the two beams simply superimpose onto the sample. Measurements of the evolution of the ratio β of the current with interferences I_{wi} over the current without interferences I_{woi} with the grating period Λ allows the deduction of the diffusion length. These two currents are those due to the probe beam contribution and, to get an accurate measurement, this beam is chopped at a frequency of a few tens or so Hz and the currents are recorded by a lock-in amplifier. Ritter *et al.* have shown that

$$\beta(\Lambda) = \frac{I_{wi}}{I_{woi}} = 1 - \frac{2\phi}{\left[1 + (2\pi L/\Lambda)^2\right]^2} \quad , \qquad (1)$$

and a fit of $\beta(\Lambda)$ gives the two parameters L and ϕ [4]. This latter parameter has only little interest and its value, always lower than one, depends on the contrast of the interference fringes, the dependence of the current with the generation rate and the ratio between the currents in the dark and under illumination. The knowledge of L is more interesting for, if they are not swept out by drift and diffusion, the slowest carriers will accumulate in a solar device and create a space charge limiting the cell efficiency. This is why the flux of the main beam must be as high as possible to be close to sun illumination in order to get an estimate of L close to that found in an operating solar device.

The MPC technique was first proposed by H. Oheda [3] and the data treatment was improved by R. Brüggemann *et al.* [9]. The basic idea is to illuminate the sample with a DC flux to which a small modulation at an angular frequency ω is added, and to record the amplitude $|\sigma_{ac}|$ and phase lag φ with respect to the alternating generation rate G_{ac} of the resulting alternating conductivity by means of a lock-in amplifier. It was shown [9, 10] that the amplitude and phase lag were due to interactions of the majority carriers with the density of localized states in the gap (DOS) and that it was possible to deduce the variations with energy of the quantity NC/μ, called MPC-DOS in the following, N being the density of states at an energy E_ω, μ the extended states mobility and C the capture coefficient of the states trapping free majority carriers, from the equations

$$\frac{N(E_\omega)C}{\mu} = \frac{2}{\pi k_B T} q G_{ac} \frac{\sin\varphi}{|\sigma_{ac}|} \quad , \qquad (2)$$

with

$$E_\omega - E_{be} = k_B T \ln\left(\frac{CN_{be}}{\omega}\right) \quad , \qquad (3)$$

in which q is the absolute value of the electronic charge, T is the temperature, k_B the Boltzmann constant, and N_{be} the equivalent density of states at the band edge located at E_{be}. It can be seen that all the quantities on the right hand side of equation (2) are known. Therefore, by varying experimentally T and ω, equations. (2)-(3) offer the possibility to scan the DOS of the material provided C, μ and N_{be} are known. One quantity of particular importance is the attempt to escape frequency $\nu = CN_{be}$ for it fixes the energy scale of the DOS spectroscopy (see equation (3)). Experimentally, to reach a large range of energy, the NC/μ spectroscopy is achieved at different temperatures with different angular frequencies. Typically, the temperature can be varied between 120 K and 450 K in 20 K steps and the frequencies range from 12 Hz to 40 kHz.

To achieve the three experiments briefly described above, one can see that we need a coherent light source, such as a laser, to create the interference pattern on the sample. Besides only a laser can provide enough flux to perform the SSPG technique with a flux as close as possible to solar illumination. However, this light must be shaped to fulfil the requirements of each technique. For the SSPC technique the flux must be constant and even adjustable to reach different values, for the SSPG technique one beam must be chopped at a low frequency and for the MPC technique the light is the addition of a DC flux, preferably adjustable, and of an AC one, adjustable too, the frequency of which should vary over several decades of frequencies. Hence, a system that can be tuned to provide all these illumination 'shapes' so as to gather the three experiments on a single bench is needed.

3. Experimental set-up

3.1 The EOM plus polarizer system

In a Pockels cell as in an EOM, containing for instance an ADP crystal, the phase shift ψ of the light can be modified according to the polarisation U applied to the crystal. In the following, we assume that the light entering the EOM is vertically polarized and, with $U = U_\pi$, it goes through the EOM with no modification. On the other hand if $U = 0$ then the exiting light is horizontally polarized.

Figure 1 explains the different axes of the system: X' and Y' are the axes of the crystal (fast and slow respectively) whereas X and Y are just the axes indicating the horizontal and vertical directions respectively. Figure 1 also explains how the EOM must be fitted on the bench to satisfy the requirements of all the experiments to be achieved, i.e. tilted 45° with respect to the vertical. We call E_M the amplitude of the light entering the EOM. With U=0 the coordinates of the electric field of the vertically polarized light entering and exiting the EOM can be written on the axes X'Y'

Journal of Physics: Conference Series **558** (2014) 012011 doi:10.1088/1742-6596/558/1/012011

$$\vec{E}_{X'Y'} = \begin{bmatrix} E_M/\sqrt{2}\cos(\Omega t') \\ E_M/\sqrt{2}\cos(\Omega t') \end{bmatrix} \quad , \qquad (4)$$

where Ω is the pulsation of the light and t' the time. If a bias U is applied to the crystal the field of the light at the exit of the EOM transforms into

$$\vec{E}_{X'Y'} = \begin{bmatrix} E_M/\sqrt{2}\cos(\Omega t') \\ E_M/\sqrt{2}\cos(\Omega t'+\psi) \end{bmatrix} \quad , \qquad (5)$$

an expression that can be re written in the XY axes after some easy calculation

$$\vec{E}_{XY} = \begin{bmatrix} E_M \sin\left(\frac{\psi}{2}\right)\sin\left(\Omega t'+\frac{\psi}{2}\right) \\ E_M \cos\left(\frac{\psi}{2}\right)\cos\left(\Omega t'+\frac{\psi}{2}\right) \end{bmatrix} \quad . \qquad (6)$$

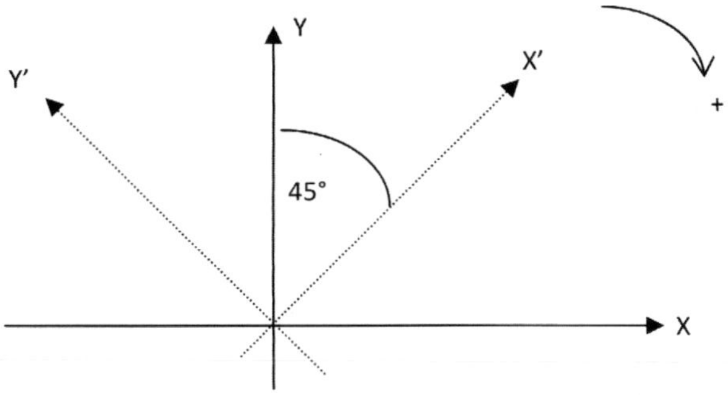

Figure 1. The different axes to be considered in the system.

Practically, the EOM and EOM driver are tuned with the following procedure. A polarizer is set at the output of the EOM, letting go through it only vertically polarized light, and the voltage of the EOM driver is fixed at a very low value U_{min} that we shall take to be equal to zero in the following. The EOM is adjusted, tilted etc.. so that the amplitude of the light after the polarizer is minimum, indicating that the light exiting the EOM is horizontally polarized. Then, the polarizer is rotated 90°, in a way such that only the horizontally polarized light can go through it and the voltage of the EOM driver is increased up to U_{max} for which the amplitude of the light going through the system EOM+polarizer is minimum again. This means that the light at the output of the EOM is vertically polarized. If we come back to the expression of the light electric field, see equation (6), it means that for $U = U_{min} = 0$ then $\psi = \pi$ and for $U = U_{max}$ $\psi = 0$, and we can deduce an expression of ψ versus U

$$\psi = \pi \frac{U_{max} - U}{U_{max}} \quad . \qquad (7)$$

The EOM driver is connected to a function generator and the final expression of U applied to the crystal is

$$U = U_{max}/2 + V_0 + V\sin(\omega t) \quad , \qquad (8)$$

the $U_{max}/2$ being fixed by the user on the EOM driver and the $V_0 + V\sin(\omega t)$ being added by the function generator after amplification. An expression of ψ is therefore

$$\psi = \frac{\pi}{2} - \pi \frac{V_0 + V\sin(\omega t)}{U_{max}} \quad . \tag{9}$$

If we assume that the polarizer is vertically positioned, letting out only the vertical component of the light electric field, we can calculate the intensity I of the light reaching the sample writing

$$E_Y = \Re e\left(\overline{E}_Y\right) = \Re e\left\{ {}^{E_M}\!/_2 \exp(j\Omega t')\left[1+\exp(j\psi)\right]\right\} \quad . \tag{10}$$

The intensity will be $I = \frac{1}{2}\overline{E}_Y\overline{E}_Y^*$ and we end with

$$I = \frac{E_M^2}{4}\left[1+\cos(\psi)\right] = \frac{E_M^2}{4}\left[1+\sin\left(\pi\frac{V_0 + V\sin(\omega t)}{U_{max}}\right)\right] \quad . \tag{11}$$

Equation (11) describes the different possibilities of flux 'shaping' by means of the EOM+polarizer system. The DC flux entering the system can be attenuated, modulated or chopped. This is illustrated in figure 2. Taking $E_M = \sqrt{2}$ and assuming that U_{max} is reached for 2.6 V, the choice of couples of (V_0, V), applied to the EOM driver, as (1.1 V, 0 V), (0.9 V, 0.12 V), (0 V, 1.3 V) gives an output flux either continuous but attenuated (diamonds), modulated (squares) or chopped (circles), respectively.

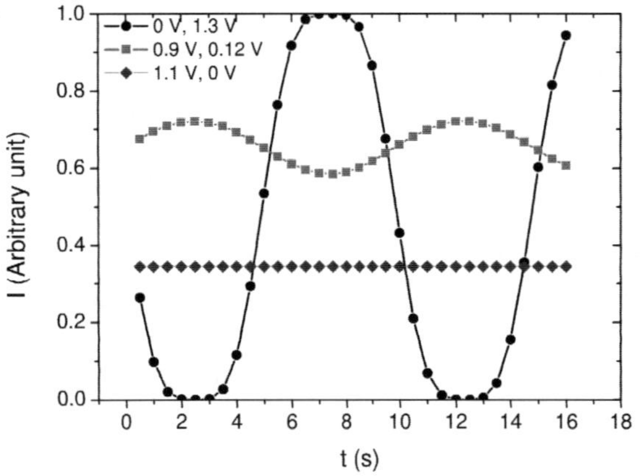

Figure 2. Intensity as function of time at the output of a system EOM+polarizer for different voltages applied to the EOM driver.

This means that the system EOM+polarizer can be used to tailor the flux impinging the sample either continuous for SSPC, modulated for MPC or chopped for SSPG, with the possibility in these two last cases to adjust the angular frequency ω of the flux variations. It can be noted that the same tailoring is possible if the polarizer is rotated 90° to let the horizontally polarized light to go through it. This just adds a phase shift of 180° to the curves displayed in figure 2 without modifying the amplitudes.

3.2 The experimental bench

Figure 3 shows a schema of the whole bench. The light source is a HeNe laser (λ = 633 nm) and the beam at its exit goes through two polarizers. The second one fixes the polarization of the light perpendicular to the plane of the figure and the global intensity of the light can be adjusted by rotating the first one. The beam is subsequently split into two equivalent beams. One, the probe beam, is attenuated and goes through the EOM+polarizer system and a second one, the main beam, is reflected toward a mobile mirror. The SSPG technique has been automated by means of a mobile mirror reflecting the main beam onto a series of nine return mirrors. Changing the choice of the return mirror changes the angle in between the two beams and therefore the grating period as depicted in the figure.

By means of the nine mirrors the grating period can be varied between 1 μm and 24 μm. For more details on this automated SSPG system see [11]. For the other experiments, SSPC and MPC, the mobile mirror is set in front of the first return mirror and a filter wheel can be rotated so as to select the illumination impinging on the sample. The two beams can be used, either at full or attenuated flux, or only one of the two beams, or even none of them (dark position) by rotating the filter wheel fitted with different neutral density filters to the suitable configuration. All the movements of the mobile mirror, rotating polarizer, and filter wheel, are computer controlled and the different configurations for SSPC, SSPG and MPC can be set one after the other automatically. The sample is fitted onto the cold finger of a cryostat, one electrode connected to a voltage source and the other one to a computer controlled switch to record the signal either on an electrometer (SSPC) or on a lock-in amplifier (SSPG, MPC). All these experiments can be thus performed at different temperatures under different conditions of flux keeping the sample in the cryostat at the same position which ensures the same state of the film for all the experiments performed one after the other.

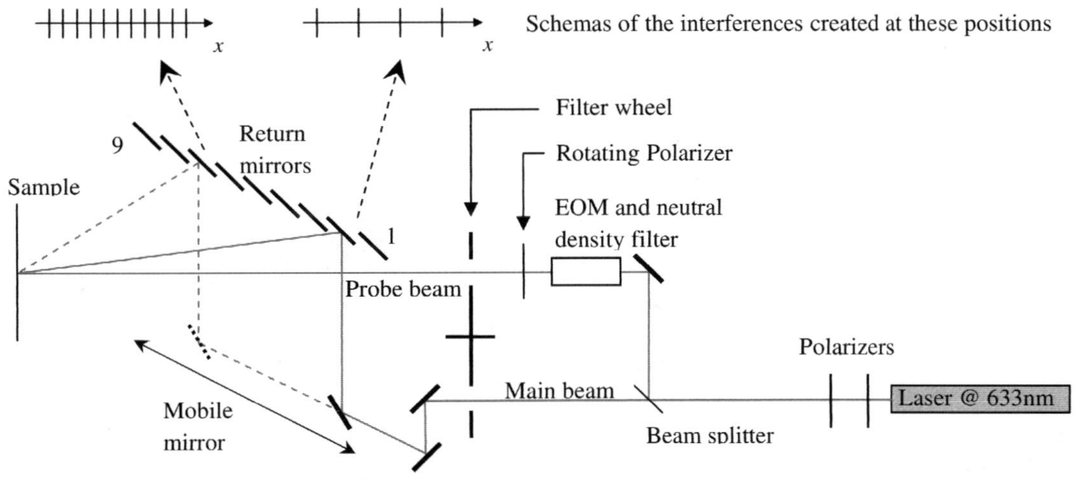

Figure 3. Schema of the experimental bench, using the EOM+polarizer system, with which SSPC, SSPG and MPC experiments can be achieved. The sample is set in a cryostat not shown here.

4. Experimental results

4.1 Hydrogenated amorphous silicon

Hydrogenated amorphous silicon (a-Si:H) was widely studied to improve its transport properties before being incorporated into solar devices and all the experiments quoted above have been applied to this material. It is not the purpose of this paper to make a review of the a-Si:H studies and some theoretical developments and basic properties of a-Si:H thin films can be found in [12 -16]. We shall briefly recall orders of magnitude of the main transport parameters found for 'standard' a-Si:H. This material, though intrinsic, is slightly n-type with an activation energy in the range of 0.6-0.8 eV. The extended states electron mobility has values in the range of 10-40 $cm^2V^{-1}s^{-1}$ [14, 15]. The mobility-lifetime product for electrons and holes at room temperature are usually in the ranges of 5×10^{-8}-5×10^{-6} cm^2V^{-1}, and 1×10^{-9}-1×10^{-8} cm^2V^{-1}, respectively, both depending on the flux and on the quality of the films [8, 12]. The capture coefficient of the conduction band tail states is of the order of 10^{-8} cm^3s^{-1} leading to an attempt to escape frequency of the order of 5×10^{11} s^{-1} [14, 16].

A 0.5 μm thick a-Si:H film was deposited onto a glass substrate by radio-frequency (13.56 MHz) powered plasma enhanced chemical vapour deposition with a power of 30 W, the silane being diluted in hydrogen (10%, 90%) at a total pressure of 2 Torr and a temperature of 190 °C. Two silver paste ohmic electrodes 1 mm apart were set on top of it. The dark conductivity and SSPC were measured each 10 K in the range 120-440 K with two different fluxes (2×10^{15} and 6×10^{16} $cm^{-2}s^{-1}$). The SSPG was measured each 20 K in the same range of temperature with a flux of 6×10^{16} $cm^{-2}s^{-1}$, the maximum flux at our disposal. The MPC spectra were recorded each 20 K in the range 110-400 K. For this last experiment the DC and AC fluxes were 8×10^{14} and 2×10^{14} $cm^{-2}s^{-1}$, respectively, and the frequencies were in the range 12 Hz-27 kHz.

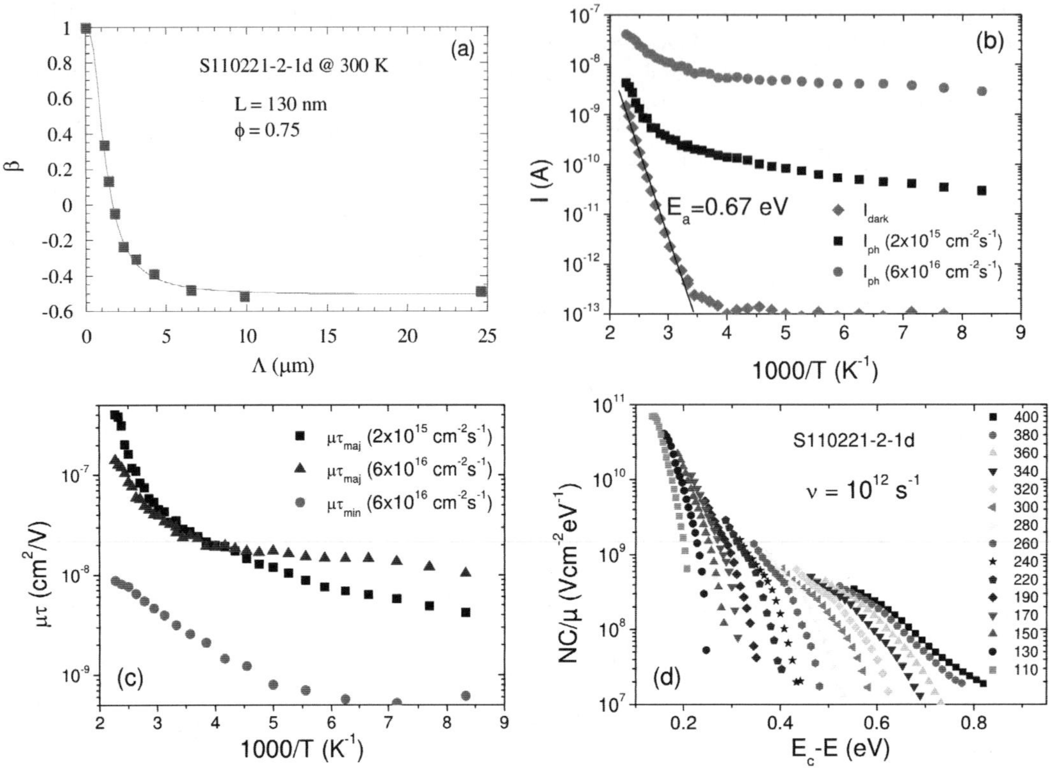

Figure 4. Results obtained on an a-Si:H thin film. (a) SSPG $\beta(\Lambda)$ curve obtained at 300 K, (b) Arrhenius plots of the dark and photo currents measured with two fluxes shown in the figure, (c) Arrhenius plots of the mobility lifetime products of the minority ($\mu\,\tau_{min}$) and majority carriers ($\mu\,\tau_{maj}$) for different fluxes, (d) MPC-DOS spectroscopy; the attempt-to-escape frequency was chosen equal to 10^{12} s^{-1} to fix the energy scale and the measurement temperatures are displayed in the figure.

We present in figure 4 several results obtained with the automated experiment described above. Figure 4a shows a curve $\beta(\Lambda)$ obtained by SSPG performed at 300 K. The diffusion length we deduce from the fitting of the $\beta(\Lambda)$ curve is of the order of 130 nm, a typical value for a-Si:H. Figure 4b displays an Arrhenius plot of the variations of the dark current (diamonds) and of two photocurrents measured respectively with fluxes of 2×10^{15} $cm^{-2}s^{-1}$ (squares) and 6×10^{16} $cm^{-2}s^{-1}$ (circles). The activation energy, E_a = 0.67 eV, deduced from the measurements in the dark, is again typical of a-Si:H. The saturation of the dark current at 10^{-13} A at low temperatures is due to the limitation of the sensitivity of the electrometer. From the SSPG and SSPC measurements we deduce the evolution with temperature of the mobility-lifetime product of the majority carriers, in this case the electrons ($\mu\,\tau_{maj}$), and of the minority carriers, here the holes ($\mu\,\tau_{min}$). Indeed, the ambipolar $\mu\,\tau$ deduced from SSPG is much smaller than that of the electrons and can be attributed without ambiguity to the minority carriers. The results are displayed in figure 4c. The saturation of the $\mu\,\tau_{min}$ values observed at low temperatures is probably due to the limitation of the experiment when dealing with very low L, i. e. lower than 50 nm [11]. Finally, in figure 4d we present the density of states (DOS) above the Fermi level deduced from MPC measurements. One can observe a 'classical' DOS made of a conduction band tail and deep states. Each spectrum displayed by a symbol was recorded at a given temperature and each symbol in a spectrum corresponds to a given frequency. The envelope of all the spectra describes the DOS. It must be recalled that the value of NC/μ is known from all the experimental quantities appearing on the right hand side of equation (2). To fix the energy scale the value of the attempt-to-escape frequency was chosen equal to $\nu = 10^{12}$ s^{-1} according to previous results [14, 16]. This value is that of the conduction band tail states and is probably different for the deep states. It is also for this value that the different band tail MPC spectra obtained at different temperatures present the best match one with the other. This property can be used to estimate the attempt-to-escape frequency for materials less studied than a-Si:H and for which ν is unknown.

4.2 Organic polymers

During the last decade bulk hetero-junction (BHJ) polymers have received a considerable attention and conversion efficiencies of solar cells larger than 12 % have been reached. In addition, the stability of the cells has been largely improved [17]. However, only few experiments dedicated to the study of transport parameters have been applied to these materials. The transport properties have been mainly studied with time-of-flight experiments and models of density of states have been proposed to explain the peculiar behavior of the drift mobility with the applied field [18, 19]. More recently photoconductivity experiments have been applied to BHJ thin films by R. A. Street *et al.* [20, 21] from the results of which a model of DOS has been derived. In addition, these authors have shown that the carrier transport could occur following a multiple trapping process as in disordered inorganic materials (e.g. a-Si:H). Therefore, the global picture describing transport processes in organic semiconductors seems very similar to that describing transport in disordered inorganic films, and, since the techniques described in this communication were successfully applied to a-Si:H, we have also tried to apply them to P3HT:PCBM thin films [22]. We are aware that the carrier generation is different (excitons instead of band to band) but in both cases electrons are generated in one material and in one band (Lowest Unoccupied Molecular Orbital) and holes in another material and another band (Highest Occupied Molecular Orbital) as in inorganic semiconductors (conduction and valence band respectively). In addition, the notion of majority and minority carrier may be slightly different in organic semiconductors than the one currently used in inorganic ones, especially the doped ones. However, as in intrinsic inorganic semiconductors (e.g. a-Si:H), a type of carrier may dominate the current because these carriers are more numerous and/or more mobile. By extension we shall still use the terms 'majority' and 'minority' carriers in the following.

The films were prepared from a mixture of powders of P3HT and PCBM in a ratio of 1:1 in weight. This mixture was dissolved in dichlorobenzene and spin-coated onto a glass substrate at a speed of 500 or 700 rpm to give a thickness of the order of 200 nm. After annealing at a temperature of 75 °C during 30 min the samples were fitted with two parallel aluminium electrodes, 0.5 mm apart, that revealed to be ohmic up to 30 V applied. All the experiments have been performed with 20 V applied. The samples were kept under vacuum (10^{-5} Torr) before and during measurements to avoid any contamination.

The measurements were performed in the range 120 K-380 K each 10 K for the SSPC technique under the same fluxes as for a-Si:H. For the SSPG and the MPC techniques, data were recorded only each 20 K. The SSPG technique was achieved with a flux of 6×10^{16} $cm^{-2}s^{-1}$. The MPC technique was achieved with a DC flux of 2×10^{15} $cm^{-2}s^{-1}$ and an AC flux four times lower. The frequencies of the modulation were varied in the range 12 Hz-8 kHz.

Among all the films tested, one, prepared at 500 rpm, exhibited very promising transport properties with a large diffusion length. The results of the different experiments are exemplified in figure 5. Figure 5a displays the variation of β vs Λ obtained at 280 K with the SSPG experiment. The symbols are the data points and the line is the fit to the data using equation (1). The agreement between both is excellent and we found L = 160 nm. In figure 5b we present the Arrhenius plots of the dark current and of two photocurrents measured with two different fluxes. From the variations of the dark current we deduced an activation energy of 0.26 eV, though it is still questionable whether this value reflects the Fermi level position in BHJ materials or not, since the mobility could be slightly activated [18].

Figure 5c displays the Arrhenius plot of the mobility-lifetime products of the minority ($\mu\tau_{min}$) and majority carriers ($\mu\tau_{maj}$) for different fluxes. It can be seen that the $\mu\tau$ deduced from the ambipolar diffusion length are much smaller than the $\mu\tau_{maj}$. This is why we have assumed that these $\mu\tau$ were those of the minority carriers. The reader may note also the limited range of temperature in which we have deduced $\mu\tau_{\rm min}$ values. Though the SSPG measurements were performed on a large range of temperature, an analysis of the data by a method proposed by Balberg *et al.* has shown that the SSPG measurements below 220 K were not reliable [23]. The variations of $\mu\tau_{\rm min}$ seem to be activated but this behavior could be a consequence of the limited range of temperature of this data. An activation could have been found also for $\mu\tau_{\rm maj}$ if the temperature range had been limited to 160-260 K. More work is needed to understand these behaviours.

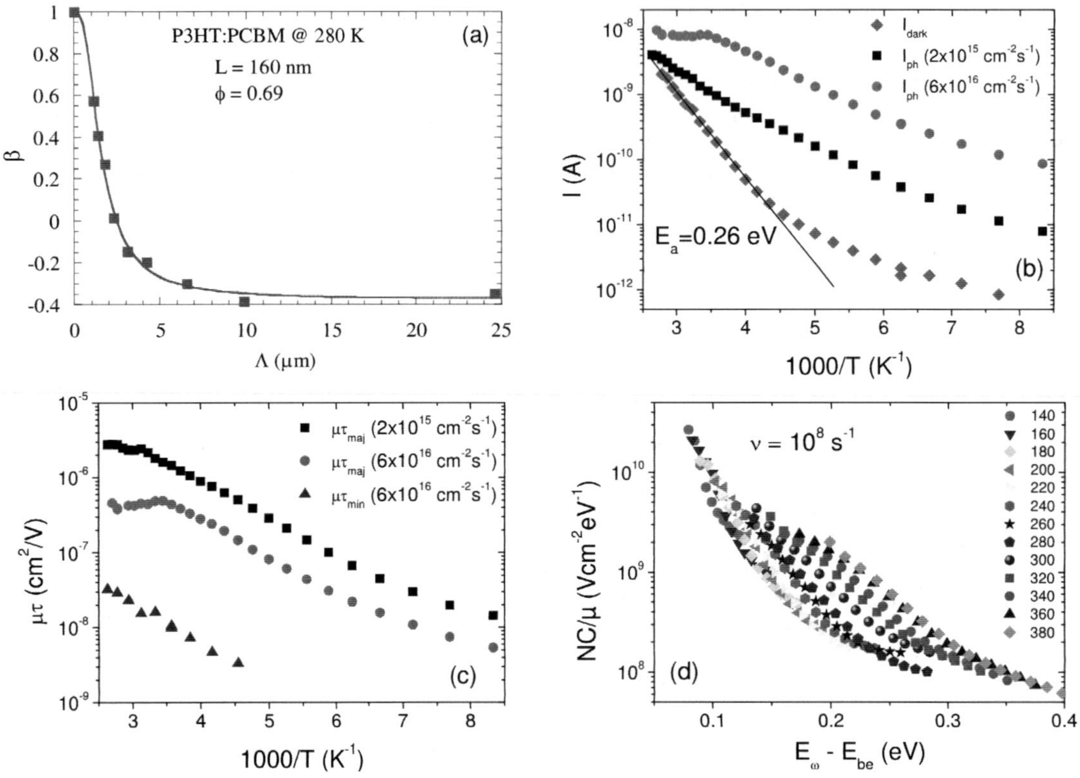

Figure 5. Results obtained on a P3HT:PCBM thin film. (a) SSPG $\beta(\Lambda)$ curve obtained at 280 K, (b) Arrhenius plots of the dark and photo currents measured with two fluxes shown in the figure, (c) Arrhenius plots of the mobility lifetime products of the minority ($\mu\tau_{\rm min}$) and majority carriers ($\mu\tau_{\rm maj}$) for different fluxes, (d) MPC-DOS spectroscopy; the attempt-to-escape frequency was chosen equal to 10^8 s^{-1} to fix the energy scale and the measurement temperatures are displayed in the figure.

Figure 5d shows the MPC-DOS spectroscopy performed on this sample. Each set of symbols corresponds to a temperature at which the frequency of the modulation was varied and each point in a set corresponds to a frequency. The energy scale was fixed from equation (3) choosing an attempt-to-escape frequency of $\nu = 10^8$ s^{-1}. This value is rather low but it is an estimate for which the MPC spectra gather at low temperature to describe a band tail and they also gather at high temperature to describe a distribution of deep states. This DOS is in fair agreement with the one proposed by Street *et al.* [20, 21]. This low value of ν would increase if we took the variations of the 'gap' of the film into account as it was the case for some other materials [24]. However, we do not know if and how the 'gap' of BHJ films varies with temperature so we have assumed a constant value for it. Besides, even if we could apply some correction to the energy scaling, the ν value would remain rather small. It may explain the rather long time of stabilization of the photocurrent or the return to dark equilibrium condition that we have observed with this sample. Indeed, a low ν value implies a rather long traffic of trapping and emission of carriers, especially emission, for the film to go from one state of equilibrium to the other, a property that was already observed for P3HT [25].

5. Conclusion

We have built a bench that allows the performing of several characterization experiments of thin photoconductive films dedicated to the solar industry. With this set up one can estimate the film transport parameters under steady or modulated illumination as well as with interferences shone onto it, all these experiments being performed with the sample in the same state (same temperature, same vacuum,..). The accuracy of these measurements was validated by experiments performed on a well known material, a-Si:H, and all the usual values of the transport parameters as well as the DOS distribution were found again. Then, we have investigated the transport properties of a blend of P3HT and PCBM. One of the advantages of this all-in-one system is that the film is never put back in the air between two subsequent experiments preventing it from any contamination. Though the transport in this organic material may be different of that in inorganic disordered semiconductors, recent publications suggest that the global picture is very similar in both films, and indeed, the results obtained on P3HT:PCBM are very close to those obtained on a-Si:H. It is clear that more investigations are still needed to fully understand all the processes implicit to the transport in organic thin film. Nevertheless, the present characterizations could be an asset in optimizing organic thin films transport properties before their incorporation in a solar device.

Acknowledgements

Many thanks to D. Tondelier and B. Geffroy from the LPICM (Palaiseau, France) for providing the P3HT:PCBM thin films. This work has been partly supported by the French National Environment and Energy Agency (ADEME) through the "Polysil" project.

References

[1] Main C 2002 *J. of Non-Cryst. Solids* **299-302,** 525
[2] Antoniadis H and Schiff E A 1991 *J. Non-Cryst.Solids* **137&138**, 435
[3] Oheda H 1981 *J. Appl. Phys.* **52**, 6693
[4] Ritter D, Zeldov E and Weiser K 1986 *Appl. Phys. Lett.* **49**, 791
[5] Haken U, Hundhausen M and Ley L 1993 *Appl. Phys. Lett.* **63**, 3063
[6] Hattori K, Koji Y, Fukuda S, Ma W, Okamoto H and Hamakawa Y 1993 *J. Appl. Phys.* **73**, 3846
[7] Longeaud C, Ventosinos F and Schmidt J A 2012 *J. Appl. Phys.* **112**, 023709.
[8] Brüggemann R 2011 in "*Advanced Characterization Techniques for Thin Film Solar Cells*", ed Abou-Ras D, Kirchartz T and Rau U, (Weinheim: Wiley)
[9] Brüggemann R, Main C, Berkin C and Reynolds J 1990 *Philos. Mag. B* **62**, 29
[10] Longeaud C and Kleider J P 1992 *Phys. Rev. B* **45**, 11672
[11] Longeaud C 2013 *Rev. Sci. Instrum.* **84**, 055101
[12] Schmidt J A and Longeaud C 2005 *Phys. Rev. B* **71**, 125208
[13] Longeaud C, Schmidt J A and Kleider J P 2006 *Phys. Rev. B* **73**, 235316
[14] Longeaud C, Schmidt J A and Koropecki R R 2006 *Phys. Rev. B* **73**, 235317
[15] Tiedje T 1984 in "*Semiconductors and Semimetals*" (21C), ed Pankove J (New-York: Academic Press) 207.
[16] Vanderhaghen R 1988 *Phys. Rev. B* **38**, 10755
[17] Scharber M C and Sariciftci N S 2013 *Prog. Polym. Sci.* **38**, 1929
[18] Bässler H 1993 *Phys. Stat. Sol. (b)* **175**, 15
[19] Novikov S V 2003 *J. Polymer Sci. B* **41**, 2584
[20] Street R A, Song K W, Northrup J E and Cowan S 2011 *Phys. Rev. B* **83**, 165207
[21] Street R A 2011 *Phys Rev. B* **84**, 075208
[22] P3HT is poly[3-hexylthiophene] and PCBM is [6,6]-phenyl C60 butyric acid methyl ester
[23] Balberg I, Delahoy A E and Weakliem H A 1988 *Appl. Phys. Lett.* **53**, 992.
[24] Luckas J, Kremers S, Krebs D, Salinga M, Wuttig M and Longeaud C 2011 *J. Appl. Phys.* **110**, 013719.
[25] Kim C H, Kisiel K, Jung J, Ulanski J, Tondelier D, Geffroy B, Bonnassieux Y and Horowitz G 2012 *Synthetic Metals* **162**, 460.

Journal of Physics: Conference Series **558** (2014) 012012 doi:10.1088/1742-6596/558/1/012012

Atmospheric pressure imaging mass spectrometry of drugs with various ablating lasers

K A Moshkunov[1], S S Alimpiev[1], A A Grechnikov[2], S M Nikifirov[3], A V Pento[1] and Ya O Simanovsky[3]

[1] General Physics Institute of RAS, Vavilov str., 38, 119991 Moscow, Russia
[2] Vernadsky Institute of Geochemistry and Analytical Chemistry of RAS, Kosygin str., 19, 119991 Moscow, Russia
[3] Advanced Energy Technologies, Novaya str., 100, 143025 Skolkovo, Russia

E-mail: moshkunov@gmail.com

Abstract. The atmospheric pressure mass spectrometric detection efficiency of organic species (tofisopam and verapamil) was measured by means of the laser ablation of dried solution drops containing known amount of the analyte. Ablated molecules were ionized by an atmospheric pressure laser plasma cell and then introduced in the TOF mass-spectrometer. The spot was formed by dripping 2 μl of solution on the stainless steel substrate and consequent drying. Then it was scanned by an intense ablating beam of various lasers (CO_2, Nd:YAG and femtosecond fiber laser) until the spot was completely eroded during the non-stop MS-analysis of ablated material. The sensitivity was defined as the ratio of the total ion current integral of the relevant mass peaks to the amount of molecules in the spot. All the tested lasers are suitable for the ablation and subsequent MS-detection of organic species in dried solution spots given enough power deposition is provided. The measured sensitivity values reach 0.1 ions/fg of tested analytes.

1. Introduction

The rapid expansion of mass spectrometry methods in the field of biology and medicine was driven by the discovery of soft non-destructive ionization methods [1] and the elaboration of high-resolution mass spectrometers [2]. Further development in the field led to the appearance of techniques that could operate with biological tissues with no preliminary procedure of preparation of probes, which cannot be studied by vacuum methods, giving rise to atmospheric pressure mass spectrometry (AP-MS).

The commonly employed atmospheric pressure method is matrix assisted laser desorption ionization MALDI [3,4]. It utilizes the ionization process by the evaporation of organic-containing matrix, which ionizes a wide variety of organic molecules. However, this requires the preparation of the sample and in the case when a study of a solid object is to be performed it can only analyse the very top layer (few microns). Surface assisted laser desorption ionization SALDI [5] analyses species in vacuum. The most basic method of ionization – electro spray [6] works at atmospheric pressure but requires the dissolution of an object in study. The ability to perform measurements in atmosphere and without sample preparation is crucial for many studies of organic tissues when depth profiling is required. Direct analysis in real time DART [7] technique allows atmospheric pressure mass analysis but is still limited to the surface layer of a sample.

The most efficient way to circumvent most of these issues is to separate the ionization from the ablation stage so that the measurement does not depend on a setup geometry, type of material and its laser absorption properties or matrix effects. Several modern techniques satisfy the need: laser assisted electro spray ionization LAESI [8], laser ablation atmospheric pressure photo ionization LAAPPI [9] for example. It utilizes laser ablation to form the flux of neutral particles, while the ionization is performed by a reaction chain driven by UV radiation or corona discharge.

The method developed in our laboratory relies on ionization by UV radiation produced by a laser plasma [10], which works with an air flow that contains molecules to be analyzed and does not depend on the way this flow was produced – either by various types of ablation, evaporation of the sample, or initially gaseous probes. The advantage of the technique is that the sensitivity of the detection does not rely on the probe type and can be calibrated for a wide range of applications. This allows the calibration of the sensitivity towards exact compounds by introducing known amount of molecules in the device. A spot produced by a drying of a drop contains well known amount of analyte. Laser ablation of it provides the required calibration stream.

However, the influence of the ablating laser parameters on the sensitivity of the method is still unclear. Wavelength, laser pulse energy and duration can affect the way the neutral flow is formed [11]. Ablation efficiency, analysed molecules fragmentation or formation of clusters affect the final amount of molecules available for ionization and subsequent mass analysis. We studied the ablation of organic-containing spots thin enough to ensure the evaporation to be governed by heating of underlying steel substrate which does not ablate itself. Three laser types were tested – surgical CO_2, Nd:YAG, and femtosecond fiber laser Fianium so that a wide range of wavelengths (355 nm – 10.6 μm) and fluences (0.065 – 1 J/cm^2) is covered.

2. Experimental

The setup consists of a time-of-flight mass spectrometer (Q-TOF reflectron MX 5311 developed by the Institute for Analytical Instrumentation of the Russian Academy of Sciences (St. Petersburg)) with attached modular atmospheric ionization device, target holder and an ablating lasers. The scheme of the atmospheric interface is presented in figure 1.

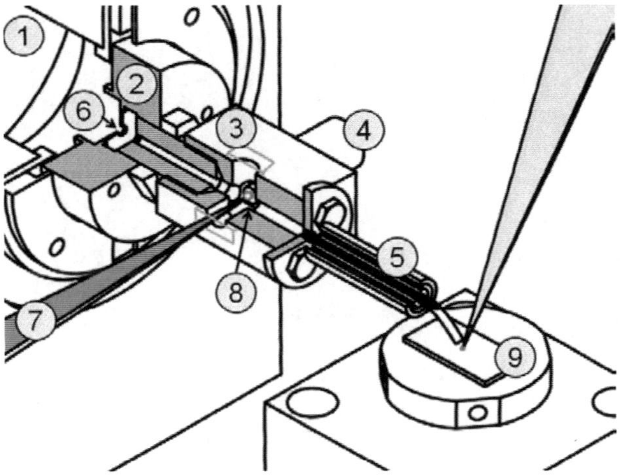

Figure 1. Atmospheric pressure interface. 1 – mass-spectrometer, 2 – diafragm holder and capillary, 3 – ionization cell, 4 – rotating metal target, 5 – heated capillary, 6 – diafragm, 7 – plasma igniting laser, 8 – plasma, 9 – sample, ablated by laser.

The sample is mounted on a two-axis scanning table with adjustable height so that samples with any height can be mounted with surface matching the firmly fixed ablating laser waist and the collection volume of the capillary. The ablated material is pumped by a capillary into an ionization chamber. The heated surface of the capillary is kept at 200° C, so that clusters dissociate into separate molecules, which are ionized by a complex chain of processes caused by the radiation of the laser ignited by plasma on a metal (tungsten) rotating target [10].

The ion stream is then conducted through the 0.3 mm diafragm that separates atmosphere from vacuum.

Though we can perform measurements on any solid target, even organic tissues, the experiment was performed on a stainless steel plate. The water-alchogol solutions containing medical drugs: one – tofisopam (1 μg/ml) and other verapamil (1 μg/ml) were dripped in 2 μl drops on the substarte and dried. The resulting spot is 1-2 mm in diameter and the average thickness of few monolayers of dried analyte. The simple estimations show that laser absorption by thin layer of analyte is negligible. The laser is dominantly absorbed by the steel substarte, which heats up and desorbs the organic overlayer.

The dried drop spots can be ablated with excessive laser power ensuring that all the material is eroded and the known amount of analyzed molecules entered the atmospheric interface. Simultaneously mass spectra are aquired and the corresponding mass peaks are monitored and integrated. In our case both analyzed species are ionized by protonation and manifest as the $M+H^+$ masses along with dimers and trimers and isotopic variations. The dominant peaks are m/z 383 and 455 for tofisopam and verapamil correspondingly. The ratio of the integrated ion signal recorded during the laser scanning of the spot to the amount of deposited molecules gives the ionization efficiency and sensitivty of the device.

While the fundamental frequency of Nd:YAG was used for plasma ignition and ionization, the ablating lasers were varied. Three lasers were tested and their parameters are presented in Table 1.

Table 1. Ablating lasers parameters.

Laser type	**Fiber**	**CO_2**	**Nd:YAG (3rd harmonic)**
Model	Fianium FP532-5	Lancet-1	RL-03/355
Wavelength	1064 nm	10.6 μm	355 nm
Average power	1.28 W	0.16 W	24 mW
Pulse duration	30 fs	0.01 s	300 ps
Pulse repetition rate	1 MHz	16 Hz	300 Hz
Pulse energy	1.28 μJ	10 mJ	80 μJ
Spot FWHM	50 μm	155 μm	100 μm
Energy density	0.065 J/cm^2	50 J/cm^2	1 J/cm^2

Mass spectra recorded during target ablation show a clear presence of tofisopam and verapamil – figure 2.

The results of dried spots scanning by three lasers and two analytes are presented in figure 3. It is a relief image of a spot containing 2 ng tofisopam or verapamil deposited on stainless steel plate. The intensity is obtained by mass spectrometric scanning of 4x4 mm frame. The images presented are the intensity of the net analyte signal plotted versus scanning coordinates (millimeters).

The mass spectra from the spots clearly indicate the presence of tofisopam and verapamil for respective scans. The noise for peaks related to tofisopam is very low and their background is negligible. A clear circle-like structure reproduces for every spot and is very characteristic. The results of the integration and sensitivity calculation are presented in table 2.

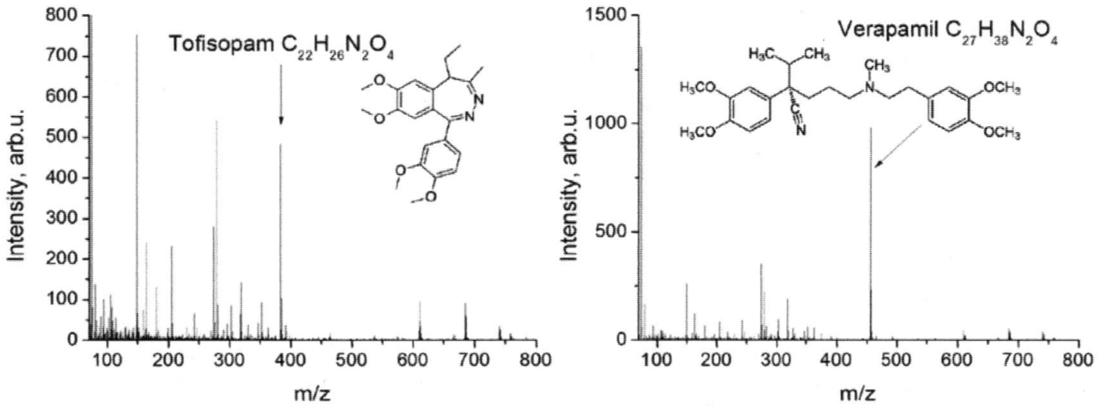

Figure 2. Characteristic mass spectra containing tofisopam and verapamil in dried spots.

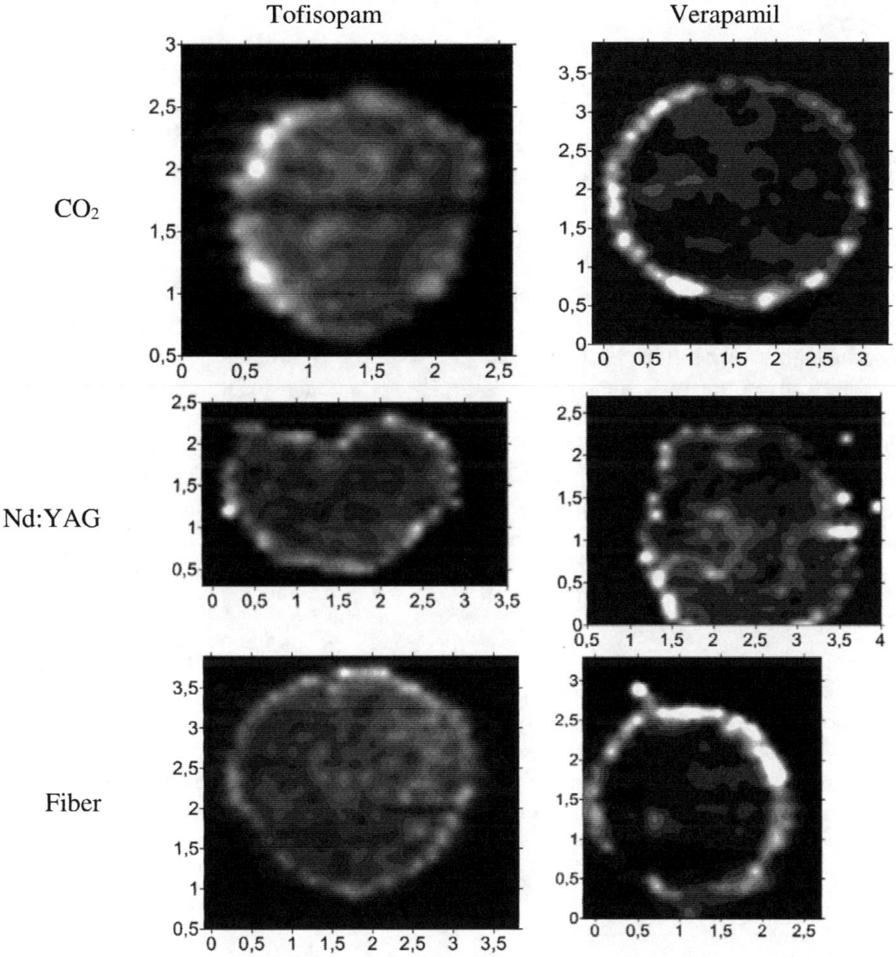

Figure 3. Mass-image (concentration distribution) of tofisopam and verapamil-containing dried spots by various ablating lasers. Image dimensions are millimetres.

Table 2. Sensitivity for tofisopam and verapamil for various ablation lasers.

Laser type	**Tofisopam, ion/fg**	**Verapamil, ion/fg**
Nd:YAG	0.110	0.004
Fiber	0.070	0.005
CO_2	0.033	0.008

4. Discussion

Recorded mass-spectra contain along with analyte-related peaks a lot of peaks that relate to molecules contained by ambient air. Nevertheless, the sensitivity of the method (table 2) is high and does not vary considerably with the laser type. The ionization occurs via a long chain of processes resulting in proton attachment to an analyte molecule. This process is often used in atmospheric ionization methods [12]. In our case, UV radiation generated by metallic laser plasma produces water ions [10] that start the reaction chain. The concentration of these proton donors is orders of magnitude higher than the concentration of analyte ions.

The mass-image of a dried spot in figure 3 exhibits a circular structure meaning that drying is accompanied with the concentration of solution additives at the periphery (liquid-solid-air border). The dried spots were barely visible before the experiment and disappeared completely after. Consecutive scans reveal very small amount of the analyte (tofisopam and verapamil) left after the first scan. Visual inspection of the stainless steel substrate reveals no damage to it and no metal ions were detected.

Several conclusions can be drawn on the basis of data in table 2. Firstly, neither average power nor laser fluence affect sensitivity considerably or there is any clear correlation. This is in accordance with the conception that the laser pulse removes all the material in the spot in few pulses. Excessive power is no more absorbed neither by the material deposited nor the temperature developed in steel substrate is enough to cause significant evaporation. Moreover, in the case of a dilute solution the evaporation process is believed to be caused not by laser absorption by the very material but rather by substrate that heats to the temperature that is enough to evaporate the organic-containing layer.

While there is no clear dependence on power parameters the sensitivity changes smoothly with wavelength. Tofisopam detection sensitivity is highest at lower wavelength and reduces with its increase and the verapamil sensitivity has a contrary dependence. The absolute values of sensitivity factor in the unknown ionization cross-section and no conclusion can be drawn from tofisopam to verapamil ratios.

5. Conclusions

Series of experiments on sensitivity calibration of atmospheric pressure mass spectrometer towards medical drugs (verapamil and tofisopam) were carried out. Known amount of molecules of the analyte was removed from the stainless steel surface by laser ablation of dried solution drops and consequently ionized by laser plasma. The resulting sensitivity reaches values 0.1 ion/fg. Investigation of the sensitivity dependence on ablating laser parameters revealed that the beam power characteristics are of low importance given the laser power is enough to ablate the thin layer of analyte from the steel substrate.

Acknowledgements

This work was supported by the Ministry of Education and Science of the Russian Federation (Contract No. 14.579.21.0020).

References

[1] Monge M E, Harris G A, Dwivedi P and. Fernández F M 2013 *Chemical reviews* **113** 2269
[2] Scigelova M and Makarov A 2006 *Proteomics* **6** 16
[3] Karas M and Hillenkamp F 1988 *Analytical chemistry* **60** 2299
[4] Laiko V V, Moyer S C and Cotter R J 2000 *Analytical Chemistry* **72** 5239
[5] Alimpiev S S, Grechnikov A A, Sunner J, Karavanskii V A, Zhabin S N, Simanovsky Ya O and Nikiforov S M 2008 *J.Chem.Phys.* **128** 014711
[6] Fenn J B, Mann M, Meng C K, Wong S F and Whitehouse C M 1989 *Science* **246** 64
[7] Cody R B, Laramée J A and Durst D H 2005 *Analytical Chemistry* **77** 2297
[8] Rezenom Y H, Dong J and Murray K K 2008 *Analyst* **133** 226
[9] Vaikkinen A, Shrestha B, Kauppila T J, Vertes A and Kostiainen R 2012 *Analytical chemistry* **84** 1630
[10] Pento A V, Nikiforov S M, Simanovsky Ya O, Grechnikov A A and Alimpiev S S 2013 *Quantum Electronics* **43** 55
[11] Vogel A and Venugopalan V2003 *Chemical reviews* **103** 577
[12] Venter A R, Douglass K A, Shelley J T, Hasman G Jr. and Honarvar E 2013 *Analytical chemistry* **86** 233

Dynamic stability of Fe under high pressure

K G Gaminchev and H Chamati

Institute of Solid State Physics, Bulgarian Academy of Sciences,
72 Tzarigradsko Chaussee, 1784 Sofia, Bulgaria

E-mail: gaminchev@issp.bas.bg

Abstract. We study the dynamic stability of bcc iron at both high pressure and temperature via Molecular Dynamics in conjunction with three different interatomic potentials constructed within the embedded-atom method. We computed the phonon dispersions, the phonon density of states, as well as the radial distribution functions. It is found that these quantities exhibit different behaviours depending on the potential used. Furthermore it is revealed that the simulated sample remains dynamically stable over a wide range of temperature and pressure for all potentials.

1. Introduction

Iron enjoys a special place in both scientific research and industrial application. It is a transition metal whose physical properties can be augmented by alloying with other chemical elements to produce materials with unique properties, such as high Tc superconductors, stainless steel, etc. Furthermore, it is the most abundant constituent of Earth's outer and inner core. It can be found in many stable crystallographic structures the temperature and the pressure phase diagram (see [1, 2, 3, 4, 5] and references therein).

Ab-initio computations put together with experimental data show that the low temperature phase (at ambient pressure) of Fe is body-centred cubic (bcc), known as α–Fe or ferrite. Increasing the temperature gradually leads to a structural phase change, at about 1185 K, from α– to γ–Fe, which appears to be face-centred cubic (fcc) or austenite. At a higher temperature, about 1667 K, iron transforms again into a stable bcc structure (δ–Fe) and above 1811 K it melts into a liquid phase. On the other hand increasing the pressure to about 13 GPa at room temperature the structure of iron changes into a hexagonal-closed packed (hcp) crystal, known as ε–Fe. Moreover there exist some experimental evidences pointing to a β–Fe phase at high-temperatures (exceeding 1500 K) and pressures (above 50 GPa) [5], although it has been thought that this phase has either an orthorhombic structure or a double hcp one (see e.g. [6] and references therein).

Using realistic computer simulations based on quantum mechanical methods to investigate the properties of materials is extremely demanding in terms of memory and/or computer time, restricting the simulations to small samples. To consider systems with very large number of atoms on a large time scale it would be suitable to employ Monte Carlo of Molecular dynamics in conjunction with (semi-)empirical interaction potentials that can be estimated with the aid of relevant experimental data sometimes supported by *ab initio* results. Beside efficiency semi-empirical models need to meet an additional criterion: reliability. First they must be able to reproduce with high accuracy various relevant experimental data, such as elastic, structural and thermal properties. Furthermore the potential must be transferable in the sense that it would predict reasonably well properties that were not used during its construction procedure. By now there exist several efficient semi-empirical potential formalisms depending on the way their corresponding expressions are derived by quantum mechanical methods. These potentials account for local density profiles of the element under study. These approaches has been successfully applied to model atomic interactions in iron and many interatomic potentials has been proposed so far.

The aim of the present paper is to study the mechanical properties of iron at high temperature and pressure with the aid an open source code for classical molecular dynamic (MD) simulation, namely LAMMPS [7]. We compute and compare some characteristic physical properties of Fe employing several interatomic potentials [8, 9, 10] (within the embedded atom potential method (EAM) [11]) designed for iron and validated by computing its fundamental physical properties. Moreover some of the potentials were checked for transferability to different atomic structural phases. Here we will turn our attention to the computation of the phonon density of states and the phonon dispersion curves along the high symmetry vectors of the bcc structure at extreme pressures and very high temperatures. Furthermore we will investigate the stability of the structure of iron under these extreme conditions.

The rest of the paper is organized as follows: In Section 2 we introduce the main theoretical models based on EAM and briefly review the computational methodology. In Section 3 we discuss the results of static equilibrium properties, as well as MD simulations. Finally we draw our conclusions in Section 4.

2. Theory and computational technique

The total energy of a monoatomic system within EAM is given by the general expression [11]

$$\mathrm{E}_{\mathrm{tot}}^{\mathrm{EAM}} = \sum_i \left[\frac{1}{2} \sum_{j \neq i} \Phi(r_{ij}) + F(\bar{\rho}_i) \right], \tag{1}$$

where $\Phi(r_{ij})$ is the pairwise interaction potential energy among atoms at sites i and j located at distance r_{ij} from each other and F represents the embedding energy of an atom i as a function of the background electron density

$$\bar{\rho}_i = \sum_{j \neq i} \rho(r_{ij}), \tag{2}$$

with $\rho(r)$ the electron density of a single atom. In general this expression may be very complex. It depends on a number of running parameters that need to be optimized by fitting the predictions of some selected properties of specific materials to their target values corresponding to experimental and/or first-principles data. In the EAM formalism, $\bar{\rho}_i$ is a linear superposition of spherically atomic electron densities. For more details the interested reader may consult the article review [12] and references therein.

The EAM potentials constructed for pure iron in references [8, 9, 10] were used to estimate some basic properties, such as the cohesive energy, the equilibrium lattice parameter and elastic constants. Moreover we used MD simulation in the isothermal-isobaric canonical ensemble (NPT) to compute the phonon dispersion curves, as well the phonon density of states (DOS) employing the technique implemented within LAMMPS according to reference [13]. To this end we used a superlattice consisting $16 \times 16 \times 16$ atoms arranged in a bcc-primitive cell. Simulations were performed under periodic boundary conditions on the three space directions, thus neglecting possible surface effects. Equations of motion were integrated using a time step $\delta t = 2 \times 10^{-15}$ s to ensure energy conservation of the trajectories during the simulation process for all potentials under study. The system was equilibrated at the desired temperature and pressure during 0.4 ns. The thermodynamic averages were computed satisfactorily over trajectories lasting 12 ps and measurements were collected every 20 MD steps. Furthermore we deduce temperature dependent elastic constants from the acoustic mode dispersion relations following the method of reference [14]. Let us recall that the elastic constants for cubic metals at finite temperatures are known to obey the following independent inequalities [15]:

$$B = \frac{1}{3}(C_{11} + 2C_{12}) > 0, \quad C' = \frac{1}{2}(C_{11} - C_{12}) > 0, \quad \text{and} \quad C_{44} > 0. \tag{3}$$

3. Results and discussions

In table 1 we present our estimates for some equilibrium (at $T = 0$ K) properties of Fe calculated within the three EAM potentials used in this study along with experimental data [16, 17, 18]. The obtained results show that all the potentials reproduce accurately well the experimental values, except for cohesive energy evaluated with the potential of reference [9] which is about 3.9 % lower. On the other hand the computed relaxed elastic constants C_{11} and C_{12} corresponding to the potentials [8, 9] are in better agreement with the experiment than that of reference [10]. As for C_{44} the closest value to experiment is the one obtained with the potential [8].

Table 1. Lattice properties of iron at equilibrium obtained with the three EAM potentials along and experimental data interpolated to 0 K.

	EAM [8]	EAM [9]	EAM [10]	Experiment
a_0 (Å)	2.8665	2.8553	2.8659	2.87 [16]
E_0 (eV)	-4.28	-4.12	-4.29	-4.28 [17]
C_{11} (GPa)	241.1	243.4	229.7	242.0 [18]
C_{12} (GPa)	146.8	145.0	135.5	146.5 [18]
C_{44} (GPa)	114.0	116.0	116.8	112.0 [18]

Table 2. Lattice properties of iron at finite temperatures obtained for the EAM potentials compared to experiment. Here $\overline{a}_V$ and ϱ ($\times 10^3$) are the average atomic volume and density, respectively.

	$T \approx 300$ K and $P \approx 0$ GPa				$T \approx 6000$ K and $P \approx 300$ GPa		
	EAM [8]	EAM [9]	EAM [10]	Exp.	EAM [8]	EAM [9]	EAM [10]
$\overline{a}_V$ (Å^3)	11.92	11.64	11.88	11.77 [19]	7.514	7.175	7.238
C_{11} (GPa)	217.7	210.1	210.7	233.1 [20]	3003	2032	1215
C_{12} (GPa)	135.7	120.8	122.9	135.4 [20]	188.6	1120	1146
C_{44} (GPa)	121.2	117.9	116.0	117.8 [20]	1960	707.7	456.7
ϱ (kg/m^3)	7.778	7.966	7.804	7.859 [21]	12.34	12.93	12.81

In table 2 we report results for the lattice properties at two specific cases corresponding to different points on the phase diagram (T, P), namely (300 K, 0 GPa) and (6000 K, 300 GPa). It is obvious that the stability conditions of equation (3) are satisfied in both cases for all potentials. Notice that at 300 K the elastic constants behaves differently for the three potentials, but they follow the same trend as experiment. In general they compare reasonably well with experimental values. At (6000 K, 300 GPa) the stability conditions (3) are fulfilled, but the elastic constants are quite different. This behaviour needs further investigation.

The phonon-dispersion curves measured during MD simulations with the three potentials are shown in figure 1 together with experimental results obtained using inelastic scattering techniques [22] and in table 3, we present our results at the-highest points in the Brillouin zone along with the corresponding experimental ones. It is clearly seen that all three potentials reproduce reasonably well the main features of the experimental curves, apart from some discrepancies at specific points. For the potential [8] this behaviour was already reported in reference [8] where the phonon dispersion curves were obtained using a different method. In the case of higher temperature and pressure however the three potentials lead to completely different behaviours. The maximum value obtained

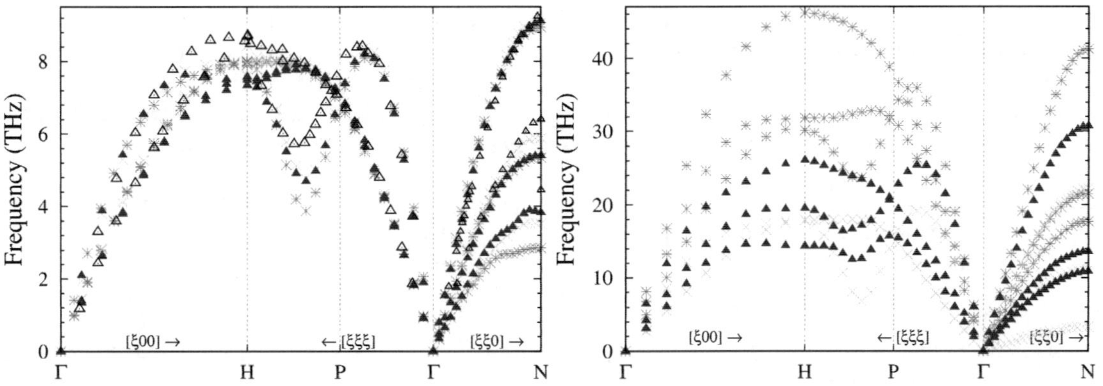

Figure 1. Comparison of the phonon-dispersion curves for iron at low temperature 300 K in the absence of pressure (left) predicted by the three EAM potentials: [8] (stars), [9] (filled triangles) and [10] (crosses). The experimental values (empty triangles) are measured at temperature 295 K [22]. The right panel shows the curves obtained at high temperature 6000 K and pressure 300 GPa with the same symbols as in the case of normal conditions.

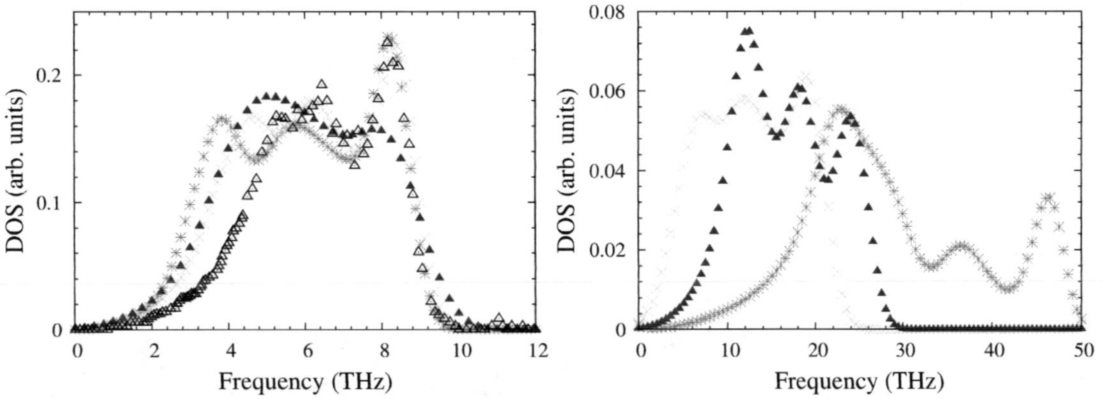

Figure 2. Phonon DOS obtained via MD simulation. Low (left) and high (right) temperature and pressure obtained with the three EAM potentials [8] (stars), [9] (filled triangles) and [10] (crosses) along with the experimental values (empty triangles) measured at temperature 300 K [23].

with the potential [8] lying on the L line in $\Gamma - \Gamma$ region is approximately 2.6 to 3 times larger than those of references [9] and [10], while the maximal value in the region $\Gamma - N$ is 1.9 times larger. In figure 2 we show the corresponding phonon DOS predicted by the three potentials. At 300 K we compare our estimates with experimental measurements obtained by nuclear resonant inelastic scattering of synchrotron radiation [23]. Notice that all EAM potentials reproduce the three main peaks of the experimental DOS. The position of the high frequency peak is found to be in a good agreement with experiment, while the lower ones are beneath those of the experimental ones in agreement with the findings of reference [24]. At high temperature (6000 K) and pressure (300 GPa) the phonon DOS for the different models show different behaviours. This unusual behaviour can be traced back to optimization procedures used to construct each potential and in the absence of experimental results we cannot conclude which potential gives the most accurate results. Despite these differences we

Journal of Physics: Conference Series **558** (2014) 012013 doi:10.1088/1742-6596/558/1/012013

can see from the behaviours of the radial distribution functions (figure 3) that the structure of the final configurations (figure 4) are indeed characteristic of stable bcc like structures, but with different degrees of order. In figure 4 we show the initial, as well as the final configurations of obtained with each potential. One can see that the structure corresponding to the potential [8] is more ordered that the two other potentials, thus confirming the behaviours of the radial distribution functions.

Table 3. Zone boundary phonon frequencies of bcc iron at "normal conditions" and at high temperature and pressure conditions (in THz). The experimental values are measured at 295 K.

	$T \approx 300$ K and $P \approx 0$ GPa				$T \approx 6000$ K and $P \approx 300$ GPa		
	EAM [8]	EAM [9]	EAM [10]	Experiment [22]	EAM [8]	EAM [9]	EAM [10]
H	8.00	7.10	8.10	8.56	46.1	26.1	17.9
N_L	8.95	8.93	9.06	9.24	41.3	30.7	22.2
N_{T_1}	2.87	3.74	3.52	4.35	17.6	10.9	3.26
N_{T_2}	5.34	5.58	5.89	6.40	21.6	13.7	11.2
P	6.96	6.67	7.02	7.20	35.9	21.0	15.4

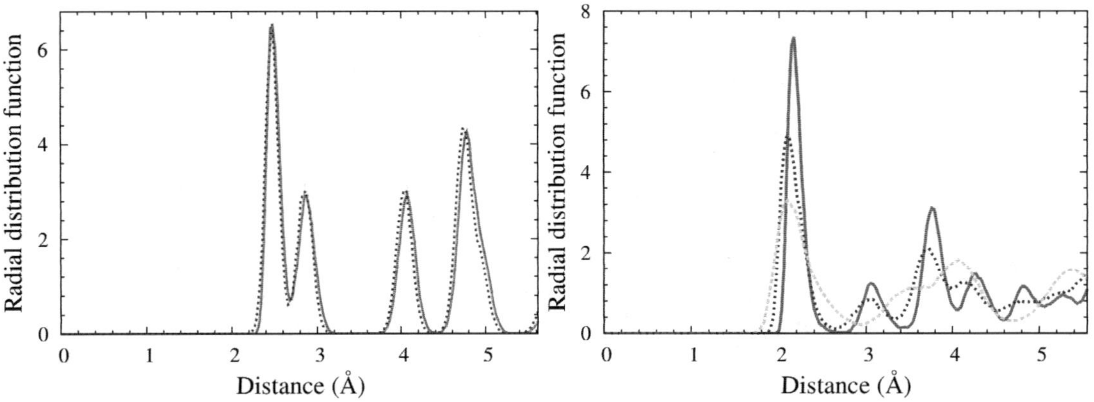

Figure 3. Radial distribution functions of Fe obtained with the EAM potentials [8] (red), [9] (green) and [10] (blue).

4. Concluding remarks

Thermodynamics and structural stability of iron are investigated using three different ineratomic potentials [8, 9, 10] within the embedded atom method. Each of the named potentials is constructed by fitting to different physical properties obtained from experimental data and/or first principles results. Minimizing the potential energy of each force field, we computed the equilibrium lattice parameter, cohesive energy and elastic stiffness coefficients. In general is it found that these potentials reproduce accurately the experimental results. Further we performed a molecular dynamics simulations to study the structural stability of iron under high temperature and pressure. To this end we have computed the phonon dispersion curves, the phonon density of states, and well as

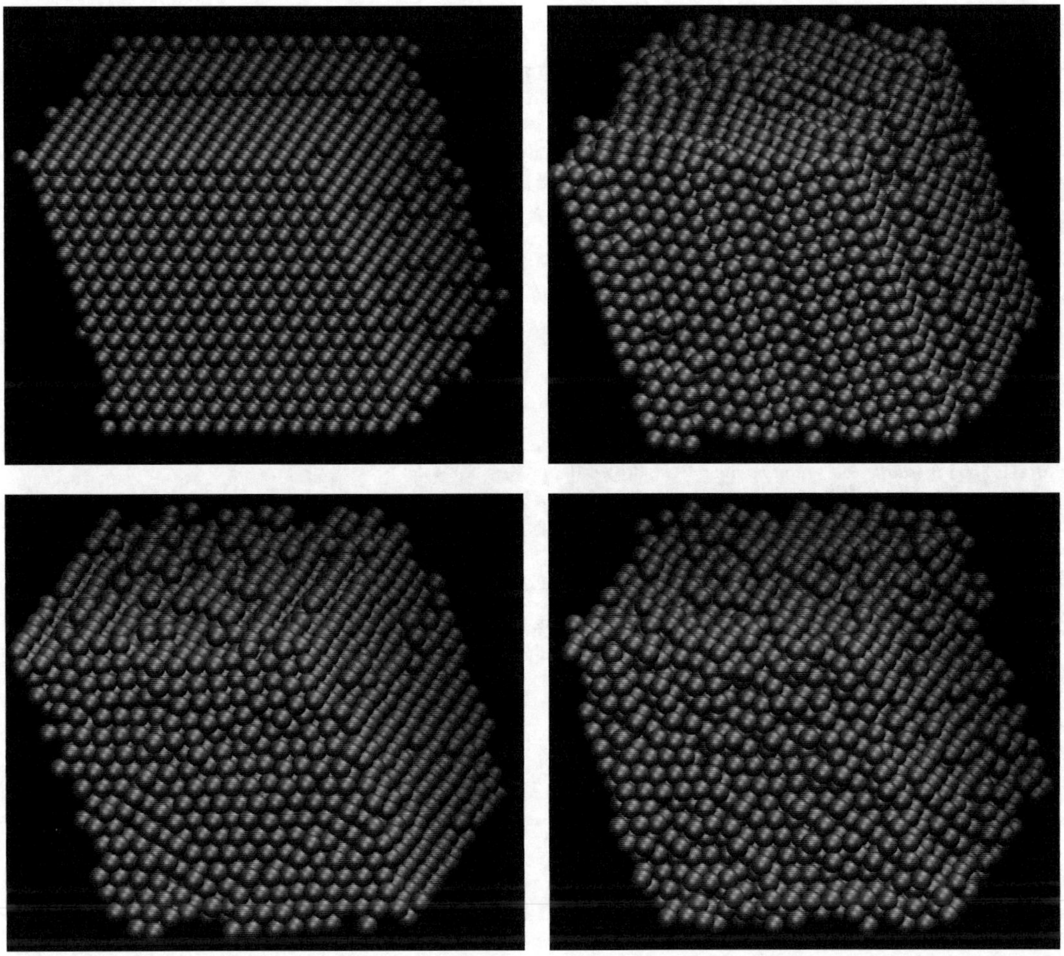

Figure 4. Initial configuration (top left) used for phonon computation at $T = 6000$ K and $P = 300$ GPa, and the final configurations at the very last time step of the simulation corresponding to the different potentials: Top right [8], bottom left [9] and bottom right [10].

the temperature dependent elastic constants. At normal conditions of temperature and pressure our results for the three potentials were found to agree very well with experimental results. To investigate the effect of the high temperature and pressure we have chosen a specific point in the phase diagram of iron, namely 6000 K and 300 GPa. Our results for the phonon dispersion curves and DOS led to completely different behaviours. This is most likely a consequence of the procedures used in the construction of each potential. Despite these discrepancies all potentials predicted stable bcc structures with different degrees of order.

Acknowledgements
Discussions with L. T. Kong are greatly acknowledged. This work was supported by the NSF of Bulgaria Grant DNTS/Germany 01/2 (03.09.2014).

References

[1] Nguyen J H and Holmes N C 2004 *Nature* **427** 339
[2] Tateno S, Hirose K, Ohishi Y and Tatsumi Y 2010 *Science* **330** 359
[3] Anderson O L 1985 *Nature* **314** 407
[4] Belonoshko A B, Ahuja R and Johansson B 2003 *Nature* **424** 1032
[5] Boehler R 2000 *Rev. Geophys.* **38** 221
[6] Rao R S, Modak P, Godwal B K and Sikka S K 1999 *Phys. Rev. B* **59** 13498
[7] Plimpton S 1995 *J. Comput. Phys.* **117** 1
[8] Chamati H, Papanicolaou N, Mishin Y and Papaconstantopoulos D 2006 *Surf. Sci.* **600** 1793
[9] Mendelev M I, Han S, Srolovitz D J, Ackland G J, Sun D Y and Asta M 2003 *Phil. Mag.* **83** 3977
[10] Zhou X W, Johnson R A and Wadley H N G 2004 *Phys. Rev. B* **69** 144113
[11] Daw M S and Baskes M I 1984 *Phys. Rev. B* **29** 6443
[12] Daw M S, Foiles S M and Baskes M I 1993 *Mater. Sci. Rep.* **9** 251
[13] Kong L T 2011 *Comput. Phys. Comm.* **182** 2201
[14] Kong L T, Li J F, Shi Q W, Huang H J and Zhao K 2012 *EPL* **97** 56004
[15] Born M 1940 *Math. Proc. Camb. Philos. Soc.* **36** 160
[16] Pearson W B 1967 *A Handbook of Lattice Spacings and Structures of Metals and Alloys* (Oxford: Pergamon Press)
[17] Kittel C 2005 *Introduction to solid state physics* 8th ed (Hoboken, NJ: Wiley)
[18] Hirth J P and Loethe J 1968 *Theory of Dislocations* (New York: Wiley)
[19] Pearson W B 1958 *A Handbook of Lattice Spacings and Structures of Metals and Alloys* (New York: Pergamon)
[20] Rayne J A and Chandrasekhar B S 1961 *Phys. Rev.* **122** 1714
[21] Simmons G and Wang H 1971 *Single Crystal Elastic Constants and Calculated Aggregate Properties: A Handbook* (Cambridge, MA: M.I.T. Press)
[22] Minkiewicz V J, Shirane G and Nathans R 1967 *Phys. Rev.* **162** 528
[23] Tsunoda Y, Kurimoto Y, Seto M, Kitao S and Yoda Y 2002 *Phys. Rev. B* **66** 214304
[24] Chamati H and Papanicolaou N 2007 *J. Optoelectron. Adv. Mater.* **9** 159

Model of nanodegradation processes in electronic equipment of NPP Kozloduy

A Popov
Sofia University, Faculty of Physics, 5 "J. Boucher" Blvd, 1164 Sofia, BG

E-mail: angelsashev@gmail.com

Abstract. From the complex studies it was proof that the main degradation processes in the three groups of elements for the extended period of time are slow; do not lead to a hopping change in basic parameters and to catastrophic failures. This gives grounds to suggest a common diffusion model , which is limited to the following: -in electronic components containing a p-n junction, is performed diffusion of residual cooper atoms, that are accumulated in the area of a spatial charge under the influence of the electric field and the local temperature, creating micro-shunt regions; -in the contactor systems whose contact surfaces are made of metal alloys under the influence of increased temperature starts decomposition of a homogeneous alloy. Conditions are created for diffusion of individual atoms to the surface, micro-phases of homogeneous atoms are formed and modify the contact resistances; -in the course of time in the insulating materials are changed the mechanisms of polarization, double bonds and dipoles are disrupting, leading to the release of carbon atoms. The latter diffuse at elevated temperatures and form conductive cords, which amend the dielectric losses and the specific resistance of the materials.

1. Introduction

A general physic model for slow degradation of electronic systems equipment with continuing resource is developed on the base of analyses of elemental diffusion processes in three separate groups of the component base: semiconductor diodes and transistors; electric contactor systems; dielectric materials.

Physical models of degradation processes in the three types of specimen are based on diffusion processes that allow accepting a general driving mechanism – diffusion of material particles, independently of their origin. There are accepted as general parameters of degradation the following factors:

- coefficient of ideality factor β for selected diodes and coefficient of conductivity h21 for selected transistors;
- microhardness H and contact resistance Rk for selected mechanical contactors;
- dielectric losses tgδ, dielectric permittivity ε and high voltage voltampere features are selected insulation materials.

2. Semiconductor diodes and transistors

Measurement of electrical parameters of semiconductor specimens is performed and analyzed by theoretical model proposed by A Popov [1] and used into methodology developed from us [2-4]. It is based on shunt areas in p-n junction of diodes and transistors in continuous operation which change their voltampere characteristics and respectively the rest of the current parameters. Diffusion and accumulation of residual impurity atoms with small covalent radius in dislocation defects that occur in

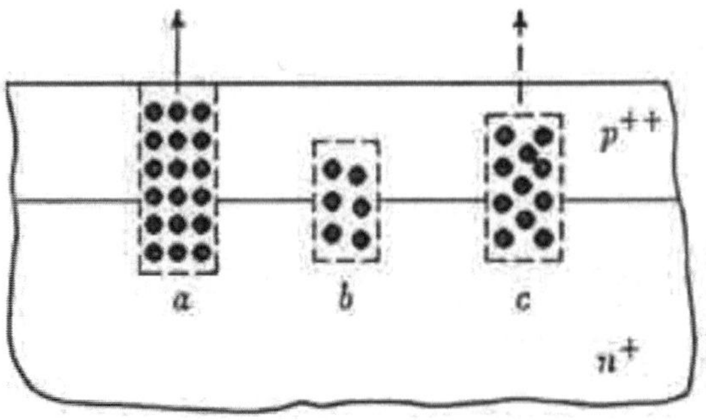

Figure 1. Possible structural defects formed in the area of p-n junction in degradation processes.

process of technological formation of active semiconductor structure are shown in cases (a, b, c in figure 1). They are considered as main reasons for formation of real shunting areas (case a). Another cases (b,c) form single deep levels and clusters. Formation of shunt areas and clusters that create deep levels into the forbidden zone influences most strongly on voltampere characteristic of p-n junction and in particular of the coefficient β (ideality coefficient) into the formula:

$$I = Io\ \exp(qU/\beta kT) \qquad (1)$$

The change of ideality coefficient shows that the tilt of volt-ampere characteristics of the p-n junction alters and respectively the current parameters of the simple diode created on the base of it are changed.

In regard to bipolar transistor created on the base of two p-n junctions (emitter and collector transitions) the current of collector transition Iko changes mostly, the ratio of electricity transition by current h21 as well as and the slope of voltampere family features.

The explanation of this degradation phenomenon can be traced to the model based on the modification of the spatial charge of the p-n junction because the formation of shunt areas after accumulation of impurity atoms on the dislocation defects. The spatial charge is presented in figure 2 in a three-dimensional version with shunt field. Ideality coefficient of this p-n junction is presented with this expression [8,9]:

$$\beta_1 = \{2[b + ch(\omega/L)]\}(b-1)^{-1} \qquad (2)$$

where $\omega = xn+xp$ is the width of the spatial charge, L- diffusion length of the carriers and b is the ratio between the mobility of electrons and holes.

Shunt areas formation and deep levels in sensitive concentrations, influencing on current parameters of diode and of transistor increase the value of the ideality factor as is shown in figure 3. In this case the relation between the width of the spatial charge and diffusion length of carriers increase (become bigger than zero) because of the spatial charge length increasing. The changes of a static current transfer constant h21 of a non-operated (upper curve) and 10-year operated (lower curve) transistor КТ310 2Б are presented in figure 4. About aging process it is proposed following mechanism:

- diffusion of impurities with a small covalent radius into dislocations intersecting p-n junction, and forming a shunt fields with a high conductivity;
- as main diffusing element are considered copper atoms contained in the synthetic semiconductor crystals in high enough concentrations, and are with significantly smaller covalent radius than that of the main atoms of the basic crystal;

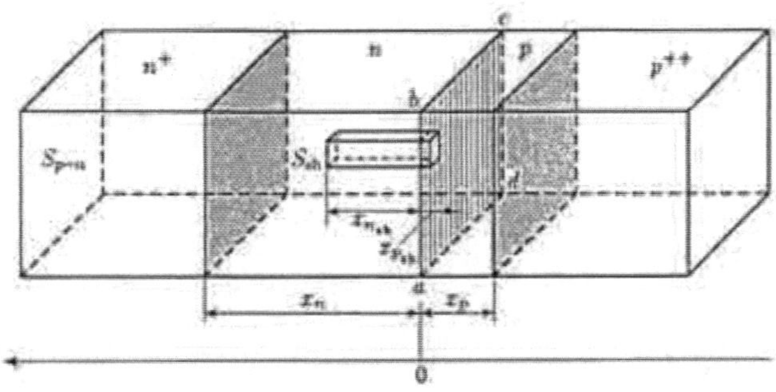

Figure 2. Three-dimensional model of shunt area in the spatial charge of p-n junction.

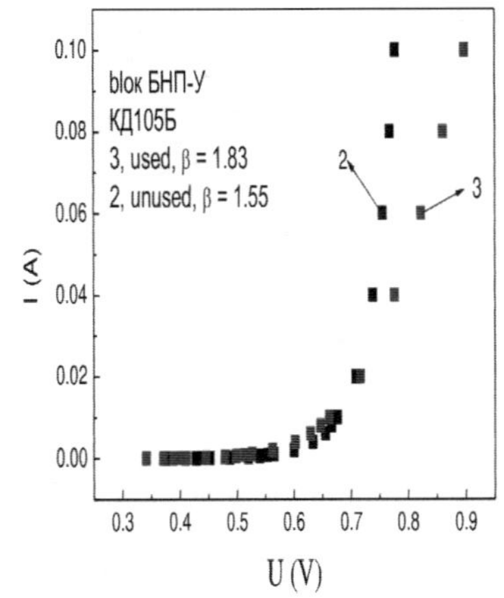

Figure 3. The ideality factor of unused and used diode.

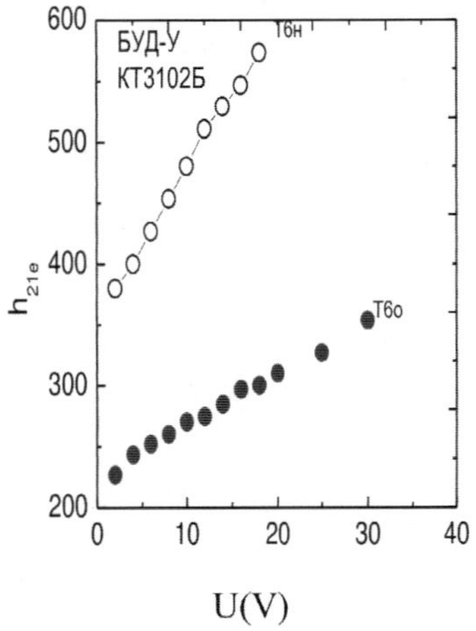

Figure 4. Current constant of unused and used transistors.

- effective coefficient of copper atoms diffusion is defined assuming that the quantity of accumulated impurities is proportionally to the operational time and temperature of p-n junction.Radius of the dislocation tube is assumed to be equal to the lattice parameter of < 10 Å. The copper diffuses into silicon as positively charged ion. The coefficient of diffusion into n-type and proper silicon is [5]:

$$D_{\text{int}} = (3.0 \pm 0.3) \times 10^{-4} \exp\left(-\frac{0.18 \mp 0.01\,\text{eV}}{\text{k}_\text{B} T} \right) \quad \text{cm}^2\text{s}^{-1} \qquad (3)$$

Positively charged copper ion form couples with negatively charged acceptor ion- boron and consequently a portion of the dissolved copper ions are stationary. The effective diffusion coefficient of copper in the silicon alloyed with boron, [6] is:

$$D_{\text{eff}} = 3 \times 10^{-4} \exp \frac{(-2090/T)}{1 + 2.584 \times 10^{-20} \exp(4990/T)(N_a/T)} \quad \text{cm}^2\text{s}^{-1} \qquad (4)$$

In p- type silicon the copper forms copper-copper pairs. The analysis shows that they are formed between atom in node and atom in internode. The time of the dissociation of the copper-copper couples is [7]:

$$\tau_{\text{Cu-pair}} = 1.47 \times 10^{-10} \exp\left(\frac{1.02 \pm 0.07 \text{ eV}}{\text{k}_\text{B} T}\right) \text{ s} \tag{5}$$

This equation gives time constant from 16 days to 9.5 years. The average value is about 8 month. Process with such a constant can be completely decisive for the slow degradation of silicon diodes and transistors. The lifetime in n-and p-silicon vary in a different manner depending on copper concentration. Influence of lifetimes and its changes on voltampere characteristics of p-n junctions is well described from the model of Stafeev [8, 9]. Taken the real values for lifetimes of electrons and p-carriers, measured into alloyed with copper n-Si and p-Si, are obtained β ≅ 2 at low alloying. At higher level of alloying where lifetimes are obtained, is measured β ≈ 3. By this model smooth growing of β could be explained with lifetimes amendment at low alloying. The large changes of β (3-7) are due to amendments of times at very high alloying with copper. From the analysis of the above models, combined with the experimentally obtained data it is built a theoretical model for the slow degradation of diodes and transistors, which are a major elemental base in electron equipment for the management and control of NPP. For diodes the analytical form of the model is represented by equation (6) and graphically presentation in figure 5:

$$\beta = (0.066 \pm 0.042) \times Years + (1.49 \pm 0.238) \tag{6}$$

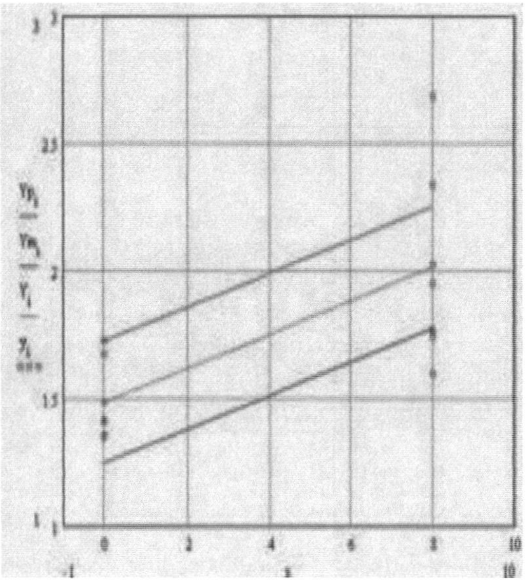

Figure 5. Graphical presentation of diodes aging as a function of operating time (Y = β, x = years).

For transistors the analytical form of the model is represented by equation (7) and graphical presentation in figure 6:

$$h_{21} = (-8.006 \pm 5.033) \times Years + (162.758 \pm 28.473) \tag{7}$$

Figure 7 shows that the relative increase of the ideality factor β of tested diodes is in the range 1.15 – 1.8. A similar picture was seen in the relative change of the transmission of the tested specimens. On the basis of the known values of the coefficient of diffusion of cooper into silicon and the dynamics of the

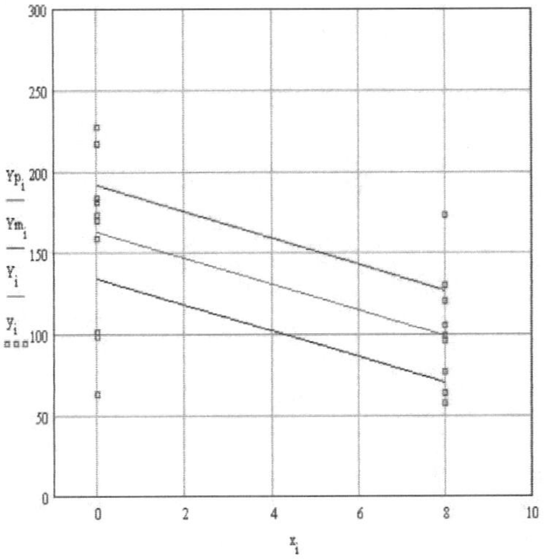

Figure 6. Graphical presentation of transistors aging as a function of the operating time (Y = h_{21}, x = years).

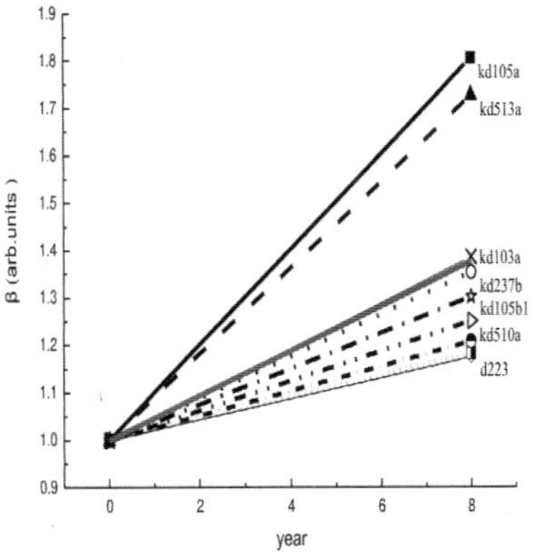

Figure 7. Relative increase of β of the tested diodes.

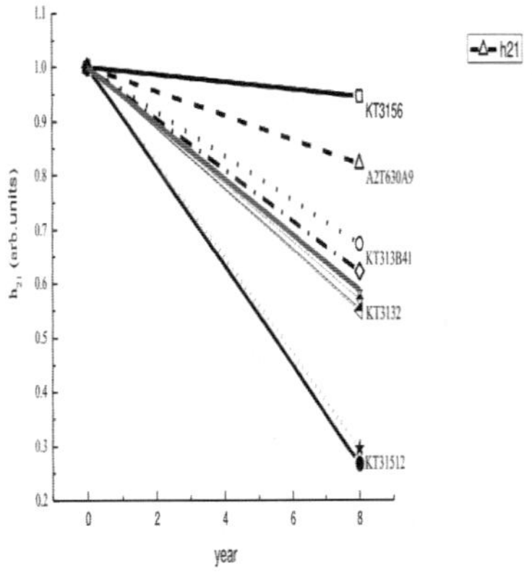

Figure 8. Arbitrary changes of h_{21} tested transistors.

formation and destruction of couples of cooper ions and precipitates were evaluated time constants of the degradation processes of about 15 days to more than 9 years.

2. Electro-contact systems

The second main group of elementary base of electronic equipment of NPP were electro contact systems, for which are studied: i) ampere-volt features of couple of pairs of contacts from each contactor for determine of contactor resistance Rk;- ii) micro-hardness H of constant surfaces of unused and used contactors independently of light and dark areas where they are detached; iii) ampere-volt characteristics for defining of surface-resistance Rs, separately of light and dark regions of contact areas (of unused and used contactors).

2.1 Measurement and analysis of contact resistance of contacts in mechanical contactor systems

The first and most important performance parameter for all contactor devices with mechanical metal contacts (or metallurgical contacts, such as thermocouples and some other control devices) is their contact resistance R_k. It is extremely sensitive to any degradation processes and its quantitative values are an indicator of whether the contactors have entered a phase of ageing or are in conditions of normal operation. Characteristic for the beginning of this degradation process is the relative increase of the contact resistance between the contacting pads, measured in closed contact. For its determination a method of comparative measurement of the residual potential difference between the contact spots in operating position of the used and unused contactors of the same type is used. Measurement is done by a four electrode method – a modified two probe method (Figure 9a). According to the equivalent scheme as shown (Figure 9b) the residual potential difference U12, registered by the voltmeter is determined by cross resistance of the contacts. Elements R_1 and R_2 represent the electric resistance of the leading rails and R_k - the aggregate contact resistance of the two contact spots $(R_k=R_{k1}+R_{k2})$:

$$R_k = R_{k1} + R_{k2} = \frac{U_{12}}{I} \tag{8}$$

The potential probes are located in close proximity to the contact spots, so that the measured potential difference *U12* reads the voltage drop mainly on a high resistive component of the near contact layers, designated with *Rk*. Reference measurements of used and unused contactors of the same type replace the absolute geometric calibration. This paragraph describes measurements done to study the contact resistance of mechanical contactors. Described measurements are non-destructive and are carried out under normal pressure on contacts: of an activated contactor for normally open contacts and of non-activated contactor for normally closed contacts and in currents comparable to the nominal ones for operation. After summarizing the results, on selected representative contact couples are modeled volt-ampere features. Furthermore, again on the selected contacts are made electric probe measurements, as well as, a study of micro-hardness and position by an X-ray microprobe. Potential probes are placed as close as possible to the contact spots, so that the measured potential difference U12 reads the voltage drop mainly on the high resistive component of the near contact layers, denoted by Rk. Measurements described are nondestructive and conducted under standard pressing on contacts: activated contactor for normal open contacts and non-activated for normal closed contacts, at nominal current operation. Measurements are compared to new and old contactors of the same type.

2.2 Measurements of micro-hardness

The study of micro-hardness contactor spots are performed using micro-hardnessmeter. Measurements are conducted at room temperature. The pressure on the intender P, could be selected in the interval 0,5-200g. By software is calculated the average micro-hardness for a given loading:

$$H = AP/d^2, \tag{9}$$

where the value of the constant A depends from the type of the selected indenter. At the same time are calculated the deviations from the average value of H, as well as the depth of indenter penetration into the specimen, d. In order to examine the profile of change in the micro-hardness as a function of the penetration depth of indenter, measurements are carried out over a wide range of load - by at least such that it is possible to measure the fingerprint, to one in which to reach a steady characteristic for a given material micro-hardness. A comparison of the profiles of unused and used specimens can show whether there have been changes in the mechanical properties of the surface layer under the influence of the operating conditions (corrosive atmosphere, high temperature, humidity, radiation). The method may also be used in carrying out the phase analysis. If the specimen is multiphase and different phases are large enough to be intendered can be measured micro-hardness of each phase. Since it depends on the composition, structure and type of processing the material, the resulting information [3], it appears useful for the study of the complex system and a contact in particular for their residual resource. This is shown in the figure 11.

Residual resource contactor elements is evaluated by comparing the slopes of the volt- ampere features of worked and non-working patterns and causes of degradation phenomena are determined by measurements of micro-hardness and phase analysis of alloys constituting the contact spots. For the

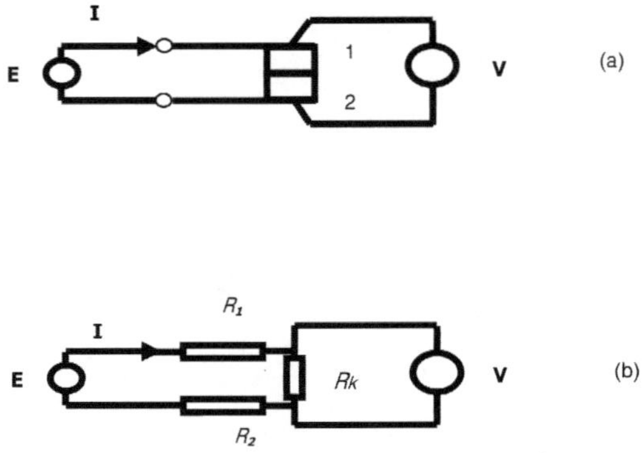

Figure 9. Schematic (a) and equivalent (b) plan of a four electrodes methodology.

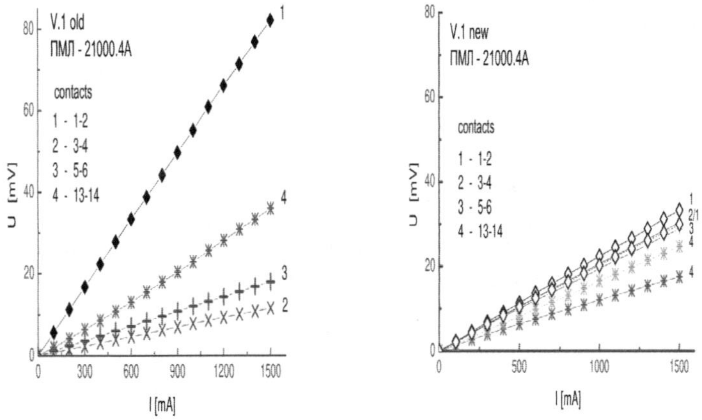

Figure 10. VA characteristics of the old (left) and the new (right) contacts.

degradation determination is used: i) linear schedule in two points, and determines the rate of degradation for a period of 8 years; ii) for a basic mechanism of aging is considered diffusion of the metal atoms in the alloy in less percentage as the formation of microphases and destructure of the contactor alloy. As the main factors facilitating this process is determined: temperature, operating current and time. Almost all unused contacts have linear volt-ampere futures and close contact resistances of several mΩ.

A large part of the used contactors are with linear volt-ampere characteristics but with higher contact resistances and they have additional working time. The average characteristics of both types are shown in figure12. Similarities in the relative degradation changes of the three investigated parameters and significantly greater number of contact sites of contactors many design variations have give the reason and an opportunity to find a unifying idea to explain the underlying processes causing degradation of contactors during operation process.

3. Insulation materials

Electronic equipment management and control of various systems of power Units № 5 and № 6 comprising insulating material is tested for the presence of degradation processes occurring in continuous use in background controlled in sense of radiation according safety standards in NPP

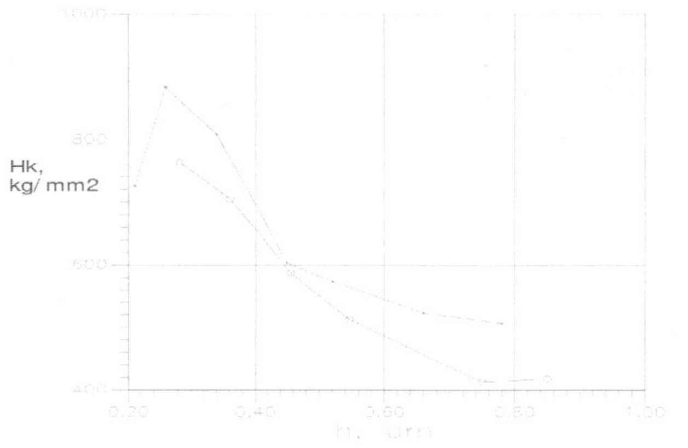

Figure 11. Microhardness dependence from the penetration depth of indenter.

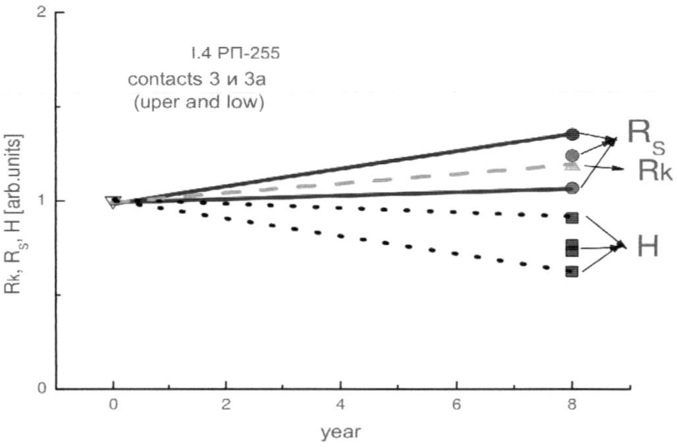

Figure 12. Relative degradation changes of the three investigated parameters for unused and used contactors.

Kozloduy by different weather conditions. Selected systems are operating in continuous mode about 10 years and they have taken separate blocks containing different dielectric materials (paint coatings, substrates for printed circuit boards, cable insulation and electrical protection elements). There were measured dielectric permittivity ε, dielectric losses factor tgδ, and specific resistivity ρ of used dielectric materials and insulators for cables of electronic equipment of NPP. The resulting values were averaged over all measured patterns and compared with available tabulated values for the specific insulating materials. The comparison is made with the table data, since it is well known that in the dielectric material was observed degradation changes in time without being subject to any real exploitation. The measurement of main parameters of the identified models is performed and analyzed by a theoretical model according to which upon degradation of dielectric and insulation materials increases the electrical conductivity at operating voltages. This is related to a change in the polarization mechanism of dielectric dipole structure or the occurrence of double injection in a dielectric, which can be considered as wideband semiconductor.

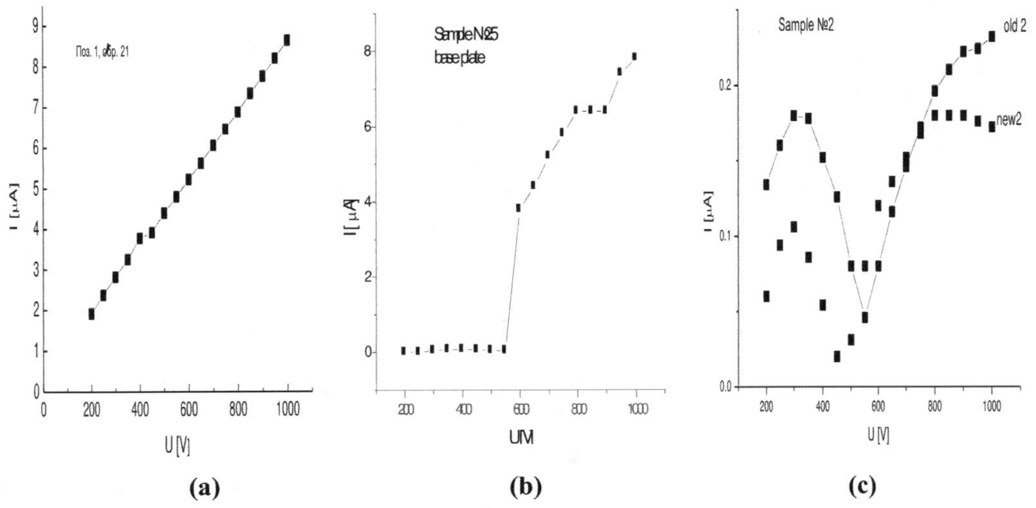

Figure 13. Three types of volt-ampere characteristics of dielectric materials.

3.1 Volt-ampere characteristics

It is examines the linearity of the volt-ampere feature of the tested specimens, as the voltage is changed from about 200 to 1000V with a step of 50V. If the ratio current/voltage is not preserved, these are symptoms of several possible processes, namely: a change of the polarizing mechanism, forming conductive cords, or the occurrence of a double injection. Received data from three types are presented in figure 13a,b,c:

a) the majority (about 80%) have no strong deviations from the linear character;

b) in another group (about 10%) there are horizontal initial sections in the volt-ampere features which are offset relative to zero of the abscissa. This is an indication of the beginning of surface leakages;

c) in a third group (about 10%) are observed sections with a negative resistance at about 500V, which suggests degradation processes associated with a double injection. For the specimens of three types of defined shape and size is possible also to determine independently the dielectric loss factor by the separate measurement of the two components of the conductivity and by bridge allows operation in two modes: -direct measurement of the dielectric loss factor; -separate measurement of the capacity (for the determination of relative permittivity) and the active impedance (for possible calculating the dielectric loss factor). Preliminary comparison of the measured values with same tabular data for the discussed materials show predominantly falling within the tolerances as is shown in for example in table 1.

Table 1. Comparison of the measured values with same tabular date of dielectric.

№	Experimental dates		Material of	Table dates	
	ε	tgδ	the samples	ε	tgδ
3	2.9 – 3	0.03	vinylite	3.2 – 4	0.01 – 0.05
14	3.6	0.13	Insulation canvas	3 – 4.5	0.02 – 0.3
15	7	0.15	bakelite	5 – 12	0.02 – 0.5
16	14	0.006	ceramics		0.0005 – 0.025

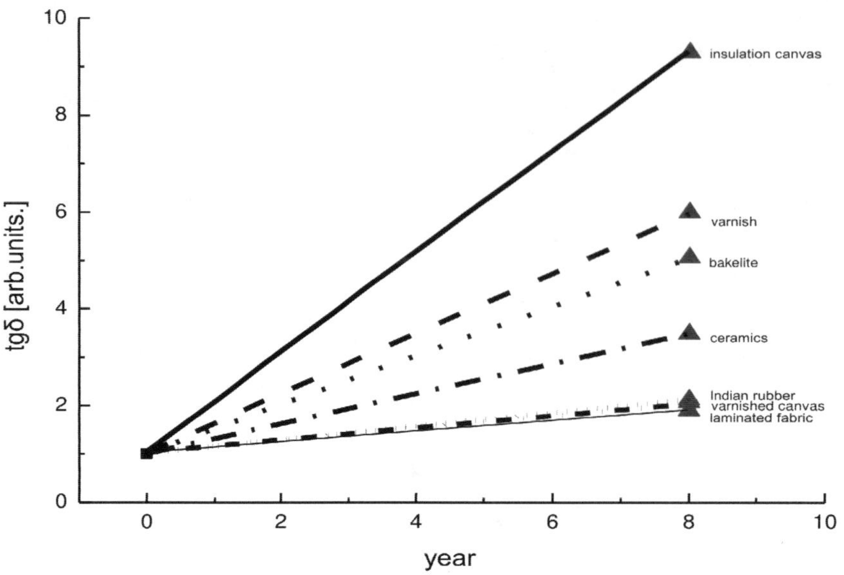

Figure 14. Grouping of dependencies of dielectric losses of different materials as a function of operating time.

Preliminary analysis of the measurements shows that the overwhelming number of cases there is an expected trend of degradation modification of the parameter, namely: reduction of dielectric permittivity and specific resistance and increase of dielectric loss factor. The fact that the relative degradation changes of the basic parameters of such a large group of diverse materials (for example is shown on figure14 for tgδ) have similar numerical intervals shows some community of processes causing degradation. This and the results of the study of the previous two groups of components (diodes and transistors and contactors) allow the construction of a general explanation of the observed changes in the parameters of electronic components and materials in the process of operation.

4. Conclusions

From the conducted studies, it is clear that the main degradation processes in the three groups of elements for an extended period of time are slow do not lead to a hopping change in basic parameters and to catastrophic failures. This gives grounds to suggest a common diffusion model, which is limited to the following: in electronic components containing a p-n junction, is performed diffusion of residual copper atoms, that are accumulated in the area of a spatial charge under the influence of the electric field and the local temperature, creating shunt channels; in the contactor systems whose contact surfaces are made of metal alloys under the influence of increased temperature starts decompositions of a homogeneous alloy. Conditions are created for diffusion of individual atoms. to the surface, micro-phases of homogeneous atoms are formed and modify the contact resistances;-in the course of time in the insulating materials are changed the mechanisms of polarization, double bonds and dipoles are disrupting, leading to the release of carbon atoms. The latter diffuse at elevated temperature and form conductive cords, which amend the dielectric losses and specific resistance of the materials.

Therefore, the total physical model of slow degradation in electronic systems can be built on diffusion processes of free atoms under the influence of local temperature and electric field. These processes are uncontrollable, their speed is slow and the time of change of the resource can be predicted by using the developed methods.

References

[1] Popov A S 1984 *Phys. Stat. Solidi* (a) **81** 669
[2] Popov A and Naydenov N 2005 *J. Opt. Adv. Mater.* **7** 1 321
[3] Naydenov N and Popov A 2006 *Mater. Sci.& Eng.* B **132** 247
[4] Naydenov N and Popov A 2007 *Conf. Proc.of J. Amer. Inst. Phys.* **899** 275
[5] Seibt M and Graff K 1988 *J. Appl. Phys.* **63** 4444
[6] Istratov A A Flink C Hiestmair H and Weber E R 1988 *Phys. Rev. Lett.* **81** 1243
[7] Istratov A A Hiestmair H Heiser T Flink C and Weber E R 1988 *Appl. Phys. Lett.* **72** 474
[8] Stafeev V I 1958 *J. Technical Physic (in Russian)* **28** 1631
[9] Stafeev V I 1959 *Solid State Physics (in Russian)* **1** 841

Unconventional phases of the alternating-spin Heisenberg chain with extra three-body exchange terms

N B Ivanov[1,2] **and J Schnack**[1]

[1]Fakultät für Physik, Universität Bielefeld, Postfach 100131, D-33501 Bielefeld, Germany
[2]ISSP, Bulgarian Academy of Sciences, Tzarigradsko chaussee 72, 1784 Sofia, Bulgaria

E-mail: nedko.ivanov@physik.uni-magdeburg.de

Abstract. The Heisenberg chain with alternating site spins $(S, \sigma) = (1, \frac{1}{2})$ defines a realistic prototype model admitting extra isotropic three-body exchange terms which drive the chain into exotic quantum phases. In this paper, we focus on the non-magnetic part of the phase diagram. Based on numerical density-matrix renormalization group and exact-diagonalization calculations, we demonstrate that the nearest-neighbor three-body interaction stabilizes (i) a critical spin liquid phase described by two Gaussian conformal theories as well as (ii) a critical nematic-like phase characterized by dominant quadrupolar S-spin fluctuations. The emergence of these phases reflects some specific features of the three-body terms such as the promotion of local collinear spin configurations and the enhanced tendency towards nearest-neighbor clustering of the spins.

1. Introduction

Competing higher-order exchange interactions have been identified as an excellent source of exotic quantum effects and phases [1]. For instance, Heisenberg models with the additional two-site biquadratic terms $(\boldsymbol{S}_i \cdot \boldsymbol{S}_j)^2$ – such as the spin-1 bilinear-biquadratic chain [2] and its higher-dimensional counterparts on square [3, 4], triangular [5, 6], and cubic [3] lattices – constitute a class of widely discussed systems supporting rich phase diagrams.

In contrast, for the time being the role of the isotropic three-body exchange terms $(\boldsymbol{S}_i \cdot \boldsymbol{S}_j)(\boldsymbol{S}_i \cdot \boldsymbol{S}_k) + h.c.$ ($|\boldsymbol{S}_i| > \frac{1}{2}$, $i \neq j, k$; $j \neq k$) remains scarcely explored. On the theoretical side, such terms have been primarily used as a tool for constructing various exactly-solvable one-dimensional (1D) models [7, 8, 9, 10]. Some special effects of the isotropic three-body exchange interactions in generic spin-S Heisenberg models have only recently been discussed [11, 12, 13, 14]. On the experimental side, more convincing experimental evidence for three-body exchange effects comes from inelastic neutron scattering measurements in the magnetic material $\mathrm{CsMn}_x\mathrm{Mg}_{1-x}\mathrm{Br}_3$ ($x = 0.28$) [15], CsMnBr_3 being known as a nearly ideal spin-$\frac{5}{2}$ Heisenberg antiferromagnetic (AFM) chain. The experimental results imply nearly equal strengths of both the biquadratic and three-site terms, which are about two orders of magnitude weaker than the principal Heisenberg coupling. In this material, the emergence of effective higher-order exchange terms was attributed to the strong magenetoelastic forces [16]. Another source of three-body exchange interactions could be the two-orbital Hubbard model, where similar terms appear in the fourth order of the strong-coupling expansion [11].

In the discussed systems, that are dominated by the Heisenberg coupling, the strengths of both higher-order terms are controlled by one and the same model parameter, so that it might be a challenge to isolate the effects resulting from three-site terms. In this connection, cold atoms in optical lattices seem to offer a promising route. For instance, as demonstrated in Ref. [17], the two-species Bose-Hubbard model in a triangular configuration can be used to generate a wide range of Hamiltonian operators, including effective three-body interactions, resulting from atomic tunnelings through different lattice paths. A general purpose of this paper is to demonstrate that some conventional alternating-spin Heisenberg systems could provide another promising route [18]. We concentrate on a prototype 1D model of this class, admitting extra three-body exchange terms, defined by the Hamiltonian

$$\mathcal{H}_{\sigma S} = \sum_{n=1}^{L} h_n \equiv \sum_{n=1}^{L} J_1 \boldsymbol{S}_{2n}\cdot(\boldsymbol{\sigma}_{2n-1}+\boldsymbol{\sigma}_{2n+1}) + J_2\left[(\boldsymbol{S}_{2n}\cdot\boldsymbol{\sigma}_{2n-1})(\boldsymbol{S}_{2n}\cdot\boldsymbol{\sigma}_{2n+1})+h.c.\right], \qquad (1)$$

where L stands for the number of elementary cells containing two different spins ($S > \sigma$). In what follows we use the standard parameterization of the coupling constants $J_1 = \cos(t)$ and $J_2 = \sin(t)$ $(0 \leq t < 2\pi)$.

In view of the numerous experimentally accessible quasi-1D spin systems described by the Heisenberg model with different pairs (S, σ) of alternating spins [19, 20, 21], the suggested model defines a simple but realistic onset to explore the impact of three-body exchange interactions. Note that for systems with $\sigma = \frac{1}{2}$ the biquadratic terms reduce to bilinear forms [22, 23, 24]. The following analysis, based on numerical density-matrix renormalization group (DMRG) and exact diagonalization (ED) calculations, is concentrated on the non-magnetic part of the ground-state (GS) phase diagram of $\mathcal{H}_{\sigma S}$ in the extreme quantum limit $(S, \sigma) = (1, \frac{1}{2})$. Keeping up to 500 states in the last (seventh) sweep ensures in most cases good convergences (up to the largest simulated system with $L = 256$) under open boundary conditions (OBC), with a discarded weight of the order of 10^{-8} or better. A detailed description of the full GS phase diagram will be presented elsewhere [25].

2. Classical states

To start with, let us briefly discuss the classical phase diagram of Eq. (1) including – apart from the standard ferromagnetic (FM) and Néel-type ferrimagnetic (FiM) phases – a highly-degenerate phase (D). These phases can be constructed from the three-spin GS configurations of the local Hamiltonians h_n. Indeed, suggesting $\boldsymbol{S}_{2n} = S(0,0,1) = \Uparrow$, one finds the GS cluster states $|\mathrm{FM}\rangle = \uparrow\Uparrow\uparrow$, $|\mathrm{FiM}\rangle = \downarrow\Uparrow\downarrow$, and $|\mathrm{D}\rangle_{L,R} = (\uparrow\Uparrow\downarrow, \downarrow\Uparrow\uparrow)$ respectively in the t intervals $(\frac{3\pi}{4}, \frac{3\pi}{2})$, $(-\frac{\pi}{2}, \frac{\pi}{4})$, and $(\frac{\pi}{4}, \frac{3\pi}{4})$. Remarkably, apart from the boundary points $t = \frac{\pi}{4}$ and $\frac{3\pi}{4}$, the classical Hamiltonian h_n supports only collinear GS spin configurations. Now, by fitting the directions of the shared σ spins, we can use these cluster states as building blocks of global (L cell) spin configurations. By construction, the obtained states correspond to local minima of the classical energy. Obviously, for $t \in (\frac{3\pi}{4}, \frac{3\pi}{2})$ $[t \in (-\frac{\pi}{2}, \frac{\pi}{4})]$, the cluster state $|\mathrm{FM}\rangle$ ($|\mathrm{FiM}\rangle$) generates a unique global configuration representing the classical FM (FiM) phase. On the contrary, for $t \in (\frac{\pi}{4}, \frac{3\pi}{4})$ there exists a manifold of degenerate optimal states based on the cluster states $|\mathrm{D}\rangle_{L,R}$ and their reversed-spin counterparts. As there are two ways to attach a new block to a given configuration, the degeneracy of the D phase is exponentially large (2^L). The established classical diagram was additionally confirmed by Monte Carlo simulations.

3. Critical spin-liquid phase (SL)

The DMRG analysis of the short-range (SR) correlations in open chains reveals in the interval $\frac{\pi}{6} \lesssim t \lesssim \frac{\pi}{2}$ a regular alternating-bond structure of the GS (and the lowest excited states) characterized by different values of the SR correlators $\langle \boldsymbol{\sigma}_{2n-1} \cdot \boldsymbol{S}_{2n} \rangle = u$ and $\langle \boldsymbol{S}_{2n} \cdot \boldsymbol{\sigma}_{2n+1} \rangle = v$

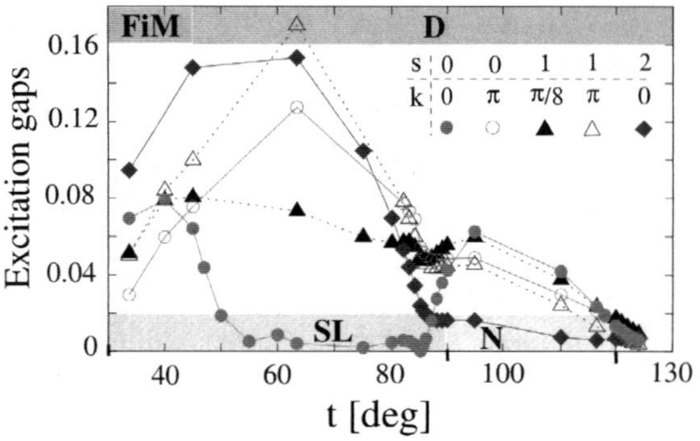

Figure 1. Numerical ED results (symbols) for the lowest excitation gaps in the $L = 8$ ring *vs* t. FiM and D are abbreviations for the ferrimagnetic (Néel type) and the 2^L-fold degenerate classical phases, respectively. SL and N denote, respectively, the spin liquid and nematic-like quantum phases discussed in the text. The lines are guides for the eye. Inset: Spins (s) and momenta (k) of the related excitations.

($u < v$). The uv (vu) "dimerized" GS $|\Psi_L\rangle$ ($|\Psi_R\rangle$) is stabilized in open chains with a σ spin on the left (right) end of the chain and corresponds to a uv (vu) dimerization of the local Hamiltonians h_n. The established uv structure of the GS is strongly pronounced in the middle of the SL region, where the values $(u, v) \approx (-1, \frac{1}{3})$ indicate the formation of nearly pure spin-$\frac{1}{2}$ states of the nearest-neighbor spins $\boldsymbol{S}_{2n} + \boldsymbol{\sigma}_{2n-1}$. Obviously, the established structural order does not violate the translational symmetry of the Hamiltonian $\mathcal{H}_{\sigma S}$ by two lattice sites.

Both types of alternating-bond states are related by the site parity operation $\mathcal{P}$: $\mathcal{P}|\Psi_{L,R}\rangle = |\Psi_{R,L}\rangle$. As the symmetry $\mathcal{P}$ is not violated in finite periodic chains, we may expect, in particular, two quasi-degenerate singlet states related to the combinations $|\Psi_{\pm}\rangle = \frac{1}{\sqrt{2}}(|\Psi_L\rangle \pm |\Psi_R\rangle)$. As seen in figure 1, a low-lying excited singlet state can be observed already in small rings ($L = 8$). A similar tendency towards doubling of the spectrum can also be identified for some triplet and quintet states in figure 2a. In fact, even in t regions with relatively large singlet gaps for small L – like the region around $t = 45°$ – the finite-size scaling (FSS) of the excitation gaps, figure 2b, supports this tendency: in particular, the FSS implies an exponentially fast (with L) doubling of the lowest singlet and triplet states. Due to strong boundary effects, the discussed doubling in the lowest part of the spectrum remains invisible in open chains up to the largest ($L = 256$) simulated system.

In figure 3a, we show DMRG results for the GS entanglement entropy $S(L, l)$ of the open alternating-spin chain at $t = 45°$. The analytical result for this quantity in critical conformally-invariant 1D systems reads [26, 27]

$$S(L, l) = \frac{c}{3\eta} \ln \left[\frac{\eta L}{\pi} \sin \left(\frac{\pi l}{L} \right) \right] + \text{const.} \qquad (2)$$

Here l is the number of unit cells in the subblock ($l = 1, \ldots, L$), c is the central charge, and $\eta = 1, 2$ for periodic boundary conditions (PBC) and OBC, respectively. The extrapolation of the numerical data for $S(L, L/2)$ *vs* $\ln(2L/\pi)$ up to $L = 256$ suggests a critical behavior with central charge $c = 1$, figure 3b. We observe two different branches of $S(L, l)$ corresponding to even and odd l. Similar even-odd oscillations have been originally reported in open spin-$\frac{1}{2}$ XXZ chains, including the isotropic limit [28]. In this work, it was clarified that the alternating part of $S(L, l)$, decaying away from the boundary with a universal power law, appears as a result of oscillations of the energy density. The latter oscillations were attributed to the tendency of

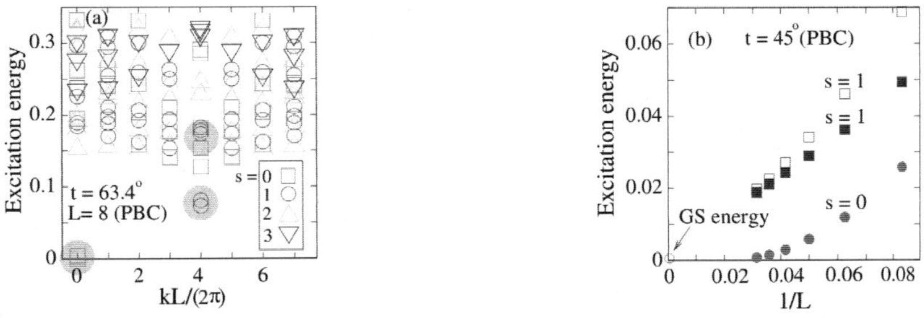

Figure 2. (a) Numerical ED results for the low-lying (spin $s = 0, 1, 2$, and 3) excitation energies of the periodic $L = 8$ chain at $t = 63.4°$. The shaded symbols correspond to the lowest quasi-degenerate pairs of singlet, triplet and quintet excitations. (b) DMRG results for the FSS of the lowest singlet and triplet modes at $t = 45°$.

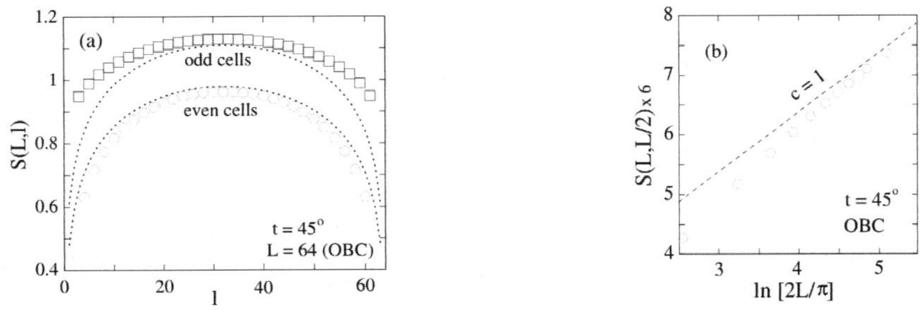

Figure 3. (a) Numerical DMRG results for the even- and odd-subblock GS entanglement entropy of the SL state ($t = 45°$) vs the number of subblock cells l. The dashed lines represent the theoretical result, Eq. (2), with $c = 1$. (b) Extrapolation of the DMRG results for $S(L, L/2)$ vs $\ln(2L/\pi)$ at $t = 45°$. The dashed line corresponds to $S(L, L/2)$ with the central charge $c = 1$.

the critical system towards formation of local singlet bonds, combined with the strong tendency of the end spins to form local singlets. To understand the even-odd cell modulation of $S(L, l)$ in the alternating spin chain, one can suggest a similar scenario, i.e., one may speculate that the even-odd cell modulation of $S(L, l)$ is related to the formation of local four-spin singlets. However, unlike for the XXZ chain, the formation of local singlet states in the alternating spin model is more complex as it includes, at least, four neighboring spins with two possible types of uv modulation.

Assuming conformal invariance, additional characteristic properties of the non-magnetic SL phase can be extracted from the FSS behavior of the GS and the lowest excited states. Numerical simulations of periodic chains are hampered by the quasi-degeneracy of the GS, so that the following analysis is performed under OBC. The expected tower of excited states related to some primary operator is defined by [29]

$$\Delta_n(L) \equiv E_n(L) - E_0(L) = \frac{\pi v_s}{L}(x_s + n) + o(L^{-1}), \tag{3}$$

where $n = 0, 1, 2, \ldots$. x_s is the universal surface exponent related to the same operator, and v_s is the sound velocity. The exponent x_s is known, in particular, for the energy states of the isotropic spin-$\frac{1}{2}$ Heisenberg chain in the m sectors ($x_s^{(m)} = m^2$, $m = 1, 2, \ldots$), where m is the z component of the total spin [30]. The Hamiltonian (1) respects the spin-rotation symmetry, so that the above asymptotic expression, when used as a fitting ansatz, has to be supplemented by

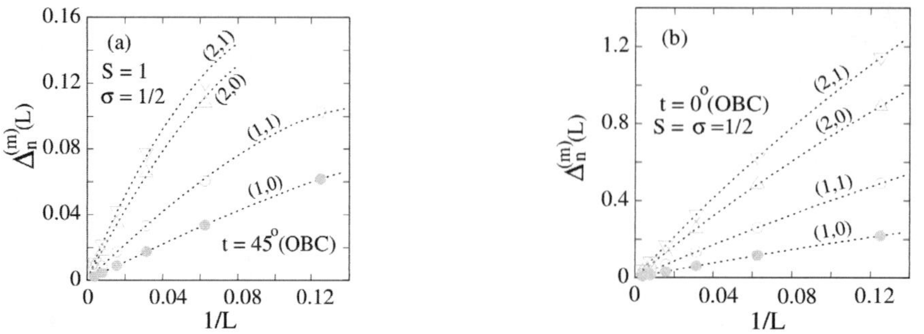

Figure 4. (a) FSS of the lowest triplet $[(m,n)=(1,0)$ and $(1,1)]$ and quintet $[(m,n)=(2,0)$ and $(2,1)]$ gaps in (a) the alternating spin $(1,\frac{1}{2})$ chain (DMRG, OBC, $t=45°$) and (b) in the uniform $(\frac{1}{2},\frac{1}{2})$ chain with $2L$ sites (DMRG, OBC, $t=0°$). The dashed lines show the best fits to the DMRG data (symbols) obtained by Eq. (4).

appropriate logarithmic terms [31].

In figure 4, we compare FSS results for the lowest two excitations in the triplet ($m=1$) and quintet ($m=2$) towers of states of the Hamiltonian $\mathcal{H}_{\sigma S}$, Eq. (1), for two cases: (i) $(S,\sigma)=(1,\frac{1}{2})$ at $t=45°$ and (ii) $(S,\sigma)=(\frac{1}{2},\frac{1}{2})$ at $t=0°$. The fit of the reduced gaps $L\Delta_n^{(m)}$ is performed by the four-parameter ansatz

$$L\Delta_n^{(m)}(L)=a_n^{(m)}+\frac{b_n^{(m)}}{\ln\left(L/\xi_n^{(m)}\right)}+\frac{c_n^{(m)}}{L}. \qquad (4)$$

For systems belonging to the Gaussian universality class – like the isotropic spin-$\frac{1}{2}$ chain in the second case – the first fitting parameter $a_n^{(m)}$ is expected to approach the exact result $a_n^{(m)}=\pi v_s(m^2+n)/2$, which gives $[m,n]\equiv a_n^{(m)}/a_0^{(1)}=m^2+n$. In fact, the performed fits for the $(S,\sigma)=(\frac{1}{2},\frac{1}{2})$ chain give the numerical estimates [1,1]=1.99, [2,0]=3.99, and [2,1]=4.96, which excellently reproduce the expected theoretical ratios. Moreover, a comparison of Eqs. (3) and (4) implies the relation $a_0^{(1)}=v_s/2$, which gives an estimate for v_s deviating only by about 0.6% from the exact result $\pi/2$. For the alternating-spin chain, similar fits give the numerical estimates $v_s=0.38$, [1,1]=2.11, [2,0]=4.44, and [2,1]=5.69. In spite of the larger deviations from the theoretical results for $[m,n]$, the observed structure of the lowest-lying part of the spectrum in the alternating-spin model remains close to the structure in the reference spin-$\frac{1}{2}$ Heisenberg chain [32].

The established one-to-one mapping of the lowest-lying excitations of both models suggests similar critical properties. Since the unit cell of the reference spin-$\frac{1}{2}$ model contains two equivalent lattice sites, under PBC this means a doubling of the spectrum and, in particular, two equivalent critical modes. This explains the discussed doubling of the lowest-lying excitations in the alternating-spin $(1,\frac{1}{2})$ ring. Thus, both the GS entanglement entropy as well as the FSS properties of the SL phase point towards a Gaussian type critical behavior. Since the alternating-spin $(1,\frac{1}{2})$ ring exhibits two equivalent critical modes, the SL state may be interpreted as a critical spin-liquid phase described by two Gaussian conformal theories associated with these modes. Similar critical phases have been studied in some exactly solvable models, including spin models with extra three-body exchange interactions. In particular, there is an exactly-solvable alternating-spin (S,σ) model [33, 34] closely related to the generic model discussed in

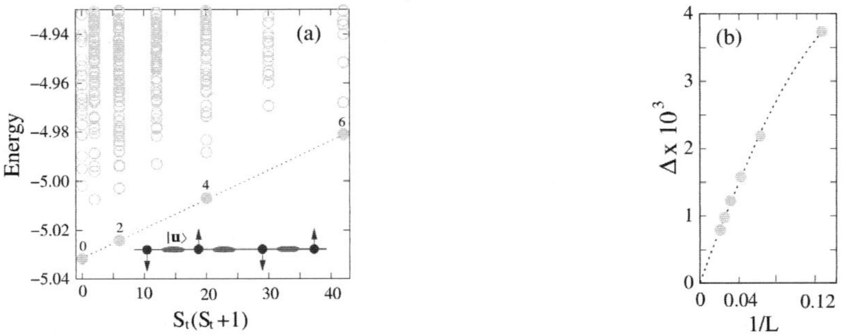

Figure 5. (a) Numerical ED results for the energy spectrum of the alternating-spin ($S = 1$, $\sigma = \frac{1}{2}$) ring ($L = 8$, $t = 110°$) *vs* $S_t(S_t + 1)$, where S_t is the total spin. The lowest multiplets in the even S_t sectors (filled circles) form a tower of states with energies $E(S) \propto S_t(S_t + 1)$. The dashed line is a guide for the eye. Inset: Cartoon of the suggested nematic-like state in the N sector. The ellipses denote the local nematic states $|\boldsymbol{u}\rangle = \sum_\alpha u^\alpha|\alpha\rangle$, where $\boldsymbol{u}$ is a unit real vector and $|\alpha\rangle$ ($\alpha = x, y, z$) is the vector basis of the spin-1 operator $\boldsymbol{S}$. (b) FSS of the lowest quintet gap ($t = 110°$, DMRG, OBC). The dashed line shows the optimal fitting ansatz $\Delta(L) = a_0 + a_1/L + a_2/L^2$.

this work at the point $t = 45°$. In fact, the difference between both models is reduced to an additional FM exchange term ($h_n^{\sigma\sigma} = J_3\boldsymbol{\sigma}_{2n-1} \cdot \boldsymbol{\sigma}_{2n+1}$, $J_3 < 0$) in the exactly solvable model. Assuming that $h_n^{\sigma\sigma}$ represents an irrelevant operator (in a renormalization group sense), it may be speculated that both models exhibit similar critical properties. In particular, for the exactly solvable (S, σ) model, it has been predicted [33] that the critical behavior can be described by an effective central charge which is the sum of the central charges related with two critical modes, t.e., $c_{eff} = 3\sigma/(\sigma + 1) + 3(S - \sigma)/(S - \sigma + 1)$. In the special case $(S, \sigma) = (1, \frac{1}{2})$ this gives $c_{eff} = 1 + 1 = 2$, which coincides with the expected critical behavior of the SL phase.

4. Nematic-like phase (N)

The numerical ED results for the low-lying excitations in the periodic $L = 8$ chain, figure (1), point towards a different non-magnetic GS in the interval $\frac{\pi}{2} \lesssim t \lesssim \frac{2\pi}{3}$. Indeed, in the vicinity of $t = \frac{\pi}{2}$ the lowest quintet excitation is strongly softened and becomes the first excited state up to $t \simeq \frac{2\pi}{3}$. DMRG results for somewhat larger periodic systems (up to $L = 28$) confirm the picture with a lowest quintet excitation [35]. Further information about the non-magnetic state can be extracted form figure 5, showing ED results for the excitation spectrum of the same system at $t = 110°$ in different total-spin (S_t) sectors. An obvious feature of the presented spectrum is the established tower of well-separated lowest multiplets containing only even S_t sectors. Furthermore, the energies in the tower scale as $E(S) \propto S_t(S_t + 1)$. The observed structure is known as a fingerprint of the spin quadrupolar (i.e., nematic) order [36], unlike the Anderson tower – a characteristic of the Néel order – containing all S_t sectors [37]. In fact, Anderson towers of states have been observed even in some finite isotropic spin-S chains and magnetic molecules [38, 39, 40], including spin-$\frac{1}{2}$ Heisenberg chains [41]. In the same spirit, we consider the specific structure in figure 5a as a fingerprint of a non-magnetic state with dominant quadrupolar spin fluctuations. The FSS of the quintet excitation gap $\Delta(L) \propto 1/L$, figure 5b, pointing towards a gapless N state, is consistent with the above suggestion. Finally, in the Inset of figure 5a we show the cartoon of a tentative nematic-like state built from the on-site nematic states $|\boldsymbol{u}\rangle$ of the spin-1 operators, which respects both the established structure of the spectrum as well as the DMRG data implying extremely weak nearest-neighbor $S\sigma$ correlations

($|\langle \boldsymbol{S}_{2n} \cdot \boldsymbol{\sigma}_{2n\pm1}\rangle| < 0.07$) in the whole region occupied by the N phase.

5. Conclusion

It is worth emphasizing that the establishment of the discussed states is directly related to some peculiarities of the dominating three-body interaction such as the maintenance of collinear classical states and the promotion of nearest-neighbor spin clustering. As with the bilinear interaction in Eq. (1), in the N phase ($J_1 < 0$) its role is to weaken the AFM nearest-neighbor $S\sigma$ correlations, while in the SL phase ($J_1 > 0$) it enhances the tendency towards the formation of spin $(S - \sigma)$ states of the neighboring spins $\boldsymbol{S}_{2n} + \sigma_{2n\pm1}$. In this sense, an analogue of the SL state in spin-S AFM chains with extra three-body terms is the recently predicted fully-dimerized (Majumdar-Ghosh type) GS [11]. Obviously, the alternating-spin systems suggest more variety in this direction. Finally, it may be expected that most of discussed effects of the three-body exchange interaction persist in higher space dimensions. We believe that the presented results will encourage future search for real systems with pronounced three-body exchange effects.

Acknowledgments

We are grateful to Jörg Ummethum for adapting his DMRG code [42, 43] to three-body exchange interactions. N.B.I. thanks Bielefeld University for its hospitality. This research was supported by the Deutsche Forschungsgemeinschaft (SCHN 615/20-1), Deutscher Academischer Austausch Dienst (Grant PPP Bulgarien 57085392), and the National Science Foundation of Bulgaria [Grant DNTS/Germany/01/2 (03.09.2014)].

References

[1] Lhuillier C and Misguich G, *Introduction to frustrated magnetism Materials, Experiments, Theory* 2011 (Springer Series in Solid State Sciences vol 164) ed C Lacroix, P Mendels and F Mila (Berlin Heidelberg: Springer Verlag) chapter 2 pp 23-41
[2] see, e.g., Läuchli A, Schmid G and Trebst S 2006 *Phys. Rev.* B **74**144426, and references therein.
[3] Harada K and Kawashima 2002, *Phys. Rev.* B **65** 052403
[4] Tóth T A, Läuchli A M, Mila F and Penc K 2012 *Phys. Rev.* B **85** 140403(R)
[5] Momoi T, Sindzingre P and Shannon N 2006 *Phys. Rev. Lett.* **97** 257204
[6] Smerald A and Shannon N 2013 *Phys. Rev.* B **88** 184430
[7] Andrei N and Johannesson H 1984 *Phys. Lett.* A **100** 108
[8] de Vega H J and Woynarovich F 1992 *J. Phys. A: Math. Gen.* **25** 4499
[9] de Vega H J, Mezincescu L and Nepomechie R I 1994 *Phys. Rev.* B **49**, 13223
[10] Ribeiro G A P and Klümper A 2008 *Nucl. Phys.* B **801** [FS] 247
[11] Michaud F, Vernay F , Manmana S R and Mila F 2012 *Phys. Rev. Lett.* **108** 127202
[12] Michaud F, Manmana S R and Mila F 2013 *Phys. Rev.* B **87** 140404(R)
[13] Michaud F and Mila F 2013 *Phys. Rev.* B **88** 094435
[14] Wang Z -Y, Furuya S C, Nakamura M and Komakura R 2013 *Phys. Rev.* B **88** 224419
[15] Falk U, Furrer A, Güdel H U and Kjems J K 1986 *Phys. Rev. Lett.* **56** 1956
[16] Falk U, Furrer A, Kjems J K and Güdel H U 1984 *Phys. Rev. Lett.* **52** 1336
[17] Pachos J K and Plenio M B 2004 *Phys. Rev. Lett.* **93** 056402
[18] As the formation of local nearest-neighbor singlet dimer states is excluded in the alternating-spin systems, one may expect a different impact of the extra three-body terms as compared to the uniform-spin Heisenberg models.
[19] Furrer A 2010 *Int. J. Mod. Phys.* B **24** 3653
[20] Furrer A and Waldmann O 2013 *Rev. Mod. Phys.* **85** 367
[21] Landee Ch P and Turnbull M M 2013 *Eur. J. Inorg. Chem.* **2013** 2266
[22] In the extreme quantum case ($S = 1$, $\sigma = \frac{1}{2}$), $\mathcal{H}_{\sigma S}$ reproduces (up to irrelevant terms) the effective Hamiltonian of an isotropic spin-$\frac{1}{2}$ diamond chain with ring interactions in the subspace where any pair of "up" and "down" plaquette spins is in a triplet state.
[23] Ivanov N B, Richter J and Schulenburg J 2009 *Phys. Rev.* B **79** 104412
[24] Ivanov N B 2009 Condensed Matter Physics **12** 435
[25] Ivanov N B Ummethum J and Schnack J (to be published)
[26] Holzhey C, Larsen F and Wilczek F 1994 *Nucl. Phys.* B **424** 443

[27] Calabrese P and Cardy J 2004 *J. Stat. Mech.* **06** P06002
[28] Laflorencie N, Sørensen E S , Chang M S and Affleck I 2006 *Phys. Rev. Lett.* **96** 100603
[29] Alcaraz F C, Barber M N, Batchelor M T, Baxter R J and Quispel G R W 1987 *J. Math. Phys.* A **20** 6397
[30] Alcaraz F C, Berganza M I and Sierra G 2011 *Phys. Rev. Lett.* **106** 201601
[31] see, e.g., Hallberg K, Wang X q G, Horsch P and Moreo A 1996 *Phys. Rev. Lett.* **76** 4955, and references therein.
[32] As may be expected, in the middle of the range occupied by the SL phase, where the doubling of the lowest singlet and triplet states with L is faster (see figure 1), the deviations of $[m, n]$ from the theoretical results are smaller. For example, at $t = 63.4°$ the same fitting procedure gives $v_s = 1.88$, $[1, 1] = 1.93$, $[2, 0] = 4.00$, and $[2, 1] = 5.35$.
[33] Aladim s R and Martins M J 1993 *J. Phys.* A **26** L529
[34] Bytsko A and Doikou A 2004 *J. Phys.* A **37** 4465
[35] Unfortunately, an extensive numerical analysis of the FSS properties of the excitation gaps is hampered by a slow convergence of the DMRG method in this region.
[36] see, e.g., Läuchli A, Domenge J C, Lhuillier C, Sindzingre P and Troyer M 2005 *Phys. Rev. Lett.* **95** 137206, and references therein.
[37] Anderson P W 1952 *Phys. Rev.* B **86** 694
[38] Schnack J and Luban M 2000 *Phys. Rev.* B **63** 014418
[39] Waldmann O 2001 *Phys. Rev.* B **65** 024424
[40] Machens A, Konstantinidis N P, Waldmann O, Schneider I and Eggert S 2013 *Phys. Rev.* B **87** 144409
[41] Eggert S and Affleck I 1992 *Phys. Rev.* B **46** 10866
[42] Ummethum J, Nehrkorn J, Mukherjee S, Ivanov N B, Stuiber S, Strassle Th, Tregenna-Piggott P L W, Mutka H, Christou G, Waldmann O and Schnack J 2012 *Phys. Rev.* B **86** 104403
[43] Schnack J and Ummethum J 2013 *Polyhedron* **66** 28

18th International School on Condensed Matter Physics IOP Publishing
Journal of Physics: Conference Series **558** (2014) 012016 doi:10.1088/1742-6596/558/1/012016

Some features of bulk melt-textured high-temperature superconductors subjected to alternating magnetic fields

P Vanderbemden[1], I Molenberg[1], P Simeonova[2] and V Lovchinov[2]

[1] University of Liège, Department of Electrical Engineering & Computer science (B28), Sart-Tilman, B-4000 Liège, Belgium
[2] Institute of Solid State Physics, Bulgarian Academy of Sciences, 72 Tzarigradsko Chaussee Blvd., 1784 Sofia, Bulgaria

E-mail : Philippe.Vanderbemden@ulg.ac.be

Abstract. Monolithic, large grain, (RE)$Ba_2Cu_3O_7$ high-temperature superconductors (where RE denotes a rare-earth ion) are known to be able to trap fields in excess of several teslas and represent thus an extremely promising competing technology for permanent magnet in several applications, e.g. in motors and generators. In any rotating machine, however, the superconducting permanent magnet is subjected to variable (transient, or alternating) parasitic magnetic fields. These magnetic fields interact with the superconductor, which yields a reduction of the remnant magnetization. In the present work we quantify these effects by analysing selected experimental data on bulk melt-textured superconductors subjected to AC fields. Our results indicate that the non-uniformity of superconducting properties in rather large samples might lead to unusual features and need to be taken into account to analyse the experimental data. We also investigate the evolution of the DC remnant magnetization of the bulk sample when it is subjected to a large number of AC magnetic field cycles, and investigate the experimental errors that result from a misorientation of the sample or a mispositioning of the Hall probe. The time-dependence of the remnant magnetization over 100000 cycles of the AC field is shown to display distinct regimes which all differ strongly from the usual decay due to magnetic relaxation.

1. Introduction

Due to their ability to carry high currents with no resistance at low temperature, superconducting materials represent a way of designing extremely promising electrical engineering applications [1-5]. In particular, (RE)$Ba_2Cu_3O_{7-\delta}$ (REBCO) superconductors (where "RE" denotes a rare-earth ion) can carry extremely large currents under high applied fields, a property that can be exploited in long length coated conductors [6] as well as in current leads [7]. When prepared in the melt-processed form, e.g. monolithic disks of a few centimeters in diameter, the so-called "single domains" contain nanometer-sized defects that can act as efficient pinning centers for flux lines [8-11]. As a consequence, these materials are able to trapped considerable magnetic flux densities and act as powerful permanent magnets [12]. The world record of trapped field in such magnets [13] has been broken very recently [14] and is now 17.6 teslas, which is far beyond the saturation magnetization of conventional ferromagnetic materials. Such large magnetic flux densities give rise to huge Lorentz forces and considerable attention needs to be given to mechanical properties [13-15]. Another important aspect is related to the use of such permanent magnets in engineering applications. In running operation, the remnant magnetization of bulk superconducting magnets is found to decrease slightly, which may be detrimental to the performances of the machine. The first origin of this decrease is magnetic relaxation, caused by a non-homogeneous magnetic flux profile (maximum in the center and zero at the sample edge) [16]. The corresponding time-dependence decay of magnetization, however, is usually logarithmic and for bulk REBCO samples at liquid nitrogen temperature (77 K), may become almost negligible after a few hours of operation [12]. The second cause of the decay is due to the interaction of the magnetized sample with transient or AC magnetic fields that may exist in the device. These time-dependent magnetic fields are due to vibrations of permanent magnets interacting with the superconductor (e.g. in magnetic bearings or magnetic levitation applications) or to variations of the rotating field generated by the stator windings (in rotating machines) [17,18]. Such parasitic alternating fields may reduce substantially the performance of the trapped field magnet [19]. In addition, the direction of the trapped field is also an important parameter, as described below.

Figure 1 shows four basic configurations of a bulk melt-textured superconductor subjected to an AC field. In this drawing, the bulk superconductor is assumed to be a cylindrical disk with the crystallographic *c*-axis perpendicular to the largest circular face (*ab* planes). The AC magnetic field, can be applied to a virgin sample (configurations 'a" and "b") or to a sample in which permanent supercurrents, flowing in the *ab* planes and generating a DC magnetization parallel to the *c*-axis (configurations "c" and "d"). The first two cases are traditional configurations in studying the AC magnetic response of a superconductor, e.g. by AC susceptibility. The AC magnetic field produces losses that, according to the Bean model in a homogeneous superconductor [16], are usually maximum when the AC currents induced by the field reach the sample center [20]. In large bulk samples, due to their relatively poor thermal conductivity [21,22], such losses may cause significant self-heating [23,24] which, in turns, has a detrimental effect on the critical current density J_c of the superconductor. In the presence of a pre-existent DC magnetization (configuration "c"), such self-heating is present and yields some decrease of the trapped flux [19], which is dependent on the cooling conditions of the superconductor [24], e.g. through its surface/volume ratio [25] or the presence of a reinforcement stainless steel ring [26]. When the AC field is applied perpendicularly to the DC magnetization (configuration "d"), the decay of trapped flux is found to be severe: this corresponds to the so-called "crossed field" configuration which may lead to the "collapse of magnetic moment" [27]. This intriguing effect, studied for more than three decades [28], has been explained by considering the modification of current distribution caused by the varying field [29-31] but it still a subject of active research, see e.g. the recent work [32] on the crossed-field effect on stacks of tapes [33]. Similarly, the behaviour of a bulk melt-textured sample subjected to a large number of AC transverse cycles is difficult to predict but plays a significant role in engineering applications.

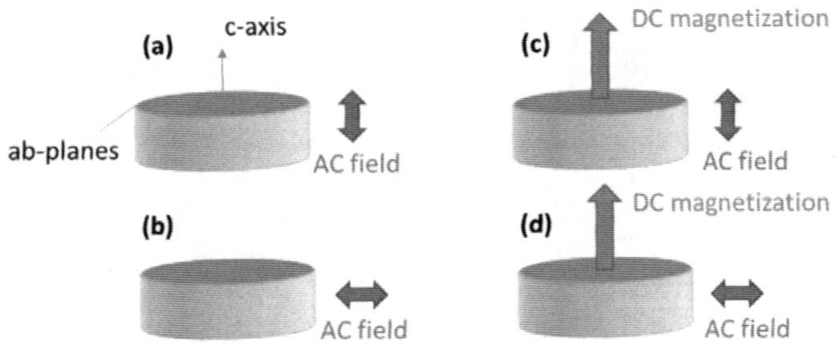

Figure 1. Schematic diagram of a bulk melt-textured superconductor, showing the *ab* planes and the *c*-axis. Four different configurations are illustrated : (a) AC field parallel to the *c*-axis, (b) AC field perpendicular to the *c*-axis, (c) sample with trapped DC magnetic moment ($\parallel c$) superimposed to a parallel AC field, (d) sample with trapped DC magnetic moment ($\parallel c$) superimposed to a transverse AC field ($\perp c$); the latter corresponds to the so-called "crossed-field" configuration.

The purpose of the present work is to analyze several situations where a bulk melt-textured sample is subjected to an AC magnetic field and to point out some features that need to be taken into account for appropriate experimental characterization of these effects or interpretation of the measured data.

2. Experiment

The experiments are carried out on bulk $DyBa_2Cu_3O_7$ single domains synthesized by a melt-growth process described in detail in an earlier work [11]. Single grain samples up to 20 mm in diameter can be routinely fabricated by this technique. The critical superconducting transition temperature of the material is $T_c \sim 89$ K and its critical current density J_c(77 K, 0.1 T) ~ 10^4 A/cm^2. Several rectangular specimens intended to AC magnetic measurements were excised from the main single domain using a wire saw. Measurements of AC magnetic properties of cubic samples (3 × 3 × 3 mm) for $H \parallel c$ were carried out as a function of temperature in an AC susceptometer based on a cryocooler [34]. The crossed-field effects were investigated at liquid nitrogen temperature (77 K) on a parallelipipedic sample of dimensions 15 × 15 × 7 mm, where the shortest dimension is parallel to the *c*-axis. The full penetration fields of this sample for $H \parallel c$ and $H \parallel ab$ were determined to be respectively $\mu_0 H_p \parallel c = 0.35$ T and $\mu_0 H_p \parallel ab = 0.25$ T. A field-cooled (FC) procedure under a constant 0.6 T background field produced by an iron-cored electromagnet was used to magnetize permanently the superconductor. The AC magnetic field was generated by a copper coil, fed either by a computer controlled power supply (quasi-static variations, frequency $f \sim 0.02$ Hz) or by a waveform synthesiser followed by an audio amplifier ($f \sim 25$ Hz). Measurements in the crossed-field configuration required to design a bespoke experimental set-up [29] adapted to large samples (~ 20 mm diameter, 5 mm thickness). Importantly, the sample was clamped firmly to prevent the strong magnetic torque (arising from the magnetic moment ∥ *c* and the field ∥ *ab*) from modifying the sample orientation. The *c*-axis component of the DC average flux density was measured through numerical integration of the induced e.m.f. across a pick-up coil closely wound around the sample, while the local flux density was measured using an AREPOC Hall probe stuck against the sample surface.

3. Results

In a first set of experiments, the AC magnetic properties of selected samples was characterized in the absence of DC magnetization (configuration "a" in figure 1). Then crossed-field effects were investigated (configuration "d" in figure 1).

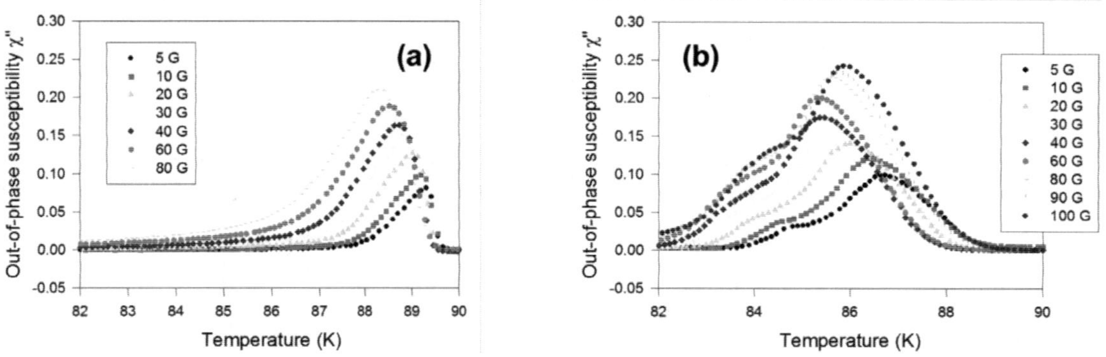

Figure 2. Out-of-phase component of the AC magnetic susceptibility χ" vs. temperature measured on two samples of bulk melt-textured $DyBa_2Cu_3O_7$ superconducting sample for several amplitudes of the AC magnetic field. The applied AC field is parallel to the *c*-axis with no DC magnetization (configuration "a" in figure 1). The amplitude ranges from 5 gauss (0.5 mT) to 80 gauss (8 mT). The frequency is fixed at 1053 Hz.

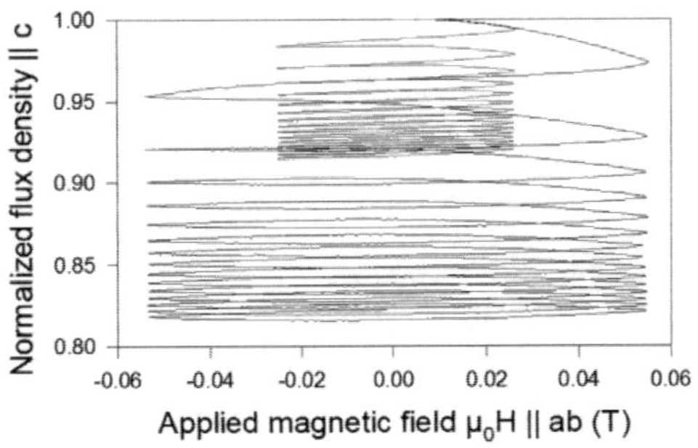

Figure 3. Decrease of the *c*-axis component of the DC remnant average flux density (normalized to its initial value) of a bulk melt-textured $DyBa_2Cu_3O_7$ superconducting sample as a function of cycles of transverse AC field (parallel to the *ab* planes) of two different amplitudes (configuration "d" in figure 1). The measurement is carried out at $T = 77$ K.

3.1. AC magnetic susceptibility in the absence of AC field

The magnetic response of melt-textured $DyBa_2Cu_3O_7$ samples was first studied as a function of temperature in order to examine the characteristics of the superconducting transition. Pick-up coils wound around the sample allow the AC magnetic susceptibility $\chi = \chi' - j\chi''$ to be determined [34,35].

Figure 3 shows the temperature dependence of the out-of-phase component χ'' at several AC field amplitudes for two selected samples extracted from same the single domain. As can be seen, sample (a) displays a well-defined single peak at the smallest field investigated and a narrow superconducting transition at $T_c \sim 89$ K. The peak is shifted to lower temperatures and its amplitude increases as the AC field is increased, as expected from the Bean model taking into account a field-dependence of the critical current density [20]. In contrast, sample (b) displays a larger superconducting transition and a well-defined shoulder that can be viewed as the convolution of two separate peaks that occur close to each other. A double peak structure was often observed in high-temperature superconductors and attributed to shielding currents circulating either within the grains (intragranular) or across the grain boundaries (intergranular). Previous works on similar melt-textured samples have shown, however, that naturally occurring grain boundaries are characterized by a very small intergranular critical current density [36] (in contrast to artificially engineered boundaries [37]). The behaviour depicted in figure 3(b) is therefore likely to be exclusively of intragranular origin. A possible explanation would be the nucleation of secondary grains of smaller size [38] or local variations of oxygen content, resulting in an inhomogeneous distribution of T_c in this particular sample. The latter hypothesis would be consistent to explain the unusual (and reproducible) feature occurring in the two peaks of the AC susceptibility : on increasing the AC field amplitude, the left peak is shifted to lower temperatures while the right main peak is shifted first to left (up to 60 G) and then to right (from 60 G to 100 G). Such a behaviour cannot be explained by the Bean model with a uniform J_c and is expected to arise from a peculiar field-dependent T_c and J_c distribution within this particular sample.

This result underlines the possible occurrence of small T_c and J_c inhomogeneities when investigating the properties of medium-size (> 1 mm) melt-textured samples. The data displayed in figure 3(b) show that such inhomogeneities can be revealed easily through AC susceptibility measurements. Therefore considerable caution need to be taken before interpreting the results; the best specimens with sharp transition – cf. figure 3(a) – should be selected carefully before proceeding to further experiments.

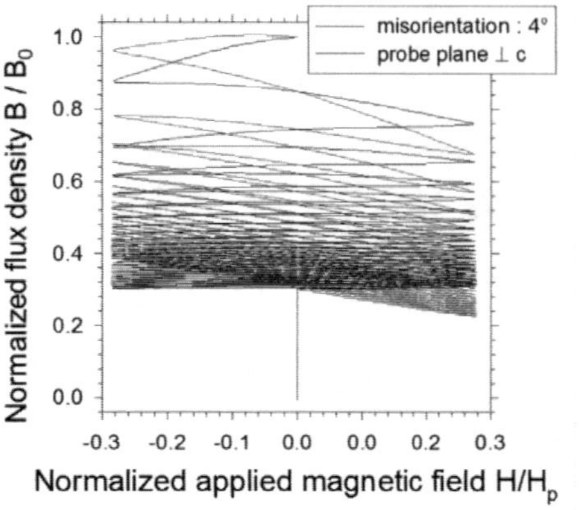

(a)

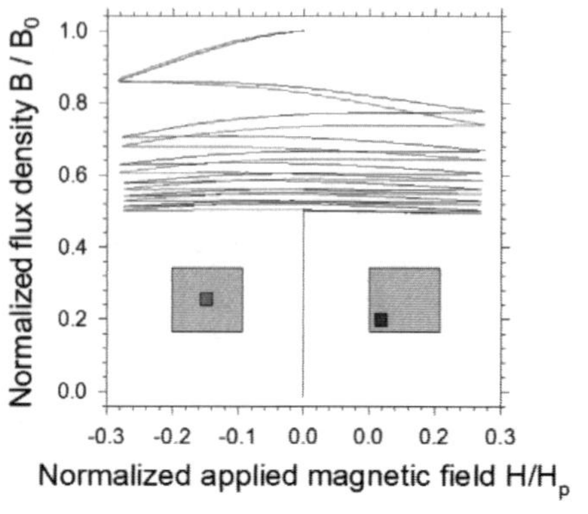

(b)

Figure 4. Two examples of crossed-field experiments on a bulk melt-textured $DyBa_2Cu_3O_7$ superconducting sample illustrating the influence of experimental parameters on the measured data. (a) Comparison between the results obtained with a correctly positioned Hall probe (blue) or when the sample is tilted at 4° (red). (b) Comparison between the results obtained with a centred Hall probe (red) or when the Hall probed is placed near the edge of the sample (blue).

3.2. AC field applied perpendicularly to pre-existing magnetization

Now we investigate the results obtained in the "crossed field" configuration. A typical result of the decay of the *c*-axis average magnetic flux density against the transverse field AC field is plotted in figure 3. This graph shows how the average flux density (probed by the sensing coil) is affected drastically when cycles of transverse field are applied to the single domain. This behaviour is in excellent agreement with other results of crossed field experiments carried out on other superconducting materials [27-31]. It should be emphasized that the DC signal plotted in figure 3 is the average flux density, which may differ from the true magnetization due to finite-size effects, as shown clearly in ref. [39]. On figure 4, the results of similar experiments are displayed, but what is plotted is the central flux density against the surface of the sample, as recorded by a miniature Hall probe stuck against the sample surface. Two experiments were carried out in order to investigate the influence of parameters of the experimental system on the measured data. On figure 4(a), the Hall probe was intentionally misoriented by a small angle (4 degrees) while in figure 4(b) the Hall probe was purposefully placed near the edge of the sample. The results are always compared to those obtained with a correctly placed Hall probe. The two set of curves plotted in figure 4(b) are very close to each other, showing that the exact positioning of the Hall probe against the surface is not a critical parameter of the experiment. Figure 4(a), however, shows that a correct Hall probe angular orientation is essential. Following similar sets of experiments carried out at different angles, it was determined in practice that the (unavoidable) experimental misorientation that can be tolerated should be 0.5° or less.

It is of interest to investigate what happens to the sample DC signal when a large number of such transverse AC field cycles are applied, and to compare the resulting decay to that caused by magnetic relaxation effects only. The result of such experiment is plotted in figure 5, where the permanently magnetized superconductor was subjected to either 10 000 or 20 000 transverse AC field cycles. After application of 10 000 cycles (white symbols), the magnetic relaxation occurs but the corresponding decay is hardly perceptible. On the contrary, the decrease of *c*-axis flux density caused by further applications of transverse field cycles (black symbols) is found to be much more significant than the

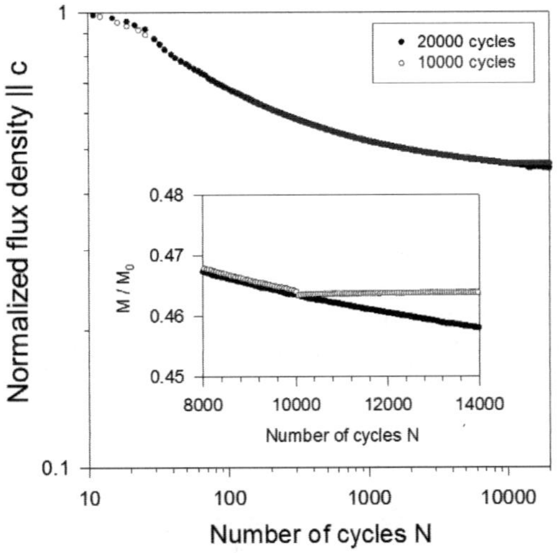

Figure 5. Log-log plot of the decrease of the *c*-axis component of the DC central flux density (normalized to its initial value) of a bulk melt-textured $DyBa_2Cu_3O_7$ superconducting sample as a function the number N of cycles of transverse AC field (parallel to the *ab* planes). The measurement is carried out at $T = 77$ K. In the two experiments, either 10 000 cycles or 20 000 cycles are applied. The inset compares the magnetization decay in the two situations.

decrease caused by magnetic relaxation (flux creep). The reason for this behaviour is also due to the modification of the flux creep regime after the transverse field cycles are applied [40]. Further investigation of such effects using field amplitudes much smaller than the full-penetration field – as is the case in rotating machines – was carried out very recently [41].

Figure 6 shows the results of the application of more than transverse 100 000 cycles. Interestingly, two distinct power law regimes are observed, respectively for $N < 500$ and $N > 5000$. The remarkable feature of the decay observed in figure 6 is the failure of fitting the experimental data by any kind of exponential curve, as evidenced from the red line in figure 6. A possible hypothesis for the faster decay for the first 500 sweeps would be an initial decrease of the critical current density J_c due to self-heating generated by the transverse field. Such magneto-thermal effects, studied numerically for superconducting cylinders of finite size [42] and analytically for infinite cylinders [24], are however unable to account for the behaviour observed in figure 6. Recent experiments [41] point out that subtraction of the extrapolated flux creep effects on the experimental data is helpful for the analysis, and demonstrate in fact that saturation of the magnetization due to AC transverse fields

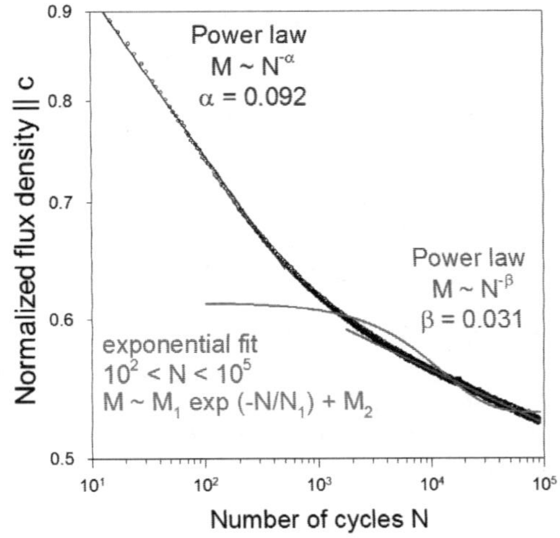

Figure 6. Log-log plot of the decrease of the *c*-axis component of the central DC flux density (normalized to its initial value) of a bulk melt-textured $DyBa_2Cu_3O_7$ superconducting sample as a function the number N of cycles of transverse AC field (parallel to the *ab* planes) for a large number of cycles. The measurement is carried out at $T = 77$ K. The blue and green lines are power law fits that are characteristic of the two regimes. The red line shows an attempt to fit the experimental data between 100 and 100000 cycles by an exponential law.

can be observed when the number of cycles is raised by a further decade. Such a result is extremely important for use of bulk melt-textured superconductors in engineering applications.

5. Conclusion

The superconducting properties of bulk melt-textured $DyBa_2Cu_3O_7$ samples have been studied under various configurations of AC magnetic fields. The results point out that the non-uniformity of superconducting properties (T_c and J_c distribution), even in millimetre size samples, may lead to unusual features in the superconducting transition. We have shown that the application of an AC field to a pre-existing magnetization requires a carefully designed experimental set-up in which the orientation of the Hall probe with respect to the sample plays a crucial role. Finally, the decay of the DC flux density was observed for a large number of AC field cycles, and the corresponding data were shown to exhibit power law behaviour. Two distinct regimes have been pointed out, but they markedly differ from flux creep effects.

Acknowledgements

We thank the University of Liège (ULg) and the Ministry of Higher Education of *Communauté Française de Belgique* for a research grant *Action de Recherches Concertées* (ARC 11/16-03). We thank Profs. M. Ausloos, H. Caps, B. Vanderheyden, as well as Dr. J.F. Fagnard and P. Laurent for many fruitful discussions. The help of Dr. J. P. Mathieu and Prof. R. Cloots for supplying the superconducting sample is also acknowledged. This work is part of a collaboration programme financed by Wallonie Bruxelles International (WBI, Belgium) and the Bulgarian Academy of Sciences (BAS, Bulgaria).

References

[1] Zhen Huang, Wei Xian, Min Zhang, Chudy M, Chen Y, Zhaoyang Zhong, Baghdadi M, Wei Wang, Spaven F, Matsuda K and Coombs T A 2013 *IEEE Trans. Appl. Supercond.* **23** 5200204
[2] Masson P J, Breschi M, Tixador P and Luongo C A 2007 *IEEE Trans. Appl. Supercond.* **17** 1533
[3] Ma G T, Lin Q X, Wang J S, Wang S Y, Deng Z G, Lu Y Y, Liu M X and Zheng J 2008 *Supercond. Sci. Technol.* **21** 065020
[4] Hsu C H, Yan Y, Hadeler O, Vertruyen B, Granados X and Coombs T A 2012 *IEEE Trans. Appl. Supercond.* **22** 7800404
[5] Werfel F N, Floegel-Delor U, Rothfeld R, Riedel T, Goebel B, Wippich D and Schirrmeister P 2012 *Supercond. Sci. Technol.* **25** 014007
[6] Obradors X and Puig T 2014 *Supercond. Sci. Technol.* **27** 044003
[7] Teshima H, Morita M and Hirano H 2006 *J. Phys.: Confer. Series* **43** 1031
[8] Moutalbi N, Noudem J G and M'chirgui A 2014 *Physica C* **503** 105
[9] Devendra Kumar N, Rajasekharan T, Muraleedharan K, Banerjee A and Seshubai V 2010 *Supercond. Sci. Technol.* **23** 105020
[10] Cardwell D A, Shi Y, Hari Babu N and Iida K 2009 *Physica C* **469** 1146
[11] Mathieu J P, Koutzarova T, Rulmont A, Fagnard J F, Laurent P, Mattivi B, Vanderbemden P, Ausloos M and Cloots R 2005 *Supercond. Sci. Technol.* **18** S136
[12] Cardwell D A *et al.* 2005 *Supercond. Sci. Technol.* **18** S173
[13] Tomita M and Murakami M 2003 *Nature* **421** 517
[14] Durrell J H *et al.* 2014 *Supercond. Sci. Technol.* **27** 082001
[15] Mathieu J P, Cano I G, Koutzarova T, Rulmont A, Vanderbemden P, Dew-Hughes D, Ausloos M and Cloots R 2004 *Supercond. Sci. Technol.* **17** 169
[16] Bean C P 1964 *Rev. Mod. Phys.* **1** 31

[17] Qiu M, Huo H K, Xu Z, Xia D, Lin L Z and Zhang G M 2005 *IEEE Trans. Appl. Supercond.* **15** 3172
[18] Matsunga K, Yamachi N, Tomita M, Murakami M and Koshizuka N 2003 *Physica C* **392-396** 723
[19] Ogawa J, Iwamoto M, Yamagishi K, Tsukamoto O, Murakami M and Tomita M 2003 *Physica C* 386 26
[20] Chen D X, Nogues J and Rao K V 1989 *Cryogenics* **29** 800
[21] Fujishiro H, Nariki S and Murakami M 2006 *Supercond. Sci. Technol.* **19** S447
[22] Marchal C, Fagnard J F, Shi Y H, Cardwell D A, Mucha J, Misiorek H, Cloots R, Vertruyen B and Vanderbemden P 2013 *Supercond. Sci. Technol.* **26** 015006
[23] Zushi Y, Asaba I, Ogawa J, Yamagishi K, Tsukamoto O, Murakami M and Tomita M 2004 *Physica C* **412-414** 708
[24] Vanderbemden P, Laurent P, Fagnard J F, Ausloos M, Hari Babu N and Cardwell D A 2010 *Supercond. Sci. Technol.* **23** 075006
[25] Laurent P, Mathieu J P, Mattivi B, Fagnard J F, Meslin S, Noudem J G, Ausloos M, Cloots R and Vanderbemden P 2005 *Supercond. Sci. Technol.* **18** 1047
[26] Kirsch S, Fagnard J F, Vanderbemden P and Vanderheyden B 2013 *Supercond. Sci. Technol.* **26** 115001
[27] Fisher L M, Kalinov A V, Savel'ev S E, Voloshin I F, Yampol'skii V A, LeBlanc M A R and Hirscher S 1997 *Physica C* **278** 169
[28] Funaki K and Yamafuji K 1982 *Jpn. J. Appl. Phys.* **21** 299
[29] Vanderbemden P, Hong Z, Coombs T A, Denis S, Ausloos M, Schwartz J, Rutel I B, Hari Babu N, Cardwell D A and Campbell A M 2007 *Phys. Rev. B* **75** 174515
[30] Vanderbemden P, Hong Z, Coombs T A, Ausloos M, Hari Babu N, Cardwell D A and Campbell A M 2007 *Supercond. Sci. Technol.* **20** S174
[31] Badía-Majós A and López C 2007 *Phys. Rev. B* **76** 054504
[32] Baghdadi M, Ruiz H S and Coombs T A 2014 *Appl. Phys. Lett.* **104** 232602
[33] Patel A, Hopkins S C and Glowacki B A 2013 *Supercond. Sci. Technol.* **26** 032001
[34] Vanderbemden P 1998 *Cryogenics* **38** 839
[35] Gömöry F 1997 *Supercond. Sci. Technol.* **10** 523
[36] Vanderbemden P, Cloots R, Ausloos M, Doyle R A, Bradley A D, Lo W, Cardwell D A and Campbell A M 1999 *IEEE Trans. Appl. Supercond.* **9** 2308
[37] Doyle R A, Bradley A D, Lo W, Cardwell D A, Campbell A M, Vanderbemden P and Cloots R 1998 *Appl. Phys. Lett.* **73** 117
[38] Laurent P, Fagnard J F, Vanderheyden B, Hari Babu N, Cardwell D A, Ausloos M and Vanderbemden P 2008 *Meas. Sci. Technol.* **19** 085705
[39] Philippe M P, Fagnard J F, Kirsch S, Xu Z, Dennis A R, Shi Y H, Cardwell D A, Vanderheyden B and Vanderbemden P 2014 *Physica C* **502** 20
[40] Voloshin I F, Fisher L M and Yampol'skiĭ V A 2010 *Low Temp. Phys.* **36** 39
[41] Fagnard J F, Kirsch S, Morita M, Teshima H, Vanderheyden B and Vanderbemden P 2014 Submitted to *Physica C*
[42] Berger K, Levêque J, Douine B, Netter D and Rezzoug A 2007 *IEEE Trans. Appl. Supercond.* **17** 3028

Neutron diffraction and RMC modeling of new amorphous molybdate system

M Fabian[1,2], E Svab[2], K Krezhov[3]

[1]Centre for Energy Research, H-1525 Budapest P.O.B. 49, Hungary
[2]Wigner Research Centre for Physics, H-1525 Budapest P.O.B. 49, Hungary
[3]Institute for Nuclear Research and Nuclear Energy, 1784 Sofia, 72 Tzarigradsko Chaussee, Bulgaria

E-mail: fabian.margit@energia.mta.hu

Abstract. The network structure of boromolybdate MoO_3-Nd_2O_3-B_2O_3 glasses has been investigated by neutron diffraction and reverse Monte Carlo (RMC) simulation technique. The partial atomic correlation functions, first neighbour atom distances and the coordination numbers have been revealed. Formation of MoO_4 (55%) and MoO_6 (45%) units was established for the binary $90MoO_3$-$10Nd_2O_3$ glass. The ternary glasses consist from MoO_4, BO_3 and BO_4 units. The relative number of BO_3 is increasing with increasing boron content. Characteristic second neighbour distributions indicate the existence of a pronounced intermediate-range ordering through MoO_4-BO_4 and MoO_4-BO_3 linkages.

1. Introduction

Rare-earth molybdate phases exhibit a great variety of important physical properties including high ion and electron conductivity of fast oxide ion conductors, non-linear optical response and luminescent properties. In contrast to crystalline molybdates, the preparation and structural information on amorphous molybdate systems is far not well known. MoO_3 is a conditional network former, it is not able to form a glass itself at even rapid cooling rates.

A systematic study to clear up the tendency for glass formation and immiscibility in the complex mixed boromolybdate glasses has been performed by Dimitriev et al. [1-5]. The phase diagram for MoO_3-$Nd(La)_2O_3$-B_2O_3 was set up [3] and the basic structural and optical characteristics were investigated by XRD, DTA, SEM, UV-VS spectra and IR spectroscopy [5]. The Nd_2O_3 is an appropriate component due to its specific optical properties but it increases the melting temperature, while by adding B_2O_3 it is possible to obtain low melting materials in a relatively wide concentration range.

In order to get a deeper insight into the network structure of MoO_3-Nd_2O_3-B_2O_3 glassy system, we have undertaken a neutron diffraction study combined with reverse Monte Carlo (RMC) computer simulation technique [6]. We expect to obtain new structural characteristics of the partial atomic pair correlation functions, especially on the first neighbour atomic distances, coordination numbers and network former units and their linkages.

2. Experimental details and methods

2.1. Samples

The amorphous materials have been prepared by a conventional melt quenching method. The nominal compositions of the studied glasses are binary $90MoO_3$-$10Nd_2O_3$ and ternary $50MoO_3$-$25Nd_2O_3$-$25B_2O_3$, $40MoO_3$-$30Nd_2O_3$-$30B_2O_3$, $20MoO_3$-$30Nd_2O_3$-$50B_2O_3$ (mol%). Commercial powder of reagent grade MoO_3, Nd_2O_3 and B_2O_3 were used as starting materials and were mixed in an alumina mortar. B_2O_3 was isotopically enriched in ^{11}B (99.6%) in order to reduce the influence of the high neutron absorption of ^{10}B present in natural boron. For the details of sample preparation we refer to [3,5].

2.2. Neutron diffraction experiments

Neutron diffraction (ND) measurements have been performed by the PSD diffractometer (λ_0=1.068 Å) [7] at the 10 MW Budapest research reactor and by the 7C2 diffractometer at the LLB-CEA-Saclay (λ_0=0.726 Å) [8], the powder specimens of about 3–4 g/each were filled in thin walled cylindrical vanadium sample holder of 8 mm and 6 mm diameter, respectively. Data were corrected for detector efficiency and background scattering. The total structure factor, *S(Q)* was calculated as described in our previous work [9]. Figure 1 shows the experimental *S(Q)* data for the investigated samples together with the results of RMC simulation (details of the RMC modeling will be discussed in the next section).

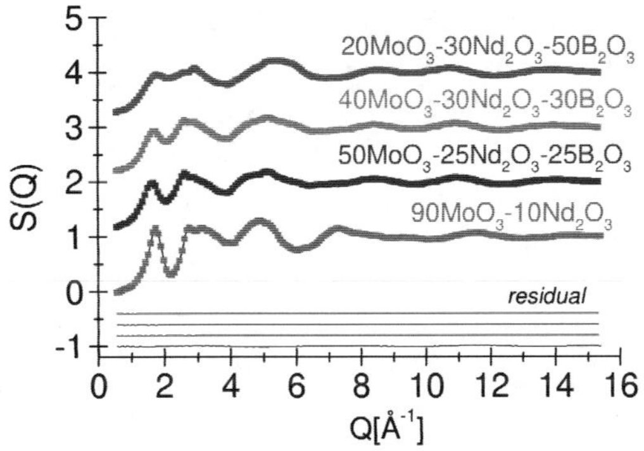

Figure 1. Structure factor of the boromolybdate glasses: experimental data (marks) and RMC simulation (solid line). The residuals of the experimental $S_{EXP}(Q)$ and calculated $S_{RMC}(Q)$ are shown at the bottom of the figure. (The curves are shifted vertically for clarity)

The overall run of the *S(Q)* data for the binary $90MoO_3$-$10Nd_2O_3$ glass is fairly different from those of the ternary glasses containing network former B_2O_3. The most striking feature is the relatively high intensity and sharp distribution of the first peak of the binary glass, which is a fingerprint of a short-range network structure, i.e. formation of well-defined units. For the ternary glasses the intensity of the first peak decreases, gets broader and slightly shifts to higher values with increasing boron content. This indicates a more complex network structure of the ternary glasses in comparison to the binary glass, as it can be expected. The second peak contains two small humps, the average main intensity changes in dependence of concentration. The third peak gets significantly broader for the ternary glasses and, for the higher order oscillations, above $Q \geq 6$ Å^{-1} phase displacement can be observed.

2.3. Reverse Monte Carlo simulation

The experimental *S(Q)* data have been simulated by the RMC method [6] using the software package [10]. The RMC minimizes the squared difference between the experimental *S(Q)* and the calculated one from a 3-dimensional atomic configuration. The RMC algorithm calculates the one-dimensional partial atomic pair correlation functions $g_{ij}(r)$, for details we refer here to our previous work [9].

In this study for the RMC starting model a disordered atomic configuration was built up with a simulation box containing 10 000 atoms. In the calculations, density values 0.0673, 0.0683, 0.0684 and 0.0692 atoms·Å^{-3} were used [11] for the $90MoO_3$-$10Nd_2O_3$, $50MoO_3$-$25Nd_2O_3$-$25B_2O_3$, $40MoO_3$-$30Nd_2O_3$-$30B_2O_3$ and $20MoO_3$-$30Nd_2O_3$-$50B_2O_3$ samples, respectively, and the corresponding RMC half-box lengths were 26.483, 26.353, 26.340 and 26.238 Å.

In the RMC simulation procedure two types of constraints were used: the minimum interatomic distances between atom pairs (cut-off distances) and connectivity constraints for Mo-O and for B-O atom pairs. Molybdenum atoms were restricted to be surrounded by 4 or 6 oxygen atoms, as suggested in previous studies [3,12]. For the B-O coordination we supposed the formation of trigonal BO_3 and tetrahedral BO_4 units, which has been reported in several works, i.e. for binary neodymium borate glasses [13], for ternary rare-earth boromolybdate MoO_3-Nd(La)$_2O_3$-B_2O_3 [3-5], for B_2O_3-Bi_2O_3-MoO_3 glass [14], and for binary alkali borate glasses [15,16]. In the RMC calculations boron atoms were forced to have 3 or 4 oxygen neighbours in the interval 1.2-1.7 Å, the upper limit was used not to overlap with Mo-O distribution.

The starting cut-off distances have been taken from the literature based on the results of crystalline compounds [17] and from glassy systems [18,19] of similar compositions. Several RMC runs have been performed by modifying slightly the cut-off distances for the various atom pairs. The convergence of the RMC simulation was good, the set of cut-off distances used in the final RMC run are the following: B-O 1.2 Å; Mo-O 1.71 Å; O-O 2.3 Å; Nd-O 2.37 Å; B-B 2.36 Å; Nd-Nd 2.65 Å; Mo-B 2.7 Å; Nd-B 3.05 Å; Mo-Mo 3.1 Å; Mo-Nd 3.45 Å.

3. Results and discussion

The calculated $S_{RMC}(Q)$s matched very well the experimental $S_{EXP}(Q)$s for all compositions, as it is illustrated in figure 1. The partial correlation functions, $g_{ij}(r)$ and coordination numbers were revealed from the RMC modeling. Here we focus on the results obtained for the network former Mo-O and B-O correlations.

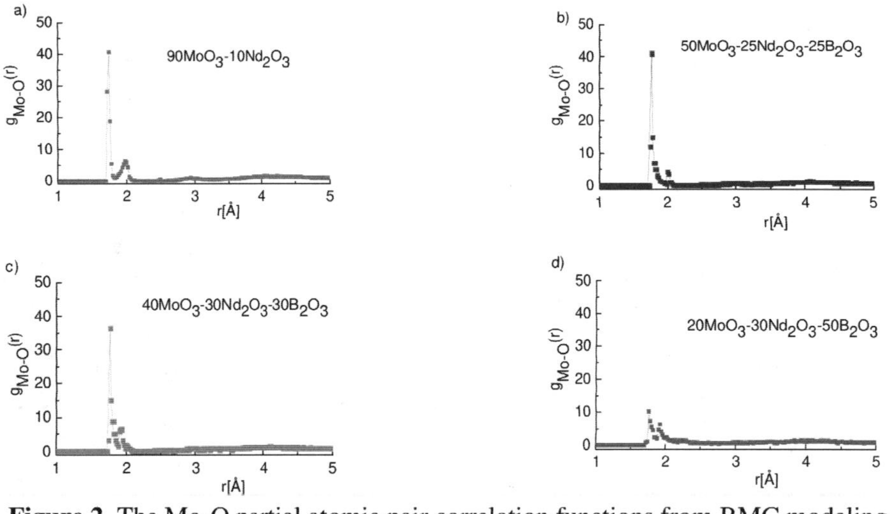

Figure 2. The Mo-O partial atomic pair correlation functions from RMC modeling.

Figure 2 shows the Mo-O correlation functions. For all compositions the first neighbour distributions show a sharp peak at 1.75±0.02 Å with a half-width value of 0.04 Å, while a significantly smaller "sub"-peak appears at 1.95±0.05 Å. For $90MoO_3$-$10Nd_2O_3$ sample the intensity of both peaks is high, and the statistics is good due to the high weighting factor, w_{Mo-O}=33.79%. The "sub"-peak at 1.98 Å is significantly broader than the first peak, the half-width value is 0.08 Å. With decreasing Mo-content the intensity of the first neighbour peak intensity decreases. The distributions for $50MoO_3$-$25Nd_2O_3$-$25B_2O_3$ and $40MoO_3$-$30Nd_2O_3$-$30B_2O_3$ samples are very similar due to the small difference in the compositions. The position of the small "sub"-peaks is slightly shifted (by 0.1 Å), which is within limit of error. For the $20MoO_3$-$30Nd_2O_3$-$50B_2O_3$ sample, even the intensity of the main Mo-O peak at 1.75 Å is fairly small, due to the small amount of molybdenum in this sample.

The average coordination number for the binary $90MoO_3$-$10Nd_2O_3$ sample is CN_{Mo-O}=4.9±0.2 atoms. If we suppose that MoO_4 and MoO_6 units are present in the network, it would mean that 55% of the Mo atoms are 4-fold oxygen coordinated, while 45% are 6-fold coordinated. Taking into consideration the $g_{Mo-O}(r)$ distribution, we can establish that the shorter distance at 1.74 Å can be attributed to the MoO_4 units, while in addition two further oxygen neighbours appear at a longer 1.98 Å distance forming MoO_6 units.

For the ternary $50MoO_3$-$25Nd_2O_3$-$25B_2O_3$ and $40MoO_3$-$30Nd_2O_3$-$30B_2O_3$ samples the average coordination number, as calculated from the RMC modeling, is CN_{Mo-O}=4.2 and 4.0±0.2 atoms, respectively. As far as, the corresponding $g_{Mo-O}(r)$ distributions contain a small "sub"-peak at 1.95 Å, we may conclude that 95% of the Mo atoms form MoO_4 units, while about 5% form MoO_6 units. For the $20MoO_3$-$30Nd_2O_3$-$50B_2O_3$ sample the average coordination number is CN_{Mo-O}=3.5±0.3 atoms, which is less than the expected 4-fold coordination. This may be the consequence of the relatively large error due to the relatively small amount of molybdenum leading to a low value of the weighting factor w_{Mo-O}=5.57%, however, distortion of the tetrahedral MoO_4 units can not be excluded.

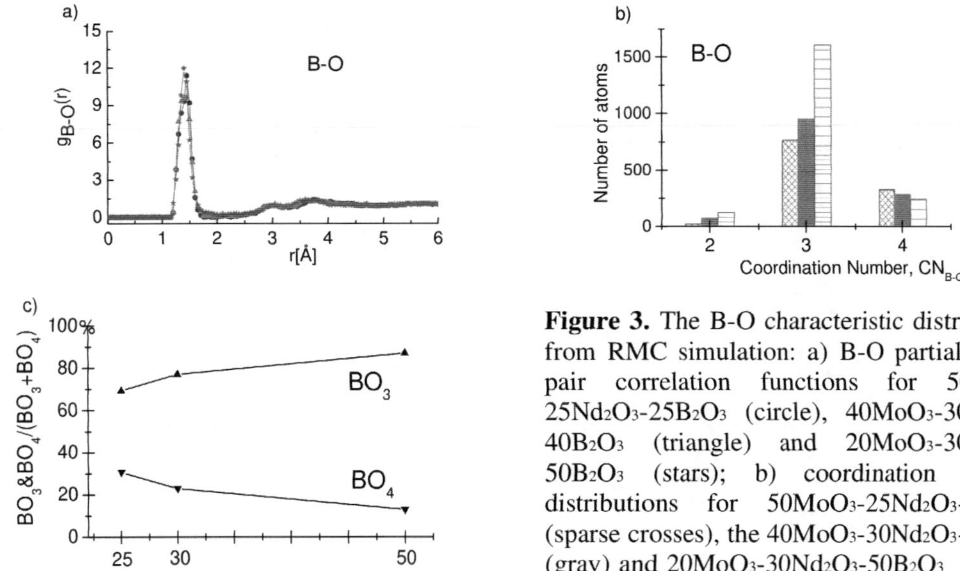

Figure 3. The B-O characteristic distributions from RMC simulation: a) B-O partial atomic pair correlation functions for $50MoO_3$-$25Nd_2O_3$-$25B_2O_3$ (circle), $40MoO_3$-$30Nd_2O_3$-$40B_2O_3$ (triangle) and $20MoO_3$-$30Nd_2O_3$-$50B_2O_3$ (stars); b) coordination number distributions for $50MoO_3$-$25Nd_2O_3$-$25B_2O_3$ (sparse crosses), the $40MoO_3$-$30Nd_2O_3$-$40B_2O_3$ (gray) and $20MoO_3$-$30Nd_2O_3$-$50B_2O_3$ (sparse) and c) fraction of BO_3 and BO_4 structural units as a function of concentration.

Figure 3 shows the characteristics for the B-O correlations. The first neighbour distributions show a relatively broad – in comparison to the Mo-O distribution - first neighbour distance at 1.40±0.02 Å with a half-width value of 0.25 Å, and they practically overlap for the three ternary MoO_3-Nd_2O_3-B_2O_3 glasses of different concentrations within limit of error as it is shown in figure 3/a. Figure 3/b illustrates the B-O coordination number distributions, which contain both 3- and 4-fold oxygen coordinated boron atoms (the small number of 2-neighbours is probably an artificial effect of RMC calculation). The average coordination numbers slightly decreases from 3.3 to 3.1 atoms with increasing boron content. The fraction of BO_3 and BO_4 to the total number of BO_3+BO_4 is an important structural characteristic. It was found that it depends on the boron concentration, as it is illustrated in figure 3/c. With increasing boron content the relative number of BO_3 units is increasing, while the number of BO_4 is decreasing. For 50MoO_3-25Nd_2O_3-25B_2O_3 sample ~70% of B-atoms are 3-coordinated, while ~30% are 4-coordinated, the average coordination number is 3.3 atoms; for 40MoO_3-30Nd_2O_3-30B_2O_3 glass ~77% of B-atoms are 3-coordinated, while ~23% are 4-coordinated, the average coordination number is 3.2 atoms; for 20MoO_3-30Nd_2O_3-50B_2O_3 sample ~87% of B-atoms are 3-coordinated, while ~13% are 4-coordinated, the average coordination number is 3.1 atoms.

In the studied system the results obtained from RMC modeling have shown significant correlations between several second neighbour atom pairs, as it is shown in figure 4. This is important, as far as the analysis of these results may give useful information on the linkage of the basic structural units forming the network structure. The linkage between the MoO_4 units leads to the Mo-Mo correlation distance at 3.2 Å. The second neighbour distances for B-B are obtained at 2.4 Å, for Mo-B at 2.8 Å, which indicate mixed linkages, and for Nd-B at 3.1 Å. These results indicate the existence of a pronounced intermediate-range ordering.

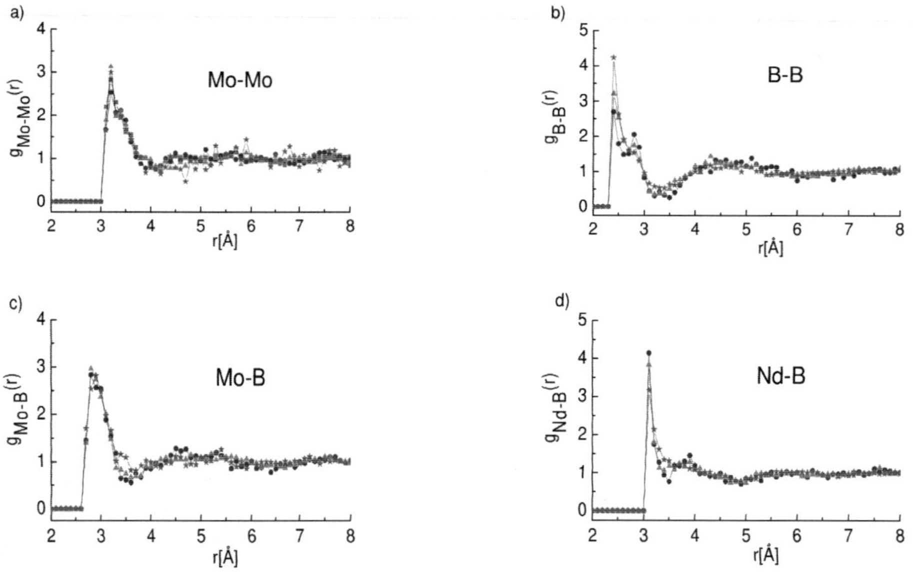

Figure 4. Second neighbour partial atomic pair correlation functions obtained from the RMC modeling for the boromolybdate glasses: a) Mo-Mo b) B-B c) Mo-B and d) Nd-B, where 90MoO_3-10Nd_2O_3 (square), 50MoO_3-25Nd_2O_3-25B_2O_3 (circle), 40MoO_3-30Nd_2O_3-40B_2O_3 (triangle) and 20MoO_3-30Nd_2O_3-50B_2O_3 (stars).

4. Conclusions

Neutron diffraction study has been performed on binary 90MoO_3-10Nd_2O_3 and ternary MoO_3-Nd_2O_3-B_2O_3 glasses prepared by melt quench technique. The network structure was modeled by reverse Monte Carlo simulation method. From the RMC modeling the partial atomic correlation functions and the coordination numbers have been revealed. Our main findings are the following:

- in the binary glass MoO_4 (55%) and MoO_6 (45%) structural units were revealed;
- in ternary system mainly MoO_4 units are present, and with decreasing MoO_3 concentration, the ratio of MoO_6 units roughly decreases;

- in the ternary glasses the B-O network is formed by BO_3 and BO_4 groups, and with increasing B_2O_3 content conversion of BO_4 to BO_3 takes place;
- in ternary system pronounced intermediate range order exists, which indicates mixed MoO_4-BO_4 and MoO_4-BO_3 linkages.

Acknowledgement - The authors are grateful to Dr. L. Alexandrov for sample preparation and for helpful discussions. The research was supported by the EU-FP7-NMI3 No.283883. Thanks to the financial support of Hungarian OTKA-PD 109384 and to the bilateral cooperation of Hungarian-Bulgarian Academy of Sciences SNK-63/2013.

References

[1] Alexandrov L, Iordanova R, Dimitriev Y, Hamda K, Ide J and Milanova M 2008 *Advanced Materials Research* **39-40** 37
[2] Aleksandrov L, Iordanova R and Dimitriev Y 2007 *Phys. Chem. Glasses: European J. Glass Sci. Technol. B.* **48** 242
[3] Aleksandrov L, Iordanova R and Dimitriev Y 2009 *Journal of Non-Crystalline Solids* **355** 2023
[4] Dimitriev Y, Iordanova R, Aleksandrov L and Kostov K L 2008 *Phys. Chem. Glasses Eur/ Glass Sci. Technol B.* **50** 218
[5] Aleksandrov L, Komatsu T, Iordanova R and Dimitriev Y 2011 *Journal of Physics and Chemistry of Solids* **72** 263
[6] McGreevy R L and Pusztai L 1988 *Mol. Simul.* **1** 359
[7] Sváb E, Mészáros Gy and Deák F 1996 *Materials Science Forum* **228** 247; http://www.bnc.hu/
[8] Ambroise J P, Bellissent R 1984 *Rev. Phys. Appl.* **19** 731; http://www-llb.cea.fr/en/
[9] Fábián M, Sváb E, Proffen Th and Veress E 2008 *Journal of Non-Crystalline Solids* **354** 3299
[10] http://www.szfki.hu/~nphys/rmc++/opening.html
[11] Alexandrov L, private communication
[12] Rada M, Rada S, Pascuta P and Culea E 2010 *Spectrochimia Acta Part A* **77** 832
[13] Burns A E, Winslow D W, Clarida W J, Affatigato M, Feller S A and Brow R K 2006 *Journal of Non-Crystalline Solids* **352** 2364
[14] Iordanova R, Aleksandrov L, Bachvarova-Nedelcheva A, Ataala M and Dimitriev Y 2011 *Journal of Non-Crystalline Solids* **357** 2663
[15] Fábián M, Sváb E, Proffen Th and Veress E 2010 *Journal of Non-Crystalline Solids* **356** 441
[16] Wright A C, Sinclair R N, Stone C E and Shaw J L 2012 *Phys. Chem.Glasses Eur. J. Glass Sci. Technol.* **B53** 191
[17] Barker R S and Evans I R 2006 *Journal of Solid State Chemistry* **179** 1918
[18] Wang H, Peng J, Chen D F and Hu Z 2006 *Solid State Sciences* **8** 1144
[19] Wang Z, Zhang L, Wang G, Song M, Huang Y and Wang G 2009 *Optical Materials* **31** 849

Electromechanical properties of biomembranes and nerves

T Heimburg*, A Blicher, L D Mosgaard and K Zecchi

The Niels Bohr Institute, University of Copenhagen, 2100 Copenhagen Ø, Denmark

E-mail: theimbu@nbi.dk

Abstract. Lipid membranes are insulators and capacitors, which can be charged by an external electric field. This phenomenon plays an important role in the field of electrophysiology, for instance when describing nerve pulse conduction. Membranes are also made of polar molecules meaning that they contain molecules with permanent electrical dipole moments. Therefore, the properties of membranes are subject to changes in trans-membrane voltage. Vice versa, mechanical forces on membranes lead to changes in the membrane potential. Associated effects are flexoelectricity, piezoelectricity, and electrostriction.

Lipid membranes can melt from an ordered to a disordered state. Due to the change of membrane dimensions associated with lipid membrane melting, electrical properties are linked to the melting transition. Melting of the membrane can induce changes in trans-membrane potential, and application of voltage can lead to a shift of the melting transition. Further, close to transitions membranes are very susceptible to piezoelectric phenomena.

We discuss these phenomena in relation with the occurrence of lipid ion channels. Close to melting transitions, lipid membranes display step-wise ion conduction events, which are indistinguishable from protein ion channels. These channels display a voltage-dependent open probability. One finds asymmetric current-voltage relations of the pure membrane very similar to those found for various protein channels. This asymmetry falsely has been considered a criterion to distinguish lipid channels from protein channels. However, we show that the asymmetry can arise from the electromechanical properties of the lipid membrane itself.

Finally, we discuss electromechanical behavior in connection with the electromechanical theory of nerve pulse transduction. It has been found experimentally that nerve pulses are related to changes in nerve thickness. Thus, during the nerve pulse a solitary mechanical pulse travels along the nerve. Due to electromechanical coupling it is unavoidable that this pulse generates a trans-membrane voltage. In the past, we have proposed that this electromechanical pulse is the origin of the action potential in nerves.

1. Introduction

Biological membranes are thin quasi-twodimensional layers mainly consisting of proteins and lipids. While research mostly focusses on the properties of individual macromolecules, e.g., on ion channel proteins or ion pumps, the total membrane possesses macroscopic cooperative features such as melting transitions and curvature fluctuations that cannot be understood on the molecular level. These properties are expressed in susceptibilities such as heat capacity, lateral compressibility, bending elasticity or capacitive susceptibility. Lipid membranes can melt from a solid to a liquid phase. In these transitions, the order of the lipids changes. Thus, the melting is associated to both, enthalpy and entropy changes. Such transitions can also be found in biological membranes under physiological conditions. As an example, a heat

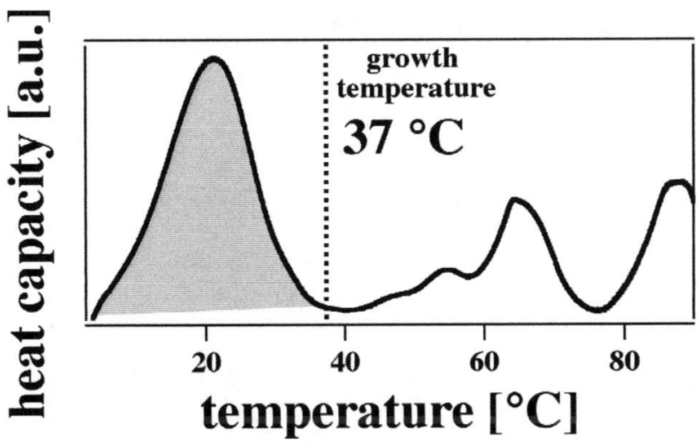

Figure 1. Heat capacity profile of E. *coli* membranes. The grey-shaded region below growth temperature represents the membrane melting transition. The peaks above growth temperature show protein unfolding. From [5].

capacity profile of E. *coli* membranes is shown in figure 1. In the melting transition, the spatial dimensions of the membrane change. For instance, upon melting the synthetic lipid dipalmitoyl phosphatidylcholine (DPPC) increases its area by about 24% and reduces its thickness by 16%. The heat absorbed in the transition is about 35 kJ/mol.

Membranes are very thin. They possess a thickness of about 5nm in their solid state. The core of the membrane is composed of hydrocarbon chains. Therefore, the membrane interior can be considered an insulator. Consequently, the biomembrane has the properties of a capacitor. Typically, the capacitance of a membrane is of the order of 1 $\mu F/cm^2$. In biological cells, the membrane is exposed to voltage differences of the order of 100 mV. Thus, the biological membrane is charged under physiological conditions.

The dimensional changes in the melting transitions have a number of consequences. Among those are [1]:

- both hydrostatic and lateral pressure changes influence the phase state of the membrane and are intrinsically coupled to heat absorption or release.
- hydrostatic and lateral pressure changes voltage across the membrane, and the charge on the membrane capacitor can change. Thus, the membrane is piezoelectric.
- voltage changes can induce membrane melting.

These features are important for various properties of biological membranes. For instance, it was shown that biomembranes slightly above a melting transition can support electromechanical solitons that resemble nerve pulses [2]. Further, in the transition one finds density fluctuations that result in the spontaneous formation of pores in the membrane [3]. These pores display open-close characteristics very similar to those reported for protein ion channels [4].

The thermodynamics of biological membranes putatively explains many properties of excitatory cells on the level of macroscopic physics rather than on the level of molecular biology. This review will introduce into some of these phenomena.

2. Membrane capacitors

The capacitance, C_m, of a planar membrane is given by

$$C_m = \varepsilon\varepsilon_0 \frac{A}{D} , \tag{1}$$

where ε_0 is the vacuum permittivity, ε is the dielectric constant, A is the membrane area and D is the membrane thickness. The charge, q, on a capacitor is given by

$$q = C_m \cdot V_m , \tag{2}$$

where V_m is the transmembrane voltage. Since in the transition the area changes by about 24%

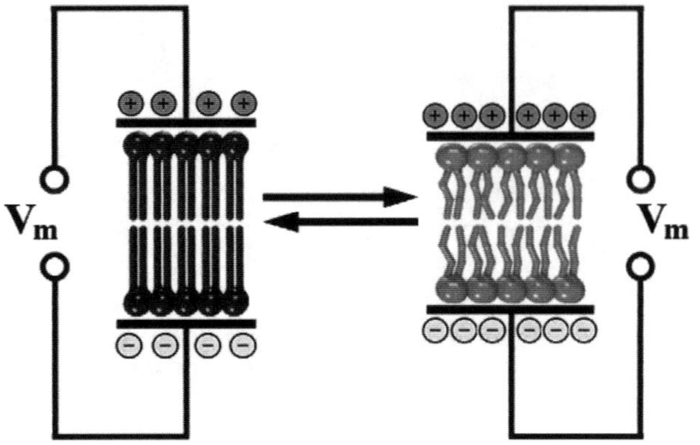

Figure 2. During the melting transition of a membrane both its area and thickness changes. This implies that the capacitance of the membrane varies as a function of phase state. It is higher in the liquid state.

and the thickness by -16%, one finds an increase in capacitance upon melting from a solid to a liquid membrane phaseof approximately 50%.

2.1. *Capacitive susceptibility*

The capacitance solely depends on the dimensions of the membrane, if $\varepsilon = constant$. However, the opposite charges on the two plates of a capacitor attract each other and generate a force on the membrane. This effect is called 'electrostriction'. If the voltage across a membrane increases, the forces on the membrane also increases. Therefore the capacitance changes as well. For a symmetric membrane, the capacitance always increases upon increasing the voltage. This effect can be taken into account by considering the capacitive susceptibility, $\hat{C}_m$:

$$\hat{C}_m = \frac{dq}{dV_m} = C_m + V_m \frac{\partial C_m}{\partial V_m} \,, \tag{3}$$

where the charge, q, is given by equation (2). The second term in this equation could be considered an excess capacitance. It assumes a maximum in the melting transition (see figure 3).

3. Fluctuations

Due to the fluctuation-dissipation theorem, all response functions (susceptibilities) are related to the mean square fluctuations of extensive variables. For instance, the heat capacity, $c_p = (\partial H/\partial T)_p$, is given by

$$c_p = \frac{\langle H^2 \rangle - \langle H \rangle^2}{kT^2} \,, \tag{4}$$

while the isothermal volume compressibility, $\kappa_T^V = -(\partial V/\partial p)_T$, is related to volume fluctuations

$$\kappa_T^V = \frac{\langle V^2 \rangle - \langle V \rangle^2}{kT} \,, \tag{5}$$

and the capacitive susceptibility is given by

$$\hat{C}_m = \frac{\langle q^2 \rangle - \langle q \rangle^2}{kT} \,. \tag{6}$$

Similarly, fluctuations in area are related to the isothermal area compressibility and fluctuations in curvature to the bending elasticity. The heat capacity assumes a maximum in the melting

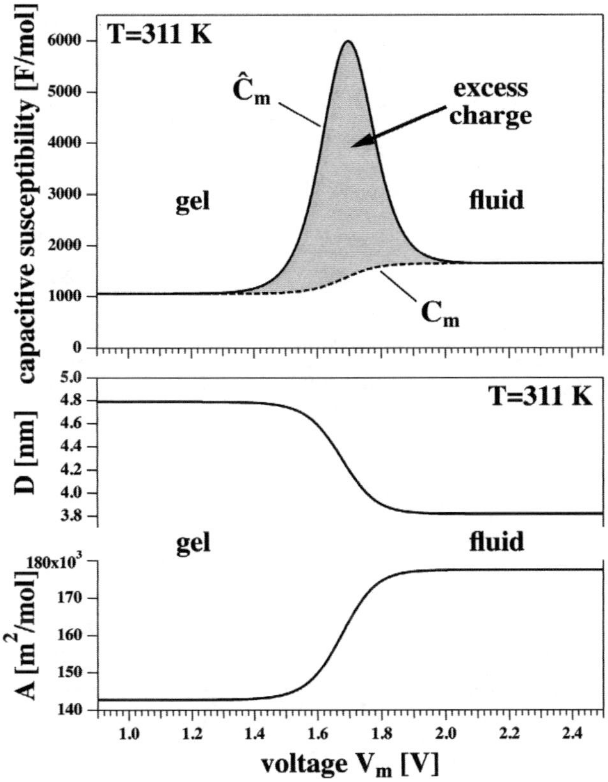

Figure 3. Increase in voltage reduces thickness and increases area. As a consequence, membranes can be moved through their melting transition by voltage changes. The capacitive susceptibility, $\hat{C}_m$, reaches a maximum at the melting temperature. The voltage-induced transition is associated to an excess charge. From [1].

transition and thus the fluctuations are at maximum. Similarly, compressibility, bending elasticity and capacitive susceptibility all assume maxima in the transition regime.

It has been shown that in melting transitions, excess volume changes are proportional to excess enthalpy changes, i.e., $\Delta V(T) = \gamma_V \Delta H(T)$. Here, γ_V is a material constant. This implies that excess volume and enthalpy fluctuations are also proportional functions. A consequence is that excess heat capacity and isothermal volume compressibility are proportional functions of temperature, pressure, etc. I.e.,

$$\Delta c_p \propto \Delta \kappa_T^V \tag{7}$$

Similarly one can directly or indirectly conclude from experiment that the excess heat capacity is proportional to other response functions of lipid membranes close to transitions [6, 7, 1], f.e.,

$$\begin{aligned} \Delta c_p &\propto \Delta \kappa_T^A \quad \text{(area compressibility)} \\ \Delta c_p &\propto \Delta \kappa_B \quad \text{(bending elasticity)} \\ \Delta c_p &\propto \Delta \hat{C}_m \quad \text{(capacitive susceptibility)} \end{aligned} \tag{8}$$

These relations are not based on first principles and should be taken as empirical correlations found to be true for membranes. The proportionality constants depend on the dimensions of the solid and liquid membrane. The heat capacity is easy to measure in a calorimeter. The other response functions can readily be calculated from the calorimetric experiment.

4. The nervous impulse

The nerve pulse consists of a propagating voltage pulse with typical velocities of 1-100 m/s that last about 1 ms. It follows that the typical dimension of a nerve pulse is about 1 mm to 10 cm.

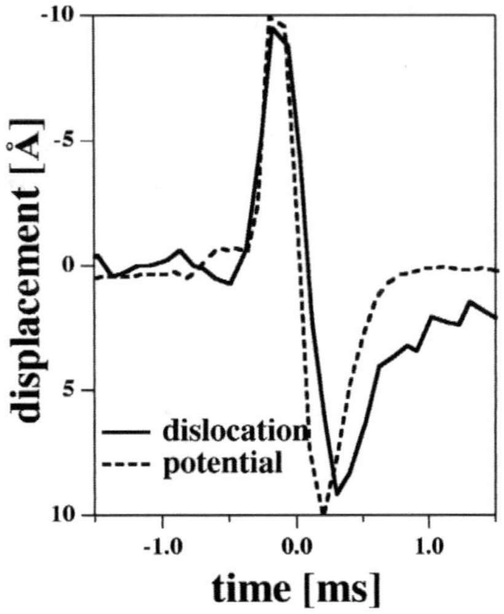

Figure 4. Thickness change of a squid axon during the action potential as a function of time (solid line). The dashed line represents the voltage change. The two functions are within error proportional. Adapted from [8].

Thus, it is of macroscopic dimension. In the biological literature, the nerve pulse is considered a purely electrical phenomenon involving capacitors (the membrane), resistors (ion channel proteins) and electrical currents (ion flows). However, during the nerve pulse one also finds changes in nerve dimensions (thickness and length [8, 9], see figure 4) and in temperature [10]. Thus, the nerve pulse should be considered a thermodynamic or hydrodynamic phenomenon. Below, we show that the nerve pulse can be seen as a localized density pulse related to the propagation of sound.

4.1. Sound velocity

The above relations (equation (8)) help to determine other membrane properties that are related to the response functions. The lateral sound velocity, c, in membranes is defined as

$$c^2 = \left(\frac{\partial \rho^A}{\partial p}\right)_S = \frac{1}{\kappa_S^A \rho^A} \tag{9}$$

Thus, it depends both on the lateral density and on the adiabatic compressibility, κ_S^A. The adiabatic compressibility is a function of frequency because it depends on the translocation of heat from the membrane to the membrane environment. The smaller the frequency, the larger is the aqueous volume that contributes as a heat reservoir and the larger is the adiabatic compressibility [11]. In the limit of zero frequency on obtains the isothermal limit and the adiabatic compressibility, κ_S^A is equal to the isothermal compressibility, κ_T^A. The frequency dependence of the sound velocity is called 'dispersion'. The sound velocity in membranes is generally higher at higher frequencies

Using the above thermodynamic relations between heat capacity and compressibility, on can calculate the low frequency sound velocity as a function of temperature (or as a function of density). Since the compressibility displays a maximum in the melting transition, the lateral sound velocity displays a minimum (shown in figure 5, left. From [2]). In this figure, small density corresponds to the liquid membrane phase while high density corresponds to the solid membrane phase. The membrane in the liquid phase is thus a spring with interesting spring properties: upon compression of the liquid phase the spring first becomes softer (in the transition) and then becomes stiffer (in the solid phase).

The lateral density of the membrane shall be given by $\rho^A = \rho_0^A + \Delta\rho^A$, where ρ_0^A is the density of the liquid membrane. The sound velocity is a non-linear function of the lateral density change,

Journal of Physics: Conference Series **558** (2014) 012018 doi:10.1088/1742-6596/558/1/012018

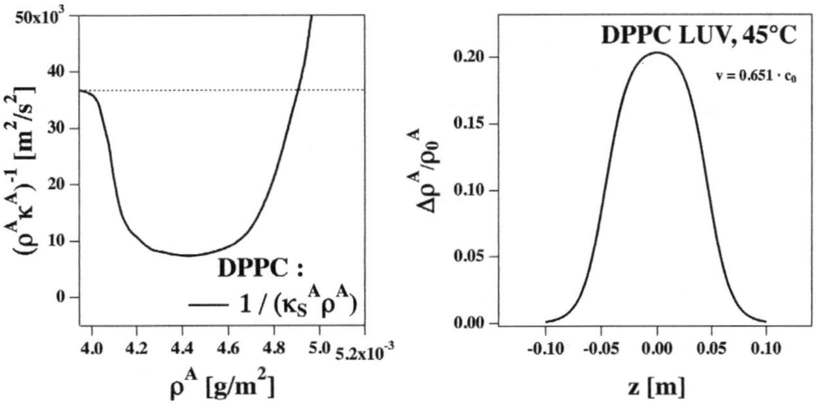

Figure 5. Left: The sound velocity in a lipid membrane close to a transition is a function of density [2]. Small density corresponds to a liquid membrane whereas high density corresponds to a solid membrane. The pronounced minimum is found in the chain melting regime. It is caused by the maximum of area fluctuations in the membrane at the transition. Right: Density soliton in a membrane cylinder using the sound velocity profile shown in the left hand panel [2]

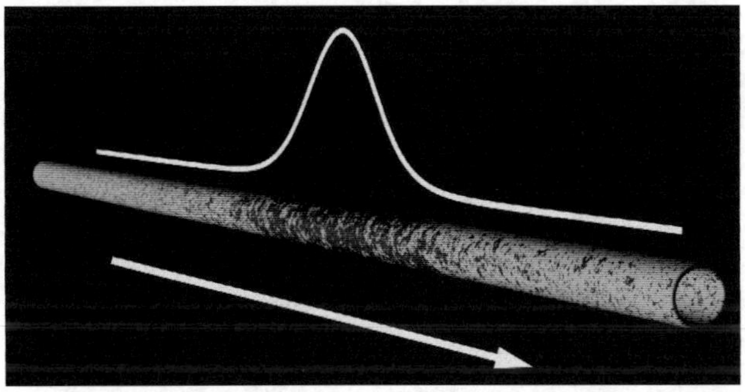

Figure 6. Schematic representation of a density soliton in a cylindrical membrane. The pulse consists of a traveling solid segment (dark shade) traveling in a liquid membrane environment.

$\Delta\rho^A$, which can be Taylor-expanded into

$$c^2 = c_0^2 + p\Delta\rho^A + q(\Delta\rho^A)^2 + ... \tag{10}$$

4.2. Solitons in nerve axons

The non-linearity of the sound velocity and the presence of dispersion give rise to the possibility of soliton propagation. Below, we show as a quasi-one-dimensional example a long cylindrical membrane comparable to the axon of a nerve. The wave equation for one-dimensional sound propagation is given by [12]

$$\frac{\partial^2}{\partial t^2}\Delta\rho = \frac{\partial}{\partial x}\left(c^2 \frac{\partial}{\partial x}\Delta\rho\right) . \tag{11}$$

By inserting equation (10) into this equation, we obtain

$$\frac{\partial^2}{\partial t^2}\Delta\rho = \frac{\partial}{\partial x}\left((c_0^2 + p\Delta\rho + q\rho^2 + ...)\frac{\partial}{\partial x}\Delta\rho\right) - h\frac{\partial^4}{\partial x^4}\Delta\rho \tag{12}$$

The second term is an ad hoc dispersion term that describes the frequency dependence of the elastic constants. Its introduction into the wave equation is justified in [2]. When inserting the parameters p and q obtained from fitting equation (10) to the experimental sound velocity profile, one finds that the above equation possesses solitary solution, i.e., localized density pulses that travel along the membrane cylinder without dissipation and without changing shape. A typical solution of equation (12) is shown in figure 5 (right). The pulse possesses a maximum amplitude and a minimum velocity when increasing the overall energy of the pulse. The maximum amplitude corresponds to the density change between liquid and solid membrane phase. Thus, the solitary pulse consists of a solid region traveling in a liquid membrane environment. This is schematically shown in figure 6.

The soliton described above shares many similarities with the nervous impulse:

- It displays a velocity similar to those of myelinated nerves.
- It is associated to transient changes in membrane thickness.
- It is associated to a reversible release and re-uptake of heat.

However, the physical principles underlying soliton propagation are very different from the mechanisms considered for nerve pulse propagation in the field of electrophysiology.

5. Ion channels

The textbook description for nerve pulse conduction is the Hodgkin-Huxley model [13]. It suggests that the nerve pulse is generated by ion currents through channel proteins. These currents charge the membrane capacitor. According to the model, channel proteins conduct ions in a voltage-dependent manner. Thus, they are considered being "voltage-gated". Combined with cable theory, this generates the possibility of propagating electrical pulses called 'action potentials'. The opening and closing of channels can be experimentally observed in electrical recordings [14]. To the contrary, in the soliton theory described above no ion channel proteins are required.

It is an interesting fact that membranes in the complete absence of proteins can form voltage-gated pores that display properties indistinguishable from protein channels [15, 4]. An example is given in figure 7 where one can see an increase in channel open-likelihood upon increase in voltage. These ion channel events result from area fluctuations in the membrane, as described by equation (8). In the melting transition, the fluctuations are large and the membrane permeability displays a maximum. Every change in a thermodynamic variable that potentially changes the membrane state can alter the permeability of the membrane [16, 3].

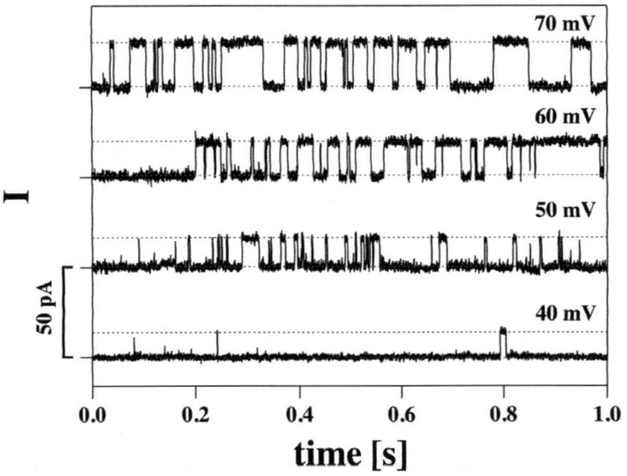

Figure 7. Quantized current events through a synthetic lipid membrane. One finds channel-like events in the complete absence of proteins. The open likelihood of of pore displays a pronounced voltage dependence. [4].

Due to the increase in channel open-probability shown in figure 7, the current-voltage relation is not linear. In particular, if the membrane displays a net polarization, V_0, in the absence of an external field, the current-voltage relation may be asymmetric and different for positive and negative voltages. A spontaneous membrane polarization could originate from an asymmetric distribution of lipids on the two sides of the membrane, or from membrane curvature. The latter effect is called 'flexoelectricity'. Its investigation was pioneered by A. G. Petrov [17]. Flexoelectricity is caused by the different dipole density on the two monolayers in curved membranes. Membrane curvature could possibly originate from slight pressure difference on the two sides of the membrane due to suction on the recording pipette. An example for an asymmetric non-linear current-voltage relation is shown in figure 8.

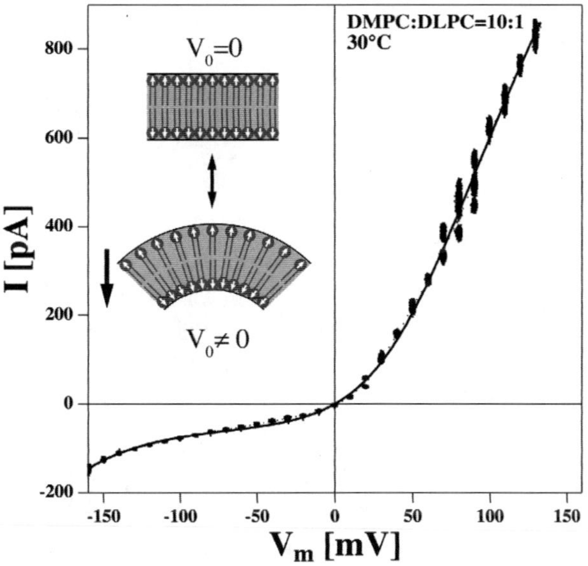

Figure 8. The current-voltage relation of the permeability of a synthetic lipid membrane is not generally symmetric even though the composition of the membrane itself is symmetric [4]. This could be caused by a permanent polarization of the membrane due to flexoelectricity [17, 18] - see insert.

In the absence of spontaneous polarization V_0 of the membrane, the electrostatic force, $\mathcal{F}$, exerted on a planar membrane by external voltage is given by

$$\mathcal{F} = \frac{1}{2}\frac{C_m V_m^2}{D} \tag{13}$$

where C_m is the membrane capacitance, V_m is the transmembrane voltage and D is the membrane thickness [1]. This force potentially reduces the thickness of the membrane [19]. The electrical work performed on the membrane by a change in thickness from D_1 to D_2 is

$$W_{el} = \int_{D_1}^{D_2} \mathcal{F} dD \equiv \alpha V_m^2 \tag{14}$$

where α is a constant. In the presence of a spontaneous polarization associated to a transmembrane voltage V_0, the electrical work instead assumes the form $W_{el} = \alpha(V_m - V_0)^2$. Since electrostatic work leads to membrane thinning, it is generally assumed that the work necessary to form a pore is proportional to the work necessary to reduce membrane thickness membrane [20, 21].

Therefore, the free energy for pore formation is given by

$$\Delta G = \Delta G_0 + \alpha(V_m - V_0)^2 \ , \tag{15}$$

where ΔG_0 is a constant.

The probability, $P_{open}(V_m)$, of finding an open pore in the membrane at a fixed voltage is given by

$$P_{open}(V_m) = \frac{K(V_m)}{1 + K(V_m)} \quad ; \quad K(V_m) = \exp\left(-\frac{\Delta G}{kT}\right) , \tag{16}$$

where $K(V_m)$ is the voltage-dependent equilibrium constant between open and closed states of a single pore.

The current-voltage relation for the lipid membrane is proportional to the likelihood of finding an open channel for a given voltage:

$$I_m = \gamma_p \cdot P_{open} \cdot V_m \tag{17}$$

where γ_p is the conductance of a single pore (or N identical pores). Eqs. 15-17 contain the theoretical description for the I-V curves of lipid channels. The solid line in figure 8 is a fit using the above description. It fits the experimental data nearly perfectly. Thus, a description based on the concept of forces induced by charging the membrane capacitor is very well able to describe experimental data of membrane permeability.

6. Summary

In this review, we summarized the evidence for electromechanical behavior of the biological membrane. The membrane can be seen as a capacitor with a spontaneous polarization. Due to forces on the capacitor, changes in transmembrane voltage can change the physical state of the membrane. E.g., it can induce membrane melting or freezing. Vice versa, lateral pressure changes in the membrane can alter the voltage on a membrane. Thus, the membrane displays piezoelectric features.

In a melting transition, the membrane displays a non-linear response to lateral pressure changes. This fact leads to the possibility of propagating density solitons in cylindrical membranes that share many similarities with the action potential in nerves. For instance, thickness and temperature changes in the nerve membrane are correctly described by the soliton approach. Further, the presence of melting transitions enhances the probability of area fluctuations in the membrane. These fluctuations lead to ion-channel-like events that are practically indistinguishable from protein ion channels. These protein channels are believed to be responsible for the nerve pulse in traditional theories. However, an electromechanical approach towards the physics of biological membranes intrinsically contains all these phenomena using the language of thermodynamics.

References

[1] Heimburg T 2012 *Biophys. J.* **103** 918–929
[2] Heimburg T and Jackson A D 2005 *Proc. Natl. Acad. Sci. USA* **102** 9790–9795
[3] Mosgaard L D and Heimburg T 2013 *Acc. Chem. Res.* **46** 2966–2976
[4] Blicher A and Heimburg T 2013 *PloS ONE* **8** e65707
[5] Heimburg T 2007 *Thermal biophysics of membranes* (Berlin, Germany: Wiley VCH)
[6] Heimburg T 1998 *Biochim. Biophys. Acta* **1415** 147–162
[7] Ebel H, Grabitz P and Heimburg T 2001 *J. Phys. Chem. B* **105** 7353–7360
[8] Iwasa K and Tasaki I 1980 *Biochem. Biophys. Research Comm.* **95** 1328–1331
[9] Iwasa K, Tasaki I and Gibbons R C 1980 *Science* **210** 338–339
[10] Ritchie J M and Keynes R D 1985 *Quart. Rev. Biophys.* **18** 451–476
[11] Mosgaard L D, Jackson A D and Heimburg T 2013 *J. Chem. Phys.* **139** 125101
[12] Sommerfeld A 1992 *Mechanik der deformierbaren Medien* (*Vorlesungen über theoretische Physik* vol 2) (Harri Deutsch)
[13] Hodgkin A L and Huxley A F 1952 *J. Physiol. London* **117** 500–544

[14] Hille B 1992 *Ionic channels of excitable membranes* (Cambridge: Cambridge University Press)
[15] Laub K R, Witschas K, Blicher A, Madsen S B, Lückhoff A and Heimburg T 2012 *Biochim. Biophys. Acta* **1818** 1–12
[16] Heimburg T 2010 *Biophys. Chem.* **150** 2–22
[17] Petrov A G 1999 *The lyotropic state of matter. Molecular physics and living matter physics.* (Amsterdam: Gordon and Breach Science Publishers)
[18] Petrov A G 2001 *Biochim. Biophys. Acta* **1561** 1–25
[19] White S H and Thompson T E 1973 *Biochim. Biophys. Acta* **323** 7–22
[20] Winterhalter M and Helfrich W 1987 *Phys. Rev. A* **36** 5874–5876
[21] Glaser R W, Leikin S L, Chernomordik L V, Pastushenko V F and Sokirko A I 1988 *Biochim. Biophys. Acta* **940** 275–287

On the critical specific heat capacity of a classical anharmonic crystal with long-range interaction

E Pisanova and S Ivanov

Department of Theoretical Physics, University of Plovdiv, 24 Tzar Assen Str., 4000 Plovdiv, Bulgaria

E-mail: katia_pisanova@abv.bg, stoil@uni-plovdiv.bg

Abstract. The bulk critical specific heat capacity of a classical anharmonic crystal with long-range interaction (decreasing at large distances r as $r^{-d-\sigma}$, where d is the space dimensionality and $0 < \sigma \leq 2$) is studied. An exact analytical expression is obtained at the upper critical dimension $d = 2\sigma$ of the system. This result depends on both the deviation from the critical point and the space dimensionality of the system, while the known asymptotic one depends only on the deviation from the critical point. For real systems (chains, thin layers, *i.e.* films and three-dimensional systems) the exact result and the asymptotic one are graphically presented and compared on the basis of the calculated relative errors. The obtained result holds true in a broader neighborhood of the critical point. The expansion of the critical region is estimated at the three real physical dimensionalities.

1. The model

The considered model describes a structural phase transition of second kind. The Hamiltonian of the model is [1]

$$H = \frac{1}{2}\sum_{\mathbf{r}}\left(\frac{P_{\mathbf{r}}^2}{m} - AQ_{\mathbf{r}}^2\right) + \frac{1}{4}\sum_{\mathbf{r},\,\mathbf{r}'}\varphi(\mathbf{r}-\mathbf{r}')(Q_{\mathbf{r}} - Q_{\mathbf{r}'})^2 + \frac{B}{4N}\left(\sum_{\mathbf{r}} Q_{\mathbf{r}}^2\right)^2, \tag{1}$$

where $P_{\mathbf{r}}$ and $Q_{\mathbf{r}}$ are the operators of displacement and momentum, respectively, of the particle of mass m at site $\mathbf{r}$ of a d-dimensional hypercubic lattice. The parameter $A = \nu_0^2 m > 0$ determines the frequency of a mode which is unstable in the harmonic approximation and the parameter $B > 0$ introduces an anharmonic interaction which is inversely proportional to the particle number N. The harmonic force constants $\varphi(\mathbf{r}-\mathbf{r}')$ which are assumed to decrease at large distances $r = |\mathbf{r}-\mathbf{r}'|$ as $r^{-d-\sigma}$, describe a short-range ($\sigma = 2$) or a long-range ($0 < \sigma < 2$) interaction.

The free energy density of the model (1), obtained by using *approximating Hamiltonian method* is [2, 3]

$$f = \frac{A^2}{B} f_0 = \frac{A^2}{B}\left(\frac{1}{2} I_{t,\lambda}^{d,\sigma}(\bar{\Delta}) - \frac{1}{4}(1+\bar{\Delta})^2\right), \tag{2}$$

where $\bar{\Delta}$ is the solution of the self-consistent equation

$$\frac{dI_{t,\lambda}^{d,\sigma}(\Delta)}{d\Delta} = 1 + \Delta. \tag{3}$$

In the thermodynamic limit $N \to \infty$ the function $I_{t,\lambda}^{d,\sigma}$ is defined by

$$I_{t,\lambda}^{d,\sigma}(\Delta) = 2t\frac{S_d}{(2\pi)^d}\int_0^{x_D} x^{d-1}\ln\left(2\sinh\left(\frac{\lambda}{2t}\sqrt{\Delta + x^\sigma}\right)\right)dx, \tag{4}$$

where $t = T/(4E_0)$ is the dimensionless temperature and $\lambda = \hbar\nu_0/(4E_0)$ is a parameter which switches on the quantum fluctuations, $E_0 = A^2/(4B)$ is the barrier height of the double-well potential in (1), $x_D = 2\pi(d/S_d)^{1/d}$ is the radius of the effective sphere replacing the Brillouin zone and $S_d = 2\pi^{d/2}/\Gamma(d/2)$ (Γ is the Euler gamma function) is the surface of the d-dimensional unit sphere. Setting $\Delta = 0$ into (3) one obtains the critical point $t_c(\lambda)$. In the disordered phase ($t > t_c$), $\bar{\Delta}$ is finite and the susceptibility of the system is $\chi = \bar{\Delta}^{-1}$.

Let us note that the equations (2) and (3) play a central role in the study of the bulk critical behavior of the model, *e.g.* for the specific heat capacity $c(T) \equiv -T\partial^2 f/\partial T^2$ from (3) and (2), we obtain

$$c(t) = -t\frac{\partial^2 f_0}{\partial t^2} = -\frac{t}{2}\left(\frac{\partial^2 I_{t,\lambda}^{d,\sigma}(\bar{\Delta})}{\partial t^2} - \left(\frac{\partial\bar{\Delta}}{\partial t}\right)^2 - (1+\bar{\Delta})\frac{\partial^2\bar{\Delta}}{\partial t^2}\right). \tag{5}$$

This model retains many fundamental properties of the real systems related to the presence of both quantum and classical fluctuations, depending on the temperature T, the quantum parameter λ, the long-range interaction exponent σ and the spatial dimensionality d. For a more complete discussion of the bulk critical behavior and the finite-size properties of the model and its generalizations, see [3] and references therein.

Exact solutions of the self-consistent equation (3), in terms of the Lambert W-function [4], have been obtained in both the quantum and the classical limits at the corresponding upper critical dimensions, $d = 3\sigma/2$ and $d = 2\sigma$, respectively [5,6]. On the basis of the solution at $d = 3\sigma/2$, an exact expressions for the bulk free energy density near the quantum critical point $T = 0$ has been obtained in [7]. The finite-size corrections to the free energy density for the pure quantum version of the model have been studied in [8].

In this paper, using the exact solution of the self-consistent equation (3) in the classical limit ($\lambda \to 0^+$) at the upper critical dimension $d = 2\sigma$ [6], we establish an exact analytical expression for the specific heat capacity of the model in terms of the Lambert W-function. For systems with real physical dimensions (chains, thin layers, *i.e.* films and three-dimensional systems), the obtained exact result is graphically presented and compared with the asymptotic one on the basis of the calculated relative error. We show that the obtained exact result for the specific heat capacity holds true in a broader neighborhood of the critical point. Besides, we give an estimate for the expansion of the critical region at the three real physical dimensionalities.

2. An exact result and its leading asymptotic behavior

For classical systems ($\lambda \to 0^+$) the self-consistent equation (3) gets the form

$$\frac{dU_{d,\sigma}(\Delta)}{d\Delta} = \frac{1}{t}(1+\Delta), \tag{6}$$

where the function U is defined by

$$U_{d,\sigma}(\Delta) = \frac{S_d}{(2\pi)^d}\int_0^{x_D} x^{d-1}\ln(\Delta + x^\sigma)dx. \tag{7}$$

For the critical temperature from (6) at $\Delta = 0$, we obtain

$$t_c = \frac{d-\sigma}{d}x_D^\sigma. \tag{8}$$

In the classical limit from (4), taking into account (6), we get

$$\frac{\partial^2 I_{t,0}^{d,\sigma}(\bar{\Delta})}{\partial t^2} = -\frac{2}{t} + \frac{1}{t}(1+\bar{\Delta})\frac{\partial \bar{\Delta}}{\partial t} + \left(\frac{\partial \bar{\Delta}}{\partial t}\right)^2 + (1+\bar{\Delta})\left(\frac{\partial^2 \bar{\Delta}}{\partial t^2}\right). \tag{9}$$

Thus, from the last equation and (5), we obtain the following expression for the specific heat capacity of the model (1) in the classical limit

$$c(t) = 1 - \frac{1}{2}(1+\bar{\Delta})\frac{\partial \bar{\Delta}}{\partial t}. \tag{10}$$

Similar result was obtained in the framework of the mean spherical model [3, p.89] which belongs to the same universality class [9].

Since in the disordered phase ($t > t_c$) near the critical point ($t \to t_c^+$) the solution $\bar{\Delta}$ of (6) decreases when t decreases and $\bar{\Delta} = 0$ in the ordered phase ($t < t_c$), then the specific heat capacity keeps its maximum value $c(t) = 1$ for all $t \leq t_c$ and the Dulong-Petit low of the classical thermodynamics holds for all $T \leq T_c$.

At the upper classical critical dimension ($d = 2\sigma$), near the classical critical point ($t \to t_c^+$), *i.e.* when $\Delta << 1$, the equation (6) can be written in the following form

$$\left(\frac{\Delta}{x_D^\sigma}\right)\ln\left(\frac{\Delta}{x_D^\sigma}\right) - x_D^\sigma\left(\frac{\Delta}{x_D^\sigma}\right) = -\epsilon, \tag{11}$$

where $\epsilon = 1 - t_c/t$ is a measure of the deviation of the critical point. From (8), for the critical temperature in this case, we get $t_c = x_D^\sigma/2$. The exact solution of (11) in terms of the Lambert W-function is [6]

$$\bar{\Delta} = x_D^\sigma \exp[x_D^\sigma + W_{-1}(-\epsilon\, e^{-x_D^\sigma})], \tag{12}$$

where $W_{-1}(x)$ is the real branch of the Lambert W-function on the interval $[-1/e, 0)$, satisfying $W_{-1}(x) \leq -1$, as $\lim_{x\to 0^-} W_{-1}(x) = -\infty$ [4]. Thus, in the neighborhood of the classical critical point ($\epsilon \to 0^+$) at the upper classical critical dimension $d = 2\sigma$, from (12) and (10), we get the following exact expression for the specific heat capacity of the model (1)

$$c(\epsilon) = 1 + \frac{1}{1 + W_{-1}(-\epsilon\, e^{-x_D^\sigma})}\left(1 - x_D^\sigma \frac{\epsilon}{W_{-1}(-\epsilon\, e^{-x_D^\sigma})}\right). \tag{13}$$

The last result shows that the specific heat capacity remains finite at $T_c(\epsilon = 0)$ but in this point its graph has a cusp. The obtained expression (13) allow us to find a critical region in which the specific heat decreases n times. For $c(\epsilon) = 1/n$, from (13) neglecting the second term in the brackets, we obtain that the endpoint of this critical region is

$$\epsilon_n = \frac{(2n-1)}{n-1}\, e^{\frac{2n-1}{1-n} + x_D^\sigma}, \qquad n > 1. \tag{14}$$

From (13), using the series in the asymptotic formula of the Lambert W-function [4] and retaining the leading term, we get the following asymptotic behavior of the specific heat capacity

$$c_{appr.} \approx 1 + \frac{1}{\ln\epsilon}. \tag{15}$$

From the other hand, the last result can be obtained from (10) using the known asymptotic behavior [6]

$$\bar{\Delta}_{appr.} \approx -x_D^\sigma \frac{\epsilon}{\ln\epsilon}. \tag{16}$$

Note that the logarithmic correction in (15) was discussed in the framework of the mean spherical model in [3].

From (15), for the endpoint $\epsilon_{n,appr.}$ of a critical region in which the asymptotic specific heat capacity decreases n times, we have

$$\epsilon_{n,appr.} = e^{\frac{n}{1-n}}, \qquad n > 1.$$

From this and (14), we get the following relation

$$\epsilon_n = \frac{2n-1}{n-1} e^{x_D^{\sigma}-1} \epsilon_{n,appr.}, \qquad n > 1 \tag{17}$$

which shows that $\epsilon_n > \epsilon_{n,appr.}$ for each $0 < \sigma \leq 2$ and gives an estimate for the expansion of the critical region. Let us note that for all n this expansion increases with σ, *i.e.* with the dimensionality of the system d.

3. Comparison between the exact result and the asymptotic one

It is easy to see from (13) and (15) that the exact result depends on both the deviation from the critical point and the dimensionality of the system, while its leading asymptotic behavior depends only on the deviation from the critical point. For systems with real physical dimensionalities (chains, films and three-dimensional systems), on figure 1 we present the obtained exact and asymptotic results for the specific heat capacity.

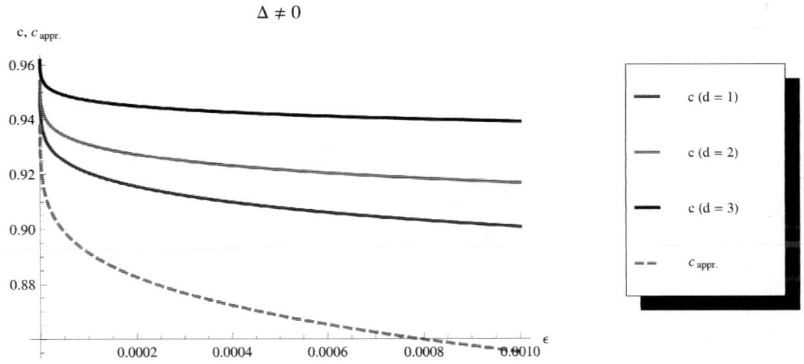

Figure 1. The dependences of the specific heats c and $c_{appr.}$ on the deviation from the critical point ϵ.

For more detailed analysis in the table 1, we give some numerical data for the computed relative errors $|1 - \bar{\Delta}_{appr.}/\bar{\Delta}|$ and $|1 - c_{appr.}/c|$ as functions of the deviation from the critical point and the dimensionality of the system.

Table 1. Percent relative errors [%].

	$\mathbf{d=1(\sigma=1/2)}$		$\mathbf{d=2(\sigma=1)}$		$\mathbf{d=3(\sigma=3/2)}$	
ϵ	$\lvert 1-\frac{\bar{\Delta}_{appr.}}{\bar{\Delta}}\rvert$	$\lvert 1-\frac{c_{appr.}}{c}\rvert$	$\lvert 1-\frac{\bar{\Delta}_{appr.}}{\bar{\Delta}}\rvert$	$\lvert 1-\frac{c_{appr.}}{c}\rvert$	$\lvert 1-\frac{\bar{\Delta}_{appr.}}{\bar{\Delta}}\rvert$	$\lvert 1-\frac{c_{appr.}}{c}\rvert$
1×10^{-7}	29.8	1.2	41.3	1.6	68.2	2.4
1×10^{-6}	33.9	1.6	47.4	2.1	78.9	3.1
1×10^{-5}	39.5	2.1	55.8	2.9	93.8	4.1
1×10^{-4}	47.5	3.1	68.2	4.2	116.0	5.8
1×10^{-3}	60.4	5.0	88.4	6.7	152.8	8.9

4. Conclusions

Using the exact solution (12) of the self-consistent equation, in the neighborhood of the classical critical point at the upper critical dimension of the system, an exact analytical expression for the specific heat capacity (13) is obtained.

It is shown that the exact specific heat capacity depends on both the deviation from the critical point $\epsilon = 1 - t_c/t$ and the long-range interaction exponent σ, *i.e.* of the space dimensionality d of the system. Its leading asymptotic behavior (15), has a logarithmic correction depending only on ϵ which is well known in the theory of the critical phenomena.

For real systems (chains, thin layers, *i.e.* films and three-dimensional systems) the specific heat capacity and its leading asymptotic behavior are graphically presented on figure 1. One can see from the graph that the specific heat capacity increases with the space dimensionality of the system. Moreover, the obtained result (13) holds true in a broader neighborhood of the critical point. The expressions (13) and (15) allow us to estimate the expansion of the critical region as a function of the spatial dimensionality of the system (17). Note that this expansion increases with dimensionality of the system d.

It is easy to see from the table 1 that at each deviation from the critical point and each dimensionality, the relative error for the specific heat capacity is less than the relative error for the inverse susceptibility.

Finally, the exact solution of the self-consistent equation can be used in investigating of the other critical thermodynamical properties of the system, *e.g.* the entropy and others.

Acknowledgments

The authors express their gratitude to Prof. N. S. Tonchev for the stimulating discussions on this topic. These research is supported by Project NI13 FMI-002.

References

[1] Schneider T, Stall E and Beck H 1975 *Physica A* **79** 201
[2] Plakida N M, Radosz A and Tonchev N S 1991 *Phase Transitions* **297** 179
[3] Brankov J G, Danchev D M and Tonchev N S 2000 The Theory of Critical Phenomena in Finite-Size Systems - Scaling and Quantum Effects *Series in Modern Condensed Matter Physics* vol 9 (Singapore: World Scientific) p 1
[4] Corlles R M, Gonnet G H, Hare D E, Jeffrey D J and Knuth D E 1996 *Advances in Comput. Math.* **5** 329
[5] Pisanova E S, Tonchev N S and Kisov Hr T 2009 *AIP Conf. Proc.* **1203** 255
[6] Pisanova E S and Ivanov S I 2013 *Bulgarian Journal of Physics* **40** 159
[7] Pisanova E S and Tonchev N S 2010 *Journal of Physics: Conference Series* **253** 012020
[8] Chamati H 1994 *Physica A* **212** 357
[9] Kadanoff L P 1971 *Proc. Intern. School of Physics "Enrico Fermi"* (New York: Academic Press)

On the statistical mechanics of shape fluctuations of nearly spherical lipid vesicle

I Bivas and N S Tonchev
Institute of Solid State Physics, Bulgarian Academy of Sciences, 72 Tzarigradsko chaussee blvd., Sofia 1784, Bulgaria

E-mail: bivas@issp.bas.bg, tonchev@issp.bas.bg

Abstract. The mechanical properties of biological membranes play an important role in the structure and the functioning of living organisms. One of the most widely used methods for determination of the bending elasticity modulus of the model lipid membranes (simplified models of the biomembranes with similar mechanical properties) is analysis of the shape fluctuations of the nearly spherical lipid vesicles. A theoretical basis of such an analysis is developed by Milner and Safran. In the present studies we analyze their results using an approach based on the Bogoljubov inequalities and the approximating Hamiltonian method. This approach is in accordance with the principles of statistical mechanics and is free of contradictions. Our considerations validate the results of Milner and Safran if the stretching elasticity K_s of the membrane tends to zero.

1. Introduction

The biological membranes are one of the most important building blocks of living matter. Their mechanical properties determine to a great extent their structure and functioning. This is the reason for the interest in these properties over the years [1, 2, 3, 4].

According to the model of Singer and Nicolson [5], the biomembrane consists of a lipid bilayer, in which integral proteins float. Evidently, in the frames of this model, the mechanical properties of the biomembrane are tightly connected with those of the lipid bilayer. In addition, the lipid bilayer is a well-defined structure that can be used in physical experiments. This is why the study of the mechanical properties of the lipid bilayers is continuously growing.

One of the most widely used methods for determination of the bending elasticity of a lipid bilayer is the analysis of the shape fluctuations of nearly spherical lipid vesicles [6, 7]. The aim of the present paper is to reconsider the theoretical basis of this analysis, proposed by Milner and Safran [6].

2. Mechanical energy stored in the shape fluctuations of a nearly spherical lipid vesicle

Following Helfrich [8], we consider a small patch of the lipid bilayer with area Δs, tension σ, and area in its tension free state Δs^{tf}. Let c_1 and c_2 be the main curvatures of the patch under consideration. If the patch is tension-free then the bending energy density g_c^0 can be written in the following form:

$$g_c^0(c_1, c_2) = \frac{1}{2}K_c(c_1 + c_2 - c_0)^2 + \overline{K}_c c_1 c_2, \tag{1}$$

where K_c is the bending elasticity of the bilayer, c_0 is the spontaneous curvature of the membrane, and $\overline{K}_c$ is the saddle splay bending elasticity. In the present paper only symmetrical membranes (with $c_0 = 0$) will be considered. The bending energy ΔG_c of the whole patch is:

$$\Delta G_c = g_c^0(c_1, c_2)\Delta s^{tf}. \tag{2}$$

If the patch is not tension free then its stretching energy density g_s is expressed via its tension σ (here we assume that σ is a constant all over the membrane [9]) as:

$$g_s = \frac{1}{2}\frac{\sigma^2}{K_s}, \tag{3}$$

where K_s is the stretching elasticity of the bilayer. The stretching energy ΔG_s of the patch is:

$$\Delta G_s = g_s \Delta s^{tf}. \tag{4}$$

Let ΔG be the total mechanical energy of of the patch:

$$\Delta G = \Delta G_c + \Delta G_s. \tag{5}$$

The total deformation energy $G(t)$ of the vesicle is obtained by integration of $\Delta G(t)$ on the vesicle surface $S(t)$:

$$G(t) = \oint_{S(t)} \Delta G(t). \tag{6}$$

In the last equation the contribution of the saddle-splay elasticity can be disregarded. This is due to the Gauss-Bonnet theorem which assures that if the topology of the vesicle does not change then the contribution of this elasticity does not depend on the shape fluctuations of the vesicle.

Let us consider a nearly spherical lipid vesicle. The volume V of the vesicle is assumed not to be fluctuating. Let R_0 be the radius of a sphere with a volume of V. Let the origin O of a laboratory reference frame be placed inside the vesicle. A point on the surface of the vesicle with polar coordinates (θ, φ) is chosen. Let $R(\theta, \varphi)$ be the modulus of the radius-vector at this point. The dimensionless quantity $u(\theta, \varphi, t)$ is defined by the equation:

$$R(\theta, \varphi, t) = R_0[1 + u(\theta, \varphi, t)], \tag{7}$$

where t is the time variable. The function $u(\theta, \varphi, t)$ is decomposed in a series as follows:

$$u(\theta, \varphi, t) = \sum_{n=0}^{n_{max}} \sum_{m=-n}^{n} u_n^m(t) Y_n^m(\theta, \varphi), \tag{8}$$

where $Y_n^m(\theta, \varphi)$ is the orthonormal basis (for simplicity chosen real) of the spherical harmonics functions [10]. A cut-off $n_{max} \sim R_0/\lambda$ is introduced in the sum, where λ is of the order of the intermolecular distance. As the harmonics with indexes $n = 1$ and $m = -1, 0, 1$ correspond to pure translation of the vesicle, the origin O can be chosen in a way that $u_1^m = 0$. Because of the requirement for volume conservation the amplitude $u_0^0(t)$ can be expressed as [6]:

$$u_0^0(t) = -\frac{1}{2\pi^{1/2}} \sum_{n=2}^{n_{max}} \sum_{m=-n}^{n} [u_n^m(t)]^2. \tag{9}$$

3. Theory of Milner and Safran

According to the theory, developed by Milner and Safran [6], the mechanical energy $G(t)$ of the vesicle is expressed as follows:

$$G(t) = \sum_{n=2}^{n_{max}} \sum_{m=-n}^{n} \frac{1}{2} K_c (n-1)(n+2)[n(n+1)+\overline{\sigma}][u_n^m(t)]^2 \tag{10}$$

where:

$$\overline{\sigma} = \frac{\sigma (R_0)^2}{K_c} \tag{11}$$

is a dimensionless parameter. The quantity σ (with dimension of tension) is considered by Milner and Safran as a Lagrange multiplier, not fluctuating with time, which ensures the mean area of the vesicle membrane to be equal to some prescribed value. The energy $G(t)$ from Eq. (10) is a sum of the energies of not interacting oscillators. This energy does not depend on the stretching elasticity K_s of the membrane. It is a function only of K_c, R_0, and σ.

In the frames of this theory, the amplitudes u_n^m of the different fluctuation modes are not time correlated. For $n \geq 2$ the time mean squares $\overline{[u_n^m(t)]^2}$ are:

$$\overline{[u_n^m(t)]^2} = \frac{kT}{K_c} \frac{1}{(n-1)(n+2)[n(n+1)+\overline{\sigma}]} \tag{12}$$

where kT is the Boltzmann factor.

4. The Model Hamiltonian

Let us denote with $H(U) \equiv H(u_2^{-2}, u_2^{-1}, \ldots, u_{n_{max}}^{n_{max}})$ the model Hamiltonian describing a fluctuating vesicle. The symbol U is used as shorthand for the real value functions $(u_2^{-2}, u_2^{-1}, \ldots, u_{n_{max}}^{n_{max}})$ that are the spherical harmonic amplitudes (see (8)), appearing in the the expansion of the vesicle shape fluctuations from the equivalent volume sphere with radius R_0. From here on the notation u_m^l is used as an abbreviation of $u_m^l(t)$. We assume that our system is described by a Hamiltonian $H(U)$, consisting in a sum of the bending $H_c(U)$ and stretching $H_s(U)$ energies:

$$H(U) = H_c(U) + H_s(U), \tag{13}$$

where

$$H_c(U) = \frac{1}{2} K_c \sum_{n=2}^{n_{max}} \sum_{m=-n}^{n} (n-1)n(n+1)(n+2)(u_n^m)^2 \tag{14}$$

and

$$H_s(U) = \frac{1}{2} \frac{[\sigma(U)]^2}{K_s} S_0^0. \tag{15}$$

In the last equation S_0^0 is the area of of the vesicle membrane when it is tension free. The membrane tension $\sigma = \sigma(U)$, which is assumed constant all over the membrane, is given by the expression:

$$\sigma(U) = K_s \frac{4\pi(R_0)^2 + \Delta S(U) - S_0^0}{S_0^0}. \tag{16}$$

The quantity $\Delta S(U)$ is the excess area of the vesicle (the difference between the area of the vesicle membrane and the area of the sphere with a volume equal to that of the vesicle). It is expressed in terms of the amplitudes u_n^m in the following way [6]:

$$\Delta S(U) = \frac{(R_0)^2}{2} \left[\sum_{n=2}^{n_{max}} \sum_{m=-n}^{n} (n-1)(n+2)(u_n^m)^2 \right]. \tag{17}$$

For further application it is convenient to present the model Hamiltonian $H(U)$ in the following equivalent form:

$$\begin{aligned} H(U) &= \frac{1}{2}K_c \sum_{n=2}^{n_{max}} \sum_{m=-n}^{n} \Bigg((n-1)(n+2) \\ &\times \left\{ n(n+1) + \sigma_0 + \frac{K_s}{2K_c}\frac{(R_0)^2}{S_0^0}\Big[\Delta S(U)\Big] \right\} (u_n^m)^2 \Bigg), \end{aligned} \tag{18}$$

where:

$$\sigma_0 = \frac{(R_0)^2}{K_c} K_s \frac{4\pi(R_0)^2 - S_0^0}{S_0^0}. \tag{19}$$

Eq. (18) gives us a hint to introduce the key quantity:

$$\hat{\sigma} = \sigma_0 + \frac{K_s}{2K_c}\frac{(R_0)^2}{S_0^0}\langle \Delta S(U) \rangle_{H(U)}, \tag{20}$$

where the symbol $\langle ... \rangle_{H(U)}$ denotes a thermodynamic average calculated with the model Hamiltonian $H(U)$:

$$\langle \dots \rangle_{H(U)} = \{Z[H(U)]\}^{-1} \int dU \dots \exp\left[-\frac{H(U)}{kT}\right], \tag{21}$$

where

$$Z[H(U)] = \int dU \exp\left[-\frac{H(U)}{kT}\right] \tag{22}$$

is the statistical sum of the model. Using Eq. (20) we can rewrite Eq. (18) in the form:

$$\begin{aligned} H(U) &= \frac{1}{2}K_c \sum_{n=2}^{n_{max}} \sum_{m=-n}^{n} \Bigg\{ (n-1)(n+2) \\ &\times \left[n(n+1) + \hat{\sigma} + \frac{K_s}{2K_c}\frac{(R_0)^2}{S_0^0}\Big(\Delta S(U) - \langle \Delta S(U) \rangle_{H(U)}\Big) \right] (u_n^m)^2 \Bigg\}. \end{aligned} \tag{23}$$

Let us note that if we drop the third term in the rectangular bracket in the rhs of Eq. (23) and replace $\overline{\sigma}$ from Eq. (11) with $\hat{\sigma}$, we obtain a result coinciding in form with the one of Milner and Safran:

$$\langle (u_n^m)^2 \rangle_{H_{MS}(U,\hat{\sigma})} = \frac{kT}{K_c} \frac{1}{(n-1)(n+2)[n(n+1)+\hat{\sigma}]}. \tag{24}$$

The relevant Hamiltonian in the calculation of the thermodynamic mean value in the lhs of Eq. (24) turns out to be the "Milner and Safran Hamiltonian" $H_{MS}(U, \hat{\sigma})$ of a system, consisting of independent harmonic oscillators, of the kind:

$$H_{MS}(U,\hat{\sigma}) = \frac{1}{2}K_c \sum_{n=2}^{n_{max}} \sum_{m=-n}^{n} \left\{ (n-1)(n+2) \Big[n(n+1) + \hat{\sigma} \Big] (u_n^m)^2 \right\}. \tag{25}$$

A sufficient condition this step to be acceptable is the smallness (in some sense) of the term

$$\frac{K_s}{2K_c}\frac{(R_0)^2}{S_0^0}\Big[\Delta S(U) - \langle \Delta S(U) \rangle_{H(U)}\Big]. \tag{26}$$

Without entering in details we note that the smallness of the above expression is equivalent to the condition for smallness of the excess area fluctuations.

To avoid confusion, we warn the reader that Eq. (24) can be obtained from Eq. (12) by the replacement of $\overline{\sigma}$ with $\hat{\sigma}$. The quantity $\overline{\sigma}$ has been introduced in the theory by hand to ensure that the mean excess area of the fluctuating membrane is equal to some prescribed value, while the quantity $\hat{\sigma}$ has a self-consistent origin depending on the model Hamiltonian itself.

It can be concluded that the approach of Milner and Safran employs the implicit assumption that the term (26) can be neglected in the model Hamiltonian $H(U)$. At this stage let us point out that an exact treatment of the thermodynamics of the model Hamiltonian $H(U)$ seems to be unrealistic due to the interaction between the modes in the the stretching energy term. Indeed, it is very hard to obtain additional results beyond the Milner and Safran ones concerning the case when modes do not interact, unless some approximation tricks or variational methods are used.

5. The approximating Hamiltonian

In this section we show that the approximating Hamiltonian $H_{app}(U, \Sigma)$ of the kind:

$$H_{app}(U, \Sigma) = \frac{1}{2} K_c \sum_{n=2}^{n_{max}} \sum_{m=-n}^{n} \{(n-1)(n+2)[n(n+1) + \Sigma](u_n^m)^2\}, \tag{27}$$

where Σ is an appropriately defined trial quantity, is the best approximation of the model Hamiltonian $H(U)$ by a Hamiltonian of a system, consisting in independent harmonic oscillators. Hereafter the problem is to obtain Σ. To this end we use the Bogoljubov variational principle [11]. In order to use it, some preliminary mathematical manipulations on the model Hamiltonian $H(U)$ need to be performed.

First, we introduce the functions $\mathcal{A}(U)$ and $\mathcal{T}(U, \sigma_0)$ in the following way:

$$\mathcal{A}(U) = \left(\frac{K_s}{2S_0^0}\right)^{1/2} \frac{(R_0)^2}{2} \sum_{n=2}^{n_{max}} \sum_{m=-n}^{n} (n-1)(n+2)(u_n^m)^2 \tag{28}$$

and

$$\mathcal{T}(U, \sigma_0) = \frac{1}{2} K_c \sum_{n=2}^{n_{max}} \sum_{m=-n}^{n} (n-1)(n+2)[n(n+1) + \sigma_0](u_n^m)^2. \tag{29}$$

Then we define a trial Hamiltonian $H_{app}(U, X, \sigma_0)$ as:

$$H_{app}(U, X, \sigma_0) = \mathcal{T}(U, \sigma_0) + 2X\mathcal{A}(U) - X^2, \tag{30}$$

where X is an arbitrary real number. The Hamiltonian $H_{app}(U, X, \sigma_0)$, obtained in this way, is linearized with respect to the squares of the amplitudes $(u_n^m)^2$. It is easy to see that:

$$0 \leq [\mathcal{A}(U) - X]^2 = H(U) - H_{app}(U, X, \sigma_0) \tag{31}$$

for arbitrary X. From Eqs. (28), (29), and (30) we obtain

$$\begin{aligned} H_{app}(U, X, \sigma_0) &= \frac{1}{2} K_c \sum_{n=2}^{n_{max}} \sum_{m=-n}^{n} \Bigg\{(n-1)(n+2) \\ &\times \left[n(n+1) + \sigma_0 + \left(\frac{2K_s}{S_0^0}\right)^{1/2} \frac{(R_0)^2}{K_c} X\right]\Bigg\}(u_n^m)^2 - (X)^2. \end{aligned} \tag{32}$$

According to the Bogolyubov inequalities (see e. g. [11]), for all X it is true that:

$$\begin{aligned}\langle H(U) - H_{app}(U, X, \sigma_0)\rangle_{H(U)} &\leq f[H(U)] - f[H_{app}(U, X, \sigma_0)] \\ &\leq \langle H(U) - H_{app}(U, X, \sigma_0)\rangle_{H_{app}(U,X,\sigma_0)}\end{aligned} \tag{33}$$

where $f[H(U)] = -kT \ln Z[H(U)]$ is the free energy of the system with Hamiltonian $H(U)$, $\langle H(U) - H_{app}(U, X, \sigma_0)\rangle_{H(U)}$ is the thermodynamic average calculated with the Hamiltonian $H(U)$, and $f[H_{app}(U, X, \sigma_0)] = -kT \ln Z[H_{app}(U, X, \sigma_0)]$ and $\langle H(U) - H_{app}(U, X, \sigma_0)\rangle_{H_{app}(U,X,\sigma_0)}$ are a free energy and a thermodynamic average, calculated with $H_{app}(U, X, \sigma_0)$. Since the residual Hamiltonian $H(U) - H_{app}(U, X, \sigma_0)$ is non-negative (see Eq. (31)), the application of the Bogolyubov inequalities (33) yields that for all X:

$$0 \leq f[H(U)] - f[H_{app}(U, X, \sigma_0)] \leq \langle[\mathcal{A}(U) - X]^2\rangle_{H_{app}(U,X,\sigma_0)}. \tag{34}$$

The best approximation of the free energy of the model system with Hamiltonian $H_{app}(U, X, \sigma_0)$ (see Eq. (32)) is given by the absolute maximum principle:

$$\max_X f[H_{app}(U, X, \sigma_0)] = f[H_{app}(U, \overline{X}, \sigma_0)]. \tag{35}$$

We note that $f[H_{app}(U, X, \sigma_0)]$ attains its maximum with respect to the parameter X at the solution of the equation:

$$\frac{\partial f[H_{app}(U, X, \sigma_0)]}{\partial X} = 0, \tag{36}$$

which yields

$$\langle \mathcal{A}(U)\rangle_{H_{app}(U,X,\sigma_0)} = X. \tag{37}$$

This is a self-consistency equation for the parameter X. It can be shown that this (self-consistency) equation has only one solution, namely $\overline{X}$. In its explicit form this equation takes the form:

$$X = \frac{kT\sigma_1}{4} \sum_{n=2}^{n_{max}} \frac{2n+1}{n(n+1) + \sigma_0 + \sigma_1 X}. \tag{38}$$

The absolute maximum condition (35) and inequalities (33) impose the following bounds on the free energy difference $f[H(U)] - f[H_{app}(U, X, \sigma_0)]$:

$$0 \leq f[H(U)] - f[H_{app}(U, \overline{X}, \sigma_0)] \leq \langle[\mathcal{A}(U) - \overline{X}]^2\rangle_{H_{app}(U,\overline{X},\sigma_0)}. \tag{39}$$

Evidently $H_{app}(U, \overline{X}, \sigma_0)$ provides the best approximation from below of the free energy of the model Hamiltonian $H(U)$ (see Eq. (23)) with the free energy of the approximating Hamiltonian

$$H_{app}(U, \overline{X}, \sigma_0) = \mathcal{T}(U, \sigma_0) + 2\overline{X}\mathcal{A}(U) - \left(\overline{X}\right)^2. \tag{40}$$

Using Eqs. (28) and (29), the approximating Hamiltonian $H_{app}(U, \overline{X}, \sigma_0)$, describing an ensemble of noninteracting oscillators, can be rewritten in the more convenient form:

$$H_{app}(U, \overline{X}, \sigma_0) = \frac{1}{2} K_c \sum_{n=2}^{n_{max}} \sum_{m=-n}^{n} (n-1)(n+2) \left[n(n+1) + \tilde{\sigma}\left(\sigma_0, \overline{X}\right)\right] (u_n^m)^2 - \left(\overline{X}\right)^2, \tag{41}$$

where

$$\tilde{\sigma}\,(\sigma_0, X) = \sigma_0 + \left(\frac{2K_s}{S_0^0}\right)^{1/2} \frac{(R_0)^2}{K_c} X \equiv \sigma_0 + \sigma_1 X \tag{42}$$

and

$$\sigma_1 = \left(\frac{2K_s}{S_0^0}\right)^{1/2} \frac{(R_0)^2}{K_c}. \tag{43}$$

The comparison of Eq. (41) with Eq. (27) shows that $\tilde{\sigma}\left(\sigma_0, \overline{X}\right)$ plays the role of Σ .

To make the respective assessments it is necessary to have the analytical expressions for the free energy term (this is not difficult because the calculations are for noninteracting harmonic oscillators) and to solve numerically the self-consistent equation in order to determine $\overline{X}$, which depends on the quantities kT, K_s, R_0, and K_c, considered as fixed values for a definite vesicle. The only free parameter in the theory remains σ_0.

Following our approach, the mean square value of the amplitude in Eq. (24) is replaced by:

$$\langle (u_n^m)^2 \rangle_{H_{app}(U,\overline{X},\sigma_0)} = \frac{kT}{K_c} \frac{1}{(n-1)(n+2)[n(n+1) + \tilde{\sigma}(\sigma_0, \overline{X})]}. \tag{44}$$

In order to obtain Eq. (44) from Eq.(24) the following replacements need to be done:

$$\langle (u_n^m)^2 \rangle_{H_{MS}(U,\hat{\sigma})} \rightarrow \langle (u_n^m)^2 \rangle_{H_{app}(U,\overline{X},\sigma_0)}, \qquad \hat{\sigma} \rightarrow \tilde{\sigma}(\sigma_0, \overline{X}). \tag{45}$$

Consequently, in our approach the term

$$H_s(U) = \frac{1}{2} \frac{[\sigma(U)]^2}{K_s} S_0^0 \tag{46}$$

ensures that the mean square amplitudes $\langle (u_n^m)^2 \rangle_{H_{app}(U,\overline{X},\sigma_0)}$ are calculated by an effective tension, appearing as solution of the following equation, deduced from Eq. (38):

$$\Sigma = \sigma_0 + \frac{kT\sigma_1^2}{4} \sum_{n=2}^{n_{max}} \frac{2n+1}{n(n+1) + \Sigma}. \tag{47}$$

The solution of this equation is $\overline{\Sigma} = \tilde{\sigma}(\sigma_0, \overline{X})$. Clearly, $\overline{X}$ is function of σ_0, which means that $\tilde{\sigma}(\sigma_0, \overline{X})$, appearing as solution of Eq. (47), depends only on σ_0. In the Milner and Safran approach the corresponding quantity $\overline{\sigma}$ is an external parameter.

In order to discuss the closeness of the free energies of the model and approximating systems, one must calculate the correlator $\langle [\mathcal{A}(U) - \overline{X}]^2 \rangle_{H_{app}(U,\overline{X},\sigma_0)}$ from Eq. (34). Since the Hamiltonian $H_{app}(U, \overline{X}, \sigma_0)$ is linear with respect to the squares of the amplitudes $(u_n^m)^2$, after some lengthly but standard calculations we obtain:

$$\langle [\mathcal{A}(U) - \overline{X}]^2 \rangle_{H_{app}(U,\overline{X},\sigma_0)} = \frac{K_s}{S_0^0} \frac{(R_0)^4}{4} \left[\frac{kT}{K_c}\right]^2 \sum_{n=2}^{n_{max}} \frac{2n+1}{[n(n+1) + \tilde{\sigma}(\sigma_0, \overline{X})]^2}. \tag{48}$$

Our further aim is to estimate the rhs of (48). Its smallness provides a sufficient condition for the replacement of the Hamiltonian $H(U)$ with $H_{app}((U, \overline{X}, \sigma_0))$.

6. Numerical estimations and discussion

If $\langle [\mathcal{A}(U) - \overline{X}]^2 \rangle_{H_{app}(U,\overline{X},\sigma_0)}$ is close enough to zero (see Eq. (39)), we can replace the model Hamiltonian $H(U)$ with the approximating one $H_{app}(U, \overline{X}, \sigma_0)$ and the model free energy $f[H(U)]$ with the approximating one $f[H_{app}(U, \overline{X}, \sigma_0)]$. To make numerical estimations, we use values of the quantities, participating in the developed by us theory that are typical for experiments where analysis of the shape fluctuations of nearly spherical lipid vesicles are carried out [7], namely: $K_s \sim 100$ erg/cm^2; $K_c \sim 10^{-12}$erg; $R_0 \sim 10^{-3}$cm; $S_0^0 \sim 4\pi(R_0)^2 \sim 1.256 \times 10^{-5}$cm^2; $\sigma_1 = 4 \times 10^9$erg$^{-0.5}$. With these values the correlator $\langle[\mathcal{A}(U) - \overline{X}]^2\rangle_{H_{app}(U,\overline{X},\sigma_0)}$ from Eq. (48), normalized by the Boltzmann factor kT, can be presented in the form:

$$\frac{\langle[\mathcal{A}(U) - \overline{X}]^2\rangle_{H_{app}(U,\overline{X},\sigma_0)}}{kT} = 2 \times 10^4 \sum_{n=2}^{n_{max}} \frac{2n+1}{[n(n+1) + \tilde{\sigma}(\sigma_0, \overline{X})]^2}. \tag{49}$$

The use of the approximating Hamiltonian is justified when the correlator is small enough. We assume that this is true when the value of the correlator is less than the Boltzmann factor kT. With these numerical values this is fulfilled when $\overline{\Sigma} \geq 2.10^4$.

If we consider lower values of the stretching elasticity K_s, the lowest value of $\overline{\Sigma}$, for which the use of the approximating Hamiltonian is justifiable, decreases. In the limiting case $K_s \to 0$ (if $\overline{\Sigma}$ is fixed), the approximating Hamiltonian can be used for all values of $\overline{\Sigma}$ satisfying the condition $\overline{\Sigma} > -6$. The last inequality is necessary in order to assure that the factors multiplying the squares $(u_n^m)^2$ of the amplitudes u_n^m of the fluctuation modes (see Eq. (41)) are finite and nonnegative.

7. Conclusion

In the present paper we show that results having the same functional dependences as those of Milner and Safran can be deduced, in accordance with the principles of statistical mechanics, by an approach based on the Bogolyubov inequalities and the approximating Hamiltonian method.

It is proved that there is a value of the dimensionless factor $\overline{\Sigma}$, related to the tension of the membrane (see Eqs. (42) and (47)), above which the application of our approach gives results that are precise enough. This value is much greater than the values appearing in the experiments. The applicability of the results for the interval of typical experimental values of this quantity remains an open question.

From Eqs. (34) and (48) it is clear that when the stretching elasticity K_s of the vesicle membrane tends to zero, keeping all the other quantities fixed, our theory becomes asymptotically exact.

References

[1] Seifert U 1997 *Adv. Phys.* **46** 13
[2] Safran S A 1999 *Adv. Phys.* **48** 395
[3] Nagle J F 2012 *Faraday Discussions* **161** 11
[4] Dimova R 2014 *Adv. Coll. Interf. Sci.* **208** 225
[5] Singer S J and Nicolson G L 1972 *Science* **175** 720
[6] Milner S T and Safran S A 1987 *Phys. Rev. A* **36** 4371
[7] Méléard P, Gerbeaud C, Pott T, Fernandez-Puente L, Bivas I, Mitov M D, Dufourcq J and Bothorel P 1997 *Biophys. J.* **92** 2616
[8] Helfrich W 1973 *Z. Naturforsh.* **28c** 693
[9] Bivas I 2002 *The Europ. Phys. J. B* **29** 317
[10] Janke-Emde-Lösch 1960 *Tafeln Höherer Funktionen* sixth ed (Stuttgart: B. G. Teubner Verlagsgeselschaft)
[11] Brankov J G, Danchev D M and Tonchev N S 2000 *Theory of Critical Phenomena in Finite-Size Systems: Scaling and Quantum Effects* (*Series in Modern Condensed Matter Physics* vol 9) (Singapore: World Scientific)

Journal of Physics: Conference Series **558** (2014) 012021 doi:10.1088/1742-6596/558/1/012021

One-dimensional dark solitons in lattices with higher-order nonlinearities

M T Primatarowa and R S Kamburova

Georgi Nadjakov Institute of Solid State Physics, Bulgarian Academy of Sciences, 1784 Sofia, Bulgaria

E-mail: `prima@issp.bas.bg`

Abstract. The existence of dark solitons in the discrete nonlinear Schrödinger equation with third- and fifth-order nonlinearities is investigated. Effects of discreteness for the homogeneous case are analyzed. Exact analytical solutions are found for wide static dark solitons in the presence of impurities. The bound soliton-defect formations can have a single minimum for attractive impurities or two minima for repulsive impurities. In contrast to the standard cubic nonlinear case, where the positions of the minima do not depend on the nonlinearity, now they are strongly influenced by the quintic nonlinearity. The model plays an important role in numerous physical systems with complicated nonlinear interactions.

1. Introduction

Phenomena associated with the interplay between nonlinearity and disorder are subject of intensive studies due to their importance for various physical systems. Nonlinearity induces the formation of spatially localized solitary waves (solitons), while the presence of impurities in inhomogeneous systems may support spatially localized impurity modes. The competition between the two different mechanism of localization leads to a complicated picture of localized states. The interaction of solitons in the continuum limit with linear [1-4] and nonlinear [5-8] point defects is widely investigated. Soliton trapping on impurities in discrete systems has been studied in [9-12]. These investigations are based on the standard nonlinear Schrödinger (NLS) equation where the nonlinearity is of third order. Considerable interest is also devoted to different variants of the NLS equation, such as the saturable NLS equation [13,14], the equation with power-low nonlinearity [6,15], and the cubic-quintic NLS equation [16-19]. The existence and stability of soliton solutions in these systems have been examined. The models appear in many physical problems of the macroscopic nonlinear dynamics of solids, nonlinear optics, Bose-Einstein condensation etc. [20]. Subject of these investigations are mostly the bright solitons. The existence of dark solitons ("holes" on a continuous wave background) in systems with various types of complicated nonlinear interactions is also important.

The present work is devoted to the study of dark solitary excitations in a nonlinear cubic-quintic lattice. The influence of the discreteness and the fifth-order nonlinearity on the soliton dynamics is analyzed. The interaction of the dark solitons with point defects in the static case is investigated.

2. Dark solitons in the homogeneous cubic-quintic lattice

We consider the dynamics of quasiparticles (e.g. phonons, excitons, magnons) with the amplitude $\alpha_n(t)$ at site n in an inhomogenious chain. When the density of the quasiparticles is high enough, their interaction should be taken into account and the system is described by the following set of nonlinear ordinary differential equations in the nearest-neighbour approximation:

$$i\frac{\partial \alpha_n}{\partial t} = -b\,(\alpha_{n+1} + \alpha_{n-1} - 2\alpha_n) + 2g\,|\alpha_n|^2\alpha_n + 3f\,|\alpha_n|^4\alpha_n + d\,\delta_{n,0}\,\alpha_n\,, \tag{1}$$

where b is the exchange interaction, g and f are the cubic and quintic nonlinear interactions, respectively. d characterizes a local linear defect of repulsion or attraction. We shall look for solutions in the form of amplitude-modulated waves

$$\alpha_n(t) = \varphi_n(t)\, e^{i(kn - \omega t)}\,, \tag{2}$$

where k and ω are the wave vector and the frequency of the carrier wave (the lattice constant equals unity) and the envelope $\varphi_n(t)$ is a real slowly varying function of the position and time. In the continuum limit $\varphi_n(t) \to \varphi(x,t)$, equation (1) transforms into the following perturbed NLS equation with cubic-quintic nonlinearities:

$$i\frac{\partial \varphi}{\partial t} = [2b\,(\cos k - 1) - \omega - d\,\delta(x)]\varphi - i2b\,\sin k\,\frac{\partial \varphi}{\partial x} - b\,\cos k\,\frac{\partial^2 \varphi}{\partial x^2} + 2g\,|\varphi|^2\varphi + 3f\,|\varphi|^4\varphi\,. \tag{3}$$

First we consider the homogeneous chain ($d = 0$). For nonvanishing boundary conditions $|\varphi| \to \varphi_0$ at $x \to \pm\infty$ equation (3) possesses a dark soliton solution of the form [21]

$$\varphi(x,t) = \varphi_0\sqrt{B}\,\sinh\xi\,(1 + B\sinh^2\xi)^{-1/2}\,, \quad \xi = \frac{x - vt}{L}$$
$$L^2 = \frac{b\,\cos k}{g\,\varphi_0^2 + 3f\,\varphi_0^4}\,, \quad B = 1 - \frac{f\,\varphi_0^2}{g + 3f\,\varphi_0^2}\,,$$
$$\omega = 2b\,(\cos k - 1) + 2g\,\varphi_0^2 + 3f\varphi_0^4\,, \quad v = 2b\,\sin k\,. \tag{4}$$

L and v are the width and the velocity of the soliton. B determines the deviation of the amplitude from its form for $f = 0$ and has to be positive ($B > 0$). The maximum intensity of the soliton is $|\varphi(\pm\infty)|^2 = \varphi_0^2$, while the minimum intensity is $|\varphi(0)|^2 = 0$.

For $f = 0$ ($B = 1$) the solution (4) turns into the well-known dark soliton of the cubic NLS equation:

$$\varphi(x,t) = \varphi_0\tanh\frac{x - vt}{L}$$
$$L^2 = \frac{b\,\cos k}{g\,\varphi_0^2}\,, \quad \omega = 2b\,(\cos k - 1) + 2g\,\varphi_0^2\,, \quad v = 2b\,\sin k\,. \tag{5}$$

When quintic nonlinearity is present ($f \neq 0$) the conditions $L > 0$ and $B > 0$ i.e.

$$\frac{b\,\cos k}{g + 3f\,\varphi_0^2} > 0\,, \quad \frac{g + 2f\,\varphi_0^2}{g + 3f\,\varphi_0^2} > 0 \tag{6}$$

have to be fulfilled. The inequalities (6) determine the range of values of the coefficients b, g, and f for which the solution (4) exists. For example if $b\,\cos k > 0$ the allowed values of f are $f > -g/3\,\varphi_0^2$ for $g > 0$ and $f > -g/2\,\varphi_0^2$ for $g < 0$. In contrast to the solution of the cubic NLS equation (5) the parameter g can have positive as well as negative values. For $g > 0$ we can choose f positive which leads to $B < 1$ or f negative ($B > 1$). For g negative we can choose only positive values of f ($B < 1$). The dependence of the quantities L and B on the quintic nonlinear coefficient f for different values of g is shown on figure 1.

Note that the soliton velocity v does not depend on the nonlinearity as in the case of bright solitons [19].

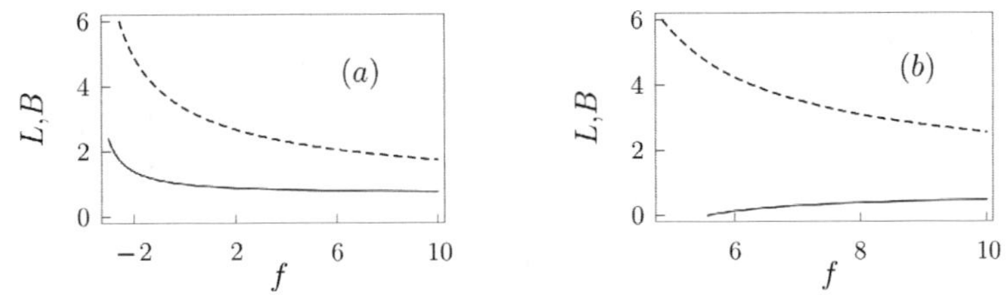

Figure 1. Dependence of the quantities L (dashed curves) and B (solid curves) on the nonlinearity coefficient f. (a): $g = 1$; (b): $g = -1$.

3. Solitons in the homogeneous quintic lattice

In this section we shall consider the case when only the quintic nonlinearity is present ($g = 0, f \neq 0$). Then $B = 2/3$ and the dark soliton solution has the form

$$\alpha_n(t) = \varphi_0 \sqrt{2} \sinh \xi \, (3 + 2 \sinh^2 \xi)^{-1/2} \, e^{i(kn - \omega t)} \,, \quad \xi = \frac{n - vt}{L}$$

$$L^2 = \frac{b \cos k}{3 f \, \varphi_0^4} \,, \qquad \omega = 2b \, (\cos k - 1) + 3 f \varphi_0^4 \,, \quad v = 2b \, \sin k \,. \tag{7}$$

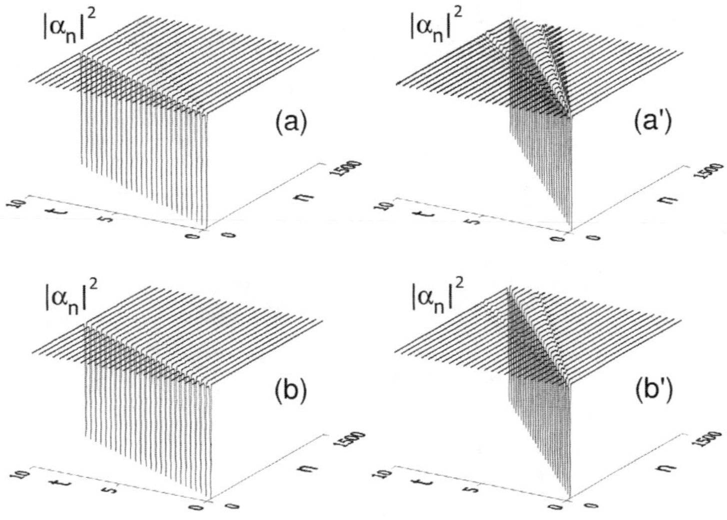

Figure 2. Propagation of dark solitons with $\varphi_0 = 0.3$, $L = 5$, $b = 1$. (a): $g = 1$, $f = -2.14$; (a$'$): $g = 1$, $f = -2.54$; (b): $g = 0$, $f = 1.57$; (b$'$): $g = 0$, $f = 1.16$. $k = 0.316$ for (a) and (b); $k = 0.787$ for (a$'$) and (b$'$). The time is in units of $100/b$.

The propagation of the dark
(velocities). For small values of k both
of the velocity the discreteness effects
background which for the cubic-quin
pure quintic one [figures 2(b$'$)]. The
parameters remain the same) as in
$L \gg 1$ is not valid any more.

figure 2 for different k values
2(a) and (b)]. With increase
is a small radiation from the
[figures 2(a$'$)] than for the
amplitude φ_0 increases (other
narrower and the condition

Journal of Physics: Conference Series **558** (2014) 012021 doi:10.1088/1742-6596/558/1/012021

We like to point out that the nonlinear terms in equation (1) can result from different models with complicated interactions. They are also a good approximation in some cases of the saturable nonlinearity of the form

$$c\,\frac{\alpha_n}{1+|\alpha_n|^2}\,, \qquad c = const\,. \tag{8}$$

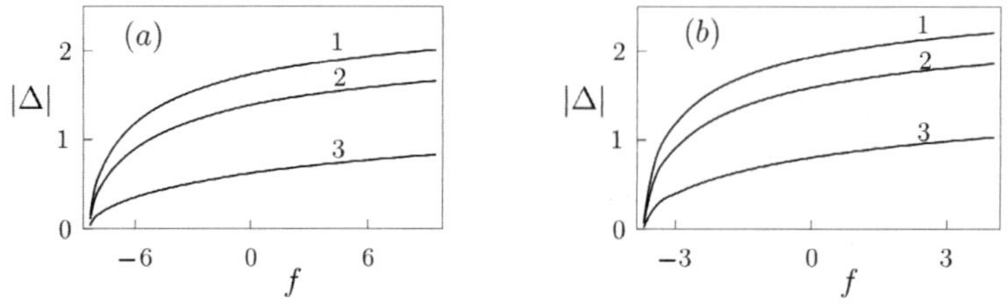

Figure 3. Dependence of Δ on the nonlinear coefficient f and the defect strength d for $b = g = 1$ and (a) $\varphi_0 = 0.2$; (b) $\varphi_0 = 0.3$. Curves 1: $|d| = 0.05$, curves 2: $|d| = 0.1$, and curves 3: $|d| = 0.5$.

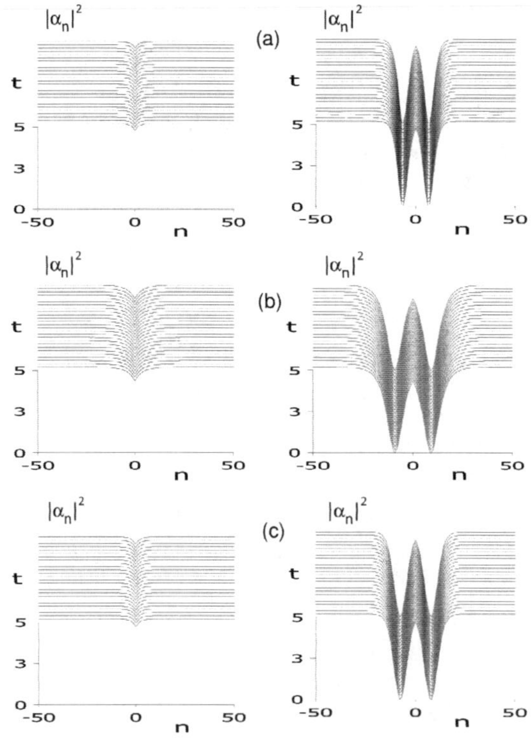

Figure 4. Evolution of the bound soliton-impurity solution (8) with $\varphi_0 = 0.3$ for (a) $g = 1$, $f = 0$; (b) $g = 1$, $f = -3$; (c) $g = 0$, $f = 3$;. The left pictures hold for attractive impurities with $d = 0.05$ ($\Delta > 0$), while the right pictures hold for repulsive impurities with $d = -0.05$ ($\Delta < 0$).

4. Interaction of dark solitons with impurities in the cubic-quintic lattice

Now we consider the inhomogeneous static case $d \neq 0$, $k = v = 0$. We found that in the continuum approximation used ($L \gg 1$) equation (1) has the following bound dark soliton-defect solution:

$$\alpha_n(t) = \varphi_0 \sqrt{B} \sinh \xi \, (1 + B \sinh^2 \xi)^{-1/2} \, e^{-i\omega t}, \quad \xi = \frac{|n|}{L} + \Delta$$

$$L^2 = \frac{b}{g\,\varphi_0^2 + 3f\,\varphi_0^4}, \quad B = 1 - \frac{f\,\varphi_0^2}{g + 3f\,\varphi_0^2}, \quad \omega = 2g\,\varphi_0^2 + 3f\varphi_0^4 \tag{9}$$

with the following relation for the determination of Δ

$$\tanh \Delta (1 + B \sinh^2 \Delta) = 2b/dL\,. \tag{10}$$

For $\Delta > 0$ the function $|\alpha_n(t)|$ has a single minimum at $n = 0$. For $\Delta < 0$ there is a maximum at $n = 0$ between two minima at $n = \pm \Delta L$.

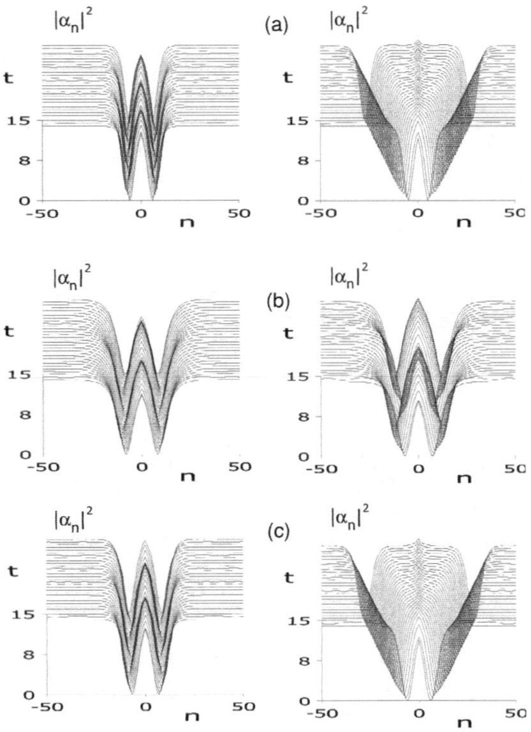

Figure 5. Evolution of the dark soliton bound on a repulsive impurity ($d = -0.05$) with initial Δ_0 and $\varphi_0 = 0.3$ for (a) $g = 1$, $f = 0$; (b) $g = 1$, $f = -3$; (c) $g = 0$, $f = 3$. The left pictures hold for $\Delta_0 = 0.9\Delta$, while the right pictures hold for $\Delta_0 = 0.8\Delta$.

Figure 3 shows the dependence of Δ on f. Similar bound soliton solutions have been obtained for point defects in the cubic NLS equation [3,22]. For $f = 0$ ($B = 1$) we obtain from (10) the simple expression for Δ

$$\sinh 2\Delta = 4b/dL\,. \tag{11}$$

The evolution of the dark soliton bound on an impurity for different values of the nonlinear coefficients is shown on figure 4. For constant amplitude φ_0 and impurity strength $|d|$ the parameters of the form (9) are $L = 3$, $|\Delta| = 1.94$ [figure 4(a)], $L = 8$, $|\Delta| = 1.18$ [figure 4(b)] and $L = 4$, $|\Delta| = 2.07$ [figure 4(c)]. The single minimum form of the bound state corresponds to attraction on the impurity (left pictures), while the two minima form corresponds to repulsion on the impurity (right pictures).

We analyze the stability of the initial form with a Δ_0 which differs from the exact value Δ. For attractive impurities ($\Delta > 0$) the solution is stable at $\Delta_0 \neq \Delta$. For repulsive impurities ($\Delta < 0$) the solution is unstable (figure 5). Its begins to oscillate for small perturbations ($\Delta_0 = 0.9\Delta$, figure 5, left pictures) or splits into two solitons which propagate with opposite velocities for larger perturbations and narrow solitons [$\Delta_0 = 0.8\Delta$, figure 5(a) and (c), right pictures]. The process depends strong on the soliton width L. For wider solutions ($L = 8$) a relative large perturbation ($\Delta_0 = 0.8\Delta$) leads still to oscillations [figure 5(b), right] in contrast to the cases with $L = 3$ [figure 5(a), right] and $L = 4$ [figure 5(c), right].

5. Conclusion

We have investigated the existence and stability of dark solitons in a discrete nonlinear chain with cubic-quintic nonlinearities. The interaction of solitons with impurities was studied as well. We have obtained analytical bound soliton-impurity solutions for wide solitons in the static case. The stability of the dark solitons is studied numerically and the influence of the different nonlinear coefficients is shown.

Acknowledgments

This work is partially supported by the EU FP7 funded project INERA (Grant agreement no: 316309).

References

[1] Kivshar Yu S, Kosevich M A and Chubykalo O A 1987 *Zh. Eksp. Teor. Fiz.* **93** 968-77
[2] Cao X D and Malomed B A 1995 *Phys. Lett.* A **206** 177-82
[3] Pushkarov D I and Atanasov R D 1990 *Phys. Lett.* A **149** 287-90
[4] Boardman A D, Bortolani V, Wallis R F, Xie K and Mehta H M 1995 *Phys. Rev.* B **52** 12736-42
[5] Burtsev S, Kaup D J and Malomed B A 1995 *Phys. Rev.* E **52** 4474-81
[6] Sukhorukov A A, Kivshar Yu S, Bang O, Rasmussen J J and Christiansen P I 2001 *Phys. Rev.* E **63** 036601(18)
[7] Kevrekidis P G, Kivshar Yu S and Kovalev A S 2003 *Phys. Rev.* E **67** 046604(8)
[8] Primatarowa M T, Kamburova R S and Stoychev K T 2007 *J. Optoel. Adv. Mat.* **9** 152-4
[9] Scharf R and Bishop A R 1991 *Phys. Rev.* A **43** 6535-44
[10] Konotop V V, Cai D, Salerno M, Bishop A R and Grnbech-Jensen N 1996 *Phys. Rev.* E **53** 6476-85
[11] Królikowski W and Kivshar Yu S 1996 *J. Opt. Soc. Am.* B **13** 876-87
[12] Palmero F, Carretero-Gonzlez R, Cuevas J, Kevrekidis P G and Krolikowski W 2008 *Phys. Rev.* E **77** 036614(11)
[13] Jakubowski M H, Steiglitz K and Squier R 1997 *Phys. Rev.* E **56** 7267-72
[14] Calvo G F, Belmonte-Beitia J and Pérez-García V M 2009 *Chaos, Solitons and Fractals* **41** 1791-8
[15] Laedke E W, Kluth O and Spatschek K H 1996 *Phys. Rev.* E **54** 4299-311
[16] Zhou C, He X T and Chen S 1992 *Phys. Rev.* A **46** 2277-85
[17] Schürmann H W 1996 *Phys. Rev.* E **54** 4312-20
[18] Tang X-Y and Shukla P K 2007 *Phys. Rev.* A **76** 013612(10)
[19] Primatarowa M T and Kamburova R S 2010 *AIP Conf. Proc.* **1203** 240-5
[20] Kartashov Y V, Malomed B A and Torner L 2011 *Rev. Mod. Phys.* **83** 247-305
[21] Pushkarov K I and Primatarowa M T 1984 *phys. stat. sol.* (b) **123** 573-84
[22] Primatarowa M T and Kamburova R S 2013 *Rom. Rep. Phys.* **65** 374-82

Soliton-impurity interaction in two Ablowitz-Ladik chains with different coupling

R S Kamburova and M T Primatarowa

Georgi Nadjakov Institute of Solid State Physics, Bulgarian Academy of Sciences, 1784 Sofia, Bulgaria

E-mail: krad@issp.bas.bg

Abstract. The interaction of solitons with point defects in a system of coupled Ablowitz-Ladik (AL) chains is studied numerically. The system is a discrete analog of coupled nonlinear Schrödinger equations. Two types of interchain coupling are investigated: one which admits reduction of the system to the standard integrable AL model (dispersive coupling) and one which couples opposite sites of the chains and does not admit reduction to the AL model (nondispersive coupling). The action of the two coupling types is additive and they can compensate each other in some cases. We have obtained that the single-peak bound soliton-defect solution (attractive impurity) is stable against perturbations, while the double-peak bound soliton-defect solution (repulsive impurity) is unstable and can be easily destroyed. Linear point defects do not influence the period of energy transfer and it is close to the period for the homogeneous case.

1. Introduction

The study of nonlinear waves is receiving much attention due to the potential application within different branches of physics from nonlinear optics to Bose-Einstein condensate. The interplay between discrete diffraction and nonlinearity leads to the formation of discrete solitons. They promise an efficient way to control switching of optical signals in a system of coupled waveguides. So waveguide-based devices have received considerable attention in literature and this field has been extensively explored theoretically and experimentally [1,2]. A discrete coupler involving two waveguides which exchange power as a result of weak overlap of their evanescent fields is the basic realization of a waveguide switch. The rate of power swapped back and forth between waveguides depends on the strength of the coupling, the degree of similarity of the waveguides and the initial pulse energy [3-7]. Waveguide arrays are particularly interesting because of their possible applications in signal processing [8-12]. All considerations regard discrete soliton switching mainly in homogenous waveguides with constant or smoothly varying coupling between them. However in practical applications the properties of inhomogeneous waveguides are more interesting and rather inevitable for switching [13-17].

Widely investigated are the standard discrete nonlinear Schrödinger (NLS) equation, as well as the completely integrable discrete Ablowitz-Ladik (AL) equation [18-20]. Although the two equations have the same linear properties and yield the same NLS equation in the continuum limit, their nonlinear properties are different. This leads to differences in the dynamics of narrow solitons (bright or dark) for the two models. Soliton solutions in two coupled discrete nonlinear chains were found and their stability was investigated in [21-24].

In the present paper we study the interaction of solitons with impurities in two Ablowitz-Ladik chains with a complicated coupling that includes linear, nonlinear and dispersive interactions between the chains.

2. The model

We shall consider two parallel chains of particles described by the following system of coupled Ablowitz-Ladik equations:

$$\begin{aligned} i\frac{\partial\alpha_n}{\partial t} &= M(\alpha_{n+1}+\alpha_{n-1})(1+\gamma|\alpha_n|^2) + [d_1(\beta_{n+1}+\beta_{n-1}) + 2d_2\beta_n](1+\gamma|\alpha_n|^2) + \varepsilon\delta_{n,n_0}\alpha_n \\ i\frac{\partial\beta_n}{\partial t} &= M(\beta_{n+1}+\beta_{n-1})(1+\gamma|\beta_n|^2) + [d_1(\alpha_{n+1}+\alpha_{n-1}) + 2d_2\alpha_n](1+\gamma|\beta_n|^2) + \varepsilon\delta_{n,n_0}\beta_n \end{aligned} \quad (1)$$

$\alpha_n(t)$ $[\beta_n(t)]$ is the amplitude of an excitation at site n of the first (second) chain, interacting with an impurity of the strength ε localized at the point n_0. M is the coupling interaction between neighboring particles in one and the same chain. The two chains are coupled to each other through the real parameters d_1 and d_2. The parameter d_2 governs the interchain coupling between opposite sites (nondispersive), while d_1 is a coupling of the AL type and allows reduction of the system to the AL model which is completely integrable. Both interactions include linear and nonlinear terms. The parameter γ determines the type of the soliton solution (bright for $\gamma > 0$ and dark for $\gamma < 0$) of the AL equation. In what follows we consider only bright solitons and set $\gamma = 1$ due to the scaling property of the AL system.

First we shall consider the homogeneous case ($\varepsilon = 0$). We briefly outline the influence of the two different couplings on the soliton properties [24]. Equation (1) can be derived from the Hamiltonian

$$\begin{aligned} H &= \sum_n [M(\alpha_n\alpha_{n-1}^* + \alpha_n^*\alpha_{n-1} + \beta_n\beta_{n-1}^* + \beta_n^*\beta_{n-1}) \\ &+ d_1(\alpha_n\beta_{n-1}^* + \alpha_n^*\beta_{n-1} + \alpha_{n-1}\beta_n^* + \alpha_{n-1}^*\beta_n) + 2d_2(\alpha_n\beta_n^* + \alpha_n^*\beta_n)] \end{aligned} \quad (2)$$

using the deformed Poisson brackets [19,20]

$$\begin{aligned} \{\alpha_n, \alpha_m^*\} &= i(1+|\alpha_n|^2)\delta_{n,m}\,, \quad \{\beta_n, \beta_m^*\} = i(1+|\beta_n|^2)\delta_{n,m} \\ \{\alpha_n, \alpha_m\} &= \{\alpha_n^*, \alpha_m^*\} = 0\,, \quad \{\beta_n, \beta_m\} = \{\beta_n^*, \beta_m^*\} = 0 \end{aligned} \quad (3)$$

and the equations of motion

$$\frac{\partial\alpha_n}{\partial t} = \{H, \alpha_n\}\,, \quad \frac{\partial\beta_n}{\partial t} = \{H, \beta_n\}\,. \quad (4)$$

The system (1) is nonintegrable but has two integrals of motion, the Hamiltonian H and the total number of particles N

$$N = \sum_n [\ln(1+|\alpha_n|^2) + \ln(1+|\beta_n|^2)]\,. \quad (5)$$

For $d_2 = 0$ and the symmetric reduction $\alpha_n(t) \equiv \beta_n(t)$ the system (1) turns in an AL equation with the well known bright soliton solution:

$$\begin{aligned} \alpha_n(t) &= \beta_n(t) = \sinh\frac{1}{L}\,\mathrm{sech}\frac{n-vt}{L}\,e^{i(kn-\omega t)} \\ v &= -2(M+d_1)L\sinh\frac{1}{L}\sin k\,, \quad \omega = 2(M+d_1)\cosh\frac{1}{L}\cos k \end{aligned} \quad (6)$$

The parameters k (wavenumber) and L (width) determine the velocity v and frequency ω of the soliton. In this case the conserved quantities have the form

$$H = 8(M + d_1)\sinh\frac{1}{L}\cos k\,, \quad N = 4/L \tag{7}$$

and it holds $\omega = \partial H/\partial N$.

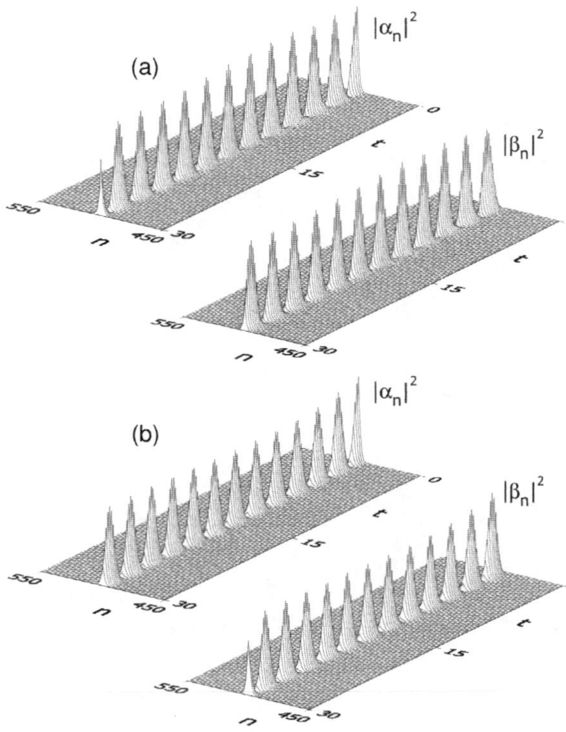

Figure 1. Switching of the bound soliton-defect solution for $M = -1$, $L = 2$, $\varepsilon = -0.32$ (attraction) and (a): $|d_1| = 0.628$, $d_2 = 0$; (b): $d_1 = 0$, $|d_2| = 0.628$. The time is in units $1/|M|$.

3. Soliton-impurity interaction

Now we consider the inhomogeneous static case $\varepsilon \neq 0$, $k = v = 0$. For $d_1 = d_2 = 0$ the system (1) decomposes into two standard AL equations which possess the following bound bright soliton-defect solution:

$$\alpha_n(t) = \beta_n(t) = \sinh\frac{1}{L}\,\mathrm{sech}\left(\frac{|n-n_0|}{L} + \Delta\right)e^{-i\omega t}\,,$$
$$\omega = 2M\cosh\frac{1}{L}\,, \quad \tanh\Delta = \varepsilon/\left(2M\sinh\frac{1}{L}\right)\,. \tag{8}$$

For $\Delta > 0$ the function $|\alpha_n(t)|$ has a single maximum at $n = n_0$ (single pick solution). For $\Delta < 0$ the function $|\alpha_n(t)|$ has two maxima at $n = n_0 \pm \Delta L$ (double pick solution).

We shall investigate the evolution of an AL bound soliton which at the initial time is launched in one of the chains

$$\alpha_n(0) = \sinh\frac{1}{L}\,\mathrm{sech}\left(\frac{|n-n_0|}{L} + \Delta\right)\,, \quad \beta_n(0) = 0 \tag{9}$$

solving numerically the system (1). The simulations are carried out for 1000 sites of each chain, $n_0 = 500$ and periodic boundary conditions. The width of the solitons is $L = 2$, which underline the discreteness of the system. For wide solitons we can use the continuum approximation and (1) turns in a system of coupled NLS equations. $M = -1$.

In the homogeneous linear case ($\varepsilon = 0$, $\gamma = 0$) the excitation will transfer from one chain to the other and back with a period $t_0 = \pi/|d|$, where d is the total linear coupling.

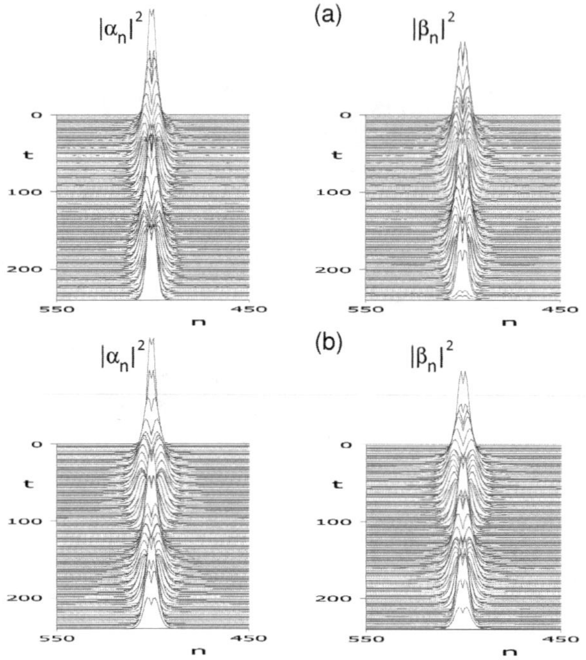

Figure 2. Switching of the bound soliton-defect solution with $\varepsilon = 0.32$ (repulsion) for (a): $|d_1| = 0.628$, $d_2 = 0$; (b): $d_1 = 0$, $|d_2| = 0.628$.

For our complicated homogeneous model ($\varepsilon = 0$, $\gamma = 1$), which is not linear we observe that an energy exchange between the two chains take place with nearly the same period

$$t_0 = \frac{\pi}{2|d_1 + d_2|} \tag{10}$$

and the energy exchange rate depends on the strength of the coupling. For small values of the coupling constants the soliton is only partially transferred. When the coupling increases the transferred rate grows. This behavior is due to the nonlinear coupling terms. We obtained that a soliton can transfer (perfect soliton switching) when the simple condition

$$4|d_1 + d_2|L^2 \gg 1 \tag{11}$$

is fulfilled.

We have observed the differences between the two types of interchain coupling. For large values of the coupling constants the soliton switching is preserve for the coupling between opposite sites of the chains d_2. For the dispersive coupling d_1 an energy exchange with the same period t_0 takes place but the soliton formation is destroyed. Obviously to achieve soliton switching $|d_1| < |M|$ has to be fulfilled. We have obtained that for wide solitons ($L \gg 1$) the dispersive character of the coupling d_1 vanish and the solitons in the two chains are stable [24].

Figure 1 shows the influence of the attractive impurity over the soliton transfer when the coupling between the two chains is of type d_1 [figure 1(a)] or d_2 [figure 1(b)]. The process does not depend on the sign of the coupling. We have obtained that the single-peak bound soliton-defect solution is stable against perturbations. The period of the energy transfer is 2.5 for both cases and is in a good agreement with the value calculated from (10).

Figures 2 and 3 show the evolution of a double pick bound soliton-defect solution (repulsive impurity) which is unstable and can be easily destroyed. It begins to oscillate for small defects (figure 2) or splits into two solitons which propagate with opposite velocities (figure 3). Figure 4 demonstrates the interplay between the two types of coupling. Their action is additive. When the constants d_1 and d_2 are of the same sign [figure 4(a)] the soliton switching is similar to that for one coupling's type but with the greater strength. Evidence for this behavior is the magnitude of the transfer period and the oscillations period. When the two couplings are of different signs the resulting coupling decreases and can be set to nearly zero as shown on figure 4(b). We observe that the soliton α_n remains in the first chain where it was initially excited and in the second chain $\beta_n \simeq 0$.

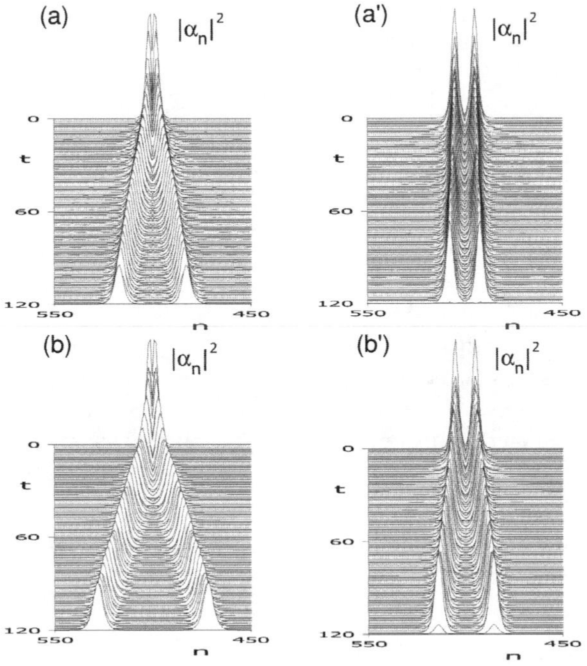

Figure 3. Splitting of the bound soliton-defect solution $|\alpha_n|$ for a repulsive defect with $\varepsilon = 0.64$ [(a), (b)]; $\varepsilon = 1.03$ [(a′), (b′)]. $|d_1| = 0.628$, $d_2 = 0$ [(a), (a′)] and $d_1 = 0$, $|d_2| = 0.628$ [(b), (b′)]. The evolution of the $|\beta_n|$ function is similar.

4. Conclusion

We have studied the interaction of discrete solitons with impurities in a system of coupled Ablowitz-Ladik chains. Two types of interchain coupling are introduced and compared. The coupling of the AL type admits reduction of the system to an AL model. Perfect soliton switching

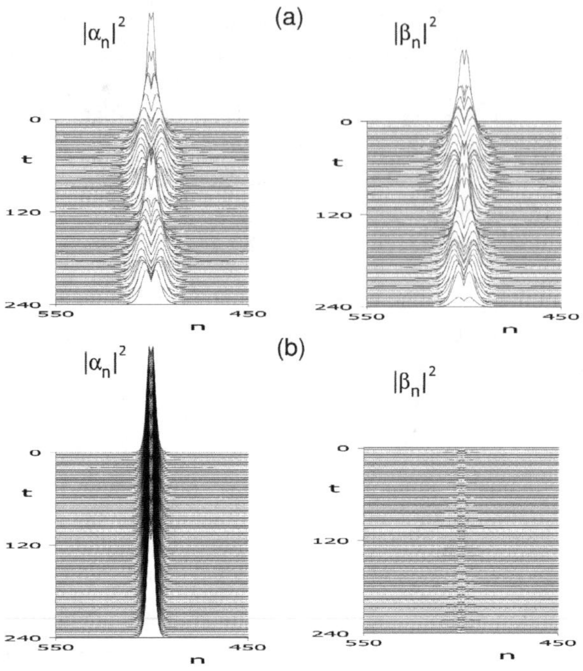

Figure 4. Time evolution of the bound soliton-defect solution for a repulsive defect with $\varepsilon = 0.32$ and (a): $d_1 = d_2 = 0.314$; (b): $d_1 = -d_2 = 0.314$.

can be obtained for large enough values of the coupling constant in both cases. In the case of dispersive coupling the soliton is destroyed when the coupling constant exceed a critical value which depends on the soliton width and the interaction between the neighboring particles in the chains. We have obtained that the single-peak bound soliton-defect solution (attractive impurity) is stable against perturbations and energy transfer period is preserved, while the double-peak bound soliton-defect solution (repulsive impurity) is unstable and can be easily destroyed.

Acknowledgments
This work is partially supported by the EU FP7 funded project INERA (Grant agreement no: 316309).

References

[1] Garanovich I L, Longhi S, Sukhorukov A A and Kivshar Y S 2012 *Phys. Rep.* **518** 1-79
[2] Joannopoulos J D, Johnson S G, Winn J N and Meade R D 2008 Photonic Crystals: Molding the Flow of Light (Prinston University Press)
[3] Jensen S M 1982 *IEEE J. Quantum Electron.* **18** 1580-3
[4] Trillo S, Wabnitz S and Stegeman G 1989 *IEEE J. Quantum Electron.* **25** 1907-16
[5] Paré C and Florjańczyk M 1990 *Phys. Rev.* A **41** 6287-95
[6] Kivshar Y S 1993 *Optics Lett.* **18** 7-9
[7] Chu P L, Peng G D and Malomed B A 1993 *Optics Lett.* **18** 328-30
[8] Aceves A B, De Angelis C, Peschel T, Muschall R, Lederer F, Trillo S and Wabnitz S 1996 *Phys. Rev.* E **53** 1172-89
[9] Christodoulides D N and Eugenieva E D 2001 *Phys. Rev. Lett.* **87** 233901(4)

Journal of Physics: Conference Series **558** (2014) 012022 doi:10.1088/1742-6596/558/1/012022

[10] Peschel U, Morandotti R, Arnold J M, Aitchison J S, Eisenberg H S, Silberberg Y, Pertsch T and Lederer F 2002 *J. Opt. Soc. Am.* A **19** 2637-44
[11] Vicencio R A, Molina M I and Kivshar Yu S 2003 *Optics Lett.* **28** 1942-4
[12] Molina M I, Vicencio R A and Kivshar Yu S 2006 *Optics Lett.* **31** 1693-5
[13] Królikowski W and Kivshar Yu S 1996 *J. Opt. Soc. Am.* B **13** 876-87
[14] Longhi S 2006 *Phys. Rev.* E **74** 026602(9)
[15] Morales-Molina L and Vicencio R A 2006 *Optics Lett.* **31** 966-8
[16] Goodman R H and Weinstein M I 2008 *Physica* D **237** 2731-60
[17] Belmonte-Beitia J, Pérez-García V M and Torres P J 2009 *Nonlin. Science* **19** 437-51
[18] Ablowitz M J and Ladik J F 1976 *J. Math. Phys.* **17** 1011-8
[19] Scharf R and Bishop A R 1991 *Phys. Rev.* A **43** 6535-44
[20] Cai D, Bishop A R and Grønbech-Jensen N 1996 *Phys. Rev.* E **53** 4131-6
[21] Bülow A, Hennig D and Gabriel H 1999 *Phys. Rev.* E **59** 2380-92
[22] Malomed B A and Yang J 2002 *Phys. Lett.* A **302** 163-70
[23] Kamburova R S and Primatarowa M T 2010 *AIP Conf. Proc.* **1203** 261-6
[24] Primatarowa M T and Kamburova R S 2012 *J. Phys.: Conf. Series* **398** 012041(6)

Surface plasmon resonance on the surface: metal - liquid crystal layer

K Zhelyazkova[1], M Petrov[2], B Katranchev[2], G Dyankov[1]

[1] Physics Department, Plovdiv University "Paisii Hilendarski", Bulgaria
[2] Institute of Solid State Physics, BAS, Sofia, Bulgaria

E-mail: katiajeliazkova@abv.bg

Abstract. Surface plasmon resonance (SPR) is widely used in different types of optical detection schemes and for light manipulation at sub-wavelength scale. Usually Kretschmann configuration is used for effective SPR excitation. There are a lot of experimental and theoretical investigations about the influence of the dielectric, adjacent to the metal, on SPR. However, till now the influence on liquid crystal layer, adjacent to the metal, is not considered to our best knowledge. The purpose of the present paper is to cover this gap. We simulate the influence of thin layer of liquid crystal, adjacent to the metal, on the SPR characteristics. For this purpose Maxwell equations are numerically solved for layer structure: prism/gold/liquid crystal/air. The light reflection spectra are calculated for chiral structure of ferroelectric smectic C (SmC*) liquid crystal layer. The angular and wavelength response are considered. Special attention is paid to SPR excitation for variation of tilt angle and different angle of incident light. The influence of SPR of the pitch length and cell thickness is also considered.

1. Introduction

Plasmonics are envisaged as a platform for future integrated photonics architectures which have the unique capability of providing sub-wavelength light confinement [1]. The functionality required for real light routing and switching applications also demand circuit elements that can be externally tuned via control mechanism, such as thermo-, electro-optic or non-linear effects. In this context, a family of tunable photonic devices has been based on the use of nematic liquid crystals [LC], inherently anisotropic materials, whose properties can be controlled by applying an electric field. Their high optically anisotropy and their extremely low power consumption have favored their use in numerous photonic platforms [2], such as tunable resonators [3], gratings [4], photonic crystals [5], or phase shifters [6], and they are also currently being explored as an efficient means of controlling nanoplasmonic resonant structures [7]. In the context of guided-wave plasmonics, LC-tunable waveguides of various designs have been proposed for low-power variable attenuators [8], phase-shifters [9], switches [10], filters [11], tunable lenses [12], beam steering [13], or modulators [14]. An important application of guided-wave plasmonics structures is sensor applications.

We focus our attention on simulation of influence of liquid crystal layer on SPR excitation which could increase the field of application. The simulation is performed for p-polarized light and for ferroelectric liquid crystal (FLC) in a chiral smectic C phase (SmC*). The SmC* structure is one in which the molecules in each layer are tilted with respect to the layer normal by an angle θ. The tilt angle projection on the layer`s plane slowly processes about a direction parallel to the layer normal resulting in a helical structure with a pitch p [15]. Our simulations are based on a theoretical model obtained by solving Maxwell equations in 4x4 matrix form for an anisotropic media [16].

We consider SmC* liquid cell which was studied in details in [17]. Regarding the structure of liquid crystal cell and molecules orientation we use the notations adopted in [17].

2. Reflection spectra at metal/ ferroelectric SmC* LC

In our simulations the physical parameters are as follows: prism (ε_p = 2.89), gold thin film with thickness d = 50nm, anisotropic layer with thickness d = 500nm and d = 1400nm, air. The permittivity of gold is according [18].

2.1. Anisotropic layer with thickness d = 500nm

To reveal the specific influence of LC layer on SPR excitation, firstly we study the plasmon resonance with usual anisotropic layer, adjacent to Au layer, with permittivity equal to LC permittivity.

2.1.1. Reference anisotropic layer with thickness d = 500 nm

This simulation clearly identifies the plasmon resonance. Figure 1 shows the reflection spectra for p-polarized light as a function of incident angle at incident wavelength λ = 632.8nm. The resonance observed at incident angle $\varphi = 66^{\circ}$ has a width of few degree, typical for surface plasmon excited at the gold/anisotropic layer interface. Observed resonance at incident angle $\varphi = 46^{\circ}$ is narrow, typically for waveguide mode propagating in the anisotropic layer. It is worthy to mention that for reference anisotropic layer with $\varepsilon_x = 2$, $\varepsilon_y = 2$, $\varepsilon_z = 3$ plasmon mode cannot be excited.

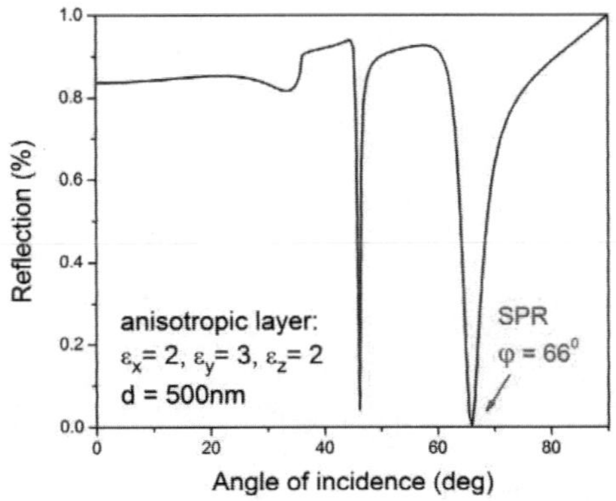

Figure 1. Reflection spectra for p-polarization light as a function of incident angle. Configuration: prism (ε_p = 2.89), Au layer with thickness d = 50nm, anisotropic layer ($\varepsilon_x = 2$, $\varepsilon_y = 3$, $\varepsilon_z = 2$) with thickness d = 500nm, air, for incident wavelength λ = 632.8nm.

2.1.2. Ferroelectric SmC LC layer with thickness d = 500nm*

Now we compare LC layer with reference anisotropic layer. The both layers have the same thickness and permittivity. The specific parameters of LC layer are pitch p = 500 nm (e.g. number of period Np = 1) and tilt angle θ (in notation of [17]). Figure 2 is analogues to figure 1 in the case of ferroelectric SmC* LC layer. The corresponding information about structure parameters is reported in the caption.

The feature of SmC* LC layer (helical structure with a pitch p) leads to plasmon resonance shift (figure 2a) compared with reference anisotropic layer (figure 1) from 66 deg to 68 deg. The other resonances in figure 2a are waveguide modes. The tilt angle $\theta = 90^{\circ}$ (figure 2b) strongly influences SPR: the plasmon resonance disappears. This can be explained simply – a molecule rotation converge the anisotropic layer with $\varepsilon_x = 2$, $\varepsilon_y = 3$, $\varepsilon_z = 2$ to layer with $\varepsilon_x = 2$, $\varepsilon_y = 2$, $\varepsilon_z = 3$ were plasmon mode cannot be excited as mention in point 2.1.1.

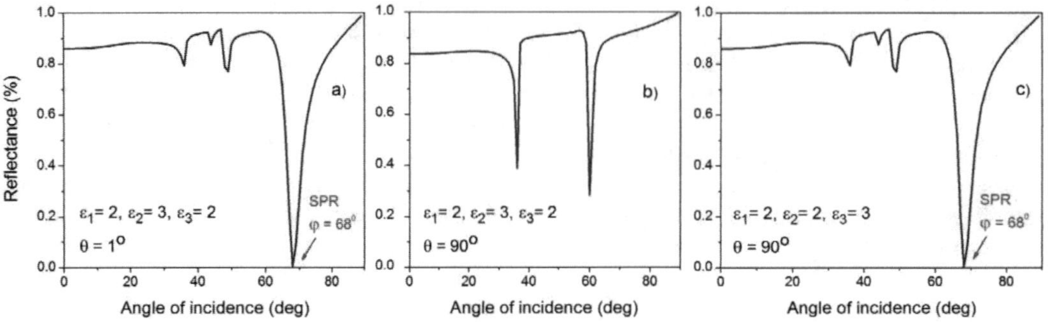

Figure 2. Reflection spectra as a function of incident angle φ for different configuration: LC ($\varepsilon_1 = 2$, $\varepsilon_2 = 3$, $\varepsilon_3 = 2$), $p = 500$nm, $Np = 1$ for tilt angle a) $\theta = 1^{\circ}$ and b) $\theta = 90^{\circ}$; c) LC ($\varepsilon_1 = 2$, $\varepsilon_2 = 2$, $\varepsilon_3 = 3$), $p = 500$nm, $Np = 1$ for tilt angle $\theta = 90^{\circ}$ and incident wavelength $\lambda = 632.8$nm.

Figure 2c shows the effect of changing the dielectric anisotropy of the LC molecule ($\varepsilon_1 = 2$, $\varepsilon_2 = 2$, $\varepsilon_3 = 3$) at tilt angle $\theta = 90^{\circ}$, which leads to case analogues to figure 2a i.e. the LC layer has the same permittivity as shown in figure 2a. For negative anisotropy ($\varepsilon_2 > \varepsilon_3$) the variation of tilt angle from 1 deg to 90 deg leads to conversion of plasmon mode into waveguide mode (see figure 2a and figure 2b).

2.1.3. Wavelength dependence of plasmon resonance at LC layer thickness d = 500 nm
The conversion of plasmon mode into waveguide mode has supposed strong dependence of SPR on the tilt angle of LC molecule. To reveal this dependence as a function of wavelength we have simulated the reflection at the angle of SPR excitation ($\varphi = 68^{\circ}$), shown in figure 3.

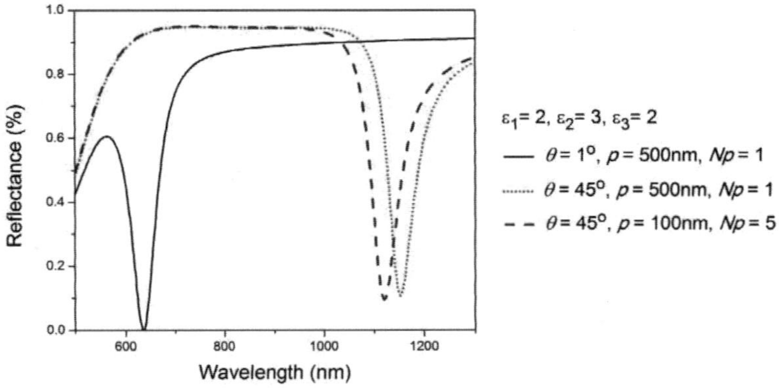

Figure 3. Reflection spectra as a function of wavelength for incident angle $\varphi = 68^{\circ}$.

Increasing the tilt angle causes a strong SPR shift to longer wavelengths (figure 3 – dot line). Pitch length increasing makes this effect smooth (figure 3 – dashed line). For correct comparison LC layer thickness is fixed, but numbers of turns are different to get different pitch.

2.2. Anisotropic layer with thickness d = 1400nm
The next step is to study the influence of thicker anisotropic layer. Keeping in mind that the penetration depth of plasmon is few hundred nanometers it is worthy to simulate LC layers with pitch

of compatible length. From common consideration one can expect that the number of waveguide modes will increase.

2.2.1. Reference anisotropic layer with thickness d = 1400nm

Following our approach from point 2.1 at thickness d=500nm, we start simulations with usual anisotropic layer. Indeed, our simulation, shown in figure 4, has confirmed the increased number of waveguide modes. The comparison of the results with figure 2 shows that the thickness of the anisotropic layer does not influence the position of the plasmon resonance.

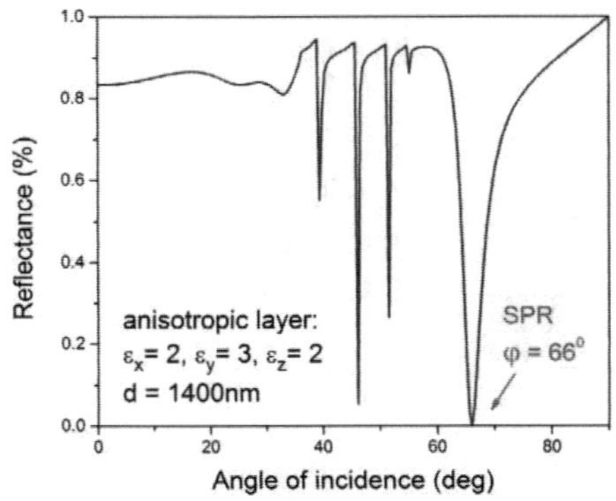

Figure 4. Reflection spectra for p-polarization light as a function of incident angle. Configuration: prism (ε_p = 2.89), Au layer with thickness d = 50nm, anisotropic layer (ε_x = 2, ε_y = 3, ε_z = 2) with thickness d = 1400nm, air, for incident wavelength λ = 632.8nm.

2.2.2. Ferroelectric SmC LC with layer thickness d = 1400nm*

Figure 5 is analogues to figure 4 in case of ferroelectric SmC* LC layer with permittivity ε_1 = 2, ε_2 = 3, ε_3 = 2, pitch p = 1400nm, number of periods Np = 1 and tilt angle as indicated in the caption of figure 5. The result shows that the existence of helical structure slightly shifts the plasmon resonance exactly at the same manner as in the case of thinner layer d=500nm.

By analogy to the case with LC layer at thickness d = 500nm, the effect of changing the dielectric anisotropy of the LC molecule (ε_1 = 2, ε_2 = 2, ε_3 = 3) at tilt angle θ = 90° is the same (figure 5a and figure 5c are similar). For negative anisotropy ($\varepsilon_2 > \varepsilon_3$) a conversion of plasmon mode into waveguide modes is observed when the tilt angle is 90 deg (see figure 5a and figure 5b), as for LC thinner layer.

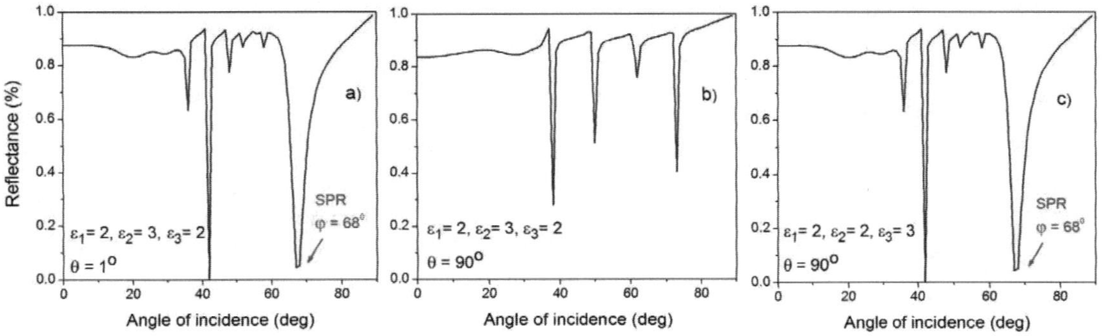

Figure 5. Reflection spectra as a function of incident angle. Configuration: LC (ε_1 = 2, ε_2 = 3, ε_3 = 2), p = 1400nm, Np = 1 for tilt angle a) θ = 1° and b) θ = 90°; c) LC (ε_1 = 2, ε_2 = 2, ε_3 = 3), p = 1400nm, Np = 1 for tilt angle θ = 90° and incident wavelength λ = 632.8nm.

2.2.3. Wavelength dependence of plasmon resonance at LC layer thickness d = 1400 nm
The wavelength dependence of SPR excitation has been simulated at the angle of SPR excitation (φ = 68°). The result is shown in figure 6. Changing the tilt angle shifts the SPR at longer wavelengths (figure 6 – dot line). The pith length influences strongly the condition of plasmon resonance (figure 6 – dashed line). Unlike the case of 500 nm this effect is much stronger – the conditions of plasmon resonance could be violated depending on the tilt angle. Observed resonances are very narrow and are not SPR. As in the case of thinner layer, here the pitch length is changed by increasing number of turns at fixed LC layer thickness.

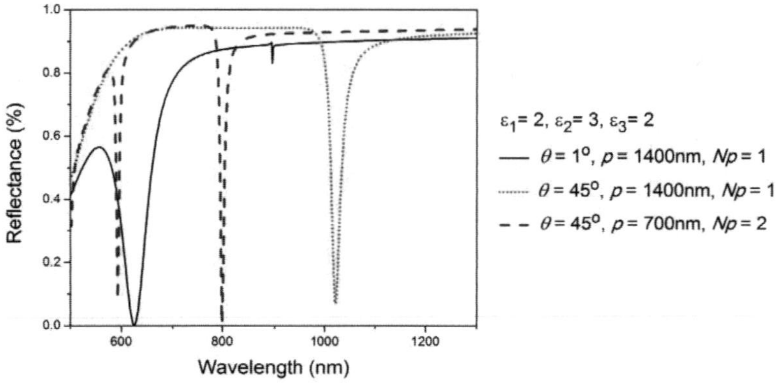

Figure 6. Reflection spectra as a function of wavelength for incident angle $\varphi = 68^{\circ}$.

The effect of tilt angle variation on SPR excitation is different for different LC layers thickness. This effect is illustrated in figure 7. The resonance shift can be used as criterion for tilt angle determination.

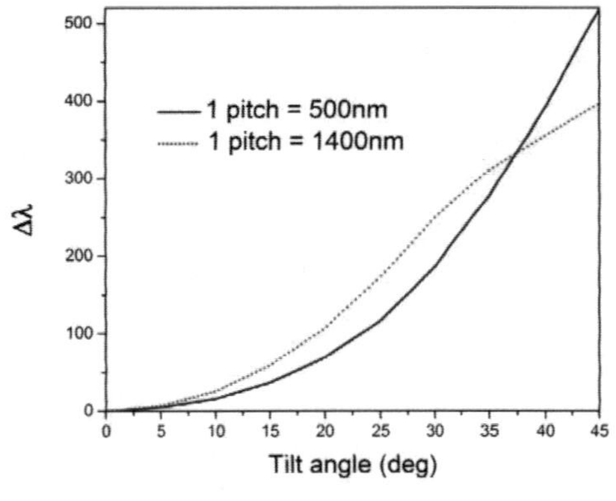

Figure 7. Plasmon resonance shift as a function of tilt angle. Configuration: LC ($\varepsilon_1 = 2$, $\varepsilon_2 = 3$, $\varepsilon_3 = 2$), p = 500nm and p = 1400nm, Np = 1 for incident angle $\varphi = 68^{\circ}$.

3. Discussion
The comparison between ordinary anisotropic layer and LC layer reveal the influence of LC layer feature on SPR: the helical structure slightly shift SPR while its influence on the waveguide modes can be neglected.

Not for all anisotropic layers SPR can be excited – the existence of evanescent field is obligatory and depends on the value of permittivity of different axes. In our geometry (following [17]) the SPR is not possible for $\varepsilon_x = 2$, $\varepsilon_y = 2$, $\varepsilon_z = 3$. This case can be accomplished for LC molecule with negative anisotropy $\varepsilon_1 = 2$, $\varepsilon_2 = 3$, $\varepsilon_2 = 2$ at tilt angle 90 deg (see figures 2a and 2b and figures 5a and 5b). This effect explains the dependence of SPR wavelength excitation on the tilt angle and can be used for LC characterization (see figures 3, 6, and 7).

The pitch strongly influences SPR excitation conditions – they can be violated as in the case of LC layer with thickness d = 1400nm (see figure 6). Probably, this is due to Bragg reflection and resonances shown in figure 6 correspond to Bragg reflection of higher order. This problem needs of more careful attention.

4. Conclusion

The reflection spectra of metal/ferroelectric SmC* LC interface as dependence of angle of incidence and wavelength has been theoretically investigated. The results show that, the tilt angle variation of LC changes the condition of SPR excitation including a conversion of plasmon mode to waveguide mode. Increasing the anisotropic layer thickness does not influence SPR, but new waveguide modes appear. SPR wavelength strongly depends on tilt angle variation. Pitch length variation causes SRR wavelength shift, the effect is more expressed for the case of thicker liquid crystal layer. Observed SPR can be used for liquid crystal characterization.

Acknowledgments

The authors acknowledge the support for this work by the department for Scientific and applied activities (SAA) within the Plovdiv University "Paisii Hilendarski" through Project YS13FF005/20.03.2013.

References

[1] Atwater H 2007 *Sci. Am.* **296** 56-83
[2] Zografopoulos D C, Asquini R, Kriezis E E, d'Alessandro A, Beccherelli R 2012 *Lab Chip* **12** 3598-610
[3] Giliardi G, Donisi D, Serpengüzel A, Beccherelli R 2009 *Opt. Lett.* **34** 3253-5
[4] Giliardi G, De Sio L, Beccherelli R, Asquini R, d'Alessandro A, Umeton C 2011 *Opt. Lett.* **36** 4755-7
[5] Tasolamprou A C, Bellini B, Zografopoulos D C, Kriezis E E, Beccherelli R 2009 *J. Eur. Opt. Soc.* **4** 09017
[6] Pfeifle J, Alloatti L, Freude W, Leuthold J, Koos C 2012 *Opt. Express* **20** 15359-76
[7] Abdulhalim I 2011 *Liq. Cryst. Today* **20** 44-60
[8] Zografopoulos D C, Beccherelli R 2013 *Plasmonics* doi: 10.1007/s11468-012-9440-7
[9] Zografopoulos D C, Beccherelli R 2013 *Photon Nanostruct.* **11** 73-84
[10] Zografopoulos D S, Beccherelli R 2013 *Appl. Phys. Lett.* **102** 101103
[11] Zhu J H, Huang X G, Tao J, Jin X P, Mei X, Zhu Y J 2011 *J. Mod. Opt.* **58** 32-7
[12] Ishii S, Kildishev A V, Shalaev V M, Drachev V P 2011 *Laser Phys.Lett.* **8** 828-32
[13] Bahramipanah M, Mirtaheri S A, Abrishamian M S 2012 *Opt. Lett.* **37** 527-9
[14] Cetin A F, Yanik A A, Mertiri A, Erramilli S, Mustecaplıoglu O E, Altug H 2012 *Appl. Phys. Lett.* **101** 121113
[15] de Gennes P G and Prost J 1993 *The Physics of Liquid Crystals* (Oxford: Clarendon Press)
[16] Berreman D W 1972 *J. Opt. Soc. Am.* **62** 502-510
[17] Abdulhalim I, Benguigui L, Weil R 1985 *J. Phys. (Paris)* **46** 815-825
[18] Johnson P and Christy R 1972 *Phys. Rev.* B **6** 4370-4379

Journal of Physics: Conference Series **558** (2014) 012024 doi:10.1088/1742-6596/558/1/012024

Liquid crystal nanocomposites produced by mixtures of hydrogen bonded achiral liquid crystals and functionalized carbon nanotubes

B Katranchev[1], M Petrov[1], E Keskinova[1], H Naradikian[1], P M Rafailov[1], U Dettlaff-Weglikowska[2] and T Spassov[3]

[1] Institute of Solid State Physics, Bulgarian Academy of Sciences, 72 Tzarigradsko Chaussee Blvd., 1784 Sofia, Bulgaria.
[2] School of Electrical Engineering, Korea University, Anam-Dong, Sungbuk-Gu, 136-713 Seoul, Korea.
[3] Faculty of Chemistry and Pharmacy, University of Sofia "St. Kl. Ohridski", 1 James Bourchier blvd.,1164 Sofia, Bulgaria.

E-mail: mpetrov@issp.bas.bg

Abstract. The liquid crystalline (LC) nature of alkyloxybenzoic acids is preserved after adding of any mesogenic or non-mesogenic compound through hydrogen bonding. However, this non-covalent interaction provokes a sizable effect on the physical properties as, e. g. melting point and mesomorphic states. In the present work we investigate nanocomposites, prepared by mixture of the eighth homologue of p-n-alkyloxybenzoic acids (8OBA) with single-walled carbon nanotubes (SWCNT) with the purpose to modify the optical properties of the liquid crystal. We exercise optical control on the LC system by inserting SWCNT specially functionalized by carboxylic groups. Since the liquid crystalline state combines order and mobility at the molecular (nanoscale) level, molecular modification can lead to different macroscopical nanocomposite symmetry. The thermal properties of the functionalized nanocomposite are confirmed by DSC analyses. The mechanism of the interaction between surface-treated nanoparticles (functionalized nanotubes) and the liquid crystal 8OBA bent-dimer molecules is briefly discussed.

1. Introduction

The nanostructured liquid crystals (LC) are of academic and industrial interest as they exhibit in many instances significantly different properties from those of distinct atoms or bulk materials [1-5]. It is possible to control the optical properties of liquid crystals by changing the molecular arrangement, inserting external agents, e.g., various non-mesogenic nanoparticles. The concentration of the nonmesogenic component, its homogeneity and its compatibility with the liquid crystal is of particular significance for the structural, optical and electrical properties of the nanocomposite. This effect is expected to be more controllable when the inserted external nanoparticles are specially functionalized. The high value of the surface-bulk ratio makes the physical properties of nanoparticles rather unique. Single-walled carbon nanotubes (SWCNT) are a special category of such nanoparticles which additionally offer a highly anisotropic impact on the LC due to their large aspect ratio [6-9]. They also exhibit a great capability for homogeneous dispersion in a liquid-crystal matrix and ability to alter its inherent bulk properties [6]. The coupling of the anisotropic elasticity of the LC medium to the optical, conductive, or magnetic properties of functionalized nanoparticles is expected to have a significant impact on the macroscopic properties of LCs designed for contemporary nanotechnology. Since the LC state combines order and mobility at the nanoscale level, even small quantities of SWCNT can effectively control the LC nanoscale assembly, which provides specific material properties.

The classical p-n-alkyloxybenzoic acid (nOBA) LC is the basic component in forming supramolecular systems. Even without complementary nonmesogenic molecules it displays a supramolecular state due to the dimeric form of its molecule, and its donor ability at contact with a nonmesogen substance predicts rich mesomorphic states [10]. On the other hand, its dimeric structure allows an induced chirality although it is an achiral LC system, i.e. it is a hopeful candidate for the realization of the long sought-after C_1 symmetry in LC systems like the bent-core molecules [11].

Recently [12,13], we investigated a series of nanocomposites, built of the hydrogen-bonded liquid crystal 7OBA and SWCNT by optical microtexture analysis and other complementary methods. The bulk interactions of the dimeric LC molecules with the SWCNT turn out to mainly govern the type (symmetry), thermal stability and chirality of the LC states induced in these nanocomposites. As a result, a cascade of phase transitions and phases not typical for pristine 7OBA are observed. The most intriguing result is the detection of a triclinic smectic C_G phase, which is chiral and biaxial, and exhibits bulk ferroelectricity. To explain this phenomenon, we assumed that the SWCNT embedded in the liquid crystal matrix, could also affect the structure of the constituent LC molecules. This is particularly important when these molecules are dimers and especially, dimers formed by linear hydrogen bonding (H-bonding) since the non-mesogenic component in the nanocomposite could provoke a conformation of the LC molecule from a closed dimer to a opened dimer (bent-dimer) or even monomer, which may lead to new electrical and thermal properties, including new phases. Recently, a smectic G phase was induced through intermolecular hydrogen bonding between mesogenic p-n-alkyloxybenzoic acids and nonmesogenic p-aminobenzonitrile [14], due to the H-bonding between the electron-rich terminal *cyano* group (as proton acceptor) and the terminal carbon of the COOH group of nOBA. This smectic G phase, however, was not interpreted as the sought-after smectic C_G phase.

The closed dimers can also be viewed as flexible bent-core-like molecules, showing similarities with phases found in large-size banana-like system [11], but exhibiting a more complex structure. A close similarity was suggested between the smectic X phase found in these substances and the C_G phase or its analogue B_7 phase in the bent-core LC [11]. These phases are examples of the typical smectic states where left- and right-handed helices are induced in an LC system whose constituent molecules are achiral. The smectic C_G phase was predicted long ago by de Gennes [15]. Being biaxial and fluid in the layers and having a macroscopic handedness due to its structure, this phase is a natural candidate to demonstrate induced helices in a system composed of achiral molecules.

In the present work we examine nanocomposites of the eighth homologue of non-chiral p-n-alkyloxybenzoic acids (8OBA) with SWCNT: 8OBA/SWCNT. In our previous studies [12] we found that for most of the SWCNT concentrations used, new chiral phases were produced and, for some specific values of cell thickness and anchoring energy, the new unique smectic C_G phase was induced. Now, we check the effect of nanotubes functionalized by carboxylic (COOH) groups (SWCNT-f). Furthermore, in the search for an adequate molecular and macroscopic model of the induction of the smectic C_G phase, our results elucidate the interactions of 8OBA with nonmesogens like SWCNT regarding the compatibility and thermal stability of the nanocomposites.

2. Experimental results and discussion

The experiments were carried out on the nanocomposites 8OBA/SWCNT and 8OBA/SWCNT-f. For their preparation 3 mg SWCNT were mixed with 1.8 g 8OBA. For the mixtures pristine and COOH-functionalized SWCNT (SWCNT-f) were used. The nanotubes were produced by Catalyzed Chemical Vapor Deposition with the following characteristics: Amorphous Carbon Content: < 3 wt%**;** Length: 5-30um; Specific Surface Area: 407 m^2/g; Electrical Conductivity: >100 S/cm; Bulk density: 0.14 g/cm^3; True density: ~2.1 g/cm^3. COOH-SWCNT contained 2.7% COOH groups.

The phases of achiral (pristine) 8OBA are isotropic (I), nematic (N), smectic C (S_C) and Crystal (Cr) (see table 1).

Table 1. Transition temperatures and enthalpies as determined by DSC for the phase transitions detected in the 8OBA, 8OBA/SWCNT and 8OBA/SWCNT-f.

8OBA		8OBA/SWCNT		8OBA/SWCNT-f	
T (°C)	ΔH (kJmol^{-1})	T (°C)	ΔH (kJmol^{-1})	T (°C)	ΔH (kJmol^{-1})
T(I-N)=141	3.2	T(I-N)=140	2.4	T(I-N)=139	2.75
T(N-S_C)=100	1.4	T(N-S_C^*)=98	1.1	T(N-S_C^*)=95	0.7
T(S_C-Cr)=89	13	T(N_r^*-C_G)=88	10.6	T(N_r^*-C_G)=87	10.9

The thermal behavior of 8OBA/SWCNT and 8OBA/SWCNT-f was examined by differential scanning calorimetry (DSC) and the results are presented in table 1. The two nanocomposites display enantiotropic phase formation. The thermal scans were recorded in both endo and exo regime. However, we base our thermal analysis and the corresponding comparisons mainly on the exo thermograms. The endo scanning was used to study the hysteresis of the first order phase transitions. An example of the thermal scanning is shown in figure 1.

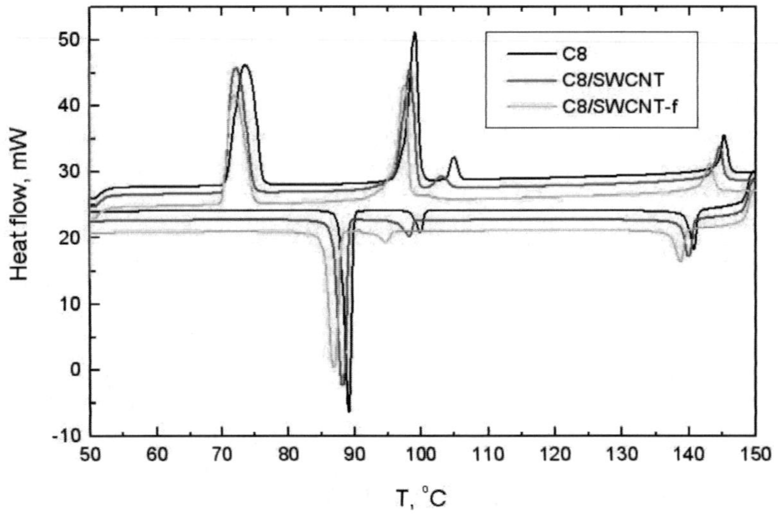

Figure 1. The DSC scans of 8OBA, 8OBA/SWCNT and 8OBA/SWCNT-f.

Using the results of the table 1, one can complete the microtexture polarization analysis, directed to the building of a qualitative model of achiral-chiral transition in 8OBA/SWCNT and 8OBA/SWCNT-f nanocomposites.

The initial liquid crystal cell (LCC) with thickness d was homogeneously orientated. The surface orientation treatment is done by usual unidirectional rubbing of the ITO layer deposited on the substrates. In figure 2 the isotropic-nematic phase transition for 8OBA/SWCNT and 8OBA/SWCNT-f marked as I-N in table 1 is presented. As is seen, the LC growth in 8OBA/SWCNT (figure 2(a)) starts by modified Maltus crosses indicating quasi-homeotropic orientation in normal N, while that of 8OBA/SWCNT-f (figure 2(b)) arises with bifocal droplets typical for the chiral states, which exist for less than one °C and then readily transform in normal nematic N. Further consecutive cooling initiates a cascade of phase transitions for both nanocomposites, capturing chiral and reentrant phases, as follows: N $\rightarrow$ chiral nematic (N^*); N^* $\rightarrow$ chiral smectic C (S_C^*); S_C^* $\rightarrow$ reentrant N^* (N^*_r); N^*_r $\rightarrow$ C_G and C_G $\rightarrow$ Cr.

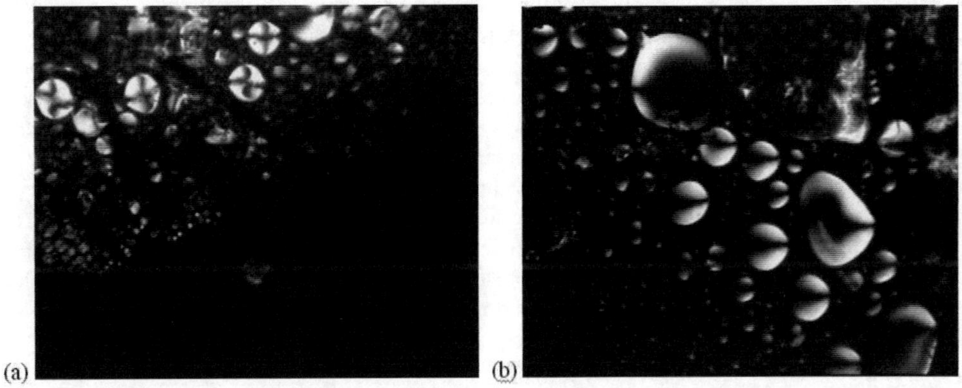

Figure 2. Textures of isotropic-nematic phase transition in unidirectionally oriented LC cells with d=8μm of: (a) 8OBA/SWCNT at T=140°C. (b) 8OBA/ SWCNT-f at T=139°C.

For clarity, we present the microtextures of these phases of 8OBA/SWCNT-f in figure 3. The nanocomposite enters at 95°C into a chiral smectic C (S_C^*) phase (in figure 3(a) the transition N $\rightarrow$ S_C^* is depicted). The textures S_C^* are characterized by typical stripes indicating the helix direction. Upon further cooling, at 93°C a reentrant chiral nematic (N_r^*) phase is reached. Figure 3(b) presents the N_r^* texture. At 87 °C we observe a transition from N_r^* to C_G (figure 3(c)). Finally, at 55°C a paramorphic solid Cr phase forms (figure 3(c)).

We identify the C_G formation by the existence of a permanent out-of-layer electric polarization P_{out} in this phase (for more details see [12]). This can be demonstrated with an electric field applied along the LC cell normal. Such application of electric field induces a second order phase transition C_G $\rightarrow$ C^* and further phase transitions to the I phase. The same phase transition can thus be a consequence of electric-field application, or temperature increase, or change in the enthalpy of the system. The requirement div$P$=0 [15] in this case allows the detection and calculation of the polarization P_{out} by the modified Kirkwood-Helfrich equation. The polarization value calculation and electric field response of this phase, using the pyro-electric and triangular dielectric methods (see the details in [12,15]), will be presented in a future work. As seen from table 1, for the two nanocomposites 8OBA/SWCNT and 8OBA/SWCNT-f the phase transition temperatures gradually decrease which is mostly pronounced for the I-N or I-N^* transitions.

Comparison of the microtexture temperature analysis with that of DSC scanning (see table 1) demonstrates that not each of the observed textural transitions is matched by a corresponding DSC peak. This indicates that some of these transitions, e.g. the N $\rightarrow$ N^*, or the S_C^* $\rightarrow$ N^*_r ones are very weak resembling second-order transitions, while others (N^* $\rightarrow$ S_C^*, N^*_r $\rightarrow$ C_G) categorically display first-order behavior.

To calculate the anchoring energy W_s in the N phase we measured the topological-defects (Neel walls) width l and using the formula $W_s=\pi^2Kd/2l^2$ [13,15], where K is the elastic constant (10^{-11} J m^{-1}) taken in the one-constant approximation [15], we found the surface energy values for the investigated different samples: $0.2\text{x}10^{-5}$ J m^{-2} (8OBA), 0.15×10^{-5} J m^{-2} (8OBA/SWCNT) and 0.1×10^{-5}J m^{-2} (8OBA/SWCNT-f). The surface energy variation from 0.2×10^{-5} J m^{-2} (8OBA) through 0.15×10^{-5} J m^{-2} (8OBA/SWCNT) to 0.1×10^{-5} J m^{-2} (8OBA/SWCNT-f) indicates that the surface anchoring reduction in (8OBA/SWCNT and (8OBA/SWCNT-f)) could take place at the expense of bulk ordering increase due to the conformability between the SWCNT-f and the 8OBA molecules. To explain this unexpected effect we propose a physical mechanism of the interaction between the molecules of

8OBA and the functionalized SWCNT, including interaction with the orienting ITO surface. Surface functionalization helps in stabilizing the dispersion, since it can prevent re-aggregation of nanotubes and also leads to coupling of SWCNT with the LC matrix. Coupling between SWCNT and the LC matrix is also very important for efficient transfer of external stress to the nanotubes. As SWCNT have some surface defects like varying diameter and bends/twists due to non-hexagonal defects, mechanical interlocking plays a role in SWCNT–LC adhesion along a SWCNT. The chemical bond, either ionic or covalent, significantly improves the interfacial interaction between matrix and filler that enables a stress transfer. The van der Waals interaction also arises from the molecular proximity and is the only mode of interaction between SWCNT and the matrix in absence of chemical bonding. Although chemical functionalization of SWCNT can improve the compatibility between SWCNT and LC, which in turn improves the mechanical properties, it may deteriorate other properties of the nanocomposites near to the surface such as electrical conductivity and increase the stability of the electrical double layer reinforcing the surface pre-tilt and decreasing the anchoring strength. This effect could be interesting for the flexo-electric response of the LC system, where strong bulk ordering and weak surface anchoring are necessary. However, for other electro-optical phenomena, where a strong anchoring is needed, a preliminary deposition of oriented SWCNT on the orienting LCC surface can be performed, as was demonstrated in [13].

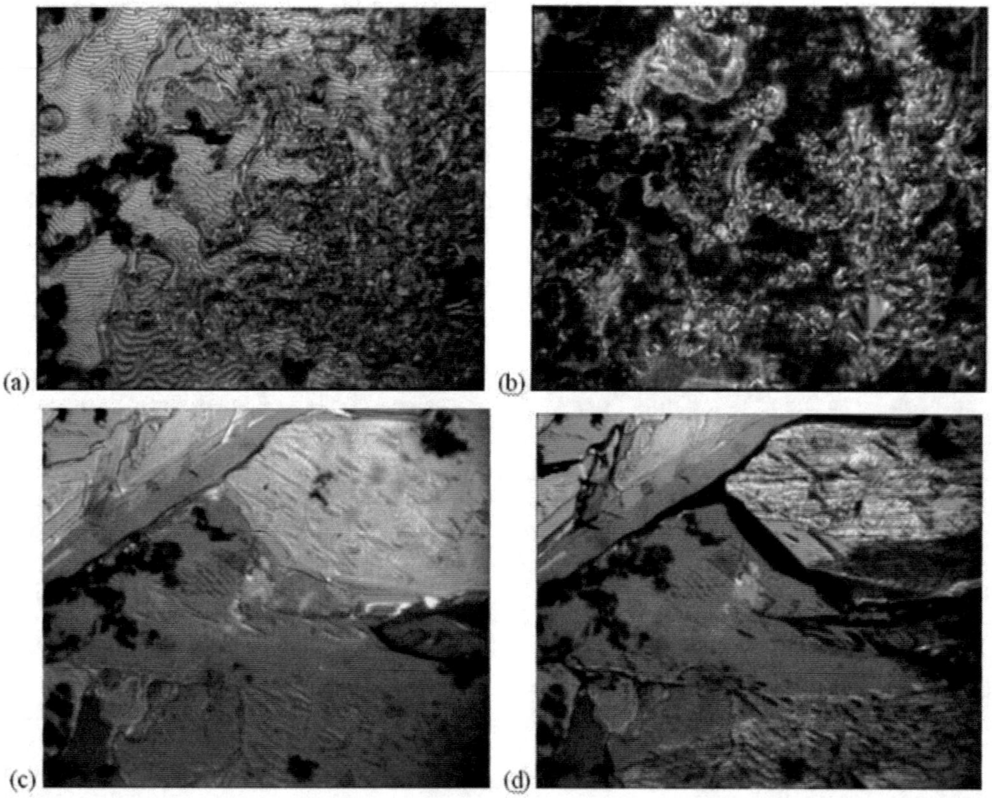

Figure 3. 8OBA/SWCNT-f textures in unidirectionally oriented LC cell with d = 8 μm: (a) Phase transition N → S_C^* at T = 95 °C. (b) N_r^* phase at T = 93 °C. (c) C_G phase at T = 87 °C. (d) Cr phase at T = 55 °C.

The chiral smectic C phase in 8OBA/SWCNT-f grows in well formed chiral lines (figure 3a) allowing control of the chiral axis and pitch value measurements. This effect and the transition to the low-temperature smectic C_G demonstrate the symmetry degradation from C_2 to C_1 as was shown in [12].

3. Conclusion

Using an optimal mixture of 8OBA with pristine and COOH-functionalized single wall carbon nanotubes (with respect to the concentration), we detect a process of chiralization of the otherwise achiral 8OBA in its pure state. The functionalized SWCNT weakly decrease the surface anchoring, but favourize the bulk ordering and the process of chiralization. The decrease of the phase transition temperature in the functionalized nanocomposite facilitates the desired effect in chiral ferroelectric liquid crystal materials concerning the isotropic-cholesteric phase transition temperature decrease. The linear decrease of the isotropic transition temperature upon electric field increase, which is equivalent to enthalpy variation, indicates a coupling of the induced out-of-layer plane polarization to the applied electric field.

Acknowledgments

This study was supported by Grant BK-04-14 from the Institute of Solid State Physics, Bulgarian Academy of Sciences.

References

[1] Hegmann T, Qi H and Marx V M, 2007 *J. Inorg. Organomet. Polym. Mater.* **17** 483
[2] Draper M, Saez J M, Cowling S J, Gai P, Heinrich B, Donnio B, Guillon D and Goodby J W 2011 *Adv. Funct. Mater.* **21** 1260
[3] Qi H and Hegmann T 2008 *J. Mater. Chem.* **18** 3288
[4] Coursault D, Grand J, Zappone B, Ayeb H, Levi G, Felidj N and Lacaze E, 2012 *Adv. Mater.* **24** 1461
[5] Taylor J W, Kurihara L K and Martinez-Miranda L J, 2012 *Appl.Phys. Lett.* **100** 173115
[6] Dierking I, Scalia G and Morales P 2005 *J. Appl. Phys.* **97** 044309
[7] Podgornov F, Suvorova A, Lapanik A and Haase W 2009 *Chem. Phys. Lett.* **479** 206
[8] Russel J M, Oh S, LaRue I, Zhou O and Samulski E T 2006 *Thin Solid Films* **509** 53
[9] Gommans H, Alldredge J W, Tashiro H, Park J, Magnuson J and Rinzler A G 2000 *J. Appl. Phys.* **88** 2509
[10] Petrov M 2010 Optical *and Electrooptical Properties of Liquid Crystals Nematic and Smectic Phases* (New York: Nova Science Publishers)
[11] Jakli A, Kruerke D, Sawade H and Heppke G 2001 *Phys. Rev. Lett.* **86** 5715
[12] Petrov M, Katranchev B, Rafailov P M, Naradikian H, Dettlaff-Weglikowska U, Keskinova E and Spassov T 2013 *Phys. Rev. E* **88** 042503
[13] Petrov M, Katranchev B, Rafailov P M, Naradikian H, Dettlaff-Weglikowska U and Keskinova E, 2013 *J. Mol. Liq.* **180**, 215
[14] Rambabu M, Pisipati V G K M., Shanmukha Kumar J V and Prasad K R S 2013 *Liquid Crystals Today* **22** 4
[15] de Gennes P G and Prost J 1993 *The Physics of Liquid Crystals* second edition (New York: Oxford University Press)

Electric field-induced optical second harmonic generation in nematic liquid crystal 5CB

S I Torgova[1], V D Shigorin[2], I A Maslyanitsyn[2], L Todorova[3], Y G Marinov[3,*], G B Hadjichristov[3] and A G Petrov[3]

[1] Lebedev Physical Institute, Russian Academy of Sciences, Leninskij pr. 53, Moscow, 119991 Russia

[2] Prokhorov General Physics Institute, Russian Academy of Sciences, ul. Vavilova 38, Moscow, 119991 Russia

[3] Georgi Nadjakov Institute of Solid State Physics, Bulgarian Academy of Sciences, 72 Tzarigradsko chaussee, Sofia 1784, Bulgaria

E-mail: ymarinov@issp.bas.bg

Abstract. Electric field-induced second harmonic generation (EFISH) was studied for the liquid crystal 4-cyano-4'-pentylbiphenyl (5CB) (a nematic phase material at room temperature). The intensity of coherent SHG from 5CB cells upon DC electric field was measured for various initial orientations of the liquid crystal. The dependence of the SHG intensity on the pump beam incidence angle was obtained in transmission geometry using sample rotation method. The experimental results (the registered light intensity in the output SHG interference patterns) were theoretically modelled and analyzed.

1. Introduction

The structure and properties of nematic liquid crystals (NLC) are extensively studied for a long time [1]. The NLC molecules are usually highly anisotropic and noncentrosymmetric and possess large hyperpolarizability. However, in the bulk phase, the NLC materials are macroscopically characterized as structures possessing inversion symmetry that prevents the even-order nonlinear optical processes. As a result, although the individual molecular hyperpolarizability of NLC is quite large, the bulk NLCs can not usually produce optical second harmonic generation (SHG) (lowest even-order nonlinear optical process). However, an external electric field (with a magnitude that is high enough) applied to NLC cells can efficiently remove the inversion centre and thus can give a rise of an electric field-induced SHG (EFISH) [2-4]. Valuable information can be obtained by varying the optical phase in EFISH experiments, e.g. by the change of the optical path. In this case, one can observe so called Maker fringes as interference patterns [5].

There are two ways to change the optical path in a NLC cell. The first is the movement of a wedge-formed sample across the incident laser beam (pump beam), usually at normal incidence. The second is the rotation of a plane-parallel sample thus changing the pump beam incidence angle. The former is known as a wedge method. Small value of a wedge angle (~1°) allows one to consider the NLC cell within the laser spotsize as a plane-parallel sample with a variable thickness and, therefore, to use well-known models. This method has been used for the studies of a number of NLCs including

* To whom any correspondence should be addressed.

MBBA, 5CB, 8CB and other cyanobiphenyls, and the temperature dependencies of the SHG intensity, as well as the coherence length, have been obtained [7-8]. However, there was no detailed analysis of the abovementioned Maker fringes.

In the present work, the dependence of EFISH intensity on the pump beam incidence angle is investigated using the rotation of a plane-parallel NLC cell. Experimental data were collected for coherent SHG from 5CB cells as measured in transmission geometry, and are linked to results from theoretical simulations responding to the NLC director orientation.

2. Experimental

The room-temperature nematic liquid crystal *p*-*n*-pentyl-*p*'-cyanobiphenyl (5CB) (from Merck) was used after rectification procedure. The 5CB cells were assembled from two parallel glass plates. The cell gap was fixed with 100 μm-thick copper spacers, also serving as electrodes. The distance between the electrodes was about 2 mm. 5CB cells with either homeotropic or planar alignment, were studied. In the first case, the inner surfaces of the glass plates of the cells were covered with a monolayer of silane in order to orient the liquid crystalline molecules homeotropically. The treatment we have used involves a deposition of silane cross-linked films [9]. For this purpose, 0.1 vol% of ODS-E silane (Chisso, Japan) was dissolved in 9:1 isopropyl alcohol/distilled water, and glass plates were dipped in the solution for several hours. The plates were then rinsed with distilled water and ODS-E was polymerized, keeping it at 110°C for one hour. Planar-oriented 5CB cells were prepared by use of glass plates covered with polyimide that was unidirectionally rubbed. Polyimide films were made by spin coating (at 2000 rpm) of a polyimide precursor 3 wt.% solution (polyimide system ZLI-2650 from Merck in cyclopentanone) onto clean glass after that baking to polymerize and followed by rubbing of the cured film with a filter paper under constant pressure of 100 g. The empty cells with the alignment layers were assembled in parallel rubbing configuration. 5CB in its isotropic phase was capillary filled into the cells, then the samples were cooled slowly to the nematic phase to achieve the desired good alignment of the nematic 5CB, as verified by polarizing microscope (Leitz DMRXP).

The prepared 5CB cells were mounted on a rotational stage driven by step motor. The SHG from the cells was studied in transmission geometry. The coherent SHG signal (highly directed SHG beam) from the 5CB cells was registered by a two-channel experimental set-up with a reference channel. In the experiments, Nd^{3+}:YAG laser was employed as a pump laser source with a wavelength of 1064 nm, pulse duration of 30 ns, and pulse energy of 0.2 mJ. Measurements were performed for both *s*- and *p*-polarization of the pump beam. The *s*-polarized SHG signal was detected as the most intense one. The diameter of the pump laser spot on the sample surface was about 1 mm and special attention has been paid to prevent the contact of the laser beam and metal electrodes. The spectrally filtered output SHG signals (at the wavelength of 532 nm) in the main and reference channels were detected by sensitive photomultipliers and processed by automated data acquisition system. An intensity calibration was carried out using a plate of *y*-cut α-SiO_2 crystal placed into the peak of Maker fringes [2]. The relative units for the measured intensity, used hereafter and given below, are 10^{-6} of the intensity of the calibration signal. SHG from empty cells has been also measured and the observed signal was negligible.

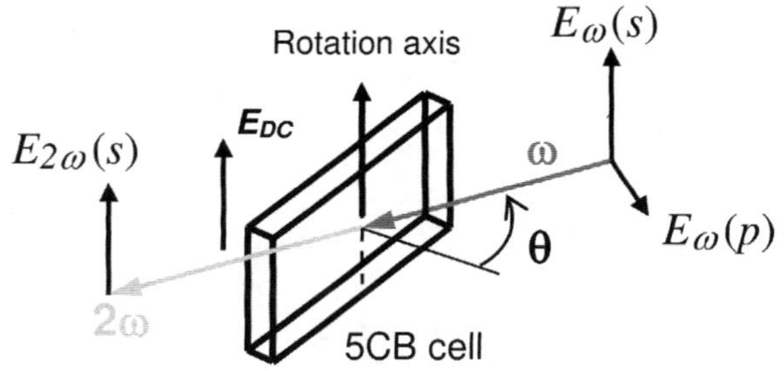

Figure 1. Geometry of the EFISH experiment performed in this work.

In the rotation experiments, the pump incidence angle θ was changed and the dependence of EFISH intensity $I_{2\omega}(\theta)$ was measured (figure 1). The rotation axis was parallel to the film plane and orthogonal to the plane of incidence. The rotation angle was changed in a step of 1°. The source of external direct current (DC) electric field was home-made high-voltage pulse generator, producing 10 μs pulses of 3 kV voltage at flat top region of the pulse and temporally synchronized with the pump laser pulse. As the external electric field was switched off, no SHG signal was detected for all cells. The experiments were carried out at room temperature (ca. 21°C).

3. Results and Discussion

Firstly, the dependencies of SHG intensity $I_{2\omega}(\theta)$ at different (s or p) polarizations of the pump beam were measured in the case of a glass cell filled with nitrobenzene (isotropic liquid having a large and well established SHG nonlinearity). The results are reported in figure 2. A single-layer model [10] and refraction index data [11] were used in the fitting procedure (generally, the method is described in [12]). The results presented exhibit two interference patterns, depending on the sum and the difference of wave vectors of the pump $\boldsymbol{k}_\omega(\theta)$ and SHG $\boldsymbol{k}_{2\omega}(\theta)$ [13]. The first pattern (the main pattern), defining the envelope curve, depends on the difference $\boldsymbol{k}_{2\omega} - 2\boldsymbol{k}_\omega$ and is a relatively slow function of θ. The second pattern depends on the sum $\boldsymbol{k}_{2\omega} + 2\boldsymbol{k}_\omega$. The amplitude of the second pattern is several times smaller than that of the first one. It should be noted also that the maximum signal for s-polarized pump is ~7 times higher than that for the p-polarized one.

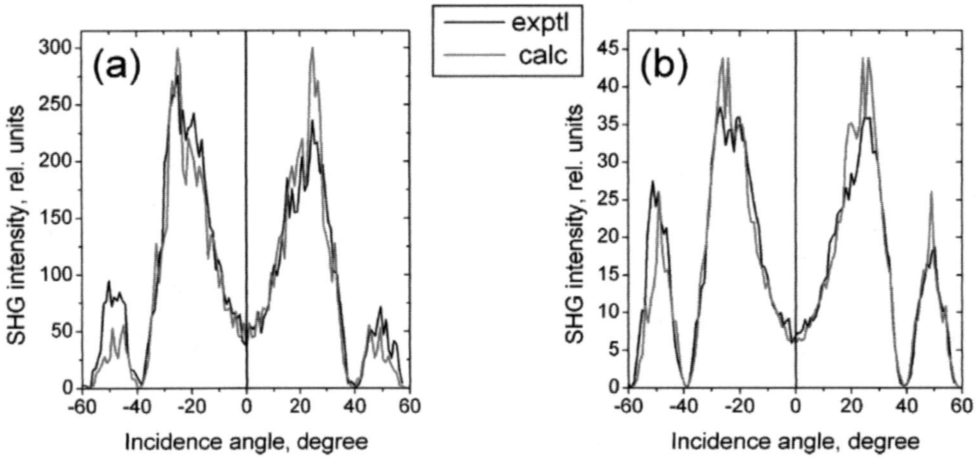

Figure 2. Angular dependence of SHG signal from pure glass cell filled with nitrobenzene: (a) s-polarized pump; (b) p-polarized pump.

The angular dependencies of SHG signal from a cell covered with polyimide, is presented in figure 3. In this case, the rubbing direction of the cell under study was parallel to the direction of the applied DC electric field. For s-polarized pump (figure 3 (a)) the frequency of Maker fringes is similar to that for nitrobenzene reported above, because the same N_e (the extraordinary refractive index) is responsible for both the pump and SHG. When the pump is p-polarized (figure 3 (b)), then it is represented with N_o (the ordinary refractive index). The difference between the wavevectors becomes significant and comparable to the wavevector sum. Now the two patterns are indistinguishable. Nevertheless there is qualitative agreement of the calculated and experimental graphs. The interference pattern exhibits an increased frequency of the oscillations and strongly reduced amplitude. The maximal SHG signal for p-polarized pump is ~180 times smaller as compared to the s-pump experiment.

The case of the same planarly-aligned (by polyimide-coated substrates) cells, but with rubbing orientation perpendicular to the direction of the applied DC electric field, is shown in figure 4. As seen, for *s*-polarized pump a new oscillation process is present. The frequency of this process is ~ 3 times higher than the main frequency and the amplitude is ~ 3 times smaller. The presence of the new process is obvious if one compares the form of the minimum in the figures 2 and 3 to that in figure 4. When the pump was *p*-polarized, then the model failed to simulate all parts of the experimental curve. It was possible to simulate either the central part (*calc2*) or the peripheral one (*calc1*). The optimal thickness values were 87 μm for *calc1* and 35 μm for *calc2*. Also the modulation depth of the experimental oscillations in peripheral parts was not 100% as predicted by the single-layer model. This indicates the failure of the single-layer model in the last case.

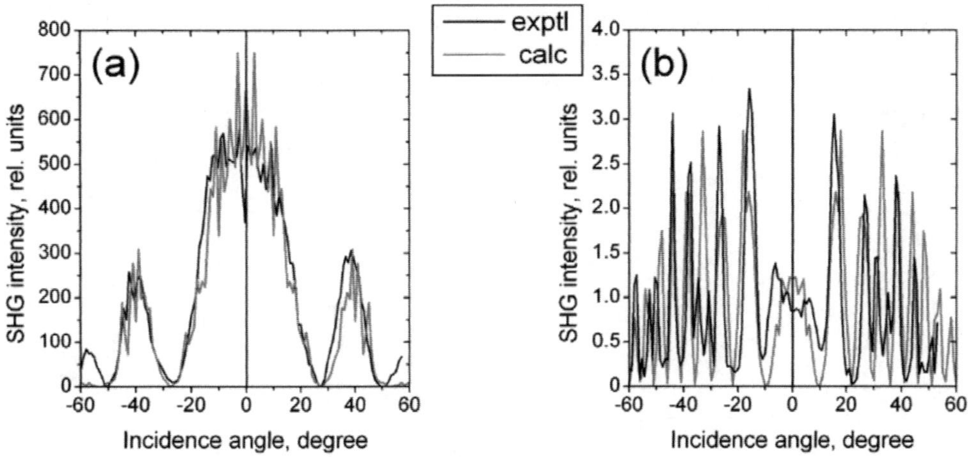

Figure 3. Angular dependencies of SHG signal from a cell covered with polyimide and filled with 5CB; the rubbing direction of the cell is parallel to the direction of the applied DC electric field: (a) *s*-polarized pump; (b) *p*-polarized pump.

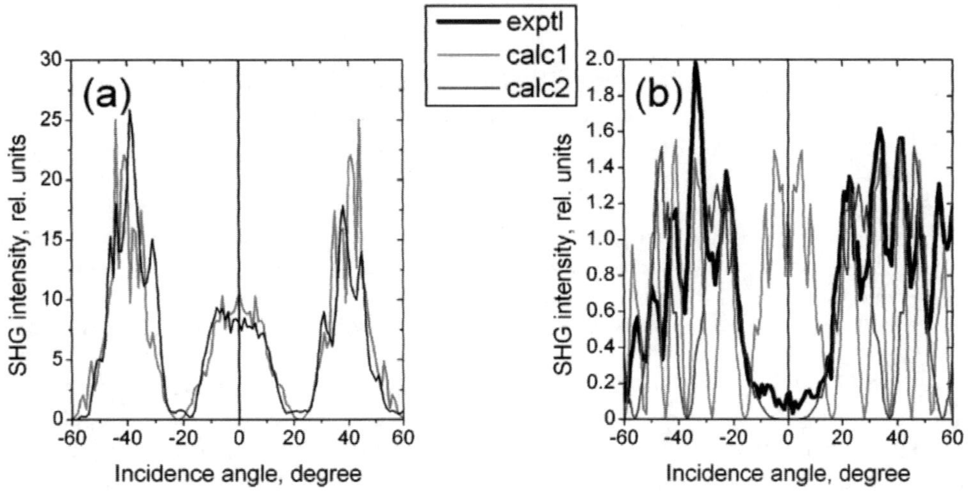

Figure 4. Angular dependencies of SHG signal from a cell covered with polyimide and filled with 5CB; the rubbing direction of the cell is perpendicular to the direction of the applied DC electric field: (a) *s*-polarized pump, (b) *p*-polarized pump.

The results for the homeotropic orientation (inner surfaces of the cell covered with silane) of the 5CB cells are shown in figure 5. For *s*-polarized pump (figure 5 (a)), the new oscillation process was also observed. Actually, within the experimental uncertainty the angular dependencies of SHG signal in this case was close to the one measured by 5CB cells with use of *s*-polarized pump and at crossed directions of the rubbing and the electric field discussed above (recall figure 4 (a)). When the pump was *p*-polarized (figure 5 (b)), then the well known phase synchronism for optical coherent process takes place, resulting in a large increase of the signal for a specific angle, so called phase matching angle. At this angle, the coherent parametric process of SHG through the NLC as a nonlinear optical medium is characterized by enhanced efficiency of the conversion from the input fundamental wave (at the wave frequency ω) to the SHG output. Furthermore, by the phase match the SHG is highly coherent, accompanied with a narrow spatial localization of the SHG beam behind the 5CB cell. Indeed, the SHG output in this case was detected as a highly collinear light beam. The optical phase matching angle is defined only by refractive index values and the wavelength of the pump. The value of this angle calculated from experiment data was in the agreement with the theoretical one. The latter was calculated using known refractive index of 5CB [11] corresponding to homeotropic orientation.

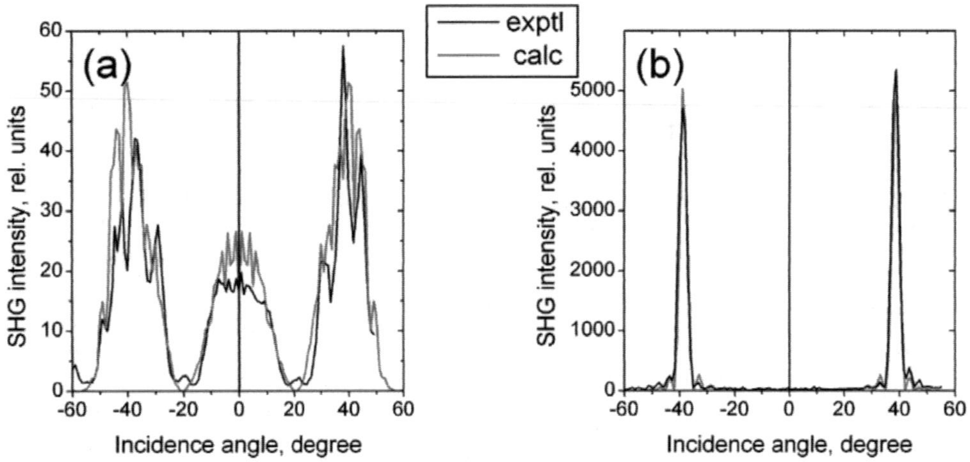

Figure 5. Angular dependencies of SHG signal from a cell covered with silane and filled with 5CB; homeotropic alignment perpendicular to the direction of the applied DC electric field: (a) *s*-polarized pump; (b) *p*-polarized pump.

Possible explanation of the fringe data can be a creation of effectively layered structure in the NLC formed under the high voltage pulses in the case when initial (without field) director orientation is not parallel to the field direction. In the homeotropic samples surface polarizations occur at both silanized (hydrophobic) surfaces due to biphilc asymmetry of 5CB molecule [14]. DC electric field normal to surface polarizations induces a component parallel to itself due to the tilt of surface director (figure 6(a)). On the other hand, in planar samples surface conditions of pretilt angle anchoring for nematic director at both surfaces lead to creation of splay deformation and developing splay flexopolarization in the bulk (figure 6(b)). However this polarization is screened due to space charges. When the electric field is applied it sweeps away the mobile charges allowing the unscreened polarization to interact with the field, and be rotated by it in one case. In most cases then opposite director orientations in upper and lower sublayers take place. The borders of the sublayers are parallel to the glass surfaces and the director orientation within the layers changes from one to another layer, leading to an effective layered structure of the sample. The optical path between these effective layers depends on the

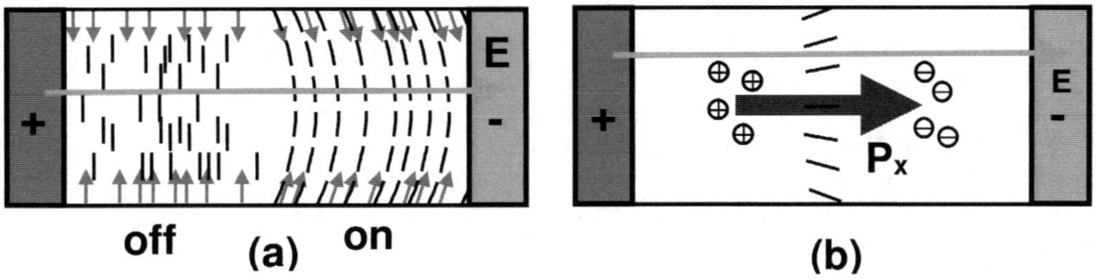

Figure 6. Flexoelectric polarization of plane nematic layers under in-plane electric field **E**: (a) Homeotropic layer. Arrows indicate surface polarization.; (b) Planar layer with free charges. Circles represent ions. The big arrow indicates flexoelectric polarization $\mathbf{P_x}$.

incidence angles. This structure could be responsible for multiple interference observed in SH amplitude modulation. Our experimental results support the conventional view of a liquid crystal (e.g., see figures 6 and 7 in references [1] and [15], respectively. The elucidation of the observed effect of variation of the EFISH intensity and angular distribution of the EFISH excitation (strongly depending on the NLC initial orientation) need further experimental and theoretical research. The efforts in this respect will be continued (work in progress).

4. Conclusions

The effect of external electric field on the coherent optical second harmonic generation in transmission was studied by different initial alignment of nematic liquid crystal (NLC), in particular 5CB. The method of sample rotation demonstrated different EFISH behavior depending on initial director orientation, as well as on the director orientation with respect to the direction of the electric field applied on the examined 5CB cells. Most probably, a layered structure of the liquid crystal under high voltage pulse is responsible for the observed variations in EFISH angular dependence. The EFISH nonlinear optical technique is highly sensitive (also sensitive to the polarization orientation of the incident laser beam) and can be sufficiently applied for optical, structural and director orientation characterization of NLC systems (including NLC mixtures, as well as various micro- and nano-scale composites, like NLC-based polymer-dispersed liquid crystals or NLC doped with nanoparticles).

Acknowledgements
This work has been carried out as a part of Bulgarian-Russian inter-academy BAS-RAS joint research project.

References

[1] Khoo I C 2007 *Liquid Crystals*. Second edition (New York: Wiley Interscience)
[2] Levine B F and Bethea C G 1975 *J. Chem. Phys.* **63** 2666
[3] Chen I H, Chu S W, Bresson F, Tien M C, Shi J W and Sun C K 2003 *Opt Lett.* **28** 1338
[4] Todorova L, Marinov Y, Hadjichristov G B, Petrov A G, Maslyanitsyn I, Torgova S, Strigazzi A, and Shigorin V 2013 *Book of Abstracts 2nd Nat .Conf. Phys. Sci.* (Sofia: Heron Press) 386
[5] Maker P D, Terhune R W, Nisenhoff , and Savage C M 1962 *Phys. Rev. Lett.* **8** 21
[6] Saha S and Wong G K 1979 *Appl. Phys. Lett.* **34** 423
[7] Saha S K and Wong G K 1979 *Opt. Commun.* **30** 119
[8] Barnik M I, Blinov L M, Dorozhkin A M and Shtykov N M 1981 *JETP* **54** 935

[9] Marinov Y, Shonova N, Naydenova S and Petrov A G 2000 *Mol. Cryst. Liq. Cryst.* **351** 411
[10] Alexandrova S, Maslyanitsyn I A and Shigorin V D 2008 *Phys.Wave Phenom.* **16** 167
[11] Li J, Wen C H, Gauza S, Lu R and Wu S T 2005 *Display Technology* **1** 5
[12] Chemla D S and Zyss J (eds) 1987 *Nonlinear Optical Properties of Organic Molecules and Crystals* (New York: Academic Press)
[13] Herman W N and Hayden L M 1995 *J. Opt. Soc. Am. B* **12** 416
[14] Petrov A G and Derzhanski A 1977 *Mol. Cryst. Liq. Cryst. Lett.* **41** 41
[15] Lavrentovich O D 2003 PRAMANA - *J. Phys.* **61** 373

Photo-stimulated electro-optic response of liquid-crystalline system with *trans-cis* photo-isomerizable agent

G B Hadjichristov[1,*], Y G Marinov[1] and C V Yelamaggad[2]
[1] Institute of Solid State Physics, Bulgarian Academy of Sciences, 72 Tzarigradsko chaussee, Sofia 1784, Bulgaria
[2] Centre for Nano and Soft Matter Sciences, Jalahalli, Bangalore 560 013, India

E-mail: georgibh@issp.bas.bg

Abstract. A rather strong photo-stimulated enhancement of photo-induced bend flexoelectric effect based on *trans-cis* photoisomerization of azo bond was found in a guest-host system formed from the nematic liquid crystal (LC) *N*-(4-methoxybenzylidene)-4-butylaniline (MBBA) as a host, and the azobenzene LC 4-hexyloxybenzoloxy-4´-cyanoazobenzene, as a guest photoactive agent at 1 wt.% concentration. Upon application of electric field, thin homeotropic layers of thickness 100 μm containing this photo-sensitized LC mixture were investigated as subjected to a relatively weak illumination with UV light (λ = 375 nm, from narrow-band light-emitting diode, LED). The stimulation of the photoactive electro-optic response of azobenzene-doped MBBA (owing to enhanced photo-induced bend flexoelectric effect driven by the photo-isomerizable dopants) was achieved by pre-resonant excitation of the photoactive agent. The degree of the effect measured is of potential interest for thin-film photoactive electro-optic applications. The UV light-induced effect in azobenzene-doped MBBA was reversible; the back (relaxation) process was stimulated by light in the blue from a LED with broadband spectrum centered at 455 nm.

1. Introduction

In the field of liquid crystal (LC) physics, flexoelectricity is well known [1, 2]. Under external electric field applied on the LC cells, the LCs exhibit flexoelectric polarization with magnitude depending on the LC material. The flexo-electric effect can be stimulated by light [3]. In the past, photo-induced dipolar flexoelectricity in homeotropic guest-host nematics upon UV illumination has been reported for a guest-host system formed by the photo-responsive azobenzene compound *p*-nitro-*p*'-diethylamino-azobenzene (D1), as a guest in a small concentration, and the nematic LC *N*-(4-methoxybenzylidene)-4-butylaniline (MBBA) as a host [4]. This UV light-induced effect related to the photo-dependent flexopolarization, labelled thereafter photoflexoelectric effect (PFE), results from *trans-cis* photoisomerization of the azo bonds of the guest compound.

The photoactive properties of azobenzene molecules are well known [5, 6]. In particular, azobenzene-doped nematics exhibit a wide variety of photo-physical and photoswitching effects [3, 7, 8], and the useful functions of this important class of synthetic photochromic LC materials have attracted much attention due to the large potential for specialty and high-technology applications [9-13]. Recently, we have investigated photosensitive guest-host systems composed of a nematic LC (as a host) mixed with the novel photoactive azobenzene compound LC 4-hexyloxybenzoloxy-4´-

* To whom any correspondence should be addressed.

cyanoazobenzene (UI-147) as a guest [14–16]. For thin (100 μm or less) homeotropic layer of such LC mixtures, the decisive role of the subsurface layer for the observed PFE was concluded. Here, we present an experimental study on PFE in thin (100 μm) homeotropic layers of nematic MBBA doped with UI-147 molecules. The enhanced photo-stimulated electro-optic response of this photosensitized nematic LC material is characterised in terms of the PFE concept [3,4,17,18], but our research is focused mainly on the photo-stimulation of the photo-physical mechanism involved that is closely related to the practical applicability of the effect and corresponding photoactive LC materials.

2. Experimental

The host compound in the investigated LC mixture was the commercial nematogen *N*-(4-methoxybenzylidene)-4-butylaniline (abbreviated as MBBA), known for a long time. The nematic-isotropic transition of this LC (from Reahim) is at 43° C. The photoactive agent in the mixture was the LC 4-hexyloxybenzoloxy-4'-cyanoazobenzene (abbreviated as UI-147, the chemical structure is given in figure 1 a). This stable (chemical- and photo-stable) azo-containing LC compound was synthesized at the Centre for Nano and Soft Matter Sciences in Jalahalli (Bangalore, India) and exhibited photophysical properties very suitable for photoactive device applications [15]. In present study, the concentration of the 'guest' UI-147 in the guest-host UI-147/MBBA molecular system was 1 wt.%. The nematic phase of UI-147 is at a relatively high temperature. However, a photoactive room-temperature LC material can be composed from UI-147 by mixing with proper room-temperature nematic LC [15], as MBBA.

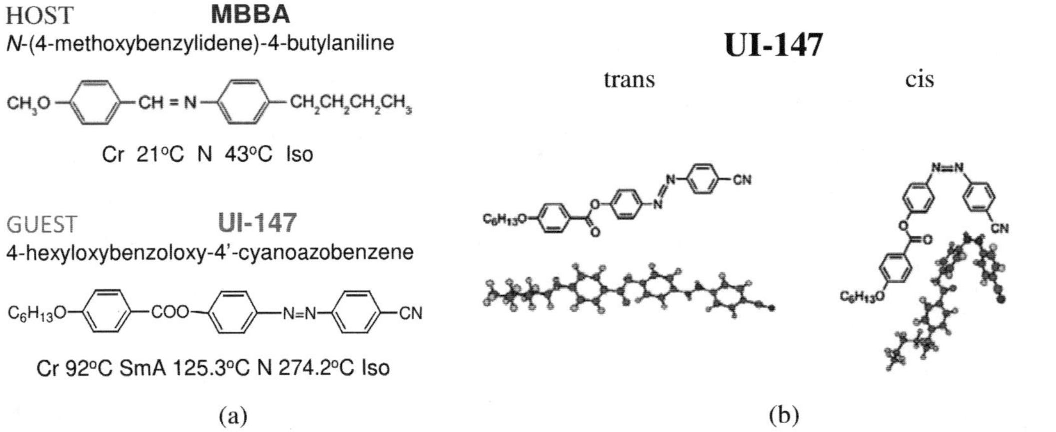

Figure 1. (a) Molecular structure of the host MBBA and azobenzene dopant UI-147. The phase transition temperatures are given. (b) *Trans* (in the left) and *cis* (in the right) conformers of UI-147.

For the investigation, homeotropic layers of UI-147/MBBA with a thickness of 100 μm were prepared. The LC mixture was sandwiched in glass cells as described in [16, 19]. Cells of dimensions 1.5 cm × 2 cm were assembled from two parallel 1 mm-thick glass plates. The copper spacers between the plates served as electrodes, with their separation being 2 mm. Initial homeotropic orientation of the LC layer was achieved using self-assembled layers of lipid lecithin [20, 21]. Identical cells, but filled with pure MBBA, were also prepared in order to be measured as reference cells in our study.

The photo-stimulated electro-optic response of the prepared UI-147/MBBA layers was studied by measurement of probe beam transmittance through the LC cells placed between crossed polarizers and subjected to a transversal DC electric field. The measurements were performed using the experimental setup schematically shown in figure 2. Linearly-polarized output of He-Ne laser (wavelength $\lambda = 633$ nm) with a power of ~ 1 mW was used as a probe beam at normal incidence on the sample. The power stability of this laser source was better than ± 1%.

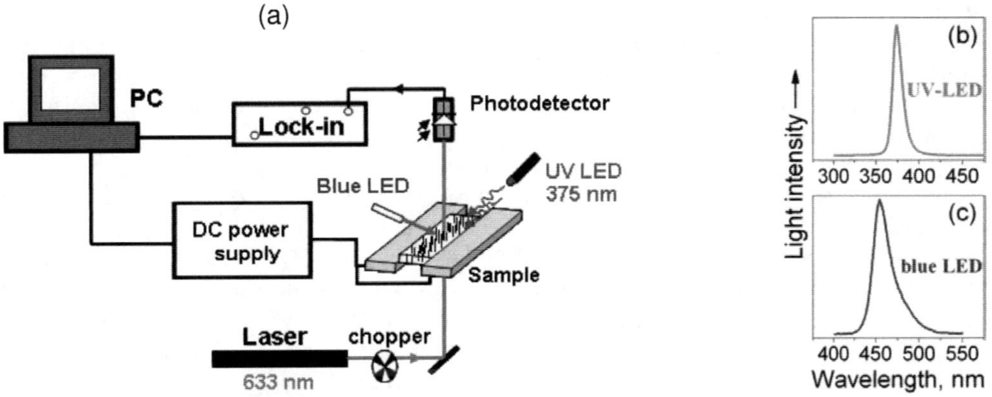

Figure 2. (a) Schematic drawing of the experimental set-up used in this work. In the right: spectral characteristics of the LEDs employed: (b) UV-LED; (c) blue LED.

The azobenzene conformers in the samples were photo-driven with two continuous light sources (light-emitting diodes, LEDs), 'UV-LED' and 'blue LED', of a relatively low optical power but highly directed unpolarized outputs. The emission from the UV-LED (type NSHU550A, Nichia Corporation, Japan) with a flux 2 mW (can type, lens window, directivity 10°) was at λ = 374.6 nm. (Since this wavelength is close to the spectral range of the visible, the light is actually UV-visible, and can be observed with naked eye). The UV-LED was narrow-band ($\Delta\lambda$ = 12 nm fwhm). The second LED (LED ENERGY curing unit, model 405-8B, Optica Laser, Bulgaria) emitted a broader spectrum ($\Delta\lambda$ = 23 nm fwhm) with a peak at 454.5 nm.

The LED illuminators were mounted on holders with micrometer motion. Thus, the UV-visible and the blue light spots of non-focused light were carefully coincided at the central zone of the LC cell, probed with He-Ne laser beam. Through a circular diaphragm, the diameter of both spots on the sample was kept fixed to extent slightly smaller than the gap between the cell electrodes. The light intensities on the sample surface was adjusted by varying the distance between the LEDs and the sample. In order to avoid the unwanted effect of so called 'surface melting' (light-induced tilted anchoring) well established for the photoactive UI-147/MBBA mixture under study [22], the light intensity from the UV-LED on the sample was kept below 3 mW/cm^2. The blue LED light (attenuated) intensity on the sample was ~ 50 mW/cm^2. Special care was taken to minimize, if not completely eliminate the effects from the unwanted stray and reflected light from various sources on the measured signal modulated by PFE.

DC voltage up to 40 V (applied between the copper electrodes of the cells) was supplied by Instek GPD-4303S Programmable Linear DC Power Supply. The same 4-channel device was employed to supply the two LEDs (DC voltage stability ± 0.01 %) and to command their output on/off functions. The DC electric field-dependent light transmittance of the LC cells was measured with a silicon photodiode as a detector, and computer-controlled lock-in amplifier (SR830 DSP, Stanford Research Systems). The PFE was also measured as depending on alternating current (AC) electric field applied on the LC cell. In this case, a sinusoidal voltage was applied and the field frequency was swept in the range from 1 Hz to 1 kHz. A computer handled the data acquisition from the lock-in instrument and also controlled the frequency sweeps.

The morphology of the samples was characterized by a transmission optical microscope (Zeiss NU-2 Universal Research Microscope). The optical absorption of the samples in the wavelength region of 300 – 800 nm, and at room temperature (ca. 22 – 24°C), was measured by means of Perkin Elmer Lambda 1050 spectrophotometer. The spectral resolution was 1 nm. During the PFE experiment the sample temperature was maintained (to within ± 0.1 °C) by a Mettler FP82 hot stage.

3. Results and Discussion

The homeotropic alignment of the prepared layers was inspected by a polarizing microscope set for crossed polarizer configuration. The high quality and extent of uniformity of the homeotropic orientation in the absence of electric field and light illumination by the LEDs was confirmed by conoscopy. Also, the lack of a 'surface melting' of the samples upon the UV-light illumination with a relatively low intensity we have employed (< 3 mW/cm^2) was verified by optical microscopy.

As known, the azobenzenes exhibit *trans* and *cis* isomers. Figure 3 (a) reports the UV-visible absorption spectrum of UI-147 where the specific spectral shapes due to the *trans* and *cis* isomers of this azobenzene are seen (the bands at 335 nm and 450 nm, respectively). Figure 3 (b) compares the optical absorption measured for the LC cell with UI-147/MBBA mixture to that with pure MBBA (both not illuminated). The effect from the UI-147 dopants can be clearly seen.

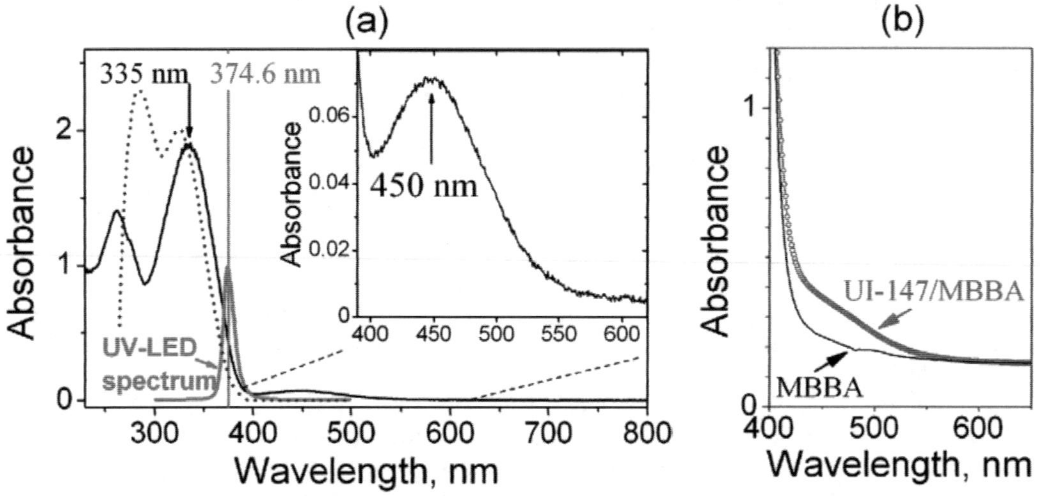

Figure 3. (a) UV-VIS absorption spectra of chloroform solutions of 0.01 % UI-147 (solid line) and 0.13mM MBBA (dotted, blue). The insert shows an expanded part of UI-147 spectrum in the blue-green region; (b) The optical absorption measured for UI-147/MBBA mixture (1 wt.% UI-147) and pure MBBA, both in 100 µm-thick cells with homeotropical alignment, not illuminated by the LEDs used here. The spectra were recorded under identical conditions.

By illumination at a wavelength lying in the absorption region, azobenzene molecules can undergo *trans-cis-trans* conformation (isomerization). This is the operation mechanism of the azobenzene-containing and azobenzene-doped LC materials for photo-control in electro-optics and photonics [3, 7–13, 23–31]. Upon illumination by UV light with appropriate wavelength and intensity, the UI-147 molecules isomerize from *trans* to *cis* form. Thus, by absorbing the UV light energy, the molecule shape of this organic compound is changed from rod-like to a bent-shaped one (see figure 1b). The corresponding change in the optical absorption after this photo-induced effect in the studied UI-147/MBBA mixture under the actual conditions of our experiment on PFE (reported below), is evident in figure 4. The resulting difference spectrum for UI-147/MBBA (see the insert in figure 4) represents the spectral absorption band of the *cis*-conformers of UI-147. Its presence indicates an efficient *trans*-to-*cis* photoisomerization of this azobenzene, even when a small part of the UV-light energy is absorbed by UI-147.

Actually, by UV-visible light 'pumping' of the UI-147 photo-agent one should consider in our case a combination of three factors: (i) the relatively low intensity of the UV-visible light incident on the sample; (ii) UI-147 excitation in the wing of its absorption band (pre-resonance condition, see figure 3 a); and (iii) the UV-light intensity loss due to the strong absorption from the host MBBA itself (figure 3b). Overlapped with the UI-147 absorption, the host MBBA absorption is crucial for the light pumping in the *trans*-form absorption band of UI-147, because this azobenzene is only 1 wt.% in the UI-147/MBBA mixture. Since the main absorption band of UI-147 is somewhat shifted to the longer wavelengths with respect to the MBBA absorption (figure 3a), the pre-resonant excitation of UI-147 becomes a favourable way for *trans*-to-*cis* photoisomerization. Upon pre-resonance conditions, the excitation wavelength is longer than the resonance wavelength but still corresponds to some small portion of the absorption spectrum. Usually, such an excitation is applied to avoid the non-desirable resonant molecular fluorescence. This approach is commonly used in resonance Raman spectroscopy (especially in various techniques of the coherent Raman spectroscopy). The pre-resonant excitation allows the loss and the other effects due to absorption to be minimized, generates much less molecular fluorescence and background noise.

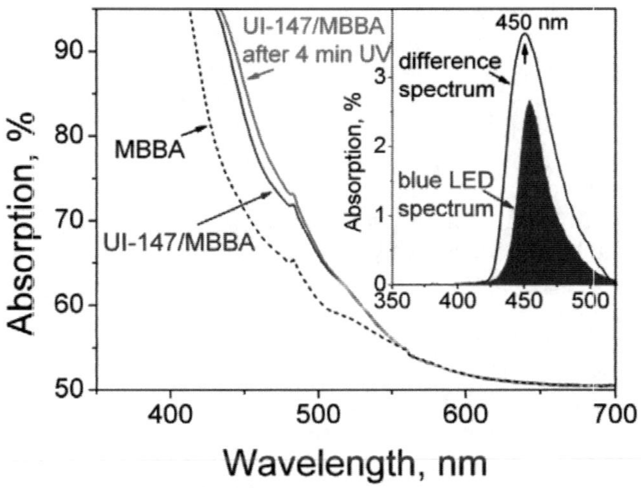

Figure 4. Absorption spectra of UI-147/MBBA recorded without (blue curve) and with (red curve) 4 min illumination with the UV-LED putted within the spectrophotometer. For the sake of comparison, the absorption spectrum of reference MBBA cell is also given (dotted). The spectra were taken under identical conditions and were corrected by the absorption recorded for an empty glass cell identical to the cells containing the LCs. The shoulders at ~ 480 nm are due to apparatus artifacts (from the lighting lamps of the spectrophotometer). In the insert is shown the difference of the UI-147/MBBA spectra (illuminated – non-illuminated UI-147/MBBA).

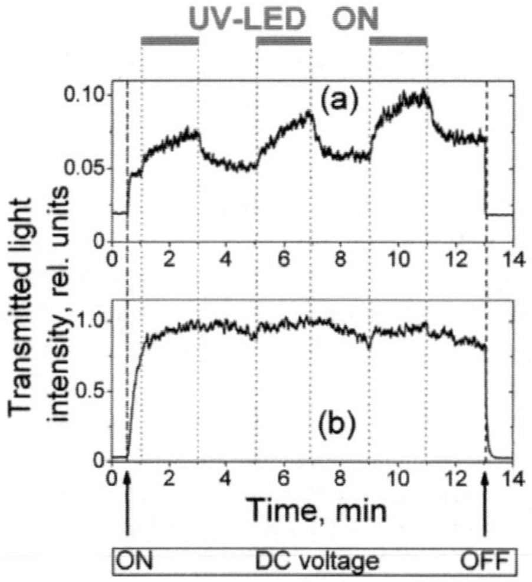

Figure 5. The time dependence of the variation of the intensity of probe He-Ne laser beam transmitted through homeotropic cell with UI-147/MBBA (a) and pure MBBA (b), both 100 μm-thick, in response to the light from UV-LED and the DC voltage ON-OFF switching. The illumination time intervals were fixed at 2 minutes alternated with equal dark periods (the periodical ON-switching of the UV-LED illuminator is indicated at the top). The UV-LED light intensity on the sample was 1.2 mW/cm^2, the DC voltage applied on the cell was 40 V, the temperature of the cell was 28 C°.

Being considerably weaker in the blue spectral region (the absorption region of the UI-147 *cis*-form), the MBBA absorption (extended up to 520 – 550 nm, figure 3 b) should not strongly affect the photo-induced *cis*-to-*trans* back-isomerization. By comparison of the difference spectrum in figure 4

Journal of Physics: Conference Series **558** (2014) 012026 doi:10.1088/1742-6596/558/1/012026

with the spectral band of our blue LED (see the insert in figure 4), one can expect (due to their optimal overlap) an efficient *cis*-to-*trans* back-photoisomerization of UI-147 azobenzene in the UI-147/MBBA cell under study. In fact, this was confirmed in our experiment and the observed chromism was fully reversible indeed. On the other hand, no change in the MBBA absorption was present upon illumination with the LEDs under the same experimental conditions.

Based on the above absorption spectral behaviors and the observed efficient interaction of the irradiation of both UV-LED and blue LED with the UI-147/MBBA under study, we probed the PFE in this photoactive guest-host system. The photoflexo-electrooptical response of homeotropic layers of UI-147/MBBA placed between crossed polarizers and subjected to in-plane electric field directed at 45° with respect to the polarizer axes orientation (figure 2) was studied by measurement of the changes in the transmitted 'probe'-light intensity by switching the 'pump'-light and electric field applied, as well as by varying the latter.

A time scan for examined UI-147/MBBA thin homeotropic layer being illuminated with the UV-LED, is shown in figure 5(a). As seen, a significant increase in the transmitted light intensity takes place. The mechanism of the registered PFE is similar to the one observed for other guest-host LC systems containing nematic LC as a host and the same photoactive azobenzene compound UI-147 as a guest [14, 15] - the near(sub)surface region rich of azo molecules is responsible for the observed PFE. The azo molecules adsorbed at the top surface and those located within the subsurface region, contribute to the photoflexo-response of the examined guest-host system. It should be noted that under the same experimental conditions, we have found no such effect in the reference MBBA cells (figure 5(b)). We have to point also out that the influence of any scattered light, light reflections, as well as possible luminescence/fluorescence from the UI-147/MBBA cells (excited by the UV-LED) on the measured transmitted light intensity, that could disturb the registered signal resulting from PFE (the transmitted probe light intensity), were negligible, as we have accurately proved in the course of the experiment.

The low intensity of the light illumination minimizes the possible thermo-induced effects. Further, owing to the low level of the incident UV-light intensity, the PFE was properly distinguished from the accompanying phenomenon 'surface-melting' of the photo-driven LC command surface. Such unwanted effect was found in identical cells with homeotropic layers of the same UI-147/MBBA mixture [22]. Upon UV light, the near(sub)surface region enriched with azo-molecules that is responsible for the PFE can be affected by the 'surface-melting' anchoring transition that disturbs the homeotropic alignment. That is why, to avoid this effect in the present experiments, the UV light intensity on the sample was reduced below the levels at which the effect of tilted anchoring occurs (~ 10 mW/cm^2 at $\lambda = 375$ nm [22]).

The characteristics of the PFE were consistent with those previously discussed in details in [4]. The magnitude of the flexoelectricity that was photo-induced in the mixture was much higher than that in the nematic host alone. Figure 6 (a) shows the DC voltage-dependent behavior (static electro-optical characteristic measured by chopped at 90 Hz probe beam), and figure 6 (b) - the photo-dependent component of the flexopolarization (the first harmonic of the modulated optical signal from transmittance of the cell) as a function of the frequency of AC electric field applied on the cell (dynamic electro-optical characteristic). The DC voltage dependences of both dark and light flexoresponses followed the theoretical predictions [4,32]. Also, a clear photo-stimulated enhancement of the amplitude of the AC electric field-excited flexo-oscillations was registered.

Figure 7 presents time courses for examined thin homeotropic layer of UI-147/MBBA with exposure to light of both UV-LED and blue LED, showing the quenching and recovery of photo-induced flexoelectric deformations of azobenzene doped MBBA by repeating cycles of alternating switching of the blue LED (the UV-LED is turned-on and remains in the ON state for the rest of the experiment). As seen, the relaxation process is accelerated by the *cis-trans* back-izsomerization of the UI-147 azo chromophores after illumination with the blue LED. The lack of noise in the time behaviors is because of the higher level of the signal due to higher intensity of the illumination with the UV-LED

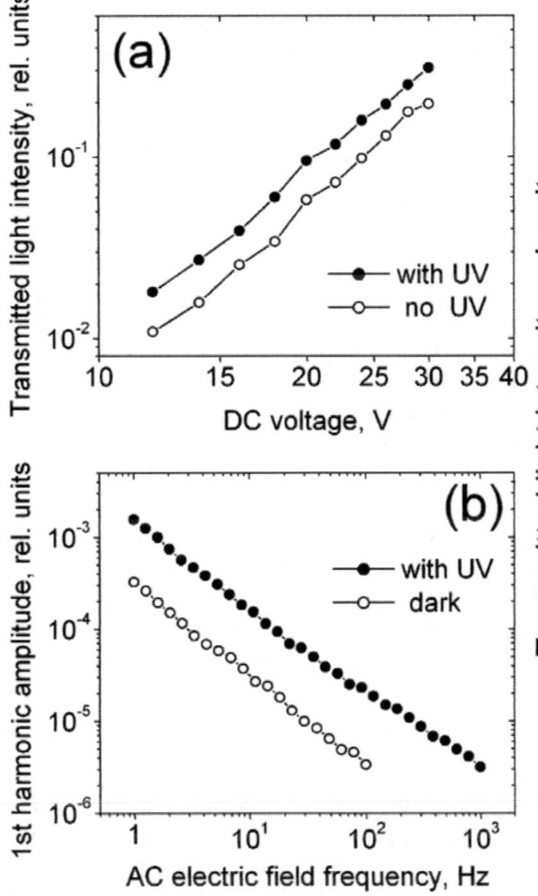

Figure 6. Log-log plot of transmitted light intensity vs DC voltage (saturated flexoresponse) (a) and the frequency dependence of the amplitude of the flexo-oscillations (the first harmonic of the modulated light signal) (b) of the layered UI-147/MBBA. The UV-LED light intensity on the sample was 1.2 mW/cm^2, the temperature of the cell was 27° C.

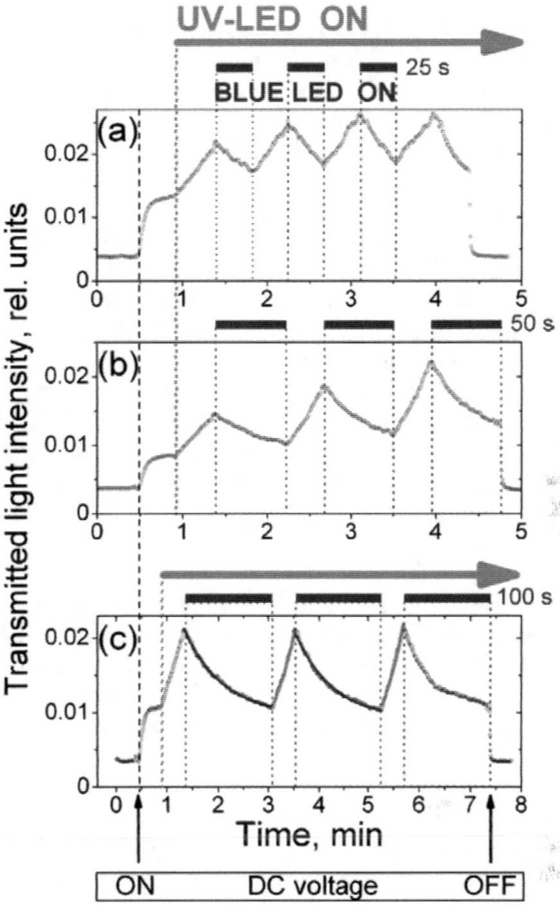

Figure 7. As in figure 5 (a), but the UV-LED lighting was steady, and the illumination with the blue LED was switched-on with periods of: 25 s (a); 50 s (b); 100 s (c), alternated with a switch-of periods of 25 s (the light switching regime is shown at the top). The lines in (c) represent the fits to the experimental data. The UV-LED light intensity on the sample was 3 mW/cm2, the DC voltage applied on the cell was 40 V, the temperature of the cell was 28 C°.

One can compare the PFE in the measured UI-147/MBBA mixture by varying the blue LED illumination interval. When the duration of the later is long enough (or the blue LED intensity is higher), the signal relaxes to the initial background level (figure 7 c), i.e. a steady dynamic balance at alternating cycles of the UV-visible irradiation occurs. The experimentally obtained temporal variation of both rise and decay parts of the PFE behavior can be exactly fitted by a simple mono-exponential function. The values of the corresponding rise and decay time constants determined in this case to be ≈ 31 s and ≈ 42 s, respectively, indicate a slow response to the illumination switching, as expected. Compared to relaxation times reported in [4] (≈ 15 s), the photo-stimulated process for the photo-responsive UI-147/MBBA material studied here, is about two times slower. It may be noted that the photoisomerization of azobenzene derivatives is a fast process, e.g. it is completed within 10 ns [33].

At the present level of UV-light intensity, the observed PFE is considerably stronger as compared to the result (12 % – 18 %) reported previously [4] for a guest-host system formed by MBBA as a host and the azobenzene dye D1 as a guest, under comparable experimental conditions (including the azobenzene concentration, the cell thickness, the applied DC voltage and temperature). The reasons could be: (i) a higher photo-induced bend flexopolarization in UI-147/MBBA than that in D1/MBBA (indeed, photoisomerization of the strongly polar UI-147 azobenzene molecules possessing a large longitudinal dipole moment and asymmetric molecular structure (polar-non polar asymmetry in the end substituents) [16], was found to be able to induce a strong PFE in nematic LCs [14, 15]); and/or (ii) stronger or more optimal UV illumination as compared with that used in [4]. Note that the UV illumination at 334 nm (Hg spectral line from a spectral lamp) employed in [4] is close to the absorption peak of the host MBBA (recall figure 3a). In our present experiment, the pre-resonance UV illumination is more favorable because in this way is avoided the strong absorption from the host MBBA (which is a loss, actually, and can bring negative effects), and in the same time provides for a sufficient photoizomerization of the guest.).

Obviously, as compared with high-performance organic photochromic molecular systems that possess very efficient electronic transition and energy transfer processes [34,35], the efficiency and speed of the photochromic function (based on photo-flexo-electro-optic response) of the azobenzene-doped photo-sensitized nematic LC studied here, are lower. Nevertheless, such azo-doped nematics (being sensitive photoactive LC materials with a wide temperature range for the nematic phase) can be still useful for practical applications (e.g., light-commanded electro-optic modulators and switches), in the case when the speed is not of crucial importance.

4. Conclusion

Here we report on enhanced photoflexoelectric effect in azobenzene-doped nematic MBBA. Through an efficient photochromic *trans-cis* photoisomerizm, the azobenzene chromophores in MBBA allow the electro-optic response of thin homeotropic layers with the doped MBBA to be efficiently controlled by UV-visible light irradiation of the layers. Based on photo-induced flexoelectric polarization, a reversible and steady photomodulation is achieved by the doped MBBA layers at alternating cycles of the UV-visible irradiation. Even at the relatively low (1 wt.%) concentration of the azo photoresponsive agent and the low-intensity illumination with UV-visible light, the *trans-cis* photoisomerization is sufficient to drive the electro-optics of the doped MBBA layers and to efficiently modulate the intensity of the light transmitted through them.

Our study shows that upon optimal illumination the photo-stimulated electro-optic response of azobenzene-doped nematics can be enhanced to a degree that can be suitable for electro-optic applications utilizing the photoactivity of the azo-dopants. To achieve an efficient *trans-cis* photoisimerization, the preresonant excitation with UV-visible light can be a reasonable alternative against the illumination by UV light at the azobenzene absorption peak. In this way, the amount of the produced *cis*-form of the photoactive dopants can be sufficient to enhance the light-induced flexoelectrooptic effect activated in the doped nematic layers. The results reported here will be useful for the practical applications of LC-based photoresponsive materials for the photo-controlled electro-optics and photonics.

Acknowledgments

The authors are very thankful to Academician Alexander Petrov from the Institute of Solid State Physics "Acad.Georgi Nadjakov", Bulgarian Academy of Sciences, and to Dr. S. Krishna Prasad from the Centre for Nano and Soft Matter Sciences, Jalahalli, Bangalore, India, for their valuable comments and suggestions. This study was carried out as a part of Indo-Bulgarian joint research project (DNTS/India-01/4/2013 "Investigation of photo-stimulation effects in nano-structured liquid crystalline systems") supported by the Department of Science and Technology, New Delhi and the Ministry of Education and Science, National Science Fund of Bulgaria.

References

[1] Petrov A G 2001 Measurements and interpretation of flexoelectricity *Physical Properties of Liquid Crystals: Nematics (EMIS Datareviews Series)* eds D A Dunmur, A Fukuda, G R Luckhurst (London: INSPEC- Institution of Electrical Engineers) chapter 5 p 251

[2] Buka A and Eber N 2013 eds *Flexoelectricity in Liquid Crystals: Theory, Experiments and Applications, Imperial College Press* (London: Imperial College Press)

[3] Hermann D S, Rudquist P, Ichimura K, Kudo K, Komitov L and Lagerwall S T 1997 *Phys. Rev. E* **55** 2857

[4] Marinov Y, Shonova N, Blinov L M and Petrov A G 1998 *Europhys. Lett.* **41** 513

[5] Natansohn A and Rochon P 2002 *Chem. Rev.* **102** 4139

[6] Yager K G and Barrett C J 2006 *J. Photochem. Photobiol. A* **182** 250

[7] Prasad S K, Nair G G, Hegde G, Sandhya K L, Rao D S S, Lobo C V and Yelamaggad C V 2005 *Phase Transitions* **78** 443

[8] Yelamaggad C V, Prasad S K and Li Q 2012 Photo-Stimulated Phase Transformations in Liquid Crystals and Their Non-Display Applications *Liquid crystals beyond displays, Chemistry, Physics, and Applications* ed Q Li (Hoboken, NJ: Wiley & Sons) chapter 5 p. 157

[9] Ikeda T and Tsutsumi O 1995 *Science* **268** 1873

[10] Tsutsumi O and Ikeda T 2002 *Mat. Sci.* **6** 563

[11] Urbas A, Klosterman J, Tondiglia V, Natarajan L, Sutherland R, Tsutsumi O, Ikeda T and Bunning T 2004 *Adv. Mater.* **16** 1453

[12] Barrett C J, Yager K G, Mamiya J and Ikeda T 2007 *Soft Matter* **3** 1249

[13] Yager K G and Barrett C J 2009 Azobenzene Chromophores in Photo-Reversible Materials *Smart Light-Responsive Materials: Azobenzene-Containing Polymers and Liquid Crystals* eds Y Zhao and T Ikeda (Boca Raton, Florida: John Wiley and Sons) chapter 1 p. 1

[14] Marinov Y G, Hadjichristov G B, Petrov A G, Sridevi S, Hiremath U S, Yelamaggad C V and Prasad S K 2010 *J. Phys. Conf. Ser.* **253** 012060

[15] Petrov A G, Marinov Y G, Hadjichristov G B, Sridevi S, Hiremath U S, Yelamaggad C V and Prasad S K 2011 *Mol. Cryst. Liq. Cryst.* **544** 3/[991]

[16] Sridevi S, Hiremath U S, Yelamaggad C V, Prasad S K, Marinov Y G, Hadjichristov G B and Petrov AG 2011 *Mater. Chem. Phys.* **130** 1329

[17] Spassova M, Petrov A G and Fendler J H 1995 *J. Phys. Chem.* **99** 15630

[18] Petrov A G 1999 *The Lyotropic State of Matter* (New York: Gordon & Breach Science Publishers)

[19] Kumar P, Marinov Y G, Hinov H P, Hiremath U S, Yelamaggad C V, Krishnamurthy K S and Petrov A G 2009 *J. Phys. Chem. B* **113** 9168

[20] Naydenova S, Petrov A G and Yarwood J 1995 *Langmuir* **11** 3435

[21] Marinov Y, Shonova N, Naydenova S and Petrov A G 2000 *Mol. Cryst. Liq. Cryst.* **351** 411

[22] Marinov Y G, Hadjichristov G B, Petrov A G, Sridevi S, Hiremath U S, Yelamaggad C V and Prasad S K 2012 *J. Phys. Conf. Ser.* **398** 012038

[23] Lim S L, Li N J, Lu J M, Ling Q D, Zhu C X, Kang E T and Neoh K G 2009 *ACS Appl. Mater. Interfaces* **1** 60

[24] Simoni F, Lucchetti L, Lucchetta D E and Francescangeli O 2001 *Opt. Express* **9** 85

[25] Sung J H, Hirano S, Tsutsumi O, Kanazawa A, Shiono T and Ikeda T 2002 *Chem. Mater.* **14** 385

[26] Nakano M, Yu Y, Shishido A, Tsutsumi O, Kanazawa A, Shiono T and Ikeda T 2003 *Mol.*

Cryst. Liq. Cryst. **398** 1

[27] Kurihara H, Shishido A, Tsutsumi O, Shiono Tand Ikeda T 2005 *Mol. Cryst. Liq. Cryst.* **441**, 171

[28] Saint-Jalm S, Miniewicz A, Karpinski P, Jarek-Mikulska U and Galewski Z 2012 *J. Molec. Liq.* **168** 21

[29] Seki T, Hara M and Nagano S 2013 Photoinduced Alignment and Structuring in Thin Films of Azobenzene-Containing Liquid Crystalline Polymers *New Frontiers in Photochromism* eds M Irie, Y Yokoyama and T Seki (Tokyo: Springer Japan) chapter 5 pp 79

[30] Son J H, Lee S H, Hong S H, Zin W C and Song J K 2013 *Liq. Cryst.* **40** 646

[31] Liu Y C, Cheng K T, Chen H F and Fuh A Y 2014 *Opt. Express* **22** 4404

[32] Derzanski A, Petrov A G and Mitov M D 1978 *J. Phys. (Paris)* **39** 273

[33] Kurihara S, Ikeda T, Sasaki T, Kim H B and Tazuke S 1990 *J. Chem. Soc. Chem. Commun.* **1990** 1751

[34] Nanasawa M 1999 Photochromism by electron transfer: photochromic viologens *Organic Photochromic and Thermochromic Compounds (Topics in Applied Chemistry, Vol. 1, Main Photochromic Families)* eds J C Crano and R J Guglielmetti (New York: Plenum Press) p. 341

[35] Irie M, Yokoyama Y and Seki T 2013 eds *New Frontiers in Photochromism* (Tokyo: Springer Japan)

Journal of Physics: Conference Series **558** (2014) 012027 doi:10.1088/1742-6596/558/1/012027

Effect of Amphotericin B antibiotic on the properties of model lipid membrane

S Kiryakova, M Dencheva-Zarkova, J Genova
Institute of Solid State Physics, Bulgarian Academy of Sciences, 72 Tzarigradsko Chaussee, Blvd., 1784 Sofia, Bulgaria

Corresponding author e-mail: sophia_kiryakova@abv.bg

Abstract. Model membranes formed from natural and synthetic lipids are an interesting object for scientific investigations due to their similarity to biological cell membrane and their simple structure with controlled composition and properties. Amphotericin B is an important polyene antifungal antibiotic, used for treatment of systemic fungal infections. It is known from the literature that the studied antibiotic has a substantial effect on the transmembrane ionic channel structures. When applied to the lipid membranes it has the tendency to create pores and in this way to affect the structure and the properties of the membrane lipid bilayer.
In this work the thermally induced shape fluctuations of giant quasi-spherical liposomes have been used to study the influence of polyene antibiotic amphotericin B on the elastic properties of model lipid membranes. It have been shown experimentally that the presence of 3 mol % of AmB in the lipid membrane reduces the bending elasticity of the lipid membrane for both studied cases: pure SOPC membrane and mixed SOPC-Cholesterol membrane.
Interaction of the amphotericin B with bilayer lipid membranes containing channels have been studied in this work. Model membranes were self-assembled using the patch-clamp and tip-dip patch clamp technique. We have found that amphotericin B is an ionophore and reduces the resistance of the lipid bilayer.

1. Introduction

Infections as a consequence of weakened immunity defences, such as from AIDS or from many cancer therapies, have renewed the scientific studies on polyene antibiotic amphotericin B (AmB). The structure of amphotericin B is given in figure 1. This antibiotic is the most common antifungal agent used to treat deep-seated mycotic infections [1].

In biological systems, AmB acts by forming ion channels that induce ion leakage across lipid membranes [2]. The formation of these channels is highly dependent on the presence of cholesterol (figure 1B) in the membrane [3]. It has been suggested that the channels are composed of AmB – cholesterol complexes made of superimposed antibiotic and cholesterol molecules. The polar groups on these molecules face toward the interior of the channel where they interact with water molecules, while the hydrophobic groups interact with the aliphatic chains of membrane phospholipids [4].

However, it has been later indicated that the mechanism of action of AmB is similar for both mammalian and fungal cells. The ion leakage that leads to the cell's death has the same effect on both types of cells because of their similar structure. Despite its toxicity this antibiotic is still irreplaceable in many micotic therapies because of its wide spectrum of application.

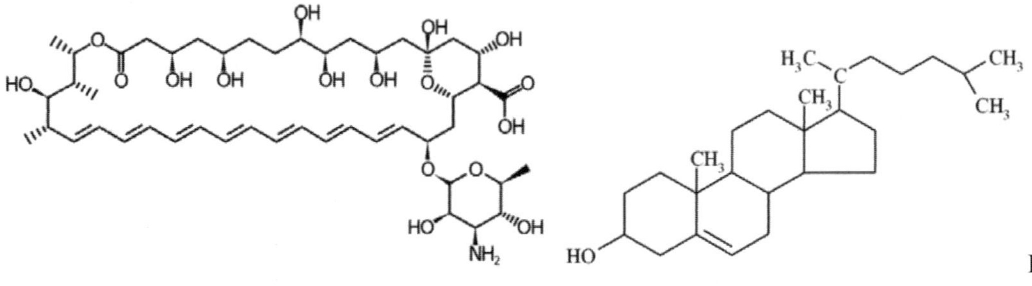

A B

Figure 1. Structure of (A)- antibiotic Amphotericin B and (B)- cholesterol.

The membranes of living cells are very complex structures composed of lipids, proteins, sugars and cholesterol. The main role of the membranes in the cell functioning is to separate the interior of the cell from the outside environment. The protoplasm inside the cells consists of many organelles (most of them membrane structures) responsible for the cell's life and functioning. They make the living cell even a more complex subject of research.

Artificially created lipid vesicles mimic the cells and their membranes. The bilayer membrane of lipid vesicles presents a simple model for investigation of membrane properties and can be formed from various lipids with controlled composition in controlled environment. With better understanding of the mechanism of AmB action on model lipid membranes we will have a clearer image on how this antibiotic interacts with living cells and their membranes.

The aim of the present study is to investigate the effect amphotericin B (AmB) and cholesterol on the properties of lipid membranes formed from natural and synthetic lipids.

2. Materials and methods

L-α-lecithin from soybean (choline content 20%), amphotericin B, HEPES, KCl (p.a.) and cholesterol (5-Cholesten-3β-ol) were obtained from Sigma. SOPC (l-stearoyl-2-oleoyl-*sn*-glycero-3-phosphocholine (C18:0/C18:1)) was purchased from Avanti Polar Lipids Inc., USA).

Methanol (p.a.), n-propanol (p.a.) and n-hexane (p.a.) were purchased from Valerus Co. All the chemicals were used without any further purification.

The vesicles for thermal shape fluctuation analysis were prepared via modified electroformation method [5]. The lipid was dissolved in chloroform 1 mg/ml. The cholesterol was dissolved in methanol 2 mg/ml. AmB was dissolved in n-propanol:distilled water (4:6, v/v). The final lipid –cholesterol solution for electroformation was prepared by mixing SOPC and cholesterol to obtain 40 mol % concentration of cholesterol in the lipid membrane. The lipid (lipid/cholesterol) solution was laid on the glasses of the electroformation cell and put under vacuum. After the entire evaporation of the solvent AmB solution was laid above the lipid layer and again put under vacuum for solvent evaporation. The electroformation cell, used for all experimental procedures contained two glasses, coated with transparent conductor, indium tin oxide (ITO; thickness of 100 ± 20 nm, resistively of 100 Ω/□) acting as electrodes and a PDMS (polydimethylsiloxane) spacer.

After the entire evaporation of the solvent the experimental cell was assembled and filled with double distilled water. A low frequency (10 Hz) sinusoidal alternative voltage was applied (1.5 V PP) to the conductive glasses overnight. This procedure leaded to the formation of vesicles, appropriate for our experiment. We have chosen giant (diameter of the order of 20-40 μm) fluctuating vesicles without any visible defects.

Thermally induced shape fluctuation method [6, 7] was used to investigate the influence of AmB and combination of AmB and cholesterol on the bending elasticity of model lipid membranes.

The samples of the fluctuating giant vesicles were observed under phase contrast microscope (Axiovert 100, Zeiss, Germany, oil immersion objective Ph3 100x magnification). The experimental equipment was improved using stroboscopic illumination, based on xenon flash lamp L6604 Hamamatsu, Japan) with damping vibration system, short light pulses (less than 3-4 μs long full width at half maximum) and high input energy (2 J) [8].

The obtained digital data was further recorded on the hard disk drive of the PC. Every second an image was acquired and recorded till the total number of images reaches a preliminary given value (about 400).

Further details on the contour determination, mean squared amplitudes calculation and fitting procedure to determine the bending elastic modulus, k_c, and the dimensionless membrane tension $\bar{\sigma}$, can be found in the articles of Genova et al. [9, 10].

For the patch-clamp experiments lipid bilayers were self-assembled at the tips of patch pipettes, using the tip-dip patch clamp technique [11], from monolayers formed by spreading L-α-lecithin from soybean in n-hexane (10 mg/ml) onto the surface of electrolyte solutions contained in Petri dishes (figure 2).

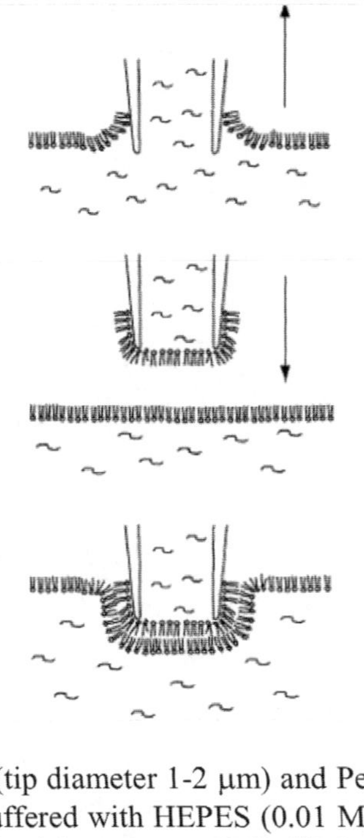

Figure 2. Scheme of lipid bilayer formation at the tip of patch pipette.

Patch pipettes (tip diameter 1-2 μm) and Petri dishes (10 cm^2 area) were filled with aqueous solutions of KCl (0.1 M) buffered with HEPES (0.01 M) at pH 7. Only bilayers with seal resistances > 1GΩ were used. AmB dissolved in n-propanol:distilled water (4:6, v/v), was added to the Petri dish after bilayer formation to a final concentration of $4x10^{-8}$ g/ml.

Solutions of cholesterol (2 mg/ml) were added to the Petri dish to final concentrations in the micromolar range.

Single AmB channel currents were monitored using a patch clamp amplifier Model 2400 (A-M Systems, Inc.) and patch clamp software. The currents were stored on a PC hard disc with 1 ms time resolution. All measurements were performed at room temperature (~22° C).

3. Results and discussion

3.1. Thermal shape fluctuations

The analysis of thermally induced shape fluctuations of giant vesicles is used to study the influence of amphotericin B and cholesterol on the bending elasticity of lipid membrane. The analysis of thermally induced shape fluctuations of giant liposomes is a classical method for studying the elastic properties of lipid membranes [6, 7]. It is based on the fact that under the Brownian motions of water molecules, bombarding the lipid membrane in water environment the lipid vesicle constantly changes its shape. Such fluctuations of the biological or model membranes are a part of the phenomena, describing the deviation of some physical properties from their equilibrium state. Strong qualification criteria for every recorded image in the image sequence and for the quality of the vesicle as a whole are imposed [10]. The white noise contribution to the fluctuation amplitudes is calculated [10] and taken into account.

The obtained experimental data for the bending elasticity modulus (k_c) for pure SOPC membrane, SOPC membrane, containing 3 mol % AmB, SOPC:Cholesterol=6:4 membrane, SOPC:Cholesterol=6:4 membrane containing 3 mol % AmB are presented in table 1.

Table 1. Bending elasticity modulus of SOPC membrane, containing AmB, cholesterol, and combination of AmB and cholesterol. The first column gives information on the studied system, the second column shows the mean weighted value of the bending elasticity modulus and the third column shows the number of vesicles over which the mean values are calculated.

System	Weighted mean value of the bending elasticity modulus $\overline{k_c} \times 10^{-19} (J)$	Number of vesicles
SOPC	1.88±0.17	6
SOPC+3mol% AmB	1.56±0.05	10
SOPC/Cholesterol=6:4	3.63±0.28	9
SOPC/Cholesterol=6:4 +3mol% AmB	3.03±0.19	12

The values of the bending elasticity modulus of each studied system are calculated as a weighted average value of about 6-12 vesicles. The value of the bending elasticity modulus of pure SOPC membrane in double distilled water, measured by the same method is given for comparison [10]. The measured bending elastic modulus for 40 mol% cholesterol in SOPC membrane is in good agreement with the values reported in the literature showing the tendency of cholesterol to increase the bending elastic modulus i.e. to stiffen the membrane [12-15]. All the experiments are performed in double distilled water environment. As it can be seen from the obtained results, the bending elasticity modulus decreases in the presence of AmB in comparison with that of pure lipid membrane for both SOPC membrane and SOPC:Cholesterol=6:4 membrane. This result is in good agreement with the theoretical predictions [16] that show that for all the admixtures, influencing the elastic modulus of lipid membranes at low concentrations in the membranes reduction of the bending elastic modulus should be observed.

3.2. Patch clamp technique

Lecithin bilayers were used to study the effect of cholesterol on AmB channels formed by several molecules self-assembled into a "bundle" structure which traverses the lipid bilayer (figure 3). Incorporation of AmB molecules into the hydrophobic core of the membrane results in the formation of molecular aggregates, which probably take the form of hydrophilic pores composed of six to nine molecules.

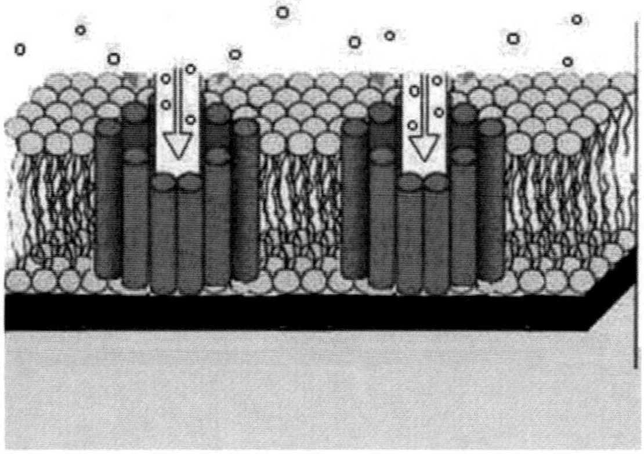

Figure 3. AmB ion channel in phosphatidylcholine bilayer model membrane.

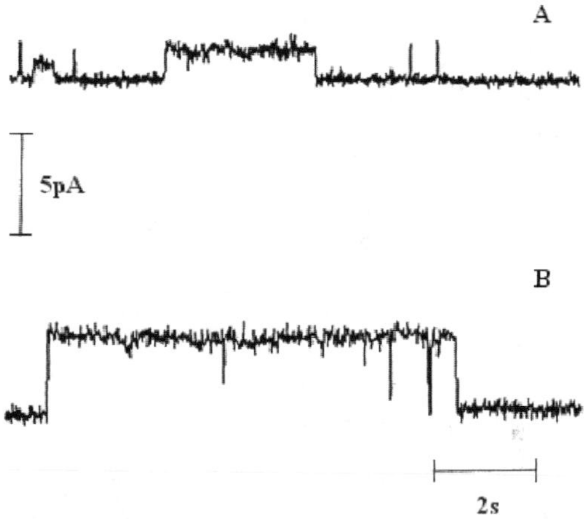

Figure 4. Records of AmB channel currents in the absence (A) and presence (B) of cholesterol in lecithin membranes (bath 100 mM KCl; pH 7.0; ±150 mV holding potential). The concentration of cholesterol is 40 molar % (B).

Figure 4 shows an example of AmB channels recorded in the absence or presence of cholesterol in lecithin membranes. In the same lipid membrane, single-channel currents appeared at ~60 min after addition of amphotericin B.

In the absence of cholesterol (figure 4A), AmB channel openings were alternated with relatively long closings. We observed that amphotericin B is an ionophore and reduces the electrical resistance of the lipid membrane. AmB and cholesterol form ion channels with millisecond dwell times (figure 4B). The volt-ampere (I–V) characteristics were linear over a range of ± 150 mV. We have shown that AmB - cholesterol complexes can produce highly conductive ion channels in phospholipid membranes [17]. Ion channels with high conductivity have great potential for application in fabrication of new types of biosensors on solid supports.

3. Conclusions

Thermally induced shape fluctuations were used to study the influence of Amphotericin B on the elastic properties of pure lipid and cholesterol-containing membranes. It was shown that in both studied systems (pure SOPC and SOPC - Cholesterol membrane) the addition of 3 mol % Amphotericin B reduces significantly the bending elasticity modulus of the membrane.

Patch-clamp technique measurements showed that in the absence of cholesterol AmB channel openings were alternated with relatively long closings. Amphotericin B reduces the electrical resistance of the lipid membrane. AmB and cholesterol interact to form ion channels in the lipid membrane with millisecond dwell times. The obtained volt-ampere characteristics were linear over a range of ± 150 mV.

Acknowledgements

This study was supported by Internal Grant VK-02-13 of The Institute of Solid State Physics, Bulgarian Academy of Sciences, Bulgaria.

References

[1] Herec M *et al* 2005 *Biochem. Pharmacology* **70** 668
[2] Bolard J 1986 *Biochim. Biophys. Acta* **864** 257
[3] Kerridge D and Whelan W 1983 *Mode of Action of Anti-fungal Agents* (Cambridge University Press, Cambridge) 343
[4] De Kruijff B and Demel R 1974 *Biochim. Biophys. Acta.* **339** 57
[5] Angelova M I, Soleau S, Méléard P, Faucon J F and Bothorel P 1992 *Prog. Coll. Pol. Sci.* **89** 127
[6] Faucon J F, Mitov M D, Méléard P, Bivas I and Bothorel P 1989 *J. Physique* **50** 2389
[7] Mitov M D, Faucon J F, Méléard P, Bivas I and Bothorel P 1992 *Adv. Supramol. Chem.* G. W. Gokel (Eds.), JAI Press Inc.: Greenwich **93**
[8] Genova J and Pavlic J 2012 *Bulg. J. Phys.* **39(1)** 65
[9] Genova J 2013 In: Iglic, A., Genova, J. (Eds.), *Adv. Planar Lipid Bilayers and Liposomes. Elsevier*, Amsterdam 1
[10] Genova J, Vitkova V and Bivas I, 2013 *Phys. Rev. E* **88** 022707
[11] Ehrlich B E, 1992 *Methods Enzymol.* **207** 463
[12] Henriksen J, Rowat A C, Ipsen J H 2004 *Eur. Biophys. J.* **33** 732
[13] Pan J, Tristram-Nagle S, and Nagle J 2009 *Phys. Rev. E* **80(2)** 021931
[14] Meleard P, Gerbeaud C, Pott T, Fernandez-Puente L, Bivas I, Mitov M D, Dufourcq J and Bothorel P 1997 *Byophys. J.* **72** 2616
[15] Genova J, Bivas I and Marinov R 2014 *Coll. Surf. A* (in press)
[16] Bivas I and Meleard P 2003 *Phys. Rev. E* **67** 012901
[17] Sykora J, Yilma S, Neely W, Vodyanoy V, 2003 *Langmuir* **19** 858

Color tunability in multilayer OLED based on DCM doped in a PVK matrix

P K Petrova, P I Ivanov and R L Tomova
Institute of Optical Materials and Technologies "Acad. J.Malinowski", Bulgarian Academy of Sciences"Acad. G. Bonchev" str., bl. 109, 1113 Sofia, Bulgaria

E-mail: petia@iomt.bas.bg; petiakl@mail.bg

Abstract. In this work, we present our achievements in color tunability in novel multilayer organic light-emitting diodes (OLEDs) based on DCM (4-(Dicyanomethylene)-2-methyl-6-[p-(dimethylamino)styryl]-4H-pyran) as red emitter doped in a composite PVK:TPD hole-transporting layer, DPVBi (4,4′-Bis(2,2-diphenylvinyl)-1,1′-biphenyl) as a separate blue emitting layer, BAlq (aluminum bis(2-methyl-8-quinolinate)-4-phenylphenolate) as hole-blocking layer and blue emitter at the same time, and $Zn(BTz)_2$ (zinc bis(2-(2-hydroxyphenyl) benzothiazole)) as yellow emitter and electron transporting layer. By modification of the OLED structure and changing the DCM doped concentration in the matrix (in the range of 0 up to 5 %) the color tunability of OLED structures has been obtained. The efficiencies, luminance and chromaticity coordinates of the fabricated OLED structures have been specified.

1. Introduction

Organic light-emitting devices (OLEDs) have received considerable attention as a promising technology for flat-panel displays [1] and solid-state lighting [2]. The successful developments of OLEDs imply the capability to obtain emission over the full visible spectrum and the generation of white emission. During the past two decades, researchers have invested a great deal of effort in developing WOLEDs. However, lifetime, performances and costs still have to be optimized to make WOLEDs commercially competitive as alternative lighting sources. Much research has been spent on finding the suitable materials to realize improvements in device architectures, to enhance OLED efficiency, and to development of new systems with low cost and low energy consumption. Various methods for generating white light have been reported [3]. Generation of white electroluminescence in WOLEDs involves simultaneous emission of light of the three primary colors (red, green and blue) or of two complementary colors (e.g. orange and blue). Combination of emitters of different colors requires proper control of energy transfer processes [4], adjustment of their relative amounts and control of their interactions on both the molecular and the device scale [5, 6]. The management of such energy transfer to ensure simultaneous light production from all the different emitting centers is a critical issue in all the methods for obtaining white light.

In this work, we report an investigation on color tunability in a novel multilayer OLED based on DCM (4-(Dicyanomethylene)-2-methyl-6-[p-(dimethylamino)styryl]-4H-pyran) as red emitter doped in a composite PVK:TPD hole-transporting layer, DPVBi [4,4′-Bis(2,2-diphenylvinyl)-1,1′-biphenyl] as a separate blue emitting layer, BAlq (aluminum bis(2-methyl-8-quinolinate)-4-phenylphenolate) as hole-blocking layer and blue emitter at the same time, and $Zn(BTz)_2$ (zinc bis(2-(2-hydroxyphenyl) benzothiazole)) as yellow emitter and electron transporting layer to achieve white emission. Different

Journal of Physics: Conference Series **558** (2014) 012028 doi:10.1088/1742-6596/558/1/012028

types of OLEDs with differing multilayer structure and doping concentrations of DCM were studied and their electroluminescent characteristics were established.

2. Experimental

We investigated the electroluminescent devices with the conventional structure: ITO/HTL/EL/M, where ITO was a transparent anode of In_2O_3:SnO_2, HTL – a hole-transporting layer, EL - emitting layers and M - a metal Al cathode. As HTL we used an in-house developed composite film of 10 wt% N,N'-bis(3-methylphenyl)-N,N'-diphenyl-benzidine (TPD) incorporated in poly(N-vinylcarbazole) (PVK) matrix. Devices with area of 1 cm^2 were prepared on commercial polyethylene terephtalate (PET) substrates coated with ITO (60 Ω/sq). The undoped and doped with DCM (from 0.15 up to 5 %) HTL (31 nm) of PVK : $TPD_{10\%}$ composite films were obtained by spin-coating from 0.75 % solution in dichloroethane at 2000 rpm. The ELs of DPVBi, BAlq and $Zn(BTz)_2$ (with total thickness of 75 nm) and metal cathode (120 nm) were deposited by thermal evaporation in vacuum better than 10^{-4} Pa at rates of 2-5 Å/s. The thickness of every sublayer included in the El layer is denoted as a superscript index ^{x}, whereas the weight percentage of DCM component in the PVK matrix is denoted as $^{y\%}$. Quartz crystal microbalance was used for an in situ monitoring of the vacuum depositions.

The PVK, TPD, DCM and BAlq were purchased from Sigma-Aldrich, ITO covered PET substrate and DPVBi from Kintec Compony, $Zn(BTz)_2$ from Yurui (Shanghai) Chemical Co. Ltd.

All measurements were performed with unpackaged devices, at room temperature and under ambient atmosphere. Electroluminescence spectra (El) and chromaticity coordinates (CIE - Commission Internationale de L'Eclairage) were recorded with an Ocean Optics HR2000+ spectrophotometer. The current-voltage (I-V) and luminance–voltage (L–V) characteristics were measured by National Instruments Data Aqusition boards controlled by Labview 7.1 written software. The luminescence (L) was measured in DC (direct current) mode and the light output was detected using a calibrated Hamamatsu silicon photodiode S2281-01. The efficiency (η_L) was calculated by equation (1) and used for quantifying the properties of the OLEDs.

$$\eta_L = L/I, \qquad (1)$$

where L is the luminescence (in cd/m^2) and I is the current density (in A/m^2).

3. Results and discussion

DPVBi is widely known to be a blue emission material in OLEDs, and normally interposed between a hole-transporting layer of NPB and an electron-transporting layer of Alq_3. DCM derivatives have been used as red doping molecules to produce OLEDs with improved efficiency and color purity. Red emission is commonly obtained by doping a red fluorescent dye into a hostmaterial that has awider band gap than that of the dopant. Generally, Alq_3 and DPVBi, or other blue emitters are used as a host

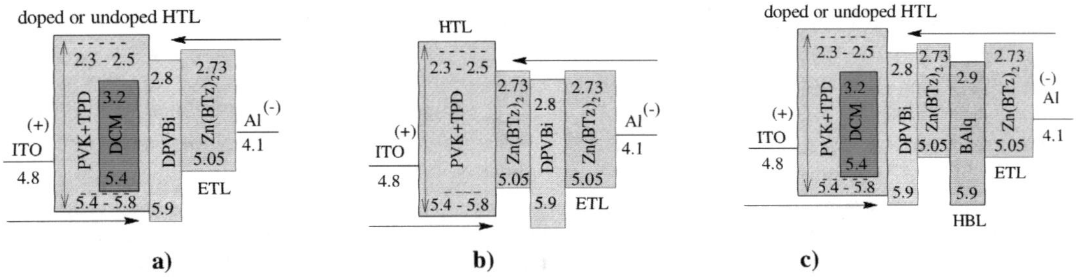

Figure 1. Schematic energy band diagrams for the investigated three types of OLED structures: a) ITO/ PVK:TPD/ $DPVBi^x$/ $Zn(BTz)_2^{75-x}$/ Al, b) ITO/ PVK:TPD/ $Zn(BTz)_2^5$/$DPVBi^5$/$Zn(BTz)_2^{65}$/ Al and c) ITO/ PVK:TPD/ $DPVBi^5$/ $Zn(BTz)_2^5$/ $BAlq_2^5$/ $Zn(BTz)_2^{60}$/ Al.

PVK:TPD, which allowed DCM to be incorporated in the PVK matrix.

To obtain white emission we combined the light emitted by separate layers irradiating in blue (DPVBi), yellow ($Zn(BTz)_2$) and red (DCM). The first layer in the investigated OLED structures is made from the composite PVK:TPD hole transporting layer which doped with DCM acts as an red light emitting layer. The red emission can be obtained either through direct excitation of the DCM fluorophores or through Förster energy transfer from the excited host matrix [7]. The second layer is of blue light-emitting material DPVBi. BAlq is added to some of structures as hole blocking and blue emitting layer at the same time. The last $Zn(BTz)_2$ layer is used as electron transporting and yellow emitting layer simultaneously.

We fabricated three types of multi-layer devices, which schematic energy band diagrams are shown in figure 1. The HOMO and LUMO values are taken from the literature [8-11].

3.1. Comparison of the different undoped structures

The first were investigated and compared three types of OLED structures with different positions and thicknesses of the color layers, as follows:

$HTL/DPVBi^{x}/Zn(BTz)_2^{75-x}$

$HTL/Zn(BTz)_2^{5}/DPVBi^{5}/Zn(BTz)_2^{65}$

$HTL/DPVBi^{5}/ Zn(BTz)_2^{5}/ BAlq_2^{5}/ Zn(BTz)_2^{60}$

The luminance–voltage curves (*L–V*), current efficiency (η_L) and El spectra of these devices are shown in figures 2 and 3, and summarized in table 1. It was found that the structure of type $HTL/DPVBi^{x}/Zn(BTz)_2^{75-x}/Al$, where x = 15 nm shows higher with 0.74 cd/A efficiency and lower with 320 cd/m^2 electroluminescent intensity in comparison with the device with 5 nm thick DPVBi layer. The current efficiencies and Luminance of structures $HTL/DPVBi^{x}/Zn(BTz)_2^{75-x}$ (shown in Fig. 2a and 2b) are better than these of the device $HTL/DPVBi^{5}/ Zn(BTz)_2^{5}/ BAlq_2^{5}/ Zn(BTz)_2^{60}$, but BAlq layer stabilize OLED performance and optimize characteristics of the structure.

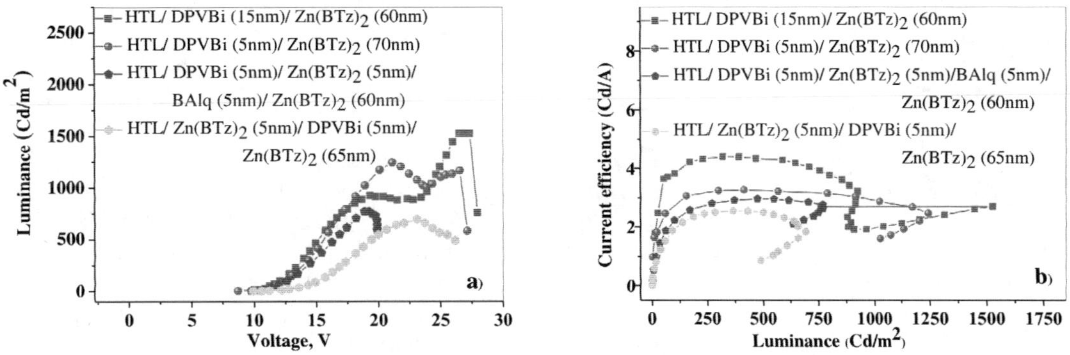

Figure 2. a) Electroluminescent intensity (L) and b) Current efficiency (η_L) of OLEDs with different architecture: $ITO/HTL/DPVBi^{x}/Zn(BTz)_2^{75-x}/Al$, $ITO/HTL/Zn(BTz)_2^{5}/DPVBi^{5}/Zn(BTz)_2^{65}/Al$ and $ITO/HTL/DPVBi^{5}/Zn(BTz)_2^{5}/BAlq_2^{5}/Zn(BTz)_2^{60}/Al$.

The device with EL structure $DPVBi^{5}/Zn(BTz)_2^{70}$ shows a broad emission band centered at wavelength of 513 nm in the green range (0.2698, 0.4435), while the $DPVBi^{15}/Zn(BTz)_2^{60}$ device demonstrates a narrow blue emission band at 497 nm with CIE (x,y) coordinates 0.2389, 0.3922. At the thinner DPVBi layer the EL intensity of the yellow emission is stronger. It can be assumed that as the recombination zone within the DPVBi layer is thinner, because of its reduced thickness, the blue color intensity is reduced. When added a BAlq layer to the basic structure $DPVBi^{5}/Zn(BTz)_2^{70}$ a broad emission band appears at 504 nm - CIE (x,y) coordinates 0.2532, 0.4225. Consequently, BAlq enhances the blue emission without reducing the yellow emission.

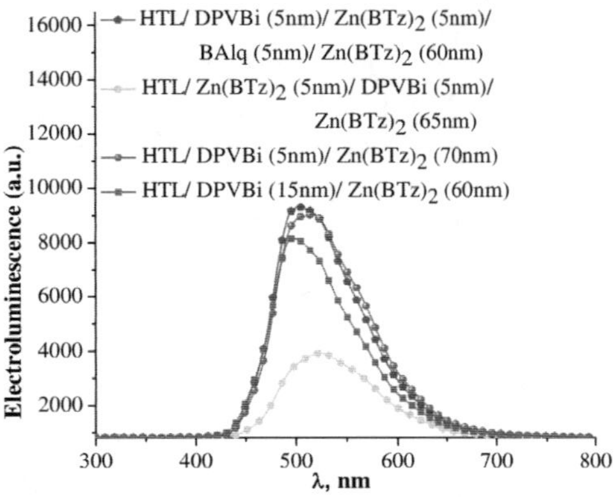

Figure 3. El spectra (at 16 V DC) of OLEDs with different architecture: ITO/HTL/DPVBi^x/Zn(BTz)$_2^{75-x}$/Al, ITO/HTL/Zn(BTz)$_2^5$/DPVBi^5/Zn(BTz)$_2^{65}$/Al, and ITO/HTL/DPVBi^5/Zn(BTz)$_2^5$/BAlq$_2^5$/Zn(BTz)$_2^{60}$/Al.

In the structure HTL/Zn(BTz)$_2^5$/DPVBi^5/Zn(BTz)$_2^{65}$ where 5 nm Zn(BTz)$_2$ layer is placed between HTL and 5 nm DPVBi an expansion of the El spectrum is observed (broad band at 522 nm and shoulder at 560 nm) and the emitted light has *CIE* (x,y) coordinates - 0.2863, 0.4551, but the electroluminescent efficiency and luminance intensity vastly reduce, see figure 2. Unfortunately this device shows the lowest current efficiency and luminance intensity, among discussed in this section structures.

Table 1. Characteristics of OLEDs with a structure: ITO/ PVK:TPD/ emitting layers/ Al. The sequences of the emitting layers are given in the first column.

Sequences of the emitting layers in OLEDs.	L_{max} (cd/m^2)	η_L (cd/A)	λ_{max} at 16V (nm)	Shoulder (nm)	CIE at 16V (x.y)
DPVBi^{15}/Zn(BTz)$_2^{60}$	929	3.21	497	524 weak	0.2389, 0.3922
DPVBi^5/Zn(BTz)$_2^{70}$	1247	2.47	513 broad		0.2698, 0.4435
Zn(BTz)$_2^5$/DPVBi^5/Zn(BTz)$_2^{65}$	695	1.85	522 broad	560	0.2863, 0.4551
DPVBi^5/ Zn(BTz)$_2^5$/ BAlq$_2^5$/ Zn(BTz)$_2^{60}$	770	2.64	504 broad		0.2532, 0.4225

3.2. Influence of DCM dye doping concentrations on the OLED performance

The second experiment was aimed to obtain a white emission via doping of the HTL with DCM in the structure HTL:DCM$^{y\%}$/DPVBi^5/Zn(BTz)$_2^5$/ BAlq$_2^5$/ Zn(BTz)$_2^{60}$. Fine tuning of the color by changing the DCM concentration in HTL at constant thickness of the layers building the basic structure was carried out. Devices with different DCM doping concentrations from 5 to 0.15 wt % were fabricated.

The characteristics of these devices are presented in figures 4 and 5, and summarized in table 2. As seen (figure 4b), the doping to 2.5% with DCM significantly increases the efficiency in comparison with the undoped structure and this doped with 5% DCM. It is interesting to note that the structures

Table 2. Characteristics of the OLEDs with structure: HTL:DCMy%/DPVBi5/Zn(BTz)25/ BAlq25/ Zn(BTz)260 by different DCM doping concentrations.

Wt% DCM doping concentrations	Lmax (cd/m2)	ηL (cd/A)	λ (nm)	λmax at 16V (nm)	Shoulder (nm)	CIE at 16V (x.y)
5% DCM	753	3.07	613	575	491, 530	0.4263, 0.4501
2.5% DCM	1065	5.22		571	493, 529, 613	0.4092, 0.4587
1.25% DCM	1090	4.62		566 broad	496, 525	0.3851, 0.4697
0.6% DCM	841	4.6		570	528, 495	0.3875, 0.4664
0.3% DCM	1340	4.70	531	557	494 weak	0.3525, 0.4678
0.15% DCM	1266	4.74	496	524	563 weak	0.2984, 0.4477

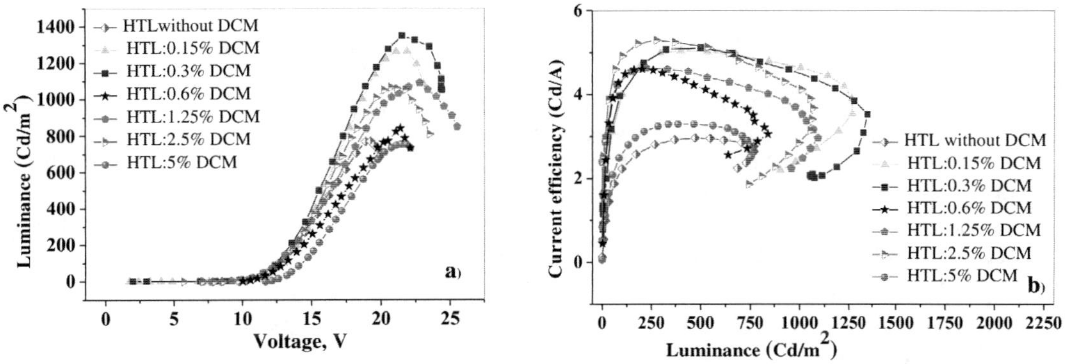

Figure 4. a) Electroluminescent intensity (L) and b) current efficiency (η_L) of the OLEDs: ITO/HTL:DCM$^{y\%}$/DPVBi^5/Zn(BTz)$_2$^{5}/ BAlq$_2$^{5}/ Zn(BTz)$_2$^{60}/Al, where y = 5%, 2.5%, 1.25%, 0.6%, 0.3% and 0.15%.

with 0.15, 0.3, 0.6 and 1.25 wt% DCM show similar current efficiencies. The luminance intensity of devices with 0.15 and 3% is strongest.

For all devices, it was observed that increasing of the DCM concentration, the intensity of the red emission increased, with respect to the intensity of the blue emission (figure 5). It can be seen that increasing DCM concentration the El spectrum of devices shifts to the red. The wavelengths of peak positions, which were in the orange-red region, are depicted in table 2.

The CIE coordinates at different DCM concentrations, given in table 2, show displacement from (0.4263, 0.4501) to (0.2984, 0.4477) with decreasing of the dopant concentration. As is apparent, the red emission predominates in the emitted light from OLEDs at all DCM concentrations, except of OLED with 0.15 wt% DCM which emission is the closest to the white light.

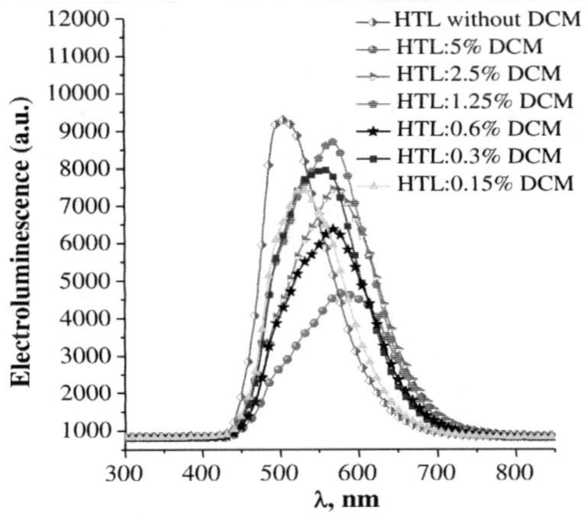

Figure 5. El spectra (at 16 V DC) of the OLEDs: ITO/HTL:DCM$^{y\%}$/DPVBi^5/Zn(BTz)$_2$^{5}/BAlq$_2$^{5}/ Zn(BTz)$_2$^{60}/Al, where y = 5%, 2.5%, 1.25%, 0.6%, 0.3% and 0.15%.

From the results obtained, it can be concluded that the optimal dopant DCM concentration is very low 0.15 - 0.3 wt%. In practice, the effective doping range requested to keep the OLED efficiency unchanged is very narrow, and limited to a few tenths of percent of the optimal concentration [12-15].

3.3. Comparing the two types of structures by equal DCM doped concentration

A comparison of the characteristics of the two types of studied structures: ITO/ HTL:DCM$^{y\%}$/ DPVBi^{15}/ Zn(BTz)$_2$^{60}/Al and ITO/ HTL:DCM$^{y\%}$/ DPVBi^5/Zn(BTz)$_2$^{5}/ BAlq$_2$^{5}/ Zn(BTz)$_2$^{60}/Al with determined optimal (0.15 and 0.3 wt%) DCM doped concentration was done. The characteristics of these devices are presented in table 3 and figures 6 and 7. As seen in the figure 6b, the efficiencies of the devices HTL:DCM$^{y\%}$/DPVBi^{5}/Zn(BTz)$_2^{5}$/BAlq$_2^{5}$/Zn(BTz)$_2^{60}$/Al are identical while these of the devices HTL:DCM$^{y\%}$/DPVBi^{15}/Zn(BTz)$_2^{60}$/Al are different each other. The efficiency of doped with 0.15 wt% DCM is 1.4-fold higher than that of this with 0.3 wt% doped DCM. The figure 6a shows clearly, that BAlq presence stabilize the recombination zone in the investigation structures and only one maximum is observed at Electroluminescence/Voltage curves of HTL:DCM$^{y\%}$/DPVBi^{5}/ Zn(BTz)$_2^{5}$/BAlq$_2^{5}$/Zn(BTz)$_2^{60}$ devices. At Electroluminescence/Voltage curves of HTL:DCM$^{y\%}$/ DPVBi^{15}/ Zn(BTz)$_2^{60}$ devices, where BAlq is missing, second maximum by increase the voltages is observed, which indicates that the recombination zone is shifted with an increase of the voltage.

Table 3. Characteristics of the OLEDs with structure: HTL:DCM$^{y\%}$/DPVBi^{15}/ Zn(BTz)$_2^{60}$ by 0.3 and 0.15 wt% DCM doping concentrations.

Wt% DCM doping concentrations	L_{max} (cd/m^2)	η_L (cd/A)	λ (nm)	λ_{max} at 16V (nm)	Shoulder (nm)	CIE at 16V (x.y)
DCM$^{0.3\%}$/DPVBi^{15}/Zn(BTz)$_2^{60}$	923 at 20V	3.05	496	529	561	0.3391 0.4601
DCM$^{0.15\%}$/DPVBi^{15}/Zn(BTz)$_2^{60}$	792 at 17V	4.16	496	525	560	0.2976 0.4364

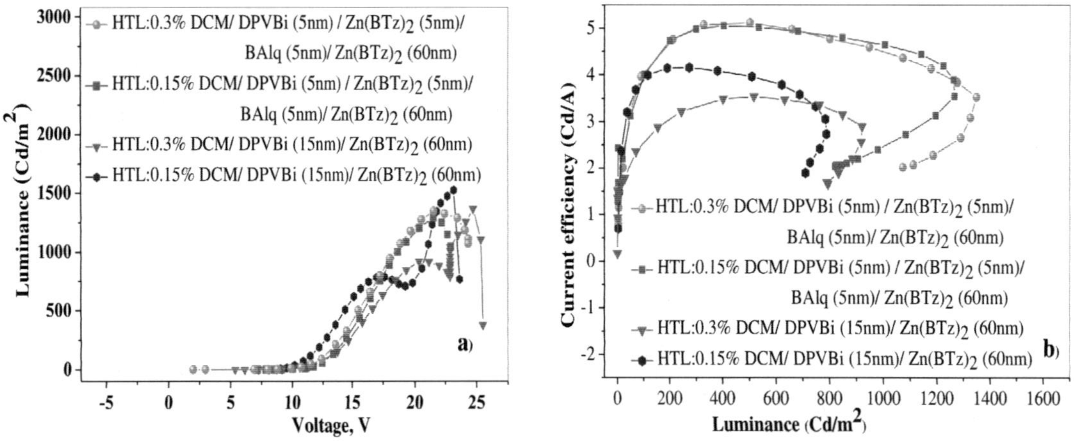

Figure 6. a) Electroluminescent intensity (L) and b) current efficiency (η_L) of the OLEDs: ITO/HTL:DCM$^{y\%}$/DPVBi^{5}/Zn(BTz)$_2^{5}$/ BAlq$_2^{5}$/Zn(BTz)$_2^{60}$/Al and ITO / HTL:DCM$^{y\%}$/ DPVBi^{15}/ Zn(BTz)$_2^{60}$/Al.

From presented in figure 7 spectra of the compared structures can be seen that the electroluminescent spectra of both devices with 0.15% DCM is very similar each other regardless of different device structures. This is also observed in the devices with 0.3% DCM. It is seen from the summarized data in table 2 and 3 that the CIE coordinates of the emitted light from OLEDs: PVK:TPD+DCM$^{0.15\%}$/DPVBi^{15}/Zn(BTz)$_2^{60}$ and PVK:TPD+DCM$^{0.15\%}$/DPVBi^{5}/Zn(BTz)$_2^{5}$/ BAlq$_2^{5}$/Zn(BTz)$_2^{60}$ is almost the same. This is also observed and at OLEDs PVK:TPD+DCM$^{0.3\%}$/ DPVBi^{15}/ Zn(BTz)$_2^{60}$ and PVK:TPD+DCM$^{0.3\%}$/DPVBi^{5}/Zn(BTz)$_2^{5}$/ BAlq$_2^{5}$/ Zn(BTz)$_2^{60}$.

It could be noted that, although BAlq was reported as a very good hole-blocking and blue-light-emission material, the results of these investigations suggested that by inserting a certain thickness of DPVBi layer (in our study 5 and 15 nm) plays an important role to modify device characteristics,

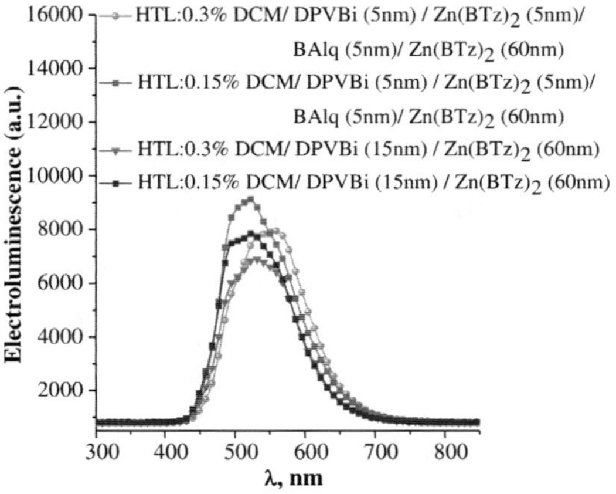

Figure 7. El spectra (at 16 V DC) of the OLEDs: ITO/HTL:DCM$^{y\%}$/DPVBi^5/Zn(BTz)$_2$^{5}/ BAlq$_2$^{5}/Zn(BTz)$_2$^{60}/Al and ITO/HTL:DCM$^{y\%}$/ DPVBi^{15}/ Zn(BTz)$_2$^{60}/Al.

which also can act as recombination-controlling layer in the multilayer devices. It offers simple way to fabricate color-tunable OLEDs by just changing the thickness of this layer.

4. Conclusion

We investigated multilayer OLEDs based on DCM as red emitter doped in a composite PVK:TPD hole-transporting layer, DPVBi as a separate blue emitting layer, BAlq as hole-blocking layer and blue emitter at the same time, and Zn(BTz)$_2$ as yellow emitter and electron transporting layer. It was shown that, via changing the OLED structure architectures and the DCM doped concentration in the matrix (in the range of 0 up to 5 %) a fine color tuning of the emitted light can be obtained. It was established that, in all of the OLED structures with doped DCM, independently of presence of the blue emitting layer, the red emission color is predominated according to the DCM concentration in the matrix. The doped OLEDs by DCM concentrations of 0.15 and 0.3 wt% show the maximum current efficiency - two times higher in comparison with undoped structure PVK:TPD/DPVBi^5/Zn(BTz)$_2$^{5}/BAlq$_2$^{5}/ Zn(BTz)$_2$^{60}.

References

[1] Nalwa H S and Rohwer L S 2003 *Handbook of Luminescence, Display Materials and Devices* **vol 1** Organic Light-Emitting Diodes (Valencia, California 91381-0751, USA: American Scientific Publishers)
[2] D'Andrade B W and Forrest S R 2004 *Adv. Mater.* **16** 1585
[3] Farinola G M and Ragni R 2011 *Chem. Soc. Rev.* **40** 3467
[4] Laquai F, Park Y-S, Kim J-J and Bascher T 2009 *Macromol. Rapid Commun.* **30** 1203
[5] Wang Q and Ma D 2010 *Chem. Soc. Rev.* **39** 2387
[6] Schwartz G, Reineke S, Rosenow T C, Walzer K and Leo K 2009 *Adv. Funct. Mater.* **19** 1319
[7] Yamashita K, Futenma J, Mori T and Mizutani T 2000 *Synth. Met.* **87** 111
[8] Weng-Chang Sh, Su Y and Liang-Wen Ji 2007 *Mater. Sci. and Eng.* **A 445–446** 509
[9] Shi M-M, Chen H-Z, Sun J-Z, Ye J and Wang M 2003 *Chem. Phys. Lett.* **381** 666
[10] Morgado J, Alccer L, Esteves M, Pires N and Gigante B 2007 *Thin Solid Films* **515** 7697
[11] Zhou X, Nimoth J, Pfeiffer M, Maennig B, Drechsel J, Werner A and Leo K 2003 *Synth. Met.*
[12] Forget S, Chenais S, Tondelier D, Geffroy B, Gozhyk I, Lebental M and Ishow E 2010 *J. Appl. Phys.* **108** 064509
[13] Yao Y S, Xiao J Wang X S, Deng Z B and Zhang B W 2006 *Adv. Funct. Materials* **16** 709
[14] Tang C W, Vanslyke S A and Chen C H 1989 *J. App. Physics* **65** 3610
[15] Chen C T *Chemistry of Materials* 2004 **16** 4389

Journal of Physics: Conference Series **558** (2014) 012029 doi:10.1088/1742-6596/558/1/012029

Electron transitions in Cr^{2+} in the aqueous solutions of $MgSO_3.6H_2O$:Cr

P Petkova[1], V Nedkov[2], J Tacheva[1], P Vasilev[1], I Ismailov[1], M. Mustafa[1] and Y. Tzoukrovsky[3]
[1]Shumen University "Konstantin Preslavsky", 115 Universitetska street, 9712 Shumen, Bulgaria
[2]Faculty of Chemistry and Pharmacy, Sofa University "St. Kliment Ohridski", 1 James Boucher Blvd., 1164 Sofia
[3]Dept. of Radiophysics and Electronics, Faculty of Physics, Sofia University "St. Kliment Ohridski", James Boucher 5, 1164 Sofia

E-mail: Petya232@abv.bg

Abstract. In this work, the absorption of the complex $[Cr(H_2O)_6]^{2+}$ is measured in the spectral region 375 – 900 nm. The aqueous solutions of $MgSO_3.6H_2O$:Cr are prepared with the concentrations 0.4%, 0.6% and 0.8%. The energies of the electron transitions in Cr^{2+} are calculated. The role of Jahn-Teller effect and the spin-orbit coupling is evaluated also. The magnetic susceptibility for chromium octahedral complex is calculated at room temperature.

1. Introduction

The cations of 3d metals in water create aqueous complexes. The creation and stability of the aqua cations are important. The number of water molecules which are connected with the metal by the direct bonds metal-oxygen determines the properties of the complex. The classical investigations, for example: mobility of the ions, ostensible radii of the hydration ions and the entropy of hydration do not give detailed information for the aqua ions. This is the reason for the investigation of their spectral properties. These properties depend on the composition and the symmetry of the surrounding medium. The aqueous solutions of 3d ions salts manifest optical activity and interesting properties in magnetic field. This fact helped us when we decided to investigate these ionic liquids. In this work, the most important accent is on the existence and influence of $[CrH_2O]_6^{2+}$ complex. Our analyses give answer of questions which are connected with the peculiarities of electron transitions in the complex and with the stability in this complex. The main aim in our work is the investigation of optical properties of complex $[CrH_2O]_6^{2+}$ in the aqueous solutions with different concentrations.

2. Experimental details

The experimental set up for the measurement of the absorption coefficient in the visible region consisted of the following: a halogen lamp with a stabilized 3H-7 rectifier, a SPM-2 monochromator, a system of quartz lenses, a polarizer, a crystal sample holder, and a Hamamatsu S2281-01 detector. The thickness of the used cuvette is d = 0.995 cm. The basic substance in the solutions is $CrCl_2.6H_2O$. The basic components in both aqueous solutions are: $Mg^{2+} + SO_3^{2-}$, $2Na^+ + 2Cl^-$ and $Na^+ + HSO_3^-$.

3. Results and Discussion

The absorption coefficient of the investigated samples has been measured to be between 375 nm and 900 nm (figure 1).

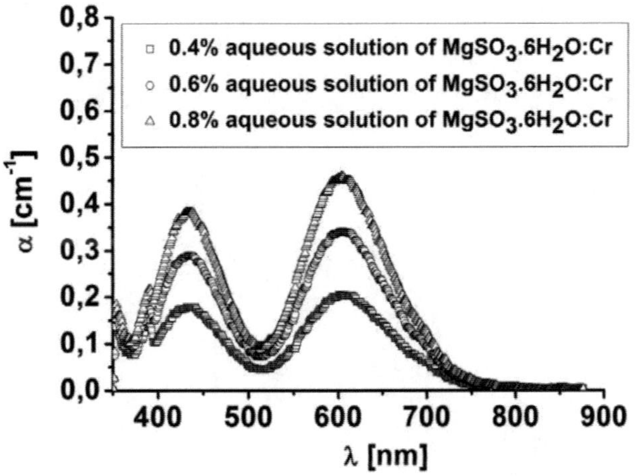

Figure 1. Absorption spectra of aqueous solutions of $MgSO_3.6H_2O$:Cr with the concentrations 0.4%, 0.6% and 0.8%.

The first derivative of the absorption coefficient at photon energy is calculated to be in the 375-700 nm spectral region. The [dα/d(hν)] determines only the number of electron transitions in a Cr^{2+} ions and it does not give an exact information about the energy position of these transitions (figure 2). This is the reason for the calculation of the second derivative of the absorption coefficient $[d^2\alpha/d(h\nu)^2]$ (figure 3). The absorption coefficient is calculated using the formula:

$$\alpha = (1/d)ln(I_0/I) \tag{1}$$

where I_0 is the intensity of the incident light, I is the intensity of the passing light and d is the sample thickness.

The electron transitions in Cr^{2+} are presented on figure 4. They are as follow: $^5D(^5E) \rightarrow {}^3H(^3E)$ (15 244 cm^{-1}) and $^3T_1(^3H) \rightarrow {}^3T_2(^3H)$ (16 639 cm^{-1}, 22 075 $cm^{-1}$and 23 364 cm^{-1}) (figure 4).

In the case of large crystal field splitting 10Dq and partially occupied shell t_{2g}, the system moment is isotropic in spite of big value of the orbital moment (figure 4). Sometimes, the t_{2g} electrons could be compared with p electrons [1]. The orbital system moment becomes anisotropic when the O_h is tetragonally distorted. The two groups of orbitals e_g and t_{2g} in O_h are splitted additionally in D_{4h} of four groups (a_{1g}, b_{1g}, b_{2g} and e_g). The groups of orbitals a_{1g}, b_{1g} and b_{2g} are non-degenerate and the orbital group (e_g) remains double degenerate. The energy of the system changes when it must realize the transition $O_h \rightarrow D_{4h}$. If our complex ML_6 (in the symmetry O_h) transforms to ML_4X_2 (in the symmetry D_4) we have to choose preliminary the deformation direction. If the ligand X determines weaker ligand field we can expect that the electrostatic repulsion between d electrons and charges of ligand X is weaker in comparison with the same repulsion along the other two coordination axes. The picture is analogious, if the M – X distances are longer than M – L distances. This is the situation of the octahedron extensed along z axis. The splitting eg → a_{1g}, b_{1g} realizes by the transition $O_h \rightarrow D_{4h}$ and we know that the energy of a_{1g} (d_z^2) is smaller than this of b_{1g} ($d_x^2 - d_y^2$). The parallel splitting $t_{2g} \rightarrow b_{2g} + e_g$ satisfies unequality e_g (d_{xz}, d_{yz}) < b_{2g}(d_{xy}). This result follows the fact that d orbitals a_{1g} and e_g

have angular part which includes the coordinate z [13]. The real orbital functions are used as a basis for the states e_g and t_{2g} [2].

The strength of the three-dimensional harmonic oscillator determines from the ratio of the real intensity and the intensity of the radiation of the electron. f = 1 for such "ideal electron". The transitions of the oscillator classify by the strength f and this connects the theory with the experiment.

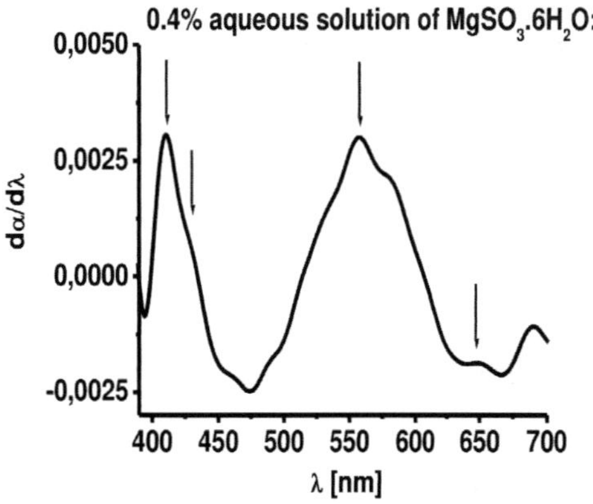

Figure 2a. The calculated first derivative of absorption coefficient to the wavelength in the spectral region 375-700 nm.

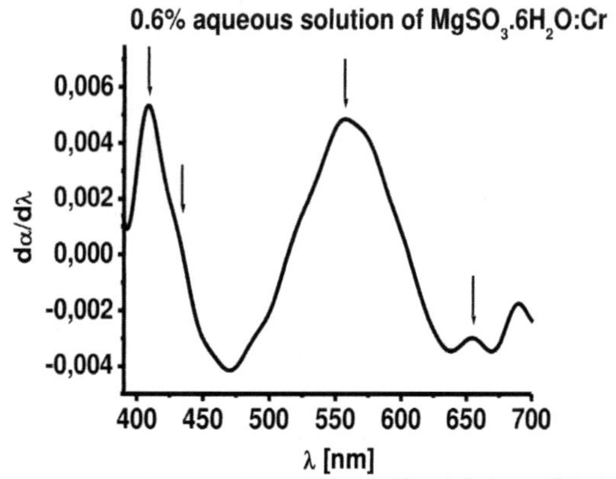

Figure 2b. The calculated first derivative of absorption coefficient to the wavelength in the spectral region 375-700 nm.

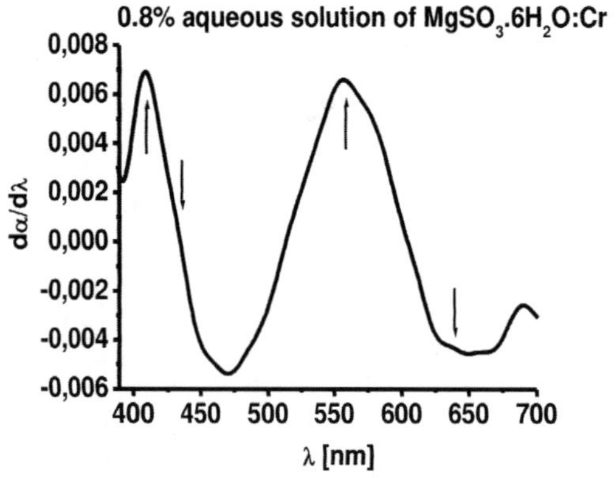

Figure 2c. The calculated first derivative of absorption coefficient to the wavelength in the spectral region 375-700 nm.

f = 1 for the allowed transition; for single symmetric band:

$$f = 4.32 * 10^{-9} \int_{\lambda_1}^{\lambda_2} \varepsilon(\lambda) d\lambda, \tag{2}$$

where ε(λ) not depends on the concentration of the solutions [3]. The next equation

$$Q_i = \left(\frac{f_i}{\nu_i * 1{,}096 * 10^{11}}\right)^{1/2} \tag{3}$$

describes the connection between the oscillator strength f_i and the transition moment of the impurity band Q_i, where ν_i is the frequency of the band centre. The strength of the oscillator $f_1 = 0.0196\text{x}10^{-5}$ for $[Cr(H_2O)_6]^{2+}$ (0,4% solution), $f_2 = 0.31\text{x}10^{-5}$ for $[Cr(H_2O)_6]^{2+}$ (0,6% solution) and $f_3 = 0.0405\text{x}10^{-5}$ at room temperature (T = 300K). This means that the transitions are spin-allowed with forbidden pairing [4]. The transition moment of the impurity bands has next three values $Q_1 = 0.0599\text{x}10^{-3}$, $Q_2 = 0.238\text{x}10^{-3}$ and $Q_3 = 0.0862\text{x}10^{-3}$.

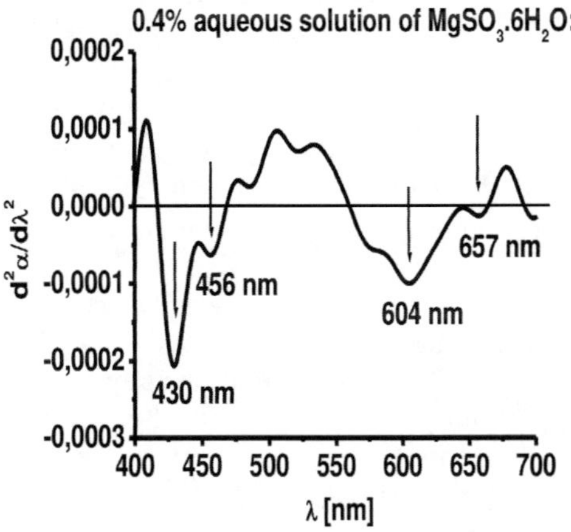

Figure 3a. The calculated second derivative of absorption coefficient to the wavelength in the spectral region 375-700 nm.

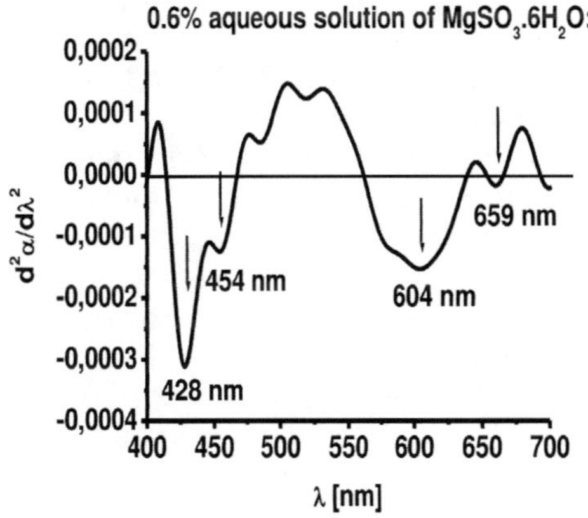

Figure 3b. The calculated second derivative of absorption coefficient to the wavelength in the spectral region 375-700 nm.

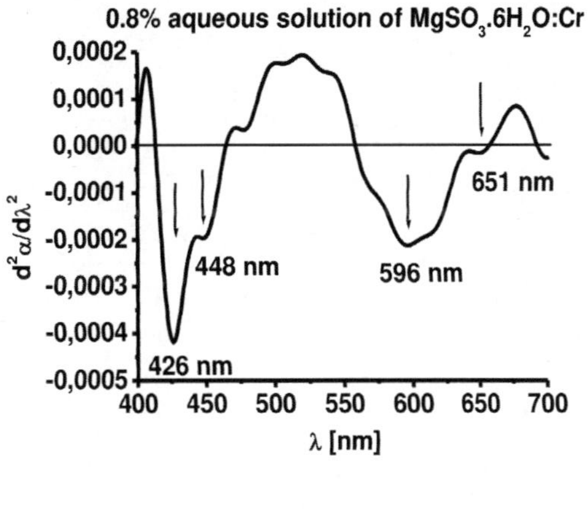

Figure 3c. The calculated second derivative of absorption coefficient to the wavelength in the spectral region 375-700 nm.

In this work, we have considered the Condon-Shortley parameters $F_2 = B + C/7$ and $F_4 = C/35$ for $[Cr(H_2O)_6]^{2+}$ [5]. These parameters are symbols which characterize the integrals of the electrons repulsion in the impurity ion. Each of the terms expressions which determines the energy dependence of these parameters, doesn't depend on the metals ion. The magnitude of the Condon-Shortley parameters changes according to the type of the impurity ion. Racah determines additionally the empirical Condon-Shortley parameters and he defines the parameter B which includes the difference between the levels by the maximum multiplicity. The second parameter C stands for the energy difference between the different multiplicity terms. The Racah parameters are determined experimentally and for Cr^{2+} ion, they have the following values: B = 810 cm^{-1} and C = 3565 cm^{-1}, therefore $\gamma = C/B = 4.40$ [6]. The Jahn-Teller effect manifests in the octahedral chromium complexes and their symmetry changes to D_{4h}. This symmetry is unstable and the manifestation of spin-orbit interaction leads to the final symmetry C_{4v}. The parameters Dq, Ds and Dt are calculated by the equations: $\Delta_1 = -3Ds + 5Dt = 1395$ cm^{-1}, $\Delta_2 = 10Dq + 4Ds + 5Dt = 5436$ cm^{-1} and $\Delta_3 = -4Ds -5Dt = 1289$ cm^{-1}. Their values are respectively 7213 cm^{-1}, 383 cm^{-1} and 49 cm^{-1}.

The symmetry of the complex is such that e_g and t_{2g} orbitals of the metal correspond only on σ anticonnecting and only on π anticonnecting orbitals [5]. The parameters $\delta\sigma$ and $\delta\pi$ are determined as follows: $\delta\sigma = \sigma_z - \sigma_{xy} = -(12/8)Ds - (15/8)Dt$, $\delta\pi = \pi_z - \pi_{xy} = -(3/2)Ds + (5/2)Dt$. The parameters σ and π are indirect indicator for the σ anticonnecting and π anticonnecting properties of the ligands. Finally, $\delta\sigma = 698$ and $\delta\pi = 483$.

It is convenient to introduce the exchange interaction J(dd) here, describing the intrashell exchange attraction between parallel spins [7]: $J(dd) = 1/14(F^2 + F^4)$ and the additional parameter describing the angular part of the multiplet splitting is $C(dd) = 1/14(9/7F^2 -5/7F^4)$. The J and C are especially useful when describing the lowest state of an l^n multiplet. The orbital quantum number l = 2 for d-electrons. According to Hund's rule this is always the state with the highest possible spin quantum number. In our case, the values of J(dd) and C(dd) are as follows: 7830 cm^{-1} and 3641 cm^{-1}. The next equations are fulfilled for our glass systems $F^2 = 49F_2$ and $F^4 = 441F_4$, where F_2 and F_4 are the Condon–Shortley parameters. The energy of levels ^5D and ^3H from the experiment is connected with F_0, $F_2 = 1319$ cm^{-1} and $F_4 = 102$ cm^{-1} as follows [8]: $E(^5D) = 6F_0 - 21F_2 - 189F_4$, $E(^3H) = 6F_0 - 17F_2 - 69F_4$. The energetic difference $E(^3H) - E(^5D) = 4F_2 + 120F_4$ is 17 516 cm^{-1}.

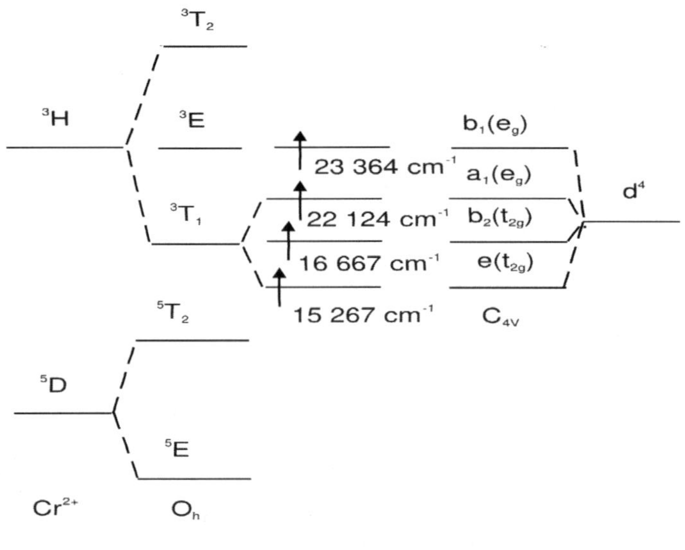

Figure 4. Energy diagram of the Cr^{2+} ion in the octahedral coordination.

The Cr^{2+} ions are paramagnetic. If the metal complexes $Cr(H_2O)^{2+}{}_6$ interact with the external magnetic field then the spin magnetic moment is expressed as

$$\mu_{eff} = [N(N+2)]^{1/2}, \quad (4)$$

where N is the number of the unpaired electrons [5]. In our case N = 4 and μ_{eff} = 4.90 (S = 2). On the other hand

$$\mu_{eff} = g_e[S(S+1)]^{1/2}, \quad (5)$$

and g_e = 2. The magnetic susceptibility can be calculated by the formula

$$\chi = \frac{N}{kT}\sum_n \left(E_n^{(1)}\right)^2, \quad (6)$$

where N = 4, k is Boltzmann constant, T = 300 K and $E_n^{(1)}$ are the values of the energy as follow $E_1^{(1)}$ = 1,89 eV, $E_2^{(1)}$ = 2,06 eV, $E_3^{(1)}$ = 2,74 eV and $E_4^{(1)}$ = 2,9 eV. The value of χ for the complexes $[Cr(H_2O)_6]^{2+}$ is χ = 3680.

4. Conclusions

The Jahn-Teller effect is stronger than the spin-orbit interaction in $[Cr(H_2O)_6]^{2+}$ complex. Our experimental results show that $\delta\sigma > \delta\pi$. Therefore, there have been observed the crystal structures order in the aqueous solutions of $MgSO_3.6H_2O$:Cr. The strength of the oscillator and the transition moment of the impurity band are greater for $[Cr(H_2O)_6]^{2+}$ in 0,6% aqueous solution.

Acknowledgements

This paper is supported by the Project BG051PO001-3.3.06-0003 "Building and steady development of PhD students, post-PhD and young scientists in the areas of the natural, technical and mathematical sciences". The Project is realized by the financial support of the Operative Program "Development of the human resources" of the European social fund of the European Union.

References

[1] Kamimura H 1956 *J. Phys. Soc. Japan* **11** 1171
[2] Hollmann N Haverkort M W Cwik M Benomar M Reuther M Tanaka A and Lorenz T 2008 *New Journal of Physics* **10** 023018 1
[3] Hartmann H Schmidtke H H 1959 *Z. Phys. Chem.* **19** 43
[4] Figgis B N and Hitchman M A 2000 *Ligand Field Theory and Its Applications* (New York: Wiley-VCH)
[5] Drago R S 1977 *Physical Methods in Chemistry* (Philadelphia, PA: Saunders W B)
[6] Tanabe Y and Sugano S 1954 *J. Phys. Soc. Japan* **9** 753
[7] Van der Marel D Sawatzky G A 1988 *Physical Review* B **37** 10674
[8] Ballhausen C J 1964 *Introduction to Ligand Field Theory* (New York:McGraw-Hill Book Company)

Synthesis, thermal and photoluminescent properties of ZnSe-based oxyfluoride glasses doped with samarium

I Kostova [1, 2*], G Okada[2], T Pashova[1], D Tonchev [1, 2] and S Kasap[2]
[1] Department of Chemical Technology and Department of Experimental Physics, Plovdiv University "P.Hilendarski", 4000 Plovdiv, Bulgaria
[2] Department of Electrical & Computer Engineering, University of Saskatchewan, Saskatoon, S7N5A9, Canada

E-mail: irena_k87@abv.bg

Abstract. Rare earth (RE) doped glasses and glass ceramic materials have recently received considerable attention because of their potential or realized applications as X-ray intensifying screens, phosphors, detectors, waveguides, lasers etc. [1]. In this work, we present a new RE doped ZnO-$ZnSe$-SrF_2-P_2O_5-B_2O_3-Sm_2O_3-SmF_3 (ZSPB) glass system synthesized by melt quenching technique. The resulting glasses were visually fully transparent and stable with glass the transition temperatures around 530°C. The thermal properties of this glass system were characterized by Modulated Differential Scanning Calorimetry (MDSC) measurements before and after annealing at 650°C. We have characterized these glasses by Raman spectroscopy and photoluminescence (PL) measurements over the UV-VIS range using light emitting diodes (LED) and laser diodes (LD) excitation sources. We have also irradiated thermally treated and non-treated glass samples by X-rays and have studied the resulting PL. We discuss the results in terms of previously reported models for Sm-doped Zn-borophosphate oxide, oxyfluoride and oxyselenide glasses.

1. Introduction

Rare earth (RE) doped glasses and glass ceramic materials have recently received considerable attention because of their potential or realized applications as X-ray intensifying screens, phosphors, detectors, waveguides, lasers etc. [1]. We have been studying ZnO based material systems including Zn-phosphate glass-ceramics [2] and Zn-B-phosphate glasses and glass ceramics [3, 4]. Most recently we included other backbone elements as fluorine and other additional metals such as Sr to this material system [5]

In various applications, it is desirable to have a stable RE-doped material system that exhibits photo- and X-ray luminescent characteristics. In the present work, our primary interest is the induced conversion of RE ions from one valence state to another towards developing a sensing medium for the measurement of dose. If such a conversion is possible, this system can be applied for high dose X-ray synchrotron-based techniques for cancer treatment such as Microbiem Radiation Therapy (MRT) as discussed previously [6]. However, there may also be additional applications of these materials such as high resolution indirect X-ray imaging [7, 8], radiation sensing, optical filters etc.

This work attempts to convert the ZnO-based oxyfluoride glass and glass ceramic systems to oxyselenide systems by including selenium components. Ciceo-Lucacel at al. have already reported the synthesis of non-doped selenium containing borophosphate glasses with possible bioactivity [7].

2. Materials and methods

2.1. Synthesis
The following reagents were used for the preparation of glass compositions: ZnO (Alfa Aesar, 99.9%), ZnSe (Alfa Aesar, 99%), SrF_2 (Alfa Aesar, 99%), P_2O_5 (Alfa Aesar, 98%), B_2O_3 (Alfa Aesar, 99.98%), Sm_2O_3 (Alfa Aesar, 99.99%), SmF_3 (Alfa Aesar, 99.99%). The synthesis was performed in a zirconia furnace Zircar 110 (with a temperature range 0 – 1750 °C) at a temperature of 1350 °C for 1 h. Seven samples of ZnO-ZnSe-SrF_2-P_2O_5-B_2O_3-Sm_2O_3-SmF_3 (ZSPBS) glass system were synthesised by conventional melt quenching technique. We varied the ZnSe content from 6.5 to 66.5 mol %.

2.2. Differential Scanning Calorimetric characterization
Differential Scanning Calorimetric (DSC) and Temperature Modulated DSC (TMDSC) experiments were performed using the TA Instruments DSC Q100 with an attached Fast Air Cooling System (FACS) at heating rates of 2 K/min, a period of 60 s and amplitude of ±1 K.

2.3. Photoluminescent Experiments
The photoluminescence (PL) spectra of synthesized glasses were measured by a fibre optic coupled CCD Aventes spectrometer AveSpec-2048 with a grating in the range 250-1100 nm. Laser diodes (LD) and Light Emitting Diodes (LED) from UV to NIR range were used as excitation sources. 3D graphs were generated using a MatLab software.

2.4. X-Ray Glass Samples Treatment
We exposed as-prepared and heat treated samples to x-rays under an x-ray tube (43855D, Faxitron) set to 110 kVp with a filament current of 3 mA. PL before and after X-ray irradiation was recorded by a spectrometer EPP2000.

2.5. Raman Spectroscopy
The Raman analysis was performed on RAM II (Bruker Optics) apparatus with a focused laser beam Nd: YAG laser (wavelength 1054 nm). Glass bulk samples were placed and scanned from 4000 cm^{-1} to 400 cm^{-1} with a resolution of 2 cm^{-1}.

3. Result and discussion
The compositions of synthesized glasses are shown in table 1. The fabricated new glass compositions are fully transparent and visually homogenous as can be seen in figure 1.

The thermal characteristics were measured by MDSC measurements. The evaluation of heat flow and heat capacity results was performed using TA Instruments Universal Analysis (UA) software. Figure 2 shows the T_g of composition 1 and the T_g of composition without any ZnSe, base sample ($ZnO_{(71.4)}$ – $SrF_{2(0.72)}$ – P_2O_5 $_{(9.69)}$ – B_2O_3 $_{(18)}$ – Sm_2O_3 $_{(0.12)}$ – SmF_2 $_{(0.12)}$) with the same glass matrix. The addition of ZnSe leads to an increase in the T_g of the glass matrix of about 20°C.

Table 1. Compositions content.

Component (mol %) / Sample №	ZnO	ZnSe	SrF_2	P_2O_5	B_2O_3	Sm_2O_3	SmF_3
1	64.9	6.5	0.72	9.69	18	0.12	0.12
2	54.9	16.5	0.72	9.69	18	0.12	0.12
3	44.9	26.5	0.72	9.69	18	0.12	0.12
4	34.9	36.5	0.72	9.69	18	0.12	0.12
5	24.9	46.5	0.72	9.69	18	0.12	0.12
6	14.9	56.5	0.72	9.69	18	0.12	0.12
7	4.9	66.5	0.72	9.69	18	0.12	0.12

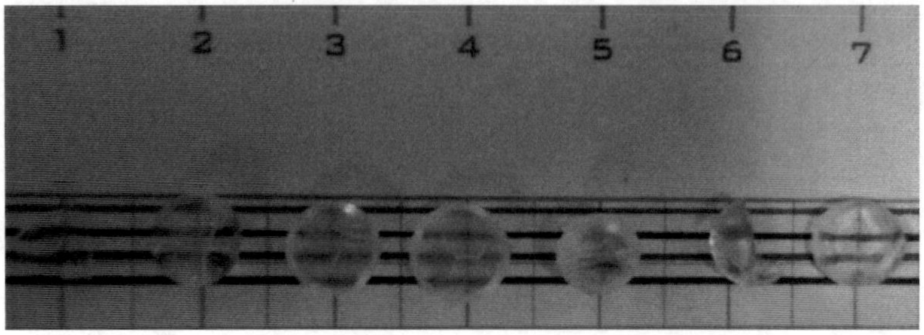

Figure 1. Glass samples (1 - 7) picture.

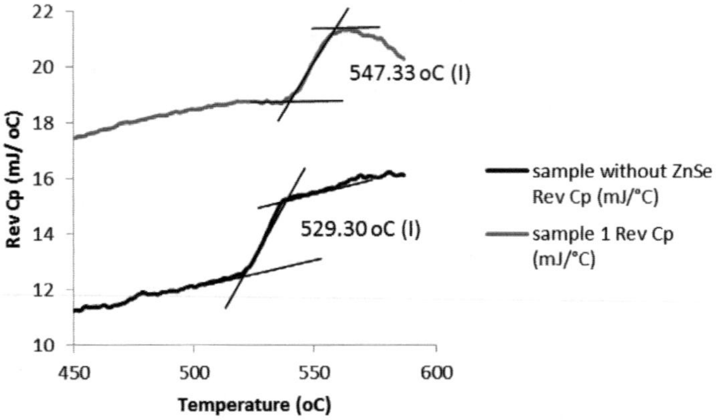

Figure 2. MDSC data of sample 1 (6.5 mol% ZnSe) and base sample (without ZnSe).

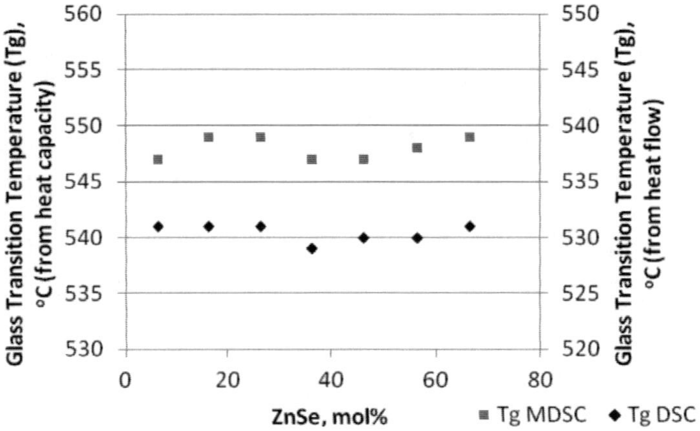

Figure 3. Glass transition temperatures dependence on zinc selenide content in the Sm-doped ZSPB glass system.

The glass transition temperatures of synthesized samples (numbers 1 to 7) are shown in figure 3. Obviously increasing the ZnSe content in the glass matrix do not significantly change the glass transition temperatures. It possibly means that oxygen and selenium are replaceable in this oxyselenide glass structure. Once selenium is included in the oxide glass system it makes the system more stable (higher T_g). Samples 1, 2, 3 and 4 were annealed at 650 °C (close to the glass crystallization temperature) for 3 hours. The MDSC data of the thermally treated samples are shown in figure 4. The results reveal a small change in T_g (about 2 °C) due to annealing. There are also some minor visual changes, e.g. samples are less transparent but are not fully crystalized with the thermal treatment process. The upper curve in figure 4 shows the T_g data of non-annealed samples for comparison.

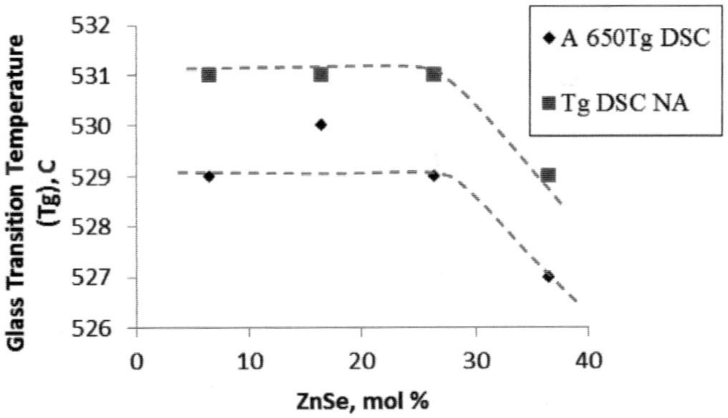

Figure 4. Glass transition temperature of samples 1, 2, 3 and 4 before and after annealing at 650 °C. The upper curve is for the non-annealed sample.

The Raman spectra of these samples are shown in figure 5. There is one strong peak at 961 cm^{-1} and two less intensive peaks at 576 and 430 cm^{-1} for all samples including those without ZnSe content. The Raman bands for Zn-phosphate glasses are 961 cm^{-1} (PO_4) stretch (non-bridging oxygen/selenium), 576 cm^{-1} is a bend mode related to the zinc-phosphate network or ZnO_4/$ZnSe_4$, and 430 cm^{-1} is a bend mode of phosphate polyhedral with a zinc modifier according to the Sandia national laboratories (USA) report [9]. The present Raman spectroscopy results are in agreement with the above findings from the MDSC analysis (figure 3) which suggest that ZnO and ZnSe based glasses are structurally identical.

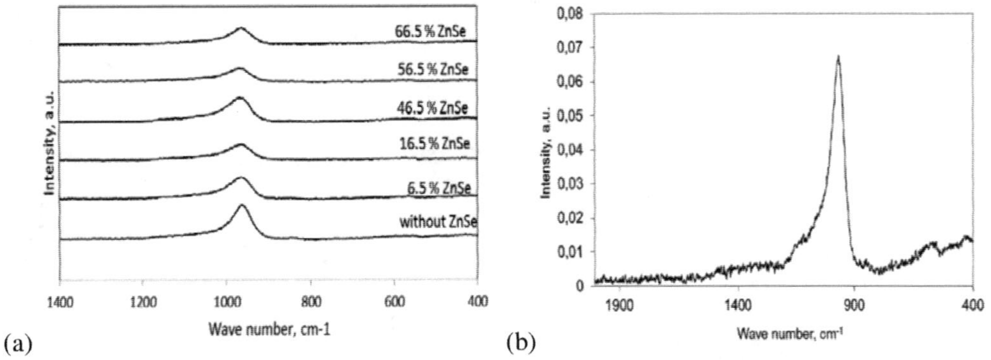

Figure 5. a/ Raman spectra of ZnSe based samples (1, 2, 5, 6 and 7) and sample without ZnSe content b/ Raman spectra of sample 5 (a typical example).

Photoluminescence experiments were performed by various LEDs with emission wavelengths of 370, 395, 405, 410, 415, 425, 435, 450, 470, 490, 505, 515, 565, 572, 590, 605, 615, and 632 nm. The results are shown in figure 6. This figure shows a 3D graph of excitation wavelengths from LEDs together with PL emission spectra for sample 1. All samples demonstrated strong photoluminescence peaks in the visible range (650 – 710 nm) due to the PL emission from Sm^{3+} ions in the glass matrix. Figure 7 presents the PL spectra before and after 60 minutes of X-ray irradiation of sample 1. The PL signal, as expected, is a typical Sm^{3+} emission spectrum. However, no conversion from Sm^{3+} to Sm^{2+} is observed after 60 min of X-ray irradiation. It can be seen from figures 8 and 9 that the PL generated by using two excitation sources (460 and 535 nm) has no PL signature from Sm^{2+} ions; even after annealing at 650 °C for 3 hours, there is no evidence of x-ray induced Sm^{3+} to Sm^{2+} conversion. Glass samples are stable under X-ray irradiation; the only effect is a minor transparent darkening.
A typical fluorescence of Sm^{3+} ions is observed with four strong peaks corresponding to transitions for all samples:

$$564\text{ nm} \leftrightarrow {}^4G_{5/2} \rightarrow {}^6H_{5/2}$$
$$600\text{ nm} \leftrightarrow {}^4G_{5/2} \rightarrow {}^6H_{7/2}$$
$$645\text{ nm} \leftrightarrow {}^4G_{5/2} \rightarrow {}^6H_{9/2}$$
$$710\text{ nm} \leftrightarrow {}^4G_{5/2} \rightarrow {}^6H_{11/2}$$

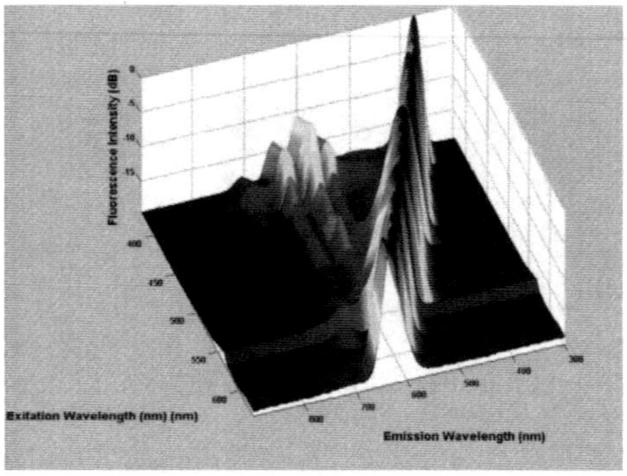

Figure 6. Excitation and emission wavelength vs. PL intensity of sample 1.

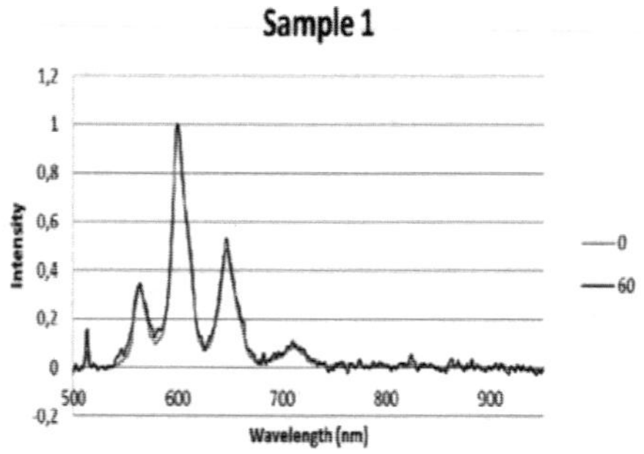

Figure 7. Photoluminescence (460 nm excitation) of sample 1 before and after 60 min X-ray irradiation. The PL arises from Sm^{3+} ions in the glass matrix.

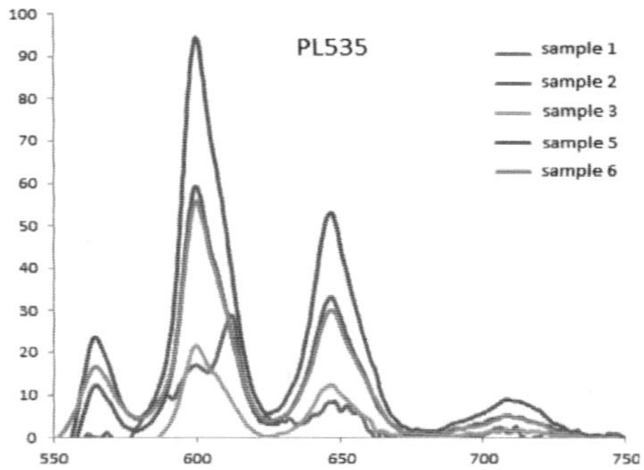

Figure 8. PL (535 nm excitation) of glass samples after 60 min X-ray irradiation.

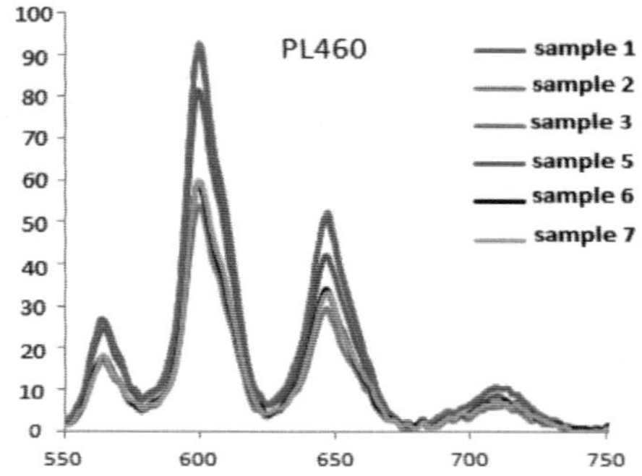

Figure 9. PL (460 nm excitation) of annealed glass samples (650 °C) after 60 min X-ray irradiation.

4. Conclusions

4.1. ZSPB glasses doped with Sm with a gradual increasing of ZnSe content from 6.5 to 66.5 mol% to replace ZnO were synthesized. These glasses are transparent, homogeneous, non-hydroscopic and nontoxic with high glass transition temperatures above 500°C.

4.2. ZnSe addition to ZnO based borophosphate glasses leads to an increase of the glass transition temperature by approximately 20°C.

4.3. Glass transition temperatures do not significantly change with the variation of ZnSe to replace ZnO in the oxyselenide glass matrix (backbone structures are identical).

4.4. The similarity of ZnO and ZnSe borophosphate glass structures is confirmed by Raman spectroscopy.

4.5. ZnSe based glasses demonstrate strong photoluminescence in the visible wavelength range due to Sm^{3+} ions in the glass matrix.

4.6. After 60 min of X-ray irradiation of annealed and non-annealed samples, the PL signal generated by using two excitation sources (460, 535 nm) shows only the Sm^{3+} spectra (no conversion of Sm^{3+} to Sm^{2+}) with some minor transparent darkening.

Acknowledgements We are grateful to Dr. Plamen Penchev and Slava Tsoneva for their help with the Raman spectroscopy measurements.

References

[1] Edgar A 2006 *Springer Handbook of Electronic and Photonic Materials* eds S.O.Kasap, P.Capper (New York, NY, USA: Springer Science) chapter 40 pp 983

[2] Stoeva Z, Patronov G, Kostova I, Chrysanthou A, Tonchev D and Kasap S 2013 *Phys. Chem. Glasses: Eur. J. Glass Sci. Technol.* **B 54** 121

[3] Kostova I, Pashova T, Patronov G, Tonchev D and Eftimov T 2013 *Scientific research of the Union of Scientists in Bulgaria-Plovdiv: series C. Natural Sciences and Humanities* ed S. Vasilev vol XVI (Plovdiv, BG: House of Scientist) pp 231

[4] Patronov G, Kostova I and Tonchev D 2013 *Bulgarian Chemical Communications* **45** *536*

[5] Tonchev D at al. 2014 *Nanoscience advances in CBRN agents detection, information and energy security* eds P. Petkov, W. Kulish, D. Tsiulyanu (Dordrecht, Nl: Springer the NATO-ASI Science for peace and security series: A – Chemistry and Biology) (accepted for publication)

[6] Belev G, Okada G, Tonchev D, Koughia C, Varoy C, Edgar A, Wysokinski1 T, Chapman D and Kasap S 2011 *Phys. Status Solidi* **C 8** 2822

[7] Ciceo-Lucacel R, Radu T, Ponta O, Simon V 2014 *Materials Science and Engineering* **C 39** 61

[8] Edgar A, Varoy C, Koughia C, Okada G, Belev G and Kasap S 2013 *Journal of Non-Crystalline Solids* **377** 124

[9] Brow R, Bencoe D, Tallant D and Kovacic L 1997 United States Department of Energy Sandia Corporation *SAND97-2391* **UC-704** 8

Journal of Physics: Conference Series **558** (2014) 012031 doi:10.1088/1742-6596/558/1/012031

Light-induced relaxation dynamics in Rh-doped $Bi_{12}TiO_{20}$ crystals

V Marinova[1], E Vlaikova[2] and E Goovaerts[3]
[1]Institute of Optical Materials and Technologies, Bulgarian Academy of Sciences, Sofia 1113, Bulgaria
[2]Institute of Solid State Physics, Bulgarian Academy of Sciences, Sofia 1764, Bulgaria
[3]Experimental Condensed Matter Physics, Department of Physics, University of Antwerpen, Universiteitsplein, B-2610, Antwerpen, Belgium

E-mail: vmarinova@iomt.bas.bg

Abstract. The lifetime of the excited charge carriers in Rh-doped BTO crystals is characterized by measuring the time-resolved photoinduced absorption (PIA) after nanosecond pulse excitation from a frequency doubled Nd:YAG laser (λ=532 nm). It was found that the Rh-addition in the BTO structure slows down the relaxation decay in comparison with non-doped BTO, which is attributed to additional trapping centers related to the rhodium dopant. The experimental curve is well fitted by a double-exponential decay which is ascribed to the presence of two different shallow traps contributing to the charge transport and recombination mechanisms in Rh-doped BTO crystal.

1. Introduction

Sillenite type photorefractive crystals (($Bi_{12}SiO_{20}$ (BSO), $Bi_{12}TiO_{20}$ (BTO)) are well known as one of the best inorganic materials for real-time image processing and related applications [1]. Due to the excellent photoconductivity and high charge carrier mobility, they found applications in dynamic information processing, interferometry, optical metrology, etc. [2]. Very recently the ability to combine sillenite crystals with liquid crystals (LCs) into an organic-inorganic hybrid structure (optically addressed light valve devices) became an attractive issue that opens further possibilities to design new nonlinear optical devices for real-time image processing and display technology [3,4]. For that reason special efforts have been made to further enhance photosensitivity and response time of the sillenites, especially in the near infrared (NIR) spectral range [5].

Generally, the modification of crystal properties by changing the stoichiometry during the crystal growth process, by reduction/oxidation treatments or by adding suitable dopant in a structure, is a well known approach. For instance, by varying the stoichiometry, Mokrushina et al. [6] found a correlation between the deficiency of oxygen in non-doped BSO crystals (grown in oxygen free atmosphere) and NIR sensitivity improvement, later on confirmed also by Raita et al [7]. Among the variety of elements transition metals ions are particularly interesting for the optimization of the properties via doping, because they can exist in different valence states after appropriate light illumination [8]. For example, we found [9] that ruthenium (Ru) is an excellent dopant in BSO structure which significantly improves the photosensitivity as well as the response speed during holographic recording in the NIR after green light pre-excitation (with optical energy close to the transition of the main photorefractive center of sillenites–the so-called antisite Bi defect –, located at 2.2 eV below the conduction

[1] To whom any correspondence should be addressed.

band (CB) at room temperature [10]). Recently [11], we have also reported that rhodium (Rh) acts as another very effective trap center in the BTO crystal structure. A response time of 0.1 s was measured during real-time holography at the NIR wavelength of 1064 nm without any preliminary treatment. When the sample was pre-exposed with 532 nm light, the detected response time reduces becomes to 0.02 s. Also, Rh is well known as a unique dopant for $BaTiO_3$ crystals, that makes it sensitive in the NIR due to the Rh^{3+} and Rh^{4+} coexistence [12]. In the present work, we report photoinduced absorption (PIA) measurements of the relaxation dynamics of the excited charge carriers in a Rh-doped BTO crystal in comparison with non-doped BTO.

2. Experimental details

Rh-doped BTO crystals were grown by the top seeded solution growth method [13]. The growth process was conducted at an atmosphere of 85% air and 15% Ar. Rhodium was introduced into the melt solution as RhO_2 and the Rh concentration of 5.2 x 10^{18} cm^{-3} in the grown crystals was determined by atomic absorption spectroscopy.

For the absorption measurements an optically polished crystal plate (thickness of 0.5 mm) was prepared. The absorption spectra were measured in the visible range using a Varian Cary 5 UV spectrophotometer.

To study the PIA and the relaxation dynamics a pump-probe technique (flash photolysis instrument LP 920) was used. The crystal plate was irradiated with single shot pulses from a frequency doubled Q-switched Nd:YAG laser (λ=532 nm, pulse duration 6 ns). A weak monochromatic probe beam (at 600 nm) transmitted through the sample was used to measure the absorption changes. The signal was detected by a photomultiplier and recorded by a high bandwidth digital storage oscilloscope. The measured time scale extends from the first few nanoseconds to the several hundred seconds after pulse photoexcitation. A xenon lamp operating in a pulsed mode was used as a source for the probe beam for short time scanning (ranges below 1 ms). Alternatively, for long-time scanning (up to 100 s) a halogen lamp was used as a probe source.

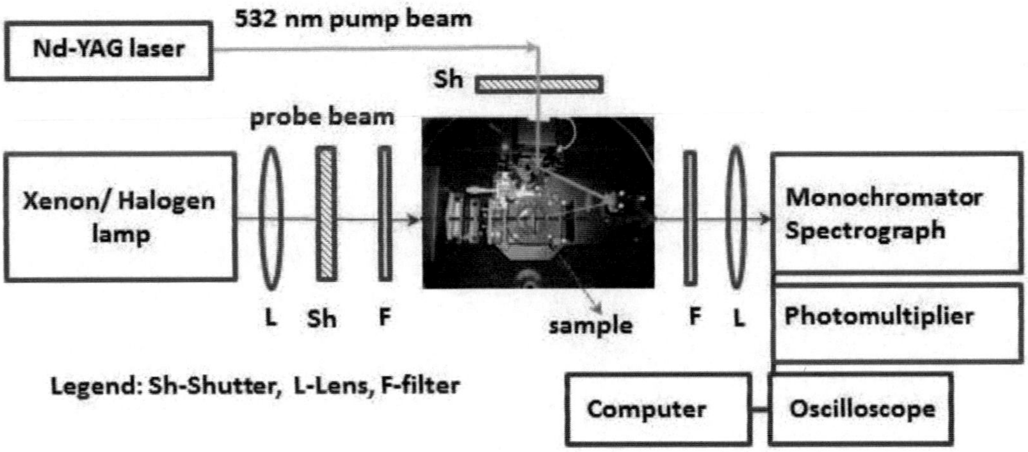

Figure 1. Schematic diagram of the flash photolysis set-up.

The crystal sample prepared in a thin plate was placed in a holder allowing near anti-collinear geometry with an angle of 5° between the pump and probe beam (figure 1). At least 10 single-shot measurements were always averaged in order to improve the signal to noise ratio. Also, we waited several minutes between each two shots, which is considered to be sufficient to complete the relaxation process.

The monitored PIA data are presented as change in the optical density (*OD*):

$$OD = -\log_{10} T = \log_{10}(I_0/I) \tag{1}$$

which is calculated from the transmittance *T*, i.e., the ratio of transmitted intensity through the sample *I* to the incident intensity I_0 of the probe light.

3. 3. Results and discussions

Figure 2 shows the absorption spectrum of a Rh-doped BTO crystal compared to the non-doped BTO. Obviously, the absorption edge is shifted to the red and NIR range by Rh addition. Moreover, a broad peak with rather low optical density is detected in the wavelength interval from 700 nm to 825 nm, which can be attributed to the increased concentration of Rh containing defects that may act as effective trap centers. By reducing the oxygen pressure during the crystal growth process, we can expect that the concentration of oxygen vacancies increases, leading to more stoichiometric defects, consequently to more Ti^{4+} vacancy defects. Such a host environment allows Rh ions to occupy energetically more suitable positions, which affects the absorption spectrum and NIR sensitivity.

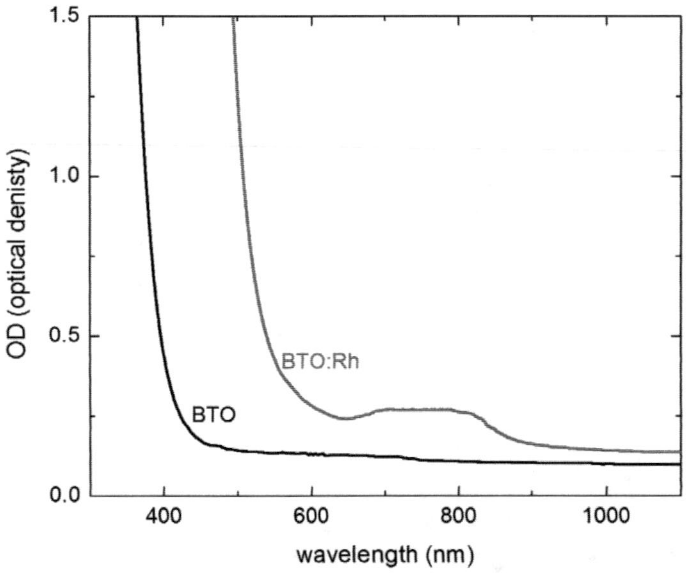

Figure 2. Optical absorption spectrum of Rh-doped BTO crystal compared to non-doped BTO.

Figure 3 shows the dark decay of the induced optical density changes for a Rh-doped BTO crystal plate measured at room temperature. A long lasting component of the relaxation process is clearly observed.
We found that the relaxation decay process for Rh-doped crystals is well fitted by a sum of two exponential decay functions:

$$OD(t) = OD_f \exp\left(-\frac{t}{\tau_f}\right) + OD_s \exp\left(-\frac{t}{\tau_s}\right) \tag{2}$$

where "*f*" and "*s*" correspond to a faster and a slower component of the decay, OD_f and OD_s denote their amplitudes, τ_f and τ_s are their time constants, respectively. The presence of these two components in the dark decay can be attributed to the existence of two different types of traps centers. We have determined a relatively fast decay component with a time constant of $\tau_{f(BTO:Rh)} = 1.1$ s and a slower one with a time constant of $\tau_{s(BTO:Rh)} =$ 13 s by fitting the BTO:Rh data using eq. (2).

Figure 4 presents the relaxation decay of a non-doped BTO crystal taken in the same experimental conditions. Again a slow exponential decay can be clearly recognized at later times in this semilog plot.

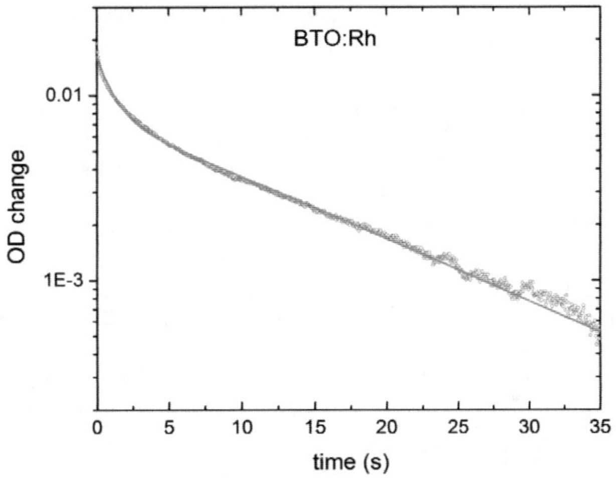

Figure 3. Time dependence of the light-induced OD change in Rh-doped BTO after ns laser pulse excitation until the full relaxation. The symbols are experimental data and the solid line is the result of the fitting procedure yielding relaxation decay time constants of τ_f = 1.1 s and τ_s =13 s at room temperature.

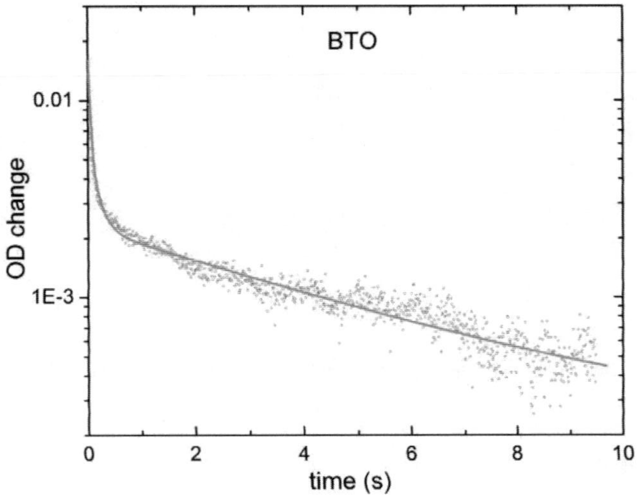

Figure 4. Time dependence of the light-induced OD change in non-doped BTO after ns laser pulse excitation until the full relaxation. The symbols are experimental data and the solid line is the result of the fitting procedure yielding relaxation decay time constants of τ_{f1}= 0.01 s, τ_{f2}= 0.15 s and τ_s= 4.5 s at room temperature.

However, for non-doped BTO the fast component is multi-exponential. Adding a single fast exponential to the slow component yields a poor fit of the data. A strong improvement was obtained with two fast components with decay times $\tau_{f1(BTO)} = 0.01$ s and $\tau_{f2(BTO)} = 0.1$ s, either of them much smaller than the value for the slow decay of $\tau_{s(BTO)} = 4.5$ s. In fact, the fast component of the decay is non-exponential which becomes more evident from measurements in shorter time scales (not shown here).

The relaxation process of non-doped BTO is completed faster in comparison with Rh-doped BTO, using the same pump intensity. Not only is the fast initial decay, relatively more important in the undoped crystal, also the decay time of the slow component is three times smaller. The slower relaxation process in the doped crystal can be attributed to the incorporation of rhodium.

The above results of non-doped BTO may be related to recently published PIA data of BTO crystals, irradiated with a high intensity laser pulse of 532 nm [14]. However the authors reported about near permanent changes in the optical properties after intense irradiation, with very slow relaxation process with decay time in order of hours. This was attributed to the existence of intermediate level traps in the forbidden band gap of the BTO crystals.

At the same time, we have recently characterized non-doped BSO crystals with rather faster relaxation decay (complete decay occurs within few ms) [15] in comparison with the non-doped BTO. The difference is supposed to be due to the different intrinsic defects and trap levels in BTO and BSO crystal structure [10].

Detection of longer relaxation times in Rh-doped BTO shows that Rh-addition introduces specific traps in BTO crystal structure. The depth of these trap levels and their concentration can contribute to slower recombination of the excited electrons in comparison with non-doped BTO. In comparison, other transition metal dopants were shown to also slow down the carrier relaxation dynamics in sillenites, with a particularly strong effect demonstrated for Ru in BSO [15] and Cr in BTO [16]. For example, we have measured relaxation decay time of $\tau_f = 3.2$ s and $\tau_s = 20$ s for Ru-doped BSO [15] and $\tau_f = 10$ s and $\tau_s \sim 10^3$ s, respectively for Cr-doped BTO [16]. Different doping behavior can be attributed to the different densities and activation energies of the relevant tapping centers, involved in the charge transport mechanism.

4. Conclusion

The dynamics of charge carriers generated by nanosecond pulse laser excitation in Rh-doped $Bi_{12}TiO_{20}$ is studied. Comparing to non-doped BTO, the rhodium addition slows down the recombination kinetics as observed by photo-induced absorption. This can be attributed to introduction of additional rhodium related defects introducing trap levels and to increased densities of the relevant trapping centers.

Acknowledgements:

This work has been performed in a bilateral collaboration project supported by the Fund for Scientific Research Flanders (FWO), Belgium, and the Bulgarian Academy of Science. Samples were grown by the Crystal Growth Laboratory, Institute of Solid State Physics, Sofia, Bulgaria.

4. References

[1] Gunter P and Huignard J P 2007 *Photorefractive materials and their applications* 3 (Berlin-Heidelberg: Springer)
[2] Frejlich J 2007 *Photorefractive materials* (New Jersey: Wiley-Interscience)
[3] Bortolozzo U, Residori S and Huignard J P 2008 *J. Phys.* D **41** 224007
[4] Bortolozzo U, Residori S and Huignard J P 2012 Materials **5** 1546
[5] Gvozdovskyy I, Shcherbin K, Evans D R and Cook G 2011 *Appl. Phys.* B **104** 883
[6] Mokrushina E V, Bryushinin M A, Kulikov V V, Petrov A A and Sokolov I A 1999 *J. Opt. Soc.* **16** 57
[7] Raita E, Kobozev O, Kamshilin A A and Prokofiev V V 2000 *Opt. Lett.* **25** 1261
[8] Ramaz R, Rakitina L, Gospodinov M and Briat B 2005 *Opt. Mat.* **27** 1547
[9] Marinova V, Liu R C, Lin S H, Chen M S, Lin Y H and Hsu K Y 2013 *Opt. Lett.* **38** 495
[10] Oberschmid R 1985 *Phys. Stat. Solidi* A **89** 263
[11] Marinova V, Liu R C, Lin S H and Hsu K Y 2011 *Opt. Lett.* **36** 1981

[12] Radoua, Delaye P, Pankrath R and Roosen G 2003 *J. Opt.* A **5** S477
[13] Sveshtarov P and Gospodinov M 1991 *J. Cryst. Growth* **113** 186
[14] Matusevich A, Tolstik A, Kisteneva M, Shandarov S, Matusevich V, Kiessling A and Kowarschik R 2008 *Appl. Phys.* B **92** 219
[15] Marinova V, Ahmad I and Goovaerts E 2010 *J Appl. Phys.* **107** 113106
[16] Marinova V and Goovaerts E 2013 *Bulg. Chem. Commun.* **45** 218

Opto-galvanic effect on degenerate magnetic states of sputtered atoms in a glow discharge

D Zhechev[1] **and V Steflekova**
Institute of Solid State Physics – BAS
72, Tzarigradsko Chaussee Blvd., 1784 Sofia, Bulgaria

E-mail: spectron@issp.bas.bg

Abstract. The opto-galvanic response of some degenerate states of sputtered atoms to linearly- and circularly polarize light is studied. On the same optical transition both time-resolved- and amplitude opto-galvanic signals are found depending on the polarizations of light absorbed. The latter induces galvanic responses differing in opto-galvanic efficiency, time-evolution and sensitivity to discharge current and laser power. The differences are ascribed to the rate constants of the decay processes, characterizing aligned and oriented atoms.

1. Introduction

The degenerate magnetic states *m* are known as the first ones excited in a coherent superposition, i.e. linearly polarized resonant light absorbed *aligns* them (coherence of the states of $\Delta m = 0, \pm 2$); circularly polarized resonant light *orients* *m*- states at either $m = 0 \rightarrow m = 1$ or $m = 0 \rightarrow m = -1$ [1]. Both *alignment* and *orientation* manifest themselves optically in linearly- or circularly polarized spontaneous emission. In a real gas discharge medium either *alignment* or *orientation* means that the resulting coherent *m*- states above appear along with other permitted *m-states*. In other words, the coherences add new spatial dispositions of valence electron. On the other hand both ionization and excitation of atoms are anisotropic in low-temperature plasma and its conductivity should depend, generally, on the magnetic states *m* too. So far, this dependence is not taken into account in modeling/calculation of the interaction atom-electron. For the first time, the relation *coherence - conductivity j* has been studied as a conductivity change Δj of the same ensemble of atoms *aligned* ↔ *non-aligned* sequentially [2-4].

Recently, a preliminary study found some different correlations *coherence (alignment/orientation) - conductivity* of some states of neon atom in a hollow cathode discharge (HCD) [5]. It means that interaction electron↔ atom (*aligned/oriented*) depend on the coherence induced.

The main objective of this study is the conductivity of an ensemble of sputtered atoms in a HCD as a function of their *alignment/orientation*. Generally, one such dependence found in detailed comparisons would represent an important result, namely, that one or other light polarization strongly influences rate constants of electron-atom interaction.

HCD is known as a reservoir of sputtered atoms, first of all [6]. One and the same ensemble of sputtered atoms that is *aligned* ↔ *oriented* in sequence is studied by using light-induced coherence. On the other hand, the resonant light absorption is the necessary condition for another effect in a gas

[1] D. Zhechev, email: spectron@issp.bas.bg

discharge, i.e. opto-galvanic (OG) one [7]. The latter is the galvanic manifestation of the redistributing action of light absorbed vs. steady state population. Ultimately, a light-induced change of gas discharge conductivity, i.e. OG signal (OGS), may be detected. The time-resolved OGS (TROGS) describes the transient processes of relaxation related to the conductivity of the medium which is absorbing. Therefore the TROGS is the most appropriate technique for precisely comparing *aligned oriented*↔ atoms in their coherent conductivity. The amplitude (OGS) (AOGS) is an averaged galvanic replica of the chopped laser beam irradiating.

This investigation is realized as a comparison of opto-galvanic signals (OGSs) induced by linearly- and circularly polarized light. OGSs on the transitions Ca I ($3p^6 3d4s\ ^3D \rightarrow 3p^6 3d4p\ ^3D^0$) ($\lambda$ = 560.13 nm), Li II transition ($1s2s\ ^3S \rightarrow 1s2p\ ^3P^o$) ($\lambda$ = 548.35 nm), Li II ($1s2s\ ^3S \rightarrow 1s2p\ ^3P^o$) ($\lambda$ = 548.57 nm) and Cd I transition ($4d^{10}5s5p\ ^3P^0 \rightarrow 4d^{10}5s7s\ ^3S$) ($\lambda$ = 325.25 nm) (designations under reference [8]) are compared.

Earlier the sputtering of Li atoms by Ne^+ ion in a HCD was modeled [9]. As for atoms of Ca they are difficult for atomization and a HCD represents the optimal medium for sputtering.

2. Experimental set-up

Two types of OGSs, i.e. time-resolved (TROGSs) and amplitude (AOGSs) are detected by using set-up in figure 1. The trademarked HCD lamps Ne/Ca-, Ne/Cd- and Ne/Li (all "Narva") were used. The discharge was produced by applying a highly stable dc voltage and was operated in a NG regime. A standard experimental scheme for TROGS detection was employed. TROGSs were induced by irradiating the HCD with an OPO system, which delivered a pulse of duration 4 ns FWHM at a repetition rate of 10 Hz and an average output power of 25 mW and were detected across R_m = 20 κΩ (C = 10 nF) with a digital oscilloscope. The RC constant was chosen low enough to detect the time-dependent signal $\Delta U(t)$. The latter represents a galvanic time-resolved replica-evolution of the peak-light perturbation. Plane- and circularly polarized light pulses irradiate the discharge in sequence and their TROGSs are compared (see Figs. 2).

AOGSs are measured by using He-Cd laser (*"KIMMON" electric co., ltd.* λ = 325 nm, 25 mW), chopper and selective nanovoltmeter "Unipan" (type 237). The values of R_m = 1 mΩ and C = 10^2 mF are used. The equality in intensity of the illuminating light polarizations is adjusted and controlled.

3. Experimental results and discussions

The analysis of the OGSs is based on the understanding that sputtered atoms which are not returned back to the cathode are thermalized by the buffers in the cathode dark space [10] and no overriding their velocity vector does not exist.

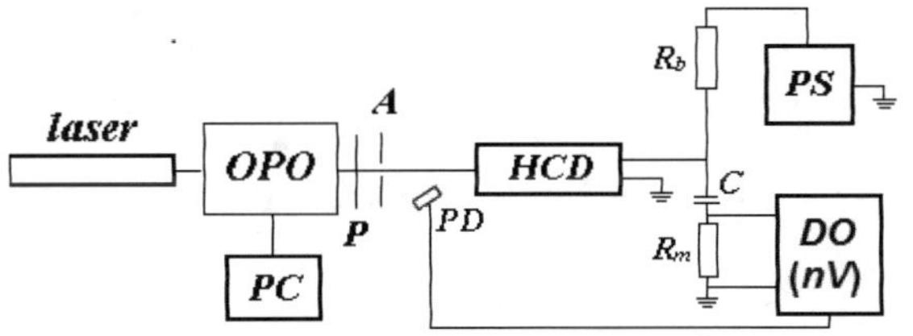

Figure 1. Experimental set-up: *HCD* – hollow cathode discharge lamp, *OPO* – optical parametric oscillator system, *P* – polarizer (linear or achromatic waveplate λ/4, 450 - 800 nm), *A* – aperture, *PD* – photo detector, *C* – decoupling capacitor 10 nF, R_b – ballast resistor, R_m = 20 κΩ – measuring resistor, *PC* – personal computer, *PS* – power supply, *DO* – digital oscilloscope, *nV* – nanovoltmeter.

Figures 2 (a ÷ c) contain certain general TROGSs regularities: *i/* the signal is composed by a fast rising peak followed by an exponential decay to either base line or a part of opposite sign, which returns to the base line; *ii/* the discharge current (dc) I extends the waveforms $\Delta U^{OG}(t)$, namely both the first peak and the rest damping part; *iii/* dc stimulates the first peak amplitude, namely $\Delta U^{OG(max)} \propto I$ on Li II transitions; on Ca I transition the inverse relationship $\Delta U^{OG(max)}(t) \propto I^{-1}$ takes place. Within these regularities the waveforms $\Delta U^{OG}(t)^{linear/circular}$ induced by linearly- or circularly polarized light are compared.

On the whole, the shapes of the two types of signals are different from one another in their parameters.

As a rule linearly polarized light induces greater galvanic perturbation $S(t)^{linear}$, where

$$S(t)^{linear/circular} = \int_0^{t1} \Delta U^{OG}(t)^{linear/circular} dt \qquad (1)$$

and t_1 [μs] – the time of integration dictated by the concrete case. The program Origin Lab is used to calculate the value of the area *S (t)*, characterizing the magnitude of the perturbation. A detailed comparative analysis of TROGSs shows the following dependences on the light polarization:

i. The inequality $|S(t)^{linear}| > |S(t)^{circular}|$ means a higher OG efficiency of linearly polarized light on irradiated transitions, namely the ratio $S(t)^{linear}/S(t)^{circular}$ varies within the limits [1.13÷1.87] in the signals in figure 2 and the relation $S(t)^{linear}/S(t)^{circular} \propto I$ takes place. Minimum values of this relation of 1.13 (at I = 1 mA) and 1.23 (1.6 mA) characterize TROGSs induced by Ca I 560.13 nm. The signals induced by Li II 548.35 nm are the closest in their OG efficiency (1.62 and 1.64). The maximum values of 1.58 (1 mA) and 1.87 (1.6 mA) characterize TROGSs induced by Li II 548.57 nm.

ii. The TROGSs in figure 2 suggest characteristic decay time T of the evolution $\Delta U^{OG}(t)$. A single-exponential approximation y = A*exp (-x/T) as in reference [11] is used here to compare T - values of each pair $\Delta U^{OG}(t)^{linear/circular}$ signals and one and the same starting t - point of the fitting is taken for each pair $a_{1,2}$, $b_{1,2}$, and $c_{1,2}$ TROGSs. The signal of lower amplitude dictates the starting point. The procedure of the best fit gives T - values (see figures 2) differing in the range of $\Delta T \in [0.96÷2.06]$ μs.
For each pair the relation $\Delta T \propto I$ take place. Despite the small values of ΔT the fact that $T^{linear} > T^{circular}$ systematically suggests that atoms perturbed by linearly polarized light relax slower than the atoms perturbed by circularly polarized light.

iii. Control TROGSs measures on the transition Ne I ($1s_5 \rightarrow 2p_9$) (640.2 nm) in Ne/Li HCD lamp give other relationships, namely $S(t)^{linear}/S(t)^{circular} \propto I^{-1}$ and the values $S(t)^{linear}/S(t)^{circular}$ = 1.94 (at I = 0.22 mA) and 1.11 (at I = 1.6 mA). Here discharge current equalizes the galvanic perturbations due to different polarizations. Another T - relation, namely $T^{circular} > T^{linear}$ characterizes TROGSs. For example, $T^{circular}$ = 26.88 μs and T^{linear} = 8.75 μs at 1.6 mA. In OG context the line Ne I 640.2 nm is unique in that it corresponds to the only transition linking the $1s_5$ and $2p_9$ states. No optical redistribution of the population Ne I p_9 $\{n_{p9}+\Delta n(t)\}$ to any Ne I s $_{2,3,4}$ states takes place. The light induced departure $\Delta n(t)$ may relax only via both spontaneous transition $2p_9 \rightarrow 1s_5$ and interaction Ne I p_9 $\{n_{p9}+\Delta n(t)\}$ + e. Therefore, the signal due to Ne I 640.2 nm line is a sensitive probe of collisional interactions. On the other hand, since the level Ne I $1s_5$ is the only in maintaining HCD at $I \leq 1$ mA [16] its disturbance is incomparable with that of the levels Ca I $3p^63d4s\ ^3D$, Li II $1s2s\ ^3S$ and Li II $1s2s\ ^3S$. Therefore, the above comparison should be regarded only as an ambiguous manifestation of TROGSs induced by linearly and circularly polarized resonant light.

In general, the understanding of the TROGS [8, 11-15] is based on the galvanic effect due to relaxation of the light-induced population transfer $\Delta n_{i\rightarrow j}$ from/to the steady state population density n_i/n_j. It is a fast process (in the 10^{-9} s time region) which is non-resolved in our experiment. Further, this perturbation relaxes $\Delta n_j(t)$ typically in the μs range by various channels dictated by the plasma

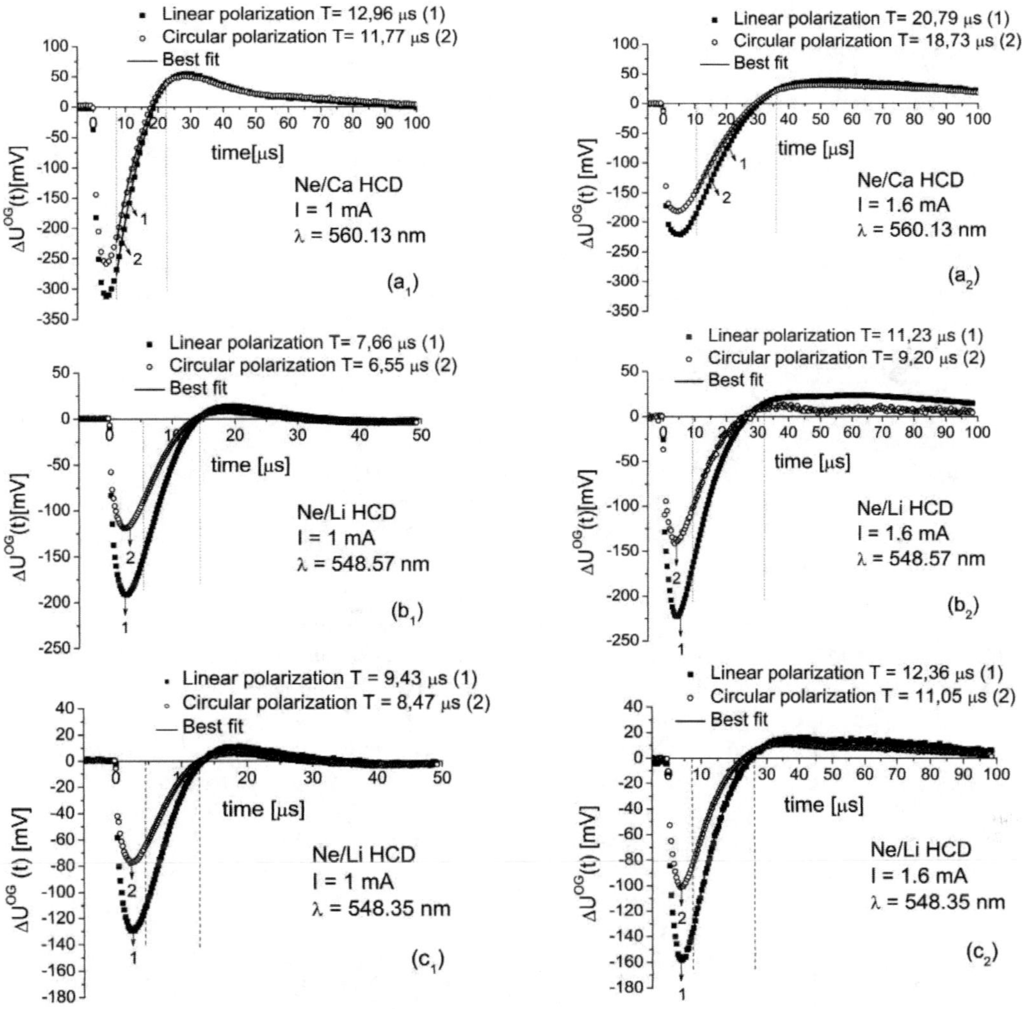

Figure 2. TROGSs induced by linear- and circular polarization of the lines Ca I ($3p^63d4s^3$D $\rightarrow$ $3p^63d4p^3$D^0) (560.13 nm), Li II ($1s2s$ ^{3}S $\rightarrow$ $1s2p$ ^{3}P°) (548.57 nm) and Li II ($1s2s$ ^{3}S $\rightarrow$ $1s2p$ ^{3}P°) (548.35 nm) in Ne/Ca- and Ne/Li HCD lamps. Broken lines point the time interval Δt [μs] of the fitting: a_1-[7.2 ÷22.4] μs; a_2-[10.4 ÷35.2] μs; b_1-[5.4÷14.4] μs; b_2-[9.2÷32] μs; c_1-[4.4÷12.6] μs; c_2-[7.2÷26] μs.

as a whole, i.e. by the collective interaction of the corresponding level with the others via the electron collisions and radiative transitions (spontaneous decay of upper state); simultaneously, the ionization also changes due to the perturbed populations $n_i - \Delta n_{ij}(t)$ and $n_j + \Delta n_{ij}(t)$. Ultimately, the contributions of these processes to the conductivity j are reduced to changes of electron number density Δn_e and/or electron drift velocity Δv_e, i.e.

$$\Delta j = c_1 \Delta n_e + c_2 \Delta v_e \tag{2}$$

These terms define also the characteristic decay time T of the evolution $\Delta U^{OG}(t)$ at light pulse short enough. Concrete processes of the above transformation are not discussed in this comparative study.

However, a specific for HCD reaction should be taken into account vs measured signals. In a HCD the dc is known to play also another specific role, namely as a factor for the atomic sputtering yield. [7, 10] The latter is a prerequisite for Penning ionization in this discharge. In our case:

$$Ne^{m}\ (^3P_2) + A \rightarrow Ne\ (^1S_0) + A^{+*} + e_{\pm 1/2}\ (A \rightarrow Ca,\ Li) \tag{3}$$

where Ne^{m} ($^3P_2$) is the metastable. The ionization (3) competes with the stepwise one and was found to influence the current/voltage curve, particularly at low dc values. [2, 3] However, of greater importance is the circumstance that the metastable spin state (magnetic number) may affect the efficiency of the process (3). The ionization cross-section for Penning collisions has been studied theoretically [16] and the author has found that in a HCD the process (3) always contributes to the conductivity. Here the population transfer $\Delta n_{i\rightarrow j}$ between states of sputtered atoms the process (3) stimulates first of all the term $c_1\Delta n_e$ in expression (2). As to the second term $c_2\Delta v_e$ its contribution may be ignored due to lower in 10^{-3} density of sputtered Ca atoms in regard to the buffers [10]. This argument is stronger in regard to Li II ions.

Really, these measurements reveal the specific galvanic contributions of the atoms when their states $\Delta m = 0, \pm 2$ or $\Delta m = \pm 1$ dominate in number the rest m- states permitted, namely aligned and oriented. The TROGSs ΔUOG(t) align and ΔUOG(t) orient describe the appearance and evolution of two specific conductivities of one and the same ensemble of atoms that are aligned ↔ oriented. Since, the processes of relaxation are the same on one and the same optical transition, the observed differences in the evolutions ΔU OG (t) linear/circular could be ascribed to different rate constants, characterizing the decay processes of the aligned/oriented ensembles of atoms.

The differences in TROGSs above are checked also on AOGSs. He-Cd laser (λ = 325.25 nm) irradiates a Ne/Cd HCD lamp. Here the signal ΔUOG is the OG effect induced by the chopped laser beam and averaged by the τ- constant of the amplifier (figure 3). Differing from one another amplitudes (ΔUOG(P) linear and ΔUOG(P) circular) and steepness ($\partial\{\Delta$UOG(P) linear)/∂P and ∂(ΔUOG(P) circular)/∂P) characterize the two types of AOGSs. Here the logical trend ΔUOG ~ P is influenced by the dc (figure 3b) and the relation ΔUOG (P) circular > ΔUOG (P) linear takes place at I = 0.7 mA and P > 0.7 mW. At least, these results confirm the differences in OG efficiency of linearly and circularly polarized light.

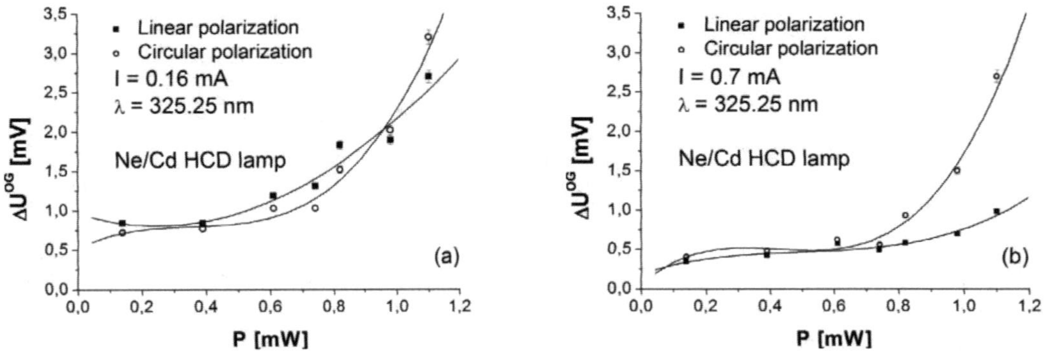

Figure 3. Amplitude OG signals induced by linear and circular polarizations of the spectral line CdI ($4d^{10}5s5p\ ^3P^0 \rightarrow 4d^{10}5s7s\ ^3S$) (325.25 nm): (a) I = 0.16 mA; (b) I = 0.7 mA.

Conclusions

Both time-resolved OGSs (TROGSs) on the transitions Ca I ($3p^63d4s^3D \rightarrow 3p^63d4p^3D^0$) (560.13 nm), Li II ($1s2s\ ^3S \rightarrow 1s2p\ ^3P°$) (548.57 nm), Li II ($1s2s\ ^3S \rightarrow 1s2p\ ^3P°$) (548.35 nm) and amplitude

(AOGSs) on the transition Cd I ($4d^{10}5s5p\ ^3P^0 \rightarrow 4d^{10}5s7s\ ^3S$) (325.25 nm) are observed to depend on the polarization of resonant light irradiating, i.e. either linear or circular.

Light polarizations induce TROGSs distinguishing in both OG efficiency and waveform $\Delta U^{OG}(t)$ parameters. As a rule linearly polarized light induces greater galvanic perturbation.

Characteristic decay times, namely $T^{linear} > T^{circular}$ of the evolution $\Delta U^{OG}(t)$ are found to depended on the dc. This *T*- relation is opposite to that on the transition Ne I ($1s_5 \rightarrow 2p_9$) (640.2 nm).

The TROGSs describe the appearance and evolution of two specific conductivities of one and the same ensemble of atoms that are aligned↔oriented, namely $\Delta U^{OG}(t)^{align}$ and $\Delta U^{OG}(t)^{orient}$. The aligned ensemble relaxes slower than oriented one.

Since the processes of relaxation are the same on one and the same optical transition the observed differences in the evolutions $\Delta U^{OG}(t)^{linear/circular}$ mean different rate constants, characterizing the decay processes of the aligned/oriented ensembles of atoms. In gas discharge plasma (self-) illuminated the interaction atom↔electron depends on the light polarization.

References

[1] Aleksandrov E B, Chaika M P and Khvostenko G I 1993 *Interference of Atomic States* (Berlin: Springer-Verlag)
[2] Zhechev D 2001 Coherent, opto-galvanic and magneto-galvanic properties of hollow cathode discharge D. Sc. Thesis (Sofia: Bulgarian Academy of Sciences)
[3] Zhechev D, Todorov G, Steflekova V, Slavov D, Polischuk V and Gorbenko A 2009 *J. Phys. B: At. Mol. Opt. Phys.* **42** 1
[4] Steflekova V, Slavov D, Zhechev D, Dimova E and Todorov G 2009 *Spectroscopy Letters* **42** 513
[5] Steflekova V 2014 *Proc. of the Summer School and International Symposium on the Physics of Ionized Gases (Belgrade, Serbia 26-29 August 2014)* p 287
[6] Zhechev D 1985 in: *Improved Hollow Cathode Lamps for Atomic Spectroscopy* ed S Caroli (London: Ellis Horwood Ltd.) p 203
[7] Barbieri B, Beverini N and Sasso A 1990 *Rev. Modern. Phys.* **62** 603
[8] National Institute of Standards and Technology - NIST Atomic Spectra Database
[9] Bundaleska N, Tomova Z, Steflekova V and Zhechev D 2008 *Vacuum* **82** 455
[10] Slezak V B, D'Accurso V and Manzano F A *J. Opt. Soc. Am. B* 1996 **13** 2701
[11] Drèze C, Demers Y and Gagné J M 1982 *J. Opt. Soc. Am.* **72** 912
[12] Gagné J M, Demers Y, Pianarosa P and Drèze 1983 C *J. de Phys.* **44** 355
[13] Mahmood S, Anwar-ul-Haq M, Riaz M and Baig M A 2005 *Eur. Phys. J. D* **36** 1
[14] Piracha N K, Feaver R, Gilani T H, Rizwan Ahmed, Ali R and Baig M A 2009 *Opt. Comm.* **282** 2532
[15] Zhechev D, Atanassova S and Todorov G 1999 *J. Phys. D: Applied Physics* **32** 598
[16] Okunevich A I 1976 *Zh. Eksp. Teor. Fiz.* **70** 899 (in Russian)

Journal of Physics: Conference Series **558** (2014) 012033 doi:10.1088/1742-6596/558/1/012033

Efficient composite broadband polarization retarders and polarization filters

E Dimova[1*], S S Ivanov[3], G Popkirov[2] and N V Vitanov[3]
[1] Institute of Solid State Physics, BAS, 72 Tzarigradsko Chaussée Blvd.,1784 Sofia, Bulgaria
[2] Central Laboratory for Solar Energy and New Energy Sources, BAS, 72 Tzarigradsko Chaussée Blvd., 1784 Sofia Bulgaria
[3] Department of Physics, St. Kliment Ohridski University of Sofia, 5 James Bourchier Blvd., 1164 Sofia, Bulgaria

*E-mail: edimova@issp.bas.bg

Abstract. A new type of broadband polarization half-wave retarder and narrowband polarization filters are described and experimentally tested. Both, the retarders and the filters are designed as composite stacks of standard optical half-wave plates, each of them twisted at specific angles. The theoretical background of the proposed optical devices was obtained by analogy with the method of composite pulses, known from the nuclear and quantum physics. We show that combining two composite filters built from different numbers and types of waveplates, the transmission spectrum is reduced from about 700 nm to about 10 nm width.We experimentally demonstrate that this method can be applied to different types of waveplates (broadband, zero-order, multiple order, etc.).

1. Introduction

Optical half-wave and quarter-wave retarders belong to the standard equipment of every optical laboratory. An optical retarder is an optical plate, made of birefringent material, which introduces a phase shift $\varphi=2\pi\delta_n L/\lambda$ of the light wave travelling through it. Here, λ is the light wavelength, δ_n is the birefringence and L is the thickness of the birefringent optical medium of the retarder. In the case of half-wave plates (HWP) the phase shift is $\varphi=(k+1)\pi$ and k=0, 1, 2, 3, ... while for the quarter-wave plates (QWP) $\varphi=(k+1/2)\pi$. The waveplates can be zero-order (k=0) or multiorder (k>0). Waveplates (WP) exhibit wavelength dispersion, due to the wavelength dependence of the phase shift and are thus manufactured for particular wavelength ranges.

Applications of waveplates are measurements with ultrashort laser pulses, e.g. terahertz time-domain spectroscopy [1, 2], microwave polarimetry [3, 4 and 5], etc. demand wider wavelength range of nearly constant phase retardation. Broadband (BB), also called achromatic, retarders are well known from the literature and are already commercially available. For instance, quarter- or half- wave plates with wide spectral range can be constructed using two or more waveplates of different birefringence [6], or waveplates with different thickness [7], or stacks of a number of equal waveplates [8, 9], twisted to specific angles, optimized to obtain nearly flat spectral response. The last mentioned case is an example that generally, a birefringent network having any required transfer function can be constructed if a proper method for the determination of the respective waveplates rotation angles is available.

Stacked composite plates can be seen as mathematically equivalent to composite pulses in quantum physics, a technique widely used in nuclear magnetic resonance (NMR) [10] and since recently, in quantum optics [11, 12]. This similarity is derived from the formal analogy between the Schroedinger equation for a two-state quantum system and the equation for the Jones polarization matrix evolution [13]. Ardavan [14] proposed to use the broad band (BB) and narrow band (NB) composite pulses of Wimperis [15] to design BB and NB WPs, respectively. In a recent work [16], the analogy between the polarization Jones vector and the quantum state vector was used to propose a calculation method to derive the twisting angles of the retarders in a stack with predefined wavelength range and polarization conversion fidelity. Recently, composite BB and NB WPs composed of stacked zero-order WPs were proposed [16] and experimentally demonstrated [17] in a double-pass setup. Both BB retarders and NB optical filters, constructed using a set of stacked standard multi-order WPs were presented in [18].

The theoretical background and calculation method used to design our polarization BB retarders and NB filters have been presented in details in [16, 18] and will not be discussed here. In the present work we will demonstrate that the wavelength range of a commercial broadband half WP can be further extended by combining it with further BB HWPs, building a composite half-wave plate. Furthermore, varying the twisting angles of the single WPs in the stack the wavelength range of the composite HWP can be even made narrower for possible applications as narrow optical polarization filter.

2. Experimental setup

2.1. Optical setup

We have used an optical breadboard to set up the experiments, cf. figure 1. The setup consisted of i) a light source, ii) the composite retarder under test, and iii) the transmission spectra recorder.

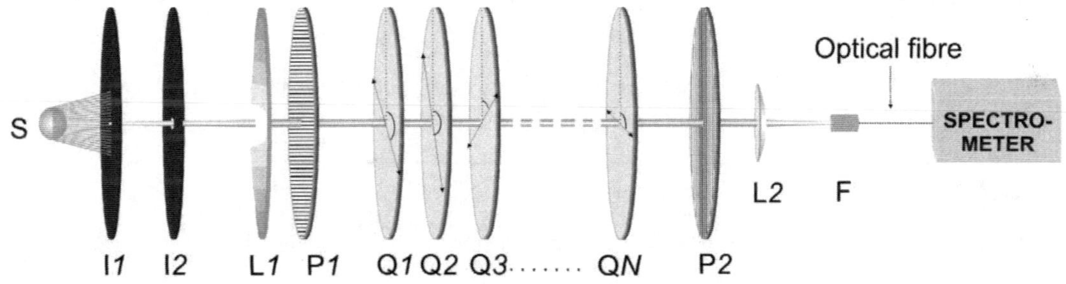

Figure 1. Experimental setup. The source S, irises I1 and I2, lens L1 and polarizer P1 form a beam of white polarized light. Polarizer P2 and lens L2 focus the analyzed beam of output light onto entrance F of an optical fibre connected to spectrometer. The investigated composite retarders are designed by stack of multiple order waveplates WPs (Qn).

As source of collimated polarized light with continuous spectrum we used a 10 W Halogen-Bellaphot (Osram) lamp, S, powered by a stabilized DC supply. For good collimation an iris I*1* with an aperture less than 0.5 mm was used, serving as a point light source, placed in the focus of a plano-convex lens L*1* (f=150 mm). A second iris diaphragm I*2* was used to collimate and thus further improve the beam so that its diameter, measured along a distance of 2 m was about 2 mm. The two polarizers P*1* and P*2* were N101-0520 (Glan-Taylor, 210-1100 nm), borrowed from a Lambda - 950 spectrophotmeter. The stack of waveplates, WP_n, to be investigated for their performance as a composite retarder was build-up of single broadband HWPs (WRM053-mica, 700-1100 nm, aperture 20 mm, Melles Griot), using RSP1 rotation mounts (Thorlabs Inc.). The optical axes of all the used waveplates could be preset to *1*° accuracy. Composite retarders consisting of up to 9 waveplates were assembled and their respective optical axes were preset to angles calculated using the model described elsewhere [16]. The waveplates were slightly tilted around the vertical axis to eliminate effects due to reflections [17].

The light passing through the composite retarder was focused by a plano-convex lens L*2* (f=20 mm) and a two-axes micro-positioner onto the optical fibre entrance "F". The later is used to feed the light to the input slit of a portable grating spectrophotometer Model AvaSpec-3648, spectral range 200 - 1100 nm, slite – 25 μm, grating – UA, controlled by AvaSoft 7.5 software.

2.2. Measurement procedure

The composite retarders were stacks of half-waveplates, assembled and adjusted according to the respective series of optical axes angle-of-rotation data delivered from the numerical calculations. For all experiments the first step was to parallel the axes of the polarizer P*1*, the analyzer P*2* and the slow axes of the waveplates which build-up the composite retarder under test. In this configuration the transmitted light spectrum was measured and stored as a reference. Thus, losses due to reflections and absorptions from all the single waveplates are taken into account. In a next step, the plates were rotated to the relative angles of their slow optical axis, according to the respective data from table 1. The analyzer P*2* was rotated to 90° with respect to its initial position. The transmission spectrum of the obtained composite retarder was recorded in respect to the reference spectrum. Thus, the measured spectrum will characterize the effective retardance of the stack of waveplates.

3. Experimental results

The theoretical model described in [16] has been applied to calculate a number of sets of rotation-angles for 3, 5, 7 and 9 BB HWPs, respectively and to construct the composite broadband retarders and composite narrowband polarization filters. Up to 10 different sets of rotation-angles for each type of composite retarders have been tested, recording their respective transmission spectra. In all cases the obtained spectra were nearly identical. Thus, we will present here only representative results for the composite retarders we have tested (see table 1).

Table 1. Calculated sets of rotation angles for construction of composite broadband polarization rotators (a) and polarization filters (b) by 3, 5, 7 and 9 BB half-wave plates (HWPs).

angles vs. waveplates	(α1, α2, α3, α4, α5, α6, α7, α8, α9)
(a)	**broadband polarization retarder**
a) 9 HWPs	(5.7, 12.5, 22.7, 34.3, 48.2, 61.5, 72.4, 81.3, 86.8)
b) 7 HWPs	(22.2, 67, 86.7, 44, 3.7, 24.5, 69.9)
c) 5 HWPs	(46.6, 0.2, 47.1, 88.9, 39.5)
d) 3 HWPs	(14.6, 44.9, 75.2)
e) 1 HWP	(45)
(b)	**polarization filter**
a) 9 HWPs	(166.4, 43, 7, 136.8, 165.2, 15.6, 43.1, 177.8, 126.7)
b) 7 HWPs	(175.6, 4.3, 170.6, 6.8, 169, 3.2, 174)
c) 5 HWPs	(7.5, 172.5, 14.2, 172.9, 8.6)
d) 3 HWPs	(14.7, 164.4, 14.7)
e) 1 HWP	(45)

3.1. Broadband composite retarders

As expected from the theory [16] and as already shown [18] a stack of identical commercial waveplates can be used effectively either as a broadband or as a narrowband composite retarder if the

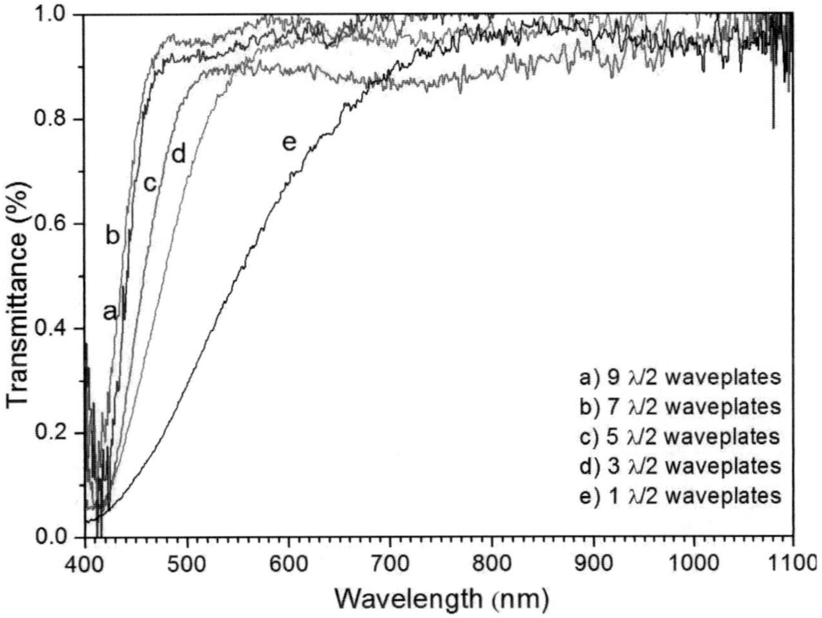

Figure 2. Transmission spectra of broadband composite retarders a), b), c) and d) assembled, respectively, by 9, 7, 5 and 3 broadband half-wave plates according to table 1(a). Each curve corresponds to the first set of angle data for the respective number of waveplates used. For comparison curve (e) presents the spectrum of one half-wave plate.

constituent WPs are arranged in a proper way. We will first show the possibility to extend the retardation of a HWP in a wider wavelength range, i.e. we will demonstrate composite BB HWPs.

Figure 2 presents the transmission spectra obtained with composite retarders assembled with different number of broadband half–wave plates, with each of them rotated to a specific angle of the slow axes according to the data from table 1 (a). It can be easily seen that with increasing number of waveplates the transmission and retardance spectra, respectively, become broader and steeper compared to the spectrum of one waveplate, cf. curves (a-e). Unfortunately, due to the experimental limitations (lamp spectrum) we were missing reliable spectral data for $\lambda \geq 1000$ nm. Thus, we do not see on the figures data for longer wavelengths.

3.2. Narrowband composite retarders

As mentioned above, according to the theoretical predictions, the wavelength range of nearly constant retardation of a composite retarder can be adjusted on demand using properly arranged stack of waveplates. In this subsection we will present representative results from a series of measurements intended to verify the possibility to produce retarders with narrower wavelength band.

Sets of rotation angles, presented in table 1(b) for composite retarders with narrow wavelength range were calculated. Although, it does not seem to be very useful for a retarder, it could be an excellent example of how a narrow band polarization optical filter can be assembled. Using the same experimental technique as described above, narrowband composite retarders were assembled with different number of broadband half-wave plates and respective angles of the slow axes from table 1(b).

Experimental transmission spectra of composite half-wave plates (HWPs) assembled with 3, 5, 7 and 9 HWPs, respectively, are shown on figure 3. On this figure the spectrum of a single half-wave plate is depicted for comparison, curve (e). The progressive narrowing of the transmission band with growing number of waveplates without significant altering of the central wavelength is clearly

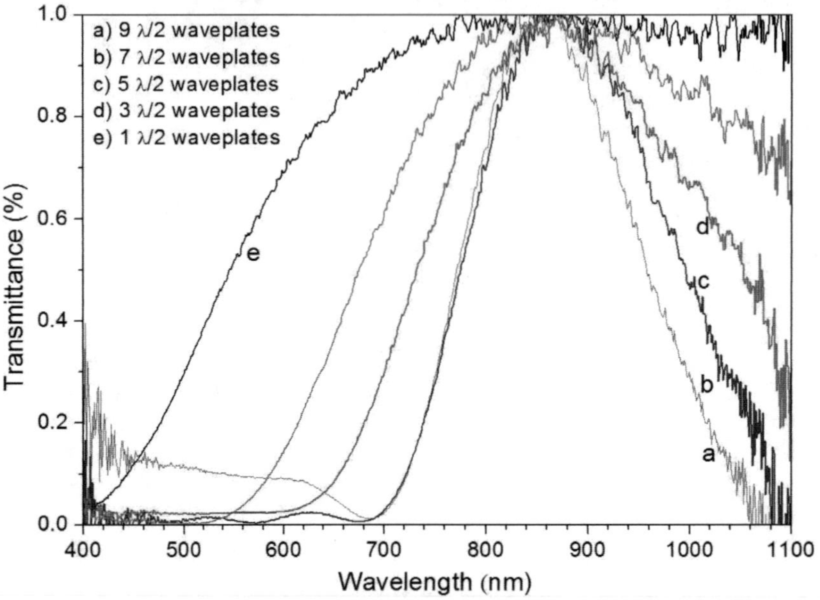

Figure 3. Transmission spectra of narrowband composite retarders assembled by 3, 5, 7, and 9 broadband half-wave plates according to table 1(a). Each curve corresponds to the first set of angle data for the respective number of waveplates used. For comparison curve (e) presents the spectrum of one half-wave plate.

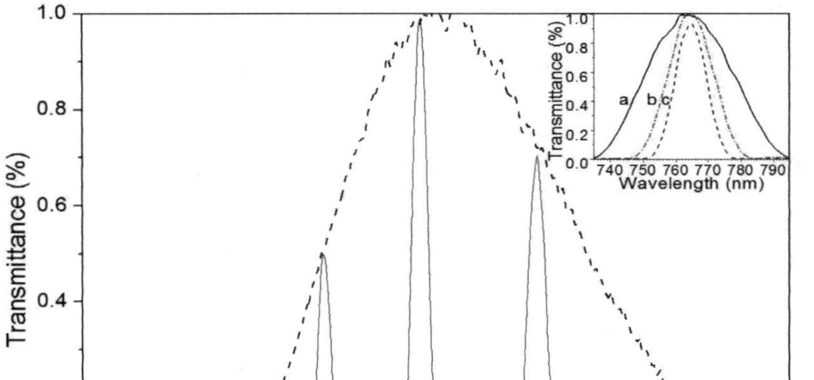

Figure 4. Transmission spectra (solid line) of a narrowband composite retarder assembled by two consecutive composite polarization filters: 9 broadband HWPs and 5 multiorder HWPs according to table 1(b). The dash curve corresponds to curve a from figure 2. The inset presents the effect of narrowing of the spectra of 1, 3 and 5 multiorder half-wave plates, curves a, b and c, respectively.

demonstrated. Thus, practical applications of composite retarders as narrowband optical filters appear to be feasible. It is also seen, that the transmission spectra show relatively low side lobes (cf. figure 3). Unfortunately, increasing the number of waveplates leads to more pronounced appearence of side lobes in the lower wavelength range.

The theoretical method used for the calculation of the rotation angles does not depend on the nature of the birefringent material. The effect of narrowing the transmission spectrum by using two consecutive composite polarization filters (CPFs) is presented in figure 4. As a result the transmission spectrum (with width of ca. 700 nm) is reduced to three peaks each with FWHM of about 10 nm. The first composite polarization filter is designed by BB HWPs with mica glass used as birefringent material while for the second composite filter a set of 5 quartz multiorder QWPs (780 nm) is used, working as HWPs at 763, 840 and 925 nm. The squeezing spectrum effect of the second CPF is presented in the inset. In both CPFs the rotation angles are taken from table 1.

Conclusion

We have demonstrated the possibility for practical realization of a new type of broadband polarization half-wave retarder plates and narrowband polarization filters, designed as composite stacks of commercial achromatic optical half-wave plates. The obtained results can be seen as further support for the calculation method described in [16].

Acknowledgments

This work was supported by the Bulgarian NSF Grant DRila-01/4 and the European Community's Seventh Frame - work Programme under grant agreement no. 270843 (iQIT).

References

[1] Grischkowsky D, Keinding S R, van Exter M and Fattinger C 1990 *J. Opt. Soc. Am.* B **7** 2006
[2] Masson J-B and Gallot G 2006 *Opt. Lett.* **31** 265
[3] Hanany S, Hubmayr J, Johnson B R, Matsumura T, Oxley P and Thibodeau M 2005 *Appl. Opt.* **44** 4666
[4] Pisano G, Savini G, Ade P A R, Hanes V and Gear W K 2006 *Appl. Opt.* **45** 6982
[5] Matsumura T, Hanany S, Ade P A R, Johnson B R, Jones T J, Jonnalagadda P and Savini G 2009 *Appl. Opt.* **48** 3614
[6] West C D and Makas A S 1949 *J. Opt. Soc. Am.* **39** 791
[7] Destriau M G and Prouteau J 1949 *J. Phys. Radium* **10** 53
[8] Pancharatnam S 1955 *Proc. Ind. Acad. Sci.* **41** 130
[9] Pancharatnam S 1955 *Proc.Ind. Acad. Sci.* **41** 137
[10] Levitt M H 1986 *Prog. Nucl. Magn. Res. Spectr.* **18** 61
[11] Häffner H, Roos C F and Blatt R 2008 Phys. Rep. **469** 155
[12] Timoney N, Elman V, Glaser S, Weiss C, Johanning M, Neuhauser W and Wunderlich C 2008 *Phys. Rev.* A **77** 052334
[13] Jones R C 1941 *J. Opt. Soc. Am.* **31** 488
[14] Ardavan A 2007 *New J. Phys.* **9** 24
[15] Wimperis S 1994 *J. Magn. Reson.* **109** 221
[16] Ivanov S S, Rangelov A A, Vitanov N V, Peters T and Halfmann T 2012 *J. Opt. Soc. Am.* A **29** 265
[17] Peters T, Ivanov S S, Englisch D, Rangelov A A, Vitanov N V and Halfmann T 2012 *Appl. Opt.* **51** 7466
[18] Dimova E St, Ivanov S S, Popkirov G St and Vitanov N V 2014 *J. Opt. Soc. Am.* A **31** 952

Journal of Physics: Conference Series **558** (2014) 012034 doi:10.1088/1742-6596/558/1/012034

Thermoluminescence properties of hausmannite Mn_3O_4 thin films induced by UV light

H Hristov[1], R Mateva[1], N Arhangelova[1], V Velev[1], Y Baneva[2], K Boubaker[3], T Larbi[4], G Moschini[5] and N M Uzunov[1, 5]
[1] Faculty of Natural Sciences, Shumen University "K. Preslavsky",
115 Universitetska str., 9712 Shumen, Bulgaria
[2] Dept. of Physics and Biophysics, Medical University,
150 Tzar Liberator str., 9000 Varna, Bulgaria
[3] Higher School of Sciences and Techniques of Tunis, University of Tunis,
63 Rue Sidi Jabeur, 5100, Mahdia, Tunisia
[4] Physics of Semiconductor Devices Uint, Faculty of Sciences of Tunis,
Tunis El Manar University, 2092 Tunis, Tunisia
[5] National Laboratories of Legnaro, INFN,
Viale dell'Universita 2, 35020 Legnaro (PD), Italy

E-mail: nikolay.uzunov@lnl.infn.it

Abstract. Thermoluminescence (TL) properties of Hausmannite thin films deposited onto glass substrates have been studied. The thin $Mn^{2+}Mn^{3+}{}_2O_4$ films were grown at 350°C on 1×2 cm^2 glass substrate by a spray pyrolysis technique. UV light from a XBO lamp was used to induce thermoluminescence in the film and the thermoluminescence properties of the Hausmannite layer were studied measuring the intensities of the TL peaks from the glow-curve deconvolution. The TL kinetic parameters of the Hausmannite were obtained from the TL glow curves analysis. The spectral emission of the $Mn^{2+}Mn^{3+}{}_2O_4$ layer has been studied using interference filters. The TL peak-fading assessment of the Hausmannite layer revealed that the peaks at 110^0C, at 142^0C and 194^0C have relatively long fading. The analysis of the thermoluminescence properties of the Hausmannite thin films deposited onto glass substrates shows that they possess very good sensitivity for UV emission and could be used for quantitative measurements of UV light.

1. Introduction

The development of thin metal-oxide layers in recent years receives an outstanding boost. On the one hand they are very interesting in terms of their fundamental physical properties, and on the other they have a wide range of applications. In particular manganese oxide thin films have attracted a great deal of attention because of their large variety of practical exploitation such as electrochemical applications, magnetic sensors and data storage based on their magnetic properties [1 – 3], catalysis of a number of processes [2, 4], supercapacitive media in electronics and microelectronics [2, 5, 6], etc. However, despite the large spectrum of studies on the properties of manganese oxide thin films, available data concerning their thermoluminescence (TL) properties are still very scarce or absent.

Spinnel Mn_3O_4 (also indicated as $Mn^{2+}Mn^{3+}{}_2O_4$) is a kind of manganese oxide naturally existing as a mineral Hausmannite. It is often obtained in a simple way by applying chemical spray pyrolysis technique [7, 8].

In this work we present the first hand results from the study on the TL emission of Hausmannite Mn_3O_4 thin film deposited onto a glass substrate using chemical spray pyrolysis technique. The TL properties of the Hausmannite thin films have been studied after irradiation with ultraviolet (UV) light. We discuss the opportunity such material to be exploited as a potential quantitative indicator, sensible to the ultraviolet light in the range of 290nm—340nm.

2. Materials and methods

Manganese oxide Mn_3O_4 layers were prepared at the Unit of Physics of Semiconductor Devices (UDPS), Tunisia. The thin films were grown at a temperature of 350°C on 1×2 cm^2 glass substrate by the spray pyrolysis technique. A schematic presentation of the process is presented in figure 1. A filtered compressed nitrogen gas was used as a gas carrier at a flow of 4 l/min. The sprayed aqueous solution, containing manganese chloride ($MnCl_2.6H_2O$) 0.1 M as precursor, was inserted at a constant flow of about 4 ml/min. The substrate temperature was fixed using a digital temperature controller with a k-type thermocouple. The distance between the nozzle and the substrate was about 27 cm. The total deposition time was maintained at 20 min. After the deposition, the coated substrates were allowed to cool down naturally to room temperature.

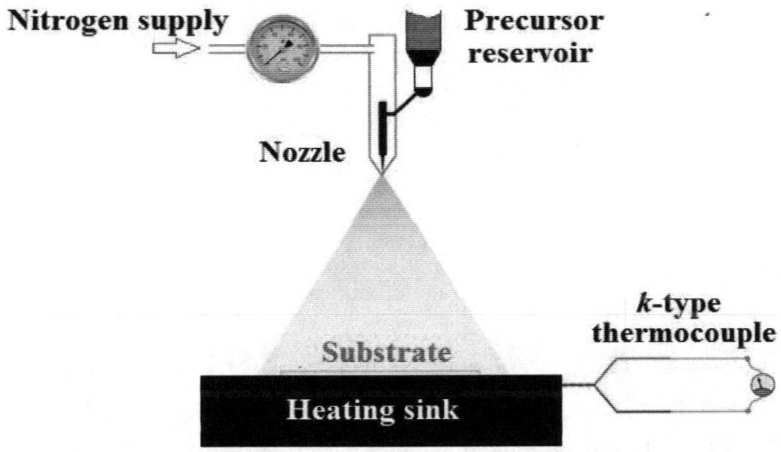

Figure 1. A schematic presentation of Mn_3O_4 thin-film deposition using spray pyrolysis technique.

To obtain the glow curves from the TL emission of the Hausmannite thin films we used a TL setup developed in the Laboratory for nuclear physics and radioecology (LNPR) at the University of Shumen [9]. The system is composed of a module for sample heating, a TL measuring module on the basis of photomultiplier tube, and a setup for sample irradiation with high-intensity large-band ultraviolet (UV) light. The TL measuring module is realized on the basis of PIC16F876 microcontroller as well as on original forward/backward feedback method, based on the so-called model predictive controllers to achieve higher temperature stability and repeatability of the analyses conducted.

The TL properties of the Hausmannite thin films have been studied after initial irradiation with a xenon short-arc lamp type XBO 75 W/2 (OSRAM), emitting light with a wavelength in the range of 250 nm $\leq \lambda \leq$ 1200 nm. The crystals have been placed at a distance of 20 cm and the time of irradiation (exposition time) was determined after studying the Hausmannite saturation curve.

The TL glow curves (GC), representing the intensity of the TL emission as a function of the linear increase of the temperature, were analyzed using glow curve decomposition analysis. Each of the peaks in the GC was treated as a composition of several glow peaks and the fitting procedure consisted in minimization of the χ^2 function varying the three parameters of each peak: the temperature, corresponding to the maximum of the peak intensity T_m; the activation energy E, and the so-called kinetic order parameter [10].

Glow curves, obtained with the TL reader were smoothed using a Savitzky-Golay filter after transferring to a personal computer. The GC data have been analyzed using a developed in the LNPR computer program implementing a minimization procedure based on conjugated gradients method.

In order to measure the fading of the GC peaks as a function of the storage time the Hausmannite thin films were irradiated with the XBO lamp for exposure time big enough to ensure full TL saturation. The irradiated samples were kept for a certain time after the irradiation in a dark storage at a temperature of 20°C. The intensity of the TL emission was measured as the net peak area of the GC peaks and the fading was presented as the peak intensity versus the storage time.

3. Experimental results

A glow curve of Mn_3O_4 thin films obtained at a heating rate of 0,16°Cs^{-1} immediately after 5 minutes of irradiation with UV light is presented with thicker black line in figure 2. Three peaks, shown with a thinner line in figure 2, have been revealed by the decomposition analysis of the glow curves: peak 1 at 110°C, peak 2 at 142°C and peak 3 at 194°C, indicated in the figure as 1, 2 and 3 respectively. Another peak with higher value of the maximum temperature position 347^{0}C, indicated as 4 in figure 2, has also been detected. In the same figure the residual is presented with a broken line and the resulting fit, which is a sum of all peaks, is presented with white line. The calculated kinetic parameters for the peaks appearing in the TL glow curves, calculated using the GC decomposition analysis, are shown in table 1.

A plot of the calculated peak intensities (the integrated counts in the corresponding peak areas from figure 2) as a function of the time of irradiation of the Hausmannite with the XBO lamp at a distance of 20 cm is shown in figure 3. From the figure it is clearly seen that a full saturation of the Mn_3O_4 film begins after 40 minutes of irradiation

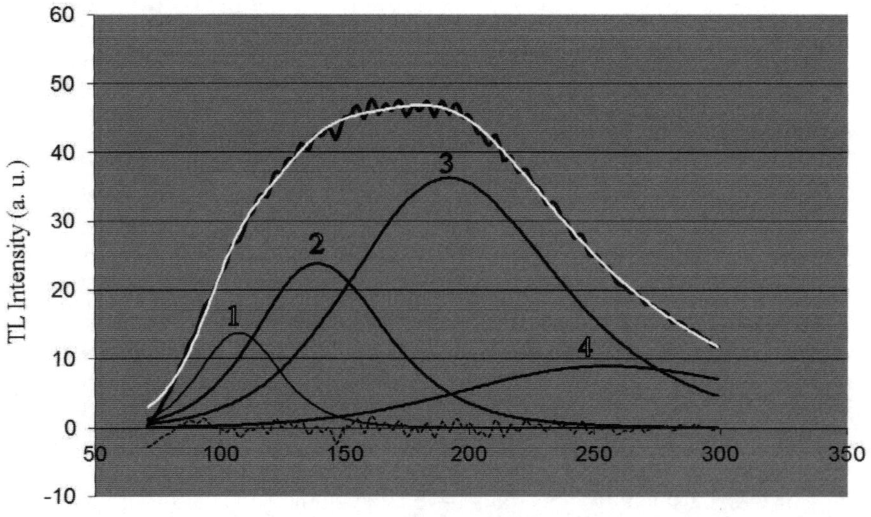

Figure 2. Glow curve of Mn_3O_4 thin film, obtained after 5 minutes irradiation with UV light.

Table 1. Calculated kinetic parameters of the peaks in the TL glow curve of Mn_3O_4 thin film.

Peak reference	Maximum temperature position (°C)	Activation energy (eV)	Kinetic order
Peak 1	110 ± 2	1.2 ± 0.2	2.0 – 0.1
Peak 2	142 ± 1.8	0.84 ± 0.09	2.0 – 0.1
Peak 3	194 ± 1.3	0.62 ± 0.08	2.0 – 0.1

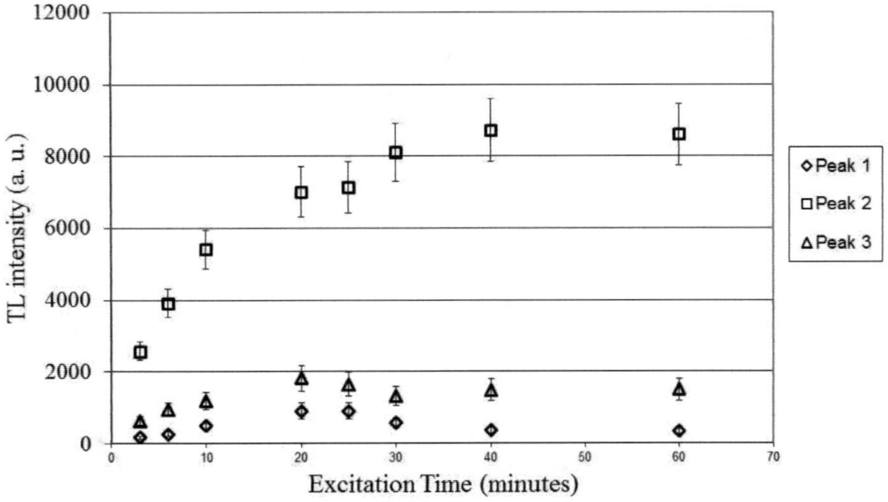

Figure 3. A plot of the peak intensities of the GC peaks of of Mn_3O_4 thin film as a function of the time of irradiation with UV light.

The dependence of the TL emission on the wavelength of the irradiating light (the spectral sensitivity) has been studied measuring the intensity of the TL glow curves for Hausmannite thin film. Interference optical filters for different wavelength were placed consecutively between the XBO lamp and the crystals during each irradiation. The Hausmannite film was exposed to irradiation from the XBO lamp for time ensuring the full saturation of the states. The calculated intensities for peak 1, peak 2 and peak 3 from the measured TL glow curves versus the wavelength of the irradiating light are shown in figure 4. The obtained thus spectral sensitivity of the material reveals a maximum at 290nm and a wavelength limit of around 340 nm.

Measurements to determine the fading of the peaks observed in fig 1 have been conducted for a maximum time period of 168 hours. To perform this kind of measurements each time the Hausmannite film was irradiated with an equal amount of light from the XBO lamp for exposure time ensuring the full saturation of the states. After the irradiation the Hausmannite film was kept for certain time period in dark place at a temperature of 20°C. A plot of the areas of the three peaks, calculated from the fits of the TL glow curves versus the storage time of the phosphors is shown in figure 5.

4. Conclusion

The quantitative assessment of the fading of Peak 1, Peak 2 and Peak 3 from fig. 2 reveals that the peaks possess low degree of fading, allowing thus stable quantitative TL measurements within a couple of weeks. The three peaks have relatively low activation energies and good count-rate statistics.

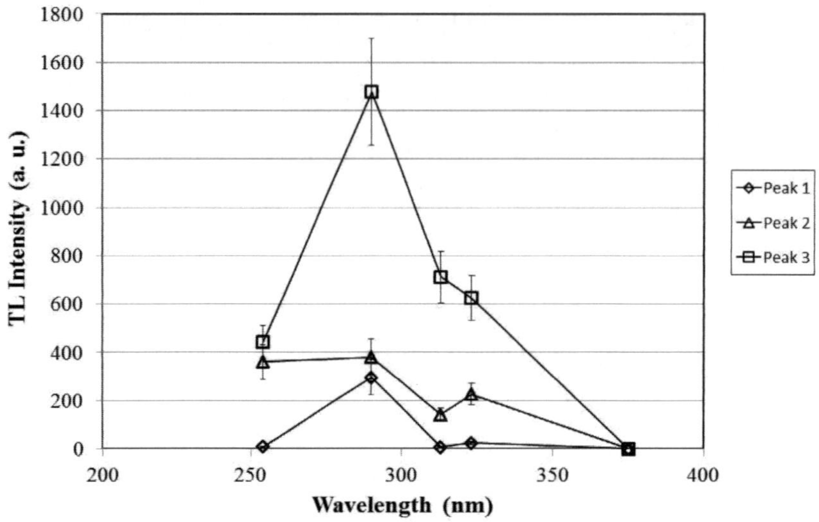

Figure 4. Spectral sensitivity of Mn_3O_4 thin film.

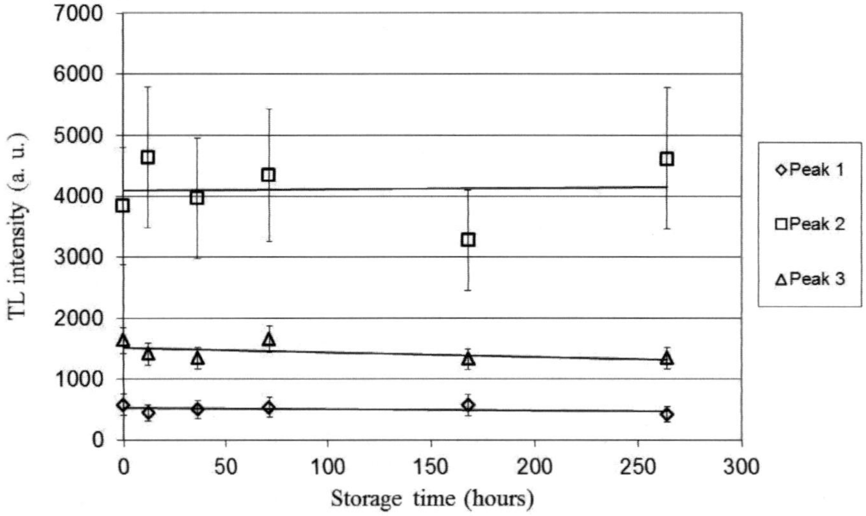

Figure 5. TL intensity of Peak 1, Peak 2 and Peak 3 as a function of the storage time. Linear least squares fits to the data are shown with solid lines.

In this connection a good reason would be the determination of dipole moment of manganese electron transitions from a metastable to the ground state as it is shown in [11] for the electrons of cobalt in the doped oxide glasses.

The assessment of the spectral sensitivity of the Hausmannite Mn_3O_4 thin film shows that the material is suitable for qualitative and quantitative measurements of UV light with a wavelength up to 350nm. Furthermore, a comparison of the intensity of the TL emission of the Mn_3O_4 thin film with other UV sensitive materials previously studied by our group [12,13] shows on the superior qualities of the Hausmannite as a UV detector. The physical characteristics of the Hausmannite studied so far suggest a reasonable opportunity to exploit such a material as an excellent media for UV measurements.

Acknowledgements
One of the authors (N Uzunov) would like to acknowledge the support received from the ICTP, Trieste, Italy and in particular the Program for Training and research in Italian Laboratories. This work has been partly supported by the projects RD-08-244 and RD-08-226/12.03.2014 of the University of Shumen, Bulgaria.

References

[1] Sarangapani S, Tilak B V, Chen C P 1996 *J. Electrochem. Soc.* **252** 3791
[2] Dubal D P, Dhawale D S Salunkhe RR Pawar S M, Fulari V J, Lokhande C D 2009 *Journal of Alloys and Compounds* **484** 218
[3] Grootendorst E, Verbeek Y, Ponce V 1995 *J. Catal.* **157** 706
[4] Yamashita Y, Mukai K, Yoshinobu J, Lippmaa M, Kinoshita T, Kawasaki M 2002 *Surf. Sci.* **514** 54
[5] Pang S C, Anderson M A, Chapman T W 2000 *J. Electrochem. Soc.* **147** 444
[6] Nam K W, Kim K B 2006 *J. Electrochem. Soc.* **153** 81
[7] Patil P S 1999 *Materials Chemistry and Physics* **59** 185
[8] Moses Ezhil Raj A, Grace Victoria S, Bena Jothy V, Ravidhas C, Wollschlager J, Suendorf M, Neumann M, Jayachandran M, Sanjeeviraja C 2010 *Appl. Surf. Sci.* **256** 2920
[9] Hristov H Y, Arhangelova N, Penev I and Uzunov N M 2008 *Annual of Konstantin Preslavski University* **Vol. XVIII B1** 30
[10] Kittis G, Gomez-Ros J and Tuyn J 1998 *J. Appl. Phys.* **31** 2636
[11] Petkova P and Vasilev P 2014 *J. Scientific and Applied Research* **5** 154
[12] Hristov H, Arhangelova N, Penev I, Bello M, Moschini G and Uzunov N 2010 *J. of Optoelectronics and Advanced Materials* **12** 892
[13] Hristov H, Baneva Y, Nedeva D, Arhangelova N, Penev I, Velev V, Moschini G, Rossi P and Uzunov N M 2012 *J. Phys.: Conf. Ser.* **398** 012043 (http://iopscience.iop.org/1742-6596/398/1/012043)

Journal of Physics: Conference Series **558** (2014) 012035 doi:10.1088/1742-6596/558/1/012035

Laser-induced electron transfer desorption/ionization of metal complexes on TiO_2 films

A A Grechnikov[1,5], V Georgieva[2], A S Borodkov[1,5], S M Nikiforov[3], Z Raicheva[2], J Lazarov[2], N Donkov[4]

[1] Vernadsky Institute of Geochemistry and Analytical Chemistry of RAS, Kosygin str., 19, 119991 Moscow, Russia,
[2] Georgi Nadjakov Institute of Solid State Physics, Bulgarian Academy of Sciences, 72 Tzarigradsko chaussee blvd., 1784 Sofia, Bulgaria,
[3] General Physics Institute of RAS, Vavilov str., 38, 119991 Moscow, Russia,
[4] Institute of Electronics, Bulgarian Academy of Sciences, 72 Tzarigradsko chaussee blvd., 1784 Sofia, Bulgaria,
[5] "Energomashtekhnika" Ltd., prospect Vernadskogo, 123, b.2, 119571 Moscow, Russia.

E-mail: grechnikov@geokhi.ru, lazarova@issp.bas.bg

Abstract. Thin titanium dioxide (TiO_2) films were studied as ion emitters for the laser-induced electron transfer desorption/ionization (LETDI) of metal complexes with organic reagents. The TiO_2 films (350 nm thick) were deposited on the silicon substrates by e-beam evaporation of TiO_2 powder. Copper complex with phthalocyanine, rhenium complex with thiocarbanilide and platinum complex with 8-quinolinethiol were studied as the test analytes. Reflectron time-of-flight mass spectrometer with the rotating ball interface was used for analysis. The analytes were applied on the surface of TiO_2 film using an electrospray deposition. All tested compounds are detected as the radical molecular ions with no fragmentation. It is found, that TiO_2 films are very stable and show good sensitivity in examined range of the analyte concentrations. The limits of detection of studied complexes were at the subfemtomole range, and the relative standard deviation was less than 10%.

1. Introduction

The analytical interest in metal-organic complexes arise from their potential use in a variety of applications of these complexes e.g. in the field of industrial photocatalysis, artificial photosynthesis, light-activated pharmaceuticals and highly efficient separation and preconcentration of metal element traces. Traditionally, investigations of such complexes have predominantly been performed using various spectroscopy methods, such as UV/visible, electron paramagnetic resonance and nuclear magnetic resonance spectroscopy. The limits of detection of spectrophotometric methods are usually at a level of not lower than 1 picomole of substance in the sample. Very few studies have employed more sensitive and specific mass-spectrometric techniques.

Recently, we introduced Laser-induced Electron Transfer Desorption/Ionization (LETDI) [1], a novel variant of Surface Assisted Laser Desorption/Ionization (SALDI) method. In SALDI, the ionized species are formed from analytes deposited on a surface of specially designed substrate, an ion emitter, which is irradiated with a laser beam at a wavelength well absorbed by the substrate [2]. The most commonly used material for SALDI is silicon, in particular, porous silicon [3], ordered nanocavity arrays [4] and amorphous silicon [5,6]. Most reported results on silicon based SALDI were obtained with basic compounds, which are detected as protonated molecules. In this case, only compounds with relatively high basicity in a gas phase (higher than 820 kJ/mol) are efficiently ionized [7]. Many important transition metal complexes have values of basicity much lower then 820 kJ/mol; therefore, they are not observed as protonated molecules by conventional SALDI. In LETDI, both substrate and analyte molecules absorb the laser radiation. This provides another mechanism for ionization of chemical compounds which is based on an electron transfer from the exited adsorbed analyte molecule to the semiconductor material. In contrast to a regular SALDI approach, chemical compounds with low basicity could be ionized by LETDI.

The previously reported results on LETDI mass spectrometry of metal complexes were obtained with silicon substrates [1,8]. In this work the titanium dioxide TiO_2 films were developed as ion emitters for the laser-induced electron transfer desorption/ionization of metal complexes. Rhenium (Re), platinum (Pt) and copper (Cu) complexes with common organic reagents were studied as test analytes.

2. Experimental

A schematic of the experimental setup is shown in figure 1.

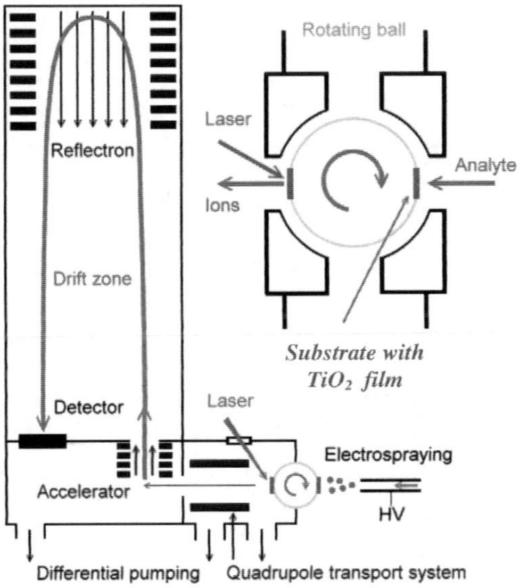

Figure 1. Schematic of the experimental setup. The instrument implements a combination of an ion source based on LETDI with a time of flight mass spectrometer. The inset shows a rotating ball interface.

The mass spectrometer was the time of flight reflectron with the orthogonal ion injection via gas-filled radiofrequency quadrupole (MX 5311, IAI of RAS, St. Petersburg). The rotating ball interface was used to couple the sample deposition chamber, where analytes were deposited on the semiconductor substrate, and the ion source, where the deposited analytes were ionized [9,10].

A diode-pumped YAG laser with near-diffraction limited beam quality, 0.35 ns pulse duration, was operated at 300 Hz. A frequency-tripling crystal yielded a 355 nm laser pulse of about 50 μJ. The laser pulse energy was varied in the range of 10-30 μJ using an attenuator. In order to more efficiently utilize the analyte, the laser focus was scanned over a 1×2 mm^2 sample deposition area using a computer-controlled, dual-mirror scanner. The laser focus moved approximately 0.1 mm between successive laser shots, and about 300 individual mass spectra were obtained from different points on the probed area of the ion emitter and were accumulated by the ADC digitizer during one "scanning cycle". The multiplier supply voltage was adjusted to provide a single-ion pulse from the ADC digitizer output that yielded an integrated signal equal to 1. Thus, the number of ions detected, for example during one raster scan, was simply obtained by integrating the output signal. The overall number of detected ions for any peak in the mass spectrum, over any number of raster scans, was similarly obtained.

The titanium dioxide TiO_2 films were deposited on the monocrystalline silicon substrates by e-beam evaporation of TiO_2 powder (initial pressure of 7×10^{-4} Pa, working pressure of 3×10^{-3} Pa and deposition rate of 50 nm/min). The results of the ellipsometric and secondary electron microscopy studies of the TiO_2 films indicated a thickness of about 350 nm, refractive index of 2.167, and a smooth and uniform surface morphology. Two substrates were installed on opposite sides of the rotating ball (see figure 1). The analytes were applied on the surface of TiO_2 film using an electrospray deposition, the ball was rotated 180°, and the analytes were ionized and desorbed by laser irradiation for the subsequent mass spectrometric analysis. All measurements were performed in the positive ion detection mode.

Copper complex with phthalocyanine was obtained commercially.

Rhenium complex with thiocarbanilide (diphenylthiourea) was synthesized according to the known procedure [11].

Platinum complex with 8-quinolinethiol was synthesized by a reaction of H_2PtCl_6 with 8-mercaptoquinoline using the microwave system Discover SP-D (CEM, Matthews NC, USA). To 10 mL of Pt solution in 2M HCl was added 1 mL of a saturated solution of 8-mercaptoquinoline in ethyl alcohol. Then, the solution was heated in the microwave system to 80°C at a rate of 12 K/min and kept for 5 min at this temperature; thereafter, the solution was cooled to room temperature, and 2 mL of chloroform was added. The mixture was allowed to separate into layers, and a chloroform fraction was taken for further analysis.

The obtained complexes were characterized by ^{1}H-NMR and UV/visible spectroscopy.

3. Results and discussion

It was found that for achieving optimum sensitivity and mass spectral performance, the TiO_2 films must first be activated by irradiating the surface by UV laser pulses to remove the adsorbed molecules. Full activation requires several (up to 10) laser pulses on each substrate spot. Activation was performed before any analysis and without analyte.

The preliminary study of threshold behavior of TiO_2 films showed that 30 mJ/cm^2 of laser intensity gives the best performance for all studied analytes. Decreasing the laser fluence leads to the dramatic decreasing of the ion signal. With the increase of laser fluence titanium containing peaks appeared in mass spectra, indicating the destruction of the film. Therefore all following experiments were carried out at 30 mJ/cm^2 of laser fluence.

A LETDI mass spectrum of the Re complex with thiocarbanilide obtained from 10^{-6} M analyte solution is shown in figure 2. The mass spectrum reflects the isotopic cluster at mass-to-charge ratio m/z 657 for the most abundant peak, with no Re-containing fragment ions observed. It is well known that natural rhenium consists of two isotopes with masses of 185 Da and 187 Da having a ratio $^{185}Re/^{187}Re = 37.4/62.6$. The characteristic Re isotope pattern considerably facilitates the identification of Re-containing ions. The measured m/z values, as well as the observed isotopic ratio, allow these peaks to be identified to correspond to the radical molecular ion isotopic cluster of the complex $ReOL_2$, where ligand L is (thiocarbanilide – H) $SN_2C_{13}H_{11}$ (structural formula of the complex is shown in the inset). Figure 3 compares the calculated and measured isotopic distribution of the molecular ion peak. It can be seen that the observed isotopic distribution agrees well with the theoretical spectra predicted for molecular radical cation $(ReOL_2)^+$. Therefore, the main ionization channel of the Re complex is electron transfer from the test molecule. In addition to Re-containing peaks, K^+ at m/z 39 and Na^+ at m/z 23 ions are observed in mass spectrum.

LETDI mass spectrum of the Pt complex with 8-quinolinethiol obtained from 10^{-6} M analyte solution is shown in figure 4. The mass spectrum is dominated by the isotopic cluster at m/z 514 and m/z 515 for the most abundant peaks. No fragment ions were detected in the mass spectrum. Pt-

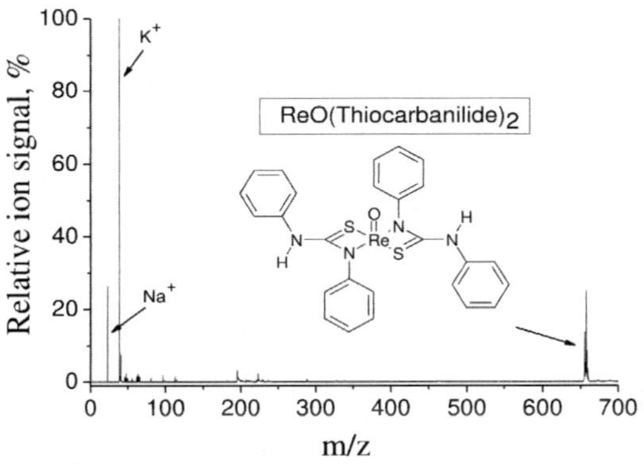

Figure 2. LETDI mass-spectrum of rhenium complex with thiocarbanilide.

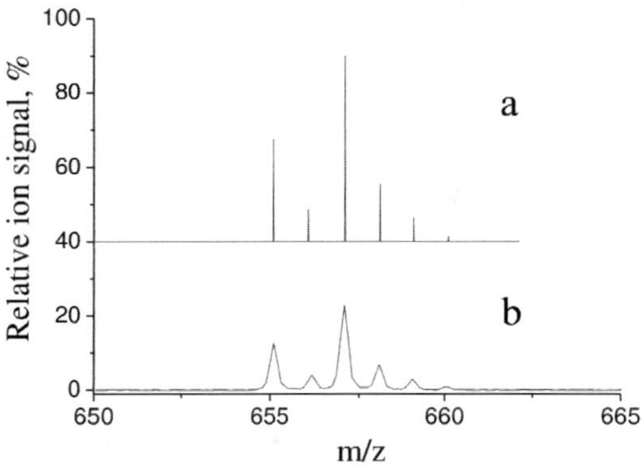

Figure 3. Calculated (a) and measured (b) isotopic distribution of the molecular ion peak $ReO(thiocarbanilide)_2^+$.

containing peaks can be easy interpreted as the molecular ion isotopic cluster of the complex PtQ_2, where ligand Q is (8-quinolinethiol – H) SNC_9H_6 (see inset of figure 4).

Figure 5 shows a LETDI mass spectrum of the Cu complex with phthalocyanine obtained from 10^{-6} M analyte solution. Similar to that of the Re and Pt complexes the molecular ion isotopic cluster at m/z 575 is observed in the mass spectrum, with no Cu-containing fragment ions detected. According to the mass spectrometric data Cu complex have composition Cu-Pht where ligand Pht is (phthalocyanine – 2H) $C_{32}H_{16}N_8$ (see inset of figure 5).

The obtained results demonstrate that LETDI on TiO_2 films is a soft ionization technique. LETDI mass spectra are "clean" with low chemical background except the peaks corresponding to the ions of sodium and potassium. However, peaks of these alkali metals are observed in the low-mass region and have not a considerable effect on the quality of the results obtained.

Additionally, experiments were carried out to compare the mass spectral characteristics of the TiO2 films to the silicon substrates. Silicon substrates were produced inside the ion source of the mass spectrometer by repeated laser irradiation according to the early reported procedure [1]. LETDI mass spectra from Si substrates found to be similar to those produced with using TiO_2 films for all tested

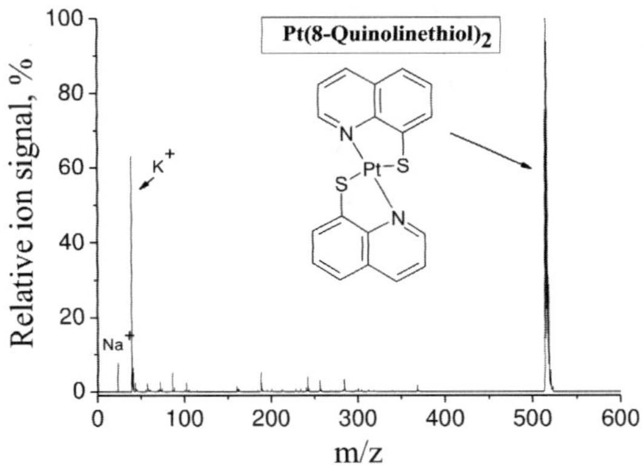

Figure 4. LETDI mass-spectrum of platinum complex with 8-quinolinethiol.

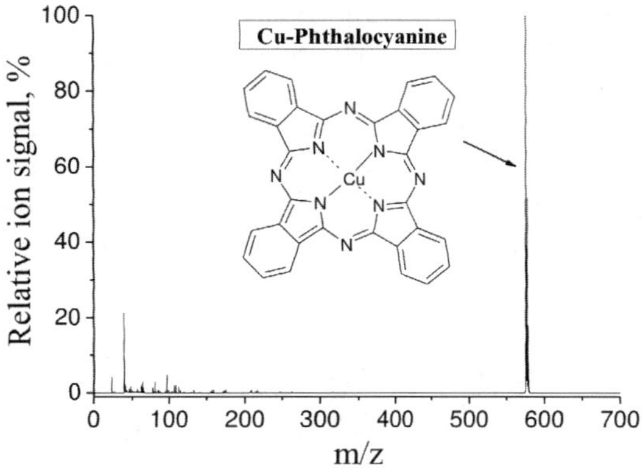

Figure 5. LETDI mass-spectrum of copper complex with phthalocyanine.

metal complexes. However, it was found that signal-to-noise ratio (S/N) obtained with using Si substrates was at least 2 times lower in comparison to S/N obtained with TiO_2 films. Another important advantage is that TiO_2 films are very stable under atmospheric conditions in contrast to Si substrates. There was no significant decrease in ionization efficiency after a few months storage of TiO_2 films in air. The surface of TiO_2 layer was not cleaned in any way, except by laser irradiation before the analysis.

To estimate the sensitivity and the limit of detection of investigated metal complexes the TiO_2 surface with known amount of deposited analyte was scanned several times until all the analyte signal vanished. The amount of analyte deposited was obtained simply from the analyte concentration, electrospray flow rate and the deposition time. The overall number of generated ions was obtained by integrating the output signal for all peaks corresponding to the metal complex including isotopes. From data obtained it was found that the limits of detection (i.e. the amount of sample needed to produce a spectrum with analyte signal-to-noise ratio $S/N \geq 3$) of studied complexes were at the subfemtomole range and the sensitivities were 20-100 ions/fmol.

The ions signal was measured for different amounts of analytes over a range of 3 orders of magnitude. The linearity of response was studied by varying the concentration of analyte solutions and the deposition time. The correlation coefficient found to be above 0.995. The relative standard deviation for five replicate analyses was less than 10%. A relatively high reproducibility is caused by several factors. Firstly, TiO_2 films were deposited using a highly reproducible method, resulting in uniformity over large areas and in high purity. Further, the surface was carefully cleaned before analyte deposition. Finally, with stable spray deposition, the rotating ball interface provided reproducible sample introduction.

The obtained results clearly demonstrate that TiO_2 films are applicable to the quantitative analysis of metal complexes with organic reagents by LETDI technique.

4. Conclusion

Thin titanium dioxide films obtained by e-beam evaporation of TiO_2 powder were investigated for the laser-induced electron transfer desorption/ionization (LETDI) of metal complexes with organic reagents. Copper complex with phthalocyanine, rhenium complex with thiocarbanilide and platinum complex with 8-quinolinethiol were studied as the test analytes. Reflectron time-of-flight mass spectrometer equipped with LETDI ion source and with the rotating ball interface was used for analysis of analyte solutions. Advantages of TiO_2 films application for LETDI of metal complexes are the following:

- high purity;
- easy and controllable method of fabrication;
- high uniformity over large areas;
- films can be deposited on any construction material;
- high stability of deposited films under atmospheric conditions;
- TiO_2 based LETDI ion source generates simple mass spectra of metal complexes, dominated by radical molecular ions. Mass spectra are "clean" with low chemical background;
- high ionization efficiencies and low limits of detection (at the subfemtomole range) of investigated metal complexes;
- chemical specificity;
- relatively high reproducibility;
- ability to analyze samples rapidly.

These advantages make TiO_2 films a potentially powerful tool for the detection of metal complexes and related compounds using LETDI technique.

Acknowledgement

The presented work was supported by the Ministry of Education and Science of the Russian Federation (Contract No. 14.579.21.0016, unique identifier RFMEFI57914X0016).

Journal of Physics: Conference Series **558** (2014) 012035 doi:10.1088/1742-6596/558/1/012035

References

[1] Grechnikov A, Nikiforov S, Strupat K and Makarov A 2013 *Anal. Bioanal. Chem.* **406** 3019
[2] Alimpiev S S, Grechnikov A A, Sunner J, Karavanskii V A, Zhabin S N, Simanovsky Ya O and Nikiforov S M 2008 *J.Chem.Phys.* **128** 014711
[3] Wei J, Buriak J M and Siuzdak G 1999 *Nature* **399** 243
[4] Finkel N H, Prevo B G, Velev O D and He L 2005 *Anal. Chem.* **77** 1088
[5] Alimpiev S S, Grechnikov A A, Sunner J, Borodkov A S, Karavanskii V A, Simanovsky Ya O and Nikiforov S M 2009 *Anal. Chem.* **81** 1255
[6] Grechnikov A A, Borodkov A S, Alimpiev S S, Nikiforov SM, Simanovskii Ya O and Karavanskii VA 2010 *J. Anal. Chem.* **65** 1504
[7] Grechnikov A A, Borodkov A S, Alimpiev S S, Nikiforov S M, Simanovsky Ya O 2013 *J. Anal. Chem.* **68** 19
[8] Borisova L V, Borodkov A S, Grechnikov A A, Ugolkova E A and Minin V V 2013 *Russ. J. Inorg. Chem.* **58** 940
[9] Alimpiev S S, Grechnikov A A, Sunner J, Karavanskii V A, Simanovsky Ya O and Nikiforov S M 2011 *Rapid Commun. Mass Spectrom.* **25** 140
[10] Grechnikov A A, Kubasov A E, Georgieva V B, Borodkov A S, Nikiforov S M, Simanovsky Ya O and Alimpiev S S 2012 *J. Phys.: Conf. Ser.* **398** 012033
[11] Borisova L V and Ermakov A N 1974 *Analytical chemistry of rhenium* (Moscow: Nauka).

Comparative analysis of electrophysical properties of ceramic tantalum pentoxide coatings, deposited by electron beam evaporation and magnetron sputtering methods.

N Donkov[1], E Mateev[1], V Safonov[2], A Zykova[2], SYakovin[3], D Kolesnikov[4], I Sudzhanskaya[4], I Goncharov[4] and V Georgieva[5]
[1]Emil Djakov Institute of Electronics Bulgarian Academy of Sciences, 72 Tzarigradsko Chaussee, 1784 Sofia, Bulgaria
[2]National Science Center "Kharkov Institute of Physics and Technology", 1 Academicheskaya Str., 61108 Kharkov, Ukraine
[3]Department of Physical Technologies Kharkov National University, 4 Svobody Sq., 61077 Kharkov, Ukraine
[4]Belgorod State National Research University, 85 Pobedy St., 308015 Belgorod, Russia
[5]Georgi Nadjakov Institute of Solid State Physics, Bulgarian Academy of Sciences, 72, Tzarigradsko Chausse, 1784 Sofia, Bulgaria

E-mail: nikolaidd@abv.bg

Abstract. Ta_2O_5 ceramic coatings have been deposited on glass substrates by e-beam evaporation and magnetron sputtering methods. For the magnetron sputtering process Ta target was used. X-ray diffraction measurements show that these coatings are amorphous. XPS survey spectra of the ceramic Ta_2O_5 coatings were obtained. All spectra consist of well-defined XPS lines of Ta 4f, 4d, 4p and 4s; O 1s; C 1s. Ta 4f doublets are typical for Ta_2O_5 coatings with two main peaks. Scanning electron microscopy and atomic force microscopy images of the e-beam evaporated and magnetron sputtered Ta_2O_5 ceramic coatings have revealed a relatively flat surface with no cracks. The dielectric properties of the tantalum pentoxide coatings have been investigated in the frequency range of 100 Hz to 1 MHz. The electrical behaviour of e-beam evaporated and magnetron sputtered Ta_2O_5 ceramic coatings have also been compared. The deposition process conditions principally effect the structure parameters and electrical properties of Ta_2O_5 ceramic coatings. The coatings deposited by different methods demonstrate the range of dielectric parameters due to the structural and stoichiometric composition changes

1. Introduction

The interest in dielectric materials and coatings applications has considerably increased in various areas of science and technology in recent years. The high dielectric parameters, band gap and low values of dielectric loss tangent are very important properties for various applications such as electronic devices, thin-film capacitors and biomedicine [1-3]. Dielectric oxides are known for their good optical properties but relatively little has been published on their electrical characteristics [4-6]. Dielectric oxide layers were proposed for double insulator thin-film electroluminescent devices [7]

Journal of Physics: Conference Series **558** (2014) 012036 doi:10.1088/1742-6596/558/1/012036

and a complete EL-device has been made with Ta_2O_5 insulators [8]. Dielectric coatings with electret properties are applied in medicine for stimulation of positive biological processes by short-range quasi-static electric field actions. Tantalum and tantalum-based compounds are promising for biomedical applications. Ta-based materials demonstrate high fracture toughness, corrosion and wear resistance, and chemical stability. The results of test on implantation of Ta in both soft and hard tissue of rats showed the good biocompatibility and osteogenesis of this metal [9]. TaC and TaN materials posses relatively high hardness due to the covalent nature of their bond [10] and demonstrate high thermal stability and superior corrosion resistance [11]. The blood compatibility of TaN films was shown to be better than that of Ta [12]. The deposition and treatment processes may change the coating's structure from amorphous one (for as-deposited films) to an orthorhombic phase, the surface topography, from smooth to nano-crystalline and improve the crystallinity of the tantalum pentoxide coatings [13]. The research of the dielectric properties of Ta_2O_5 ceramic coatings deposited on glass substrates by means of e-beam evaporation and magnetron sputtering methods was presented.

2. Materials and methods

The study of Ta_2O_5 ceramic coatings structure and properties which effect on electrophysical parameters was performed. The substrates were glass samples. The main parameters of the process were described in our previous study [14]. The evaporation process was carried out at initial vacuum of 9.3×10^{-4} Pa, operational vacuum of 4×10^{-3} Pa, anode current of 50 mA and calculated evaporation power of 350 W. The deposition rate under these conditions was 30 nm/min. For magnetron sputtering process Ta target was used. The substrates were displaced far from the magnetron axes to avoid bombardment with ions accelerated by the static electric field. The operating gas was chosen between O_2 (ensuring the highest possible degree of oxidation, but having the lowest deposition rate), Ar (ensuring the highest deposition rate, but the oxidation is only via residual gas of the vacuum system – high vacuum – not UHV) and mixture of 50% O_2 and 50% Ar – the intermediate case. Typical operating pressure was about 4 Pa and operating power was 80 W. The bias at such conditions was about 300 V. The e-beam evaporated (EB) and magnetron sputtered (MS) Ta_2O_5 ceramic coatings structure and composition were investigated by means of photoelectron spectroscopy XPS and X-ray diffraction XRD methods. X-ray diffraction profiles of Ta_2O_5 were observed by means of diffraction device "DRON-3" with filtered Cu-Ku radiation. X-ray photoelectron measurements have been carried out on the ESCALAB MkII (VG Scientific) electron spectrometer at a base pressure in the analysis chamber of 5×10^{-8} Pa using anode AlKα X-ray source with excitation energies of 1486.6 eV (160 W). The spectra are recorded at the total instrumental resolution (as it was measured with the FWHM of $Ag3d_{5/2}$ photoelectron line) of 1.18 eV for AlKα excitation source. The energy scale has been calibrated by normalizing the C1s line of adsorbed adventitious hydrocarbons to 285.0 eV. The processing of the measured spectra includes a subtraction of X-ray satellites and Shirley-type background. The peak positions and areas are evaluated by a symmetrical Gaussian-Lorentzian curve fitting. The relative concentrations of the different chemical species are determined based on normalization of the peak areas to their photo ionization cross-sections, calculated by Scofield [15].

The surface topography and roughness parameters of the tantalum pentoxide ceramic coatings were evaluated by scanning electron microscopy (SEM) and atomic force microscopy (AFM) methods. The thickness parameters were obtained from the SEM cross-section measurements. The analysis of capacity, dielectric constant, parameters of dielectric loss tangent depending on the frequency of electromagnetic field for oxide ceramic coatings deposited in the different sputtering regimes was made by LCR-meter BR-2876. The research of near-surface electric field distributions and surface potentials were made. The correlation of surface characteristics and field distributions was analyzed.

3. Results and discussion

The surface properties of e-beam evaporated and magnetron sputtered Ta_2O_5 coatings were investigated. The SEM surface images show no cracks on the surface of the tantalum pentoxide

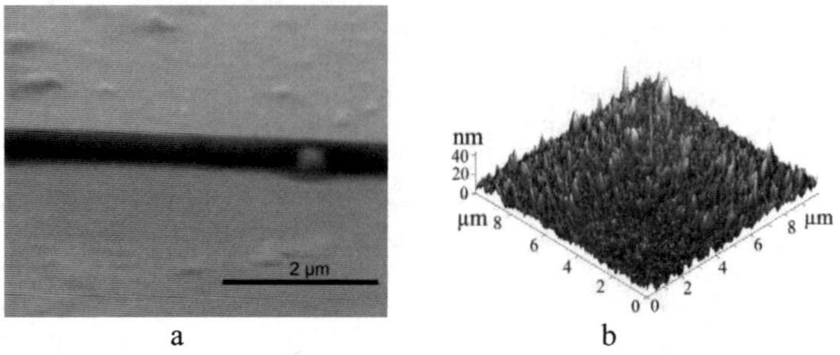

Figure 1. Surface topography of the e-beam evaporated Ta_2O_5 coatings by SEM (a) and AFM (b).

ceramic coatings in a good agreement with Atomic Force Microscopy (AFM) measurements, which show relatively flat surfaces (figure 1).

The structure of Ta_2O_5 coatings was investigated by means of the XRD method. X-ray diffraction profiles of as-deposited Ta_2O_5 coatings .demonstrated an amorphous nature for both e-beam evaporated and magnetron sputtered Ta_2O_5 coatings, no peaks were observed. XPS survey spectra of the e-beam evaporated and magnetron sputtered Ta_2O_5 films were obtained. All spectra consist of well-defined XPS lines of Ta 4f, 4d, 4p and 4s; O 1s; C 1s. All binding energies of the high-resolution spectra were calibrated with the C 1s binding energy of 285.0 eV. Ta 4f doublets are typical for Ta_2O_5 coatings and have two peaks: Ta 4f7/2 at ~ 26.2 for e-beam evaporated and 26.5 eV for magnetron sputtered layers and Ta 4f5/2 whose binding energy higher by 1.9 eV. The Ta 4f lines of the films deposited agree well with the Ta 4f doublet representative of the Ta-0 bond in Ta_2O_5. The Ta/O ratio estimated from the spectra was in the range 0.41 for magnetron sputtered and 0.37 for e-beam evaporated Ta_2O_5 coatings (table 1). The results demonstrate that magnetron-sputtering process improves the surface atomic concentration and the stoichiometry of the Ta_2O_5 films (Figure 2). The O1s spectra further support this assumption. The O 1s peaks of the deposited layers are centered at binding energies of 530.8 eV for e-beam evaporated and 530.6 eV for magnetron sputtered coatings.

Table 1. Surface atomic concentrations of Ta_2O_5 coatings (at.%).

	XPS spectra lines		
Type of coating	O1s	Ta4f	Ta4f/ Os1
Ta_2O_5 MS	7.6	14.2	0.41
Ta_2O_5 EB	7.4	14.1	0.37

The dielectric properties of tantalum pentoxide coatings were characterized by LCR-meter BR-2876 controlled by computer. Frequencies were chosen in the range from 100 Hz to 1 MHz with an oscillation voltage of 1V. The analysis of capacity, dielectric constant ε and dielectric loss tgδ measurements as a function of the frequency for e-beam evaporated and magnetron sputtered Ta_2O_5 ceramic coatings was made (figure 3).

Figure 3 shows the dielectric constant ε and the dielectric loss tgδ of e-beam evaporated in comparison with magnetron sputtered Ta_2O_5 ceramic coatings. The dielectric constant is calculated on the basis of the formula for a parallel plate capacitor: $\varepsilon = dC/\varepsilon_0 S$, where C/S is the capacitance per unit area and d is the thickness of the layer.

The analysis of dielectric constant ε and dielectric loss tgδ measurements depending on the frequency for e-beam evaporated and magnetron sputtered Ta_2O_5 ceramic coatings shows dielectric constant, which are steadily decrease at higher frequencies and losses, which are increasing at lower

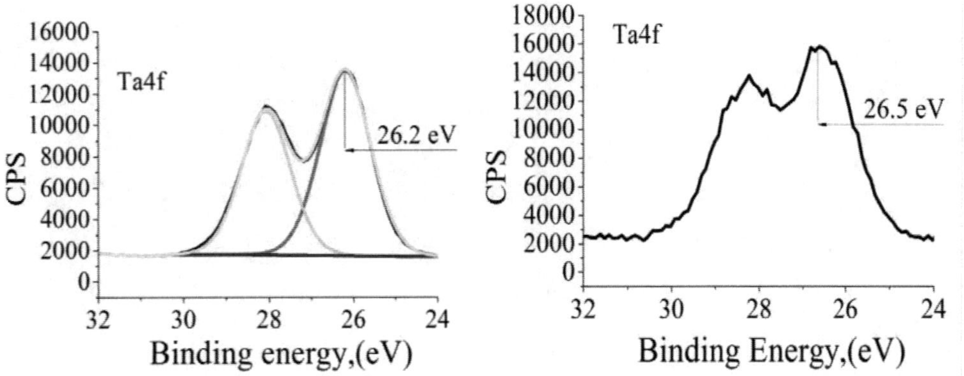

Figure 2. High-resolution XPS spectra of Ta 4f and composition of the e-beam evaporated (a) and magnetron sputtered (b) Ta_2O_5 coatings.

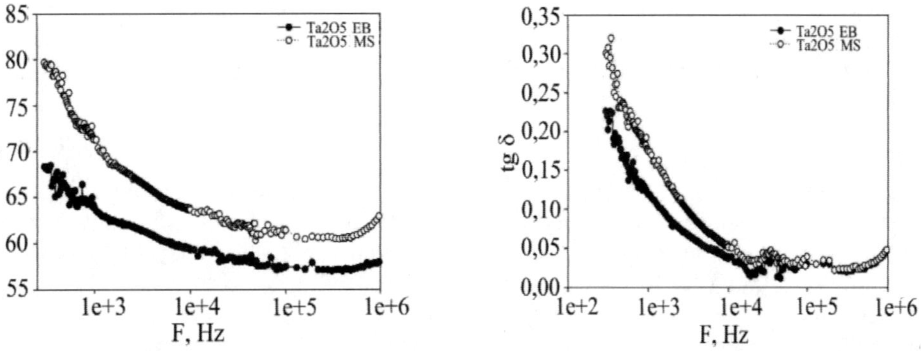

Figure 3. Dielectric constant ε (a) and dielectric loss tgδ (b) measurements as a function of the frequency for e-beam evaporated and magnetron sputtered Ta_2O_5 ceramic coatings

processes. The coatings demonstrate high dielectric parameters: the dielectric loss tgδ < 1 for all Ta_2O_5 ceramic coatings. The difference in apparent dielectric constant can be explained by the different structure and composition parameters of tantalum pentoxide ceramic coatings deposited by e-beam evaporation and magnetron sputtering methods. A dielectric constant obtained for e-beam evaporated layers are substantially lower than for magnetron sputtered, due to the degradation of the structural properties and the impurities penetration during evaporation process. Electrical properties are structurally sensitive and structural variation leads to significant changes of electrophysical characteristics [13, 16, 17].

The surface was observed by scanning microprobe method, phase contrast of coating's surface scan was analyzed and surface charge distribution by Kelvin probe method was evaluated. The method was used for analysis of electric field and charge distribution on the surface of tantalum pentoxide coatings deposited by different processes. The method realized by two stages scanning of surface relief and probe measurements at resonance frequency with the constant bias between probe and surface 2V. The images were formed as a result of probe oscillation phase changing (figure 4).

The results show that surface relief has no significant differences in the case of e-beam evaporated and magnetron sputtered Ta_2O_5 ceramic coatings. The surface potential was in the range 3 – 5 mV. The process of deposition changes the electric state of surfaces in the case of both e-beam evaporated and magnetron sputtered coatings. The effect of constant electric fields on the cell's behaviour was previously reported [18]. Osteoblast and osteoclast cells migrate in opposite directions

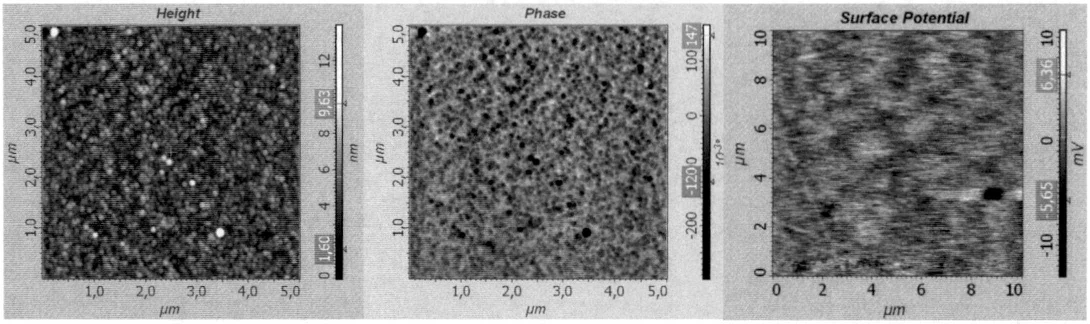

a b c

Figure 4. Surface scans of surface relief (a), phase contrast (b) and surface potential (c) of e-beam evaporated Ta_2O_5 ceramic coatings.

by action of a constant electric field. The variation of the electric field values allows to control the cell response in vitro tests [19]. The changes in the cell membranes potential took place at the potential values about 10 mV. Therefore, the surface potential of Ta_2O_5 ceramic coatings deposited by both e-beam evaporation and magnetron sputtering methods can effect on further cell/material response on cellular level.

4. Conclusion

The tantalum pentoxide ceramic coatings have been deposited by e-beam evaporation and magnetron sputtering process in the study. The coatings were transparent and amorphous and the surface shown no cracks. The Ta 4f lines of the deposited coatings agree well with the Ta 4f doublet representative of the Ta-0 bond in Ta_2O_5. The Ta/O ratio estimated from the spectra was in the range 0.41 for magnetron sputtered and 0.37 for e-beam evaporated Ta_2O_5 coatings. The results demonstrate that magnetron sputtering process improves the surface atomic concentration and the stoichiometry of the Ta_2O_5 layers. The coatings demonstrate high dielectric parameters: the dielectric loss $tg\delta < 1$ for all Ta_2O_5 ceramic coatings. Electrically the e-beam evaporated tantalum pentoxide coatings have a smaller value of dielectric constants compared with magnetron sputtered. This is probably due to the more perfect structure and composition of the magnetron sputtered coatings. The results show that the electro physical properties are strongly influenced by the coating's deposition parameters. The deposition process conditions principally effect on the structure properties and electrical behavior of the e-beam evaporated and magnetron sputtered Ta_2O_5 coatings. The coatings deposited by different methods demonstrate the range of dielectric parameters due to the structural and stoichiometric composition changes.

Acknowledgment

The work was supported under a program for international scientific collaboration between the Bulgarian Academy of Science and the National Academy of Science of Ukraine, research program of National Academy of Science of Ukraine № 24-04-14 and program of Russian Fund of Fundamental Research №14-02-90457.

References

[1] Robertson J 2004 *Eur. Phys. J. Appl. Phys.* **28** 265
[2] Miyazaki S 2001 *J. Vac. Sci. Technol.* B **19** 2212
[3] Guha S, Cartier E, Bojarczuk N A, Bruley J, Gignac L and Karasinski 2001 *J Appl. Phys. Lett.* **90** 512
[4] Vanbesien K, De Visschere P, Smet P F and Poelman D 2006 *Thin Solid Films* **514** 323
[5] Minami T, Shirai T, Nakatani T and Miyata T 2000 *Jpn. J. Appl. Phys.* **39** 524

[6] Kavanagh Y, Alam M J and Cameron D C 2004 *Thin Solid Films* **447/448** 85
[7] Minami T, Kobayashi Y, Shirai T, Miyata T and Suzuki T 2002 *Jpn. J. Appl. Phys.,* **41** 478
[8] Minami T 2003 *Solid State Electron.* **47** 2237
[9] Matsuno H, Yokoyama A, Watari F, Uo M and Kawasaki T 2001 *Biomaterials* **22** 1253
[10] Miyazaki T, Kim H M, Kokubo T, Ohtsuki C and Nakamura T 2002 *Biomaterials* **23** 827
[11] Georgiev G, Fescheschiev N, Popov D and Uzunov Z 1989 *Vacuum* **36** 595
[12] Leng X Y, Sun H, Yang P, Chen J Y, Wang J, Wan G J, Huang N, Tian X B, Wang L P and Chu P K 2001 *Thin Solid Films* **398/399** 471
[13] Uthanna S, Jagadeesh C S V, Sreedhara R P and Mohan R G 2008 *J. Phys.: Conf. Series* **114** 01203
[14] Donkov N, Mateev E, Zykova A, Safonov V and Luk'yanchenko V 2010 *J. Phys.: Conf. Series* **223** 012030
[15] Scofield J H 1976 *J.Electron Spectroscopy and Related Phenomena* **8** 129-37.
[16] Atanassova E., Kalitzova M, Zollo G, Paskaleva A., Peeva A., Georgieva M and Vitali G 2003 *Thin Solid Films* **426** 191-99
[17] Jagadeesh C S V, Chandrasekhar M, Mohan R G and Uthanna S 2009 *J. Mater Sci-Mater. Electron.* **20** 295.
[18] Ferrier J, Ross S M, Kanehisa J and Aubin J E 1986 *J. Cell Physiol.* **129** 283
[19] Qui Q, Sayer M, Kawaja M, Shyen X and Davies J E 1998 *J. Biomed. Mater. Res.* **42** 117

Study of quartz crystal microbalance NO_2 sensor coated with sputtered indium tin oxide film

V Georgieva[1], M Aleksandrova[2], P Stefanov[3], A Grechnikov[4], V Gadjanova[1], T Dilova[3] and Ts Angelov[1]
[1] Georgi Nadjakov Institute of Solid State Physics, Bulgarian Academy of Sciences, 72, Tzarigradsko Chausse Blvd., 1784, Sofia, Bulgaria
[2] Technical university of Sofia, Department of Microelectronics, Kliment Ohridski blvd, No 8, bl. 1, 1000, Sofia, Bulgaria
[3] Institute of General and Inorganic Chemistry, Bulgarian Academy of Sciences, "Acad. Georgi Bonchev" str.bl. 11, 1113, Sofia, Bulgaria
[4] Vernadsky Institute of Geochemistry and Analytical Chemistry of RAS, Kosygin str., 19, 119991, Moscow, Russia

E-mail: lazarova@issp.bas.bg

Abstract. A study of NO_2 gas sorption ability of thin indium tin oxide (ITO) deposited on 16 MHz quartz crystal microbalance (QCM) is presented. ITO films are grown by RF sputtering of indium/tin target with weight proportion 95:5 in oxygen environment. The ITO films have been characterized by X-ray photoelectron spectroscopy measurements. The ITO surface composition in atomic % is defined to be: In-40.6%, Sn-4.3% and O-55%. The thickness and refractive index of the films are determined by ellipsometric method. The frequency shift of QCM-ITO is measured at different NO_2 concentrations. The QCM-ITO system becomes sensitive at NO_2 concentration $\geq$ 500 ppm. The sorbed mass for each concentration is calculated according the Sauerbrey equation. The results indicated that the 1.09 ng of the gas is sorbed into 150 nm thick ITO film at 500 ppm NO_2 concentration. When the NO_2 concentration increases 10 times the calculated loaded mass is 5.46 ng. The sorption process of the gas molecules is defined as reversible. The velocity of sorbtion /desorption processes are studied, too. The QCM coated with thin ITO films can be successfully used as gas sensors for detecting NO_2 in the air at room temperature.

1. Introduction

In the recent years one of the most important problems in the surrounding world is the increasing concentration of gas pollutants in the atmosphere. Among the hazardous gasses, the nitrogen dioxide (NO_2) can be extremely harmful for the human health, causing different senses irritations [1]. This is the reason, the researchers in this field to investigate intensively the possibilities for development of highly sensitive, fast, reversible, low temperature and low cost NO_2 detector.

It is well known that the quartz crystal microbalance (QCM) type sensors can react to the mass changes, respectively to the concentration changes, even in the nanogram range, in the case when suitable sensitive coating is deposited on the resonator electrodes [2,3]. Most typical application for this purpose have been found to be the metal oxide thin films, like WO_3, TiO_2, etc. [4,5]. Several papers have been reported in the literature, concerning deposition of different materials for effectively detection of NO_2. $Ge_{33}Se_{67}$ chalcogenide glass was thermal evaporated and concentration of 1.48×10^{-10} mole of NO_2 molecules was detected [6]. Another used material is poly (3,4-ethylenedioxythiophene) (PEDOT) electrochemically deposited on the gold electrodes of the resonator, working at 35°C [7].

Indium tin oxide (ITO) also seems to be appropriate for gas detection. The available results show ITO films used for methanol, ethanol and ammonia vapors [8,9,10]. The principle of detection relies on the resistance change whit the vapors concentration changing and can be activated at optimum temperature in the range 150°C - 450°C. Hot wire semiconductor sensor with high surface area ITO film was prepared by the method of thermal oxidation for iso-butane detection at optimum temperature higher than 120°C [11]. Regarding NO_2 measurement, nanostructured Indium Tin Oxide (ITO) thin film has been produced by RF magnetron sputtering and concentrations lower than 50 ppm can be detected at high temperature of ~ 320°C. Again the sensor is resistive-type [12].

QCM principle in combination with ITO adsorbing film for NO and NO_x detection has been reported in [13,14]. It has been established that introduction of hydrogen during sputtering of ITO results in higher conductivity and more oxygen vacancies in the ITO films, which is ascribed as the key mechanism for sensing [13]. In [14] the acceptable mechanism of the NO_x adsorption in the deposited ITO film on QCM has not been established. By the authors' knowledge no available results about fabrication of reversible QCM sensor with sputtered ITO film from non-oxide target for application in nitrogen dioxide sensor working at room temperature are reported in the literature.

The aim of this work was to prepare and investigate the QCM-ITO structure for precision NO2 detection at different gas concentrations. The sensitive layers were deposited by RF sputtering of indium/tin target in oxygen environment.

2. Experimental

2.1. ITO films preparation and methods of characterization

ITO films were prepared by radio frequency sputtering in reactive atmosphere, using 3 inch non - oxidized target with In:Sn in a weight proportion of 90:5. The base pressure in the system was approximately 8.10^{-6} Torr. The partial pressure of the oxygen was set to 1.10^{-4} Torr and the total after introduction of the sputtering argon gas was adjusted to $1.5.10^{-2}$ Torr and was maintained constant during the process of deposition. The sputtering voltage was 0.85 kV and the plasma current of 0.16 A was measured (plasma power 136 W). The produced layers are neither thermally treated during deposition nor after that. In order to control the thickness and the composition of the film, together with the resonator, control pieces of glass and silicon were put on the holder, close to the target sample. The control samples were preliminary cleaned in detergent solution consisting of distilled water, ammonia and hydrogen peroxide in ratio 3:1:1. For the silicon piece additional treatment in water solution of hydrofluoric acid was performed for the native silicon dioxide removing. At this process' parameters a deposition rate of about 4 nm/min was achieved. The as-deposited ITO films were studied by ellipsometric methods and X-ray Photoemission Spectroscopy (XPS), in order to define their morphology, refractive index, thickness, surface state and composition. For this purpose a control silicon wafer, placed next to the resonator structure during sputtering. A Rudolph 435 ellipsometer at a wavelength of 632.8 nm was used. The XPS measurements are performed in ESCALAB Mk II (VG Scientific) electron spectrometer with aluminium radiation (hn = 1486.6 eV). X-Ray diffraction spectra are collected on Bruker D8 Advance diffractometer using CuK radiation and SolX detector.

2.2. Sensor QCM – ITO structure

The ITO films were deposited on gold electrodes of quartz resonator by r.f. sputtering on both sides subsequently. An AT-cut approximately 16 MHz QCM with 8 mm diameter of the quartz plate and 4 mm gold electrodes was used as substrate. The ITO layer was deposited only on the gold electrode, using shadow stencil mask with low thermal capacity. The resonance frequency (f) and dynamic equivalent parameters of the QCM-ITO structure were measured during the investigation.

2.3. Sensing investigation of the QCM – ITO structure

The sensing properties of QCM-ITO were measured on home made laboratory set up. The methodology and the schematic diagram of the set up were described in detail in [15]. The test gas used in experiments was 1% NO_2 diluted in synthetic air. The synthetic air was used as carrier gas too. The flows of both the test and carrier gases were measured and controlled by mass flow controllers. By changing their ratio different concentration of the NO_2 in camera was created. The QCM was installed inside the camera on special holder. The temperature was measured by a Pt-thermosensor located next to the sample. All measurements were performed at room temperature (T) maintained with an accuracy of ±0.2°C. A frequency counter connected to the QCM and joined to the computer for data recording the QCM frequency. Thereby, the frequency change as a function of time was measured. The measurement process can be conditionally divided into three stages:

- purging the camera with QCM-ITO structure with carrier gas until the frequency reaches a constant value (f_o),
- switching on the NO_2 gas flow and reaching the frequency saturation value (f),
- switching off the NO_2 gas flow and purging camera with carrier gas flow until reaching the f_o.

By measuring the time–frequency characteristics (FTCs) of the QCM-ITO at different NO_2 concentrations the sensitivity was determined. Experiments with NO_2 concentrations in the gas flow from 50ppm to 5000 ppm were carried out.

3. Results and discussion

3.1. Characterization of ITO thin films

The value of the refractive index is n=2.42 and the thickness is 150 nm as calculated from the ellipsometric measurements carried out at λ=632.8 nm. XPS measurements (figure 1) show that the surface composition of the ITO film is: O-55.1 at%, In–40.6 at% and Sn–4.3 at.%, which means that the film is indium rich. The O/In+Sn intensity ratio is 1.23 suggesting that the lower oxygen concentration probably results in more oxygen vacancies thus leading to creation of large number of active sites on the surface. The XPS spectra of the ITO film show In $3d_{5/2}$ and In $3d_{3/2}$ peaks at 444.4 and 452.0 eV, respectively. The binding energies of the photoelectron Sn $3d_{5/2}$, and Sn $3d_{3/2}$ peaks are 486.4 and 494.8 eV, respectively. These values reveal that the valence states of In and Sn are mainly +3 and +4, respectively, which is in consistent with the data for a $(In_2O_3)_{0.90}(SnO_2)_{0.1}$ phase of ITO oxide.

The 150nm thin films were deposited on both sides of the electrodes of the resonators. The QCM – ITO parameters of the structure were measured on every step of the experiments to ensure that the sensing response to NO_2 was the actual films' behaviour. The initial basic parameter such as motional resistance (R_s) was measured to be 9.74 ohm and quality factor (Q) of 40 323 was calculated. The parameters of the loaded QCM were suitable for gas sensing studies.

3.2. Gas sensing properties of ITO films to NO_2

The sorption ability of the ITO films was studied changing the NO_2 concentration in the range of 10 ppm - 5000 ppm. It was observed that QCM system became sensitive at NO_2 concentration ≥500 ppm. Response was detected at concentrations higher than 500 ppm up to 5000 ppm.

Frequency – time characteristics (FTCs) of the QCM-ITO structure under investigation were measured at 500 ppm, 1000ppm, 2500 ppm and 5000 ppm NO_2 concentrations. Figure 2 and figure 3 show the NO_2 response of ITO films for 1000ppm and 5000ppm respectively. At 1000 ppm the frequency change is linear for the sorption and desorption processes. The times for reaching saturation and recovering of the system are comparable (respectively 5.9 min. and 5.6 min.). Based on the sorption/desorption behaviour at this concentration, it is possible to suggest a mechanism of surface sorption. Increasing the NO_2 concentration leads to change in the mechanism of the sorption/desorption process.

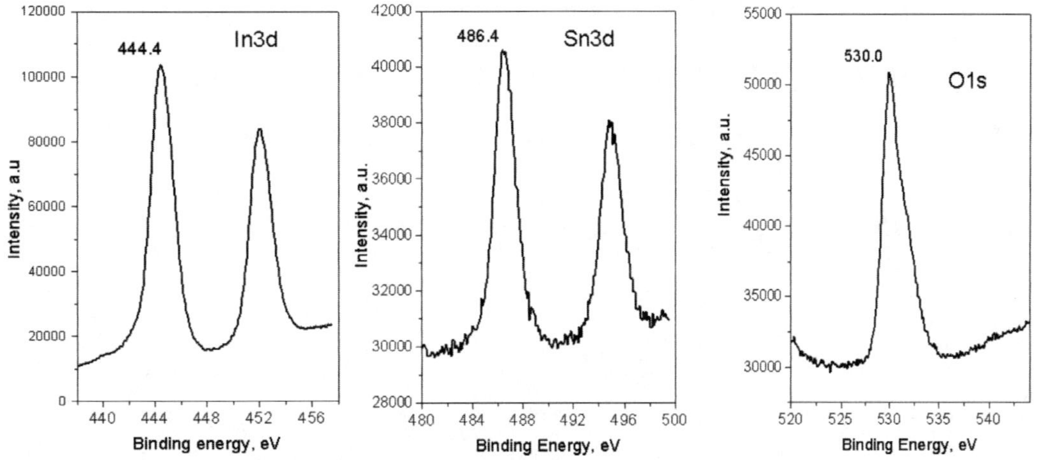

Figure 1. XPS spectra of an ITO film.

Typical FTC at high NO_2 concentration (5000 ppm) is shown in figure 3. Two stages can be observed during the change in the frequency of the FTC – a fast one and a slow one. The recovering (unloading) of the QCM goes off in the same manner. At the beginning of the sorption a sharp decreasing in the frequency was measured (f_{s1}), afterward slightly decrease of the frequency was detected until saturation was reached. This effect can be explained by sorption in depth of the film.

During the desorption process a similar effect was observed – a fast frequency recovering f_{d1} and slow reaching of the starting frequency f_{d2}. The comparison of the data shown in figures 2 and 3 demonstrates the resulting effect of the NO_2 concentration on the sorption and desorption processes.

The parameters of the described processes are summarized in table 1, where t_s and t_d are the total times for sorption and desorption at given constant NO_2 concentration, respectively. The times for

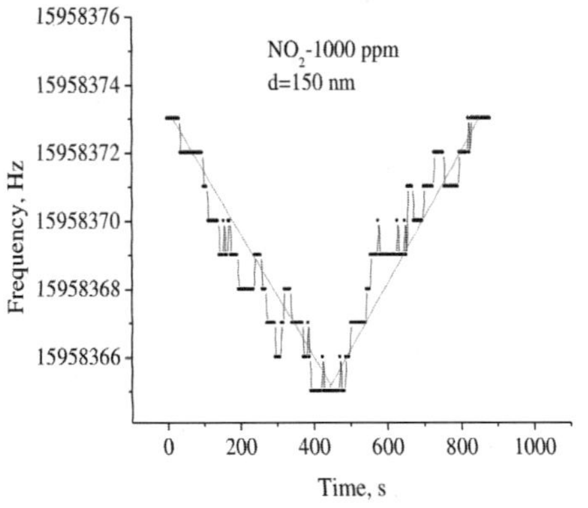

Figure 2. Frequency-time characteristic of ITO-QCM at 1000 ppm NO_2 concentration, thickness of ITO-150 nm.

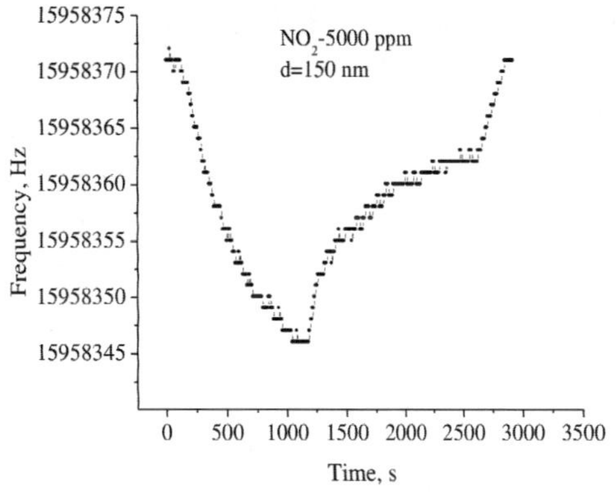

Figure 3. Frequency-time characteristic of ITO-QCM at 5000 ppm NO_2 concentration, thickness of ITO-150 nm.

reaching saturation of the system increase from 4.2 min to 14.9 min with NO_2 concentration increasing from 500 ppm to 5000 ppm. The same behaviour can be observed during the recovering process, where the increase is from 2.6 min to 23.4 min. The sorption and desorption rates (V_s and V_d respectively), defined as a frequency change with the time, shows an insignificantly increase of V_s from 1.2 Hz/min to 1.7 Hz/min and decreasing of the desorption rate V_d from 1.9 Hz/min to 0.9 Hz/min almost twice in the investigated concentration range.

Table 1. Sorption and desorption parameters of ITO (150nm) - QCM structure.

NO_2, ppm	t_s, min	t_d, min	V_s, Hz/min	V_d, Hz/min
500	4.2	2.6	1.2	1.9
1000	5.9	5.6	1.4	1.4
2500	8.9	14.5	1.7	1.0
5000	14.9	23.4	1.7	0.9

Based on the measured FTCs, the sorbed mass for each NO_2 concentration was calculated. The quantitative relationship between the change in frequency of an AT-cut piezoelectric crystal (f_o) and the mass change (Δm) caused by mass loading on the crystal surface was given by Sauerbrey equation [16]:

$$\Delta f = -\ 2.26.10^6\ f^2_o\ \Delta m/s, \tag{1}$$

where f_o is the basic frequency of the unloaded piezoelectric crystal in MHz, Δm is the change of the mass on the surface in gram, and s is the surface area of the electrode in cm^2. The surface of the electrode in the experiment was 0.1256 cm^2. The resonant frequency shift (Δf) can be calculated as ($\Delta f = f - f_o$), where f is the frequency of the structure measured at saturation. Thereby using Sauerbrey's equation the sorbed mass on the resonator surface for each NO_2 concentration can be calculated. Figure 4 demonstrates the frequency shift and corresponding sorbed mass increase with increasing in NO_2 concentration. The obtained dependence for investigated structure QCM-ITO has almost linear characters. Tenfold increase in the concentration from 500ppm to 5000ppm leads to three times higher sorbed mass from 1.9 ng to 5.46 ng.

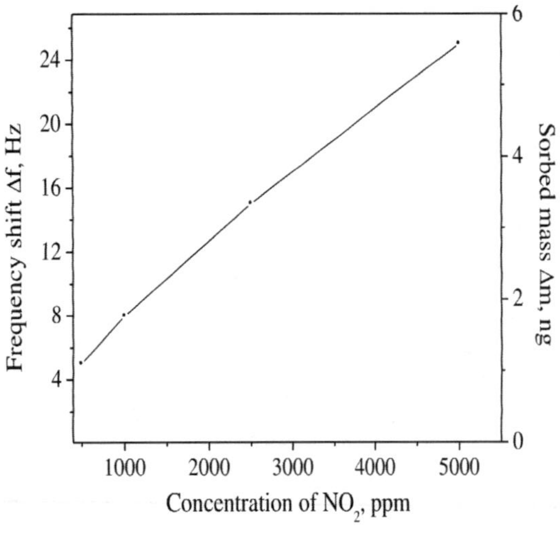

Figure 4. Dependence of the NO_2 concentration on the ITO-QCM frequency shift and mass loading. The thickness of ITO is 150 nm.

4. Conclusion

The QCM-ITO structure was realized. The ITO films were deposited by r.f. sputtering. The film properties such as: refractive index, morphology and composition were defined. The gas-sensing properties of the ITO to NO_2 in the air were studied. The ITO films were found to be sensitive to NO_2 concentrations above 500ppm. The sorption/desorption processes of NO_2 were estimated. The QCM-ITO structure is promising for NO_2 detection. The work demonstrated that the realized QCM-ITO structure could be used as gas sensor for NO_2 in the air. Further investigations were requires in aim of improving gas-sensing properties.

Acknowledgement

The work is supported partly by the bilateral project (2012-2014) ISSP, BAS and GEOKHI and Project ISSP-BAS - № 05-13/ 13.04.2013.

References

[1] Nieding G, Krekeler H, Fuchs R, Wagner M and Koppenhagen K 1973 *Internationales Archiv für Arbeitsmedizin* **31** 61
[2] Montagut Y, García J V, Jiménez Y, March C, Montoya Á and Arnau A 2011 QCM Technology in Biosensors *Biosensors – Emerging Materials and Applications* ed P A Serra (Rijeka, Croatia: InTech) chapter 9 pp 153-178
[3] Vashist S K and Vashist P 2011 *Journal of Sensors* **2011** 1
[4] Boyadjiev S and Rassovska M 2007 *International Scientific Conference Electronics ET'07, (Sozopol, Bulgaria, 19–21 September)* p121
[5] Georgieva V, Spassov L, Mogilevsky A, Grechnikov A, Donkov N, Yordanov Ts and Gadjanova V 2009 *Journal of Optoelectronics and Advanced Materials* **11** 1367
[6] Chen P, Mitkova Maria, Tenne D A , Wolf K, Georgieva V and Vergov L 2012 *Thin Solid Films* **525** 141
[7] Fort A, Innocenti M, Foresti M L, Mugnaini M, Pasquini I, Pigani L, Rocchi S and Vignoli V 2009 *3rd International Workshop on Advances in sensors and Interfaces (Trani, Italy, 25-26 June 2009)* p 41
[8] Vaishnav V S, Patel P D and Patel N G 2005 *Thin Solid Films* **487** 277
[9] Vaishnav V S, Patel P D and Patel N G 2005 *Thin Solid Films* **490** 94
[10] Lin Ch, Chen H, Chen T, Huang Ch, Hsu Ch, Liu R and Liu W 2011 *Sensors and Actuators B* **160** 1481
[11] Lee S, Lee Y, Shim Ch, Choi N, Joo B, Song K, Huh J and Lee D 2003 *Sensors and Actuators B* **93** 31
[12] Vijayalakshmi K, Ravidhas C, Pillay V V and Krishna D G 2011 *Thin Solid Films* **519** 3378
[13] Zhang J, Hu J Q, Zhu F R, Gong H and O'Shea S J 2002 *Sensors and Actuators B* **87** 159
[14] Zhang J, Hu J Q, Zhu F R, Gong H and O'Shea S J 2003 *Sensors* **3** 404
[15] Georgieva V, Stefanov P, Spassov L, Raicheva Z, Atanassov M, Tincheva T, Manolov E and Vergov L 2009 *Journal of Optoelectronics and Advanced Materials* **11** 1363
[16] Sauerbrey G 1959 *J. Physics* **155** 206

Electrical characterization of In-N codoped p-type ZnO films grown by chemical methods

M Duta[1,*], S Simeonov[2], D Spasov[2], S Mihaiu[1], M Anastasescu[1], A Szekeres[2] M Gartner[1] and M Zaharescu[1]
[1] Institute of Physical Chemistry "Ilie Murgulescu", Romanian Academy, 202 Splaiul Independentei, 060021 Bucharest, Romania,
[2] Institute of Solid State Physics, Bulgarian Academy of Sciences, 72 Tzarigradsko Chaussee, 1784 Sofia, Bulgaria

E-mail: maria_duta@yahoo.com

Abstract. Preliminary results of the studies of electrical properties of ZnO films, codoped with donor and acceptor pair In-N are reported. The films were grown on Si substrates by chemical route procedures involving a sol-gel followed by hydrothermal treatment at 500°C. For electrical characterization, current-voltage and capacitance-voltage measurements were conducted on MIS structures with embedded ZnO(In,N) films. The estimated doping concentration in the studied films is higher than $1\text{x}10^{16}$ cm^{-3} corresponding to a semiconductor with good conductivity. However, the obtained large values of the differential specific resistivity (ρ = $3.2\text{x}10^5$ Ωcm) suggest that the nitrogen acceptors in these films are compensated by some kind of donor-type defects.

1. Introduction

Zinc oxide is a transition metal oxide semiconductor with wide band gap (3.37 eV) and large exciton binding energy (~60 mV), good transparency, high electron mobility and strong room-temperature luminescence [1-3]. These properties determine its feasibility in emerging applications for transparent electrodes in liquid crystal displays, in energy-saving or heat-protecting windows, and in electronics as thin-film transistors and light-emitting diodes [3-5].

Zinc oxide can be obtained in a large variety of morphologies, function of the deposition method and technological conditions [6]. Among these, we count vapor-phase deposition [7], pulsed laser deposition [8], or metallic-organic chemical vapor deposition [9]. However, physical deposition techniques have a series of disadvantages such as complex equipment and process, as well as involving higher temperatures. Chemical methods for obtaining thin films such as chemical bath deposition [10-12] or sol-gel method [13] are more accessible, economical and can be more easily up-scaled to industry [14]. The use of a ZnO seed layer, especially when promoting 1D formations, is well documented in the literature [15, 16] and promotes the nucleation and subsequent growth of continuous, homogenous ZnO films [17].

While ZnO as a material has been widely investigated, the current trend is the obtaining of p-type conduction through different doping mechanisms [18,19]. This may prove difficult due to the self

[*] To whom any correspondence should be addressed.

compensating effect, deep acceptor level and low solubility of acceptor dopant ions [4]. Codoping with a donor-acceptor pair (group III - group IV metal ions) has been reported to lead to p-type conduction in ZnO, with improved electrical properties as compared to monodoping [20].

In this work we report preliminary results of the studies of electrical properties of ZnO films, codoped with donor and acceptor pair In-N, further denoted as ZnO(In,N) films, grown by chemical route procedures involving a sol-gel and a hydrothermal chemical procedure. For electrical characterization, MIS structures with embedded ZnO(In,N) films were prepared and from their current-voltage (I-V) and capacitance-voltage (C-V) characteristics electrical parameters such as conductivity, mobility, carrier and doping concentrations were estimated and compared with Hall measurement data.

2. Experimental details

The studied ZnO(In,N) films were deposited on single-crystal p-type Si substrates (with specific resistivity of 19÷29 Ωcm), by applying a chemical method in two steps.

The first stage is aimed at obtaining an undoped ZnO seed player through dipping the substrates in a 0.5 M solution of zinc acetate dihydrate [$Zn(CH_3COO)_2x2H_2O$] in ethanol, to which triethanolamine was added as chelating agent. The seed layer was thermally treated at 400°C for 1 hour.

The second stage consisted of the hydrothermal deposition of the codoped layer. For this purpose, an aqueous solution (0.05 M) of zinc nitrate tetrahydrate [$Zn(NO_3)_2x4H_2O$] and hexamethylene-tetramine [$(CH_2)_6N_4$] with a ratio of 2:1 was prepared. In order to obtain 1 at. % indium concentration in the film, indium nitrate [$In(NO_3)_3$] was added to this solution. The resulted solution was treated at 90°C for 2 hours in a teflon autoclave. Finally, the codoped ZnO films were cleaned with hot water, dried at room temperature and subjected to heat treatment at 500°C for 1 hour in air. After the final technological step the film thickness was approximately 150 nm, as determined from ellipsometry.

For the electrical measurements, MIS structures with the ZnO(In,N) films were formed by vacuum evaporation of Al contacts, as on the film surface Al dots (contact area S=2.8x10^{-3} cm^2) were deposited through a metal mask while continuous Al film was deposited on the Si wafer backside. The room temperature current-voltage (I-V) characteristics were measured with a cycle sequence beginning from 0 V toward negative or positive voltages applied to the Al-dot contact on the film surface followed by a voltage reversal toward zero applied voltage. The room temperature 1 MHz capacitance-voltage (C-V) measurements were carried out by a digital LCR meter E7-12.

Room temperature Hall effect measurements were done using the van der Pauw method, on a HMS-5000 equipment from Ecopia.

Spectroscopic ellipsometry (SE) measurements were performed by VASE-Woollam equipment at an angle of incidence of 70°, with 5 nm step, in the UV-VIS-NIR spectral range at room temperature.

3. Results and discussion

Typical I-V characteristics of a MIS structure with ZnO(In,N) film is shown in figure 1. When negative voltage is applied to Al-dot contact on the film surface the current is several times higher than that measured when positive voltage is applied.

At high negative voltages the current density is 50-60 times higher than that for positive voltages with the same magnitude. As the Si substrate is p-type, such current behavior is a direct evidence of the appearance of p^+ - p junction at the ZnO(In,N) - Si interface. At negative voltages a hysteresis is observed because the current at given negative voltages is predominantly higher during voltage sweep toward the zero voltage in comparison with the current during the voltage sweep from zero to negative voltages. Furthermore, when the applied voltage describes a sweep from zero to highest voltage and back to zero, a residual current continues to flow with different time-constants and diminishing amplitude for several hours. These observations show that during the current flow some of holes are captured at deep levels in the ZnO(In,N) film and their release interferes with the current in the subsequent voltage sweep.

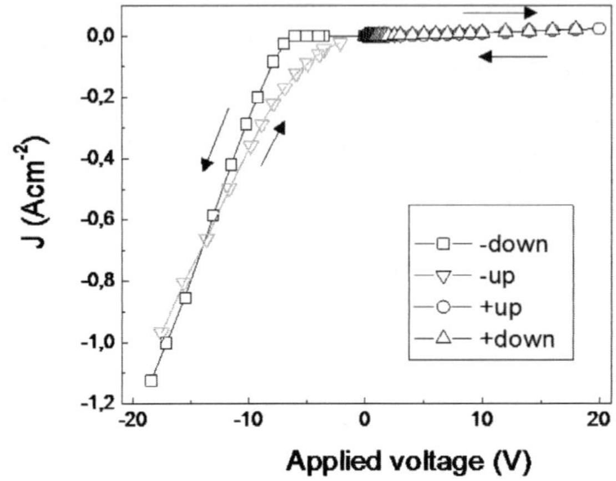

Figure 1. Room temperature I-V characteristics of a MIS structure with ZnO(In,N) film. The sweep directions are shown with arrows.

The specific resistivity ρ of 3.2x10^5 Ωcm for ZnO(In,N) film is calculated from the I-V characteristics (figure 1) at negative voltages corresponding to an electric field of 10^6 V/cm.

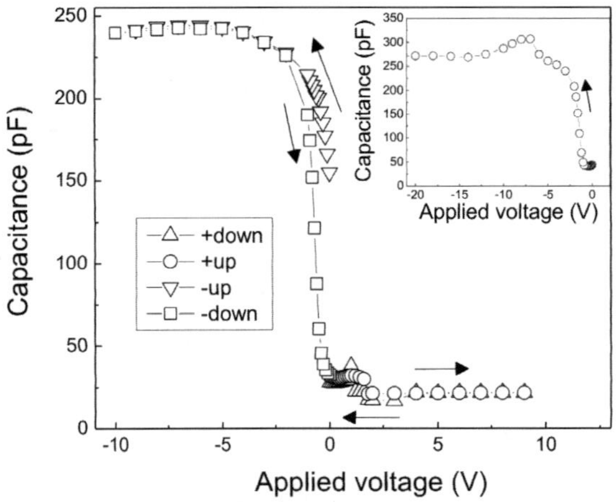

Figure 2. C-V characteristics of MIS structure with ZnO(In,N) film at 1 MHz. The sweep directions are shown by arrows. The inset shows another C-V characteristics for the same ZnO(In,N) film but recorded for other Al-dot contact on the film surface.

The charge density can be estimated from the capacitance values in the depletion region (from -5 to 0 V) of the C-V characteristics. By measuring several MIS structures using Al-dot contacts at different places on the film surface a non-homogenous charge distribution in the film bulk was observed, resulted in different shape and capacitance values. This is illustrated in the inset in figure 2.

Replotting the C-V curves from figure 2 as S^2/C^2 versus applied voltage in the depletion region (not shown here), the density of acceptor charge N_A can be calculated from the slope B of the linear part of this dependence:

$$B = 2/(qN_A\varepsilon_{ZnO}) \tag{1}$$

where q is the electric charge (q=1.6021x10^{-19} C) and ε_{ZnO} is the dielectric permittivity of the film.

The density of acceptor charge obtained in depletion region close to 0 V in "+down" sweep direction is in the range of 10^{14} – 10^{15} cm^{-3} as it can be expected for Si substrates with specific resistivity in the 19-29 Ωcm range. For the depletion region near -5 V in "-down" sweep direction, the charge density value is above 10^{16} cm^{-3}. As this value is determined by the charge of diffused holes from the film, i.e. from the p^+ side of the p^+ - p (ZnO-Si) junction, its value can be used for an estimation of net acceptor concentration, N_A, in the ZnO(In,N) film. Therefore, the p-type doping level in the ZnO(In,N) film is in the order of 10^{16} – 10^{17} cm^{-3}. Taking into account this net acceptor concentration (~ 2x10^{16} cm^{-3}), the effective hole mobility in these ZnO films is estimated to be 9.83x10^{-4} cm^{2}V^{-1}s^{-1}. Such low mobility values are a consequence of some kind of hopping conductivity via deep levels in semiconductor materials. Moreover, since the I-V characteristics revealed a high resistivity ZnO(In,N) film (ρ = 3.2x10^5 Ωcm), it suggests that the nitrogen acceptors in the film are compensated by some kind of donor defects. As it has been shown, the use of a combined sol-gel and hydrothermal deposition method leads to the formation of randomly oriented ZnO nanorods [21, 22]. This could contribute to the low mobility most likely due to interface scattering between 1D structures.

From figure 2 it is seen that there is a hysteresis making a whole voltage sweep cycle and thus, the zero voltage capacitance values are different for negative and positive sweep directions. This effect is caused by carriers being captured at deep traps in the film. While remaining at zero voltage position, the return of the capacitance value to its initial level requires prolonged time, which is an indication of large time constants of these traps. Because of that, several hours are needed for complete discharging of the traps. This is partly illustrated in figure 3, where discharge of the deep levels with time after removing the applied voltage is given. From this plot it is evident that the release of captured charges in the ZnO film takes place with different time-constants in the order of minutes.

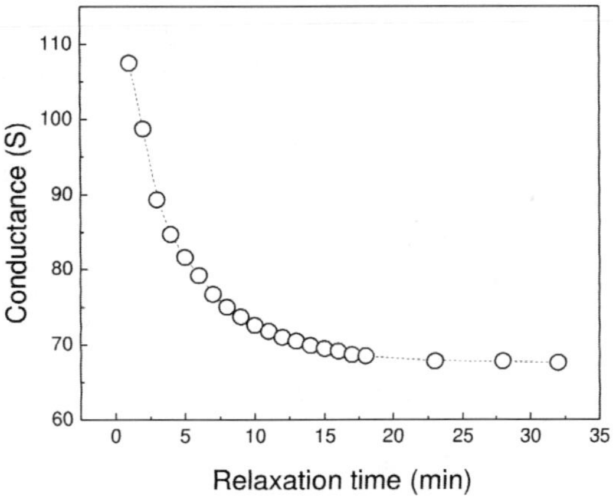

Figure 3. Discharging of deep levels in time, at zero voltage.

For a comprehensive electrical study of the codoped films, Hall effect measurement was also attempted. In order to avoid the influence of silicon substrate on the results, we have deposited a ZnO(In,N) film on glass substrate using the same chemical procedure parameters. The film thickness obtained was approximately 290 nm.

The Hall effect measurement has confirmed good semiconducting properties of the ZnO film, while also exhibiting p-type conduction. The p-type characteristic of the ZnO(In,N) film was proved by the positive sign of the Hall coefficient ($R_H = +29.7\ cm^3C^{-1}$). A high carrier concentration of $2.1 \times 10^{17}\ cm^{-3}$ was determined, which is well correlated with the value of $10^{16} - 10^{17}\ cm^{-3}$ obtained through C-V analysis. The resistivity was determined to be $1.8 \times 10^2\ \Omega cm$ which is comparable to values reported in the literature for Hall effect measurement [23]. From this, a mobility of $0.2\ cm^2V^{-1}s^{-1}$ is obtained. In contrary to that, the analysis of the I-V and C-V characteristics revealed hole mobility values in ZnO(In,N) films with approximately two orders of magnitude lower and resistivity values with three orders of magnitude higher than those from the Hall effect measurements. These differences could be attributed to the different method principles: during the Hall measurement the current flows parallel to the film surface between 4 contact points, while at the I-V and C-V measurements the current is perpendicular to the Al-dot area, as it flows through the p-type ZnO/p-type Si structure. Since non-homogenous charge distribution in the film bulk is registered (figure 2), this may lead also to different values for the same parameter as determined through the different techniques. Furthermore, as was discussed above, during I-V measurements holes are trapped in ZnO(In,N) film creating an internal electric field which decreases the external electric field of the applied voltage and, thus decreases the effective mobility. The establishment of real causes of the observed discrepancy needs further investigations.

4. Conclusions

It has been established that codoping with In and N of ZnO films and using the combined sol-gel and hydrothermal deposition method for film deposition yields a comparatively high net acceptor concentration of the order of 10^{16}-$10^{17}\ cm^{-3}$. By I-V and C-V measurements, deep levels are registered in the investigated ZnO(In,N) films. Because of these deep levels the effective mobility of holes in these films is low and it is in the order of $10^{-3}\ cm^2V^{-1}s^{-1}$. Hall effect measurements have confirmed the p-type conduction of the film. While carrier concentration values were similar, resistivity and mobility determined through the two methods differed, due to the method principle of measurement. Thus, a non-homogeneous carrier distribution in the film was evidenced.

Acknowledgements

The authors acknowledge with thanks the financial support of this work in the frames of Project INERA under the contract REGPOT 316309.

References

[1] Yan Z, Zhang X, Liu Y, Zhou X and Liang J 2011 *J. Mater. Sci.* **46** 2392
[2] Clatot J, Campet G, Zeinert A, Labrugere C and Rougier A 2011 *Appl. Surf. Sci.* **257** 5181
[3] Mbamara U S, Akinwunmi O O, Obiajunwa E I, Ojo I A O and Ajayi B 2012 *Journal of Modern Physics* **3** 652
[4] Yuan N, Fan L, Li J and Wang X 2007 *Appl. Surf. Sci.* **253** 4990
[5] Ma A M, Gupta M, Chowdhury F R, Shen M, Bothe K, Shankar K, Tsui Y and Barlage W 2012 *Solid-State Electronics* **76** 104
[6] Chen X, Man A, Ng C, Djurisic A B, Ling C C and Chan W K 2012 *Thin Solid Films* **520** 2656
[7] Zhao Q, Willander M and Klason P 2007 *Appl. Phys.* A **88** 27
[8] Huang M, Feick H, Tran N, Weber E, Yang P and Wu Y 2001 *Adv. Mater.* **13** 113
[9] Sun Y, Ashfold M and Fuge M 2004 *Chem. Phys. Lett.* **396** 21
[10] Jacobs K, Balitsky D, Armand P and Papet P 2010 *Solid State Sci.* **12** 333
[11] Yang L L, Zhao Q X and Willander M J 2009 *J. Alloys Compd.* **469** 623
[12] Le H Q and Chua S J 2011 *J. Phys. D: Appl. Phys.* **44** 125104
[13] Zaharescu et al 2014 *Thin Solid Films* DOI: 10.1016/j.tsf.2014.02.090
[14] Urgess Z N, Oluwafemi O S and Botha J R 2012 *Mater. Lett.* **79** 266
[15] Yin Y T, Que W X and Han C H 2010 *J. Sol-Gel Sci. Technol.* **53** 605

Journal of Physics: Conference Series **558** (2014) 012038 doi:10.1088/1742-6596/558/1/012038

[16] Cha S N, Song B G, Jang J E, Han I T, Ha H J, Hong J P, Kang D J and Kim J M 2008 *Nanotechnology* **19** 235601
[17] Wen X, Wu W, Ding Y and Wang Z L 2012 *J. Mater. Chem.* **22** 9469
[18] Bian J M, Li X M, Gao X D, Yu W D and Chen L D 2004 *Appl. Phys Lett.* **84** 541
[19] Yang T H and Wu J M 2012 *Acta Materialia* **60** 3310
[20] Chen L L, Lu J G, Ye Z Z, Lin Y M, Zhao B H, Ye Y M, Li J S and Zhu L P 2005 *Appl. Phys. Lett.* **87** 252106
[21] Kim K H, Utashiro K, Abe Y and Kawamura M 2014 *Int. J. Electrochem. Sci.* **9** 2080
[22] Feng X, Feng L, Jin M, Zhai J, Jiang L and Zhu D 2004 *J. Am. Chem. Soc.* **126** 62
[23] Chen L L, Ye Z Z, Lu J G and Chu P K, 2006 *Appl. Phys. Lett.* **89** 252113

Journal of Physics: Conference Series **558** (2014) 012039 doi:10.1088/1742-6596/558/1/012039

Chemosorption hydrogen gas sensor based on MOSFET with optical activation

T V Stoyanova[1], V V Tomaev[1,3], N D Stoyanov[2], S K Andreev[4]
[1]Depatment of Physics, National university of mineral resources, Saint-Petersburg, V.O. 21-ya liniya, 2, Russia
[2]LED Microsensor NT, LLC, Saint-Petersburg, Politechnicheskaya 26, Russia,
[3]Depatment of Chemistry, State University, Saint-Petersburg, Petrodvoretc, Universitetskii prosp. 26 Russia
[4]Institute of Solid State Physics, Bulgarian Academy of Sciences, Sofia, 72 Tzarigradsko Chaussee Blvd., Bulgaria

E-mail: Stoyanova_TV@list.ru

Abstract. In this paper we present an approach which allows to increase the sensitivity of the sensors at room temperature. Field effect transistors with SnO_2 gate in combination with optical activation of the adsorption centres on the surface were utilized. Studies of photosensitivity were carried out using a set of LEDs in the visible spectral range by the method of impedance spectroscopy. Optical activation of the adsorption centres in the SnO_2 gate leads to the additional accumulation of positive charges on top of the gate H^+. Positive charge caused broadening of the channel in MOSFET sensor, resulting to lower real component of the impedance. These results show that optical activation allows to detect significantly lower concentrations of the hydrogen contained gases.

1. Introduction

Gas detectors and sensors based on chemisorption are widely used in practically all branches of industry, transport and medicine. In some industries without gas detectors and sensors it is impossible to provide the required level of human safety, including ecology and technogenic safety.
According to the principle of the chemical sensitivity, sensors can be divided into two large groups:

1. Chemically sensitive optical sensors. Their selective sensitivity to certain chemicals is based on characteristic absorption of electromagnetic radiation in a gaseous environment. Main advantage of the optical sensors is that measured chemicals do not affect the sensor module. However, they cannot be used for measurement of many important chemicals, including hydrogen and oxygen due to the lack of strong enough absorption bands.

2. Chemically sensitive sensors based on direct effects of the environment on the sensor module. In this group sensors, chemical or physico-chemical reactions in the volume or on the surface of the sensitive element lead to changes in the electrical parameters.

Such types are catalytic and electrochemical sensors. They have good sensitivity but poor performance (speed of response), high operating temperature (up to +400 - +450 °C) and short lifetime of the working cell.

Created later adsorption sensors have significantly better performance and lifetime. Generally in such sensors catalytic layers based on Pd, Pt, or SnO_2, deposited on ceramic or silicon are used.

The main mechanism of detection in such sensors is based on registration of changes in the electrical parameters during adsorption of molecules of the measuring environment on the surface of the sensor (change in conductivity or capacitance of the sensing element). The mechanism of the hydrogen sensitivity in the Schottky diodes with palladium contacts was described by Ludstrem [1, 2]. In Schottky diodes the dark current and capacitance are proportional to the concentration of hydrogen in the environment. In the n-type structures current increases, in the p-type structures, current decreases [3]. Pd layer catalyzes the dissociation of hydrogen-containing compounds to ions of hydrogen and other radicals. The selectivity of the sensor to the hydrogen is due to the higher solubility of hydrogen in the palladium in comparison to the other substances.

At room temperature, a small change of the electric characteristics is observed that not provide the required sensitivity. Effect of hydrogen influence on the electrical characteristics of the diodes and MOS structures with Pd or Pt layers increases with increasing temperature primarily due to the more active dissociation of the molecules. In the modern miniature adsorption sensors a heater is integrated in the sensor cell.

SnO_2 also catalyzes the dissociation of hydrogen-containing compounds. There are two main mechanisms of change of the concentration of charged carriers in the surface layer of the semiconductor due to gas adsorption:

1. Chemisorption of gases in the charged form which causes near-surface band bending of the semiconductor, described in the Schottky model [4]. This process leads to a change in the charged carrier concentration in the surface region of the semiconductor. Acceptor gases (with a large electron affinity) during chemisorption lead to negative charging of the surface layer. Their surface electronic states are located below the Fermi level of the semiconductor. Acceptor gases are O_2, Cl_2, F_2, NO_2, etc.. In donor gases, surface electronic states are located above the Fermi level of the semiconductor, so they lead to positive charging. Donor gases are H_2, CO, NH_3 and others. They reduce the resistance of the SnO_2 layers.

2. Direct interaction of gases with electrically active defects on the surface, which also changes the concentration of charged carriers in the semiconductor. High operating temperature of the sensors significantly limits the scope of applications. In this paper we present an approach which allows to increase the sensitivity of the sensors at room temperature, using field effect transistors with SnO_2 gate (figure 1) in combination with the optical activation of the adsorption centers on the surface.

2. Materials and methods

Field-effect transistors with a tin dioxide gate were fabricated in the Institute of Solid State Physics, Sofia. Their sensitivity to hydrogen-contained gases was investigation [6 - 8]. Dielectric SiO_2 layer with 20 nm thickness was deposited on a silicon substrate. Source and drain areas were formed by diffusion of donors through the windows in the dielectric layer. A layer of tin dioxide with thickness of 60 nm and resistance of 10 to 100 ohms / square was deposited on the gate dielectric layer.

The influence of the NH_3 concentration in the range of 19 - 52,000 ppm on the drain current at different voltages on the drain and gate was investigated. Using of pulse mode of operation allowed to obtain repeatable results. Measurements have shown that the change in drain current in the presence of ammonia is strongly dependent on the drain-source voltage V_D. This fact can not be explained only by change in the work function in SnO_2 layer. Model, which allowed not only to explain qualitatively but also to calculate the current change in the presence of NH_3 was developed [8].

When the drain-source voltage V_D and a positive gate voltage V_G are applied, a conductive n-channel in silicon under the gate is formed (figure 2). The potential distribution in the channel and, accordingly, the transverse field E_n and E_p depend on the ratio of voltages V_D and V_G.

In the presence of moisture atoms, on the surface of SnO_2 take place dissociation of NH_3+H_2O to NH_4^+, positive ions and negative ions OH^-. Due to the influence of the distribution of the transverse field, positive ions migrate toward the source (to a negative area), and the negative ions migrate toward the drain.

The redistribution of ions changes the effective value of the gate voltage. Positive ions NH_4^+ make

Figure 1. MOSFET with SnO_2 gate.

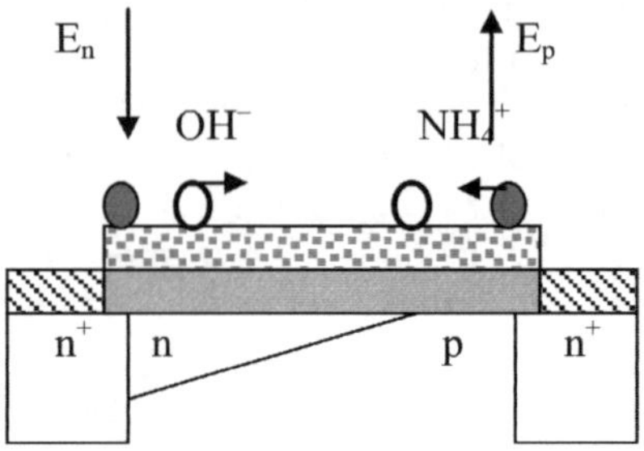

Figure 2. Migration of NH_4^+ and OH^- ions on the surface of tin dioxide.

an additional positive contribution ΔV_P in the gate potential V_G. Similarly, negative ions OH^- give an additional negative contribution ΔV_n in the gate potential V_G. Thus, the total change in the gate potential ΔV_G determined by the influence of two types of ions, and can have both positive and negative sign:

$$\Delta V_G = \Delta V_p - \Delta V_n. \quad (1)$$

Created sensor structure based on a field effect transistor with a SnO_2 gate demonstrated sensitivity to hydrogen-containing gas at room temperature. This structure was used in carrying out experiments on the optical activation of adsorption centers, with the purpose to further improve the sensitivity and selectivity. These experiments were made in St. Petersburg.

Activated adsorption (requires activation energy) is usually accelerated by heating, but the concentration of adsorption centres can be also increased by illumination of the surface with light of appropriate wavelength. Illumination of the structures affects both the adsorption level of equilibrium and the kinetics of adsorption. Experiments [5] had shown that illuminated semiconductor showed a significant increase in the adsorption ability of the adsorbent, as was detected by a pressure drop in the adsorption volume. Maximum sensitivity with respect to the adsorption properties of semiconductor correspond to the maximum internal photoelectric effect.

Studies of photosensitivity were carried out using a set of LEDs in the visible spectral range by the method of impedance spectroscopy which is based on the study of the electrical response of a system imposed by electrical field with variable frequency.

Impedance measurements were carried out using impedancemeter Z-500P. The device allows to explore the properties of electrically conductive materials by recording the impedance spectra (complex resistance to the sinusoid current) at a constant polarizing voltage. Impedancemeter measures the real Z' and imaginary Z" components of the complex resistance when AC (sinusoidal) voltage with an amplitude value of 10 mV was applied to the sample. The voltage was applied between the drain and source of the MOSFET. The gate voltage was not applied. Activation of the adsorption centers in the SnO_2 gate was provided by electromagnetic radiation with wavelengths in the visible range. Red, yellow, green, blue LEDs were used as a source of electromagnetic radiation (table 1).

The measurement results were obtained in the form of hodograph curves of the impedance - plots of $\underline{Z}(\omega)$ in the coordinates Z' and -Z", corresponding to the resistive and capacitive properties of the system, respectively (figure 3).

As a rule, the impedance spectrum of polycrystalline materials in the complex plane has two semicircles, indicating the contribution to the overall conductivity of the material volume of the grains and their boundaries.

Table 1. Main parameters of the LEDs used in the experiment.

Parameter	LED model			
	LEDis625-R5-A30	LEDis590-R5-A30	LEDis505-R5-A30	LEDis470-R5-A30
Color	Red	Yellow	Green	Blue
Wavelength, nm	623-657	588-592	503-507	467-473
Brightness (20 mA), mCd	3500-5500	3500-5500	6500-8500	2500-4500
Electrical power, mW	35-65	35-65	60-95	60-95
Forward voltage, V	1.8-2.2	1.8-2.2	3.0-3.2	3.0-3.2

The presence of only one circle indicates a high degree of homogeneity of the sample. In the analysis of the experimental results it is common to use an approximation of equivalent circuits.

Obtained frequency dependences of Z' and Z" are compared to the known characteristics of the basic impedance model circuits. Received during the experiments results (figure 4) can be associated with a parallel circuit consisting of resistor R_p and capacitor capacity C_p. The equation describing the graph can be written as:

$$\left(Z' - \frac{R_p}{2}\right)^2 + Z''^2 = \left(\frac{R_p}{2}\right)^2, \tag{2}$$

that is a circle with the coordinates (Z' and -Z"). The point at which the arc intersects the axis (on the right site) corresponds to the active impedance Z' in the "volume" of the sensor.

The dependence of the complex dark impedance versus frequency has two main regions: low frequency (less than 10 kHz) and high frequency (more than 20 kHz) (figure 5 (a)). In the low-frequency region the main contribution to the complex impedance makes the active component Z'. When electromagnetic radiation from the visible LEDs exposes the surface of the MOSFET structure, at high frequencies both real Z' and the imaginary Z" components of the complex impedance region remain without change.

But at low frequencies we recorded significant changes in the components of the impedance. Significant reduction of the real component was obtained, while the imaginary component increased. Hodograph curves shifted up along the axis Z", which corresponds to a change in capacitance of the system. The complex impedance increased about 2 times for the blue and green LEDs and a little less for the red and yellow ones.

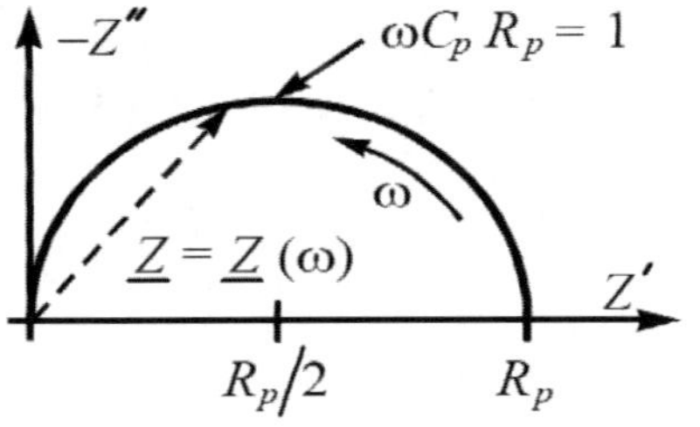

Figure 3. Hodograph impedance corresponding to a parallel circuit connection capacity C_p and resistance R_p.

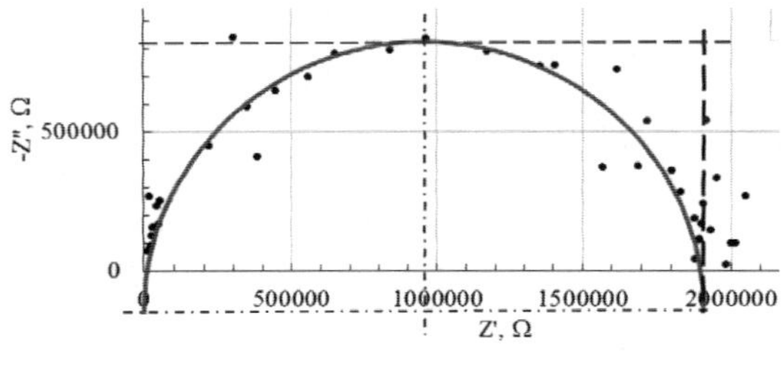

Figure 4. Hodograph of the optical cell with no lighting.

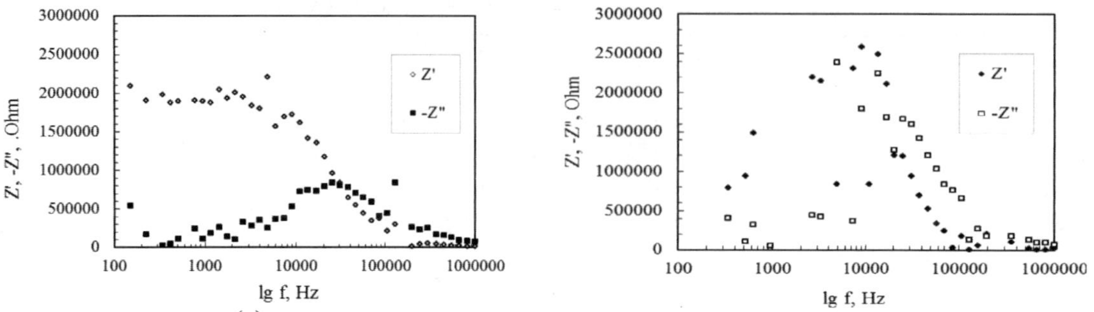

(a) (b)

Figure 5. Dependence of the complex impedance vs. frequency without radiation (a) and with radiation from yellow LED (λ=590 nm) (b).

This difference may be explained by differences in the energy of the LED radiation (table 1). In the future experiments effect of radiation in ultraviolet and infrared range will be investigated.

Optical activation of the adsorption centres in the SnO_2 gate leads to the additional accumulation of positive charges on top of the gate H^+. Positive charge caused broadening of the channel in MOSFET sensor, resulting to lower real component of the impedance. These results show that optical activation allows to detect significant lower concentrations of the hydrogen contained gases.

3. Conclusion

Studies have shown the ability to create sensors for hydrogen and hydrogen contained gases, operating at room temperature with a sufficiently low threshold. This result was achieved using field effect transistor with a tin dioxide gate in combination with original technique of optical activation of the adsorption centers.

References

[1] Lundström I, Shivaraman S, Svensson C and Lundkvist L 1975 *Applied Physics Letters* **26** 55
[2] Lundstrom I and Soderberg D 1981 *Sensors and Actuators* **2** 105
[3] Poteat T L and Lalevic B 1982 *IEEE Trans Electron Devices* **29** 123
[4] Gopel W 1985 *Technisches Messen* **52** 92
[5] Wolkenstein F F 1953 *Uspekhi Fizicheskikh Nauk* **L** 253
[6] Andreev S K, Popova L I, Gueorguiev V K and Stoyanov N D 1994 *Sensors and Actuators* **B 18** 457
[7] Popova L I, Andreev S K, Gueorguiev V K and Stoyanov N D 1994 *Sensors and Actuators* **B 18** 543
[8] Andreev S K, Popova L I, Gueorguiev V K and Stoyanov N D 1994 *Sensors and Actuators* **B 18** 540

Low pressure and time storage influences on the electrets stability of HDPE composite films

T Yovcheva[1], A Viraneva[1], M Galikhanov[2]
[1]Department of Experimental Physics, University of Plovdiv, 24 Tzar Assen Str., 4000 Plovdiv, Bulgaria
[2]Kazan National Research Technological University, 68 "Karl Marx" Kazan Republic of Tatarstan, Russian Federation

E-mail: yovchevat@gmail.com

Abstract. The influence of both low pressure and time storage on the surface potential decay of high density polyethylene (HDPE) composite films with different weight concentrations of the zeolite particles - 0 wt.%, 2 wt.% and 6 wt.% were studied. The samples were charged in a corona discharge by means of a corona triode system for 1 minute under room conditions. Positive or negative 5kV voltage was applied to the corona electrode and 1 kV voltage of the same polarity as that of the corona electrode was applied to the grid. After charging, the electret surface potential was measured by the method of the vibrating electrode with compensation. Two groups of tests were performed. In the first group after charging, the electrets were placed into a vacuum chamber where the pressure was reduced step by step in the range from 1000 mbar to 0.1 mbar. At each step the samples were stored for 1 minute. Then the electrets were removed from the vacuum chamber, the surface potential was measured again and the normalized surface potential was calculated. It was established that the low pressure had led to the surface potential decay of the electrets. The influence of the low pressure was analyzed by the equation that describes processes of desorption from the electret surface accompanied with surface diffusion. In the second group after charging, the electret surface potential was measured with the time of storage for two months. The experimental results obtained show a significant change in the electret behaviour of the composite films after the incorporation of zeolite particles with different concentration into the HDPE matrix. It was established that the surface potential decay depends on the corona polarity and the particle concentration.

1. Introduction

Considerable scientific interest has been shown in the surface potential decay of corona charged polymeric materials such as polypropylene (PP), polyethylene (PE), polyethyleneterephthalate (PET) and etc. that has been stimulated by industry demand related to the adjustment of Xerox-photography processes, assessment of cable insulation, and evaluation of electrostatic risks or monitoring of aged insulators [1-3]. Over the last few years, however, together with these materials another relatively new direction has been developed. It is the investigation of different composite electret films with specific properties obtained by the incorporation of different particles into polymer matrices [4]. Polymer zeolite composites have attracted wide interest because of enhancing polymer properties and extending their applications [5]. In [6] composites of low-density polyethylene (LDPE) and zeolite H-b particles at different zeolite contents up to 10 wt% were investigated. It was established that after incorporating zeolite H-b particles, the polyethylene composite films exhibit considerable degradation in mechanical properties and they have shown a good dispersion of zeolite particles in the PE matrix. Lopes and co-authors [7, 8] have investigated dielectric properties of composite polymers of α-PVDF doped with different NaY zeolite particles. It was established that dielectric properties and conductivity can be tailored by zeolite inclusion in polymer matrix and the presence of the zeolite particles increases ac conductivity of the matrix.

In the literature, however, there are no data available about the influence of the incorporation of zeolite particles on the charge decay of electret polymer films at different conditions. The aim of the present paper is to investigate the influence of both low pressure and time storage on the surface potential decay of HDPE composite films with different concentrations of the zeolite particles.

2. Experimental

2.1. Preparation of the composite films

The high density polyethylene (HDPE) brand PE2NT11-285D and the zeolite particles with a density of 2.4 g/cm3 and a particle diameter of less than 50 μm were used. Mixing of the polymer with the zeolite particles had been carried out on the mixer (Brabender Plastograph EC Plus) for 7 minutes at a temperature of (170 ± 5) oC. Polymer composite samples were prepared as 250 μm films under a hydraulic press at a temperature of (170 ± 5) oC and holding time of 5 min.

2.2. Corona treatment and surface potential measurement

Samples of 30 mm diameter were cut from the HDPE composite films. Charging of the samples in a corona discharge was carried out by means of a conventional corona triode system (figure 1) consisting of a grounded plate electrode, a corona electrode and a grid placed between them. The distance between the corona electrode and the grid was 10 mm and the distance between the grid and the grounded plate electrode was 3 mm. The samples were charged for 1 minute under room conditions. The voltage of the corona electrode was ±5 kV. The electret surface potential was limited by the grid potential, which was set at 1 kV with the same polarity as the corona electrode voltage.

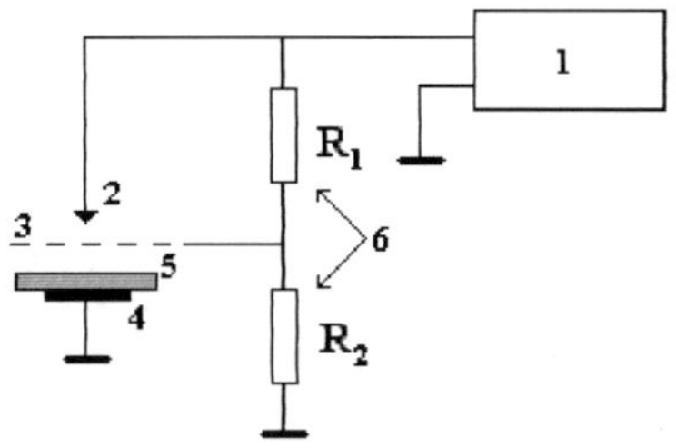

Figure 1. Scheme for obtaining electrets: 1. high voltage source; 2. corona electrode; 3. grid; 4. plate grounded electrode; 5. sample on a metal pad; 6. voltage divider.

The electret surface potential was measured by the method of the vibrating electrode with compensation [1] and the estimated error was better than 5%. After charging, the initial surface potential V_0 was measured and the electrets were divided in two groups according to the kind of treatment – different low pressure and time storage.

2.3. Low pressure treatment

After charging the samples, the initial electret surface potential V_0 was measured and then the electrets were placed into a vacuum chamber. The pressure in the vacuum chamber was reduced step by step in the range from 1000 mbar to 0.1 mbar. At each step the samples were held at respective constant pressure for 1 minute. After that the electrets were removed from the vacuum chamber, the surface potential V was measured again and the normalized surface potential V/V_0 was calculated.

3. Results and discussion

3.1. Time storage influence on electrets surface potential decay

The dependences of the normalized surface potentials V/V_0 on the storage time for positively and negatively charged HDPE composite films with different weight concentrations of the zeolite particles (0 wt.%, 2 wt.% and 6 wt.%) had been studied for two months. These dependences are presented in figures 2 and 3 respectively. The surface potential was measured once a week except for the first 20 days when the charge was rapidly decaying. Steady state values of the surface potential were established for all samples investigated after 30 days storage time.

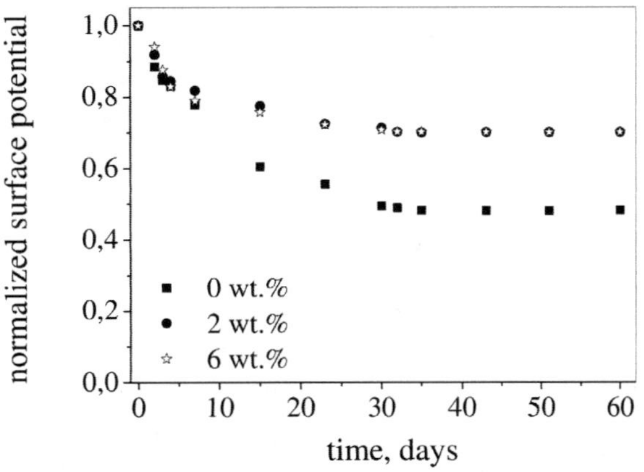

Figure 2. Time dependences of the normalized surface potential for positively charged HDPE composite films with different concentration of the zeolite particles.

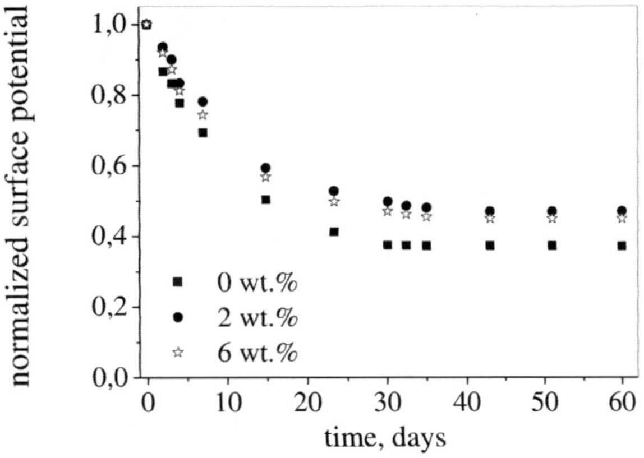

Figure 3. Time dependences of the normalized surface potential for negatively charged HDPE composite films with different concentration of the zeolite particles.

Each point in the figures is a mean value from 6 samples. The calculated standard deviation was better than 5% from the mean value with confidence level 95%.

The results illustrated in figures 2 and 3 demonstrate the following peculiarities:

- For all investigated samples the normalized surface potential are initially sharp decaying exponentially for the first 10 days, then are slowly decreasing to the 30th day and are practically stabilized to the 60th day.

The value of the surface potential of electrets depends on the amount of trapped charges in the different localized surface states of the samples. In the initial period of time after the corona charging, the surface potential rapidly decreases due to the release of the weakly captured charges from the shallow energy states. Then the surface potential stabilizes to a steady state value caused by the tightly captured charges in the deep energy traps.

- The steady state values of the normalized surface potential for positively charged HDPE composites are higher than those for negatively charged ones independently of the concentration of the zeolite particles incorporation in the HDPE matrix;
- The normalized surface potential of HDPE composite films with different concentrations of the zeolite particles (2 wt.% and 6 wt.%) decays slower than the one of the pure HDPE films independently of the corona polarity.

Therefore, the incorporation of the zeolite particles in the HDPE matrix independently of the concentration leads to an increase of the electret surface potential value and to obtaining the stable electrets.

For all samples the experimental results fitted very well to an exponential dependence of the type:

$$y = y_\infty + (1 - y_\infty)\exp\left(-\frac{t}{\tau}\right), \tag{1}$$

where y = V/V0 is the normalized surface potential, y∞ = (V/V0)st is the steady state level which is reached for t >>τ and τ is a relaxation time. Equation (1) is analogous to an equation characterizing the kinetics of adsorption-desorption processes [10]. In our case, these processes are not parallel but in series. Firstly, during the corona discharge, an adsorption takes place, as a result of which the accumulated surface charge determines V0 at t = 0. Subsequently, the release of the captured charges from the energy states takes place and its development is described by the second term in equation (1). Above a certain value of the time (t >>τ), the steady state level of the normalized potential y_∞ can be determined. Table 1 shows two obtained parameters y∞ and τ by fitting to the equation (1) and the determination factor R2 for all the samples as well as experimental values of the steady state level $(V/V_0)_{st}^{\exp}$ measured on 60th day.

The very good coincidence between the values of the parameter y∞ and the experimental measured normalized surface potential on 60th day $(V/V_0)_{st}^{\exp}$ as well as the determination factor R2

Table 1. Values of y_∞, τ, R^2 and $(V/V_0)_{st}^{\exp}$ for all investigated HDPE composite films.

Concentration of the zeolite	Corona polarity	y_∞	τ	R^2	$(V/V_0)_{st}^{\exp}$
0 wt.%	Positive	0.47	11.6	0.99	0.48
	Negative	0.36	10.0	1.00	0.37
2 wt.%	Positive	0.71	7.2	0.97	0.70
	Negative	0.46	11.4	1.00	0.47
6 wt.%	Positive	0.71	5.8	0.98	0.70
	Negative	0.44	10.4	1.00	0.45

that is very close to 1 for all the samples prove an equation 1 true. It is cleaned that relatively weakly-trapped charged is released over tens of days as the relaxation time τ is around 10 for all samples. Moreover the relaxation time does not depend of the zeolite concentrations for negatively charged HDPE composite films while it decrease with increasing of the zeolite concentrations for positively charged HDPE composite films.

3.2. *Low pressure influence on electrets surface potential decay*

The dependences of the normalized surface potential V/V_0 on the low pressure for positively and negatively charged HDPE composite films with different weight concentrations of the zeolite particles (0 wt.%, 2 wt.% and 6 wt.%) have been studied. Dependences of normalized surface potential on normalized pressure for HDPE composite electrets charged in positive or in negative corona are presented in figures 4 and 5.

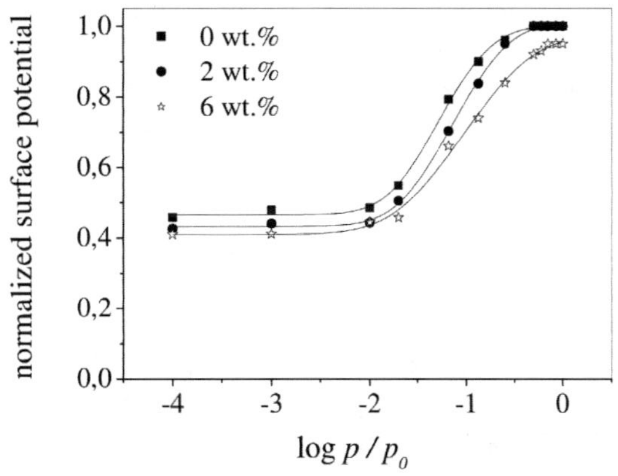

Figure 4. Dependences of the normalized surface potential on the normalized pressure for positively charged HDPE composite films with different concentration of the zeolite particles.

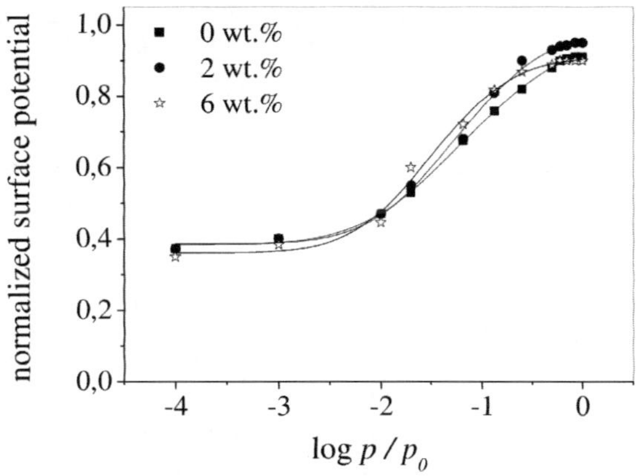

Figure 5. Dependences of the normalized surface potential on the normalized pressure for negatively charged HDPE composite films with different concentration of the zeolite particles.

The symbol V_0 in the figures marks the initial value of the surface potential measured just after electrets charging, and the symbol p_0 marks the atmospheric pressure. Each point in the figures is a mean value from 6 samples. The calculated standard deviation was better than 10% from the mean value with confidence level 90%.

The results presented in figures 4 and 5 show that:

- Independently of the concentration of zeolite particles three parts are observed in each curve. There are two regions of high and low pressure where the surface potential is constant. For each curve a relatively narrow region of pressures exists where a sharp decay of surface potential occurs. A similar behavior has also been observed in [9].
- The steady state values of the normalized surface potential at a pressure of 0.1mbar for the samples charged in a positive corona are higher than those for the samples charged in a negative corona independently of the zeolite particle concentrations.
- The steady state values of the normalized surface potential at a pressure of 0.1mbar do not depend of the incorporation of zeolite particles in HDPE matrix.

The results obtained (figures 4 and 5) were analyzed by the equation that describes the processes of desorption from the electrets surface accompanied with surface diffusion. This equation (2) was used and described earlier in [11]:

$$\theta = a + \frac{1}{2} b\left(1 + erf\left(\frac{x-c}{\sqrt{2}d}\right)\right). \tag{2}$$

Here $\theta = V/V_0$ is the normalized surface potential, $x = \log p/p_0$ and a, b, c and d are parameters. The parameter a is the minimum value of the normalized surface potential, b is the difference between the maximum and the minimum values of the normalized surface potential, c is the midpoint of the pressure range where the surface potential decay occurs and d is the half-width of the range of sharp potential decay.

Values of the parameters a, b, c and d for HDPE composites charged in a positive or a negative corona to an initial surface potential value of 1000 V are presented in table 2.

The data obtained in table 2 for HDPE/zeolite composite materials could have practical benefits in its use under conditions of reduction pressures. If you know the initial surface potential value V_0 of a certain HDPE/zeolite composite material, data from table 2 could be used to determine the pressure range in which there will be a potential sharp decay - from parameter d, and the minimum value of the surface potential to which it will decrease under reduced pressure - from parameter a.

Table 2. Values of the parameters *a*, *b*, *c* and *d* obtained by fitting to the equation 2.

Concentration of the zeolite	Corona polarity	a	b	c	d
0 wt.%	Positive	0.50±0.01	0.47±0.01	-1.28±0.02	0.42±0.02
	Negative	0.42±0.01	0.48±0.02	-1.49±0.05	0.63±0.07
2 wt.%	Positive	0.47±0.01	0.51±0.01	-1.14±0.02	0.46±0.02
	Negative	0.42±0.01	0.52±0.02	-1.26±0.04	0.68±0.05
6 wt.%	Positive	0.44±0.01	0.50±0.02	-1.03±0.04	0.58±0.05
	Negative	0.39±0.01	0.46±0.01	-1.29±0.04	0.69±0.05

4. Conclusion

Steady state values of the surface potential of the HDPE/zeolite composite electrets were established for all investigated samples after 30 days storage time. Therefore, this composite electrets material will be good for practical applications after 1 month when their electrical charge is stabilized.

Our experimental data show that the low pressure at which HDPE/zeolite composite electrets have been stored should be the important factor determining electrets charge decay. This is extremely important, and should be taken into account in the study of the stability of electrets sensors and devices that are used at room temperature and function under real conditions involving various low pressure levels.

References

[1] Sessler G and Gerhard-Multhaupt R 1999 *Electrets* 3rd Ed. (Laplacian: Press Morgan Hill California)
[2] Pouillès V, Lebey T and Castelan P 1996 *J. Appl. Phys.* **79** 8620
[3] Koch D and Molinié P 2008 *J. Elecrtrostat.* **66** 467
[4] Galikhanov M, Borisova A and Deberdeev R 2005 *Russian J. Appl. Chem.* **78** 820
[5] Zhang Y, Xu C, He Y, Wang X, Xing F, Qiu H, Liu Y, Ma D, Lin T and Gao J 2011 *J. Biomater. Sci. Polym. Ed.* **22** 809
[6] Huang B, Guo Y, Zhang T, Zhang X and Guo L 2013 *Packag. Technol. Sci.* **26** 1
[7] Lopes A, Gonçalves R, Costa C, Fonseca A, Botelho G, Neves I and Lanceros-Mendez S 2012 *J. Nanosci. and Nanotech.* **12** 6804
[8] Lopes A, Costa C, Sabater R, Neves I, Gomez Ribelles J and Lanceros-Méndez S 2013 *Solide State Ionics* **235** 42
[9] Viraneva A, Yovcheva T and Mekishev G 2012 *Polypropylene* (InTech Open Access Publisher: Rijeka Croatia)
[10] Lazneva E, 1990 *Laser desorption* (LGU Press: Leningrad) in Russian
[11] Viraneva A, Yovcheva T, Pisanova E, Gencheva E and Mekishev G 2010 *AIP Conf. Proceed.* **1203** 382

Optical properties of polyimide layers prepared by vacuum deposition in the presence of gas

V Strijkova and G Danev
Institute of Optical Materials and Technologies "Acad. J. Malinowski", Bulgarian Academy of Sciences, Acad. G. Bonchev Str., Bl. 109, 1113 Sofia, Bulgaria

E-mail: vily@iomt.bas.bg

Abstract. Thin polyimide (PI) layers were prepared by simultaneous vacuum evaporation of the polyimide precursors pyromellitic dianhydride (PMDA) and 4,4'-oxydianiline (ODA). After deposition the layers were thermally treated 1 hour at 170°C followed by 1 hour at 250°C. Precursors were deposited in a residual atmosphere of different gases - argon, nitrogen, helium and air at a pressure 10^{-2} Pa. Conditions for fabrication of PI layers with reproducible composition, uniform and smooth surface, without mechanical and chemical defects have been established. It was found that the preparation conditions do not significantly affect the optical properties of the layers. The PI layers before and after thermal treatments are transparent over 85% in the studied spectral range (from 350 to 900 nm). The thermal treatment leads to a bathochromic shift, in a consequence of the transmission edge shift to the longer wavelengths, for all the samples, independently of the deposition conditions in the vacuum chamber. The average refractive indices at 633 nm varied from 1.67 towards 1.72 depending on the deposition conditions and thermal treatment.

1. Introduction

In the last few years, the interest in the field of thin organic layers has grown dramatically due to their successful application in optical and electronic devices, such as light emitting diodes or field effect transistors [1]. Organic and polymer thin layers will play a pivotal role in the development of next-generation electronic devices, passivation coatings, and chemical and biological sensors. The superior dielectric and thermal performances of polyimides (PI) have led to their widespread use for electronic applications (as an insulator, barrier layer or capsulation layer) [2-6]. PI layers are used in photovoltaic (PV) cells due to its dielectric properties, but also have requirements for their optical properties [7]. Polyimide-based materials are expected to have potential applications in the field of organic optical waveguides [8]. PI layers are conventionally prepared from polyamic acid solution (the precursor of PI) [9, 10] by coating on a substrate and transformation to PI through thermal or chemical dehydration. One of the used methods for PI layer preparation is physical vapour deposition (PVD) of precursors, like diamines and dianhydriedes [11] followed by a thermal treatment.

It is known that there is a relationship between the conditions of preparation and structure of the layers on the one hand, and their structure and physical and chemical properties on the other. Our previous studies have shown that the introduction of gas into a vacuum chamber leads to a change in micro-mechanical and electrical properties of polyimide layers [12].

FTIR spectroscopy summarized that presence of an inert gas (regardless of its type) results in an increase of the degree of imidization [12]. The layer produced in nitrogen atmosphere is with the highest degree of imidization. The degree of imidization decreases around 10% for layers produced in the presence of air compared with the layers produced in the presence of nitrogen [12].

The aim of this study is to investigate the optical properties of thin PI layers, obtained by PVD depending on the residual gas in the vacuum chamber – helium, argon, nitrogen and air. The results obtained could lead to a better understanding of the relationship between preparation conditions, structure, and optical properties of thin PI layers.

2. Experimental

2.1. Sample preparation

The vacuum deposited polyimide (VDP) layers with thicknesses of 300 nm and 1000 nm were prepared on soda-lime-glass plates. The vacuum chamber was evacuated by oil diffusion pump to a pressure of < $7x10^{-4}$ Pa. The precursors used pyromellitic dianhydride (PMDA) and 4,4'-oxydianiline (ODA), were evaporated from two independent thermally-heated Knudsen type evaporation sources and deposition rates from 0.2 to 2 A/sec, the latter being carefully controlled by quartz oscillators. Thus, the optimal ratio in the flux of 1:1 for the ODA:PMDA vapours was ensured [13]. The process of maintenance of a definite permanent residual pressure of the inert gas in the vacuum chamber was performed by the introduction of the corresponding gas by a mass flow controller which is computer-controlled via information obtained from the vacuum meter. The precursors condensed on the substrates in the presence of gas atmosphere at vacuum 10^{-3} and 10^{-2} Pa. Thus the mixed (precursors-gas) layers were built-up. The co-deposited precursor layers were transformed into PI layers by a two - step thermal treatment at strictly controlled temperatures (1 hour at 170°C followed by 1 hour at 250°C).The goal of the thermal treatment is to accelerate the polycondensation solid state reaction between the monomers, leading to the PI formation [6].

2.2. Methods of investigation

Thickness of the PI layers was measured using profilometer Talystep.

Surface structure and morphology of vacuum deposited PI layers was examined with a scanning electron microscope (SEM) Philips 515 mode secondary electron image (SEI). To increase the contrast of the image, the sample surface was tilted relative to the primary electron beam. The angle between the normal of the sample surface and the direction of the scanning beam was altered from 0 to 40 °.

UV-Vis spectra of PI layers deposited in presence of different gas before and after thermal treatment were investigated by UV-Vis-NIR spectrophotometer Cary. The refractive index values of the different layers were determined by Swanepoel method [14] through a program developed in our institute by Konstantinov and associates [15].

3. Results

By means of SEM the surface morphology of the layers deposited in presence of different gas has been investigated. Figure 1 presents the surface morphology of PI layer deposited in an atmosphere without gas and vacuum with pressure (p) of 10^{-4} Pa. In figure 2 the surface morphology on PI layer obtained in a residual of gas He with p ~ 10^{-2} Pa is presented. The PI layers deposited by varying residual gas atmosphere (air, argon, nitrogen) have similar surface morphology.

Figures 1 and 2 show, that thin PI layers remain homogeneous, devoid of micro-cracks and defects independently of the residual gas. The surface of the PI layers is smooth, without noticeable peculiar defects or granular structure. Therefore it can be concluded that there was no influence of the different gas atmosphere on the surface morphology of the different samples.

Figures 3 and 4 present the cross section of the above PI layers - deposited in an atmosphere without presence of gas in the vacuum chamber and in a residual of gas helium, respectively.

As seen in the both pictures, without gas (vacuum 10^{-4} Pa) and in the presence of helium, the observed growth of the columns from the substrate to the surface, determines the presence of free volume in the layers.

Figure 1. SEM microphotographs of the surface morphology of VDP 1000 nm thick PI layer deposited without presence of gas in the vacuum chamber (vacuum 10-4 Pa).

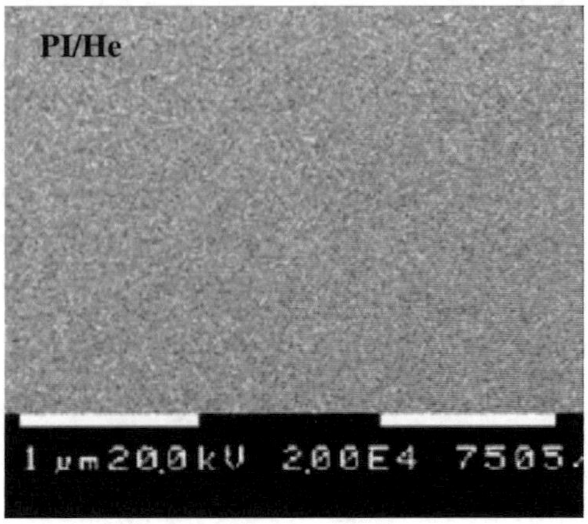

Figure 2. SEM microphotographs of the surface morphology of VDP 1000 nm thick PI layer deposited in the presence of helium in the vacuum chamber.

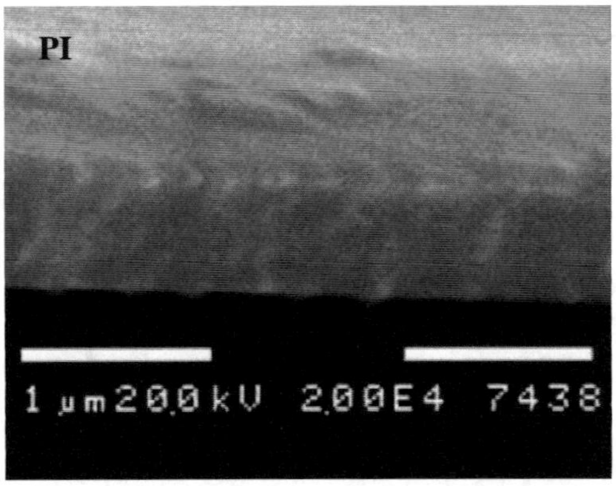

Figure 3. SEM microphotographs of the cross section of VDP 1000 nm thick PI layers deposited without gas in the vacuum chamber (vacuum 10-4 Pa).

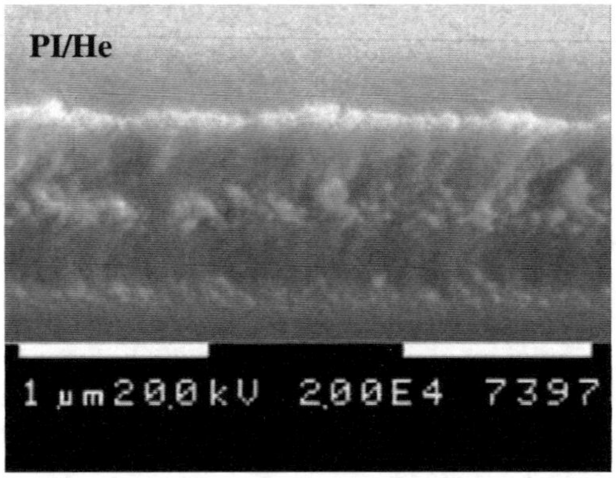

Figure 4. SEM microphotographs of the cross section of VDP 1000 nm thick PI layers deposited in the presence of helium in the vacuum chamber.

The layers obtained in an atmosphere of helium introduced in the vacuum chamber have narrower columns. The layers are built up of fine grain structure compared to the layer deposited in an atmosphere without gas.

Figures 5 and 6 show UV-Vis spectra of PI layers deposited in presence of different gases before and after thermal treatment, respectively. As seen after the thermal treatment all spectra are displaced to a long wavelength region. The effect of bathochromic shift, as well as the colorization of PI layers are due to the charge transfer complexes that occur between the donor and acceptor units in the polyimide chain [16]. The color of the samples changes from colorless before thermal treatment, to light brown – after treatment for 1 h at 170 ° C and 1 h at 250 ° C.

The layers possess a high transmission (>80%) in the range of 440–900 nm. Our experiments show that the shape of the curves and the optical parameter values do not depend on the gas in the vacuum chamber, when the PI layer is formed.

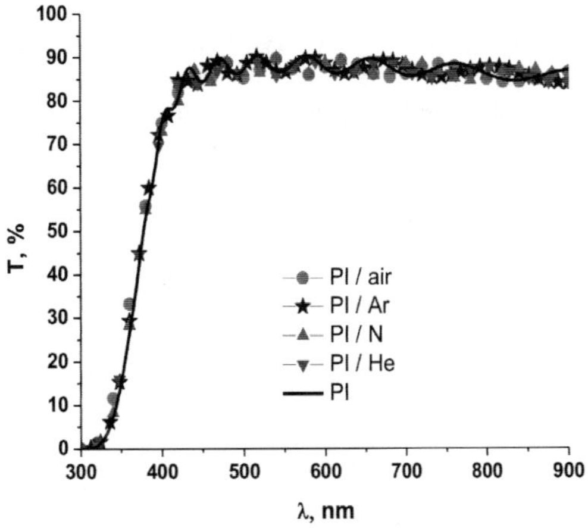

Figure 5. Optical spectra of PI layers deposited in presence of different gas in the vacuum chamber before thermal treatment.

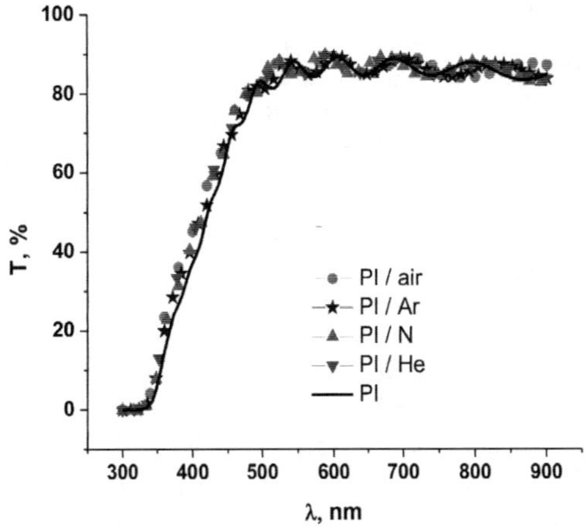

Figure 6. Optical spectra of PI layers deposited in presence of different gas in the vacuum chamber after thermal treatment.

Figure 7 shows the refractive index distribution of PI layers deposited in the presence of different gases after thermal treatment and figure 8 shows the diagram of refractive index values depending on the residual gas.

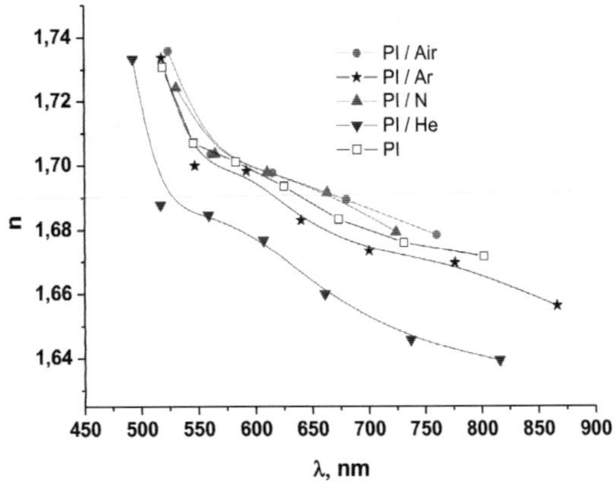

Figure 7. Refractive index distribution of PI layers deposited in presence of different gas in the vacuum chamber.

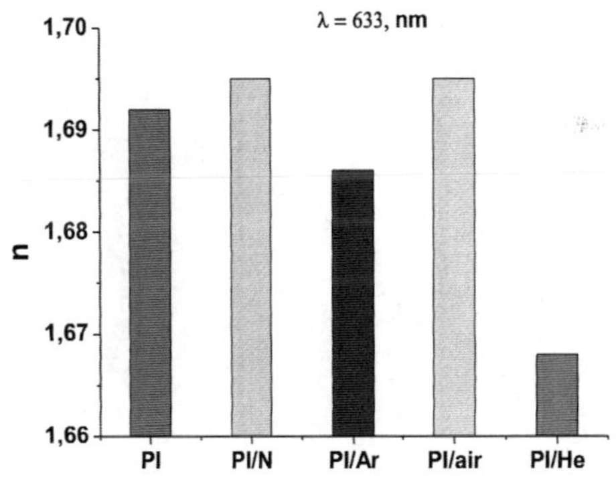

Figure 8. Diagram of the refractive index values of PI layers depending on the residual gas in the vacuum chamber.

The results obtained have shown that there is a small difference in the refractive index value of the different layers with the exception of these obtained in the residual atmosphere of helium. The other layers have refractive index values comparable to that of the layer deposited in an atmosphere without gas and vacuum with pressure of 10^{-4} Pa.

4. Conclusion

The study was associated with the possibility of modifying the properties of a thin vacuum deposited PI layers by varying the conditions of their preparation. The surface of the PI layers is smooth, without noticeable peculiar defects or granular structure. There was no influence of the different gas atmosphere on the surface morphology of the different samples. The PI layers display a high transmission (>80%) in the studied range independently of the residual gas. The most significant changes in the refractive index values were found for polyimide layers obtained in the presence of helium atmosphere. The optical properties of these VDP PI layers make them suitable for various applications in the photovoltaic industry or in the field of organic optical waveguides.

References

[1] Peisert H, Knupfer M, Schwieger T, Fuentes G, Olligs D, Fink J and Schmidt Th 2003 *J. Appl. Phys.* **93** 9683
[2] Hatton B, Landskorn K, Hunks W, Bennet M, Shukaris D, Perovic D and Ozin G 2006 *Materials Today* **9** 22
[3] Maex K 2003 *J. Appl. Phys* **93** 8793
[4] Hoofman R 2005 *Microelectron. Eng.* **80** 337
[5] Hougham G, Tesoro G, Shaw J, 1994 *Macromolecules* **27** 3642
[6] Strunskus Y and Grunze M 1994 *Polyimides - fundamentals and applications* eds. M. Crosh and K. Mittal (New York: Marcel Dekker, Inc) p187
[7] French R, Rodrı´guez-Parada J, Yang M, Derryberry R and Pfeiffenberger N, 2011 *Solar Energy Materials & Solar Cells* **95** 2077
[8] Xie S Zhang Z and Wei W 2007 *Journal of the Korean Physical Society* **51** 1536
[9] Sroog C, Endrey A, AbramoV, Berr C, Edwards W and Olivier K 1965. *J. Polym. Sci.* A **3** 1373
[10] Edwards W 1965 US Patent 3 (to E.I. du Pont de Nemours and Co.) 179
[11] Salem J, Sequeda F, Duran J, Lee W and Yang R 1986*J. Vac. Sci. Technol.* **A 4** 369
[12] Strijkova V and Danev G 2013 *Bulg. Chem. Communications: Conf. Series* B **45** 185
[13] Spassova E 2003 *Vacuum* **70** 551
[14] Swanepoel R. 1983 *J. Phys. E: Sci. Instrum.* **16** 1214
[15] Private communication
[16] Gordina T, Kotov B, Kolninov O and Pravednikov A Charje *Transfer Complexes of Gertain Aromatic Polyimides,* NASA Technical Memorandum, NASA'TM-7797

Preparation and characterization of mesoporous Nb_2O_5 films for sensing applications

K Lazarova, B Georgieva, M Spasova and T Babeva
Institute of Optical Materials and Technologies "Acad. J. Malinowski", Bulgarian Academy of Sciences, Acad. G. Bonchev str., bl. 109, 1113 Sofia, Bulgaria

E-mail: babeva@iomt.bas.bg

Abstract. The possibility of combining sol–gel synthesis with molecular self-assembly and phase separation is explored to produce Nb_2O_5 films with tailored pore structures at the nanometer scale. Thin Nb_2O_5 film are fabricated by spin coating of a Nb sol prepared by sonocatalytic method using $NbCl_5$ and ethanol mixed with structure directing agent Pluronic PE 6800 in different concentrations which acts as a supra molecular pore template. Post deposition treatment is applied for film consolidation and pore formation. Surface morphology and structure of the films are studied by Transmission Electron Microscopy and Selected Area Electron Diffraction. The optical properties (refractive index and extinction coefficient) along with the thickness of the films are determined from reflectance spectra of films deposited on silicon substrates. The overall porosity of the films and the amount of adsorbed acetone vapors are quantified by means of Bruggeman effective medium approximation using already determined optical constants. The sensing properties of the samples are studied by measuring reflectance spectra prior to and after exposure to acetone vapors with controlled concentrations. Furthermore, the potential of using the studied mesoporous Nb_2O_5 films for chemical sensing with optical read-out is demonstrated and discussed.

1. Introduction

Mesoporous materials have attracted the scientific interest due to their remarkable properties and potential applications in many advanced areas such as drug delivery systems, sensors, catalysis, photovoltaic cells, and fuel cells [1]. Among several advantages, mesoporous materials offer possibilities of functionalization and a high active surface area due to their controlled porosity in 2-50 nm range and narrow pore size distribution. Besides mesoporous oxide thin film is a promising matrix for embedding isolated or interacting metal nanoparticles modifying in this way its optical, electronic and catalytic properties [2].

One attractive application of mesoporous films is for building blocks of vapor responsive Bragg stacks [3-5]. The idea behind the detection is the change of refractive index and / or thickness of the layers in the Bragg stack due to condensation of vapors in the pores which results in a shift of the high reflectance band and consequent change of stacks color [6,7]. Because the high optical contrast (difference in refractive indices of the layers in the stack) is essential for proper operation of the device most commonly a combination of SiO_2 and TiO_2 mesoporous thin films is used in the stack [4,5,8]. It will be a real advantage if material with higher refractive index is used instead of TiO_2. Thus a higher optical contrast could be achieved enabling even an omnidirectional reflectance to be realized [9]. Recently we have shown that Nb_2O_5 thin films with tunable refractive index and good optical properties could be produced by simple sol-gel method [10]. Besides, Nb_2O_5 exhibits the highest refractive index among the widely used TiO_2 and Ta_2O_5 sol-gel films [11]. Despite there are few reports for fabrication of mesoporous Nb_2O_5 films [12 and refs 27, 29, 56, 172-175, 238-240 cited herein] a lack of optical characterization exists.

In this paper we report studies of sol-gel derived mesoporous Nb_2O_5 thin films prepared by evaporation-induced self-assembly method using Pluronic PE6800 as a molecular pore template. Optical properties and thickness of the films are calculated from measured reflectance spectra. The refractive index curves are further used for determination of overall porosity of the films using Bruggeman effective medium approximation. The sensing properties of the films are studied by measuring reflectance spectra prior to and after exposure to acetone vapors. The potential application of films as sensitive elements of optical vapour sensors is explored.

2. Experimental details

Thin mesostructured Nb_2O_5 films were prepared through an evaporation-induced self-assembly method using spin-coating of Nb sol mixed with inorganic template Pluronic PE 6800 (BASF) in volume ratio of 5:1. The Nb sol was prepared by a sonocatalytic method using 0.400g $NbCl_5$ (99%, Aldrich) as a precursor, 8.3 ml ethanol (98%, Sigma-Aldrich) and 0.17 ml distilled water [10,13]. Instead of stirring, the solution was subjected to a sonication for 30 min and aged for 24 h at ambient conditions prior to deposition. For deposition of mesoporous Nb_2O_5 thin films mixtures of Nb sol and template in volume ratio 5:1 were prepared. Three different batch solutions of Pluronic in ethanol with weight concentration of 1.5, 3 and 5 wt.% were used in the mixtures. An amount of 0.3 ml was dropped on the pre-cleaned Si substrates and spun at a rate of 2500 rpm for 30 s. The template is eliminated by annealing at 320 °C for 30 min at 5 °C / min acceleration. Dense Nb_2O_5 thin films were also prepared for comparison. Purposely 0.3 ml of Nb sol was spun coated at 2500 rpm for 30 s on Si substrate. The annealing conditions were the same as for the mesoporous films.

The surface morphology and structure of the films were studied by Transmission Electron Microscopy (TEM) and Selected Area Electron Diffraction (SAED) using HRTEM JEOL JEM 2100 (Japan) microscope. The optical properties (refractive index (n) and extinction coefficient (k)) along with the thickness (d) of the films were determined from reflectance spectra of the films measured at normal light incidence by UV-VIS-NIR spectrophotometer Cary 05E (Varian, Australia) using non-linear curve fitting method [10]. The experimental errors for n, k and d are 0.005, 0.003 and 2 nm, respectively. The vapor sensing measurements on films were implemented in a Cary 05E spectrophotometer equipped with a homemade bubbler system for generation of vapors from liquids with controlled concentrations [14].

3. Results and discussions

3.1. Morphology study of Nb_2O_5 thin films

Figure 1 shows typical TEM images of the samples with different concentrations of Pluronic from 1.5 to 3 to 5 wt.% (figure 1(b), 1(c) and 1(d), respectively) which are compared to dense Nb_2O_5 thin films shown in figure 1(a). It is seen that the addition of template leads to formation of pores arranged non periodically. The size of the pores gradually increases with concentration of Pluronic. The SAED diffraction (not shown here) had proven that all samples were amorphous.

3.2. Optical properties and porosity of Nb_2O_5 thin films

The measured reflectance spectra (R) of both dense and mesoporous Nb_2O_5 films deposited using different concentrations of Pluronic are presented in figure 2 (a). The reflectance spectrum of uncoated silicon substrate is also shown. Comparison of R spectra of films and bare substrate shows that reflectance of the films decreases with increasing of template concentration. Besides, the reflectance minimum shifts towards longer wavelengths. This means that the growth of pore size manifest itself in increase in optical thickness (the product of refractive index and physical thickness) of the films.

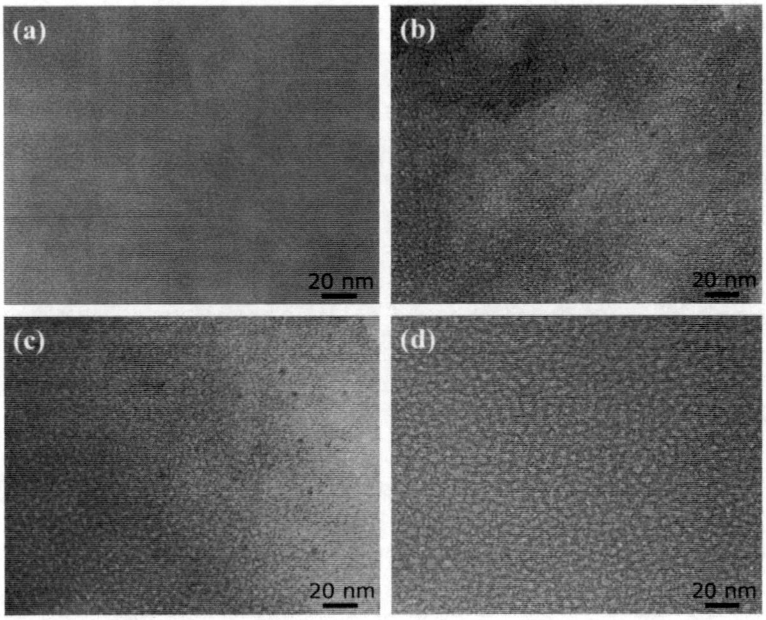

Figure 1. TEM images of dense Nb_2O_5 film (a) and porous Nb_2O_5 films consisting of Nb sol and Pluronic in the same volume ratio (5:1) but with different concentrations of Pluronic: 1.5 wt.% (b), 3 wt.% (c) and 5 wt.% (d).

In order to clarify the optical thickness dependence of R the measured spectra are used further for calculation of n, k and d of the films. The calculation procedure is explained in details elsewhere [10]. Briefly, a Wemple-diDomenico single oscillator model [15] and an exponential decay function are used for the description of wavelength dependences of n and k, respectively. The models parameters along with the thickness of the films are determined through a non-linear minimization of the discrepancies between the measured and calculated R spectra. The refractive index curves of the films with different Pluronic concentration (i.e different pore sizes) along with dispersion curve of dense Nb_2O_5 film are presented in figure 2(b). All curves show normal dispersion (decrease in n with wavelength) that is typical for transparent or slightly absorbing materials. With increasing the pore size a significant decrease of refractive index is observed. The values of n at wavelength of 600 nm decrease from 2.14 for the dense film to 1.88, 1.77 and 1.53 for porous films with Pluronic concentration of 1.5, 3 and 5 wt.%, respectively (table 1). Simultaneously there is no change in the extinction coefficient (table 1). Interestingly an increase in thickness is observed: d changes from 37 nm for the dense film to 47, 57 and 85 nm for porous films with different concentrations of Pluronic (table 1). We may conclude that the annealing conditions are appropriately chosen so that no contraction of pores to take place.

In order to quantify the degree of porosity of the studied films we used Bruggeman effective medium approximation (BEMA) [16]. The idea is to regard the mesoporous film as an effective medium consisting of dense Nb_2O_5 and voids filled with air. The effective dielectric constant $\boldsymbol{\varepsilon}_e$ depends on the dielectric constants of the phases presented and their volume fractions:

$$f_d \frac{\varepsilon_d - \varepsilon_e}{\varepsilon_d + 2\varepsilon_e} + f_{air} \frac{\varepsilon_{air} - \varepsilon_e}{\varepsilon_{air} + 2\varepsilon_e} = 0,\ f_d + f_{air} = 1, \quad (1)$$

ε_d and ε_{air} are the dielectric constants of dense Nb_2O_5 and air, respectively, and $f_d = V_d/V_{tot}$ and $f_{air} = V_{air}/V_{tot}$ are their volume fractions (V_d and V_{air} being the volumes occupied by dense Nb_2O_5 and air, respectively, $V_{tot} = V_d + V_{air}$ is the total volume of the film).

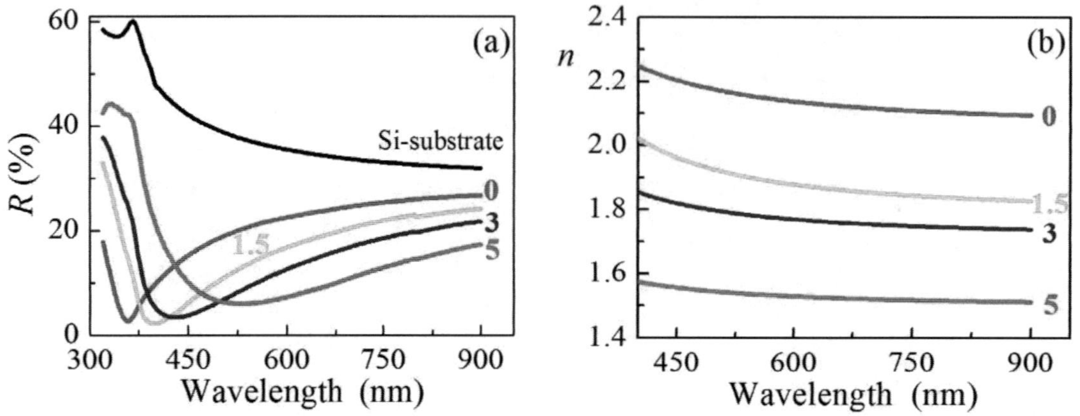

Figure 2. Reflectance spectra (a) and refractive index (b) of Nb_2O_5 films deposited on Si-substrate using Pluronic with different concentrations indicated by the numbers in the figures and the same volume ratio of Nb sol and Pluronic equal to 5:1.

For clarity, it is important to note that the two parameters ε and n are related as follows: $\varepsilon = n^2$. Thus using the already determined values of refractive index n_d and n_e, (ε_d and ε_e, respectively) (figure 2 (b)) and considering that $\varepsilon_{air}=1$, the voids volume fractions f_{air} are calculated from eq.1 and presented in table 1. As can be expected the increase of the pore size leads to significant increase of the free volume fraction that reaches value of 51 % for the film with the highest concentration of Pluronic (table 1).

Table 1. Refractive index (n), extinction coefficient (k), thickness (d), volume fractions of air (f_{air}) and acetone (f_{ac}) in % calculated from eq. 1 and 2 respectively, relative reflectance change $\Delta R_{ac}/R_{Ar}$ and refractive index change Δn_{ac} after exposure to acetone vapors as a function of Pluronic concentration.

Pluronic, wt. %	n (600 nm)	k (600 nm)	d, nm	f_{air}, %	$\Delta R_{ac}/R_{Ar}$, %	$\Delta n_{ac} = n_{ac} - n_{Ar}$	f_{ac}, %
0	2.14	0.020	37	0	0	0	0
1.5	1.88	0.018	47	22	1.2	0.0034	0.9
3	1.77	0.018	57	31	2.3	0.011	2.8
5	1.53	0.021	85	51	2.7	0.007	1.4

3.3. Vapor sensing experiment

Reflectance spectra of the samples are measured prior to and after exposure of the films to acetone vapors at relative pressure $p/p_0=1$ (p_0 is the pressure of saturated vapors at zero degrees). The generation of acetone vapors from liquid and the control of their concentrations are realized by homemade bubbler system explained in details in [14]. Before running the spectra all films are exposed to dry Ar flow with rate of 500 sccm for 10 min in order to empty the pores. The values of relative reflectance changes $\Delta R_{ac}/R_{Ar}$ (where $\Delta R_{ac} = R_{ac} - R_{Ar}$, R_{ac} and R_{Ar} are reflectance spectra after acetone and argon exposure, respectively) as a function of Pluronic concentration are presented in figure 3 (a) and in table 1. No change of R is observed for dense Nb_2O_5. For porous films a gradual increase of $\Delta R_{ac}/R_{Ar}$ with Pluronic concentration is obtained. Considering that R_{ac} is a function of n, k and d of the films we can write ΔR_{ac} as follows:

$$\Delta R_{ac} = \frac{\partial R_{ac}}{\partial n}\Delta n_{ac} + \frac{\partial R_{ac}}{\partial k}\Delta k_{ac} + \frac{\partial R_{ac}}{\partial d}\Delta d_{ac}, \qquad (2)$$

where $\partial R_{ac}/\partial n$, $\partial R_{ac}/\partial k, \partial R_{ac}/\partial d$, are the partial derivatives of R_{ac} with respect to n, k and d and $\Delta n_{ac,}$ Δk_{ac} and Δd_{ac} are the absolute changes in n, k and d of the films due to exposure to acetone vapors, respectively.

The exposure of the porous films to acetone results in vapor condensation in the pores and replacement of the air inside with acetone with higher refractive index. As a result the effective refractive index of the films increases, i.e $\Delta n_{ac} > 0$. We assume no changes in k and d, i.e Δk_{ac}=0 and Δd_{ac}=0. The values of Δn_{ac} are calculated through equation 2 using the measured values of ΔR_{ac} (table 1) and numerically derived partial derivatives. Figure 3 (b) and table 1 show Δn_{ac} as a function of Pluronic concentration. It is seen that the change of n is the strongest at Pluronic concentration of 3 wt.%.

After acetone adsorption the exposed films consist of three phases: dense Nb_2O_5, acetone and air with respective dielectric functions ε_d, ε_{ac} and ε_{air} and volume fractions f_d, f_{ac} and f_{air}. Then BEMA can be written in the form:

$$f_d\frac{\varepsilon_d-\varepsilon_e}{\varepsilon_d+2\varepsilon_e}+f_{air}\frac{\varepsilon_{air}-\varepsilon_e}{\varepsilon_{air}+2\varepsilon_e}+f_{ac}\frac{\varepsilon_{ac}-\varepsilon_e}{\varepsilon_{ac}+2\varepsilon_e} = 0, \qquad f_d + f_{air} + f_{ac} = 1 \qquad (3)$$

The values of the volume fraction of acetone vapors in the film are calculated using equation 3 and are presented in table 1. Surprisingly, the highest value of f_{ac} is obtained for porous film with Pluronic concentration of 3 wt.%. For higher and lower concentration of Pluronic the amount of adsorbed acetone vapors is smaller.

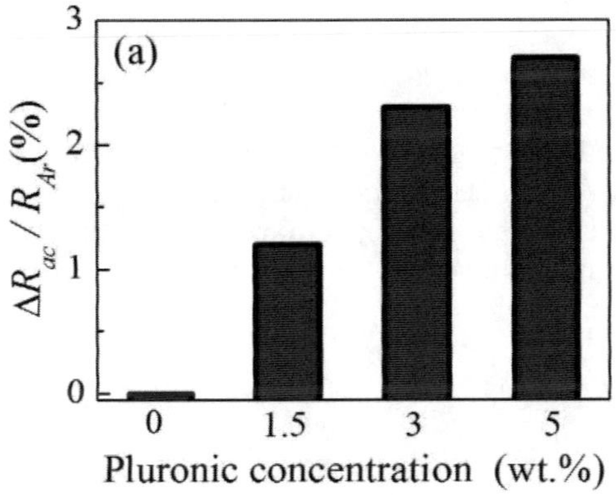

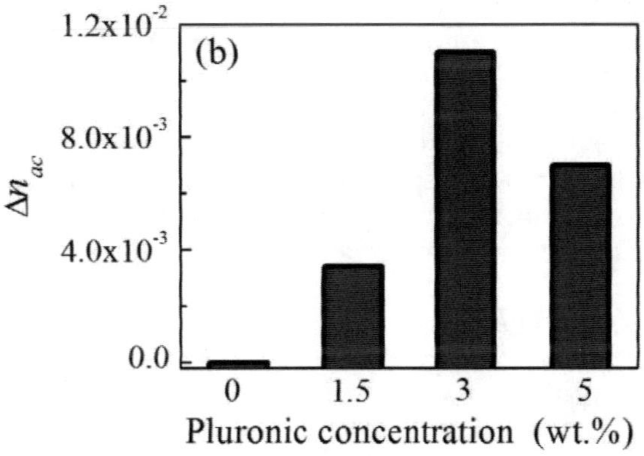

Figure 3. Relative reflectance change $\Delta R_{ac}/R_{Ar}$ (a) and refractive index change Δn_{ac} (b) after exposure to acetone vapors as a function of Pluronic concentration.

Previously we have shown that the strength of the optical response is determined by the interplay between surface hydrophobicity and tension, refractive index and molecular size of the analyte [17]. Considering that the refractive index and molecular size are the same for all studied samples we can assume that the surface hydrophobicity and tension mainly influence the adsorption strength. It may happen these two parameters to vary with the pore size maximizing its positive impact on adsorption at particular pore size. Another possible reason for the peak in adsorption is the interconnectivity of the pores that could be the highest for the sample with 3wt.% Pluronic. We suppose that the periodical arrangement of the pores will improve further the optical response.

As we note in the introduction part a possible application of the studied mesoporous films is in responsive Bragg stacks [18,19]. Bragg stacks are multilayered systems comprising high and low refractive index materials with quarter-wave optical thicknesses ($nd = \lambda_c/4$) arranged in a periodic manner. Due to the interference phenomena, Bragg stacks exhibit a band of high reflectance that is centered at λ_c, and it is highly sensitive to changes in optical thickness of the constituent films. If vapor sensitive materials are used as building blocks, then the position of the reflectance band can be used as an indication of the concentration of the detected analyte [18,19].

For achieving high sensitivity of vapor responsive Bragg stacks all layers across the stack should be accessible to the detected analyte. Our previous studies have shown that for Bragg stack comprising alternatively deposited MEL-type zeolites films and dense Nb_2O_5 films only five layers are accessible to the analyte (acetone in the current case) [14]. We expect that the implementation of mesoporous Nb_2O_5 films in Bragg stack instead of dense ones will improve the effective penetration of the analyte through the layers resulting in increased sensitivity and simultaneously preserving the high optical contrast. These experiments are ongoing in our group.

4. Conclusions

It is shown that the combination of sol–gel synthesis with phase separation is successful approach for deposition of mesoporous Nb_2O_5 films with enhanced porosity. The fabrication of thin Nb_2O_5 films with tailored pore structure at the nanometer scale through evaporation-induced self-assembly method is demonstrated. Purposely a mixture of Nb sol and Pluronic PE 6800 in different concentrations is spun coated on silicon substrate and subjected to post deposition annealing at 320 ^{o}C for 30 min at 5 ^{o}C/min acceleration of the temperature for pore formation and film consolidation. An increase in film thickness from 37 to 85 nm is observed when Pluronic concentration changes from 0 to 5 wt.% thus confirming the appropriateness of the choice of the annealing conditions ensuring no pore contraction to take place. Simultaneously a decrease of refractive index from 2.14 to 1.53 is obtained due to the higher free volume fraction originating from the increased pore size.

An improvement of sensing properties of mesoporous films is demonstrated as compared to the dense films. The strongest response of Nb_2O_5 films towards acetone vapors is achieved for films with Pluronic concentration of 3 wt.%. For higher and lower concentrations of Pluronic the amount of adsorbed acetone vapors is smaller. Two possible reasons are discussed: i) the interconnectivity of the pores that maximizes at certain pore size and ii) a specific interplay between surface hydrophobicity and tension that favours acetone adsorption at a particular Pluronic concentration / pore size.

Acknowledgement

The authors are grateful to Dr. Daniela Karashanova from IOMT-BAS for the help with TEM investigations.

References

[1] Zhu Y P, Ren T Z and Yuan Z Y 2014 *New J. Chem.* **38** 1905
[2] Fuertes M C, Marchena M, Marchi M C, Wolosiuk A and Soler-Illia G 2009 *Small* **5** 272
[3] Soler-Illia G, Angelome P C, Fuertes M C, Calvo A, Wolosiuk A, Zelcer A, Bellino M G and Martınez E D 2011 *J. Sol-Gel Sci. Technol.* **57** 299
[4] Ghazzal M N, Deparis O, Errachid A, Kebaili H, Simonis P, Eloy P, Vigneron J P, Coninck J and Gaigne E M 2012 *J. Mater. Chem.* **22** 25302
[5] Fuertes M C, López-Alcaraz F J, Marchi M C, Troiani H E, Luca V, Míguez H and Soler-Illia G 2007 *Adv. Funct. Mater.* **17** 1247

[6] Calvo M E, Colodrero S, Hidalgo N, Lozano G, Lopez-Lopez C, Sanchez-Sobrado O and Mıguez H 2011 *Energy Environ. Sci.* **4** 4800
[7] Bonifacio L D, Lotsch B V, Puzzo D P, Scotognella F and Ozin G A 2009 *Adv. Mater.* **21** 1641
[8] Choi S Y, Mamak M, Freymann G, Chopra N and Ozin G A 2006 *Nano Letters* **6** 2456
[9] Kim S H and Hwangbo C K 2002 *Appl. Opt.* **41** 3187
[10] Lazarova K, Vasileva M, Marinov G and Babeva T 2014 *Optics & Laser Technology* **58** 114
[11] Babeva T, Lazarova K, Vasileva M, Gospodinov B and Dikova J 2013 *Bulg. Chem. Comm.* **45** 28
[12] Gu D and Schuth F 2014 *Chem. Soc. Rev.* **43** 313
[13] Arfsten N J and Gavlas J F 2004 US patent 6811901 B1
[14] Lazarova K, Awala H, Thomas S, Vasileva M, Mintova S and Babeva T 2014 *Sensors* **14** 12207
[15] Wemple S and Di Domenico M 1971 *Phys. Rev. B* **3** 1338
[16] Bruggeman D A G 1935 *Ann. Phys.* **24** 636
[17] Babeva T, Todorov R, Gospodinov B, Malinowski N, Fallah J and Mintova S 2012 *J. Mater. Chem.* **22** 18136
[18] Calvo M and Mıguez H. 2013 in *Responsive Photonic Nanostructures: Smart Nanoscale Optical Materials*, ed Y Yin (Cambridge, UK: The Royal Society of Chemistry) pp 1–21.
[19] Ghazzal M N, Deparis O, De Coninck J and Gaigneaux E M 2013 *J. Mater. Chem. C* **1** 6202

Effect of the substituents on the photophysical, electrochemical and electroluminescence properties of OLED dopant Iridium bis(2-phenylbenzothiozolato-N,C2′)(acetylacetonate)

P Ivanov, R Tomova* and P Petrova

Institute of Optical Materials and Technologies „Acad. J. Malinowski", Bulgarian Academy of Sciences, Acad. G. Bonchev str. bl. 109, 1113 Sofia, BG

E-mail: reni@iomt.bas.bg

Abstract. The effect of two substituents: clorine and 1,3-diphenylpropane-1,3-dionate, placed on different position in the molecule of Iridium (III) bis(2-phenylbenzothiozolato-N,C2′)-(acetylacetonate) (bt)2Ir(acac), on its electrochemical behaviour, photophysical and electroluminescence properties were investigated. Three complexes (bt)2Ir(acac), Iridium (III) bis[2-(4-chlorophenyl)benzothiazolato-N,C2']-acetylacetonate (Clbt)2Ir(acac), in which the Cl atom was introduced on the 4-position in the benzothiazole ring, and the new Iridium (III) bis[2-phenylbenzothiazolato-N,C2']-(1,3-diphenylpropane-1,3-dionate) (bt)2Ir(dbm), where ancillary acetylacetonate ligand was replaced by 1,3-diphenylpropane-1,3-dionate, were synthesized and characterised by 1H-NMR and elemental analysis. The HOMO/LUMO energy levels of the complexes were determined by cyclic voltammetry (CV) and their properties were established by UV-Visible and fluorescence spectroscopy. The application of (Clbt)2Ir(acac), (bt)2Ir(bsm) and (bt)2Ir(acac) as dopants in hole transporting layer (HTL) of Organic light-emitting diodes(OLEDs). It was found that with respect to the reference (bt)2Ir(acac): both LUMO and HOMO of the substituted complexes were shifted to more positive values accordingly with 0.23 and 0.19 eV for (Clbt)2Ir(acac) and 0.14 and 0.12 eV for (bt)2Ir(dbm). OLEDs doped with 1 w% of the complexes irradiated the warm white light with Commission internationale de l'éclairage (CIE) coordinates: 0.24;0.38 for (Clbt)2Ir(acac), 0.30;0.44 for (bt)2Ir(acac) and 0.28;0.46 for (bt)2Ir(dbm). Devices doped with 10 w% of all complexes irradiated in the yellow orange region of the spectrum.

1. Introduction

Heavy metal complexes are of exceptional interest due to their applications in OLEDs. Efficient intersystem crossing caused by heavy metal ions in these complexes removes the spin-forbidden nature of the radiative relaxation of the triplet excited state, thus, both the electrogenerated singlet and triplet excited states contribute to light emission and can realize a theoretical internal quantum efficiency of nearly 100% [1, 2]. Among them, cyclometalated iridium(III) complexes are the most promising because of their quasi-octahedral geometry permitted introducing of specific ligands in a controlled manner, good stability, high photoluminescence (PL) quantum yields and short triplet state lifetimes. The yellow-emitting Ir(III) complexes are of great interest in view of the fact that in combination with blue emitters, they can be used for fabrication of WOLEDs. The design and synthesis of new organic ligands for Ir complexes is highly desirable in order to improve the efficiency and the color purity [3]. The photophysical properties of Ir phosphors can be tuned by varying the ligands [4], its substituent groups [2, 5, 6] and/or ancillary ligands [7, 8]. Such complexes are normally used as emitting guests in blends with host materials, where hole-transporting hosts include carbazole derivatives such as CBP (4, 4′-*N*,*N*′-dicarbazole)biphenyl and PVK [poly(9-vinylcarbazole)] [1, 9-11] or polyfluorenes [12, 13]. Researchers from Dupont de Nemours et al. have synthesized and characterized a series of iridium (III) complexes with fluorinated 2-arylpyridines and showed that the emissive colors of the materials can be finely tuned by systematic control of the nature and position of the substituents of the ligands [14]. 2-Phenylbenzo[d]thiazole (bt) is a general ligand framework to construct Ir(III) complexes such as yellow $(bt)_2Ir(acac)$, which is particularly promising to develop two-color white OLEDs with simple architectures [16, 17]. Up to now, many $(bt)_2Ir(acac)$ derivatives by structure modifications on bt ligand with adjustable color, high luminescent efficiency and electroluminescent performances have been synthesized and investigated [4, 17-18].

These research activities inspired us to initiate a study to investigate the role played from two different substituents chlorine and 1-Phenyl-eicosane-1,3-dione placed on two different position in $(bt)_2Ir(acac)$: in benzothiazole ligand and instead of ancillary acetylacetonate ligand.

2. Experimental

2.1. Synthes of Iridium complex

The cyclometalated Ir(iii) complexes used in this study (figure 1) have been synthesized according a method reported by Nonoyama [19]. The details of the synthesis, the ^{1}H-NMR spectra and elemental analysis of the complexes are described elsewhere [20, 21].

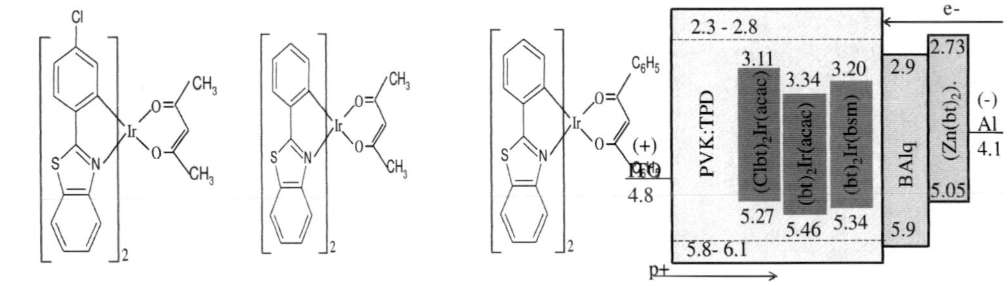

$(Clbt)_2Ir(acac)$ $(bt)_2Ir(acac)$ $(bt)_2Ir(dbm)$

Figure 1. Chemical structure of used Ir(III) complexes and energy diagram of investigated OLED.

2.2. *OLED fabrication*

We investigated $(Clbt)_2Ir(acac)$, $(bt)_2Ir(acac)$ and $(bt)_2Ir(dbm)$ as a dopants into a HTL in multilayered OLED structure: ITO/doped-HTL/EL/ETL/M. ITO was a transparent anode of In_2O_3:SnO_2, M - a metallic Al cathode, HTL - 4,4'-Bis(9H-carbazol-9-yl)biphenyl (CBP) incorporated in poly(N-vinylcarbazole) (PVK) matrix, EL - electroluminescent layer of Bis(8-hydroxy-2-methylquinoline)-(4-phenylphenoxy) aluminum (BAlq) and ETL - electron-transporting layer of Bis[2-(2-benzothiazoly)phenolato]zinc(II) $(Zn(bt)_2)$. This structure was chosen in order of obtaining a white light by mixing of red emission from the dopants and blue emission from the EL of BAlq. Devices with area 1cm^2 were prepared on commercial polyethylene terephtalate (PET) substrates coated with ITO (40Ω/sq). The layer (30nm) of PVK:$CBP_{10\%}$(relatively to PVK) + Ir(III) complex composite films were prepared by spin-coating from 0.75% solution in dichloroethane at 2000 rpm. All organic layers: BAlq (40nm), $Zn(bt)_2$ (35nm) and the Al cathode (100nm) were deposited in succession by thermal evaporation in vacuum better than 10^{-4} Pa at rates 2-5 Å/s, without breaking vacuum. The layers thicknesses were controlled with quartz crystal microbalance sensor, positioned near the PET/ITO substrate.

2.3. Instruments and measurements

The UV and fluorescence spectra of the complexes in solutions were recorded on a Thermo Spectronic Unicam 500 spectrophotometer and a Varian Cary Eclipse fluorescence spectrophotometer, respectively. All measurements of OLED structures were performed at room temperature and under ambient atmosphere, without any encapsulation. The electroluminescent spectra (El) and CIE coordinates were obtained by Ocean Optics HR2000+ spectrometer. The current-voltage (I-V) curves were measured by programmable with Labview power supply. The luminescence (L) was determined in DC (direct current) mode and the light output was detected using a calibrated Hamamatsu silicon photodiode S2281-01. The electroluminescent efficiency (η_L) were calculated by equations (1) and used for quantifying the properties of the OLEDs.

$$\eta_L = L/I,\ \text{cd/A}, \qquad (1)$$

where L is the luminance (in cd/m^2), I is current density (in A/m^2).

The CV measurements were done on air in 0.001M solution of Ir complexes in acetonitryle in presence 0.1M $LiClO_4$ as supporting electrolyte, on air, in standard three-electrode electrochemical cell with platinum plates as working and counter electrodes and Ag/AgCl - as reference electrode, using potentiostat Gamry PC 3.

3. **Result and discussion**

3.1 Cyclic voltammetric measurements

CV was employed to investigate the electrochemical behaviour of the Ir complexes and estimation of their ionization potentials (Ip) and electron affinities (Ea) in accordance with the theory proposed by Pommerehne et al. [22].

$$I_p = E_{1/2}^{\text{ox}} + 4.8 \qquad (2)$$

$$E_a = E_{1/2}^{\text{red}} + 4.8 \qquad (3)$$

$$E_g = I_p - E_a, \qquad (4)$$

where: $E_{1/2}^{\text{ox}}$ and $E_{1/2}^{\text{red}}$ are electrochemical potentials over ferrocenium/ferrocene (Fc^+/Fc) reference (vs. Ag/AgCl), and 4.8 eV is the value of Fc with respect to zero vacuum level. I_p and E_a may be regarded as the HOMO and the LUMO and the electrochemical gap between them as the band gap (E_g) of the complex. E_g may be evaluated from the long-wavelength absorption edge pursuant to the theory reported by Burrows et al. [23] using the equation (5):

$$E_g = h.f = h.c\ /\ \lambda = 1241/\lambda, \qquad (5)$$

where: h is Planck's constant; c is the speed of light and λ is the wavelength in nm.

During the cathodic CV scan (anodic scan was compromised by the presence of oxygen), shown in figure 2, all complexes exhibited two reversible reduction waves with very close potentials (0.98 - 1.22 V vs Ag/AgCl). The first of them was taken for estimation of E_a and LUMO level of the complexes. The data of absorption-edges and electrochemical potentials $E_{1/2}^{\text{red}}$, determined from figures 3 and 2, and E_g, E_a and I_p, calculated according to Eqs. (5), (3) and (4), are summarized in table 1. Both substitutions: of H with electron withdrawing Cl in benzothiazole ligand and of two CH_3 groups with two electron withdrawing phenyl groups in ancillary acetilacetonate ligand, shifted the reduction potentials of $(Clbt)_2Ir(acac)$ and $(bt)_2Ir(dbm)$ to more negative values, in comparison with the base $(bt)_2Ir(acac)$, with 0.24 V and 0.15 V respectively. As, the values of absorption-edges were not significantly different from each other, the optical bandgaps were almost the same and as a result, both the LUMO and the HOMO of the complexes were shifted to more positive values with 0.23 and 0.19 eV for $(Clbt)_2Ir(acac)$ and 0.14 and 0.12 eV for $(bt)_2Ir(dbm)$.

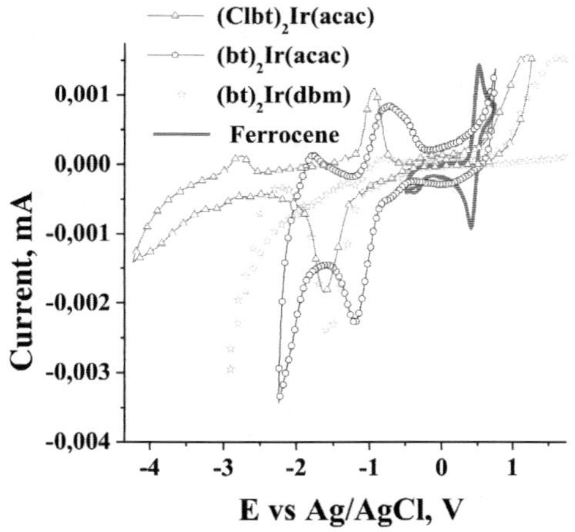

Figure 2. Reduction CV waves of 0.001 M Ir complexes and Ferrocene in acetonitryle and 0.1 M $LiClO_4$, at scan rate 100 mV/min.

Evidently the both substituents brought only a small difference to the HOMO and LUMO levels and the electronic behaviour of these iridium complexes

Table 1. Electrochemical data.

Complex	$-E_{1/2}^{red}$ vs. Ag/AgCl(V)	$-E_{1/2}^{red}$ vs. Fc/Fc$^+$(V)	Absorption edge(nm)	Eg (eV)	Ip(-HOMO) (eV)	Ea(-LUMO) (eV)
$(Clbt)_2Ir(acac)$	1.22	1.69	570	2.16	5.27	3.11
$(bt)_2Ir(acac)$	0.98	1.46	585	2.12	5.46	3.34
$(bt)_2Ir(dbm)$	1.13	1.60	580	2.14	5.34	3.20

3.2 UV-VIS absorption and photoluminescence

The UV-VIS absorption (curves 1-3) and PL spectra (curves 4-6) of three Ir(III) complexes in CH_2Cl_2 are shown in figure 3 and all data are summarized in table 2. The absorption spectra could be divided into two regions. The region below 360 nm contains bands for electronic transitions, which lead to excitations in the bt ligand. Most probably, the band around 330 nm is due to transition to lowest singlet ligand centered (^{1}LC) excited state. This is consistent with publications about similar complexes published before [4]. The shoulders above 360 nm are assigned to singlet and triplet metal-to-ligand charge transfer bands (^{1}MLCT, ^{3}MLCT), respectively. Effective spin-orbit coupling (SOC) in these complexes lead to mixing of the singlet and triplet states making the band for resonant transition $^1S_0 \rightarrow ^3$MLCT detectable in absorption spectrum (around 480 nm). The bands for ^{1}MLCT transition are centered at 440 nm. Under excitation at 330 nm light, $(Clbt)_2Ir(acac)$, $(bt)_2Ir(acac)$ and $(bt)_2Ir(dbm)$ exhibited strong yellow or orange phosphorescence with maximum emission peaks at 552, 558, 558 nm and shoulders at 590, 593, 600 nm, respectively, which could be attributed to predominantly ^{3}MLCT state. It is seen that the Pl spectrum of $(Clbt)_2Ir(acac)$ was blue-shifted with 6 nm with respect to the Pl spectrum of the $(bt)_2Ir(acac)$, that is typical for derivates containing electron withdrawing substituent in benzothiazole ring [24]. The Pl spectrum of $(bt)_2Ir(dbm)$ was almost the same with this of $(bt)_2Ir(acac)$, except that the shoulder at 600 nm was red shifted with 7 nm and more pronounced. The PVK and CBP (curves 7 and 8) emitted a blue light with peaks at 367 and 385 nm respectively.

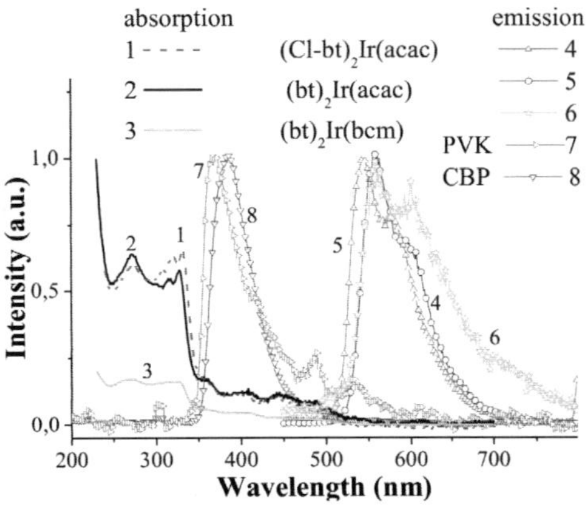

Figure 3. UV-VIS absorption and PL emission spectra of Ir complexes, PVK and CBP in CH_2Cl_2.

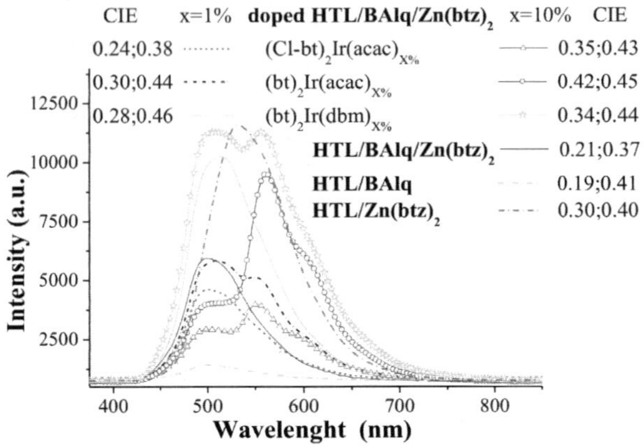

Figure 4. El spectra and CIE coordinates of devices doped with 1 and 10 w% of the dopants at DC 16 V.

The MLCT absorption bands of three Ir(III) complexes ranged from 360 nm to 480 nm, exhibiting good spectral overlap with the PL emission bands of PVK and CBP, which is a good prerequisite for the efficient Förster or Dexter energy transfer from the host (PVK:CBP) to the iridium complex guests would be expected.

Table 2. Photophysical data of $(Clbt)_2Ir(acac)$, $(bt)_2Ir(acac)$ and $(bt)_2Ir(dbm)$ in Ar-saturated CH_2Cl_2.

Complex	Absorbtion, λ[nm] (lgε)*	Emission λ_{max}/ *shoulder*[nm]
$(Clbt)_2Ir(acac)$	272(3.89), 320(3.92), 331(3.94),4 00(3.23), 442(3.18)	552/590
$(bt)_2Ir(acac)$	272(3.93), 315(3.86), 327(3.88), 07(3.21), 442(3.19), 489(3.03)	558/593
$(bt)_2Ir(dbm)$	269(4.23), 315(4.21), 327(4.21), 362(3.75), 407(3.64), 445(3.45), 492(3.25)	558/600

* ε is the molar absorption coefficient (molar absorptivity)

2.2. Electroluminescence

The electroluminescence (El) spectra of devices doped with three investigated complexes, at two different concentrations and theirs CIE coordinates, taken at 16 V DC, are shown in figure 3. For comparison, the El spectra of devices with undoped HTL and EL of BAlq, Zn(bt)2 and BAlq/Zn(bt)2 are presented as well. It is seen that the device with the EL of BAlq emitted at 496 nm, and that with Zn(bt)2 at 524 nm. The shape of the El spectra of OLED with successively evaporated layers of BAlq/Zn(bt)2 was just the same as that of the OLED with EL of BAlq. In our opinion, it is due to the fact that in both cases the recombination zone is near the HTL/BAlq interface. As the HOMOZn(bt)2 level (5,05 eV) is lower than the HOMOBAlq level (5,90 eV) (figure 1), Zn(bt)2 did not stop the holes, which flow unimpeded from the Zn(bt)2 layer. This is why, in the BAlq/Zn(bt)2 structure, Zn(bt)2 acted as an ETL only. The El spectra of the doped devices did not include any Zn(bt)2, PVK and CBP emission. The addition of iridium complexes into HTML caused the red shifting of the EL spectra of devices, in comparison with undoped device. As can be seen from the CIE coordinates and colour diagram, shown in set, at the low concentration (1 w%) of the dopants all three devices irradiated the warm white light (0.24;0.38 for (Clbt)2Ir(acac), 0.30;0.44 for (bt)2Ir(acac) and 0.28;0.46 for (bt)2Ir(dbm)), close to the absolute white (0.33, 0.33). At the higher concentration (10 w%) of the dopants all devices irradiated in the yellow orange region of the spectrum (0.35;0.43 for (Clbt)2Ir(acac), 0.42;0.45 for (bt)2Ir(acac) and 0.34;0.44 for (bt)2Ir(dbm)) and theirs El spectra were basically the sum of predominantly emissions of the dopants (at: 551 nm for $(Clbt)_2Ir(acac)$, 560 nm for $(bt)_2Ir(acac)$ and 556 nm for $(bt)_2Ir(dbm)$) and BAlq (at 496 nm).

The L-I-V characteristics and El efficiencies of base OLED structure HTL/BAlq/$Zn(bt)_2$ and doped with 1 and 10 w% of Ir complexes devices are shown in figures 5a, 5b and 5c. Inside in the figures are presented the voltages provided the luminance 500 cd/m^2 ($V^{w\%}_{500cd/m2}$) and the current density 200 A/m^2 ($V^{w\%}_{200A/m2}$), and El efficiencies at the luminance 500 cd/m^2 ($\eta^{w\%}_{500cd/m2}$).

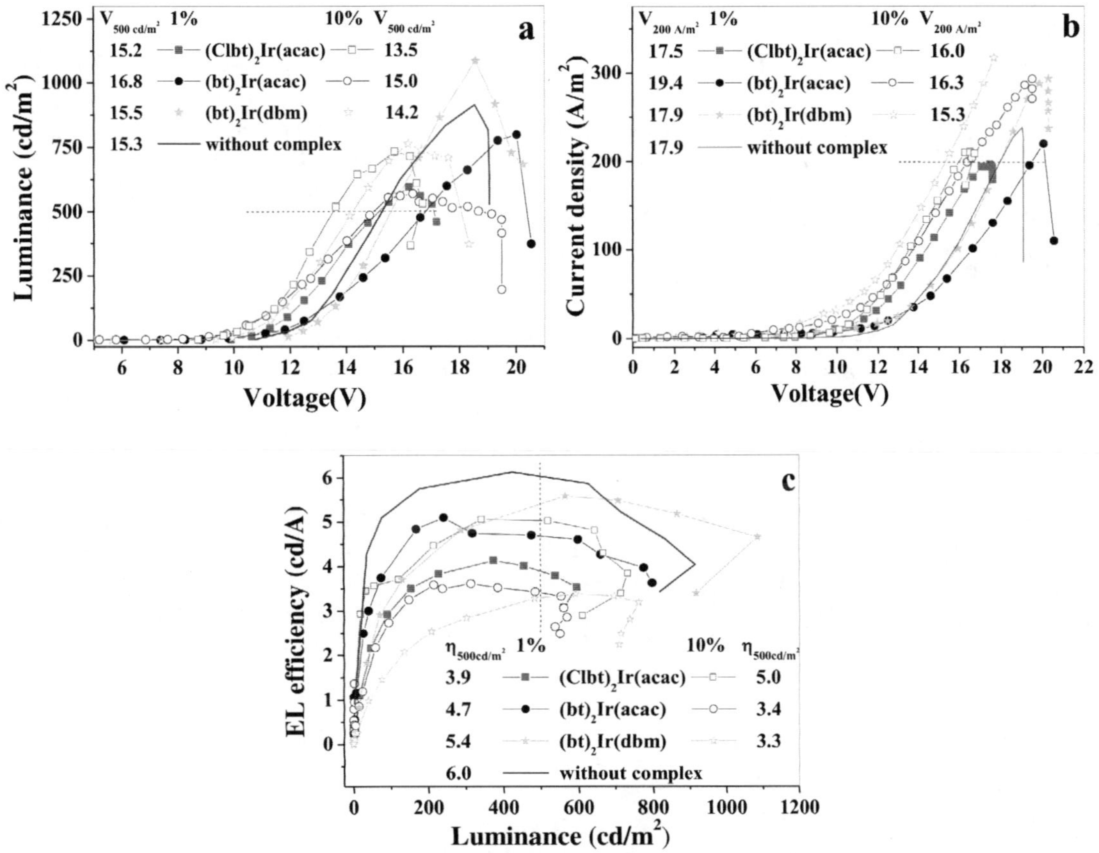

Figure 5. Luminance-voltage (a), Current density-voltage (b) and Electroluminescent efficiency – luminance (c) characteristics for undoped and doped with 1 and 10 w% of the complexes devices.

It is seen that the adding of all three dopants to the HTL of the base device reduced $V^{w\%}_{500cd/m2}$ and $V^{w\%}_{200A/m2}$ at both concentrations. The only exception was the device doped with 1 w% $(bt)_2Ir(acac)$, in which there was rise of $V^{1\%}_{500cd/m2}$ and $V^{1\%}_{200A/m2}$ in comparison with undoped device. At the same time the El efficiencies for all doped devices, at the both concentrations, were lower than this of the base structure. All devices doped with 10w% showed lower efficiencies than these of devices doped with 1w%, except device doped with $(Clbt)_2Ir(acac)$, at which $\eta^{10\%}{}_{500cd/m2} > \eta^{1\%}{}_{500cd/m2}$.

Electroluminescence in phosphorescent OLED comes either from the energy transfer from the host to the guest molecules, as described previously, or from the charge trapping by the guest molecules, where the dopant acts as a trap site, which can capture a charged carrier. The increasing of current densities and decreasing of OLED efficiency with increasing of the dopant concentration on the one hand and the absence of any PVK and BCP emission in El spectra of doped devices, even at low concentrations of the dopant, on the other hand, showed the charge trapping in the Ir complex, rather than the energy transfer, as the dominant mechanism in investigated OLEDs. Since the holes injected from the anode fall directly on the HOMO level of the dopant without any difficulties (figure 1), the trapping of electrons on the LUMO level is the process determining the intensity of electroluminescent emission of devices. Therefore the lower LUMO levels, respectively shallower electrons traps, of both substituted complex in comparison with this of reference $(bt)_2Ir(acac)$, are the reason for higher electroluminescence of devices with $(Clbt)_2Ir(acac)$ and $(bt)_2Ir(dbm)$.

4. Conclusion

We investigated the effect of Cl and 1,3-diphenylpropane-1,3-dionate as substituents placed in $(bt)_2Ir(acac)$. The present results show that the incorporation of an electron-withdrawing substituent either on the benzothiazole ring of the 2-phenylbenzothiazole ligand or in ancillary ligand, only has a subtle tuning effect on the energy bandgap so that the phosphorescence of substituted complexes is still located in the same orange range as that of the parent $(bt)_2Ir(acac)$. Both substituents shifted the HOMO/LUMO levels of the complexes to more positive values and thereby enhance the electroluminescence of OLEDs.

References

[1] Bauer R, Finkenzeller W, Bogner U, Thompson M and Yersin H, 2008 *Org. Electron.* **9** 641
[2] Wang R, Deng L, Zhang T and Li J 2012 *Dalton Trans.* **41** 6833
[3] Seo J, Seung Lee C, Kim Y and KimY 2008 *Thin Solid Films* **517** 1346
[4] Lamansky S, Djurovich P, Murphy D, Razzaq F, Lee H, Adachi C, Burrows P, Forrest S and Thompson M 2001 *J. Am. Chem. Soc.* **12** 4304
[5] Zhang L, Li L, Shi L and Li W 2009 *Opt. Mater.* **31** 905
[6] Choi J, Jung C, Kwon J, Cho H, Lee J, Lee J-Ik, Chu H and Hwang D 2009 *Synth. Met.* **159** 1517
[7] Liu Z, Bian Z, Ming L, Ding F, Shen H, Nie D and Huang C 2008 *Org. Electron.* **9** 171
[8] Hu J, Zhang J, Shih H, Jiang X, Sun P and Cheng C 2008 *J. Organomet. Chem.* **693** 2798
[9] Tavasli M, Moore T, Zheng Y, Bryce M, Fox M, Griffiths G, Jankus V, Al-Attar H and Monkman A 2012 *J. Mater. Chem.* **22** 6419
[10] Kawamura Y, Goushi K, Brooks J, Brown J, Sasabe H and Adachi C 2005 *Appl. Phys. Lett.* **86** 071104
[11] Adachi C, Baldo M, Forrest S, Lamanski S, Thompson M and Kwong R 2001 *Appl. Phys. Lett.* **78** 1622
[12] Ha S and Noh Y 2013 *Jpn. J. Appl. Phys.* **52** 10MB11
[13] Chang J, An J and Im Ch 2005 *J. of Korean Phys. Soc.* **47** 1028
[14] Grushin V, Herron N, Lecloux D, Marshall W, Petrov V and Wang Y 2001 *Chem. Commun. (Cambridge)* 1494
[15] Chen P, Xie W, Li J, Guan T, Duan Y, Zhao Y, Zhang L and Li B 2007 *Appl. Phys. Lett.* **91** 023505
[16] Zhao B, Su Z, Li W, Chu B, Jin F, Yan X, Zhang F, Fan D, Zhang T, Gao Y and Wang J 2012 *Appl. Phys. Lett.* **101** 053310
[17] Liu D, Ren H, Deng L and Zhang T 2013 *ACS Appl. Mater. Inter.* **5** 493744
[18] Li M, Zeng H, Meng Y, Sun H, Liu S, Lu Z, Huang Y and Pu X, 2011 *Dalton Trans.* **40** 7153

[19] Nonoyama M 1974 *Bull. Chem. Soc. Jpn.* **47** 767
[20] Ivanov P, Tomova R, Petrova P, Stanimirov S and Petkov I 2012 *J. Phys. Conf. Ser.* **398** 012052
[21] Ivanov P, Tomova R, Petrova P, Stanimirov S and Petkov I 2014 *J. Phys. Conf. Ser.* **514** 012038
[22] Pommerehne J, Vestweber H, Guss W, Mahrt R, Bässler H, Porsch M and Daub J 1995 *Adv. Mater.* **7** 551
[23] Burrows P, Chen Z, Bulovic V, McCarty D and Forrest S 1996 *J. Appl. Phys.* **79** 7991
[24] Adachi C, Baldo M, Forrest S and Thompson M 2000 *Appl. Phys. Lett.* **78** 1704

Journal of Physics: Conference Series **558** (2014) 012044 doi:10.1088/1742-6596/558/1/012044

Ion implantation induced modification of ta-C films

M Sandulov, M Berova and T Tsvetkova*

Institute of Solid State Physics, Bulgarian Academy of Sciences, 72 Tzarigradsko Chaussee, 1784 Sofia, Bulgaria

*E-mail: tania_tsvetkova@yahoo.co.uk

Abstract. Thin film samples (d~40nm) of tetrahedral amorphous carbon (ta-C), deposited by filtered cathodic vacuum arc (FCVA), have been implanted with N^+ at ion energy E = 20 keV and ion dose $D = 3.10^{14}$ cm^{-2}. The induced structural modification of the implanted material results in a considerable change of its optical properties, best manifested by a significant shift of the optical absorption edge to lower photon energies as obtained from optical transmission measurements. This shift is accompanied by a considerable increase of the absorption coefficient (photo-darkening effect) in the measured photon energy range (0.5÷3.0 eV). These effects could be attributed both to additional defect introduction and increased graphitization, as confirmed by IR and Raman measurements. The optical contrast thus obtained (between implanted and unimplanted film material) could be made use of for information archiving, in the area of high-density optical data storage while using focused ion nano-beams.

1. Introduction

Thin film materials of tetrahedral amorphous carbon (ta-C) have attracted great interest for a long time from both scientific and industrial perspectives. The term tetrahedral is used to describe amorphous carbon films with a large percentage of *sp*3 bonding (up to 87%). The films are manufactured using a variety of techniques, including filtered cathodic vacuum arc (FCVA), pulsed laser deposition (PLD) and mass selected ion beam deposition (MSIBD) [1-4]. The high *sp*3 content in the films results in unique properties that include extreme hardness (~70 GPa), chemical inertness, high electrical resistivity, and wide optical band gap [5–8]. Other important factors which make the films an attractive material for coatings include a smooth surface and low friction, thermal stability and transparency over a wide spectral range. These properties also offer advantages as compared to another wide optical bandgap material – silicon carbide (SiC) – for uses in nano-scale optical data recording for archival information storage using focused ion beams (FIB) techniques, where SiC thin films have found useful applications recently [9-16].

In the case of polycrystalline silicon carbide (pc-SiC) thin films, ion bombardment is used to amorphise areas of the films by computer operated FIB systems, thus creating useful optical contrast between non-irradiated polycrystalline areas and the irradiated amorphous areas, which can be further used for nano-scale optical data recording for archival information storage [9-11]. In the case of hydrogenated amorphous silicon carbide (a-SiC:H) films, computer operated FIB systems are used to both introduce irradiation defects and additionally chemically modify the amorphous structure of the films, thus reducing their optical bandgap in even a more effective manner for the useful creation of optical contrast between implanted and non-implanted areas of the film material for applications in nano-scale optical data recording [12-16].

In both polycrystalline and amorphous SiC film materials, a considerable part in the creation of useful optical contrast between irradiated and non-irradiated areas of the films is played by the transformation of substantial part of the present diamond-like (sp3) carbon bonds, before the irradiation, into graphite-like (sp2) carbon bonds, as a result of it [11,16]. It is expected, that a similar mechanism of the carbon bonds transformation would result when applying ion bombardment with different ions, e.g. nitrogen ions (N^+), in ta-C films, so that to achieve useful optical contrast between irradiated and non-irradiated areas of the films as reported in a previous publication [17], which could possibly be of interest for further uses in archival information storage.

2. Experimental

Thin ta-C films (d~40 nm) were deposited on Corning glass substrates using a commercial FCVA system (Commonwealth Scientific Corporation). Carbon plasma is produced from the arc spot on the cathode, 99.999% pure graphite in high vacuum. Cathodic arcs are prolific generators of highly ionized carbon plasmas. With the FCVA technique, the plasma stream is steered through a magnetic filter to eliminate neutral particles generated at the cathode. At the filter exit, the fully ionized plasma, consisting of carbon ions and electrons, streams towards the substrate. The films were deposited at room temperature with an arc current of 120 A under floating conditions.

Ion implantation of N^+ was carried out at room temperature (RT) using a commercial broad-beam ion implanter. The ion-beam intensity was I~2 μA/cm^2, the ion energy was E=20 keV, and the ion dose used was D=3.10^{14} cm^{-2}. SRIM simmulation programme [18] was used to determine the projected range Rp = 29 nm and the struggle ΔRp = 10 nm of the implanted N^+ ions into the ta-C film samples (d = 40 nm).

Atomic force microscope (AFM) - Dimension 3000 Digital Instruments - was used to analyze the topography of the N^+ ion implanted samples.

The structural modifications of the ta-C films by the N^+ ion implantation were monitored by Raman spectroscopy using Horiba Jobin-Yvon LabRam confocal Raman microscope fitted with an argon ion laser. The spectra were taken using a 100× objective lens, an excitation wavelength (λ) of 532 nm and a spot size of ~ 0.70 μm at 10% of total laser power. The laser exposure time was 30 s and the collected spectra were averaged three times with a dwell time of 1 s.

Fourier Transform Infrared (FTIR) spectroscopy measurements were done by Shimadzu FTIR Spectrophotometer IRPrestige-21 in the spectral range 350-4000 cm^{-1} in reflectance mode (specular reflectance attachment), Al mirror. The samples were measured in absorbance.

3. Results and discussion

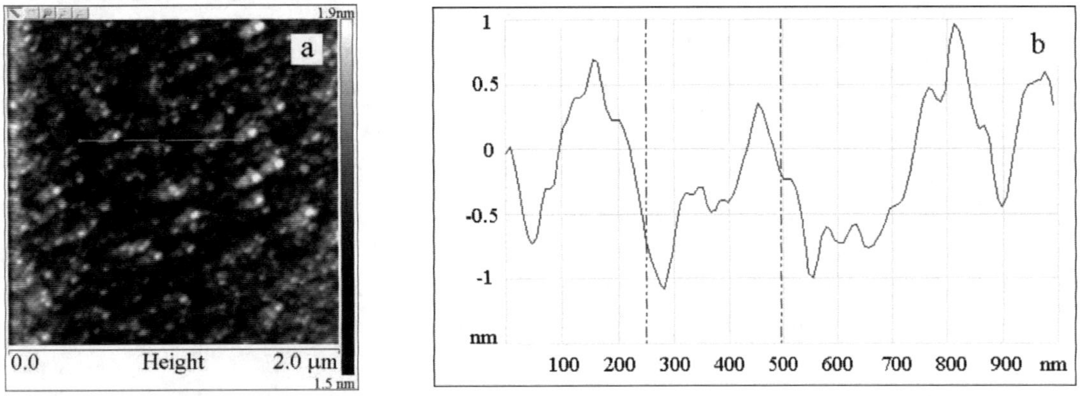

Figure 1. AFM results on N^+ ion implanted ta-C sample: (a) Topography image; (b) RMS roughness is less than 2nm at different areas of the samples.

Journal of Physics: Conference Series **558** (2014) 012044 doi:10.1088/1742-6596/558/1/012044

3.1. AFM spectroscopy results on ta-C samples implanted with N^+ broad-beam implantation The N^+ ion implanted samples topography was studied by AFM spectroscopy both before and after the ion implantation. In both cases surface roughness (RMS) is less than 2 nm, as shown in figure1 above.

3.2. Raman spectroscopy results on ta-C samples implanted with N^+ broad-beam implantation
The useful optical contrast, between irradiated and non-irradiated areas of the films, which was reported in a previous publication [17], which could possibly be of interest for further uses in archival information storage, is expected to be resulting from the ion beam induced structural modification leading to increased optical absorption in the visible range of the spectrum.

The possible mechanism is presumed to be that the nitrogen ion irradiation, with relatively low doses ($D \sim 3\times10^{14}$ cm^{-2}), induces structural modifications by increased defects introduction and converting sp3-bonded carbons into sp2-bonded ones in ta-C films, as confirmed by the Raman results shown in figure 2. The Raman spectra show a distinctive change after the N^+ implantation, namely the ratios of D and G bands [I (D)/I (G)] decreases drastically, so that in some cases is leading to complete disappearance of the D band. The *D* peak arises from aromatic rings. *I(D)/I(G)* will increase with increasing disorder in graphite with the presence of such rings. However, for more disorder, clusters decrease in number become smaller and more distorted, until they open up [19]. Analysis of the G band enhancement implies clustering of the sp2-bonded carbon aromatic rings with implanting N ions.

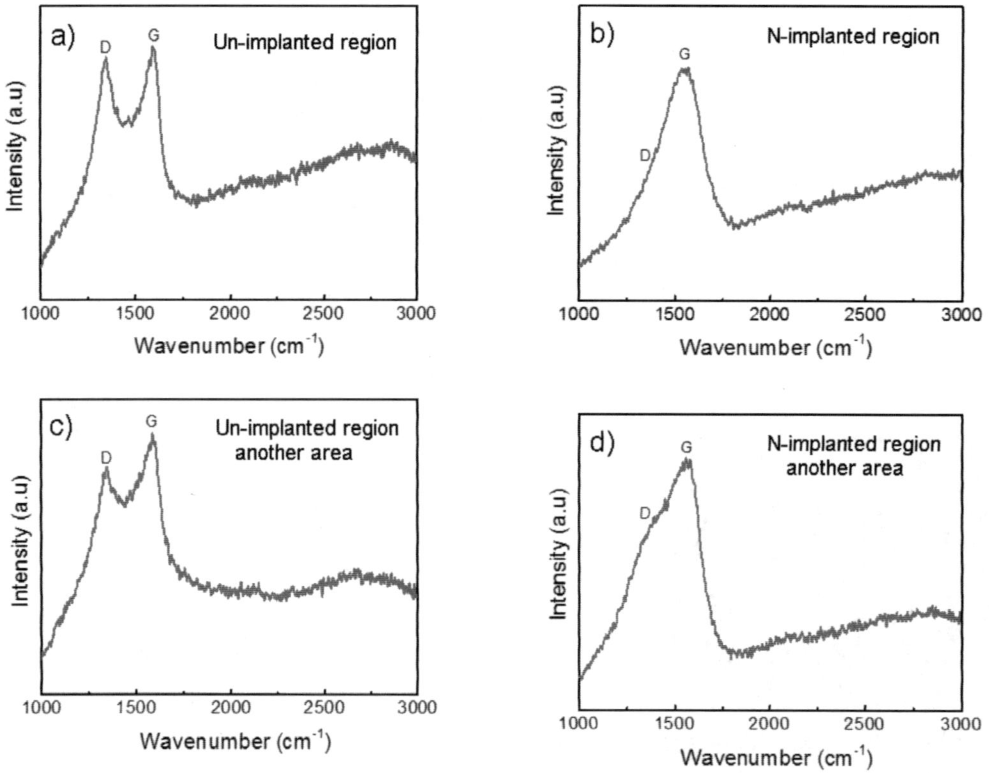

Figure 2. Raman results on ta-C samples: unimplanted (a,c) and N^+ ion implanted (b,d).

3.3. Infra-red (IR) spectroscopy results on ta-C samples implanted with N^+ broad-beam implantation

The ion-implanted ta-C samples with N^+ broad-beam implantation with a dose $D = 3\times10^{14}$ cm^{-2} were studied by infra-red (IR) spectroscopy in the range 1000÷2000 cm^{-1}, where C-C bond vibrations are expected at about 1358-1569 cm^{-1} [20]. The obtained IR spectroscopy results are displayed in figure 3, where the sp^2 C-configuration at ~1550 cm^{-1} is shown to be slightly increasing as a result of the N^+ ion implantation, confirming the Raman data presented in figure 2.

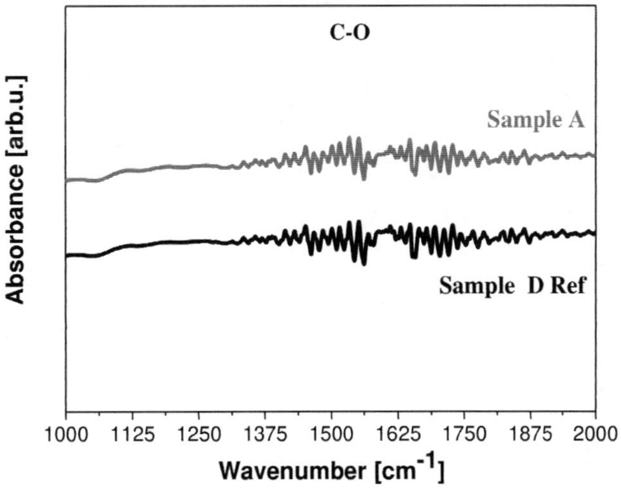

Figure 3. IR results on ta-C samples: unimplanted (D) and N^+ ion implanted (A).

4.Conclusion

$N+^+$ broad-beam ion implantation in ta-C samples was carried out at room temperature at a relatively low dose, which has been shown previously to lead to a useful optical contrast between irradiated and non-irradiated areas of the films, which could possibly be of interest for further uses in archival information storage. This optical contrast has been confirmed to be resulting from the ion beam induced structural modification, including increased defect intriduction and enhanced graphitisation, i.e. sp3 to sp2 conversion of the C-C bonds, leading to increased optical absorption in the visible range of the spectrum.

Acknowledgements

The authors would like to thank Dr. K.Nagareddy from the University of Exeter, UK, for the help with the AFM and Raman measurements. The help of Dr. K.Gesheva from the Bulgarian Academy of Sciences, Bulgaria, with the measurement of the IR spectra is also gratefully acknowledged.

References

[1] McKenzie D, Muller D and Pailthorpe B A 1991 *Phys. Rev. Lett.* **67** 773
[2] Fallon P J, Veerasamy VS, Davis CA, Robertson J, Amaratunga G A J, Milne W I and Koskinen J 1993 *Phys. Rev.* B **48** 4777
[3] Friedmann T A, McCarty KF, Barbour J C, Siegal M P and Dibble D C 1996 *Appl. Phys. Lett.* **68** 1643
[4] Lifshitz Y, Kasi S R and Rabalais J W 1989 *Phys. Rev. Lett.* **62** 1290
[5] Silva S R P, Xu S, Tay B K, Tan H S, Scheibe H J, Chhowalla M and Milne W I 1996, *Thin Solid Films* **290–291** 317

Journal of Physics: Conference Series **558** (2014) 012044 doi:10.1088/1742-6596/558/1/012044

[6] Veerasamy V S, Yuan J, Amaratunga G, Milne W I, Gilkes K W R, Weiler M and Brown L *Phys. Rev. B* 1993 **48** 17954
[7] Robertson J 1991 *Solid State Chem.* **21** 199
[8] Xu S, Flynn D, Tay B K, Prawer S, Nugent K W, Silva S R P, Lifshitz Y and Milne W I 1997 *Philos. Mag. B* **76** 351
[9] Ruttensperger B, Krötz G, Müller G, Derst G and Kalbitzer S 1991 *J. Non-Cryst. Solids* **137-138** 635
[10] Müller G 1993 *Nucl Instrum Methods B* **80-81** 957
[11] Kalbitzer S 2004 *Nucl Instrum Methods B* **218** 343
[12] Tsvetkova T 1996 Ion Beam Modification of Amorphous Silicon-Carbon Alloys *Beam Processing of Advanced Materials*, ed J Singh, S Copley and J Mazumder (ASM International, Materials Park) pp 207-2018
[13] Tsvetkova T, Takahashi S, Zayats A, Dawson P, Turner R, Bischoff L, Angelov O and Dimova-Malinovska D 2005 *Vacuum* **79** 94
[14] Tsvetkova T, Takahashi S, Zayats A, Dawson P, Turner R, Bischoff L, Angelov O and Dimova-Malinovska D 2005 *Vacuum* **79** 100
[15] Takahashi S, Dawson P, Zayats AV, Bischoff L, Angelov O, Dimova-Malinovska D, Tsvetkova T and Townsend P 2007 *J. Phys. D: Appl Phys.* **40** 7492
[16] Tsvetkova T, Sellin P, Carius R, Dimova-Malinovska D and Angelov O 2007 *J. Optoel. Adv. Mater.* **9** 375
[17] Sandulov M, Berova M, Tsvetkova T and Zuk J 2014 *Acta Physica Polonica* in press
Ziegler J F, Biersack J P and Littmark U 1985 *The Stopping and Range of Ions in Matter* vol 1 (New York: Pergamon)
[18] Ferrari A C and Robertson J 2000 *Physical Review B* **61** 14095
[19] Ferrari A C, Rodil S E and Robertson J 2003 *Physical Review B* **67** 155306

Spectroscopic studies of SiO_x films irradiated with high energy electrons

V Dzhurkov[1], D Nesheva[1*], M Scepanovic[2], N Nedev[3], S Kaschieva[1], S N Dmitriev[4] and Z Popovic[2]

[1]Institute of Solid State Physics, Bulgarian Academy of Sciences, 72 Tzarigradsko Chaussee Blvd, 1784 Sofia, Bulgaria
[2] Center for Solid State Physics and New Materials, Institute of Physics, University of Belgrade, Pregrevica 118, Belgrade 11080, Serbia
[3]Institute of Engineering, Autonomous University of Baja California, Benito Juarez Blvd. s/n, C.P.21280, Mexicali, B.C., Mexico
[4]Joint Institute for Nuclear Research, Flerov Laboratory of Nuclear Reactions Dubna, Moscow region 141980, Russia

E-mail: nesheva@issp.bas.bg

Abstract. The effect of 20 MeV electron irradiation on the vibrational properties of homogeneous SiO_x films and SiO_x films containing amorphous or crystalline Si nanoparticles is studied. Layers with $x = 1.15$ and 1.3 and two different thicknesses, 200 nm and 1000 nm, were deposited on crystalline silicon substrates and annealed at 250 °C for stabilization. No nanoparticle growth occurs at this temperature. A part of the films were annealed in an inert atmosphere at 700 °C or 1000 °C for 60 min in order to grow amorphous (a-Si) or crystalline (nc-Si) silicon nanoparticles, respectively. Samples from all types of the films were irradiated with 20 MeV electrons at close to room temperature and a fluence of 2.4×10^{14} el.cm^{-2}. The effect of the irradiation was investigated by means of Fourier Transform Infrared Absorption (FTIR) and Raman scattering spectroscopy. The FTIR results show no appreciable changes in the composition of the homogeneous SiO_x films and the oxygen content in the matrix of a-Si-SiO_x composite films. The electron irradiation causes an increase of the intensity of the light scattered from the pure silicon phase. The results obtained are discussed in terms of disorder decrease in the Si nanoparticles stimulated by MeV electron beam irradiation.

1. Introduction

The radiation induced changes of electrical properties of solid-state materials and devices are well-known phenomena. Such changes may cause an electronic device or system to fail and therefore during the last few decades considerable activity has taken place to explore such alterations in various types of semiconductor devices and in particular in those based on metal-oxide-silicon structures [1]. The nature of the interaction between the impinging particle and the target depends on the particle mass, charge, and kinetic energy. Target properties of importance are mass, charge, impurity and density. When electrons pass through matter, several possible processes occur, including ionization in which moving particle loses energy in small steps through interactions with the electrons in the material through which it passes. Once the particle loses enough energy, such that it no longer has sufficient energy to excite an electron, then it may lose energy by nuclear collisions. Further, creation of X-rays from Bremsstrahlung (electron deceleration) and elastic scattering from nuclear and electronic interactions take place. The path of an electron is zigzagged i.e. it is not well defined and an energy of MeV is high enough for electrons to penetrate through the whole Si - SiO_2 samples and to create radiation defects (electrons and holes) in the oxide, at the Si - SiO_2 interface and in the Si wafers [2, 3]. Annealing of ion-irradiation created defects has also been observed [4].

After creating defects the intense electron beam irradiation of solid state dielectrics and semiconductors can further form nanoclusters. The interest in formation of such nanosized particles in silicon oxide films is conditioned by the prospect of development of new functional materials and devices suitable for usage in intense radiation fields in spacecraft, medicine, nuclear power plants, etc. It has been reported that the electron beam in the transmission and scanning electron microscopes (TEMs and SEMs) can induce formation of nanostructures [5, 6]. Formation of Si nanocrystals (NCs) in a SiO_2 thin film matrix at 560 °C by irradiation of a high intensity convergent electron beam in an ultrahigh vacuum field emission transmission electron microscope using acceleration voltage of 100 kV has been reported in [7]. For Ge it was demonstrated that nanocluster formation occurs at room temperature under irradiation with 400 kV electrons and a fluence higher than 3.75×10^{22} el.cm^{-2}. [8]. The authors have also demonstrated that ripening takes place at temperatures far below the activation threshold of the corresponding thermal process, which is about 1000 K. Phase separation in SiO_x films [9] as well as defect production and annealing in high energy ion-irradiated Si nanocrystals [10] have also been investigated under irradiation with high-energy (50 MeV) ion beams. To the best of our knowledge, the information on the effect of MeV electrons on the lattice structure of Si nanoparticles (crystallization/amorphization, level of disorder) is exiguous.

In this work SiOx films with x = 1.15 and 1.3 were deposited on crystalline silicon substrates and then annealed at high temperatures to prepare composite films containing amorphous or crystalline Si nanoparticles in a SiOy matrix; y = 1.8 - 1.9 or 2, respectively. Both types of films were irradiated with 20 MeV electrons in vacuum and the effect of the film irradiation on the vibrational properties of the film matrix and Si nanoparticles was investigated by means of Fourier Transform Infrared (FTIR) absorption and Raman scattering spectroscopy, respectively. The obtained results are discussed in terms of MeV electron beam stimulated ordering of the Si nanoparticles.

2. Experimental details

SiOx layers with an initial composition of $x = 1.15$ and 1.3 and thickness of 200 and 1000 nm were prepared by thermal evaporation of SiO at a vacuum of 1 × 10−3 Pa on p-type (100) c-Si substrates ((7.2 – 10) Ω×cm) maintained at room temperature. The film composition has been set by the applied preparation conditions and proven by means of Rutherford Back Scattering [11]. The SiO evaporation was carried out from a tantalum crucible situated at the base of a molybdenum cylindrical screen [11]. The film thickness and deposition rate were monitored by a calibrated quartz microbalance system. Prior to film deposition the silicon wafers were cleaned chemically using a standard procedure for the microelectronics industry. All as-deposited layers were annealed at 250 oC for 30 min in an Ar atmosphere to keep them stable at room conditions. No formation of pure silicon phase occurs during this annealing procedure [11]. In order to grow Si nanoparticles an additional furnace annealing at 700 °C in Ar or 1000 °C in N_2 atmosphere for 60 min was carried out. Previous high resolution electron microscopy and Raman scattering measurements have shown the formation of amorphous silicon (a-Si) nanoparticles in films annealed at 700 °C and Si nanocrystals (nc-Si) in films annealed at 1000 °C [12, 13].

Both homogeneous SiO_x and composite films (a(nc)-Si-SiOx) containing nanosized amorphous or crystalline particles were irradiated with 20 MeV electrons at approximately 30 °C. Electron irradiation with a fluence of 2.4×1014 el.cm^{-2} was carried out in Microtron MT-25 in FLNR, Joint Institute for Nuclear Research, Dubna. No significant temperature increase was observed during irradiation.

The Raman scattering measurements of non-irradiated and electron irradiated films were performed in the backscattering geometry at room temperature in the air using a TriVista T557 triple spectrometer with a 900/900/1800 grooves/mm gratings combination. The spectrometer was equipped with a confocal microscope and a nitrogen-cooled charge coupled device detector. Two lines of a mixed Ar^+/Kr^+ gas laser with wavelengths of 514.5 or 488 nm and output power of about 50 mW were used for excitation. Fourier Transform infrared Transmission (FTIR) spectra of the films were measured using a Perkin Elmer Spectrum One FTIR spectrometer in transmission mode in the 450 – 4000 cm^{-1} range with a resolution of 2 cm^{-1}.

3. Results and Discussion

Experimental FTIR spectra of non-irradiated and 20 MeV electron irradiated 200 and 1000 nm thick SiO_x films with $x = 1.3$ annealed at 250 °C (not containing Si nanoparticles) are shown in figure 1. The stretching vibration band of the 1000 nm films is significantly broader than that of the 200 nm ones which is most likely due to some compositional variations in the film depth. Similar IR spectra were recorded from the $x = 1.15$ films. It is known that the main band in the IR spectra of SiO_2 is observed at ~ 1080 cm^{-1}. It is due to an asymmetric stretching vibration of the Si–O–Si bridge and consists of one transverse optical (TO) mode at 1080 cm^{-1} and others longitudinal optical (LO) modes in the range 1200 - 1300 cm^{-1} [14, 15]. With a decrease of the oxygen content ($x < 2$) the position of the peak is shifted to lower frequencies ('red' shift) with respect to silica, the characteristic LO shoulder at the high-frequency side of the band disappears and the full-width at half maximum (FWHM) of the band increases. The band position (~ 1014 cm^{-1}) seen in figure 1 is in good agreement with our previous results for $x = 1.3$ films [11]. No significant electron beam induced alterations in the band position and shape are observed which indicates that the film composition has not been changed upon the applied electron irradiation of 2.4×10^{14} $el.cm^{-2}$. Hence based on the IR transmission data one can conclude that though the energy of irradiating electrons is rather high no appreciable electron beam induced phase separation occurs in the homogeneous SiO_x films. This observation differs from the result reported in [9] showing that it takes a fluence of about 10^{14} $ions.cm^{-2}$ 50 MeV Cu ions to obtain a Si-O-Si infrared absorption spectrum characteristic for a Si-SiO_2 composite film structure. Rather, our result is in line with the data of other authors for keV electrons in TEMs showing that high electron fluencies and an increased film temperature (above 500 °C) are required during irradiation [7, 8] to cause phase separation in Si-rich silicon oxide films and formation of pure silicon phase.

Figure 2 shows IR transmission spectra of a homogeneous x = 1.3 film annealed at 250 °C and a pair of non-irradiated and 20 MeV electron irradiated composite films with the same initial composition containing a-Si nanoparticles. The film thickness is 200 nm. As expected [11] the furnace annealing at 700 °C causes both a "blue" shift of the stretching vibrational band from ~ 1014 cm^{-1} to ~ 1060 cm^{-1} and a FWHM decrease; the shoulder at ~ 1200 cm^{-1} is also better seen. This indicates an increase in the oxygen content of the matrix due to formation of pure silicon phase but the oxygen content in the matrix is still below the stoichiometric proportion. No appreciable changes in the band position and FWHM related to the electron irradiation are observed in figure 2. The $x = 1.15$ films showed the same behavior thus implying that the electron irradiation does not cause further phase separation in the composite films and does not create a high defect density in the matrix which should result in FWHM increase.

The described IR transmission data give information mainly about the composition and the lattice disorder in the silicon oxide films but they do not provide information about the relatively small amount of pure silicon phase formed due to phase separation in SiO_x films. Raman spectroscopy can provide a fast and nondestructive method to detect the existence of silicon phase and to determine whether silicon particles are amorphous or crystalline [16, 17]. The ''finite size effects,'' which destroy the full translation symmetry of a crystalline material, result in a "red shift" and low-frequency asymmetric broadening of the most intensive silicon Raman band at about 520 cm^{-1} [18].

All experimental Raman spectra presented in this paper have been normalized with respect to the band located between 930 and 1030 cm^{-1} after subtraction of a linear baseline. This band was assigned to multi-phonon scattering coming from the c-Si silicon substrate [19, 20] and it was assumed that its intensity had almost the same value for the films with a given thickness and composition (which

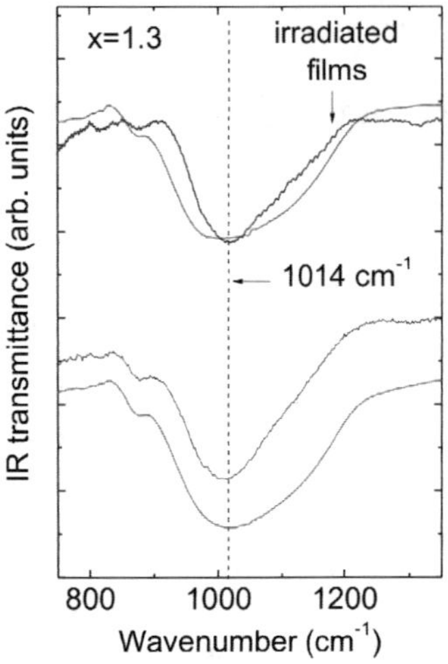

Figure 1. IR transmission spectra of pairs of 200 nm (the spectra with narrower band) and 1000 nm thick SiO_x films with $x = 1.3$. All films were annealed at 250 °C for 30 min in argon and a film from each pair was then irradiated with 20 MeV electrons.

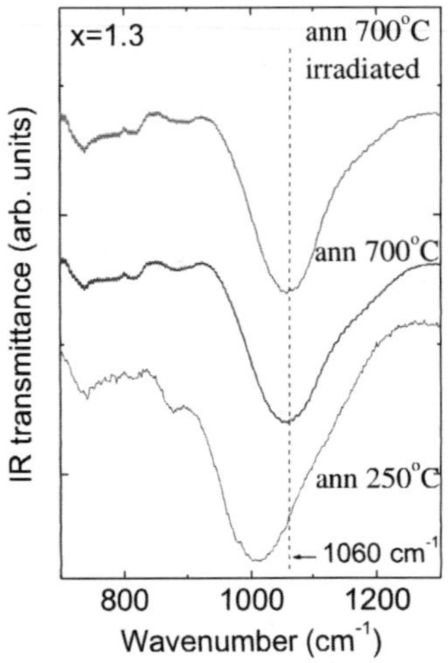

Figure 2. IR transmission spectra of a homogeneous $x = 1.3$ film and a pair of non-irradiated and 20 MeV electron irradiated composite films with the same initial composition but furnace annealed at 700 °C for 60 min and containing a-Si nanoparticles.

determine the exciting light absorption). Normalized Raman spectra of three non-irradiated 200 nm thick films with $x = 1.3$ annealed at different temperatures are shown in figure 3 on an enlarged scale applied to reveal better the low intense features in the spectra. The inset in figure 3 depicts the same spectra in a scale in which only a very strong band situated at ~ 521 cm^{-1} is well seen; this band can be assigned to Raman scattering from 1TO phonons of the c-Si substrate. A low intensity feature at around 300 cm^{-1} also can be ascribed to the substrate, whereas the weak band at ~ 430 cm^{-1} can be related [21] to scattering in amorphous SiO_2. It is seen from figure 3 that only bands originating from the crystalline Si substrate are well resolved in the spectrum of the homogeneous SiO_x film annealed at 250 °C. In the spectrum of the film annealed at 700 °C two additional broad bands, centered at ~ 150 cm^{-1} and 480 cm^{-1} are observed. These bands are typical for a-Si [16, 17] and indicate existence of amorphous silicon phase in the film. A weak asymmetry at the low frequencies side of the c-Si band at ~ 521 cm^{-1} is observed in the spectrum of the film annealed at 1000 °C. It can be related to the presence of Si NCs whose formation has been proven by TEM investigations [12, 13].

Figure 4 depicts normalized Raman spectra of three pairs of $x = 1.3$ non-irradiated and MeV

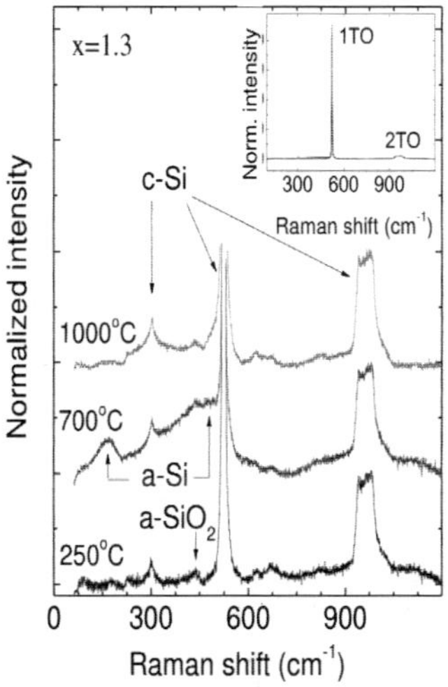

Figure 3. Normalized Raman spectra of three non-irradiated $x = 1.3$ films annealed at different temperatures denoted in the figure. The spectra of the films annealed at 700 °C and 1000 °C are vertically shifted for clarity. Inset: The same spectra shown in a 10 times compressed scale.

electron irradiated films annealed at the temperature denoted in the corresponding figure (a), (b) and (c). It is seen that the electron irradiation did not cause appreciable changes in the Raman spectrum of the homogeneous film annealed at 250 °C (figure 4 (a)), i.e. no bands due to scattering from pure Si phase in the SiO_x film appeared. This observation is in agreement with the FTIR result (figure 1) and confirms that there was no electron induced phase separation in homogeneous films. However the irradiation of the composite film annealed at 700 °C (figure 4 (b)), containing amorphous silicon phase, has caused an increase of the light scattering from this phase. Namely, the intensity of the band at ~ 480 cm^{-1} is significantly higher than the intensity of the multiphonon scattering from the c-Si substrate in the Raman spectrum of the electron irradiated sample. Figure 5 shows normalized Raman spectra of three MeV electron irradiated $x = 1.15$ films annealed at different temperatures. Again, no new features in the 250 °C spectrum are seen whereas the intensity of the band at ~ 480 cm^{-1} is significantly higher than the intensity of the c-Si multiphonon scattering band in the 700°C spectrum. In the Raman spectra of the films annealed at 1000 °C (figure 4 (c) and figure 5) for both compositions the asymmetry of the 521 cm^{-1} c-Si band at the low frequencies side is better distinguished after MeV electron irradiation.

Our previous studies [11] as well as the results in figure 2 have shown that the phase separation process in the films annealed at 700 °C is not completed and a-Si nanoparticles in a SiO_y ($y < 2$) matrix are formed. The observed electron irradiation increase of the scattering from the a-Si phase could be ascribed to an increase of its amount upon irradiation. Should this be the case an increase of the oxygen content y in the SiO_y matrix would take place, which was not observed. In fact the IR transmission (figure 2) as well as the Raman scattering spectra (figures 4 and 5) of the homogeneous films from both compositions have not given indications for electron induced phase separation. No appreciable change in the matrix oxygen content and appearance of Raman bands characteristic for a-Si have been observed. Therefore we suppose that the applied electron irradiation does not cause formation of Si nanoclusters. Our previous investigations have shown silicon nanoparticle formation in $Si\text{-}SiO_2$ structures upon 20 MeV electron irradiation and a fluence of 1.2×10^{15} el.cm^{-2} [2, 22]. Perhaps the lack of nanoparticle formation in this study is due to the relative low irradiation dose (2.4×10^{14} el.cm^{-2}) applied; maybe it is less than a required critical dose and experiments with higher doses are planned.

The furnace annealing at 1000 °C causes a complete phase separation in SiO_x films (and formation of Si NCs in a SiO_2 matrix) and no electron stimulated phase separation is expected in such films. The observed low frequency asymmetry increase of the 1TO c-Si band in the electron irradiated films with

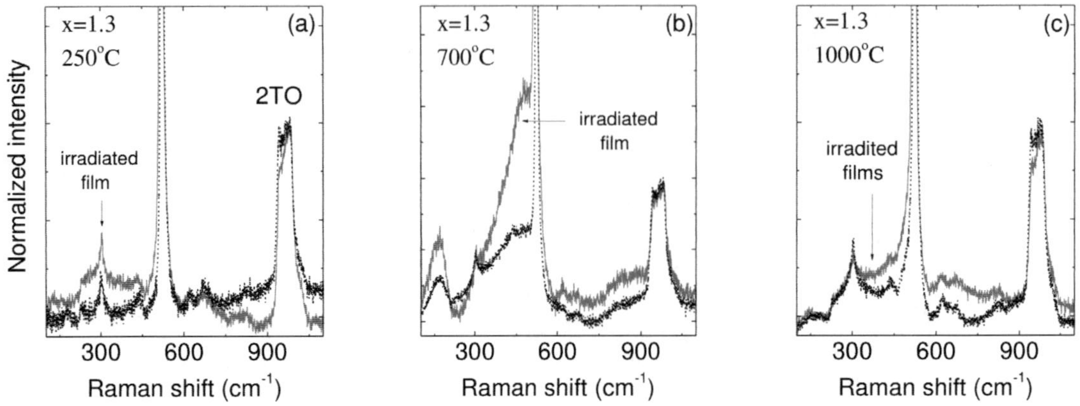

Figure 4. Normalized Raman spectra of three pairs of $x = 1.3$ films. Each pair was annealed at the temperature denoted in the corresponding figure and then a sample of each pair was 20 MeV electron irradiated.

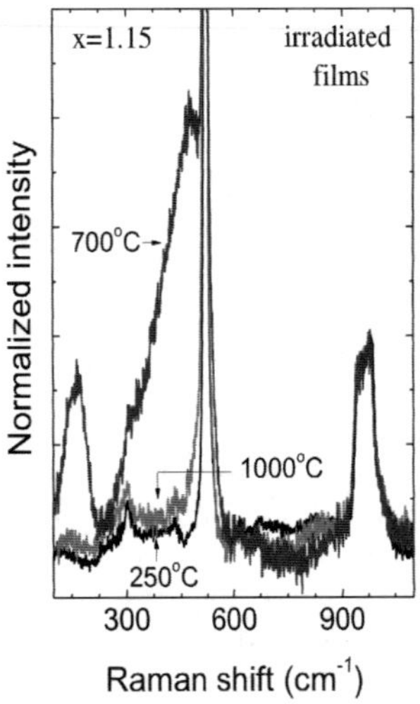

Figure 5. Normalized Raman spectra of three 20 MeV electron irradiated $x = 1.15$ films annealed at different temperatures denoted in the figure.

Si NCs and the a-Si Raman bands increase give us basis to assume that the electron irradiation reduces the structural disorder in the both types of nanoparticles. This could happen: (i) if an electron induced annealing takes place or (ii) due to a nanoparticle size increase caused by electron stimulated Ostwald ripening (small particles dissolve and redeposit onto larger particles).

The radiation effect on the nanoparticle crystallinity is an important point. The results on irradiation with electrons having energy < 1 MeV speak about rather weak effects. Very high fluencies (400 keV electrons of up to 1.7×10^{24} el.cm^{-2}) were required to amorphize 5 nm Si NCs [23]. Crystallization of Ge nanoclusters in a SiO_2 matrix has been observed under irradiation with 400 keV electrons at room temperature and a fluence above 4×10^4 C.cm^{-2} (2.5×10^{23} el.cm^{-2}) but ripening has been observed at temperatures far below the activation threshold of the corresponding thermal process, which is about 730 °C [8]. It has been suggested that the Ge diffusion in SiO_2 necessary for the cluster formation is a secondary process and that it is induced by the displacement of O and Si atoms due to electron nucleus collision [8]. However, it has been observed that the irradiation of Si-SiO_2 samples with high energy electrons (> 10 MeV) caused radiation assisted annealing at relative low dose ($3\text{x}10^{16}$

el.cm^{-2}) [2, 4]. This observation implies that the irradiation with > 10 MeV electrons at room temperature stimulates processes that normally occur in the Si-SiO_2 structures at high temperature (> 800 °C). Although the detailed mechanism is still unclear this result gives some support to the above assumption for electron stimulated annealing. Nevertheless, at this stage we cannot distinguish between the electrons induced nanoparticle annealing and Ostwald ripening process. High resolution TEM investigations on 1000 °C annealed films are planned whose results can favor one of the assumptions.

4. Conclusions

Homogeneous SiO_x films with $x = 1.15$ and 1.3 as well as composite films containing amorphous or crystalline Si nanoparticles in a SiO_y matrix ($y = 1.8 - 2$) were irradiated at room temperature with 20 MeV electrons and a fluence of 2.4×10^{14} el.cm^{-2}. The Fourier transform infrared absorption data have shown that no appreciable electron induced changes in the composition of both the homogeneous films and the oxide matrix of the composite films occurred thus indicating that no phase separation was initiated by the electron irradiation. It has been concluded that the applied electron irradiation does not cause formation of Si nanoclusters in the films studied; higher fluences are required.

A significant increase of the Raman scattering from the pure silicon phase has been seen in the films containing amorphous Si nanoparticles. In the electron irradiated films with Si NCs the low frequency asymmetry of the 1TO c-Si Raman band, caused by the finite size effect in NCs, also increased. The observed intensity increase has been related to disorder decrease in the Si nanoparticles stimulated by the MeV electron irradiation. It has been assumed that either an annealing or a nanoparticle size increase caused by Ostwald ripening take place during electron irradiation.

Acknowledgements: This work is partially supported by Grant № 41/30.01.2013/p.17 from JINR, Dubna, Russia. The financial support of the Serbian Academy of Sciences and Arts is also gratefully acknowledged.

References

[1] Fourches N T and Nenoi M 2012 *Current Topics in Ionizing Radiation Research* **32** 741 (Rijeka: InTech)

[2] Kaschieva S and Dmitriev S N 2010 *Radiation Defects in Ion Implanted and/of High-Energy Irradiated MOS Structures* (New York: Nova Science Publishers)

[3] Zatsepin A F, Kaschieva S, Dmitriev S N and Buntov E A 2008 *Nucl. Instr. Meth. in Phys. Res.* B **266** 5027

[4] Zatsepin A F, Kaschieva S, Biryukov D Yu, Dmitriev S N and Buntov E A 2009 *Technical Physics* **54** 323; *Original Russian text published* 2009 *in Zhurnal Tekhnicheskoi Fiziki* **79** 155

[5] Wang L M, Wang S X, Ewing R C, Meldrum A, Birtcher R C, Newcomer P P, Weber W J and Matzke H 2000 *Mater. Sci. Eng.* A 286

[6] Shibata M, Nitta Y, Fujita K and Ichikawa M 1998 *Appl. Phys. Lett.* **73** 2179

[7] Takeguchi M, Tanaka M and K Furuya 1999 *Appl. Surf. Sci.* **146** 257

[8] Klimenkov M, Matz W, Nepijko S A and Lehmann M 2001 *Nucl. Instr. Meth. in Phys. Res.* B **179** 209

[9] Arnoldbik W M, Tomozeiu N, van Hattum E D, Lof R W, Vredenberg A M and Habraken F H P M 2005 *Phys. Rev.* B **71** 125329

[10] Pacifici D, Moreira E C, Franzò G, Martorino V, Priolo F and Iacona F 2002 *Phys. Rev.* B **65** 144109

[11] Nesheva D, Bineva I, Levi Z, Aneva Z, Merdzhanova T and Pivin J C 2003 *Vacuum* **68** 1

[12] Nesheva D, Raptis C, Perakis A, Bineva I, Aneva Z, Levi Z, Alexandrova S and Hofmeister H 2002 *J. Appl. Phys.* **92** 4678

[13] Curiel M A, Nedev N, Nesheva D, Soares J, Haasch R, Sardela M, Valdez B, Sankaran B, Manolov E, Bineva I and Petrov I 2010 *Mat. Sci. Eng.* B **174** 132
[14] Schumann L, Lehmann A, Sobotta H, Riede W, Teschner U and Hubner K 1982 *Phys. Stat. Solidi* **110** 69
[15] Tsu D V, Lucovsky G and Davidson B N 1989 *Phys. Rev.* B **40** 1795
[16] Iqbal Z and Veprek S 1982 *J. Phys.* C **15** 377
[17] Nesheva D, Raptis C and Levi Z 1998 *Phys. Rev.* B **58** 7913
[18] Campbell I H and Fauchet P M 1986 *Solid State Commun.* **58** 739
[19] Temple P A and Hathaway C E 1973 *Phys. Rev.* B **7** 3685
[20] Borowicz P, Latek M, Rzodkiewicz W, Łaszcz A, Czerwinski A and Ratajczak J 2012 *Adv. Nat. Sci.: Nanosci. Nanotechnol.* **3** 045003
[21] Rahmani A, Benoit M and Benoit C 2003 *Phys. Rev.* B **68** 184202
[22] Kaschieva S, Gushterov A, Angelov Ch and Dmitriev S N 2012 *J. Phys.:Conf. Series* **356** 012005
[23] Kachurin G A, Ruault M O, Gutakovsky A K, Kaitasov O, Yanovskaya S G, Zhuravlev K S and Bernas H 1999 *Nucl. Instr. and Meth. in Phys. Res* B **147** 356

Electrical conductivity and optical properties of tellurium-rich Ge-Sb-Te films

V Dzhurkov[1], S Fefelov[2], D Arsova[1], D Nesheva[1] and L Kazakova[2]
[1] G. Nadjakov Institute of Solid State Physics, Bulgarian Academy of Sciences, 72 Tzarigradsko Chaussee blvd., 1784 Sofia, Bulgaria
[2] Ioffe Physical - Technical Institute, Russian Academy of Sciences, 26 Polytehnicheskaya str., 194021 S-Petersburg, Russia

E-mail: valeri.dzhurkov@issp.bas.bg

Abstract. The Ge-Sb-Te based (GST) films are intensively studied because they are suitable for reversible phase-change storage media. $Ge_2Sb_2Te_5$ films are most frequently used for memory applications while Te enriched GST films are also promising but not well studied. In this work Te enriched GST films with two different compositions and various thicknesses (120 – 750 nm) were prepared by thermal evaporation of previously synthesized $Ge_{15}Sb_5Te_{80}$ and $Ge_{15}Sb_{15}Te_{70}$ glasses. Optical transmission and electrical conductivity measurements in sandwich and planar contact configuration were carried out. Temperature measurements of the planar electrical conductivity were performed in the range 77 - 300 K. Values of around 0.7 eV have been determined for the optical band gap of both compositions; the film refractive index is in the range 3.8 - 4.5. The sandwich conductivity is in the interval $(2.5 – 5)\times10^{-4}$ $(\Omega.cm)^{-1}$ while the planar conductivity is around 10^4 times greater. The obtained results are discussed in terms of existence of an amorphous volume part as well as a crystalline surface layer in the films.

1. Introduction

Chalcogenide films from the Ge-Sb-Te system are intensively studied because they are very suitable for reversible phase-change storage media. Phase-change applications utilize differences in the optical and electrical proprieties of the material in crystalline and amorphous state. Optical storage applications are based on small differences in the reflectivity [1], while electronic applications stand on large (more than 10^3) change in the electrical conductivity [2]. The transition from amorphous (low conductive) to crystalline (high conductive) state in chalcogenide films by applying electrical pulses was reported at first in Ref. [3] and patented by S. Ovshinsky in 1966. This transition is known as “switching effect”.

The Ge - Sb -Te (GST) films with stoichiometric compositions lying on the $GeTe$-Sb_2Te_3

pseudobinary line in the Ge-Sb-Te system are widely used nowadays for memory applications and the most explored among them is $Ge_2Sb_2Te_5$ (GST225). However the GST films with non-stoichiometric compositions, especially Te enriched ones are not studied in detail. It has been reported that the thermal stability of Te-enriched GST glasses with constant amount of Te depends on the Sb concentration [4] and a phase-separation has been observed during the crystallization process [5]. Further studies on new GST materials will give possibility to improve the parameters of the already developed memory elements and will provide additional information about the nature of the switching effect.

Recent current-voltage (I - V) measurements we performed on $Ge_{15}Sb_5Te_{80}$ films have shown typical switching effect [6]. When the applied voltage reaches a threshold voltage U_{th} an electrical breakdown occurs - the voltage on the film decreases abruptly and the film conductivity arises of up to 10^4 times. In addition some oscillations have been observed in the high conductive part of the I - V curve which are not discussed in the available literature related to the switching effect in chalcogenide films. The obtained results are promising for phase change memory applications of these films as well as of Te enriched films with similar compositions and the knowledge of their optical and electrical properties is important for such applications.

In the present work films from two $Ge_{15}Sb_5Te_{80}$ and $Ge_{15}Sb_{15}Te_{70}$ glassy compositions with various thicknesses are produced by thermal evaporation in vacuum. Both compositions are rich in Te in comparison with GST225; the atomic percentage of the over-stoichiometric Te is more than 50%. These compositions have not been studied previously by other researchers and have been chosen because an easier crystallization is expected than in GST225. Data about films' optical band gap, index of refraction and electrical conductivity are reported and the structural homogeneity of the films is discussed.

2. Experimental

The $Ge_{15}Sb_5Te_{80}$ and $Ge_{15}Sb_{15}Te_{70}$ glassy alloys were synthesized by melt quenching technique using elemental Ge, Sb and Te with 5N purity, sealed in evacuated (~ 10^{-3} Pa) quartz ampoules. The ampoules were heated in a furnace at 950 ^{0}C for 20 h with continuous rocking to ensure homogeneity. The melts were rapidly cooled inside the furnace. The obtained glasses were powdered and deposited onto glass substrates at a rate of 3 - 4 nm/sec using thermal vacuum evaporation technique. The material source temperature was ~ 600 ^{0}C and the temperature of the substrate ~ 30 ^{0}C. The thickness of the films was in the range 100 - 750 nm and was controlled in situ by a MIKI FFV quartz microbalance system. The films area was 1 cm^2.

Optical transmission (*t*) spectra were measured in the 800 – 2500 nm spectral region using a precise spectrophotometer Perkin Elmer Lambda 1050 NB. The dispersion of the refractive index (*n*) and the film thickness (*d*) of the thickest films were determined in the low absorption range by the well-known method of Swanepoel. The absorption coefficient (α) of the films was calculated in the strong absorption range and used for determination of the optical band gap. For the thickest films α was determined using the Swanepoel's approach. Since no good interference was observed for the thinnest films their absorption coefficient has been evaluated directly from the experimental transmission spectrum in the region of low transmission ($t < 10$ %) using the relation:

$$ln(\alpha) = (d^{-1}) \times ln[(1-R)^2/t] \quad (1)$$

where the film thickness (*d*) is in centimeters and the reflectance $R = 0.35$ [7].

Two types of DC electrical measurements were performed. Current-voltage characteristics were measured in sandwich geometry and at different points of the film surface. Sputtered Au layer was used as bottom electrode while the top contact was realized by pressing a gold needle mounted on a two axis stage. The area of the top Au contact was ~ 10^{-4} cm^2 and the distance between the points explored was ~1 mm. The measurements were carried out in current controlled mode; details about the advantages of this mode are given in Ref. [6, 8]. Dark current (I_d) temperature dependences were measured in the temperature range 77- 300 K at two heating rates (0.05 and 0.1 K/sec) by means of a Keithley 6487 picoammeter. For these measurements the films were provided with planar Au contacts with a length of 8 - 10 mm and 1 - 1.5 mm spaced, prepared on the top film surface by cathode sputtering. The contacts showed Ohmic-like behavior in the range 0 - 10 V and a voltage of 0.1 V was applied. From the data obtained in the planar geometry room temperature conductivity of the films was calculated using the total film thickness.

3. Results and Discussion

3.1. Optical properties

In the strong absorption range the absorption coefficient (α) of semiconductors follows the relation:

$$(\alpha E) = B\left(E - E_g\right)^m \tag{2}$$

where B is a parameter depending on the transition probability, E is the photon energy and m is an index, which characterizes the transition processes: normally $m = 1/2$ for allowed direct electron transitions and $m = 2$ – for indirect transitions.

A simple presentation [9] can be used to determine the transition type; the ratio of (αE) to its first derivative $(\alpha E)^I$ is examined with respect to the photon energy E in accordance with the relation:

$$(\alpha E)/(\alpha E)^I = 1\,/\,m\left(E - E_g\right) \tag{3}$$

It is seen that provided the dependence $(\alpha E)/(\alpha E)^I$ versus E is linear the index m can be found from the slope of the line. Our experimental data showed good linearly following relation (3) and values of 1.6 and 1.8 were calculated for the $Ge_{15}Sb_5Te_{80}$ and $Ge_{15}Sb_{15}Te_{70}$ films, respectively. Both values are nearer to $m = 2$ rather than to $m = 0.5$ and therefore a conclusion for indirect optical transitions for both compositions has been made. It has to be pointed out that for GST films most authors consider indirect optical band gap [11].

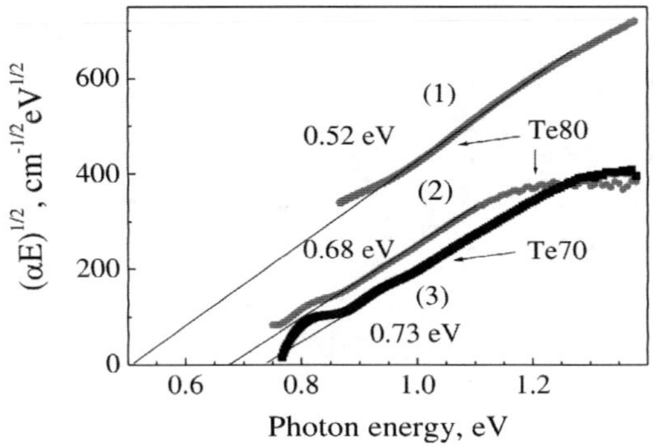

Figure 1. Spectral dependences of the optical absorption for two $Ge_{15}Sb_5Te_{80}$ films with $d = 120$ nm (1) and 750 nm (2) and a $Ge_{15}Sb_{15}Te_{70}$ film, $d = 750$ nm (3).

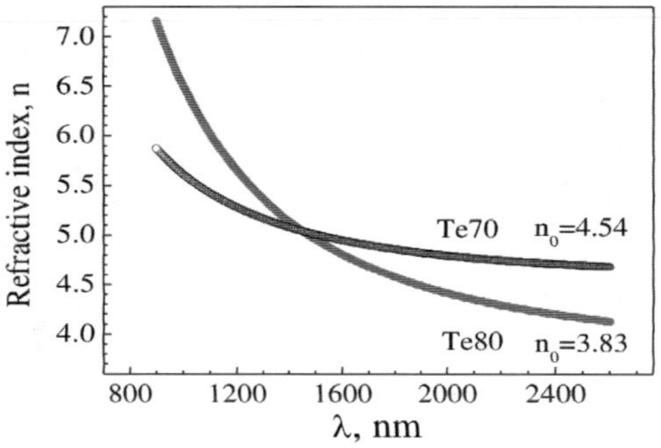

Figure 2. Spectral dependences of the refractive indexes of 750 nm films from both compositions.

The spectral dependences of $(\alpha E)^{1/2}$ for films with $d = 750$ nm from both compositions are presented in figure 1. Using the Tauc approach [10] values of 0.73 eV and 0.68 eV have been obtained for the optical band gap energy (E_g) of the 750 nm thick Te70 and Te80 films, respectively. They are close to the E_g values reported for GST225 films, prepared by using various deposition techniques [9, 12, 13] as well as to the data published for other GST films [14, 15]. The results obtained indicate small E_g decrease with increasing Te content and a better pronounced decrease when the film thickness decreases. A value of 0.52 eV has been obtained for the optical band gap of the 120 nm $Ge_{15}Sb_5Te_{80}$ film. The observed E_g diminish with the thickness decrease may be connected with an increase of Te content or/and an increase of defect density. Such decrease has already been observed on other chalcogenide films [9]. It has been related to an increase of unsaturated bonds existing in an amorphous material with the thickness decrease which produce localized states in the band gap of amorphous solids. Probably a defect density increase is the reason for our observation, as well, since the increase of the Te content in the source material of 10 at.% causes a considerably smaller E_g decrease.

Dispersion curves of the refractive index n are presented in figure 2 for films from both compositions with $d = 750$ nm, n_0 represents the refractive index value at $\lambda \rightarrow \infty$. High n values have been obtained for both compositions which are typical for amorphous GST films [2]. However these values are significantly lower than the refractive indexes of crystalline (face centered cubic structure) films [9, 12] and indicate that the main part of the volume of the 750 nm thick films is amorphous.

3.2. Electrical properties

The room temperature conductivity values of the investigated GST films with different thicknesses obtained from both sandwich and planar contact configuration measurements are summarized in table 1. In the sandwich structures an initial conductivity (σ-sandwich) was measured using a current pulse with $I_{max} = 6.5$ μA at which the corresponding value of the measured voltage is a few tenth of a volt and no switching effect occurred [6, 8]. The σ-sandwich value defines the conductivity in the high-resistive state of the film.

The results for σ-sandwich are similar for the films of both compositions and lie in the range $(2 - 5)\times10^{-4}$ $(\Omega.cm)^{-1}$. The conductivity variations with the thicknesses are comparable with those observed when measuring σ-sandwich at different points on the surface of a given sample. The σ-sandwich values are comparable with the conductivity of amorphous GST films reported in the literature [2 and references therein, 9, 12] and indicate, in agreement with the optical data, that the main part of the films is amorphous. Surprisingly the σ-planar values obtained for the 400 - 750 nm films of both compositions are more than 10^4 times higher than the conductivity determined in sandwich contact configuration. The planar conductivity of the films with the higher Te content is higher and an appreciable increase with the thickness decrease is observed.

Table 1. Thickness and conductivity of Ge-Sb-Te films.

Composition	Thickness, d (nm)	σ -planar $(\Omega.cm)^{-1}$	σ -sandwich $(\Omega.cm)^{-1}$
$Ge_{15}Sb_{15}Te_{70}$	350	7.4	-
	500	6.5	$2.5*10^{-4}$
	750	1.5	$3.0*10^{-4}$
$Ge_{15}Sb_5Te_{80}$	120	75	$5.0*10^{-4}$
	400	19	-
	750	-	$2.5*10^{-4}$

The temperature dependences of the planar conductivity of $Ge_{15}Sb_5Te_{80}$ films with $d = 120$-500 nm are shown in figure 3 in a semi-logarithmic plot. An increase of the conductivity with increasing temperature is observed. This is typical behavior for semiconductors. The dark current activation energy (E_a) was evaluated by applying the well known formula in the high temperature range:

$$\sigma = \sigma_0 exp\left(-\frac{E_a}{k_B T}\right) \qquad (4)$$

In this formula k_B is the Boltzman constant and T is temperature. The obtained activation energies for all films are rather low (less than 0.1 eV) when compared with the value typical for GST225 films (E_a ~ 0.37 eV for modern phase change memory devices under low bias [16]). For example for as-prepared $Ge_{15}Sb_5Te_{80}$ with $d = 400$ nm $E_a = 0.08$ eV (figure 1, curve 2) and it decreases after annealing the sample at 473K for 45 min in an Ar atmosphere (figure 1, curve 4); this decrease is accompanied with a conductivity increase of about 40%. For the thinnest as-deposited film E_a is very low (~ 0.01 eV) (figure 1, curve 1).

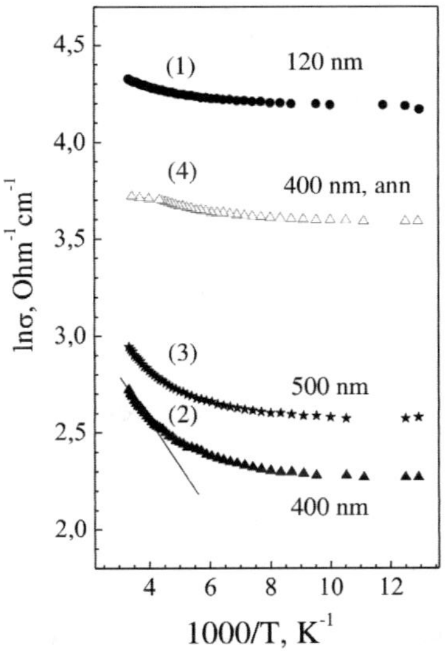

Figure 3. Temperature dependences of DC conductivity of as-deposited $Ge_{15}Sb_5Te_{80}$ films with three different thicknesses (curves 1 - 3). Curve (4) was measured after annealing of the 400 nm film at 473 K for 45 min in an Ar atmosphere.

The observed big differences between the conductivities measured in planar and perpendicular directions, as well as the rather low E_a values obtained for the planar conductivity imply existence of a highly conductive layer at the film surface i.e. presence of a significant in depth variation in the film structure or composition. As known the transition of the GST films from amorphous to crystalline state causes a conductivity increase of around 4 - 5 orders of magnitude [10, 17]. The room temperature conductivity of face centered cubic crystalline GST films is ~10 $(\Omega.cm)^{-1}$ while for the films with hexagonal structure it is around 10^3 $(\Omega.cm)^{-1}$ [10]. As seen from table 1 and figure 3 the planar conductivity of our films is in the range 1 - 100 $(\Omega.cm)^{-1}$. Based on this result and keeping in mind the conclusions taken from the optical and sandwich electrical measurements for an amorphous structure of the main part of the films' volume, one can assume existence of a thin crystalline layer at the film surface. This is not surprising since the films are enriched in Te and surface crystallization can be easily induced due to some surface contamination, specific surface defects or local increase of Te at.%. We have not found in the literature information about existence of such crystalline layer on top of Te enriched GST amorphous films. It should be mentioned that an increase of around 50% in σ-planar has been registered after a 10 month storage of the films at standard room conditions. The observed conductivity increase with the film thickness decrease could be related to the usage of the total film thickness in the σ-planar calculations. Most likely the thickness of the crystalline surface layer is similar for all films and its real conductivity is higher than those listed in table 1. As shown in [17] the crystallization of GST films of four different compositions starts at ~ 380 K and hence the increase of the $d = 400$ nm film conductivity after annealing at 473K (figure 3) can be related to complete film crystallization.

4. Conclusions

Tellurium enriched GST films with two different compositions and various thicknesses (120 – 750nm) were prepared by thermal evaporation of previously synthesized $Ge_{15}Sb_5Te_{80}$ and $Ge_{15}Sb_{15}Te_{70}$ glasses

Journal of Physics: Conference Series **558** (2014) 012046 doi:10.1088/1742-6596/558/1/012046

and data about their optical characteristics and electrical conductivity have been obtained. Based on the optical transmission spectra a conclusion for indirect optical transitions for both compositions has been made and values of around 0.68 eV and 0.73 eV have been determined for the optical band gap of Te80 and Te 70 films, respectively. A considerable optical band gap decrease has been observed with decreasing film thickness which has been related to a defect density increase. The refractive index values, in the range 3.8 - 4.5, have indicated that the main part of the film volume is amorphous. This conclusion has been supported by the results obtained for the sandwich conductivity (2.5 - 5.0)$\times 10^{-4}$ $(\Omega.\text{cm})^{-1}$ of the films which is close to the data reported by other author for amorphous films of similar compositions. Much higher (~ 10^4 times) planar conductivity has been measured which has been assigned to existence of a crystalline layer at the film surface. The structural non-homogeneity of the films may be the reason for the oscillations observed in the high conductive part of the I - V curve.

References

[1] Zhou G 2001 *Mater. Sci. Eng.* A **73** 304
[2] Popescu M 2005 *J. Ovonic Reseach* **1** 69
[3] Kolomiets B T and Lebedev E A 1963 *Radiotech. and Electron.* **8** 2097 (in Russian)
[4] Zaluska A, Zaluski L and Lasocka M 1980 *J. Thermal Anal.* **19** 257
[5] Belhadji M, Ziani N and Mostefa M 2005 *Chinese J. Phys.* **43** 95
[6] Fefelov S, Kazakova L, Kozyukhin S, Tsendin K, Arsova D and Pamukchieva V 2014 *Technical Physics* **59** 546
[7] V Pamukchieva and A Szekeres 2008 *Opt. Mat.* **30** 1088
[8] Almasov N, Bogoslovskiy N, Korobova N, Kozyukhin S, Fefelov S, Kazakova L, Jakovlev S, Tsendin K and Guseinov N 2012 *J. Non-Cryst. Solids* **358** 3299
[9] Fang M, Li Q and Gan F 2004 *Chinese Opt. Lett.* **2** 177
[10] B S Lee, J R Abelson, S G Bishop, D H Kang, B Ki Cheong and Ki B Kim 2005 *J. Appl. Phys.* **97** 093509
[11] Tauc J, Grigorovichi R and Vancu A 1966 *Phys. Stat. Solidi* **15** 627
[12] Němec P, Moreac A, Nazabal V, Pavlišta M, Přikryl J and Frumar M 2009 *J. Appl. Phys.* **106** 103509
[13] Kumar S, Singh D, Shandhu S and Thangaraj R 2012 *Appl. Surf. Sci.* **258** 7406
[14] Kumar S, Singh D and Thangaraj R 2013 *Thin Solid Films* **531** 577
[15] Abd El-Rahman A A, Hafez F M, Ahamd F and Elokr M M 2012 *Chalcog. Lett.* **9** 41
[16] Nardone M, Simon M, Karpov I V and Karpov V G 2012 *J. Appl. Phys.* **112** 071101
[17] Morales-Sanchez E, Prokhorov E F, Gonzalez-Hernandez J and Mendoza-Galvan A 2005 *Thin Solid Films* **471** 243

Outdoor degradation of thin film amorphous silicon based PV modules

M Berov[1], P Ivanov[1], N Tuytuyndziev[1] and P Vitanov[1]
[1]Central Laboratory of Solar Energy and New Energy Sources, 72 Tzarigradsko shaussee, 1784 Sofia, Bulgaria

E-mail: vitanov@phys.bas.bg

Abstract. One of the main problems in thin film silicon based modules is the deterioration of their performance upon exposure to light. The presented work focuses on a methodology for evaluation of thin-film photovoltaic module degradation behavior under real operating conditions. The outdoor degradation of double junction a-Si:H/a-Si:H modules was investigated using automated measurement setup for a period of two years. A deterioration of the module's maximum power was observed due to the well known Staebler-Wronski effect, which main causes are the decrease of open circuit voltage and the fill factor of the module. The obtained results can be correlated to the technology and construction of the thin film silicon based modules.

1. Introduction

One of the main problems in thin film silicon based modules is the deterioration of their performance upon exposure to light. The degradation of the electric parameters of the a-Si thin film modules over the time is observed upon exposure to solar radiation and in conditions of electric current flow which is manifested as decreasing of module power and, respectively, its efficiency. When the amorphous silicon layer is exposed to light or when an electric current is injected into the layer, a degradation effect takes place. The weaker bonds within the amorphous silicon network are broken and, thereby, new dangling bonds are created. Usually, after a time of about 1000 hours of light exposure, amorphous silicon layers "stabilize". It is, however, often observed that degradation still continues, albeit at a lower rate, even after 1000 hours of light exposure.

In practice, for present state of the art the p-i-n amorphous silicon solar cells one finds degradation in the range 10% to 30% for cells of i-layer thickness between 200 nm and 400 nm and cells of thicker i-layers have higher relative degradation. The use of tandem (double junction) solar cells is one way to reduce the effect of light-induced degradation because the top cell in a tandem structure can be kept very thin, and therefore shows little degradation.

The presented work focuses on a methodology for evaluation of thin-layer photovoltaic module degradation behaviour under real operating conditions. An outdoor measurements of six double junctions a-Si:H/a-Si:H modules have been carried out at the Central Laboratory of Solar Energy and New Energy Sources, located in Sofia, Bulgaria. The degradation of module performance, initially due to the well-known light-induced degradation described by Staebler and Wronski [1] has been investigated. The module performance parameters have been determined by outdoor measurements for a long period of time (two years) under varying conditions of solar irradiance and air temperature in order to estimate the degradation effects.

2. Experimental set-ups

2.1 Selection of module types

The measurements are focused on thin film modules, in particular, p-i-n - type double junction (a-Si:H/a-Si:H) structure. They are rated by the manufacturer at 40Wp. Typical thickness of the i -layer of the thin top cell in a tandem structure is 100-120 nm and for the bottom cell i – layer is 400-500 nm. The modules are glass/glass laminates of size 1245x635mm, and 0.79 m^2 module area.

Six PV modules, labelled M 1 to M 6, have been deployed outdoors at the beginning of June 2008. During the whole measurement period the modules were mounted on a fixed south facing open rack at a fixed 45° tilt all the year round.

2.2 Outdoor electrical characterizations.

A major problem in thin-film photovoltaic modules is the degradation of their parameters with time at operation in real conditions. In amorphous semiconductors under solar radiation and at electric current flow degradation of the fundamental electric parameters with time is observed, which is manifested as a reduction of module power and, respectively, its conversion factor. The investigation of the performance of single- and multi-junction a-Si thin film modules at outdoor conditions are the object of many papers [2, 3, 4 and 5]. In real conditions test the modules are put under the continuous influence of solar radiation and temperature.

The especially designed experimental setup which was developed for to perform long-term electrical measurements is presented in figure 1. The unit MUX-1 switches over the modules from standby generation state in the maximum power point, to measurement mode for tracing the I-V curves of the modules, at a specified interval (i.e. every 30 minutes). In I –V curve tracing mode, the modules are consequently connected by unit MUX-2 for measurement by the Agilent 34970A measurement system. The data acquisition software is based on the LABVIEW package for laboratory automation.

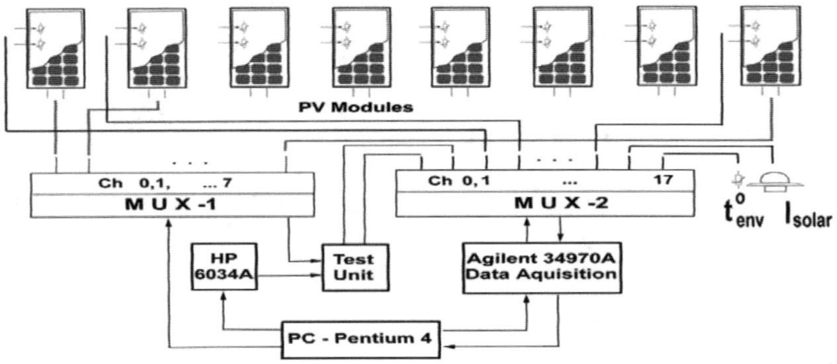

Figure 1: Automated measurement setup.

The accumulated experimental data are processed with especially developed software. In order to process the data correctly filters of random fluctuations have been employed. Usually the parameters of solar radiation are measured and then one proceeds with the measurement of the modules load characteristics. This approach allows obtaining various functional dependencies by processing the results of the database. For example:

- The generated power P_{max} as a function of time, using solar radiation and ambient temperature as a parameter;
- The short circuit current I_{sc} as a function of time at fixed values of solar radiation and for a certain temperature range;
- The degradation of the fill factor (FF) with time, FF being determined at a fixed value of solar radiation and fixed temperature range.

An advantage of the proposed methodology provides an objective measurement and processing of large amounts of data. Measurements are made in real conditions, following all environmental parameters (solar radiation, air temperature, spectral density of solar radiation). The PV modules are kept in generation mode and the current magnitude can be set. The developed measurement system allows assessing the degradation at different types of thin-film PV modules experimentally.

3. Results and discussion

A set of 6 thin film amorphous Si modules (M1...M6) are measured for a two-year period. With our developed software, the dependence of electrical parameters with time can be plotted, so that one can see the different degradation effects. In figure 2 the degradation of P_{max} for the all modules can be seen. One can choose the desired ambient temperature and irradiance range using the sliders on the left side of the figure.

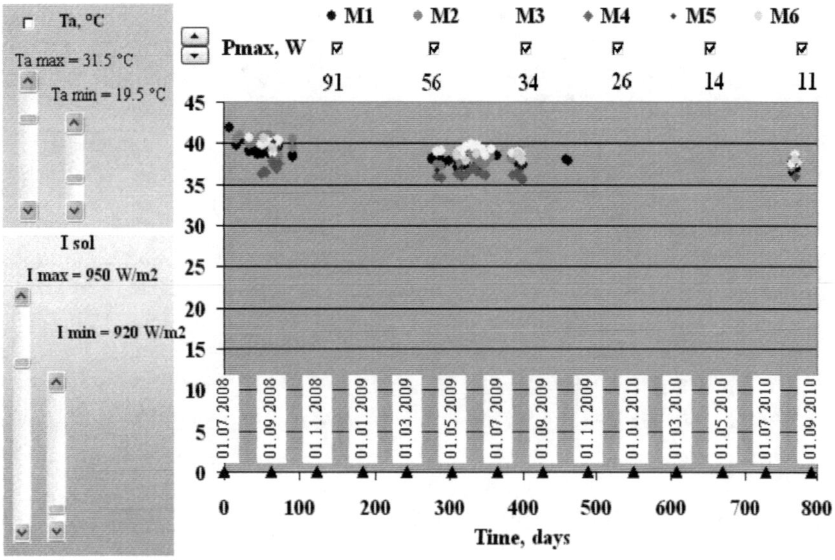

Figure 2. Degradation of P_{max}.

It should be noted that the degradation of P_{max} is more pronounced in the first 120 days. This cannot be seen for module M4, as it was exposed to solar radiation 6 months before it was connected with the other 5 modules to the measurement system. From the collected data during the investigated long- term period it is possible to estimate the deterioration of the P_{max} as a function of the time. The light- induced degradation of P_{max} at different solar radiation is shown in figure 3 and table 1. The most severe degradation was observed after the first 3 months of outdoor exposure.

From the accumulated in the database results one can assess maximum power output P_{max}

deterioration at various solar radiation values. The light- induced degradation of Pmax at different light intensities is shown in figure 3 and table 1. After the first 3 months of outdoor exposure, the rate of degradation for year is between 2 to 5%.

The power output degradation due to illumination is the manifestation of the Staeble-Wronski effect. The drop in performance is ascribed to the creation of additional metastable defects in the absorber layer, which act as extra trapping and recombination centres. As a result of trapping, the space charge distribution in the intrinsic a-Si:H is changes in such a way that internal electric field

Table 1. Measurements of module M2 have been made each year on July 31st. The air temperature and the time of measuring are the same.

Radiation (W/m^2)	P_{max} (W)	After 1 year 2009 (W)	After 2 years 2010 (W)
1000 (13:30 h)	45	41	39.3
900 (12 h)	41.3	38.1	35.8
750 (11 h)	37.8	33.7	32.9
300 (9 h)	15.7	14.2	13.8

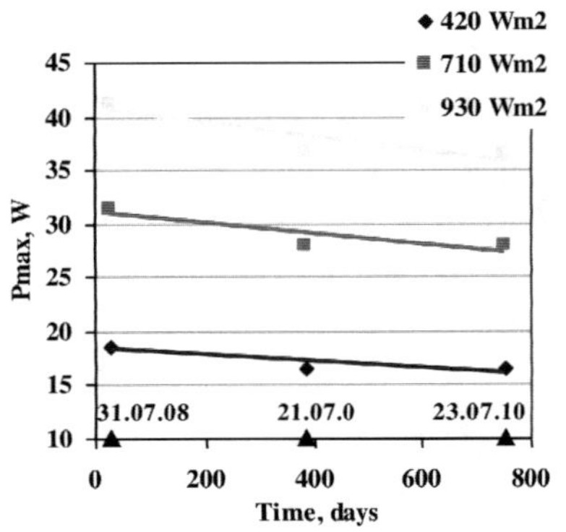

Figure 3. P_{max} at different light intensities.

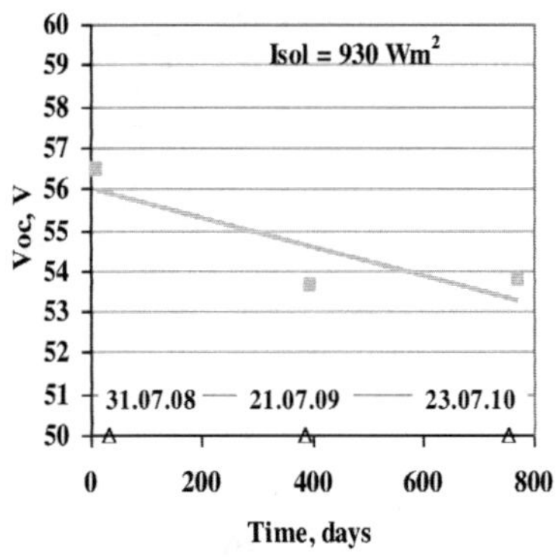

Figure 4. Degradation of open-circuit voltage V_{oc}.

The power output degradation due to illumination is the manifestation of the Staebler-Wronski effect. The drop in performance is ascribed to the creation of additional metastable defects in the absorber layer, which act as extra trapping and recombination centres. As a result of trapping, the space charge distribution in the intrinsic a-Si:H is changes in such a way that internal electric field across this layer is distorted. This leads to a lower carrier drift and thus to a lower carrier collection. One way to minimize this effect is the use of multijunction structure. By multijunction structure a total absorption is obtained that is the same or even higher than a single junction with comparable thickness. The measured modules are double junction and for this reason after the initial degradation the performance of solar cells stabilizes.

The experimental data about the stability of the electrical parameters demonstrate a decrease of the open circuit voltage V_{oc} shown in figure 4. The same behaviour is observed under different intensities of solar radiation. It is most probably that the defects and microvoids are forming conductive paths causing increased leakage current. The open-circuit voltage V_{oc} depends on the value of photo-generated current density, but it depends also on the diode reverse saturation current J_o, also [6]. Therefore, V_{oc} can decrease if the diode factor n increases (indicating a diode of a lower quality) [6]. It should be noted that short circuit current I_{sc} does not change upon the same conditions, even it values increase – figure 5 (start day 1.07.2008). The measurements of I_{sc} in August 2008 and the same month in 2010 showed that this parameter remains constant value (figure 6).

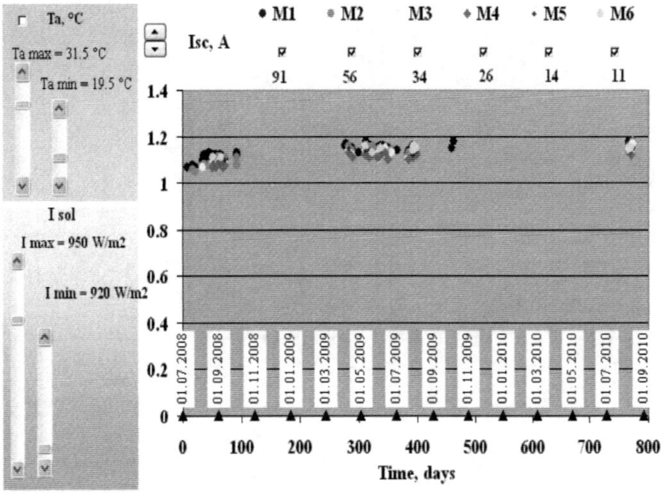

Figure 5. Degradation of I_{sc}

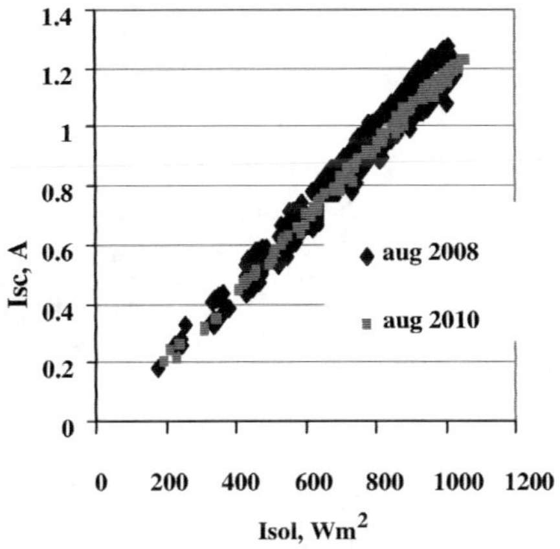

Figure 6. Short circuit current vs. irradiance for module M1.

The most significant degradation is obtained for the value of the fill factor FF shown in figure 7. The measured value is a very sensitive experimental indication of the quality of a solar cell and it should be used as a complementary quantity, in addition to V_{oc}, to characterize solar cell quality.

The difference of the fill factors between the initial and the degraded state can be explained by the effect of light induced degradation. The most common effects encountered are:

- collection problems due to the low i-layer quality or to defects at the p/i and i/n interfaces;
- high series resistance due to unsatisfactory contact layers;
- low parallel (shunt) resistance;
- high diode reverse current due to leakage paths and bad quality diodes.

One possible reason for this is the decrease in carrier collection caused by light induced defects which leads to degradation of the FF.

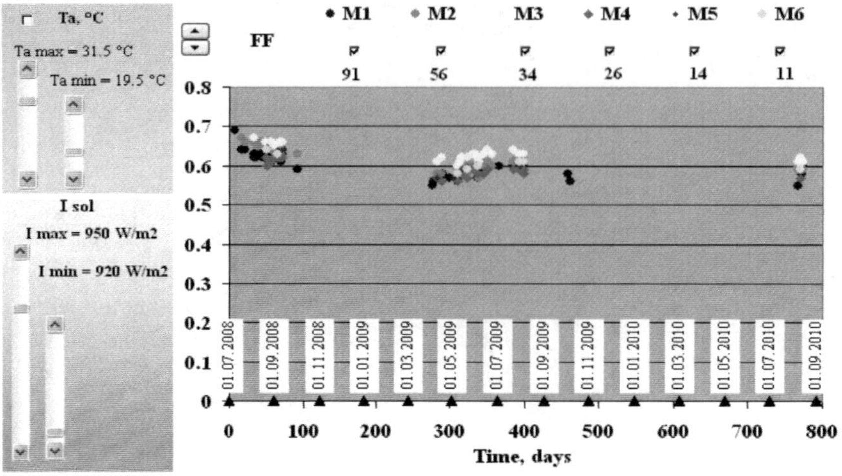

Figure 7. Degradation of FF.

Shunt resistance R_{sh} represents the ohmic shunts through the diode due to pinholes and defects within the i-layer. Figure 8 shows R_{sh} (shunt resistance) for M1 determined from I-V characteristics using the one-diode model. The values in blue show the initial values of R_{sh} at different irradiance, and in red are the values after 2 years.

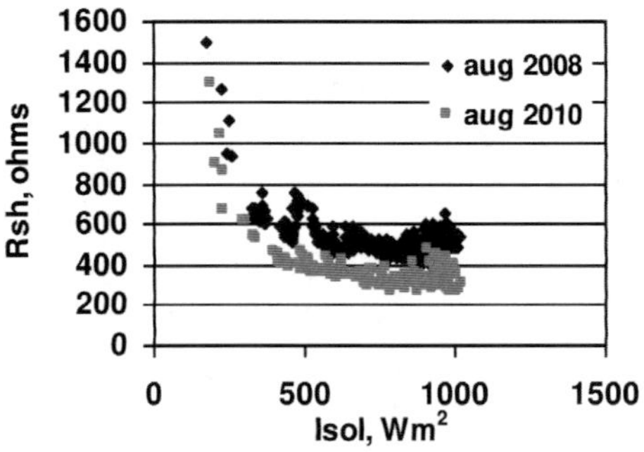

Figure 8. Shunt resistance R_{sh} vs solar irradiance. I-V curves have been taken down in August 2008 and August 2010, respectively.

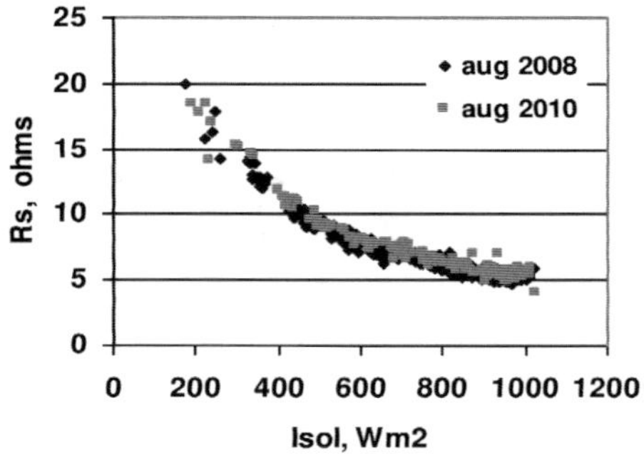

Figure 9. Series resistance R_s vs solar irradiance. I-V curves have been taken down in August 2008 and August 2010, respectively.

It can be seen that R_{sh} changes only slightly in the irradiance range 400 W/m2– 1000 W/m2, R_{sh} degrades by 50% after 2 years of outdoor exposure. One explanation is that the defects and microvoids formation directly affect shunt resistance. Most likely, this is caused by generated defects in the i-layer, which influence the diode quality factor. The shunt resistance R_{sh} have no effect on short circuit current, but lead to strong decrease in U_{oc}, which is clearly seen in our case.

Series resistance R_s is calculated from the load characteristics of module M1, applying the one diode model. Usually, R_s represents the contact resistance due to the TCO- layer and the insufficiently

Journal of Physics: Conference Series **558** (2014) 012047 doi:10.1088/1742-6596/558/1/012047

doped p- and n-layers. The estimated value reveals that R_s depends on the solar irradiance level, but does not show light-induced degradation effect – figure 9).

4. Conclusions

The behaviour of double junction (a-Si:H/a-Si:H) PV modules at outdoor conditions has been investigated for two years and deterioration of their performance upon exposure to sunlight observed.

It was found that the U_{oc} is decreasing during this period. No change in I_{sc} was observed. Using the one diode model, R_{sh} and R_s were estimated. It was found that R_{sh} reduces its value with 50% during these two years while no significant changes have been observed for R_s. The methodology used on the experimental data in the database, enables us to study the degradation effect of module's performance, and to correlate the results with the production technology and construction of the thin-film silicon modules.

5. References

[1] Staebler D I and Wronski C R 1977 *Applied Physics Letters* **31** 292
[2] Radue C and van Dyk E 2009 *Physica B* **404** 4449
[3] van Dyk E, Audouard A, Meyer L and Woolard D 2007 *Solar Energy Materials & Solar Cells* **91** 167
[4] Nikolaeva-Dimitrova M, Skoczek A, Kenny R and Dunlop E 2008 *Proceedings of the 23th European PV Solar Energy Conference (Valencia, 01-05 September 2008)* pp 2079-2083
[5] Nikolaeva-Dimitrova M, Robert Kenny R, Dunlop E and Pravettoni M 2010 *Prog. Photovolt. Res. Appl.* **18** 311
[6] Shade H and Bouree J 2010 Module fabrication and performance *Thin-Film Silicon Solar Cells*, ed. A Shah (Lausanne: EPFL Press) chapter 6 pp 269-368
[7] Wronski C R, Pearce J M, Koval R J, Ferlauto A S and Collins R S *World Climate & Energy (Rio de Janeiro, January 6-11 2002)* p 64

Spectral sensitivity dependence on the details of preparation for P3HT:PCBM bulk heterojunction solar cells

M Sendova-Vassileva, G Popkirov and P Vitanov
Central Laboratory of Solar Energy and New Energy Sources, Bulgarian Academy of Sciences, 72 Tzarigradsko Chausse, 1784 Sofia, Bulgaria

E-mail: marushka@phys.bas.bg

Abstract. The performance of bulk heterojunction P3HT:PCBM polymer solar cells is studied using current-voltage and spectral sensitivity measurements. Front transparent contacts of either ITO or a sputtered multilayer TiO_2/Ag/TiO_2 transparent conductive electrode are applied. Thin PEDOT:PSS and P3HT:PCBM polymer layers are deposited on these substrates by spin coating. The degradation of the solar cells is studied by subjecting them to UV irradiation. The obtained spectral sensitivity curves are analyzed in comparison with the optical absorption spectra of the transparent electrodes and active layers employed. It is observed that there is a pronounced dependence of the spectral sensitivity curves on the thermal treatment and UV irradiation applied to the solar cells. The performance of the multilayer TiO_2/Ag/TiO_2 transparent conductive electrodes is also studied in comparison with that of the ITO ones.

1. Introduction

The design and study of solar cells employing polymer materials is a field of intense activity in recent years. This is due to the results up to now, which promise easy to produce, low cost, non-toxic photovoltaic devices. One of the properties very important for long serving structures like solar cells is their stability. This challenge is especially significant for polymer solar cells. There are many ways in which they can degrade connected with the different layers comprising them [1]. In this contribution we present data on the degradation of encapsulated P3HT:PCBM bulk heterojunction solar cells under concentrated UV radiation from a solar simulator. This type of treatment is studied less often. The concentration is needed in order to speed up the degradation as the exposure of the studied cells to real conditions of solar irradiation for more than 30 hours failed to produce a noticeable deterioration of their performance. Besides, the filtered UV radiation does not increase noticeably the temperature of the cells and thus excludes heating effects. The aim of this contribution is to study the effects of degradation using the solar cell spectral sensitivity spectra.

As explained in [2] the multilayer transparent conductive films TiO_2/Ag/TiO_2 are studied because they avoid the use of expensive indium, can be thinner and more transparent for the same conductivity and can be deposited at lower temperature. According to the authors of [3] a thin amorphous TiO_x layer can play the role of electron collection and hole blocking layer and contribute to the air stability of the solar cells. In this report TiO_2/Ag/TiO_2 films are experimented as front contacts in P3HT:PCBM bulk heterojunction polymer solar cells and as a potential intermediate layers in tandem polymer solar cells. Immediately on top of the transparent conductive film PEDOT:PSS hole conducting layer is deposited. If the combination works the electrons and holes will recombine at the interface between the two films.

2. Experimental

The samples studied in this work are deposited as described previously [2, 4]. The solar cells consist of the following layers: front transparent contact, PEDOT:PSS, bulk heterojunction P3HT:PCBM, back contact of Ag deposited by sputtering without annealing in the vacuum chamber.

The front transparent contact is either ITO used as purchased or TiO_2/Ag/TiO_2 multilayer contact deposited in our laboratory. The method used for forming the contacts is RF magnetron sputtering. The equipment allows the deposition of layers using three targets at the same time. The multilayer contacts are formed on glass through a shadow mask by depositing of TiO_2, Ag, TiO_2 consecutively. The TiO_2 layers are 25 nm thick as measured by ellipsometer. The Ag layer has granular structure which is discussed in more detail below. The pattern of the front transparent contacts on the glass is similar to the one of the purchased ITO covered glass which is used as standard substrate and front contact. However, the active area of the cells in the case of our shadow mask is 3.75 mm^2 instead of 4.5 mm^2.

The active area of the samples is encapsulated and electrical connection legs are mounted as described in [3]. Post-production annealing of some of the samples is performed by heating the already encapsulated cells on a hot plate at 150°C for 10 min.

Current-voltage (J-V) measurements upon illumination by a halogen lamp are carried out by a computer controlled home-made potentiostat. The spectral response measurements are performed on a computer controlled set-up including a halogen lamp, a DK-240 monochromator, a focusing lens and a chopper as light source, and a home-made current amplifier and Brookdeal 401 lock-in amplifier for the photocurrent measurement. The illumination spectrum is corrected using a calibrated Si diode from Hamamatzu. The optical transmission of individual layers is measured by a Shimadzu UV-3600 double beam spectrometer.

UV degradation of the cells is performed with a solar simulator modified with a UV-pass-filter and a lens to concentrate the illumination intensity on the sample 70 times. Samples are irradiated for intervals from 2 to 18 hours and their I-V curves are measured periodically.

3. Results and discussion

3.1. Effect of degradation and annealing

Figure 1 shows typical current-voltage characteristics of the solar cells with ITO front contact as deposited, after post-deposition annealing and after 18 hours of UV degradation of an annealed sample. It is obvious that post-deposition annealing improves significantly the parameters of the polymer solar cells as already reported in [4]. Figure 2 demonstrates the evolution of the main parameters of the solar cells during the UV irradiation. It is the short circuit current (I_{sc}) that gets the greatest reduction of its values followed by an almost 50% decrease of the open circuit voltage (U_{oc}).

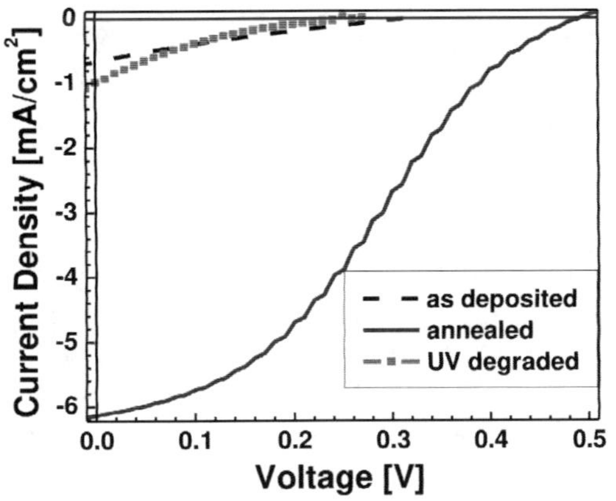

Figure 1. J-V characteristics of polymer solar cells with ITO front contact - as deposited, after annealing and after 18 hours UV degradation of the annealed cell.

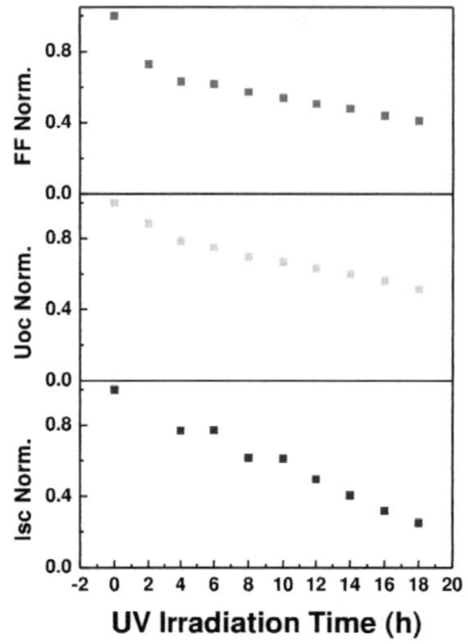

Figure 2. Evolution of J-V parameters with the time of UV irradiation.

Consequently, the fill factor (FF) falls by more than 50% as well. Concentrated UV irradiation affects the cells strongly and as seen in figure 1, the solar cell parameters return to values similar to the ones before post-production annealing.

In an attempt to learn more about the details of the degradation process the spectral response curves of as deposited, annealed and UV degraded (18 h) solar cells are measured. Typical spectra are shown in figure 3. Figure 4 shows the absorption spectra of the bulk heterojunction P3HT:PCBM layer as deposited without any treatment and annealed for 30 min at four different temperatures as indicated in the figure. More details about the interpretation of the dependence of the absorption spectra on the annealing temperature are given in [5].

The optical absorption of the bulk heterojunction P3HT:PCBM layer in the visible part of the spectrum is due to the polymer P3HT [6]. It demonstrates three main absorption peaks: 517 nm attributed to π-π absorption and the 552 and 602 nm ones - due to vibronic absorption in P3HT in the

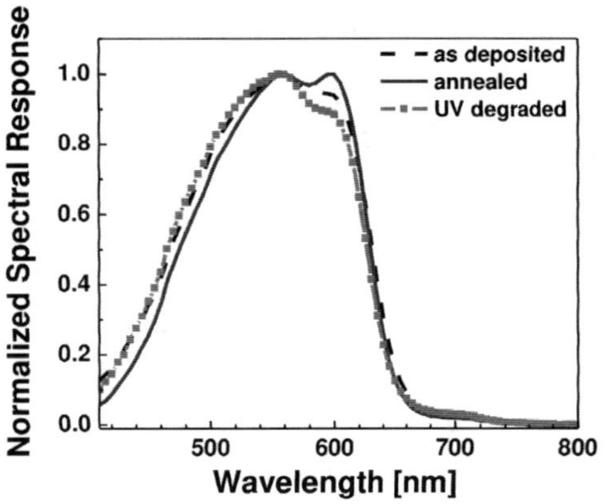

Figure 3. Normalized spectral sensitivity curves of polymer solar cells with ITO front contact - as deposited, after annealing and after 18 hours UV degradation of an annealed cell.

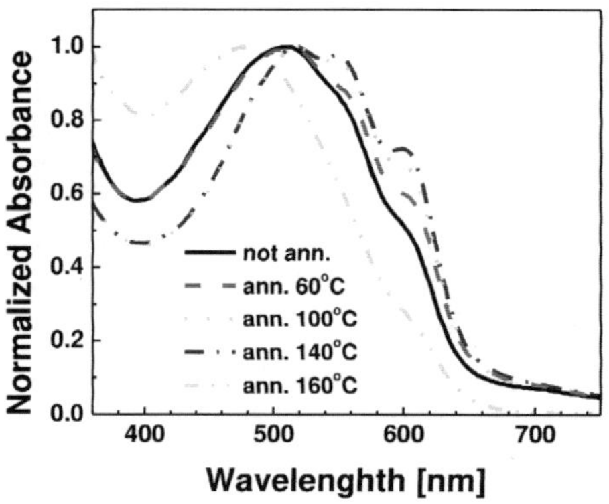

Figure 4. Normalized absorption of P3HT:PCBM layers annealed at different temperatures in Ar as indicated in the figure.

solid state. The 602 nm band is connected to interchain interaction and has a higher intensity for higher degree of interchain order. These features are clearly visible in the absorption spectra in figure 4 and are best expressed in the films annealed at 140°C.

The spectral response curves in figure 3 exhibit clearly visible the bands at 552 and 602 nm. The absorption peak at 517 nm (cf. figure 4), however, cannot be distinguished as a peak in the spectral response. Perhaps, as in inorganic solar cells the separation and extraction of carriers is hindered for higher energy photons. The difference between the spectral response of the three kinds of samples studied is in the relative intensity of the 602 nm vibronic band. Post-deposition annealing sharpens that band and presumably increases the interchain order of the polymer. This can be correlated with the improvement of the solar cell parameters which has been observed by a number of authors [7, 8]. Of course this improvement may have other reasons connected with the contacts as well [9]. The fact that this band weakens after the UV degradation process could be an indication that the fall in the values of the main solar cell parameters can be connected in part to chemical processes occurring in the active layer under the influence of the high energy UV radiation and deteriorating the interchain order.

3.2. TiO_2/Ag/TiO_2 front transparent contacts

The experiment of applying the multilayer TiO_2/Ag/TiO_2 transparent film as front contact aims to find out whether a functioning solar cell could be formed using it and to compare that cell with the one with ITO front contact. A positive result would allow the preparation of a tandem organic solar cell in which the combination of the TiO_2/Ag/TiO_2 multilayer with PEDOT:PSS hole transporting layer could be used as an intermediate layer between the two subcells. If the structure works the electrons and holes will recombine at the interface between these two layers.

As mentioned in the Experimental section the multilayer TiO_2/Ag/TiO_2 transparent film consists of two 25 nm TiO_2 dielectric layers and a granular Ag layer sandwiched between them. The size, shape and density of the metal granules were investigated with AFM on specially prepared samples. Silver was deposited on a polished silicon wafer under the same conditions that were used in the multilayer. In figure 5 an AFM image of this layer is presented.

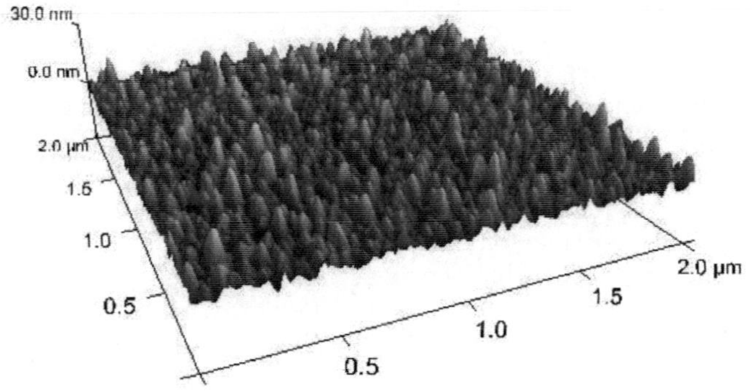

Figure 5. 3-D AFM image of metallic granules.

Analysis of the AFM data have show that the metal Ag layer has granular structure with grain height of 10-20 nm and width of about 80 nm.

Figure 6 compares the transparency of the ITO front contact and the multilayer sputtered TiO_2/Ag/TiO_2 contact experimented in this work. It can be seen that the multilayer contact has comparable transparency to the ITO one. The reason for the fall in transparency in the 400-500 nm range has not been clarified but it is outside the absorption range of the active layer of the solar cell and would not affect its performance significantly. The conductivities of the two transparent contacts are very similar as well, 20 and 17 Ω/□, respectively. The $TiO_2/Ag/TiO_2$ layer demonstrates metal-like electron conductivity and its optical transmission falls above 700 nm due to the reflection connected with free carriers.

Figure 7 shows the quantum efficiency curve measured on two typical solar cells differing only in the type of front contact. The curves are given in relative units but their values can be compared. It can be seen that the shape of the quantum efficiency curves are very similar but the values of QE are smaller for the multilayer $TiO_2/Ag/TiO_2$ transparent front contact. Possible reasons are the high contact resistance or the unfavourable for the hole transport mutual valence band (HOMO) energy positions of PEDOT:PSS and TiO_2 [10]. However, experiments indicate that the $TiO_2/Ag/TiO_2$ multilayers behave more like degenerate semiconductor or metal than as a wide bandgap semiconductor.

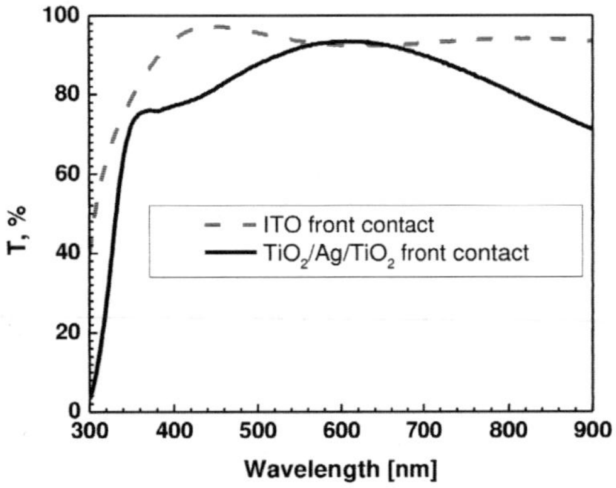

Figure 6. Optical transmission of the two different front transparent contacts used.

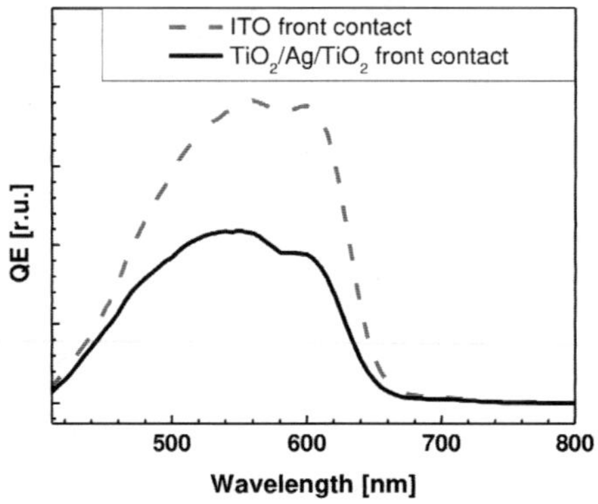

Figure 7. Relative quantum efficiency of the solar cells with different transparent front contacts, as shown in the figure.

Still our results show that there is a functioning solar cell with the new $TiO_2/Ag/TiO_2$ front transparent contact. Its quantum efficiency is lower than that of the ITO containing one and further experiments are needed to optimize this solar cell architecture.

4. Conclusion

The effect of concentrated UV radiation on bulk heterojunction P3HT:PCBM polymers solar cells has been studied. This kind of irradiation is interesting because it speeds up the solar cell degradation and separates the effect of light from that of heat. The latter often recovers some of the degradation. Concentrated UV irradiation strongly affects both the short circuit current and the open circuit voltage. The study using the spectral sensitivity curves of the solar cells shows that the degradation possibly results from the loss of polymer interchain order in the active layer of the cells.

The sputtered $TiO_2/Ag/TiO_2$ front transparent contact exhibits good transparency and lateral conductivity. Functional solar cells are realized using that contact instead of ITO. The shape of the quantum efficiency curves of the polymer solar cells is not affected by its use in the place of the ITO front electrode. However the current density is reduced. This is most probably due to high contact resistance or an energy barrier at the interface.

Acknowledgement. The authors are grateful to V. Gancheva, G. Grancharov from Institute of Polymers, Bulgarian Academy of Sciences, E. Lazarova, Ch. Dikov form Central Laboratory of Solar Energy and New Energy Sources, Bulgarian Academy of Sciences, for layer deposition, V. Bozhilov and E. Bubev from University of Chemical Ttechnology and Metallurgy, Sofia for UV irradiation and O. Milushev for technical help.

References

[1] Grossirod N, Kroon J M, Andriessen R and Blom P W M 2012 *Org. Electron.* **13** 432
[2] Sendova-Vassileva M, Dikov Ch, Popkirov G, Lazarova E, Gancheva V, Grancharov G, Tsocheva D, Mokreva P and Vitanov P 2013 *J. Phys.: Conf. Series* **514** 012018
[3] Kuwabara T, Nakayama T, Uozumi K, Yamaguchi T and Takahashi K 2008 *Sol. Energ. Mat. Sol. C.* **92** 1476
[4] Sendova-Vassileva M, Popkirov G, Vitanov P, Dikov Ch, Gancheva V, Tsocheva D and Mokreva P 2012 *J. Phys.: Conf. Series* **398** 012049
[5] Sendova-Vassileva M, Bakardjieva V, Ivanova T, Lazarova E, Gancheva V, Tsocheva D, Mokreva P, Terlemezyan L and Vitanov P 2013 *CR Acad Bulg Sci* **10** 1393
[6] Li P, Chen L J, Niu G X, Zhang T, Xiang J, Cai L, Hu Y, Zhang Y J, Wan K M and Song Q L 2014 *Sol. Energ. Mat. Sol. C.* **125** 96
[7] Dennler G, Scharber M C and Brabec C J 2009 *Adv. Mater.* **21** 1323
[8] Li G, Shrotriya V, Huang J, Yao Y, Moriarty T, Emery K and Yang Y 2005 *Nat. Mater.* **4** 864
[9] Ahlswede E, Hanisch J. and Powalla M 2007 *Appl. Phys. Lett.* **90** 063513
[10] Kim J Y, Lee K, Coates N E, Moses D, Nguyen T-Q, Dante M and Heeger A J 2007 *Science* **317** 222

Deposition by liquid epitaxy and study of the properties of nano-heteroepitaxial structures with quantum dots for high efficient solar cells

I I Maronchuk[1*], I E Maronchuk[1], D D Sankovitch[1], K Lovchinov[2], D Dimova-Malinovska[2*]
[1] Laboratory of Applied Physics and Nanotechnologies for the Energy Application, Sevastopol National University of Nuclear Energy and Industry, str. Kurtchatova, 7, 299033 Sevastopol, Russia
[2] Central Laboratory of Solar Energy and New Energy Sources, Academy of Sciences, bulv. Tzarigradskao chaussee, 72, 1784 Sofia, Bulgaria

E-mail: doriana@phys.bas.bg, igimar@mail.ru

Abstract. The paper presents experimental results showing the possibility of liquid-phase epitaxy with a pulse-cooled substrate nano-heteroepitaxial structures with Ge quantum dots and GaP matrix for production of single-junction solar cells working under sun light concentration. The preparation of the surface of Si wafers suitable for growing of these structures is described. The optimal technological conditions for growing of the of GaP buffer layer on the Si wafer from supersaturated solutions of melts Sn are applied. Structures with Ge quantum dots are grown at different conditions and study of their structural and optical properties is performed. The surface morphology of the grown materials is studied by AFM microscopy. The photoluminescence spectra of nano-heteroepitaxial structures with Ge quantum dots are presented and analyzed.

1. Introduction

Currently 80% of solar cells converting directly (not concentrated) solar radiation efficiency ranging from 12% to 22% are based on monocrystalline, polycrystalline or amorphous silicon. Provided photovoltaics growth at 30% per year until 2030, the need for silicon will increase by more than 200 times that is a problem, both technological and environmental. In this regard the development of solar cells and modules working under high concentration of the solar radiation using single junction solar cells on the basis of nano-hetero-epitaxial (NHE) structures with quantum dots (QDs) arrays offer the possibility for further increasing of the efficiency. The idea launched by A. Nozik [1] to achieve the thermodynamic limit of photovoltaic conversion efficiency (till 84%) using both, short and long wave range of the solar spectrum by applying QDs of narrow gap semiconductors has been confirmed theoretically

Different methods of deposition of high efficiency solar cells using nano-heterojunctions (NHJ) structures with QDs have been experimented. Attempt to use molecular beam epitaxy (MBE) and chemical vapor deposition from organometallic compounds (MOCVD-technology) in the process of preparation of solar cells with high efficiency on the basis of nano-heterojunctions (NHJ) with QDs, where the matrix material used is GaAs, and the QDs material is InAs have been unsuccessful [2].

In this work the method of liquid phase epitaxy (LPE) in the regime of impulse cooling and heating (ICH) of the Si substrate is applied to deposit Ge QDs arrays without elastic - tense "wetting" layer in between QD thus assuring conditions for their formation close to the equilibrium and growing of the structures with reduced carrier recombination and solar cells with efficiency > 35% [3].

It is known that the morphology and the composition of the substrate affect significantly the process of growth of NHE structures with QDs, because it determines the parameters of the crystal lattice of the surface layer and thus the quality of the NHE structures with QDs deposited by LPE method using ICH of the substrate [4]. Therefore, high-quality surface preparation of the substrates plays an important role in applied technology LPE.

The aim of this paper is formation of Ge QDs during the LPE process of crystallization with applying procedure with ICH of the substrate and study of the surface morphology and photoluminescence of p-n nanostructures based on A3B5 semiconductors (GaP) on Si substrates obtained in the process of multi-layer epitaxial growth. Using Si substrates which are less expensive that A_3B_5 semiconductors offers advantages concerning decreasing the cost of the solar cells.

2. Experimental

Deposition of the NHE structure with QDs was performed by the LPE method with impulse cooling and heating of the substrate. Special equipment for growing of multilayer NHJ structures with horizontal supplier has been constructed. In this method the front surface of the substrate is stand in a close contact with the saturated melt solution heated to the temperature T. A heat sink which temperature, THS, which is lower than the temperature of the substrate by the amount ΔT = T –THS is situated on the back surface of the substrate. After some time, which depends on the duration of the cooling impulse τ ≈ 10-3 ÷ 10-1 sec, a heat sink takes the temperature of the substrate and nanolayer with QDs arrays is crystallized on the front surface of the substrate [3].

Due to the different in lattice constants of the substrate material (Si) and the material of QDs (Ge), in the initial stage a continuous "wetting" layer by the Stranski-Krastanov's mechanism is formed. With increasing of the thickness of the growing layer the periodic stresses appears at the interface with the substrate. During the process of growing the formation of the Ge QDs takes place in the sites with minimum mechanical stress, i.e. in the positions with perfect structure [5]. The density of the crystallized QDs at optimum conditions is limited mainly by the period of the perfect crystalline structure in the interface.

Growing of NHE structures with of QDs of narrow-gap material-Ge, with a surface density < 1012 cm-2 on the wider band gap material Si single crystal substrate than used in [2] is performed at 450-500°C temperature of the oven in atmosphere of highly purified hydrogen. Growing of QDs arrays is completed when the temperature of the heat sink becomes equal to the substrate temperature. After that the heat sink is removed and heater with temperature TH above the substrate temperature Ts by an amount ΔTH = TH - Ts. is placed on the back surface of the substrate. The heating impulse leads to the dissolution of the nano-sized "wetting" layer situated between the QDs on the front surface of the substrate. The preliminary experiments show that removing part of the "wetting" layer results in formation of NHE structures with QDs with significantly reduced concentration of generation-recombination centers and in formation of the QDs arrays with parameters close to the "ideal" in the case when the heating impulse amplitude and duration are about 0.3-0.7 of the corresponding values of the cooling impulse. The heating impulse, ΔTH, not only leads to removing the part of the "wetting" layer, but also eliminates the islands of fused quantum dots, where the areas with QDs are slicked in the islands in the place with misfit dislocations, thus it promotes their separation and formation of the QDs with less variation in the size. The process of removing of the part of the "wetting" layer between QDs is terminated when the temperature of the heater becomes equal to the substrate temperature. After that the arrays of the QDs is covered by nano-layer of the matrix material (spacer layer) - GaP. The thickness of the spacer layer should provide tunneling of charge carriers from one QDs arrays located parallel to the surface of the substrate to the other QDs arrays – the vertically coupled QDs.

Location of the superlattices with vertically coupled QDs in the p-n junction, leads to the separation of the generated in QDs carriers by the electrostatic field of the p-n junction.

Multiple repeating steps of growing of QDs arrays and formation of the spacer layers of different type conductivity lead to creation of multi-layer p-n NHE structures with QDs. Monocrystalline Si wafers n-type conductivity with the crystallographic orientation (111), sheet resistance of 20 Ω/□ and diameter of 40 mm are applied as substrates. The solvent used is Sn of high purity. As the matrix material wide-gap A3B5 semiconductor – GaP, is used and narrow-gap semiconductor Ge - for the QDs formation.

The equipment for deposition of NHE structures is designed and manufactured. It consists of a heating furnace, a quartz reactor with a working cassette and snap, a temperature regulator, gas supplier system and vacuum pumps. Heating (to maximum 1200 oC) and cooling are performed with rate of 30oC/min and 10oC/min, respectively. A special graphite cassette-holder is constructed for growing NHE structures from melt-solution on substrate with different diameter and thicknesses.

The surface morphology of the samples is studied by AFM microscope type CMM-2000 (Proton—Zelenograd, Russia) with soft cantilever tip MSCT (Veeco, USA). The spectra of photoluminescence of the grown nano-heteroepitaxial structures with QDs are measured using spectral equipment with monochromator type MDP-4.

3. Results and discussion

The surface morphology of the Si wafers used for deposition NHE structures is studied before and after dynamic chemical treatment. Figure 1 shows the AFM picture of the pristine silicon substrate before chemical and dynamical processes of preparation. It is observed that the initial silicon wafer has a number of defects - the average surface roughness of the scanned area of 3.7 m length was 3.2 nm and the maximum difference in texture surface - 30 nm. Scratches obtained after the mechanical and chemical polishing are seen. The three-dimensional picture (figure 1 b) reveals the depth of scratches in some places as high as 40 nm and a length of more than 400 nm.

The preparation of the Si wafer for deposition of the NHE structures is carried out by applying the chemical-dynamic treatment using two standard procedures for removing the damaged layer as a result of cutting and the chemical mechanical polishing Deshu's [7] and RCA methods [8]. Improvement of the surface morphology of the original Si wafer is achieved.

Figure 2 demonstrates the AFM image of the Si wafer after the chemical-dynamic treatment applying the Deshu's method [6]. The polished Si wafer is degreased, the oxide and damaged layers - removed. It is clearly seen from the AFM image shown in figure 2 a the lack of scratches and the presence of small pits formed during the etching of the material, the maximum size of which is about 300 nm and a depth - 8 nm. The wafer treatment is carried out immediately before starting the process of epitaxy.

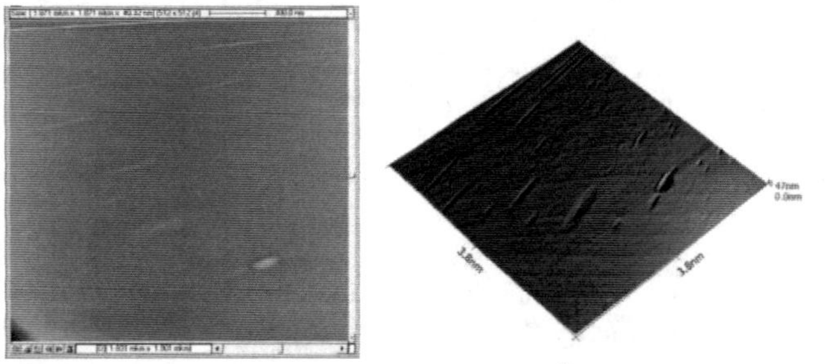

a b

Figure 1. AFM images of the pristine Si substrate: (a) surface view and (b) three-dimensional view. The marker in (a) corresponds to 200 nm.

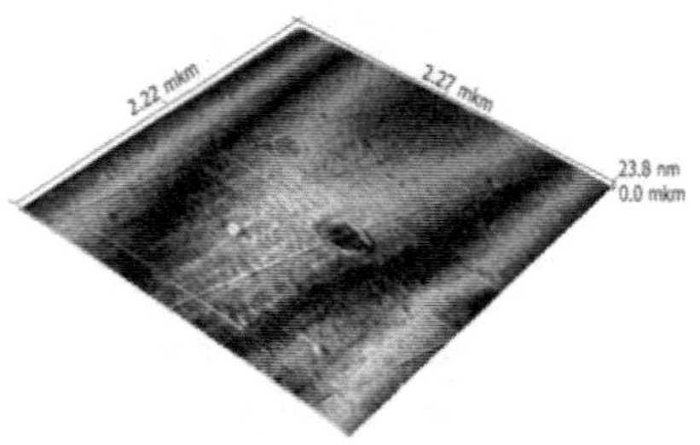

Figure 2. The AFM image (2.22 μm x 2.27 μm) of the Si wafer after treatment by the Deshu's method.

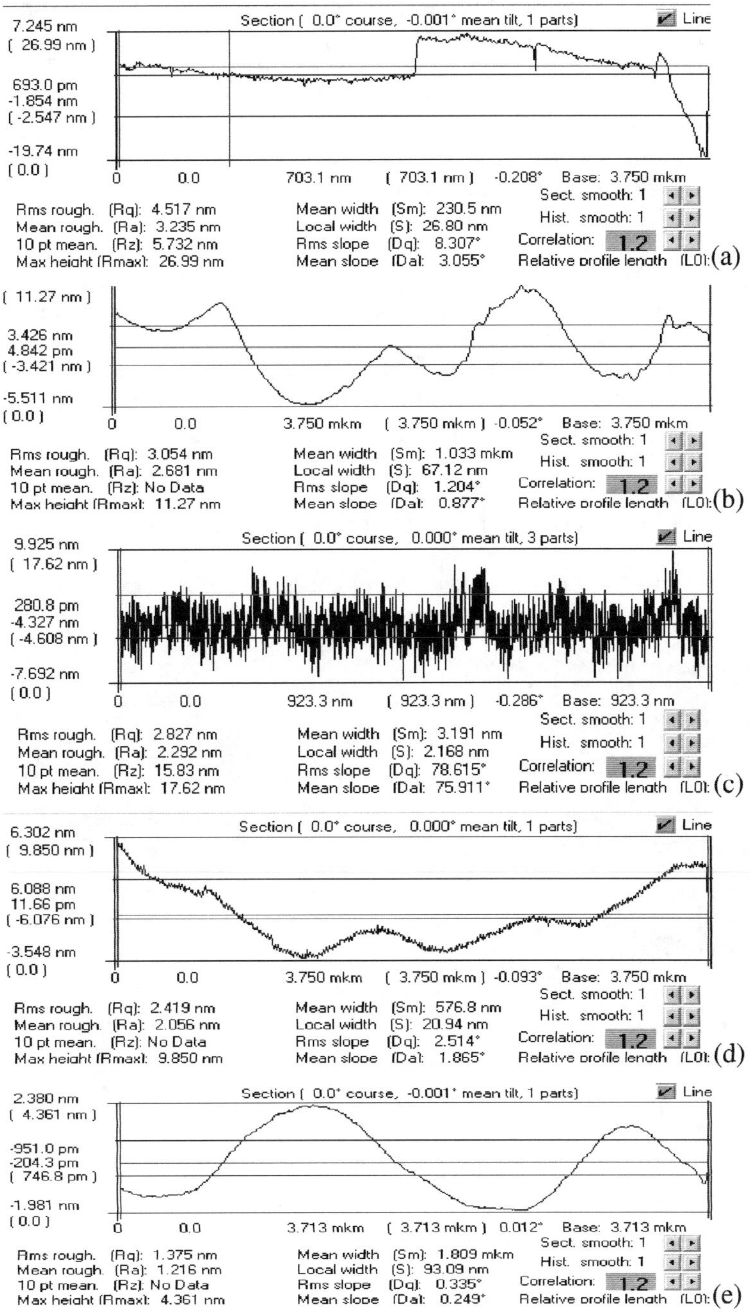

Figure 3. Surface topography of the subatrate of Si Si substrates before treatment (a), after treatment by chemical-dynamical method RCA (b), after treatment by Deshus's chemical-dynamical method (c), with growth buffer layer after Deshus's method treatment and (d) and with grown buffer layer after RCA treatment (e). The profiles are taken on a segment of 3.7 μm.

Figure 3 shows the surface topography of the Si substrates before treatment (a), after treatment by chemical-dynamical method RCA (b), after treatment by Deshus's chemical-dynamical method (c), after Deshus's method treatment with growth buffer layer and (d) and after RCA treatment with grown buffer layer (e). It is seen that the maximum difference in high of the surface features of 11 nm and 18 nm and the average roughness of 2.6 nm and 2.4 nm in the linear segment of 3.7 µm microns are obtained after treatment by RCA (figure 3 b) and by Deshus's method (figure 3 c), respectively. Both techniques of chemical and dynamic treatment significantly improved the surface morphology of the original Si substrate, the average surface roughness of the samples obtained are practically at the same level within the error of the measurement method, and the maximum difference in the high of the surface features is in favor of techniques RCA. From the figures 3 c and d is observed that the average surface roughness is 2.0 nm and 1.2 nm and difference in the surface texture is 10 nm and 4 nm after formation of the buffer layer on the substrates treated by Deshus's and RCA method, respectively. The results demonstrate that obtained qualities of the buffer layers after RCA treatment is better and can be considered to be suitable for growing of high-quality NHE structures with QDs on Si substrates.

During the process of LPE the optimum technological conditions for epitaxial growth of the buffer Si layer from supersaturated solutions of melt Sn have been established.

A thick GaP layer doped with Sn is deposited on the obtained buffer layer. For growing of the Ge QDs 1 and 3 cooling impulses with difference in the temperature of the heat sick and substrate, $\Delta TH = 5^{o}C$ are applied.

The study by AFM demonstrates that the surface roughness of the structure with grown Ge QDs is in the range of 1 ÷ 5Å. The preliminary study shows that the obtained after one and three pulses are

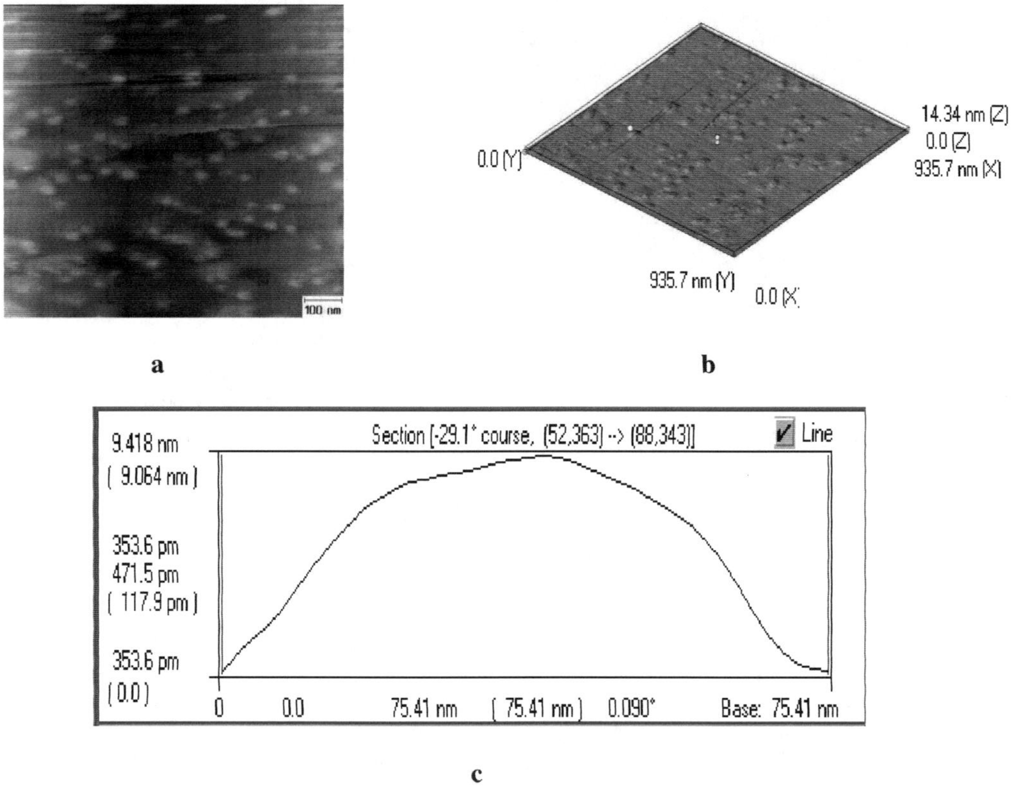

Figure 4. AFM surface view (a), three dimensional image (b) and profile (c) of the Ge QDs in the matrix of GaP on the (111) Si substrate obtained after applying of 3 impulses. The marker in (a) corresponds to 100 nm.

small in size and the QDs with sharp edges corresponding to the crystallographic orientation of the original Si substrate are not seen. Figure 4 shows typical two- and three-dimensional images and the profile, respectively, of the Ge QDs grown after applying of three pulses of cold heat sink. The average height of the QDs is about 8 nm -10 nm and an average width is about 40 nm - 60 nm. The formation of the Ge QDs can be explained by the presence of the stress during the growing of the Ge nano-grains due to the mismatch in the lattice constants of GaP and Ge (3.7%), which results in formation of Ge QDs with $d \leq 10$nm.

The optical properties of the NHE structures with QDs are investigated on the basis of the photoluminescence spectra. The study of the n-p NHE junction structures with Ge QDs is performed. For comparison the PL spectra of the n-p GaP structures without Ge QDs are presented as well.

Two types of NHE structures are fabricated and their photoluminescence (PL) spectra are measured and compared between them and with PL of the p-n GaP red diode grown on Si substrate. Structure 1 consists of Si substrate treated by RCA method, a buffer layer of GaP doped with Sn, n-p superlatices consisting of Ge QDs arrays separated by spacer layer of n-type GaP (doped with Te) and Ge QDs arrays separated by spacer layer p-type GaP (doped with Zn), and p-type GaP 80 nm thick layer. In between the superlatices are deposited:

- n-type GaP 20 nm thick layer to the n- type superlatice;
- p-type GaP 20 nm thick layer to the p- type superlatice.

Structure 2 is different from the structure 1 by the lack of 20 nm n- and p- type GaP layers between the superlatices with Ge QDs.

Figure 5 demonstrates the PL spectra of the structures 1 excited by two laser radiation with λ = 4880Å (curves 1 in figure 5 (a) and λ = 5145Å (curve 1 in figure 5 ((b)). For comparison the PL spectra of the p-n GaP red diode without Ge QDs are shown as well (curves 2 in figures 5 (a) and (b)). The PL spectrum of the structure 1 with Ge QDs (Figure 5 (a) excited by laser with λ = 4880Å shows a broad band with a maximum at 1.49 eV. The PL maximum position of the p-n diode without Ge QDs is observed at higher energy – 1.77 eV. The intensity of the PL in the structures with Ge QDs is stronger than the intensity of PL of the junction without QDs. Under excitation with laser radiation with the larger wave length, λ = 5145Å (figure 5 b), the position of the PL maximuma does not change however the intensity of maximum the PL band of the structure 1 with Ge QDs decreases about 3 times and it become less intensive than the intensity of PL of the n-p GaP structures without QDs. The

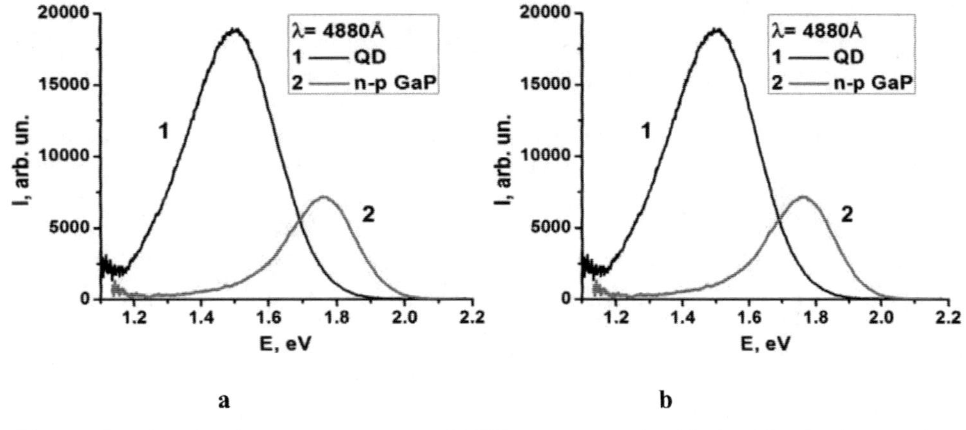

Figure 5. Photoluminescence spectra of NHE structures 1 with Ge QDs in the matrix of GaP (a) under excitation with wave lengths: λ=4880Å (a) and λ=5145Å (b) (curves 1). For comparison the PL spectra of the n-p GaP structures without Ge QDs are shown as well (curves 2).

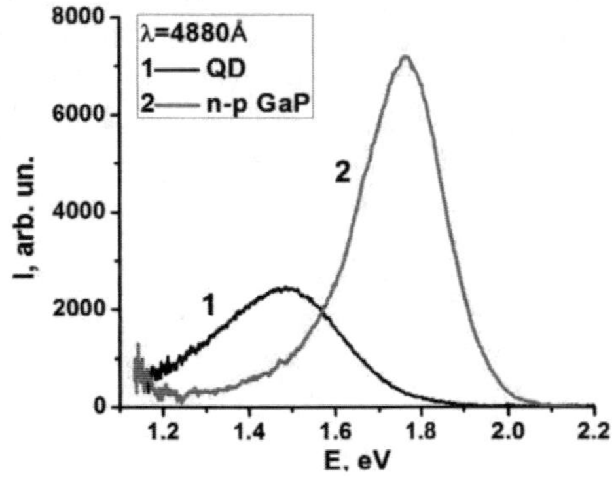

Figure 6. Photoluminescence spectra of NHE structures 2 with Ge QDs, without GaP layer between the superlatices measured under excitation with wave lengths λ=4880Å (curve 1). For comparison the PL spectra of the n-p GaP structures without Ge QDs are shown as well (curves 2).

ratio of the intensity of the maximum of PL bands under excitation with λ = 4880Å is about 2.5 and under excitation with λ = 5145Å - about 0.86. The observed dependence of the PL intensity on the energy of the exciting radiation can be explained by the effect of excitation splitting of the high-energy quant of the radiation. In this case of the presence of the quantum dots, two or more electron-hole pairs can be generated when of the photon energy exceeds about two time the energy of formation of electron-hole pairs (Nozick's effect) [1].

Figure 6 shows the PL spectrum of the structure 2 measured under 4880 nm laser excitation. PL band due to the Ge QD is observed at the same energy 1.49 eV however its intensity is lower about 6 times than PL intensity of the structure 1 with 20 nm GaP layer between the superlatice. The ratio in the intensity of the PL of the structure with QDs and that of the p-n GaP diode without QDs is about 0.35. This observation demonstrates that the 20 nm GaP layer between the superlatices with Ge QDs leads to decreasing in the nonradiative recombination of nonequilibrium carriers. It is possible to conclude that to increase the radiative recombination in the NHE structures it is necessary to create arrays of QDs not in the depletion region (structure 2) but in the volume of p- n junction as it is in the case of the structure 1 which is more favorable case for application into solar cells. Applying the effective mass approximation using the Schrödinger equation at different values of the size of the Ge QDs is possible to calculate the energy spectrum of charge carriers in Ge QDs between the GaP layers:

$$E_n = \frac{\pi^2 \hbar^2}{2m^* d^2}\left(n - arctg\sqrt{\frac{U_0}{E_n}} - 1\right)^2 \qquad (1)$$

where En is the energy of the mini-gap of the QDs material with quantum number n, measured from the bottom of the conduction band (for electrons) or valence band (for holes); U0 - potential barriers for the carriers; d - the width of the quantum hole (pit) m * - effective mass of the charge carriers.

If the size of the Ge QDs is assumed to be d $\leq$ 13 nm the calculated values for the energy are comparable with the energy of the PL peak position in figures 5and 6 (curves 1). The slight difference in the size of the Ge QDs between the values of the size of the Ge QDs (~10 nm) observed in the AFM images and the used for the theoretically estimation by eq.1 (d $\leq$13 nm) could be explained with not perfect control of the temperature during the growing of the superlattices Ge QDs/GaP layers. It should be noted that our results about the size and density of the Ge QDs reported here are close to the obtained and with the published by the other authors [9].

4. Conclusions

Equipment has been designed and used for LPE growing of hetero-epitaxial structures with Ge quantum dots from molten solution in p-n junction GaP solar cells on Si substrate. The technological conditions for preparation of the surface of Si substrate suitable for growing of buffer layer from supersaturated solutions of melt Sn are established and applied for deposition of NHE structure with Ge QDs. The Ge QDs are grown by the method of LPE with applying of impulse cooling and heating of the substrate. The study of the structural properties by AFM of the substrate after applying different treatment reveals the morphology, surface roughness, the linear dimensions of quantum dots and their density. It is observed that the structure with Ge quantum dots has more intensive bands of PL than that without QDs due to the increasing of the radiative recombination of the carriers, e.g. in the presence of the Ge QDs the nonradiative recombination is reduced. It is possible to conclusion that in order to increase the radiative recombination in multilayer p-n NHE structures it is necessary to create arrays of QDs in the volume regions with p-and n- GaP. The work is in progress to apply the method of LPE growing of the nano-heteroepitaxial A3B5 structures with Ge QDs in p-n junction GaP on Si substrates for high efficient low cost solar cells.

Acknowledgement
The work has been supported by the NSF- of Bulgaria - grant DNTS/Ukraine - 01/0005.

References
[1] Nozik A 2007 *Technology review* **March/April** 49
[2] Alfelov I 1998 *Physics of Semiconductors* (in Russian) **32** 3
[3] Maronchuk I E, Kulyutkina T F and Maronchuk I I *Method of growth epitaxial nano hetero-junction structures with quantum dots array-Patent Ukraine* UA № 94699 Class C 30B 19/00, C 30B 29/00, H 01L 21/20. 20.09.2010-10.06.2011 bulletin №5
[4] Ledencov N N 1998 *Physics of Semiconductors (in Russian)* **32** 385
[5] Dubrovsky V G 2009 *Theory and formation epitaxial structure* (Moscow: MIR) (in Russian) p. 350
[6] Maronchuk I E, Maronchuk I I, Kulyutkina T F and Buikovsky S 2012 *Nanosystem, nanomaterials, nanotechnology: Scientific proceedings (in Russian)* **10** 77
[7] Shwartz B and Robbins H 1976 *Journal Electrochemical Society* **1906** 123
[8] Kern W 1990 *Journal Electrochemical Society* **137** 1887
[9] Kuhler U 1991 *Surface Science* **248** 321

X-ray investigations of proton irradiation induced structure development in polyester fibers

V Velev[1], A Popov[2], N Arhangelova[1], P Kyurkchiev[1], S Stefanova[1], P Sartori[3] and N Uzunov[1,3]
[1]Konstantin Preslavsky University, 115 Universitetska Street, 9700 Shumen, Bulgaria
[2]University "Prof. Dr. Assen Zlatarov", 1 "Prof. Jakimov" Street, 8000 Burgas, Bulgaria
[3] National Laboratories of Legnaro, INFN, Viale dell'Universita 2, 35020 Legnaro (PD), Italy

E-mail: v.velev@shu-bg.net

Abstract. The role of proton irradiation on the structural changes in partially crystalline poly (ethylene terephthalate) (PET) fibres has been investigated using a wide-angle X-ray scattering (WAXS). Experimental data about the influence of the exposure dose and the irradiation conditions in the processes of destruction and crosslinking in the irradiated objects have been obtained. The structural changes and the packing density of the macromolecular chain segments of the poly (ethylene terephthalate) filaments irradiated with different doses have been analyzed. A mechanism of the structural reorganization initiated by the irradiation has been supposed on the basis of the changes in the geometry of the intensity distribution of the X-ray diffraction patterns of the studied objects.

1. Introduction

Poly(ethylene terephthalate) (PET) is a representative of the linear polyesters crystallizable thermoplastic polymers. PET is characterized by relatively high glass transition temperature, high thermal stability and low crystallization rate. Moreover, the widespread use of PET is due to its ability to crystallize in conditions of deformation under temperatures and mechanical stresses applied during various treatments. PET can easily turn into uni- and biaxial oriented state with different qualitative properties. Due to the above-mentioned, PET is a widely used in many technological and industrial areas such as composite materials for the aerospace and space satellite industry, but mainly for the production of thin films and fibers.

The properties of PET filaments are highly dependent on the super molecular structure of the polymer in the oriented state. This, in turn, depends on the type of polymer, the initial isotropic super molecular structure, and the conditions of its restructuring in oriented. Ion beam irradiation is widely used in the polymer technology in order to change and improve the chemical and physical properties of the polymer materials in the recent years [1 - 12]. Proton irradiation of flexible-chain polymers, including PET is quite common in the practice [6 - 9].

Considering the wide application of PET it is interesting and very important from a technological and scientific point of view to study the influence of proton irradiation on the structure and properties of PET filaments.

In the present work is studied the influence of proton irradiation with low power and different exposure durations on the structure and properties of undrawn partially crystalline uniaxially oriented

PET. X-ray diffraction method was used for determination of the of object's ability for reorganization of the super molecular structure.

2. Experimental

2.1. Materials

The initial characteristics of the investigated PET fibres were as follows:

speed of fibrillate 4110 *m/min;*
number of single filaments in the complex thread 32;
diameter of a single fibre 11,0 μm;
degree of crystallinity α = 36,9 %;
birefringence $\Delta n.10^3 = 5{,}82$.

The studied samples are produced on industrial spinning installation Furnet (France). The presented parameters indicate that the investigated PET fibers are produced at a relatively high spinning speed, possess significant preliminary orientation and respectively degree of crystallinity.

2.2. Methods

2.2.1. Irradiation of the PET samples with protons. PET filaments have been fixed on flat plates with central holes (diameter of 1 *cm*) in the form of a layer with a thickness consistent with the range of the used for the irradiation proton beam. Plates with the samples were mounted on the outside wall of a circular holder (figure 1). The sample holder is placed in a chamber in which is achieved a high vacuum (figure 2). During the measurement the holder is rotated thereby are irradiated all positions of the holder loaded with samples. Samples were irradiated with protons using the accelerator facilities AN 2000 at the National Laboratories of Legnaro, INFN, Italy. Proton beams with energy of 2 *MeV* obtained by Van de Graaff accelerator have been used to irradiate the polymers' samples. The proton beam passed through a beam diffuser was collimated to a diameter of 4 *mm*. The collimated beam gone through the samples was collected for measuring of the total beam charge.

With purpose to prevent the samples from excessive heating the beam current was kept small enough with value of 5 *nA*. For each sample the total beam charge was measured and the number of protons irradiated the sample was calculated.

Figure 1. The circular holder and attached to it plates with the PET filaments.

Figure 2. Vacuum camera together with the measuring circular holder.

2.2.2. Wide-angle X-ray scattering. Due to the precise geometric and structural sensitivity of X-ray diffraction methods, the angular deviation of the diffraction reflections gives perfect information about the packing density of the structural elements on the atomic-molecular level. Knowing very well the crystallography of PET, the alteration in the geometry of the intensity distribution of the diffracted X-ray radiation can adequately visualize the repacking of the macromolecular chain segments in the volume of the crystal polymer phase, the transient zones of the crystallites and its mesophases. Furthermore, the changes in the intensity ratios, profile characteristics and angular deviation of the diffraction reflections provide information concerning to the crystalline phase perfection.

The structure of the non-irradiated fibers was studied by wide- angle X-ray scattering (WAXS) using a diffractometer HZG 4 (Freiberger Präzisionsmechanik, Germany) and Ni-filtered Cu K_α radiation with a wavelength of 1,5418 $\AA$. Equatorial scattering was monitored in transmission mode.

Structure of the irradiated fibers was studied by wide-angle X-ray scattering Diffractometer URD - 6 (under license of SIEMES) of the company "Freiberger Präzisionsmechanik" (Freiburg im Breisgau, Baden-Württemberg, Germany). Used is β-filtered with Ni-filter Cu K_α radiation with the same wavelength as mentioned above.

3. Results and discussion

The main effects of proton irradiation of polymer objects, in particular PET fibers might be destructive and crosslinking effects as the crosslinking effects are secondary after ionization and radicalization. The intensity distribution of the diffracted X-ray radiation of the non-irradiated PET fibers is present on the figure 3. The diffraction patterns obtained after the proton irradiation of the studied objects for 20 *s* (5.10^{12} p/cm^2), 40 *s* (10^{13} p/cm^2) and 2040 *s* (5.10^{14} p/cm^2) are shown on figure 4. The comparison between the intensity distribution of the source structure of PET fibers (figure 3) and the proton irradiated for 20 *s* (figure 4) shows partially crystalline structure formed during the fibre formation process.

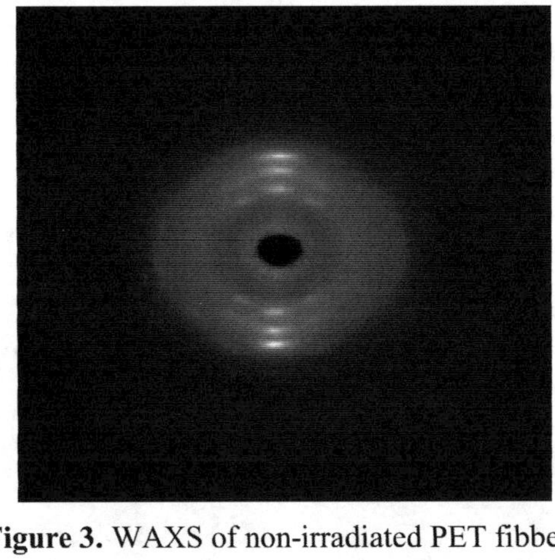

Figure 3. WAXS of non-irradiated PET fibbers.

The observed impact of the irradiation on the geometry of the intensity distribution of the diffracted X-ray radiation in the objects diffraction patterns consists of angular deviation of intensity changes in the range 10 $^0 2\theta$ - 30 $^0 2\theta$. In the diffraction patterns obtained after irradiation for 40 *s* and 2040 *s* (figure 4) can be formulated the following basic changes.

Main trend in the observed diffraction pattern of the PET fibers exposed for 40 *s* is the displacement of the diffraction peaks to the smaller diffraction angles. Observed effect is due to the increase in the distances as well between the folds as well as between the planes of the folding in the crystalline phase of the polymer.

The widening between the folds is larger, which can be a result from reasons as follows:

- the folds stretching, twisting of segments and distancing of the stretched during the fibrillate chains added contribution to increasing of the degree of crystallinity [13];
- improvement of the stacking of the macromolecular chain segments in the transition zones;
- formation of crystals with stretched chains.

Due to the different densities of the crystals with stretched chains and crystals with folded chains could be formed crystallites or a micelle structure with a lower physical density of the crystal, respectively, distances between the chains. Furthermore, the angular deviations can be a result of widening and multiple splitting of peaks, which in turn may be due to the morphology of the crystalline phase formed in different sizes and dispersion, type, quality and perfection.

The trend observed in the diffraction pattern of the PET fibers irradiated with 5.10^{14} p/cm^2 (figure 4) is the offset of the diffraction peaks towards larger diffraction angles. In this diffraction spectra (lasting 2040 *s* exposure) sufficiently precise information indicating the possibility of realization of recrystallizations processes resulting from the relaxation after the segmental destruction occurred.

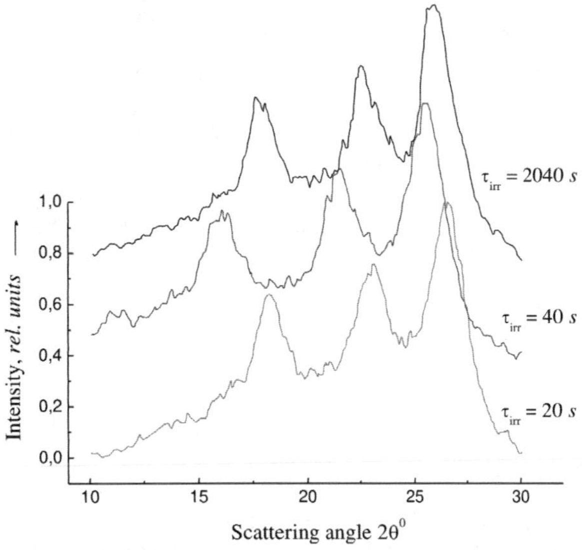

Figure 4. WAXS diffraction patterns of PET fibers irradiated with protons for 20, 40 and 2040 *s*. The curves are shifted vertically for clarity.

Interesting is that the proton irradiation during 2040 *s* returned to middle position and narrowed somewhat, again in an intermediate position the distances between the peaks. Observed effects are evidence of completely increasing of the studied between planar distances in the elementary crystal cell. On one hand such an extension of the cell (decreasing of the polymer mass density) can be associated with the logical improving the arrangement of the segments in the cell. Moreover, the displacement of the small angular diffraction peaks is larger than the shift of the peaks formed at larger angles. On the other hand, the third longest exposure causes a slight positional shift of the multiple peaks back to the right to the larger angles, as well as bringing them closer to each other.

This is the opposite effect to the observed at exposures 20 *s* and 40 *s*, but to a lesser extent. The complex summary variation in the position of the diffraction peaks may be interpreted as steps of the improvement of the polymer crystalline phase. In the starting structure of this type of samples having a sufficiently large preliminary orientation exist the so called frozen micro stresses.

It can be assumed that the destruction over a certain level of the blocking and supporting the tension segments, fixing the momentary raw structures, does not improve the structural perfection but leads to distortion of the achieved order. The clarification of such ambiguities requires further planning of multi factorial extended experiment that also includes other analytical methods, as is the purpose of our future work.

4. Conclusions

On the basis of the experiments of proton irradiation of the studied polymer objects and subsequent analyzes can be made conclusions about the relationship between the radiation rate and the changes of the fibers super molecular structure organization as follows:

The low energy proton irradiation with low power and beam current of 5 *nA* during relatively short exposition to irradiation, (respectively the absorbed dose) can cause noticeable structuring effect in the studied PET fibers.

Through examination of the angular deviations, changes in the intensity ratios and the profile of specific diffraction reflections of PET are defined the changes in the segmental packing in the amorphous and crystalline phase of the polymer.

The complex summary variation in the position of the diffraction peaks may be interpreted as steps of the improvement of the polymer crystalline phase.

It is supposed that the major modifying and structuring effects of the proton irradiation are ionization, radicalization, destructions, recombination and cross linking.

The complete clarification of the existing ambiguities requires further planning of multi factorial extended experiment that also includes other analytical methods, as is the purpose of our future work.

Acknowledgement

Part of the present work has been supported by Grant 08-226/12.03.2014 from Konstantin Preslavsky University, Shumen.

References

[1] Abu Saleh S and Eyal Y 2003 *Nuclear Instruments and Methods in Physics Research Section B: Beam Interactions with Materials and Atoms* **208** 137

[2] Hnatowicz V 2003 *Nuclear Instruments and Methods in Physics Research Section B: Beam Interactions with Materials and Atoms* **209** 145

[3] Zhu Z and Kelley M J 2004 *Applied Surface Science* **231–232** 302

[4] Awasthi K, Kulshrestha V, Avasthi D K and Vijay Y K 2010 *Radiation Measurements* **45** 850

[5] Ra S-J, Jung M-H, Kil J-K and Kim K-R 2011 *Journal of the Korean Physical Society* **59** 703

[6] Velev V, Popov A, Arhangelova N, Kyurkchiev P, Hristov H, Bello M, Moschini G and Uzunov N 2014 *J. Sci. and Appl. Res.* in press

[7] Mallick B, Patel T, Behera R C, Sarangi S N, Sahu S N and Choudhury RK 2006 *Nucl. Instr. Meth. in Phys. Res.* B **248** 305

[8] Abdesselam M, Stoquert JP, Chami S, Djebara M, Chami AC and Siad M 2009 *Nucl. Instr. and Meth. in Phys. Res.* B **267** 108

[9] Kumar V, Sonkawade R G, Chakarvarti S K, Singh P and Dhaliwal A S 2012 *Radiation Physics and Chemistry* **81** 652

[10] Kumar V, Ali Y, Sonkawade RG and Dhaliwal AS 2012 *Nucl. Instr.and Meth. in Phys. Res.* B **287** 10

[11] Vasiljeva I V, Mjakin S V, Makarov A V, Krasovsky A N and Varlamov AV 2006, *Applied Surface Science* **252** 24 8768

[12] Burillo G, Tenorio L, Bucio E, Adem E and Lopez G P, 2007 *Radiation Physics and Chemistry,* **76** 1728

[13] Kilian et. al 1960 *Koll. Zeitschrift* **172** 166

WAXS studies of heat - mechanically modified amorphous PET fibers. Role of the tensile stress

V Velev[1], A Popov[2], P Kyurkchiev[1], L Veleva[3] and M Pencheva[1]
[1]Konstantin Preslavsky University, 115 Universitetska Street, 9700 Shumen, Bulgaria
[2]University "Prof. Dr. Assen Zlatarov", 1 "Prof. Jakimov" Street, 8000 Burgas, Bulgaria
[3] Paul Scherrer Institute, Nuclear energy and safety department, 5232, Villigen, Switzerland

E-mail: v.velev@shu-bg.net

Abstract. The present work is devoted to the investigation of the structure developments in as-spun amorphous poly (ethylene terephthalate) (PET) filaments occurred as a result of heat mechanically modification. The degree of crystallinity of the untreated samples was 1,7 %.
The thermal deformation experiments were carried out under isothermal conditions. PET yarn was annealed during 10 *min* at constant temperature of 80^0C after which the sample is subjected to a well-defined constant tensile stress for 120 *s* at the same temperature. The mechanical load is gravitationally in the range from 0 *MPa* to 30 *MPa* and with increment step of 3 *MPa*. Using of wide angle X-ray scattering (WAXS) were investigated the structural rearrangements in the studied samples caused by the fibers treatments. Dependences between the strain force values and the running in the specimen's structure development are established. And in particular, it was found that a small increase of the tensile stress from 3 *MPa* to 6 *MPa* leads to a massive increase in the fibers degree of crystallinity with more than 33%.

1. Introduction

The flexible chain polymers are frequently used in different technological and commercial fields, but most often for the production of fibre-forming polymeric materials, films, etc. In the case of undrawn polymeric filaments, the final structure is determined mainly from the melt spinning conditions and of the subsequent heat-mechanically treatments, too. Depending on the forming conditions in the fibers are formed areas with an increased order of the macromolecular segments, meso- and crystalline phases with different perfection which under appropriate conditions can be converted into crystal nucleus, and they are the so-called semi- or quasi-crystalline nuclei. Another consequence of the filaments formation is the formation of regions with frozen stresses in them. The fibers formation is accompanied by the alternative processes of the macromolecular segments orientation as well as from the destruction, tangling, untangling, stress relaxation, deformation, etc. The above-mentioned processes have a direct influence on the physical and in particular on the mechanical properties of the prepared fibers. At the same time the optimal realization of the high temperature orientation modification of polymer filaments remains complex and still insufficiently studied process. Therefore the study of the structural reorganization mechanisms during uniaxial drawing needs of special attention. There are described investigations of the relationships between treatments of PET with

Journal of Physics: Conference Series **558** (2014) 012051 doi:10.1088/1742-6596/558/1/012051

different initial structures and the caused phase and structure evolution [1-5]. The effects of the strain force, strain rate and temperature on the structure development of PET filaments have been well studied and reported by a number of researchers [6-18] including the authors of the present article. It is studied the influence of the treatment and production conditions on the fibers deformation behavior [6, 7] mechanical properties [8, 9, 14] and spinning process [17].

Despite the large number of carried out investigations PET remains an interesting research object, intensively studied in the recent years. The complex effects of the preliminary fiber orientation and combined tension stress with thermal treatments in the temperature region just above the glass transition temperature on the deformation behaviour and super molecular structure development in uniaxially oriented PET remains not fully clarified.

The aim of the present study is the structural reorganization behaviour of amorphous PET fibers caused by combined thermal and mechanical treatments at constant temperature of 80^0C and mechanical tensile loading with values from 0 *MPa* to 30 *MPa*.

2. Experimental

2.1. Materials

Amorphous PET as-spun multifilament yarns were used as precursor samples. The initial characteristics of the studied PET fibers were as follows:

speed of fibrillate 1150 *m/min;*
number of single fibers in yearn 32;
diameter of a single fiber 44 μm;
degree of crystallinity α = 1,7 %;
birefringence $\Delta n.10^3$ = 3,58.

The filaments are produced by melt spinning on the industrial spinning machine of the company "Furnet" (France).

2.2. Methods

2.2.1. Simultaneous heat-mechanical modification. The filaments simultaneous heat-mechanical modification (SHMM) was performed using a specialized device designed and produced in the author's laboratory. Thermal deformation experiment consisted of annealing of the studied yarn during ten minutes at the required temperature of 80 ^0C followed by the application of tensile stress for 120 seconds. The samples annealing temperature is closely above the filaments glass transition temperature, determined in our previous work of 74 ^0C. The tensile stress values are in the interval from 0 *MPa* to 30 *MPa* with increment step of 3 *MPa*.

2.2.2. Wide-angle X-ray scattering. Structure of the treated PET fibers was studied by wide-angle X-ray scattering (WAXS) using apparatus Diffractometer URD - 6 (under license of SIEMES) of the company "Freiberger Präzisionsmechanik" (Freiburg im Breisgau, Baden-Württemberg, Germany). Used is β-filtered with Ni-filter Cu K_α radiation with a wavelength λ = 1,5418$\AA$.

3. Results and discussion

It is known that the wide-angle X-ray scattering method is geometric - structural sensitive therefore it was selected for the study of the treatments induced structural changes in the investigated PET objects. Due to the steric factor determined by the monomer composition, PET have narrow and hard defined options for mutual packing of the macromolecular chain segments. This determines his poor polymorphism and opportunities for the formation of various transient structural states and mesophases in improving of the quality and the packing density of the macromolecular segments in the crystallization process. The only known for now triclinic crystal modification of PET and the lack of stable previously known pseudo crystalline states and mesophases determine in fact the complete absence of its polymorphism.

However, as is noted above, depending on the conditions, and other specific characteristics of the crystallization processes, the quality and perfection of the crystalline phase can be found in many structural states. Right from the beginning of the emergence of some distant three-dimensional order in separate micro local areas in the polymer volume the specifics of the registered X-ray diffraction pattern reflecting the geometry package of the macromolecular segments on atomic - molecular (crystallographic) level reflect any change and it is criterion for its evaluation.

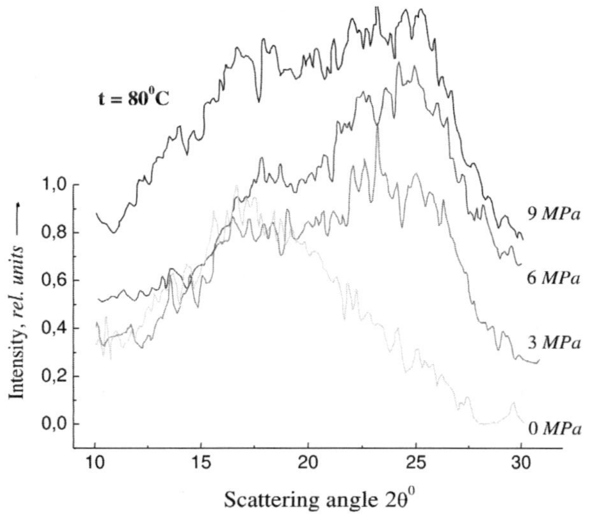

Figure 1. WAXS curves of amorphous PET yarns heat mechanically modified at isothermal conditions and under tensile stress with different values.

All major structural characteristics of that a finely-crystalline, not morphological, crystallographic elementary level is described by the angular positions and deviations, intensity ratios and profile changes in X-ray diffraction reflections in the angle range from 16 $^0 2\theta$ to 28 $^0 2\theta$.

At thermo-mechanically modified (TMM) objects the infrastructure of the intensity distribution of diffracted X-ray looks more similar to the known highly oriented or well-crystallized objects therefore the structure of TMM samples can be assumed and proved more easily, precise and unambiguously. At heat mechanically-initiated destruction the supporting factor "free volume" also exist around the overstressed in the state before destroying macromolecular segments. Comparing the Kilian's model [19] with the received diffraction patterns can be suggested the structural changes in their mutual pickings of the macromolecular chain segments in the heat-mechanically modified PET fibers.

The obtained wide-angle X-ray scattering curves of TMM PET fibers under tensile stress values from 0 *MPa* to 9 *MPa*, from 12 *MPa* to 21 *MPa* and from 24 *MPa* to 30 *MPa* are presented respectively in figures 1, 2 and 3.

The observed effects of the SHMM on the geometry of the intensity distribution of the diffracted X-ray in the objects diffraction patterns (figures 1-3) is expressed in angular deviations of intensity changes in the angular range from 10 $^0 2\theta$ to 30 $^0 2\theta$. The starting amorphous sample of undrawn fibers shows on the diffraction pattern increased intensity about 16,5 $^0 2\theta$ (figure 1).

This is the reflex 010 which reflects the width of the macromolecules folds, which is a thermodynamic dependent value and is relatively stable over a wide range of conditions. Application of tensile stress of 3 *MPa* leads to the occurrence of intensity peak (figure 1) at an angle of about

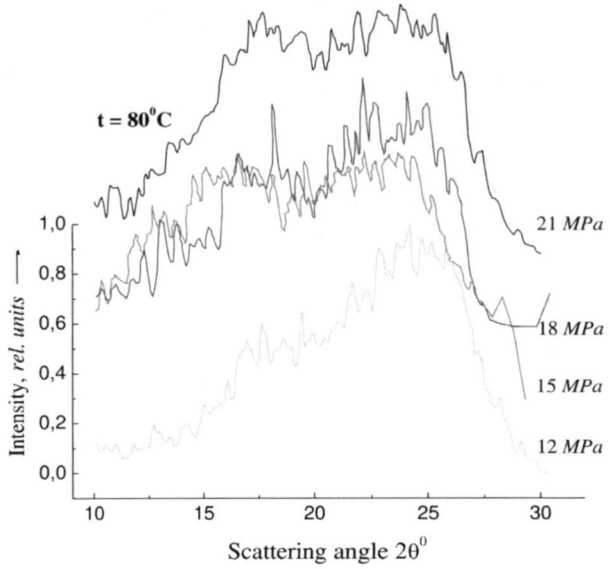

Figure 2. WAXS curves of amorphous PET yarns heat mechanically modified at temperature of 80^0C and under tensile stress with values of 12, 15, 18 and 21 *MPa*.

23,7 $^0 2\theta$ (reflex 100) reflecting the distance between the planes on folding of the macromolecular chain segments equaling to about 3,5 Å. At the same time there is an intensity increase and on the positions of the diffraction reflections from the type of 1-1-1, 1-10 and 011.

The tensile stress increasing to 6 *MPa* enhances this effect (figure 1) and further load increase equalizing the intensity throughout the observed range of 16-28 $^0 2\theta$. The last two effects are related to improving the quality and packing density of the macromolecular chain segments and the order disturbing of the distance between the folding planes. These effects appear consecutively two more times with the next tensile stress increasing up to value of 24 *MPa* (figure 2).

As can be seen from figure 3 the increase in gravitational load up to 30 *MPa* did not causes significant changes in the course of the diffraction curves that shows rather the absence of further structural reorganization than the development of the above mentioned processes.

In conclusion it can be said that TMM has a major impact on the structural reorganization of PET in solid phase state. Depending on the objects and conditions of the thermo-mechanical treatments significant redistribution of the intensity of the scattered and diffracted X-ray radiation is observed.

The registered effects may be explained by an increase in the width of the folds, respectively, decrease the distance between the planes of the folding of the chain segments. From a geometric standpoint, these two effects are interrelated, and even highly mutually dependent. There is observed a marked increase in the orientation and overall degree of the objects crystallinity, which is a confirmation of the suggested mechanism of TM modification. Registered effects are evidence of the crystallization of the studied PET fibers and especially of the impact of the strain stress on the crystallization process. Tensile stress is the factor which prevents the destruction of the quasi-crystalline nuclei and turns them into crystal nuclei thus starting the crystallization.

In addition to the above described studies it was examined and the deformation behavior of the PET fibers during the thermal deformation experiments [18]. The obtained results show large increase of the final yarns length with only minor increase of the stress level increasing from 3 *MPa* to 6 *MPa*.

Furthermore, with purpose to determine the effect of TMM on the structural changes in the studied PET objects the heat mechanically treated amorphous samples were tested by differential scanning

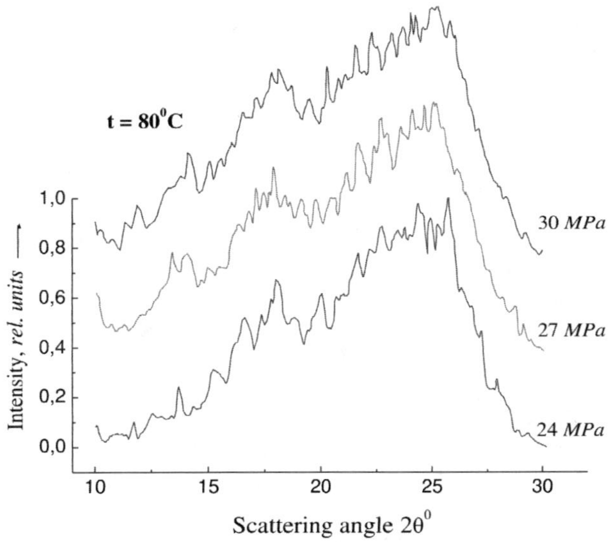

Figure 3. WAXS curves of amorphous PET yarns heat mechanically modified at temperature of 80^0C and under tensile stress with values of 24, 27 and 30 *MPa*.

calorimetry (DSC), too. On the basis of the obtained DSC data it was calculated and the degree of crystallinity of the studied PET fibers [18].

Degree of crystallinity of the PET filaments heat treated without applied strain stress (0 *MPa*) marks negligible growth from 1,7 % to 2,0 %. But only the insignificant increase of the tensile stress from 3 *MPa* to 6 *MPa* leads to a massive increase in the samples degree of crystallinity with more than thirty percents reaching value of 34,7%.

Further more, at loadings above 9 *MPa* both the filaments crystallinity and final bundle length almost no changed and reach to a plateau up to the load value of 30 *MPa.* The above mentioned additional results confirm the established by X-ray diffraction experimental fact for the decisive role of the tensile stress in the process of crystallization of the tested amorphous PET yarns.

4. Conclusions

It is investigated the structure developments in as-spun amorphous poly (ethylene terephthalate) filaments occurred as a result of heat mechanically modification at isothermal conditions in rubbery state.

Using wide-angle X-ray scattering it was found that the heat mechanically modification has a major impact on the structural reorganization of PET in solid phase state and leads to the crystallization of the studied objects.

The effect of the annealing temperature of 80 ^0C on the fibers degree of crystallinity is minor; In contrast, the applied tensile stress strongly influences the structure development of the studied specimens.

It is established that only the insignificant tensile stress increasing leads to a massive increase in the fibers degree of crystallinity with more than thirty percents.

Remains interesting and unclear the question about the role of the superposition annealing temperature/tensile stress in the temperature range between 80^0C and 100^0C on the structure development of preliminary and additionally oriented PET materials and objects and such experiments are coming.

Acknowledgement
Part of the present work has been supported by Grant 08-226/12.03.2014 from Konstantin Preslavsky University, Shumen.

References
[1] Todorov L V and Viana J C 2008 *International J. Material Forming* **1** 661
[2] Göschel U 2005 *Handbook of Thermoplastic Polymers: Homopolymers, Copolymers, Blends, and Composites,* Weinheim: Wiley-Vch Verlag GmbH **Ch. 6**
[3] Todorov L V, Martins C I and Viana J C 2012 *J.of Appl. Polymer Sci.* **124** 470
[4] Todorov L V, Martins C I and Viana J C 2011 *J.of Appl. Polym. Sci.* **120** 1253
[5] Bouti S, Kieffel Y, Pohlink K, Hiver J M, Dahoun A and Etienne S 2009 *Journal of Physics, Conf. Series: Materials Science and Engineering* **5,** doi:10.1088/1757-899X/5/1/012006
[6] Desai P and Abhiraman A S 1986 *J. of Polym. Sci. Part C: Polymer Letters* **24** 135
[7] Betchev Ch 1995 *Polym. Testing* **14** 163
[8] Rudolf A, Geršak J and Smole M S 2012 *Textile Research Journal* **82** 161
[9] Rudolf A and Geršak J 2012 *Textile Research Journal* **82** 148
[10] Velev V, Androsch R, Dimov T, Popov A and Radusch H-J 2010 *J. of Optoelectronics and Advanced Materials* **12** 958
[11] Haji A, Rahbar R S and Kalantari B 2012 *Materials Research* **15** 554
[12] Velev V, Popov A, Androsch R, Dimov T, Angelov T and Radusch H-J 2012 *Journal of Physics: Conference Series* **398** 012055
[13] Vibhor J 2009 *A thesis presented to the academic faculty,* Georgia Institute of Technology
[14] Haji A and Rahbar R S 2012 *Chemical Industry & Chemical Engineering Quarterly* **18** 233
[15] Murthy N S and Grubb D T 2003 *Journal of Polymer Science: Part B: Polymer Physics* **41** 1538
[16] Minqiao R, Zhiying Z, Shizhen W, Jia W and Xiao C 2006 *Journal of Polymer Research* **13** 9
[17] Ziabicki A and Jarecki L 2007 *Journal of Applied Polymer Science* **105** 215
[18] Velev V, Popov A and Bogdanov B 2011 *Thermodynamics - Systems in Equilibrium and Non-Equilibrium* ed Moreno-Pirajan J C (Rijeka: INTECH) **Ch.5** 89
[19] Kilian et. al 1960 *Koll. Zeitschrift* **172** 166

Journal of Physics: Conference Series **558** (2014) 012052 doi:10.1088/1742-6596/558/1/012052

Optical modeling of bulk-heterojunction organic solar cells based on squarine dye as electron donor

S Kitova[1], D Stoyanova[1], J Dikova[1], M Kandinska[2], AVasilev[2] and S Angelova[3]
[1]Institute for Optical Materials and Technologies "Acad. J. Malinowski", Bulgarian Academy of Sciences, "Acad. G. Bonchev" str. Bl. 109, 1113 Sofia, Bulgaria
[2]Sofia University "St. Kliment Ohridski, Faculty of Chemistry and Pharmacy, 1 James Bourchier Blvd., 1164 Sofia, Bulgaria
[3]Institute of Organic Chemistry with Centre of Phytochemistry, Bulgarian Academy of Sciences, "Acad. G. Bonchev" str. Bl. 9, 1113 Sofia, Bulgaria

E-mail: skitova@iomt.bas.bg

Abstract. The potentiality of a squarine dye (Sq1) for using as electron donor component in bulk heterojunction organic solar cells (BHJ) has been studied from the optical point of view. The soluble n-type fullerene, (6,6)-phenyl C_{61} butyric acid methyl ester ($PC_{61}MB$) was chosen as acceptor. Optical modelling based on transfer matrix method was carried out to predict and improve photovoltaic performance of a BHJ device with blended Sq1/$PC_{61}MB$ active layer. The dependence of the absorption and the calculated maximum short circuit photocurrent (J_{sc}^{max}) on the thickness of the active layer (d_{act}), was investigated for two weight ratios of Sq1 and $PC_{61}MB$. Thus, the optimal d_{act} was calculated to be about 100 nm, which provides an efficient overlapping of the total absorption with solar spectrum in the range between 580 and 900 nm. Besides, it is found that the insertion of ZnO or C_{60} spacer layer shifts J_{sc}^{max} peak to lower d_{act} and significantly enhances J_{sc}^{max} for active layers with d_{act} < 50 nm, which is mainly due to improved light absorption by a factor of 5 to 10. Simultaneously, for d_{act} <100 nm the optical effect of inserted PEDOT:PSS hole transporting layer is negligible.

1. Introduction

During the past years the organic photovoltaics (OPVs) are subject of continuously growing interest as promising alternative for high cost crystalline silicon cells (c-Si), which still dominate the market. In the most successful devices, a soluble n-type fullerene, (6,6)-phenyl C_{61} butyric acid methyl ester ($PC_{61}MB$) as acceptor, is blended with the donor from conjugated p-type polymers to form bulk-heterojunction (BHJ) cells [1]. An attraction of these devices, which are processed from solution, is the ability to deposit active layer precursors over large areas in a single step, using techniques ranging from inkjet printing to spin-coating. Small molecule donors offer the same facile processing however are easier to synthesize in many variants and purify, are intrinsically monodisperse, and are often environmentally more stable [2]. The disadvantage of lower solubility of this class of organic semiconductors can be overcome by functionalizing of their molecular structure. This approach has been successfully applied to different dyes and very encouraging results about the performance of devices developed on this base has been achieved [3].

Squaraine dyes have been the subject of many recent investigations owing to their unique photochemical and photophysical properties [4]. Principal attractions of this dye class include facile synthetic access, a wide variety of possible structural motifs and high thermal and chemical stabilities

[5]. Squaraines are the products of condensations of electron-rich molecules with squaric acid [6-7]. Their structure can be either symmetric or nonsymmetric. The nonsymmetry depends on the synthetic route employed and has enabled the preparation of a wide variety of functional structures, which have been exploited in, for example, photoconductive materials, light emitting field-effect transistors, IR absorber in optical disks, and dye-sensitized solar cells [4-8]. Squaraine dyes are promising molecular donor active materials for BHJ solar cells due to their strong absorption coefficients up to $3x10^5$ cm^{-1} that can be extended from the green to the near infrared region [4]. The application of symmetric squaraine dyes in BHJ cells was recently reported by Pagani, Facchetti and co-workers [4, 5]. The authors demonstrated the fabrication and initial characterization of BHJ solar cells based on alkyl- and alkenyl-functionalized squaraine dyes as donors, with $PC_{61}BM$ and $PC_{71}BM$ as acceptors. The results are promising regarding power conversion efficiencies (PCEs) reported to date for ambient solution-processed small molecular OPVs. Based on that, we choose this approach to develop new squarine dyes suitable for application as electron donors in BHJ organic solar cells.

As a first step in our investigations, in the present work a symmetrical n-hexyl substituted squarine dye 2-(5-(((2,2-diphenylhydrazono)methyl)-1-hexyl-1H-pyrrol-2-yl)-4-(5-(2,2-diphenyl-hydrazono)-ylidenemethyl)-1-hexyl-2H-pyrrol-1-ium)-3-oxocyclobut-1-enolate was synthesized by a modified by us method, cited above. The structure and electronic properties of the dye, which we labeled as Sq1, were studied by means of density functional theory (DFT) calculations. Further, optical modeling based on transfer matrix method (TMM) was performed to predict and improve photovoltaic performance of a bulk heterojunction cell with active layer from a solid-state blend of Sq1 and the fullerene derivative $PC_{61}BM$. The influence of PEDOT:PSS hole transporting layer as well as optical spacers from ZnO and C_{60} layers on the optical performance of the model OPVs was also studied.

2. Experimental

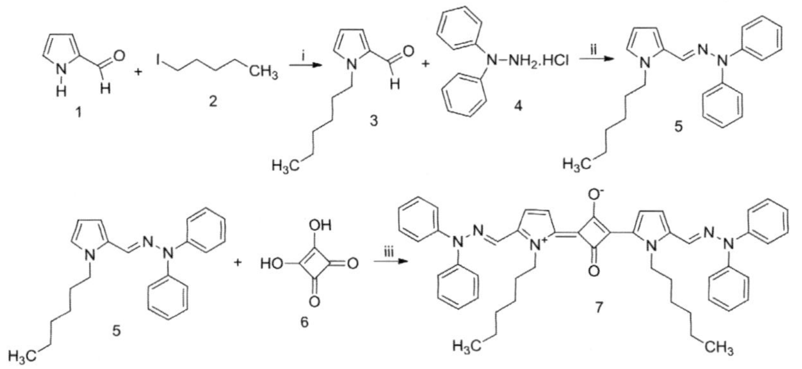

Figure 1. Synthesis of the squarine dye Sq1. Reactants, conditions and yields: i) anhydrous DMF, NaH, 48 h at r.t., 77%; ii) DIPEA, anhydrous EtOH, 96%; iii) BuOH/toluene = 1:1, cat. 4-(dimethylamino) pyridine, 2 h reflux, 65%.

The synthesis of the target squaraine dye Sq1 (indicated as 7 in figure 1) was performed via a modified by us previously reported procedure [4, 5]. The use of the sterically hindered base DIPEA instead of imidazole [4] increased the reaction yield probably due to avoid of possible side reactions. Another improvement is the addition of catalytic amounts of 4-(dimethylamino)pyridine. Double decrease of the reaction time and increase of the reaction yield have been observed. Following the purification procedure by recrystallization from iso-propanol we obtained the target dye Sq1 as golden needles as was described in [4]. The dye was purified by flash column chromatography with subsequent recrystallization and its structure was elucidated by elemental analysis and Nuclear magnetic resonance (NMR) spectroscopy.

The molecular ground-state geometry of Sq1 compound was optimized using Becke's three-parameter hybrid exchange-correlation functional B3LYP [9] with double-ζ polarized basis set, 6-31+G(d,p). Local minima were verified by establishing that all vibrational frequencies are real. All calculations were performed using Gaussian 09 [10].

Single dye films were prepared from solution of the compound in chlorobenzene by spin coating. In the preliminary experiments the concentration of the solution, the spin rate and duration were varied in order to obtain films with thickness, suitable for optical measurements. Thus, 45 nm films were deposited on substrates from solution containing 12 mg ml^{-1} Sq1, spin coated at 1200 rpm for 180 s. Blended films were prepared also by spin coating from chlorobenzene solutions of 1:1 and 1:3 (weight ratios) mixture of Sq1 and $PC_{61}MB$. 80 nm thick PEDOT:PSS films were also obtained by spin-coating on the substrates at 5000 rpm and heated at 120°C for 20 min. under argon atmosphere. $PC_{61}BM$ and PEDOT:PSS were commercially available. In all cases preliminary carefully cleaned BK7 optical glass plates and Si wafers were used as substrates.

Optical constants of the Sq1 dye and blended Sq1/$PC_{61}MB$ films were determined with high accuracy on the basis of three spectrophotometric measurements at normal incidence of light - transmittance T and reflectances R_f and R_m of the both kinds of films, deposited on transparent (BK7) and opaque (Si wafers) substrates, respectively [11].

Optical modeling of a BHJ device stack was carried out using the transfer matrix formalism, based on the Fresnel formulas for the several interfaces occurring in the cell. This approach has been explained in great detail in the literature [12, 13]. In our work a Matlab script developed by Burkhard and Hoke, which treats the case of normal incidence of light, was applied [14]. The absorption distribution for each wavelength over the film thickness has been calculated in the wavelength range 350 – 900 nm. Multiplying with the light of the AM 1.5 sun spectrum and integration over the film thickness results in the total number of photons, absorbed for each wavelength. Lastly, a summation over all wavelength results in the total number of absorbed photons. The possible maximum current density J_{sc}^{max} was calculated assuming that each absorbed photon results in a collected electron i.e. the internal quantum efficiency, IQE, equals one [14].

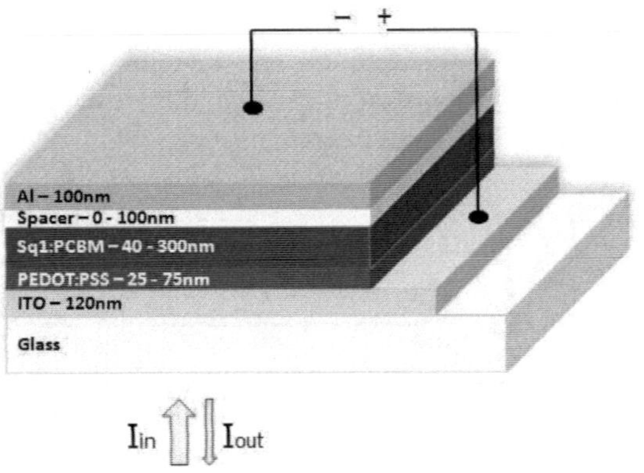

Figure 2. Scheme of a model BHJ organic solar cell.

For optical modeling we have chosen a BHJ device structure, used in [5], for studing the same squarine dye. The architecture of our model cell is presented in figure 2. As seen, the cell consists of ITO (120 nm)/PEDOT:PSS (x nm)/blend Sq1/$PC_{61}MB$ (1:1 or 1:3 wt:wt) (y nm) /LiF (0.6 nm)/Al (120 nm). The anode represented glass substrates, pre-coated with indium tin oxide (ITO), modified by spin-coated polyethylenedioxythiophene/polystyrenesulphonate (PEDOT:PSS) hole transporting layer. The cathode consisted of LiF capped with Al. Very thin LiF films were inserted between active layer and Al surface to improve the ohmic contacts for electron collection. At the next step of modeling LiF was substituted with ZnO or C_{60} spacer layer. We have used literature data for the optical constants values of ITO, Al, LiF [15] and ZnO [16]. The optical constants of C_{60} films were determined experimentaly by applying the same procedure as for Sq1 dye. For the purpose 25 nm thick C_{60} films were thermally evaporated in vacuum (10^{-4} Pa).

3. Results

Possible molecular structure of our Sq1 dye and the distributions of the frontier orbitals are shown in figure 3. The highest occupied molecular orbital (HOMO) is spread over the whole length of the conjugated chain, which is one of the prerequisites for achieving high power conversion efficiency. The lowest occupied molecular orbital (LUMO) is localized predominantly on the central four membered ring and the pyrrole rings. Neither HOMO nor LUMO are spread over the electronically inactive aliphatic chains, attached to the backbone. A detailed analysis of electronic transitions behavior and molecular orbital (MO) energy diagram of the dye studied under the framework of DFT and its time-dependent version is in progress.

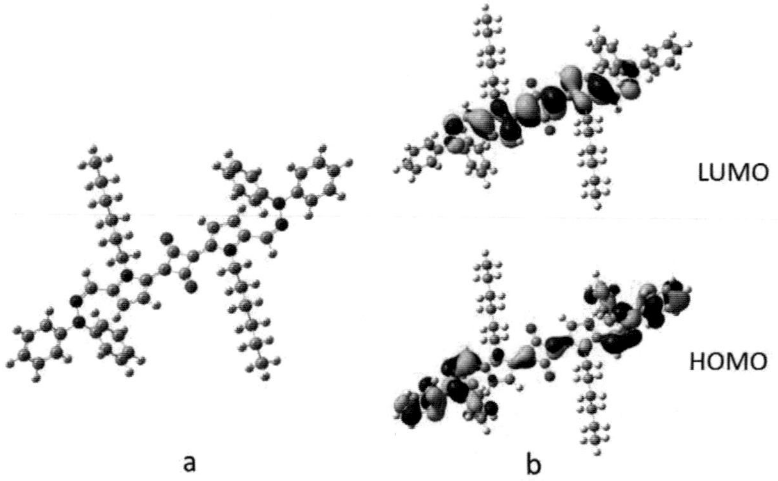

Figure 3. B3LYP/6-31+G(d,p) optimized geometry (a) and isocontours plots (isovalue = 0.02 a.u.) of the B3LYP/6-31G(d,p) calculated frontier orbitals of Sq1 (b); Color code indicates different phases of MO.

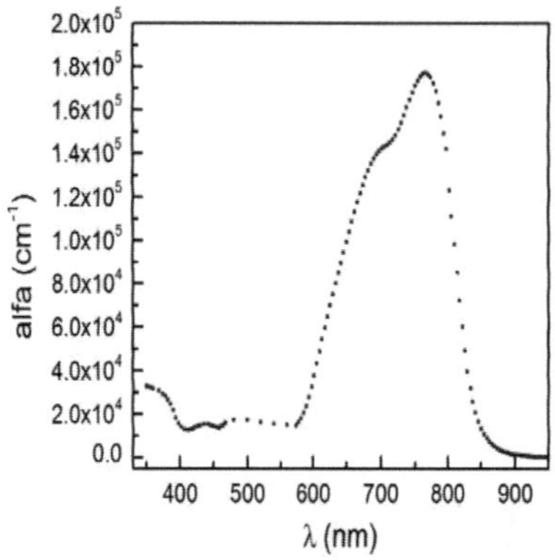

Figure 4. Absorption coefficient profile ($\alpha=4\pi k/\lambda$) of 45 nm thick dye film.

Figure 4 shows the absorption coefficient α determined for 45 nm thick Sq1 dye film. It is seen that the absorption area of the film covers not very broad spectral region from 590 to 840 nm with two strong typical for squarines absorption peaks of about 1.8 x 10^5 cm^{-1} and 1.4x10^5 cm^{-1} at λ = 766 nm and 695 nm, correspondingly. The results obtained are in a good accordance with the data given in [5]

Journal of Physics: Conference Series **558** (2014) 012052 doi:10.1088/1742-6596/558/1/012052

for the identical dye.As it was already mentioned, the optical simulation was performed for a standard BHJ cell structure, where the light enters through the glass substrate, sequentially passing the layers of ITO, PEDOT:PSS, Sq1/PC_{61}MB blend layer and LiF (see figure 2). Then the light is reflected back from the Al electrode and finally leaves the solar cell partly at the front again. We have found that the presence of LiF film has no influence on the optical model results, probably due to its very low thickness of 0.6 nm.

Figure 5 presents the spectral dependence of total absorption within the Sq1/PC_{61}MB (1:3 wt.ratio) active layer for a set of different layer thicknesses. For comparison, the photon flux of a standard AM1.5 solar spectrum is shown as a gray solid line. The blue light is mainly absorbed by PC_{61}MB while red and near infrared only by Sq1 dye. It is seen that the absorption of blended film overlaps the solar spectrum in broad range from 400 to 900 nm.

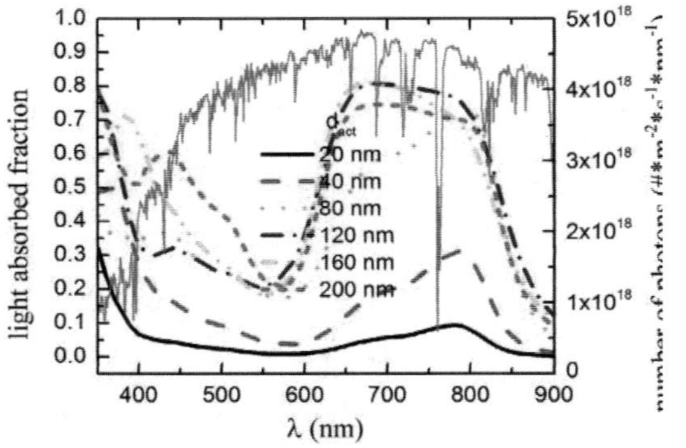

Figure 5. Fraction of incident light absorbed within the 1:3 dye/PC_{61}BM layer of the solar cell for several layer thickness. $d_{PEDOT:PSS}$ =75 nm. For comparison, the photon flux of a standard AM1.5 solar spectrum is shown as a gray solid line (right axis).

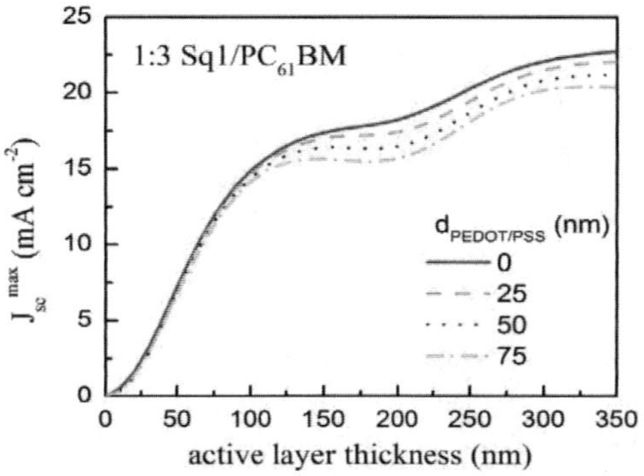

Figure 6. Calculated J_{sc}^{max} under 100 mW cm^{-2} AM 1.5 spectral illumination as a function of active layer thickness (IQE =1) for OPV device with indicated thickness of PEDOT:PSS hole transport layer. The same results were obtained for device with 1:1 dye/PC_{61}BM.

Assuming an IQE=1, the maximum short circuit current density J_{sc}^{max} under AM 1.5 illumination in dependence of 1:3 Sq1/PC_{61}MB layer thickness is depicted in figure 6 for several thicknesses of the hole transporting layer. Due to the interference effects in the device structure J_{sc}^{max} does not follow a straight line, but shows a local maximum at about 100 and 300 nm. Further, it is established that for $d_{act} < 100$ nm the optical effect of the PEDOT:PSS hole transporting layer is negligible. In general, the photocurrent is higher for a larger d_{act}, but due to the limited exciton diffusion length and limited charge mobility in the organic layer the real photocurrent decreases with increasing d_{act}. Indeed, the experimental results obtained in [4] showed that measured J_{sc} is low in a 140 nm thick layer, but by reducing its thickness down to 30 nm, OPV performance can be enhanced and J_{sc} increased from 1.85 to 3.10 mA cm^{-2} [4]. In particular, for the same squarine dye the best OPV performance and the highest J_{sc} has been measured at 35 nm thick Sq/PC_{61}MB (1:3 wt. ratio) layer [5]. As can be seen from figures 5 and 6, however the fraction of incident light absorption within our active layer and the calculated J_{sc}^{max} for this thickness are very low.

It is known that insertion of a transparent film between the active layer and the metal electrode can redistribute the optical electric field in the OPV stack, and then influence the device performance. This is referred to as the optical spacer effect [17]. It has been found that in polymer solar cells with conventional structure, an optical spacer of ZnO with optimal thickness leads to a significant increase of the photocurrent when the active layer thickness is below 60 nm [17]. In figure 7 the calculated J_{sc}^{max} as a function of d_{act} for devices with different thicknesses of ZnO spacer layer is presented.

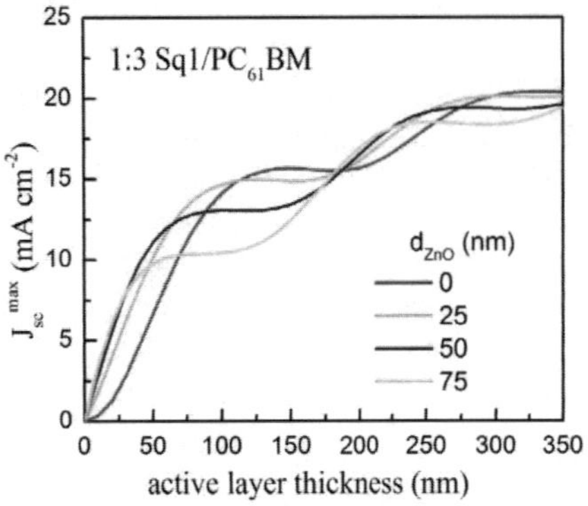

Figure 7. Calculated J_{sc}^{max} as a function of active layer thickness (IQE =1) for OPV device with indicated thickness of ZnO spacer layer. $d_{PEDOT:PSS}$ =75 nm.

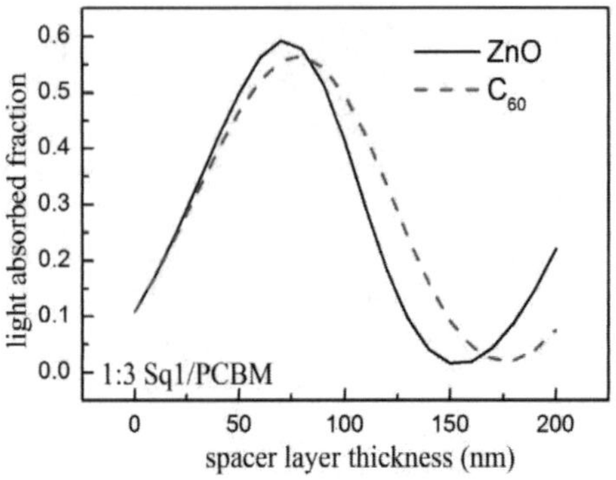

Figure 8. The variation of absorption within 35 nm thick active layer as a function of thickness of ZnO and C_{60} optical spacer layers. $d_{PEDOT:PSS}$ =75 nm.

The results demonstrate that the increasing of ZnO layer thickness shifts the J_{sc}^{max} peak to lower d_{act} and significantly enhances J_{sc}^{max} at active layers thicknesses below 50 nm. Besides, our calculations showed that fullerene C_{60} layer has analogous effect. As seen from figure 8 the presence of both spacer layers causes the increasing of the absorption in the active layer by a factor of 5 to 10. While ZnO film is completely transparent in the visible range of light, C_{60} exhibits some light absorption, which explains the obtained smaller light absorption improvement.

Finally, it should be mentioned that experimental studies of a real device with Sq1/PC_{61}BM active layer are in progress and will be forthcoming in a separate paper.

4. Conclusions

Optical modeling based on TMM has been performed to investigate the performance of OPVs consisting of glass/ITO/PEDOT:PSS/Sq1:PC_{61}BM/spacer/Al. The dependence of the absorption and J_{sc}^{max} on the active layer thickness is studied for two weight ratios of Sq1 and PCBM. The results show that the total absorption within the Sq1/PC_{61}MB blend active layer overlaps the solar spectrum in a broad range from 400 to 900 nm. Due to the interference effects in the device structure, J_{sc}^{max} shows a local maximum at active layer thickness of about 100 and 300 nm. However, because of the limited exciton diffusion length and charge mobility in the organic layer of practical importance are thicknesses below 60 - 70 nm, which has been proved experimentally. For these thicknesses of the Sq1/PCBM active layer the optical effect of inserted PEDOT:PSS hole transporting layer is negligible. On the contrary, the insertion in the device stack of ZnO or C_{60} spacer layer with thicknesses up to 50 nm shifts J_{sc}^{max} peak to lower active layer thicknesses. Moreover, a significant increase of J_{sc}^{max} for active layers thinner than 50 nm is found, due to the improved light absorption by a factor of 5 to 10.

References

[1] Dennler G, Scharber M and Brabec C 2009 *Adv. Mater.* **21** 1323
[2] Walker B, Kim C and Nguyen T 2011 *Chemistry of Materials* **23** 470
[3] Xin Xo, Wei G, Wang S, Zimmerman G, Renshaw Ch, Thompson M, and Forrest S 2012 *Adv. Mater.* **24** 1956
[4] Silvestri F, Irwin M, Beverina L, Facchetti A, Pagani G and Marks T J. 2008 *Am. Chem. Soc.* **130** 17640
[5] Bagnis D, Beverina L, Huang H, Silvestri F, Yao Y, Yan H, Pagani G, Marks T and Facchetti A 2010 *J. Am. Chem. Soc.* **132** 4074

[6] Choi H, Baik C, Kang S, Ko J, Kang M, Nazeeruddin M and Graetzel M 2008 *Angew. Chem.* **120** 333
[7] Li C, Yum J, Moon S, Herrmann A, Eickemeyer F, Pschirer N, Erk P, Schçneboom J, Muellen K, Graetzel M and Nazeeruddin M 2008 *Chem. Sus. Chem.* **1** 615
[8] Chen X, Guo J, Peng X, Guo M, Xu Y, Shi L, Liang C, Wang L, Gao Y, Sun S and Cai S 2005 *J. Photochem. Photobiol. A* **171** 231
[9] Becke A 1993 *Chem. Phys* **98** 5648
[10] Gaussian 09 program, Revision C.01, Gaussian, Inc., Wallingford CT, 2010
[11] Babeva Tz, Kitova S and Konstantinov I 2001 *Appl. Opt.* **40** 2682
[12] Pettersson L, Roman L and Inganä O 1999 *J. Appl. Phys.* **86** 487
[13] Peumans P, Yakimov A and Forrest R 2003 *J. Appl. Phys.* **93** 3693
[14] Burkhard G, Hoke E and McGehee M 2010 *Adv. Mater.* **22** 3293
[15] Smith D, Shiles E and Inokuti M 1985 *Handbook of Optical constants of Solids* (New York: Academic) p 377
[16] Liu Y, Hsieh J and Tung S 2006 *Thin Solid Films* **510** 32
[17] Gilot J, Barbu I, Wienk M and Janssen R 2007 *Appl. Phys.Lett.* **91** 113520

Journal of Physics: Conference Series **558** (2014) 012053 doi:10.1088/1742-6596/558/1/012053

Magnetic properties of fibrous Fe–Al–O prepared by electrospinning

V Lovchinov[1], P Simeonova[1], Ph Vanderbemden[2] and K Krezhov[3]
[1] Institute of Solid State Physics, Bulgarian Academy of Sciences,
72 Tzarigradsko Chaussee Blvd., 1784 Sofia, Bulgaria
[2] University of Liège, S.U.P.R.A.T.E.C.S., Department of Electrical Engineering and Computer Science B28, Sart-Tilman, Liège
[3] Institute for Nuclear Research and Nuclear Energy, Bulgarian Academy of Sciences
72 Tzarigradsko chaussee blvd, 1784 Sofia, Bulgaria

E-mail: lovcinov@issp.bas.bg

Abstract. Recently, we obtained a new Fe-Al-O fibrous material by applying the novel route of electrospinning. Here we report on the properties of the as prepared material established by PPMS magnetometry and X-ray diffraction. X-ray diffractograms revealed the coexistence of the orthorhombic multiferroic phase $FeAlO_3$ along with traces of unreacted impurities. The structure of the $FeAlO_3$ phase is described in the orthorhombic $Pna2_1$ space group and is of great current interest since the compound features piezoelectricity, ferrimagnetism and magnetoelectric effects at low temperatures. In contrast to known bulk material obtained by classical routes the new fibrous material shows a lower value of saturation magnetization and the magnetic transition occurs at a much lower temperature. In the discussion an attempt is made to disclose the reasons for the observed differences.

1. Introduction

In the last decade a lot of efforts have been made for obtaining monophase samples of $FeAlO_3$.This compound undergoes a ferromagnetic phase transition at about 280 K with strong 180° cation-anion-cation super-exchange antiferromagnetic interactions. Moreover $FeAlO_3$ is multi- ferroic and at low temperature it shows an extremely high magnetic anisotropy [1,2]. These properties as well as its relatively simple crystal structure make it a perspective material for important technological applications. Additionally, studying the properties could help for clarification of the reason for appearance of magnetoelectric coupling in these compounds. Another important reason for the interest towards $FeAlO_3$ is that most magnetoelectric materials are based on heavy metal oxides, mainly lead (Pb) and bismuth (Bi). That is why $FeAlO_3$ is perspective as lead – free multi-ferroic material, showing piezoelectrisity and ferrimagnetism at low temperature [1-4]. As is known, in recent years the search of alternative materials with substantial technological impact grows constantly by taking, however, into account their environmental impact to remain acceptably low due to environmental, social and public health reasons [4]. Therefore, the fabrication of single-phase lead-free multi-ferroics (such as

$FeAlO_3$) with appropriate properties for the technological application is a provocative challenge for physicists and material scientists.

The standard routes of synthesis of orthorhombic $FeAlO_3$ are complicated due to the very narrow range of stoichiometry and thermodynamic stability of this phase in the binary equilibrium diagram $Fe_2O_3 - Al_2O_3$. Different experimental approaches were applied so far for synthesis of single-phase $FeAlO_3$ such as solid state reaction, sol-gel, co-precipitation and high-energy ball-milling techniques. Recently, we obtained a new fibrous material of Fe-Al-O by applying the novel approach of electrospinning [5]. For the first time it was demonstrated that this method provides the possibility to prepare fibrous Fe-Al-O non-woven material. However, the resulting long fine fibres of multiferroic $FeAlO_3$.were mixed with $FeAl_2O_4$. This necessitated searching for new precursors and related conditions for the post-heat treatments.

In this study, further investigations with modified blend solutions for preparation of Fe-Al-O fibres are described. The new material was characterized through X-ray powder diffraction and AC susceptibility measurements.

2. Experimental details

Details of the electrospinning technique and conditions for preparation of the Fe-Al-O material as non-woven mat were reported in [5]. Iron nitrate based precursor was used and the temperature for non-isothermal calcinations was 800 °C. The present fibrous material was obtained by electrospinning of a different liquid precursor, which contains a mixture of blend aqueous solution of aluminum acetate $AlOH(CH_3COO)_2.H_2O$ and iron citrate $C_{12}H_{10}Fe_3O_{14}$, dissolved in deionised water in equal molar concentrations. A 10,0 wt% aqueous solution of water soluble polymer polyethileneoxide-PEO is added in weight ratio 1 : 1 (POE : Fe-Al precursor) in order to assist the electrospinning process. The actual working parameters were as follows: high voltage – 15 kV, working distance - 20 cm, flow rate - 0,3 ml/h and electrospinning time of 10 hours. The samples were deposited onto fused quartz substrates, which thereafter facilitate the calcinating procedure at 1100°C for 2 hours, preceded by stepwise water and PEO removal at 120°C or 360°C, which was accomplished completely for about 6 hours.

The new Fe-Al-O material thus obtained was morphologically inspected using a scanning electron microscope Jeol SEM T-200. In order to amplify the contrast during the SEM observations, the samples were beforehand covered with vacuum deposited carbon and gold films. The phase composition of the thermally processed mats was verified from X-ray powder diffraction (XRD) data collected with a Philips diffractometer, using Cu $K\alpha$ radiation. Temperature dependences of magnetization were measured using a physical properties measuring system (PPMS, Quantum Design) in the temperature range from 2 K to 300 K and applied magnetic field of H_{DC} = 1kOe.

3. Results and discussion

The SEM examinations on the microstructure of Fe-Al-O non-woven mats are illustrated by the images in figure 1. The SEM micrographs revealed morphologically homogeneous fibrous materials composed of long fibers with nearly regular cross sections.

However, the fibres cross-section size is of the order of 30-50 nm for the Fe-citrate based material and is substantially reduced in comparison with the previously estimated cross-section of about 1μm for the Fe-nitrate based material. The micrographs evidenced as well that by following the newly developed procedure the resulting material was of better quality in terms of both morphological and phase quality. Here is to note that the Fe-Al-O non-woven material reported earlier [5] had two major drawbacks: 1) the orthorhombic $FeAlO_3$ phase was of small amount at the expense of a greater quantity of spinel phase $FeAl_2O_4$; 2) the woven material featured a number of undesirable topological defects such as broken – **A**, fused fibres **B** or knots **a** (figure 1 left panel).

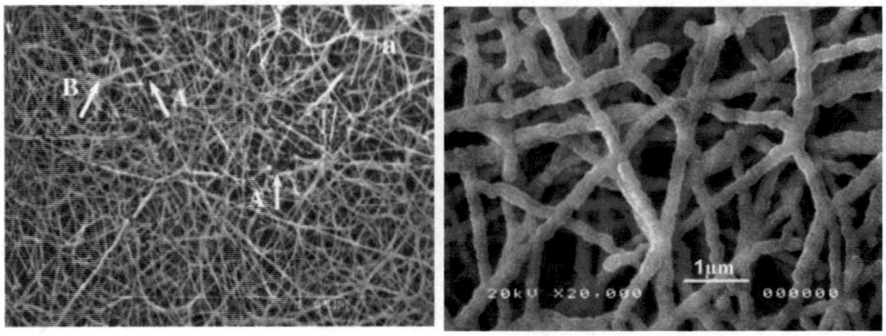

Figure 1. SEM images of Fe-Al-O non-woven mats obtained by electrospinning using Fe-nitrate based precursor (left -magnification 2000) and Fe-citrate based precursor (right- magnification 20000).

Figure 2 shows the XRD patterns of Fe-Al-O non-woven mats obtained earlier (left) and at present (right).

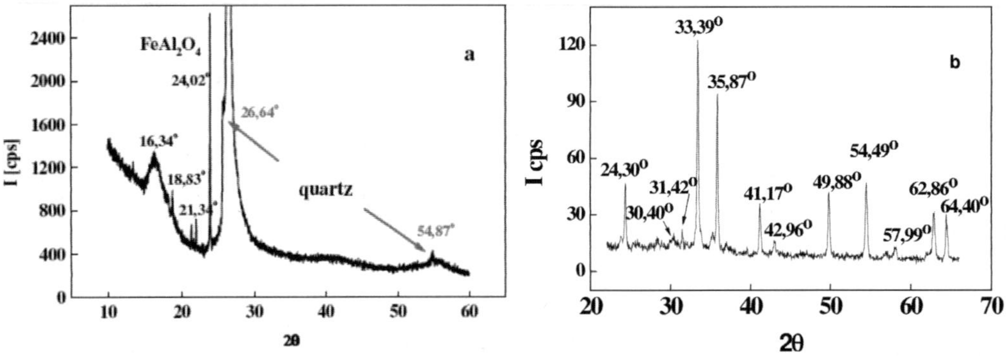

Figure 2. Room temperature XRD pattern of Fe-Al-O non-woven mats obtained by electrospinning using ferric nitrate based precursor (left) and ferric citrate based precursor (right).

In figure 2 (left) the peaks at 2θ=26,64° and 2θ=54,87° belong to the quartz substrate, the sharp intensive peak at 2θ=24,02° belongs to the $FeAl_2O_4$ phase whereas the peak at 2θ=18,83° corresponds to the $FeAlO_3$ phase. The orthorhombic perovskite $FeAlO_3$ is reported to crystallize in the non-centrosymmetric space group *Pna*21 space group $Pna2_1$. The cations are distributed among four types of sites labelled Fe1, Fe2, Al1 and Al2. The three first sites are octahedral and the last one is tetrahedral. The octahedra share edges whereas the nearly regular tetrahedra share oxygen at the corners. There is site disorder among the cationic sites (in fact all sites are occupied by both Fe^{3+} and Al^{3+} ions) and site occupancies do not vary with lowering of temperature. The compound presents a classical Néel ferrimagnetism as established by neutron diffraction [6]. The Néel sublattices **A** = Fe1 + Al1 and **B** = Fe2 (+ Al2) are responsible for the strongly antiferromagnetic ~ 180° bonds Fe1 - O – Fe2 and Fe2 - 0 – Al2. The direction of easy magnetization is *a*-axis and *c* is the polar axis.

Figure 3 shows a typical plot of inverse susceptibility versus temperature. The curve consists of a linear Curie–Weiss-type region at sufficiently high temperatures (above 200 K). The zero field cooled (ZFC) and field cooled (FC) magnetization (M) versus temperature plots are also displayed (zero-field cooling protocol with H_{DC} = 1kOe applied field). A typical Hopkinson-type maximum exists at about

28K in the ZFC curve and the ZFC and FC curves deviate substantially (such a behavior is often called a spin-glass-like effect). Such behavior has been observed for a number of manganites in the low temperature range [7]. This effect could be a result of different factors—a kind of magnetic disorder, magnetic frustration, the existence of small magnetic particles or clusters, etc. In all cases the main physical reason for it is that the system is energetically degenerate, and the main physical parameter responsible for the effect is magnetic anisotropy. However, more extended and detailed investigations are necessary for a better understanding this effect.

It can be summarized that the magnetic measurements on cooling the samples reveal a magnetic transition (T_N of about 125 K, taken from the differential maximum of the ZFC M (T)) that can be attributed to a transition from the paramagnetic to the ferrimagnetic state. While the spontaneous magnetization is still close to the value of about 0.4 μB/ Fe^{3+} as estimated by Bouree et al. [6] for microcrystalline powder of $FeAlO_3$, the magnetic transition temperature is substantially lower than the values ranging from 250K to 280K reported by different authors for polycrystalline substances of various average grain size produced by commonly used techniques and could be related to the nanosized diameter of the fibres.

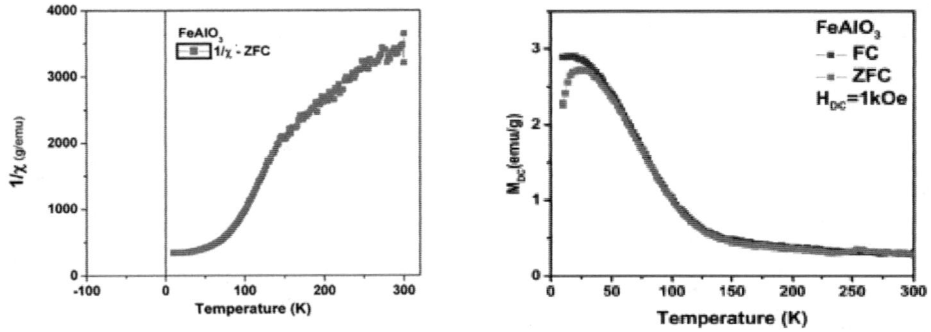

Figure 3. Non-woven mat Fe-Al-O from ferric citrate based precursor: (Left) Temperature dependence of the inverse magnetic susceptibility $1/\chi$; (Right) ZFC and FC thermomagnetic curves in a field of 1 kOe.

4. Conclusion

It was confirmed that the electrospinning technique with successive steps of calcination of the product provides a reproducible route for fabricating of Fe-Al-O non-woven mats. A new precursor based on ferric citrate was developed and new appropriate conditions for calcinations were implemented successfully for the fabrication of Fe-Al-O substance composed of long filaments of uniform size ranging in diameter between 30 and 50 nm that are more than an order of magnitude thinner than those previously reported [5]. The new material is with much less topological defects, improved morphology and a higher content of the orthorhombic $FeAlO_3$ phase. Nevertheless, it shows a lower value of saturation magnetization and the magnetic transition occurs at a much lower temperature as compared to bulk $FeAlO_3$ material obtained by classical routes.

Acknowledgments
This study is supported by WBI – Belgium in the frame-work of joint collaboration research program with BAS, Bulgaria. The authors would like to thank Assoc. Prof. K. Starbova and Assoc. Prof. N. Starbov for samples preparation and SEM visualization and Dr. K. Buchkov for help in magnetization measurements.

Journal of Physics: Conference Series **558** (2014) 012053 doi:10.1088/1742-6596/558/1/012053

References

[1] Schieber M, Frankel R B, Blum N A and Foner S 1967 *J.of Appl.Phys* .**38** 1282
[2] Devaux X, Rousset A, Broto J M, Rakoto H and Askenazy S 1990 *J.Mat. Science Letters* **9** 371
[3] Cotica L F,deMedeiros S N, Santos I A, Paesano A,daCunha JBM, Venet M, Garcia D, Eiras JM
2006 *Ferroelectrics* **338** 241
[4] Cotica L F, Santos I A, Venet M, Garcia D, Eiras J M, CoelhoA A, 2008m *Solid StateComm* **147** 123
[5] Starbov N, Starbova K, Vanderbemden Ph, Simeonova P and Lovchinov V 2012 *J. Phys.: Conf. Series* **398** 012050
[6] Bourée F, Baudour Jr, Elbadraoui E, Musso J,Laurent C and Rousset A ,1996 *Acta Cryst* **B52** 217
[7] Tzankov D, Kovacheva D, Krezhov K, Púzniak R, Wiśniewski A, Sváb E, Mikhov M, 2005 *J. Phys.: Condens. Matter* **17** 4319

Journal of Physics: Conference Series **558** (2014) 012054 doi:10.1088/1742-6596/558/1/012054

Electrical parameters of thin nanoscale SiO_x layers grown on plasma hydrogenated silicon

S Alexandrova[1], A Szekeres[2], E Halova[1] and N Kojuharova[1]

[1]Department of Applied Physics, Technical University, 8 Kl. Ohridski Blvd., 1797 Sofia, Bulgaria
[2]Institute of Solid State Physics, 72 Tzarigradsko Chaussee, 1784 Sofia, Bulgaria

E-mail: ehalova@tu-sofia.bg

Abstract. In the present paper results are presented on electrical characterization of the interface Si/SiO_x, formed by oxidation on Si wafers, previously exposed to rf hydrogen plasma. As a tool of investigations multiple frequency *C-V* and *G-V* measurements are applied. The data analysis was performed using two-frequency method to extract generalized frequency independent *C-V* characteristic. Interface trap densities were evaluated from the generalized *C-V* data by comparison with theoretical data for an ideal interface. A set of localized states, acting as interface traps, was found that characterize the interface of Si to substoichiometric SiO_x, layer with $x < 2$. The dielectric constant of the oxides was calculated from the capacitance in accumulation of the generalized *C-V* curves. The thickness and the refractive index of the oxide layers were obtained from ellipsometric data analysis assuming the oxide-Si substrate as single layer system. From the data for the dielectric constant and refractive index suggestion is made that the grown oxides on hydrogenated Si contain voids thus reducing the dielectric constant. Correlation with oxide mechanical stress is found.

1. Introduction

Contemporary semiconductor electronics is still largely dominated by the silicon technology. Reaching Si CMOS limits, great efforts have been exerted to search for a replacement of the SiO_2 by other dielectric layer with higher dielectric constants, to design different gate structures or even to develop entirely different materials for non-Si based electronic elements. However, still no reasonable candidate has been found to replace the thermally grown SiO_2 with its high dielectric quality and perfect interfaces. Si will continue to be the main substrate due to its overall excellent properties. Beyond for CMOS digital applications, Si/SiO_2 has been used for applications in various structures from solar cells to graphene films and carbon nanotubes, new 2D geometry, double gate devices, ballistic nanotube transistors etc. Detailed discussion on future developments of nanodevices can be found in the recent reviews [1, 2, 3]. It can be concluded that at least in the next decade Si/SiO_2 will remain as an important building block.

Small sized silicon oxide in nanodevices obtained either by thermal oxidation of Si or by some deposition technique is usually non stoichiometric SiO_x. Thermal oxidation is conducted at lower temperatures compromising film growth rate and film quality. The interface to the Si can reveal lower quality, higher gate leakage and interface trap density. Improvement can be achieved by incorporation of hydrogen. Heating in hydrogen containing atmosphere has been widely used the past decades due to different beneficial effects such as stabilization of device characteristics through gettering of impurities and saturation of defects. Applied to silicon oxide layers in the nanometric scale, hydrogenation can have a beneficial impact for improving the defect and leakage characteristics.

Our previous studies have shown that hydrogenation of Si before thermal oxidation has a potential for obtaining higher interface quality to ultrathin oxides with thickness around and below 15 nm [4]. An optical study using micro-Raman and spectral ellipsometry revealed the possibility to control the mechanical stress in the Si below the interface [5]. The possible future application of the thin films grown on the hydrogenated Si is conditional upon the studying of their electrical properties. Express and well developed technique for characterization is the capacitance-voltage *C-V* method. Combined with conductance-voltage *G-V* measurements taken at different frequencies it can deliver valuable information about interface quality.

The aim of the present paper is electrical characterization of the Si/SiO_x interface, formed on Si wafers exposed to rf hydrogen plasma prior oxidation. During plasma hydrogenation the Si wafers could be heated up to 300ºC. The SiO_x layers with thickness below 10 nm were formed by thermal oxidation at 850ºC. Thickness was ellipsometrically controlled.

For electrical characterization purposes the Si/SiO_x structures are incorporated in a MOS device. As a tool of investigation multiple frequency *C-V* and *G-V* measurement techniques are applied.

2. Sample preparation

The structures used in this study were MOS capacitors formed on 5-10 Ohm.cm *n*-type (111)-oriented single-crystal Si wafers. Oxides with a thickness below 10 nm were grown on rf plasma hydrogenated Si by thermal oxidation at 850ºC in dry O_2 ambient. The Si wafers were cleaned using a standard RCA procedure (H_2SO_4/H_2O_2 solution followed by a dip in diluted HF and a rinse in de-ionized H_2O). Some of the substrates were subsequently hydrogenated by exposure to a rf hydrogen plasma in a planar plasma unit. The rf generator (13,56 MHz) was capacitively coupled to the reactive chamber with an input power of 15 W. The hydrogen gas pressure was 133 Pa. The substrates were kept on the lower electrode. Part of the substrates was hydrogenated without heating and for another part the temperature was 300ºC. Duration of the plasma exposure was 15 min.

The oxidation was performed in same oxidation run for all substrates, the plasma exposed and those with treated in RCA only. The oxides grown on RCA cleaned samples served for comparison, to reveal the effect of hydrogenation and possible advantages of the proposed plasma technology as well.

For electrical characterization $1{,}96\times10^{-3}$ cm^2 Al dots on the oxide surface and Al continuous film on the Si back-side were deposited by vacuum evaporation to form MOS structures. Information about the concentrations of the electrically active defects was gained from analysis of the frequency dispersion of the *C-V* and *G-V* characteristics taken at room temperature in the frequency range 1 Hz-300 kHz. The measurement unit was a Precision Component Analyzer Wayne-Kerr 6425. The interface trap density D_{it} was evaluated by comparison of the experimental and ideal theoretical *C-V* characteristics. For that purpose a two frequency method has been applied [6] developed to explore thin and/or leaky oxides. The MOS capacitor is modelled as a three-element circuit. By this technique a generalized frequency independent C-V curve is evaluated.

Using the measured *C-V* and *G-V* curves at two frequencies in the range 1Hz - 300 kHz the dissipation *D* is calculated from

$$D = \frac{G}{\omega C} \tag{1}$$

where *G* is the parallel conductance and *C* is the capacitance measured at the different frequencies. Using the method presented by Yang and Hu [6] the generalized capacitance C´ is:

$$C' = \frac{f_1^2 C_1(1+D_1^2) - f_2^2 C_2(1+D_2^2)}{f_1^2 - f_2^2} \tag{2}$$

where C_1 and D_1 refer to values measured at the frequency f_1 and C_2 and D_2 refer to values measured at the frequency f_2. where C_1 and D_1 refer to values measured at the frequency f_1 and C_2 and D_2 refer to values measured at the frequency f_2.

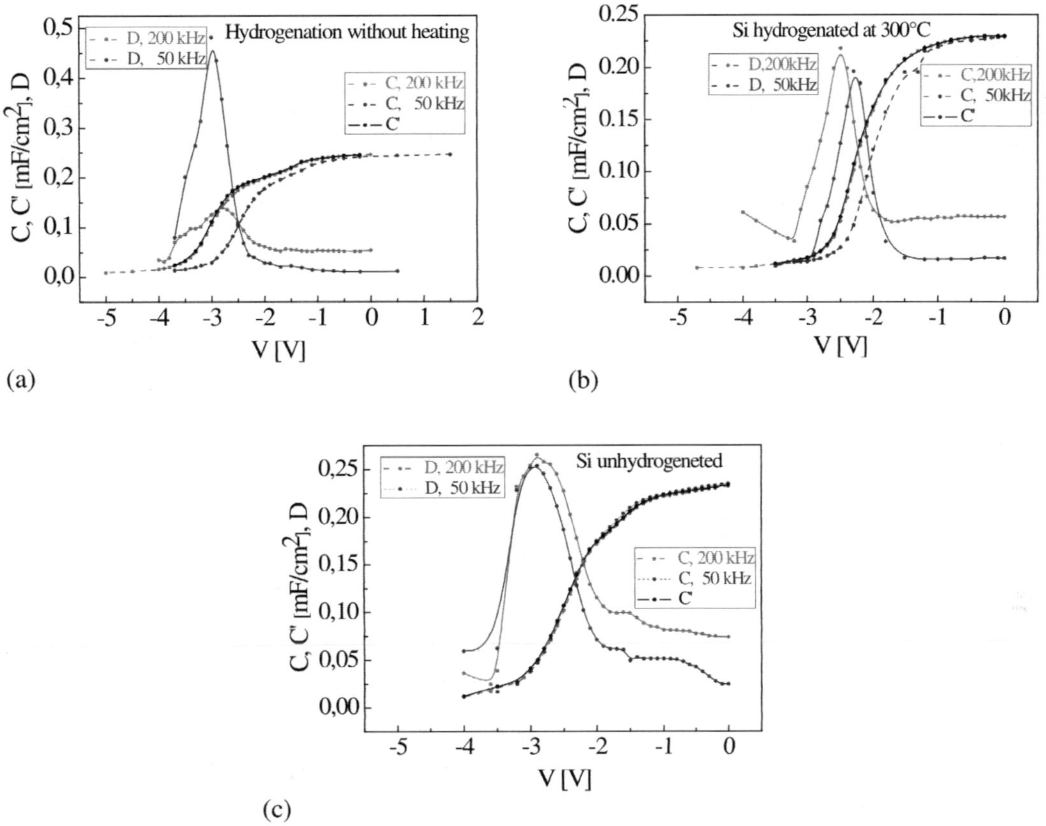

Figure 1. *C-V* and *D-V* curves at different frequencies and generalized *C-V* characteristics
a) Si hydrogenated without heating; b) Si hydrogenated at 300°C; c) Si unhydrogenated.

Additional information about the oxide and interface region to the Si was gained using a Rudolf Research spectroscopic ellipsometer, in the wavelength range 280-820 nm, at an incidence angle of 70°. The accuracy of the thickness determination was ± 0.2 nm. The refractive index was calculated assuming the oxide-Si as a single oxide layer system. The mechanical stress across the oxide was evaluated from analysis of the spectra gained by micro-Raman spectroscopy (RS).

3. Results and Discussion

The analysis of the electrical characteristics has revealed that the MOS structures with SiO_x layer grown on hydrogenated Si indicate shifts and shapes of the measured curves that refer to small oxide charge and low leakage currents, i.e. high quality dielectric layer. Applying two-frequency method to the *C-V* and *G-V* characteristics taken at a variety of frequencies [6], a generalized *C-V* curve for every sample was generated. In these curves the series resistance and the leakage through the oxide layer is accounted for. The generalized *C-V* curves are displaced in figure 1.

The shape of the curves for oxides on Si hydrogenated without heating (figure 1a) shows variations typical for high density of interface traps, even higher than the oxide on unhydrogenated Si (figure 1c). Hydrogenation of the Si wafer at 300°C ensures oxide with *C-V* and *G-V* curves with more regular shape (figure 1b).

From comparison of the generalized *C-V* curves with the theoretical *C-V* curves, calculated for idealized case without presence of interface traps, the energetic spectra of the interface traps over Si bandgap have been calculated. The results for the oxides grown on hydrogenated and unhydrogenated Si wafers are given in figure 2. The spectra show the presence of interface traps with localized energy levels in the Si bandgap. It can be seen that the spectrum of the oxide grown on unhydrogenated substrate reveals a set of energetical levels in the Si bandgap and a steep increase near the conduction band edge with relatively high density. It should be remembered that this oxide has not been subjected to any thermal treatment for trap concentration decrease. The oxide grown on hydrogenated Si shows decrease of the density near midgap. Localized interface levels appear similarly to oxide on unhydrogenated Si and approximately at same energetical positions in the band. Results concerned to the oxide grown on a Si substrate that has been heated up to 300°C during plasma hydrogenation

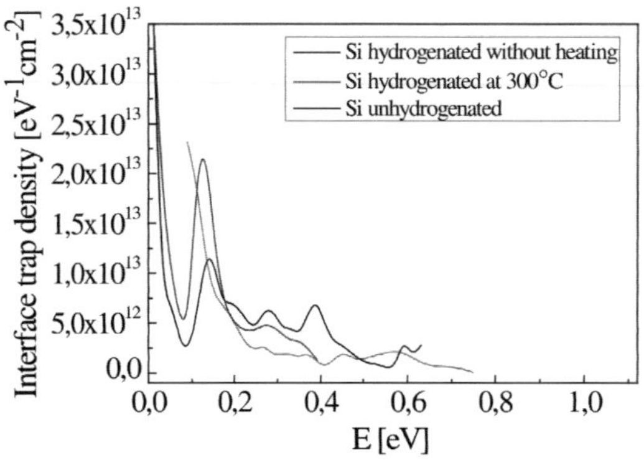

Figure 2. Interface trap density spectra.

are also presented. It can be seen that in this case the spectrum flattens, localized states have disappeared, the density of the traps is reduced. The explanation could be in the combined annealing effect of the increased temperature and hydrogen incorporated during plasma hydrogenation which serves to repair damaged chemical bonds in the interface region. Hydrogen supply only, as evident from the spectrum of the oxide grown on Si hydrogenated without heating, although showing beneficial effect, is still not sufficient to bring the density of the traps to substantially low level. Localized trap levels in Si bandgap can be associated to the interface region where transition from the crystalline Si to the amorphous silicon oxide occurs. Most probably the oxide layer is not stoichiometric SiO_2 because of the oxidation temperature that is much below the condition for growing oxide with complete oxidation of the Si. The growth mechanism of the oxide, and consequently, its structure, is influenced by the presence of hydrogen in the hydrogenated Si top layer. For that reason in the interface spectra appears a set of localized states due to deformed bonds, not necessarily the single levels that are usually contributed to well defined dangling bonds at the interface of Si with high temperature SiO_2.

Data from ellipsometric study also indicates that the grown oxides should be defined as SiO_x, where $x < 2$. The refractive index values at 633 nm for the oxides are over 1.46 which is typical for stoichiometric SiO_2. Taking the thickness of the oxide layers from the ellipsometric data, the dielectric constant of the oxide was calculated from the capacitance in strong accumulation of the generalized *C-*

V curves. The results for the refractive index and the dielectric constant of the oxides are summarized in the table.

Table. Parameters of oxides grown on Si with different pre-oxidation treatment.

Si wafer pre-oxidation treatment	Refractive index n	Oxide thickness d_{ox} [nm]	Dielectric constant ε	σ x10^8 [N/m^2]	ρ/M [mol/cm^{-3}]
Plasma hydrogenation: unheated	1,620	13,60	3,76	1,94	0,0476
Plasma hydrogenation: 300°C	1,478	13,20	3,42	1,80	0,0381
Without hydrogenation	1,468	14,12	3,73	2,48	0,0374
SiO_2	1,460	-	3,90	-	0,0368
SiO	1,960	-	5,00	-	0,2760

For comparison, in the table data for ε and n is given for stoichiometric SiO_2 and silicon monoxide SiO [7, 8, 9]. However, oxide parameters can depend on oxide thickness and be different from those for bulk relaxed material. The parameters for oxides on hydrogenated Si can differ also from those of oxides grown on unhydrogenated Si before oxidation, which is evident from the table.

It can be expected that the dielectric constant ε should fall between the values of 3,9 for stoichiometric SiO_2 and 5 for silicon monoxide. However, our results are below this interval. The refractive index data reveal higher values as compared to stoichiometric SiO_2 and substantially lower values than SiO monoxide. This implies that the oxide composition can be characterized rather as substoichiometric silicon dioxide, which becomes clearly evident for the oxides on plasma hydrogenated Si wafers being more pronounced for hydrogenation without heating. From the data for the dielectric constant also further indications can be made about the structure of the oxides. It can be suggested that the films contain voids with $\varepsilon = 1$, which lowers the dielectric constant. The voids have been proposed to be created in Si top layer during plasma hydrogenation which persist even after oxidation [4]. Inclusion of voids in oxide network would imply smaller film density. Next, the film density can be extracted from the refractive index by using the Lorentz-Lorenz relation:

$$\frac{n^2-1}{n^2+2}=\frac{4\pi}{3}N\alpha\frac{\rho}{M} \qquad (3)$$

where n is the index of refraction, M is the molecular mass, ρ is the mass density, N is Avogadro's number 6,022x10^{23} mol^{-1} and α is the total polarizability of the molecule. However, without specific composition data for the film, calculation of its density is not possible. Since the composition of the investigated oxides is expected to be SiO_x, where x varies from 1 to 2, the molecular mass of the film is between 44,08 g/mol for silicon monoxide and 60,08 g/mol for SiO_2 and the density between 2,13 g/cm^{-3} and 2,2 g/cm^{-3}, respectively. Some insight to the dielectric constant dependence on the structure and composition of the oxide can be gained from the ρ/M ratio. Substituting the data for the refractive index n and taking the total polarizability $\alpha = \alpha_e$ (electron polarizability) = 2,95x10^{-24} cm^3 [10] results for ρ/M are obtained and included in the table. Again it can be concluded that the oxides are substoichiometric silicon oxides with x < 2. The ratio ρ/M is higher than the value for relaxed SiO_2 but much smaller that for SiO. It should be remembered that the oxide parameters, dielectric constant, refractive index, mass density and interface trap density depend not only on composition, but on thickness as well [11, 12] For oxides grown thermally on Si at 1000°C [11] with similar thickness as in the present study, the density amounts to about 4 g/cm^{-3} corresponding ρ/M = 0,0666 for SiO_2 composition, as suggested by the authors.

Further indication on the composition and structure of the oxides comes from the data for the mechanical stress over the oxide as gained from micro-Raman analysis. The stress generated from oxidation process only for the oxides on RCA treated Si shows relatively high value. Preoxidation hydrogenation of the Si obviously leads to relaxation of the stress mainly through creation of voids retained after oxidation.

4. Conclusions

Electrical characterization of silicon oxides grown on rf plasma hydrogenated Si was conducted. Interface trap densities were evaluated from *C-V* and *G-V* measurements taken at different frequency applying two-frequency technique. A set of localized states acting as interface traps was found that characterize the interface of Si to substoichiometric SiO_x layer. For oxide grown on Si hydrogenated at increased wafer temperature most of the traps are annealed out. The traps should be due to deformed bonds, not necessarily the single levels that are usually contributed dangling Si bonds at the interface of Si with high temperature SiO_2. The dielectric constants were extracted from the generalized *C-V* curves. Suggestion is made that the grown oxides on hydrogenated Si contain voids thus reducing the dielectric constant. Correlation is found with the results from ellipsometrically obtained refractive index and its relation to oxide mass density. The mechanical stress on the oxide estimated from micro-Raman evaluation shows decrease due to presence of voids in the oxide. Conclusion is made that the oxides grown on plasma hydrogenated Si can be characterized as substoichiometric SiO_x layers.

References

[1] Saraswat K 2010 *How far can we push Si CMOS?* http://www.ohio.edu/people/starzykj/network/class/ee516/slides/Future%20Devices.pdf
[2] Iwai H *Ultimate CMOS scaling* 2013 Korean International Summer School on Nanoelectronics 2–5 July 2013 Daejon, Korea http://www.iwailab.ep.titech.ac.jp/pdf/iwaironbun/201307nanokiss.pdf
[3] Balestra F 2014 *Beyond CMOS Nanodevices* 1 ed. (New York: Wiley & Sons - Interscience)
[4] Alexandrova S, Szekeres A and Halova E 2010 *IOP Conf. Series: Materials Science and Engineering* **15** 012037
[5] Alexandrova S, Szekeres A and Valcheva E 2012 *Phisica status solidi (c)* **9** 2203
[6] Yang K J and Hu Ch 1999 *IEEE Trans. Electron Dev.* **46** 1500
[7] Hass G and Salzberg C D 1954 *J. Opt. Soc. Am.* **44** 181
[8] Philipp H R 1971 *J. Phys. Chem. Solids* **32** 1935
[9] Gupta T 2009 *Dielectric Materials in Copper Interconnect Technology* **Chapter 2** p 67 (New York: Springer Verlag)
[10] Lasaga A C and Cygan R T 1982 *American Mineralogist* **67** 328
[11] Rzodkiewicz W and Panas A 2009 *ACTA PHYSICA POLONICA A* **116** S92
[12] Alexandrova S and Szekeres A 2010 *Solid State Phenomena* **159** 163

Effect of post-synthesis acid activation of TiO_2 nanofilms on the photocatalytic efficiency under visible light

I Stambolova[1,*], V Blaskov[1], M Shipochka[1], A Eliyas[2] and S Vassilev[3]

[1] – Institute of General and Inorganic Chemistry, Acad. G. Bonchev St., bl. 11, Bulgarian Academy of Sciences, Sofia 1113, Bulgaria
[2] – Institute of Catalysis, Acad. G. Bonchev St., bl. 11, Bulgarian Academy of Sciences, Sofia 1113, Bulgaria
[3] – Institute of Electrochemistry and Energy Systems, Acad. G. Bonchev St., bl.10, Bulgarian Academy of Sciences, Sofia 1113, Bulgaria

E-mail: stambolova@yahoo.com

Abstract. Nanosized TiO_2 films were deposited by spray pyrolysis and thermally treated at 400°C. Then the films were dipped in 1M aqueous solution of HCl. The activated samples were divided into two parts – one part was dried (A) and another was annealed (AT) in air. The photocatalytic degradation of Reactive Black (RB5) textile dye under visible light was tested. The following instrumental methods: X-ray diffraction (XRD), Raman spectroscopy and X-ray photoelectron spectroscopy (XPS) were applied for the phase and surface characterization of obtained samples. According to Raman and XRD analyses all films are anatase. The XRD showed that activated films are better crystallized than non activated TiO_2 film. The presence of chlorine at 200.3 eV was registered for acid activated samples by X-Ray photoelectron spectroscopy. The acidic activated films exhibited higher rate of dye photodegradation than that of the reference TiO_2 sample. The photocatalytic efficiency decreases in the order A > AT> non activated TiO_2 films. The degradation rate constant for acid activated films is two times higher than those of the reference film. The hydroxyl content in TiO_2 acidic activated films is greater than that of the non- activated films, which results in significant increase in the photocatalytic activity. In addition, the presence of chlorine may also lead to enhancement in efficiency.

1. Introduction

The uncontrolled release of colored waste water contaminated with dyes from textile, paper, rubber and plastic industries is resulting in serious environmental pollution [1]. Recent studies have shown that heterogeneous semiconductor photocatalysis can be an alternative method to the conventional ones (flocculation, sedimentation, absorption etc.) for the removal of dye pollutants from water [2]. The heterogeneous photocatalytic process consists of making use of the near ultraviolet band of the solar spectrum to photoexcite a semiconductor catalyst in contact with water in the presence of oxygen. Oxidizing species (hydroxyl radicals) which attack oxidizable contaminants are generated producing CO_2, H_2O and diluted inorganic acids. Titania exhibits outstanding decomposition ability for many organic compounds induced by illumination with UV light [3]. Various methods are available for the preparation of nanostructured TiO_2-based photocatalytic films, such as electrochemical method [4], chemical vapour deposition (CVD) [5] as well sol gel methods [6]. Nevertheless, most of them are complicated, expensive and not suitable for practical production of high-effective catalyst. Among the chemical films deposition methods the spray pyrolysis is promising for industrial applications, due the possibility to obtain thin films with desired morphology and thickness on large-area substrates [7].

There are many approaches, proposed to increase the photocatalytic activity such as doping with transition metals, non metals [8], mixing with other semiconductor oxides [9]. Recently the acid treatment is applied to enhance the photocatalytic performance of TiO_2 powders and films, but the studies are quite scarce in the available literature. Liang et. al. have revealed the positive effect of HCl treatment on the activity of TiO_2 thin films [10]. Another scientific groups have improved the photocatalytic properties of TiO_2 powders by treatment in different acids [11, 12].

The aim of this study is to investigate the effect of the HCl activation of sprayed TiO_2 film on the surface features and the rate of photocatalytic degradation of real textile di-azo dye under Visible light.

2. Experimental details

2.1. Preparation of the films

The aluminium foil plates (75 x 25mm, thickness 0.3 mm) were used for the deposition of thin TiO_2 films. The substrates were cleaned successively in hot ethanol and acetone. The solution of $TiCl_4$ (0,4 M) was diluted with a mixture of isopropanol and butyl carbitol ($C_4H_9OC_2H_4OC_2H_4OH$). The aerosol of precursor solution was generated by pneumatic glass nebulizer to the heated substrate at 300-400°C. All samples were thermally treated at 400°C (label T). One part of T films was dipped in 1M aqueous solution of HCl. After acidic activation the samples were divided into two parts – one part was dried at 100°C (label A) and another was annealed at 400°C in air (label AT).

2.2 Films characterization

The phase composition of the samples was studied by X-ray diffraction (XRD) with CuK_α-radiation (Philips PW 1050 apparatus). The composition and electronic properties of the films were investigated by X-ray photoelectron spectroscopy (XPS). The measurements were performed in VG ESCALAB II electron spectrometer using AlK_α radiation with energy of 1486.6 eV. The binding energies were determined with an accuracy of ±0.1 eV utilizing the C1s line at 285.0 eV (from an adventitious carbon) as a reference. The chemical composition of the films was investigated on the basis of peak areas and binding energies of O1s, Ti2p and Cl2p photoelectron peaks (after linear subtraction of the background and Scofield's photoionization cross-sections). The Raman spectra were obtained using a LabRAMHR visible single spectrometer.

2.3. Photocatalytic tests

The photocatalytic experiments were conducted with powerful visible-light irradiation source (TUNGSRAM 500W lamp). The TiO_2 coated substrate was placed into the vessel and covered completely with 150 ml solution of Reactive Black 5 dye with 20 ppm concentration and stirred continuously with a magnetic stirrer at a constant magnetic stirring rate (400 rpm) under oxidative conditions at room temperature.The photocatalytic degradation was evaluated by taking aliquot of the solution and measuring the residual concentration using spectrophotometer UV-1600PC in the wavelength range from 200 to 800 nm at regular time intervals.

3. Results and discussion

We have applied Raman spectroscopy for more precise phase analysis, because this technique is more surface sensitive, while XRD is applied for bulk analysis. The Raman spectra of non activated and acid activated TiO_2 films have shown the presence of anatase phase (figure 1).

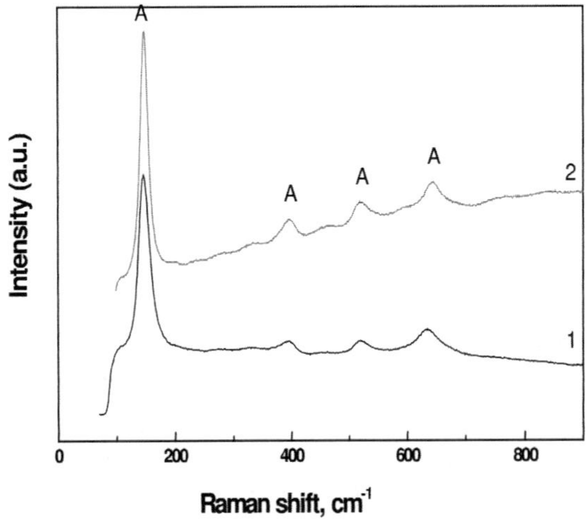

Figure 1. Raman spectra of non activated (1) and acid activated (2) TiO_2 film.

According to the literature the main bands of anatase monocrystal were registered at 144, 197, 320, 399, 519, 639 cm^{-1} [13]. It has to note that the anatase main bands of the T films are broadened and high frequency shifted with respect to the values of bulk crystal due to the lack of long range order, which is typical for nanocrystalline materials [14]. The X-ray diffraction analyses have proved that the main phase in the films is the anatase and the acid modification leads to a better crystallization of A and AT films. The crystallites size of T, A and AT films are 20 nm, 18 and 17 nm, respectively.

According to the XPS analyses the films are non-stoichiometric and on their surface are present hydroxyl groups, which were proved by the deconvolution of the O1s peaks.

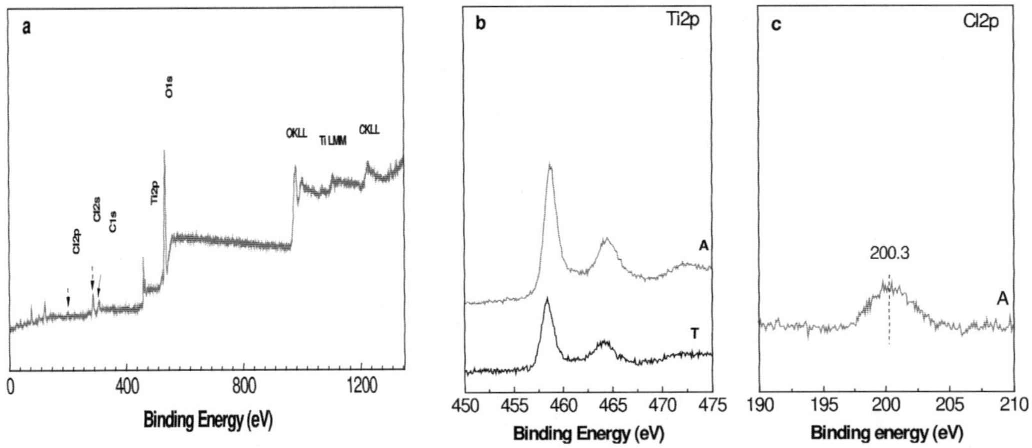

Figure 2. (a) XPS survey spectrum of acid treated TiO_2 films; (b) Ti2p core level spectra of the films T, A and AT; (c) Cl2p photoelectron spectrum of acid treated TiO_2 films.

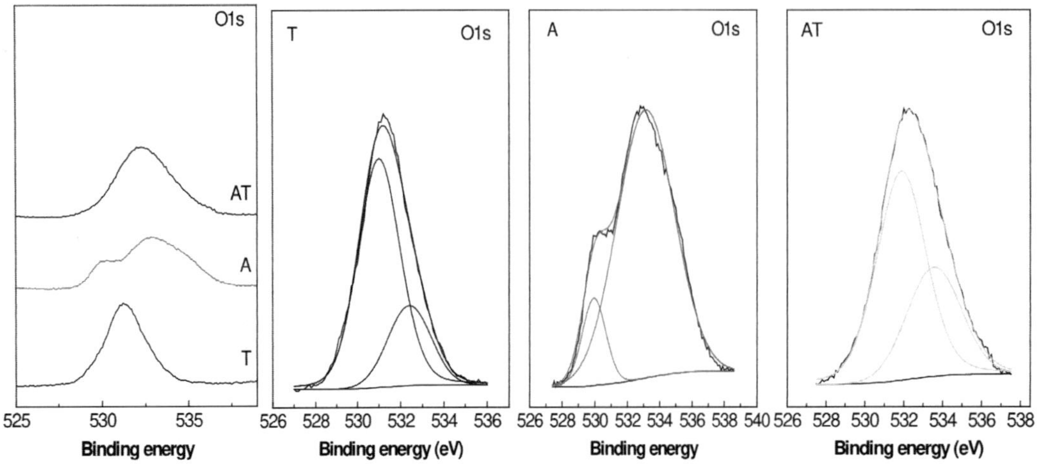

Figure 3. Deconvolution of the O1s photoelectron spectra of non activated (T), acid treated (A) and acid activated and thermally treated at 400°C (AT) films.

Figure 2 shows the XPS survey spectrum, Ti2p core level spectrum of non treated and acid treated TiO_2 films and Cl2p photoelectron spectrum of acid treated TiO_2 films. The Ti2p spectra of sample T and A are sharp and symmetric and have a maximums at 458.4 eV and 458.8 eV, respectively, typical for TiO_2. For sample AT the concentration of titanium strongly reduces. The presence of chlorine at 200.3 eV was registered for acid activated samples (figure 2-c). Figure 3 shows the deconvolution of the O1s photoelectron spectra of the TiO_2 films, obtained by different treatments The O1s peaks are wide and asymmetric and deconvoluted by Lorentzian–Gaussian curve fitting into two components [15]. The first component with the lower binding energy is attributed to O^{2-} ions in the TiO_2 lattice and the second one with the higher binding energy is ascribed to oxygen atoms in hydroxyl groups. The shape of O1s peaks for not activated films and acid activated films is different, showing that the amount of adsorbed OH^- groups is increased for samples A. After temperature treatment at 400°C amount of OH^- groups is decreased (sample AT).

The reaction kinetics is represented by plotting the natural logarithm of the concentration ratio as a function of the illumination time. The experimental results show that the photocatalytic decoloration of the dye with the films under UV-light irradiation follows a pseudo first order reaction kinetics, expressed by the equation:

$$\ln(C/C_0) = -kt \qquad (1)$$

where C_0 is the initial concentration of the dye, C is the concentration of the dye after irradiation during a selected time interval t and k is the rate constant.

The photocatalytic studies have proved that the kinetic curves for RB5 degradation follow the pseudofirst order kinetics of the Langmuir - Hinshelwood model as confirmed by plotting of ln(C/Co) ***vs*** reaction time *t*. (figure 4). A straight line with R^2 (correlation coefficient) higher than 0.9 is obtained. It can be seen that the activated TiO_2 films show higher photocatalytic activity than non activated films. The thermal treatment after acid activation has strong influence on the photocatalytic efficiency. Samples A possess faster rate of dye degradation than AT films. The photocatalytic behavior of the acid activated films can be explain as follows: (i) Surface of the TiO_2 films is protonized as a result of the treatment with HCl. The amount of OH groups on the surface considerably increase (as was proved by XPS analysis) in order to maintain the equilibrium OH/H^+ and the photocatalytic degradation of the dye increases [11]. (ii) the presence of Cl^- on the surface of activated TiO_2 films leads to more efficient degradation of the dye in the visible region due to the red shift. Xu et al. [16] have found that the doping of chlorine resulted in red shift of absorption and higher acidity, which were the reasons for higher photocatalytic activity.

Journal of Physics: Conference Series **558** (2014) 012055 doi:10.1088/1742-6596/558/1/012055

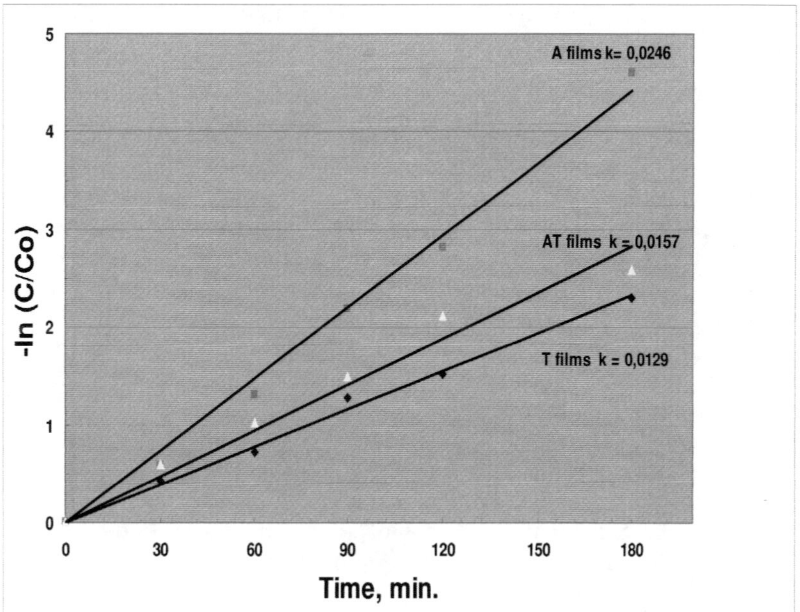

Figure 4. Kinetic curves of RB5 degradation for TiO_2 films, non activated (T), activated in HCl and dried at 100°C(A), activated in HCl and treated at 400°C (AT).

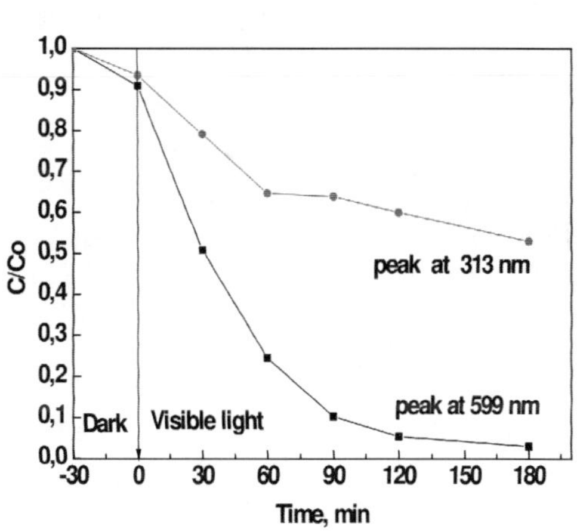

Figure 5. Concentration changes *vs* time of illumination for samples A, for the peak at 599 nm (corresponding to azo-bond) and at 313 nm (corresponding to naphtalene-ring).

The photocatalytic measurements have shown that the adsorption peaks corresponding to the N=N bond (at 599 nm) and to naphthalene rings (at 313 nm) decrease with the time (figure 5).
The peak at 313 nm not only diminishes, but it is shifted to lower values with the time, due to the destruction of naphthalene structure. The plot C/C_0 versus time (where C_0 and C are initial absorbance

and absorbance of the photolyzed solution at time t, respectively) displayed a considerably degradation rate for azo groups. The stability of the unsaturated six-membered ring leads to slower rate of degradation.

4. Conclusions

The effect of HCl acid activation of TiO_2 nanosize sprayed films on their photocatalytic efficiency was investigated. One part of the samples was dried (A) and another was annealed (AT) in air. The photocatalytic degradation of Reactive Black (RB5) textile dye under visible light was tested. The activated films are better crystallized than non activated TiO_2 films. The acidic activated films exhibited higher rate of dye photodegradation than reference TiO_2 sample. This photocatalytic behavior can be explain as follows: (i) Surface of the TiO_2 films is protonized as a result of the activation with HCl. The amount of OH groups on the surface considerably increase (as was proved by XPS analysis) in order to maintain the equilibrium OH/H^+ and the photocatalytic degradation of the dye increases and (ii) the presence of chlorine on the surface of activated TiO_2 films leads to more efficient degradation of the dye in the visible region due to the red shift.

References

[1] Pekakis PA, Xekoukoulotakis N P and Mantzavinos D 2006 *Water Research* **40** 127
[2] Hoffmann H, Martin S, Choi W and Bahnemann D 1995 *Chem. Rev.* **95** 69
[3] Stambolova I, Shipochka M, Blaskov V, Loukanov A and Vassilev S 2012 *J. Photochem.Photobiol. B* **117** 19
[4] Wang S, Wu X, Qin W and Jiang Z 2007 *Electrochim. Acta* 53 1883
[5] Kim BH, Lee JY, Choa Y H, Higuchi M and Mizutani N 2004 *Mater. Sci. Eng. B* **107** 289
[6] Kaneva N, Stambolova I, Blaskov V, Dimitriev Y, Bojinova A and Dushkin C 2012 *Surf. Coat. Technol.* **207** 5
[7] Stambolova I, Blaskov V and Dimitrov O, 2013 *Rev. Rom. Chem.* **58** 839
[8] Sun H, Wang S, Ming H, Ang M and Li Q 2010 *Chem. Eng. J.* **162** 437
[9] Enesca A, Isac L, Andronic L, Perniu D and Duta A 2014 *Appl. Catal. B* **47** 175
[10] Liang J, Feng Y, Liang G, Zwang J, X. Yan X and Jin Z 2003 *J. Mater. Sci. Lett.* **22** 1503
[11] Colon C, Sanchez-Espana J M, Hidalgo M C and Navia J A 2006 *J. Photochem. Photobiol. A* **179** 20
[12] Cheng B, Yu J-G, Zhang X-Q and Zhao X-J 2003 *Mater. Sci. Lett.* **22** 967
[13] Chaves A, Katiyan K S and Porto S P S 1974 *Phys. Rev.* **10** 3522
[14] Bersani D, Lottici P and Montenero A 2000 *J. Mater. Sci.* **35** 4301
[15] Liang J, Feng Y, Liang G Ji, Z. Wang J, Yan X and Jin Z 2003 *J. Mater. Sci. Lett.* **22** 1503
[16] Xu H, Zheng Z, Zhang L, Zhang H and Deng F 2008 *J. Solid State Chem.* **181** 2516

Effect of nanodiamond modification of siloxane surfaces on stem cell behaviour

M Keremidarska[1], T Hikov[2], E Radeva[2], L Pramatarova[2] and N Krasteva[1*]

[1]*Institute of Biophysics and Biomedical Engineering, Bulgarian Academy of Sciences, Acad. Georgi Bonchev Str., Block 21, 1113 Sofia, Bulgaria*
[2]*Institute of Solid State Physics, Bulgarian Academy of Sciences, 72 Tzarigradsko Chaussee, Blvd., 1784 Sofia, Bulgaria*

E-mail: nataly@bio21.bas.bg

Abstract. Mesenchymal stem cells (MSCs) hold a great promise for use in many cell therapies and tissue engineering due to their remarkable potential to replicate indefinitely and differentiate into various cell types. Many efforts have been put to study the factors controlling stem cell differentiation. However, still little knowledge has been gained to what extent biomaterials properties influence stem cell adhesion, growth and differentiation. Research utilizing bone marrow-derived MSCs has concentrated on development of specific materials which can enhance specific differentiation of stem cells e.g. osteogenic and chondrogenic.
In the present work we have modified an organosilane, hexamethyldisiloxane (HMDS), with detonation nanodiamond (DND) particles aiming to improve adhesion, growth and osteodifferentiation of rat mesenchymal stem cells. HMDS/DND films were deposited on cover glass using two approaches: premixing of both compounds, followed by plasma polymerization (PP) and PP of HMDS followed by plasma deposition of DND particles. We did not observe however an increase in rMSCs adhesion and growth on DND-modified PPHMDS surfaces compared to unmodified PPHMDS. When we studied alkaline phosphatase (ALP) activity, which is a major sign for early osteodifferentiation, we found the highest ALP activity on the PPHMDS/DND material, prepared by consequent deposition while on the other composite material ALP activity was the lowest. These results suggested that DND-modified materials were able to control osteodifferention in MSCs depending on the deposition approach. Modification of HMDS with DND particles by consequent plasma deposition seems to be a promising approach to produce biomaterials capable to guide stem cell differentiation toward osteoblasts and thus to be used in bone tissue engineering.

1. Introduction

About 40 years ago Friedenstein described for the first time spindle shaped stromal cells in bone marrow that proliferated and form colonies [1]. Cells were named colony-forming unit-fibroblasts (CFU-F) but later different authors used different synonyms such as mesenchymal stem cells (MSCs), skeletal stem cells [2], stromal stem cells, mesenchymal progenitor cells, etc. to define almost the same cell population [3]. Under defined *in vitro* conditions MSCs are able to differentiate into multiple cell types, e.g. osteoblasts, chondroblasts, adipocytes. Easy isolation and multipotent properties of MSCs make them an attractive choice for different clinical applications [4].

Trauma-related skeletal deformities and disorders are among the most widely spread diseases in the world - over 40 % of the population suffers from them. Many of the patients require bone graft substitutes to repair the injury or defect [5]. An ideal bone-graft substitute must provide biomaterials and growth factors for osteoinduction and progenitor cells for osteogenesis. Until recently the growth factors and cytokines were considered as the most potent factors directing *in vitro* differentiation of MSCs. However differentiation driven by the substrate would be a better proposition in terms of implanting tissue-engineered constructs to elicit a specific cellular response [6]. Synthetic materials offer a greater degree of control of material's properties, reproducibility and scalability over naturally occurring materials. Organosilanes, a class of synthetic polymers, meet with a number of indicators the requirements of medical practice, including excellent inertness, smoothness, stability and biocompatability. They have a rare combination of properties that makes them unique: high hydrophobicity, solubility in non-polar solvents, high chain flexibility, low cohesive energy, very low surface tension, good thermal stability and resistance to UV. In particular, PPHMDS has the advantages to be easily processed and fabricated by the method of plasma polymerization and its properties can be readily controlled by varying process parameters. However, the low mechanical stability and high hydrophobicity are serious disadvantages for bone mimicry [7]. Modification with nanodiamonds (NDs) seems to be promising in improving mechanical and cell-contacting properties of siloxanes due to the superior mechanical features and good biocompatibility of nanodiamonds [8]. ND particles can be deposited onto polymer surface or can be implicated into monomer structure. Here, we have compared both methods of modification of PPHMDS with detonation generated ND particles and how both methods affect surface properties and consequently, the stem cell behaviour. Although there are some papers concerning effect of nanodiamond coatings on osteoblast adhesion [9] there is currently a lack of knowledge about how nanodiamond-modified materials can control stem cell adhesion and the subsequent related osteodifferentiation.

2. Materials and methods

2.1. Synthesis and modification of PPHMDS surfaces

Three groups of plasma polymer films were prepared for the purposes of this study: plasma polymerized hexamethyldisiloxane (PPHMDS) and two polymer composites of PPHMDS and DND, named as DND1 and DND2. Preparation and characterization of the materials are described in [10]. As control surfaces cover glasses (CG) were used.

2.2. Cell culture

Rat mesenchymal stem cells were isolated from bone marrow of femurs and tibias of 4-week-old male rats according the centrifugation method of Dobson et al. [11]. Cells were cultured under conditions previously described in [12].

2.3. Cell adhesion assay

To assess initial cell adhesion rat mesenchymal stem cells were seeded in a concentration of 2.10^4 cells/ml onto tested materials, precoated for 30 minutes with 20 μg/ml fibronectin (Sigma, Germany) and were incubated for 2 hours in serum-free media. Phase-contrast micrographs of the attached cells were taken by Zeiss Axiovert 25 inverted microscope (Germany). The pictures were further used for measuring the cell spreading area by image analysis software (Image J).

2.4. Cell growth assay

Evaluation of cell proliferation ability of rMSCs was performed by measuring the activity of the cytoplasmic enzyme lactate dehydrogenase (LDH) with LDH assay (Roche Diagnostic, Germany). Cells were seeded in initial concentration of 2.10^4 cells/ml and cultured for one week on the films in media supplemented with 10% foetal bovine serum (Sigma, Germany). At the first and the seventh day cells were treated according to the manufacturer's instructions. LDH activity was assessed by measuring the optical density with a plate reader at wavelength λ=492 nm.

2.5. Alkaline phosphatase activity assay

At the 1st and 7th day of incubation of the cells on the materials the levels of expression of the enzyme alkaline phosphatase (ALP) were evaluated. For this aim cells were rinsed with Tris buffered saline (TBS) and lysed with Triton X-100. 50 µl of cell lysates were mixed with 100 µl enzyme substrate-p-nitrophenylphosphate (Sigma, Germany) and incubated at 37° C for 2 hours. Optical density was measured at wavelength λ =405 nm.

2.6. Statistical analysis

All experiments were conducted in triplicate. The results are presented as mean values ± standard deviation (SD). For statistical evaluation the Student's test has been applied. The significance was determined at $p \leq 0.05$.

3. Results and discussion

3.1. Characterization of PPHMDS surfaces

Engineering a successful biomaterial to establish optimal interactions with cells for good osteointegration requires knowledge about materials and how their properties influence cell behaviour [13]. Surface topography and wettability are the major materials' factors affecting cell-biomaterial interactions [14]. Therefore, we have characterized first PPHMDS surfaces in respect to their surface roughness and hydrophilicity by AFM and goniometry, respectively and then have studied stem cell behaviour. AFM measurements (data not shown here) revealed that nanodiamond modification increased the surface roughness of the materials, especially in the case of DND2 samples. Water contact angle data showed that the samples, obtained by consequent deposition of PPHMDS and DND were hydrophilic while the other PPHMDS/DND films were hydrophobic with WCA similar to pure PPHMDS film (figure1). In summary, both composite films have different roughness and wettability depending on the method of preparation.

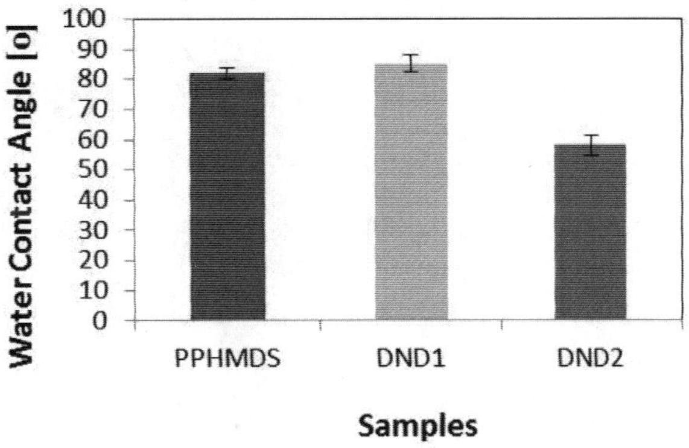

Figure 1. Water contact angle of the polymeric films.

3.2. Initial cell adhesion

Since cell processes such as adhesion, proliferation and differentiation are directly influenced by material's properties they can be used to assess the effect of material modification on cell behaviour. The initial cellular events taking place on biomaterials surface mimic to a certain extent the natural adhesive interactions of cells with the extracellular matrix [15]. In fact, living cells do not interact directly with foreign materials. Upon contact of a biomaterial with physiological fluids *in vivo* or culture medium *in vitro* the first event that happened is adsorption of proteins such as fibronectin (FN), vitronectin (VN), fibrinogen (FG) on followed by attachment of the cells to the adsorbed protein layer. In order to mimic physiological conditions we have pre-adsorbed our materials with FN which is one of the most important proteins mediating cell adhesion [16]. Representative phase contrast images of living rMSCs incubated for 2 hours on various substrates are shown in figure 2. Surprisingly, well spread cells with polygonal morphology, suggesting better interactions with material, were observed mainly on unmodified PPHMDS films. Although the largest cells were observed on DND2 surfaces, in general the cells on DND-modified materials were mainly round with many protrusions.

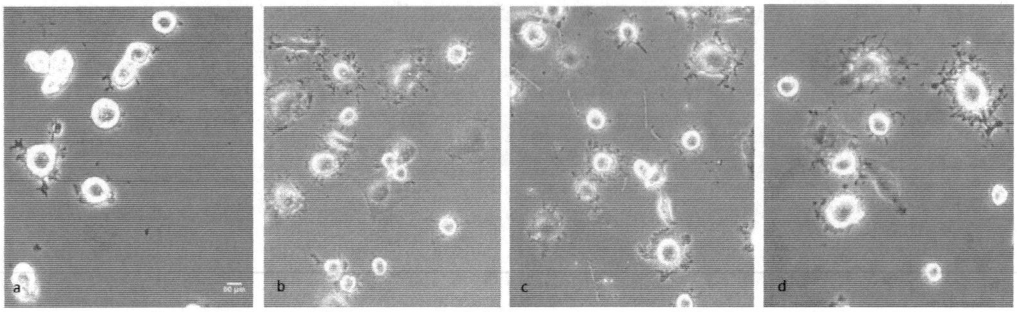

Figure 2. Phase contrast micrographs of rMSCs after 2 hours incubation on FN coated a)CG, b)PPHMDS, c)DND1 and d) DND2. Mag. 20x.

These results can be explained with adsorption and organization of FN. It is well known that fibronectin molecules adsorb preferentially on hydrophobic surfaces but usually they are tightly connected to the material and their cell binding domains stay hidden for the cells [17]. Therefore it is not uncommon that cells on the most hydrophobic DND1 showed the smallest average spreading area as it can be seen from figure 3.

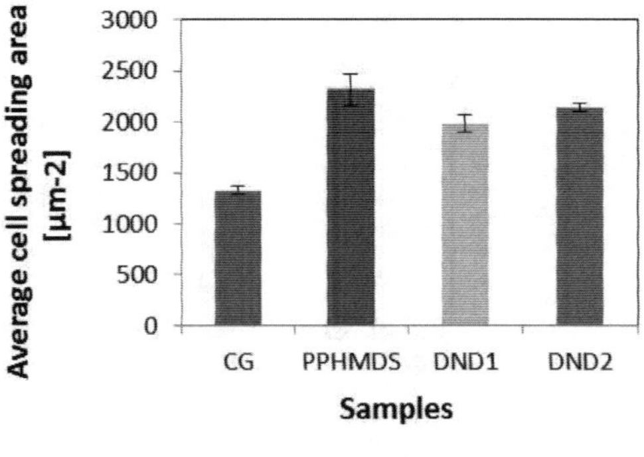

Figure 3. Image J measurement of cell spreading area of rMSCs after 2 hours incubation on FN coated polymeric surfaces.

Opposite, on hydrophilic surfaces FN undergoes greater extension of its dimmer providing more binding sites for the mesenchymal stem cells and thus strength the cell-material interactions. In this context moderately hydrophilic DND2 should provide better substrate for cell adhesion. Moreover, DND2 surface is rougher than the other materials and there is data in literature about improved stem cell attachment on rough surfaces [18]. However our results from overall cell morphology and analysis of cell spreading area surprisingly demonstrated that the hydrophobic and the smoothest PPHMDS favours better the cell binding and spreading of rMSCs compared to both DND-modified surfaces. Therefore not only hydrophobicity and surface roughness but also more surface properties must be taken into account to explain differences in stem cell behaviour on DND-modified and unmodified PHMDS surfaces.

3.3. Cell growth

Further, we have studied cell growth and differentiation in order to make connection between cell adhesion and growth and function of stem cells. We observed an increase in cell growth over the entire culture period (7 days) on all surfaces tested meaning that DND modification did not adversely affect rat MSCs. Among siloxane surfaces the highest proliferative ability in rMSCs at the 1st day was found to induce DND1 while at the 7th day - PPHMDS films (figure 4). However, compared to the control surfaces (cover glasses) the cell growth on siloxane and DND-modified siloxane surfaces was suppressed.

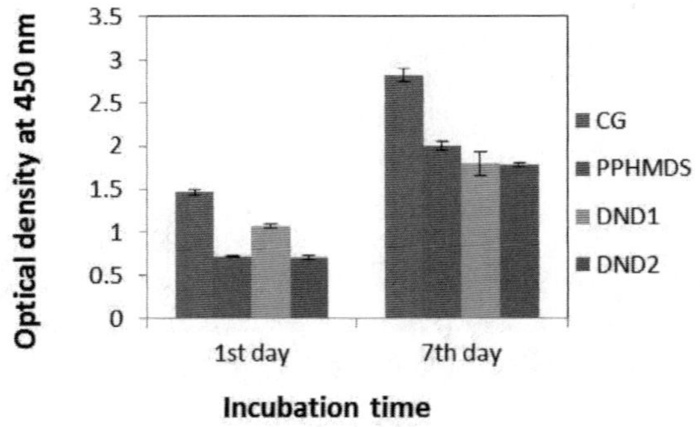

Figure 4. LDH assay of rMSCs on polymeric films.

3.4. Alkaline phosphatase activity

To fully characterize rMSCs behaviour on PPHMDS surfaces we measured ALP activity on days 1 and 7. Although rMSCs demonstrated comparatively low expression of the enzyme at the first day with insignificant differences between materials at day 7 dramatic increase in ALP activity was detected. As it can be seen in figure 5 the cells, except these, cultured on DND1, showed almost three times higher levels of the alkaline phosphatase. The strongest ALP activity in rMSCs was provoked by DND2 materials. Keeping in mind that activity of this enzyme is among the major markers for early osteodifferentiation [19], these results suggest that DND2 managed guide the stem cells toward osteogenesis.

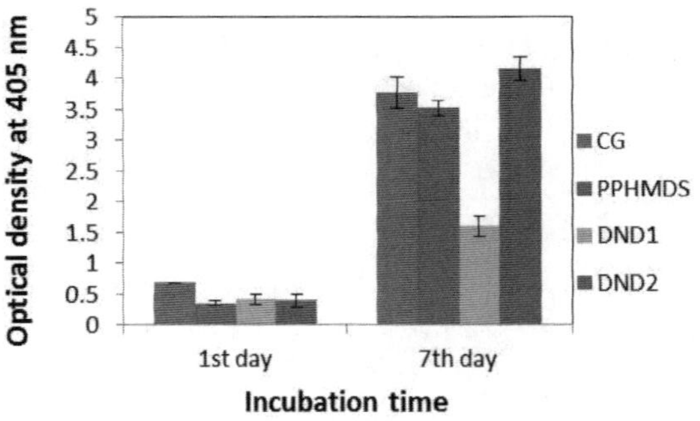

Figure 5. ALP assay of rMSCs on polymeric films.

4. Conclusions

Among the materials studied in this work unmodified PPHMDS films supported the greatest adhesion and proliferation of rMSCs. However, the most significant indication for osteodifferentiation stem cells exhibited after cultivation on the composite films obtained by consequent PP of HMDS and NDs. Therefore, we can conclude that for bone tissue engineering applications modification of polymeric films with nanodiamond particles by consequent deposition of the polymer and nanoparticles should be the method of choice because it induced in greatest degree differentiation of mesenchymal stem cells toward osteoblast cells.

Acknowledgements

This work was supported by Bulgarian National Scientific Research Fund project №: DFNI_B01_18/2012.

References

[1] Friedenstein AJ, Chailakhyan RK, Latsinik NV, Panasyuk AF and Keiliss-Borok IV. 1974 *Transplantation* **4** 331
[2] Bianco P 2011 *Endocrinology* **152** 2957
[3] Väänänen HK 2005 *Ann. Med.* **7** 469
[4] Ding DC, Shyu W-C and Linn S-Z. 2011 *Cell Transplant.* **1** 5
[5] Ramakrishnaiah Siddappa Cellular and molecular prerequisites for bone tissue engineering 2007 PhD thesis University of Twente, Enschede, The Netherlands
[6] Curran JM, Chen R and Hunt JA . 2006 *Biomaterials* **27** 4783
[7] Colas A and Curtis J 2005 *Silicone biomaterials: History and chemistry & Medical applications of silicones* (Biomaterials Science) ed B Ratner (Elsevier) chapter 2 and 7
[8] Thalhammer A, Edgington RJ, Cingolani LA, Schoepfer R and Jackman RB. 2010 *Biomaterials* **31** 2097
[9] Yang L, Sheldon BW and Thomas JW. 2009 *J. Biomed. Mat. Res.* A **91** 548
[10] Keremidarska M, Radeva E, Eleršič K, Iglič A, Pramatarova L and Krasteva N. Plasma deposited composite coatings to control biological response of osteoblast-like MG-63 cells 2014 *IOP: Conference series Preprint*
[11] Dobson KR, Reading L, Haberey M, Marine X and Scutt A. 1999 *Calcif. Tissue Int.* **65** 411
[12] Keremidarska M, Radeva E, Pramatarova L and Krasteva N. 2014 *Compt. Rend. Acad. Bulg.Sci.* **67** 1113
[13] Tirrell M, Kokkoli E and Biesalski M. 2002 *Surf. Sci.* **500** 61
[14] Ponsonnet L, Reybier K, Jaffrezic N, Comte V, Lagneau C, Lissac M and Martelet C. 2003 *Mat. Sci. Eng.* C **23** 551
[15] Salmeron-Sanchez M and Altankov G. 2010 *Cell-protein-material interactions in tissue engineering* (Tissue engineering) ed D Eberli (InTech) chapter 4 pp 77-102
[16] Yamagata M, Suzuki S, Akiyama SK, Yamada KM and Kimata K. 1989 *J. Biol. Chem.* **264** 8012
[17] Bergkvist M, Carlsson J and Oscarsson S. 2003 *J. Biomed. Mater. Res.* A **64** 349
[18] Deligianni D, Katsala ND, Koutsoukos PG and Missirilis YF. 2001 *Biomaterials* **22** 87
[19] Birmingham E, Niebur GL, McHugh PE, Shaw G, Barry FP and McNamara LM. 2012 *Eur. Cell Mater.* **23** 13

Plasma deposited composite coatings to control biological response of osteoblast-like MG-63 cells

M Keremidarska[1], E Radeva[2], K Eleršič[3], A Iglič[4], L Pramatarova[2] and N Krasteva[1*]

[1]*Institute of Biophysics and Biomedical Engineering, Bulgarian Academy of Sciences, Acad. G. Bonchev Str., Block 21, 1113 Sofia, Bulgaria*
[2]*Institute of Solid State Physics, Bulgarian Academy of Sciences 72, Tzarigradsko Chaussee, Blvd., 1784 Sofia, Bulgaria*
[3]*Institute J. Stefan, Jamova 39, 1000 Ljubljana, Slovenia*
[4]*Laboratory of Biophysics, Faculty of Electrical Engineering and Laboratory of Clinical Biophysics, Faculty of Medicine, University of Ljubljana, Trzaska 25, Slovenia*

E-mail: nataly@bio21.bas.bg

Abstract. The successful osseointegration of a bone implant is greatly dependent on its ability to support cellular adhesion and functions. Deposition of thin composite coatings onto the implant surface is a promising approach to improve interactions with cells without compromising implant bulk properties. In this work, we have developed composite coatings, based on hexamethyldisiloxane (HMDS) and detonation nanodiamond (DND) particles and have studied adhesion, growth and function of osteoblast-like MG-63 cells. PPHMDS/DND composites are of interest for orthopedics because they combine superior mechanical properties and good biocompatibility of DND with high adherence of HMDS to different substrata including glass, metals and plastics. We have used two approaches of the implementation of DND particles into a polymer matrix: pre-mixture of both components followed by plasma polymerization and layer-by-layer deposition of HMDS and DND particles and found that the deposition approach affects significantly the surface properties of the resulting layers and cell behaviour. The composite, prepared by subsequent deposition of monomer and DND particles was hydrophilic, with a rougher surface and MG-63 cells demonstrated better spreading, growth and function compared to the other composite which was hydrophobic with a smooth surface similarly to unmodified polymer. Thus, by varying the deposition approach, different PPHMDS/DND composite coatings, enhancing or inhibiting osteoblast adhesion and functions, can be obtained. In addition, the effect of fibronectin pre-adsorption was studied and was found to increase greatly MG-63 cell spreading.

1. Introduction

Biomaterials for bone regeneration should support cellular adhesion and growth, maintain cellular differentiation and functions and also should have mechanical properties similar to the native bone [1-4]. Since materials designed as bone implants differ in their capability to attach cells many surface modification techniques, aiming to enhance and accelerate anchorage of bone cells to the implant surface are under investigation. An approach to improve surface properties of the implants without compromising the overall structure of the material is deposition of thin coatings onto the implant surface. Plasma deposition is a cost effective and economic method for the preparation of thin and regularly structured layers on bone implants that has been gaining importance for the last several years as a tool to modify a material surface. A major advantage of this technique is that by simple adjusting of the process parameters from one monomer different polymers could be obtained [5]. Hexamethyldisiloxane is a widely used monomer in plasma polymerization due to its volatility, cheap availability and relatively non-toxic nature. By implementation of nanodiamond particles into a HMDS matrix during plasma polymerization it is possible to obtain highly dense, pinhole-free and well-adherent films on a variety of substrates like conventional polymers, glass and metals. Nanodiamonds have received great attention in recent years because they exhibit many properties that are attractive for orthopedic implants, such as extremely high hardness, chemical and thermal resistance, high fracture toughness and low coefficient of friction [6]. Moreover, due to their non-toxicity, high biocompatibility and feasibility to be modified with different macromolecules nanodiamonds can be used to improve interactions with bone tissue cells [7, 8].

Here we have compared the ability of two DND/PPHMDS composite layers, obtained by different procedures to support cell adhesion, growth and function by MG-63 cells. This cell line is a commonly used model for osteoblast cells because they express many of the characteristics of normal osteoblasts. We have correlated biological response with the surface properties of composite materials to provide an insight toward controlling osteoblast behavior by manipulating diamond-containing composite surfaces. Although there are several investigations on adhesion of osteoblast cells and cytotoxicity of ND coatings the biological response of osteoblasts toward diamond coatings is still not clear and has not been sufficiently studied to date.

2. Materials and methods

2.1. Preparation of plasma polymer films

Three groups of plasma polymer films were prepared for the purposes of this study: plasma polymerized hexamethyldisiloxane (PPHMDS) and two polymer composites of PPHMDS and DND, named as DND1 and DND2. As control surfaces cover glasses (CG) were used.

PPHMDS films were prepared by plasma polymerization (PP) as previously described [9].

ND particles used in PPHMDS/DND composites were synthesized by detonation of carbon-containing explosive mixtures as previously described in [9] and were characterized by FTIR-spectroscopy, oxidative titration, and pH-measurement as described in [10]. For preparation of DND1 films 0,03 g DND powder was added to 30 ml HMDS and the suspension was treated ultrasonically for 15 min, then continuously stirred on a magnetic stirrer at 20° C and 275 rpm immediately before deposition aiming to homogenize well DND particles in the monomer. The deposition of PPHMDS/DND films on the CG was performed at the following parameters: current density 0.04 mA/cm^2, monomer flow rate of 5 revolutions in the valve for 15 minutes.

DND2 films were prepared by consequent deposition of PPHMDS as a first layer on CG at a current density 0,08 mA/cm^2 and monomer flow rate 10 l/h for 15 min, followed by plasma deposition of ethanol suspension of DND particles on the PPHMDS film at a current density 0,16 mA/cm^2 and monomer flow rate 10 l/h for 10 min. DND suspension was prepared by dispersion of DND powder in ethanol (C_2H_5OH), followed by ultrasonic treatment for 15 min and intensive stirring on the magnet stir at 20° C and 275 rpm. After the deposition the samples were washed with deionized water and air dried. All films were deposited onto commercially available cover glasses (Menzel Glaesser) in a vacuum chamber with a system for glow discharge previously described [9].

2.2. Surface characterization of plasma polymer films

The morphologies of the films were observed under an atomic force microscope (AFM, Solver PRO, NT-MDT, Russia). To obtain information about the wettability of samples water contact angles were measured through the sessile drop method using the Easy Drop analyzer (FM40, Kruss, Germany).

2.3. Cell culture
A human osteosarcoma cell line, MG-63 (CRL-1427, American Type Culture Collection), passage 23 to 26, was used for cell experiments cultivated under the conditions previously described [11].

2.4. Initial cell adhesion assay
MG-63 cells with a concentration of 2.10^4 cells/ml were seeded on polymeric films, individually placed in 24-well cell culture plates (Nunc, Denmark). Before cell seeding half of the materials were pre-adsorbed with 20 μg/ml fibronectin (Sigma, Germany) for 30 min at RT. Cells were incubated for 2 hours at 37° C in humidified air in serum-free culture medium (DMEM, Sigma). At the end of incubation non-adherent cells were removed by washing with PBS and phase-contrast picture of adherent cells were taken. The pictures were further used for counting the cell number and measuring the cell area by Image J software.

2.5. Cell proliferation assay
Cell proliferation was determined at the 1st and the 7th day of incubation using a modified lactate dehydrogenase assay (LDH, Roche Diagnostics, Germany), following the manufacturer's instructions.

2.6. Alkaline Phosphatase Activity assay
To measure ALP activity cells on polymer films were washed with TBS and lysated with 5 mM $MgCl_2$/0,5%Trton X100. The cell lysates were incubated with p-nitrophenylphosphate (Sigma, Germany) at alkaline conditions (pH=10) for 2 hours and the reaction was stopped with 0,5N NaOH. The amount of p-nitrophenol liberated was measured by microplate reader at 405 nm.

2.7. Statistical Analysis
Cell experiments were conducted in triplicate and the results are presented as mean values ± standard deviation (SD). For statistical evaluation the student's test has been applied. The significance was determined at $p \leq 0.05$.

3. Results and discussion

3.1. Characterization of plasma polymer films
Our results showed that modification of HMDS with DND particles by two different approaches of plasma polymerization resulted in different surface properties of the obtained composites which in turn affected cell behaviour. DND2 films prepared by subsequent deposition of monomer and particles had higher surface roughness and hydrophilicity compared to the other composite layers (DND1) obtained by pre-mixing of monomer with the nanodiamond. AFM scans of the sample surfaces, shown in figure 1, revealed that the surface of DND2 is much rougher compared to PPHMDS and DND2 films probably because DND particles aggregate during their deposition onto PPHMDS layer and thus making the surfaces rough while DND1 composite has similar to PPHMDS smooth surface.

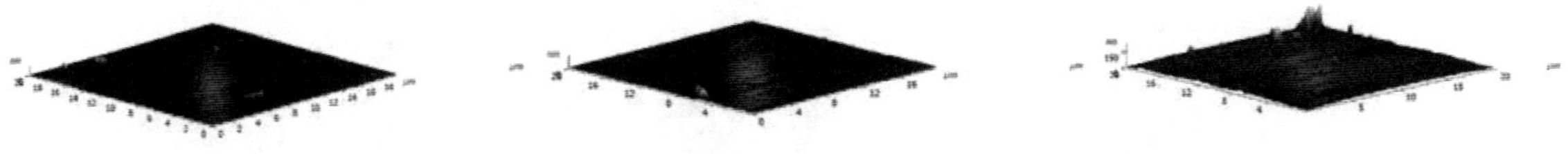

Figure 1. AFM 3-D images of polymeric and composite films deposited on cover glasses.

Water contact angle (WCA) measurements, summarized in table 1, show that both composite layers differed significantly in respect to their hydrophilicity. While DND1 is hydrophobic similarly to unmodified PPHMDS, DND2 is moderately hydrophilic which is most probably due to the fact that the nanodiamonds are in ethanol, rendering them hydrophilic and resulting in a hydrophilic surface.

Sample – name	WCA [°]
PPHMDS	81±2
PPHMDS/DND1	84±2
PPHMDS/DND2	58±2

Table 1. Water contact angle (WCA) of plasma films deposited on cover glass.

Based on the literature, we assumed that DND2 would be preferred by osteoblasts, because rougher surfaces have been shown to promote the attachment, growth and differentiation of bone cells [12, 13]. On the other hand, it is well known that mammalian cells interact better with hydrophilic, rather than hydrophobic surfaces. Therefore higher hydrophilicity of DND2 suggested better interactions with cells. It is also believed that the hydrophobicity does not affect directly cells, but by changing the biological activity of the adsorbed adhesive proteins [14]. Since FN has been isolated and identified as a main protein responsible for the adhesion of cells *in vitro*, such variations in the biological activity of a material can be explained by the different conformation of adsorbed FN. On hydrophobic surfaces cells cannot reorganize adsorbed FN, because FN is bound very strongly to the material while on hydrophilic materials FN is loosely bound, its cell binding domains are expressed resulting in more cell contacts.

3.2. Cell adhesion

We have focused mainly on the initial adhesion of cells, estimating cell morphology, the number of attached cells and cell spreading area. In addition, we have studied cell growth and function which are other indicators of biocompatibility of the material. Phase-contrast micrographs showed that at the 2nd hour of incubation the majority of MG-63 cells on plain films remained rounded and small in size which is indication of poor cell adhesion (figure 2). On DND2 osteoblasts seem to be more spread out compared to the other materials. When materials were pre-coated with FN cell adhesion improves signifficantly as can be concluded from the polygonal morphology of cells. The largest cells were observed on PPHMDS.

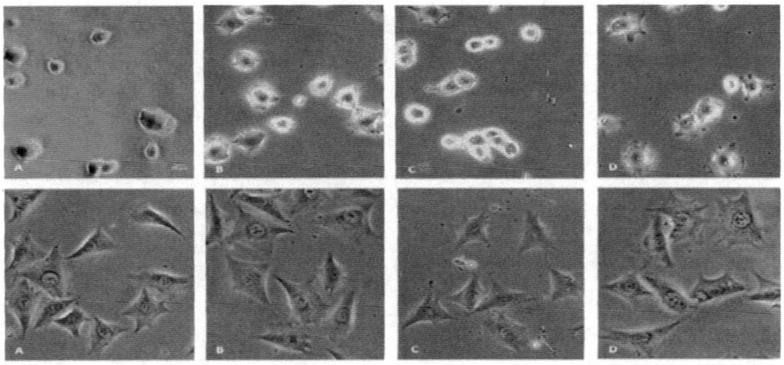

Figure 2. Overall morphology of MG63 cells cultured for 2 hours on plain (upper panel) and FN pre-coated (low panel) cover glass (A), PPHMDS (B), PPHMDS/DND1 (C) and PPHMDS/DND2 (D), bar 50 μm.

Results from counting the number of attached cells demonstrated that plasma polymerized films varied in their ability to support cell attachment. As is shown in figure 3 the greatest number of attached MG-63 cells were found on DND2, in both groups, plain and FN-coated. Surprisingly, pre-adsorption with FN did not improve cell adhesion on materials. Only on controls (CG) and DND1 FN pre-adsorption increased the number of attached cells while on PPHMDS and DND2 even reducing in cell attachment was observed. Cell attachment characterized only partially cell-material interactions. More information can be received from cell spreading, which is expression of an active physiological process. The extent of cell spreading was quantified by measuring the average cell area following digital image analysis of phase-contrast pictures. MG-63 cells demonstrated reduced spreading on DND1 compared to PPHMDS and DND2 ($p < 0.05$) in both, plain and FN-coated materials (figure 4).

On plain surfaces the highest cell spreading area was estimated on DND2 suggesting that the second method of modification of the polymer with DND stimulated better spontaneous spreading of MG-63 cells.

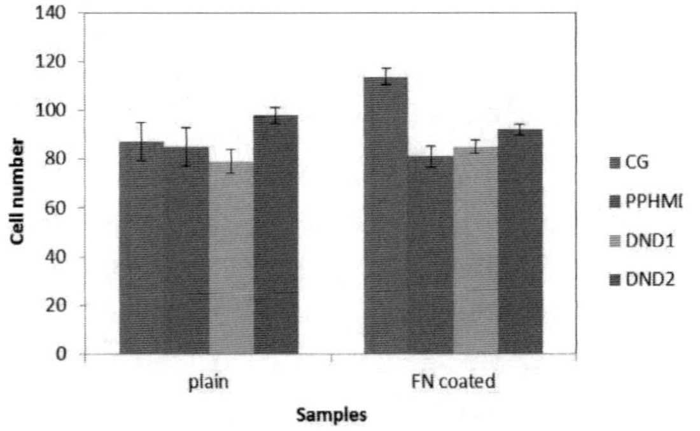

Figure 3. Adhesion of MG-63 cells to different substrata after 2 hours of incubation.

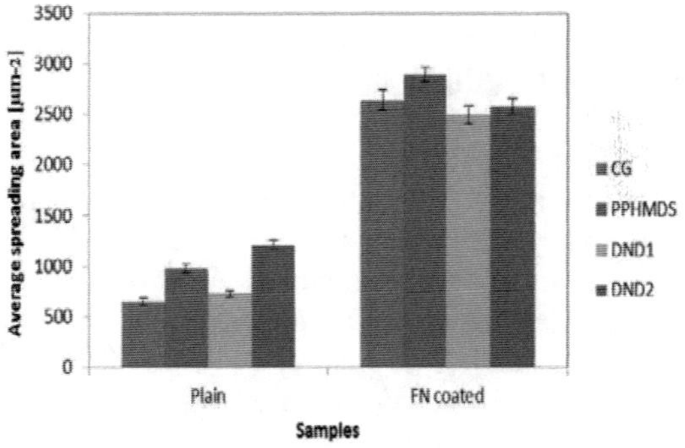

Figure 4. Average spreading area of MG-63 cells cultured for 2 hours on different substrata.

After coating with FN the spreading area on all materials significantly increased, assuming that the materials adsorbed and organized well FN, which in turn promotes cell attachment and spreading. However, the highest average spreading area of MG-63 cells was found to be on PPHMDS films despite their hydrophobicity.

3.3. Cell proliferation

Another parameter characterizing cell behaviour in contact with biomaterials is cell growth. In general, MG-63 cells grew well on all tested materials, because an increase in their LDH activity from the 1st to the 7th day was observed (figure 5). The highest cell proliferation rates were found on both DND composites compared to unmodified PPHMDS and control CG. There was no apparent difference in cell growth on the two composite layers although the cell adhesion and spreading differed significantly.

3.4. Alkaline phosphatase activity

Cell function was investigated by measuring the alkaline phosphatase activity of MG-63 cells cultured for 7 days on different plasma polymer films. Alkaline phosphatase (ALP) is a membrane–anchored enzyme, widely used as a marker for early osteogenic differentiation. It enhances the local concentration of inorganic phosphate and thus participates in the initiation of the mineralization of the bones. As can be seen from figure 6 MG-63 cells expressed low levels of alkaline phosphatase activity on all studied materials which did not increase significantly after one week incubation. On the first day, the enzyme activity was highest in the control cells (CG) and the lowest in DND2 however the differences were not statistically significant. On day 7, the cells plated on DND-modified materials showed higher ALP-activity compared to the unmodified polymer and control. The highest enzyme activity was measured in MG-63 cell incubated on DND2 materials suggesting that the second method of modification induced spontaneous osteogenic differentiation of cells.

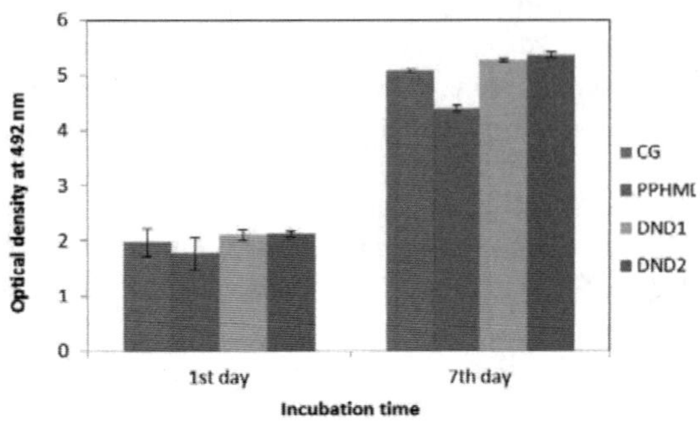

Figure 5. Cell proliferation of MG-63 cells cultured for 7 days on different surfaces.

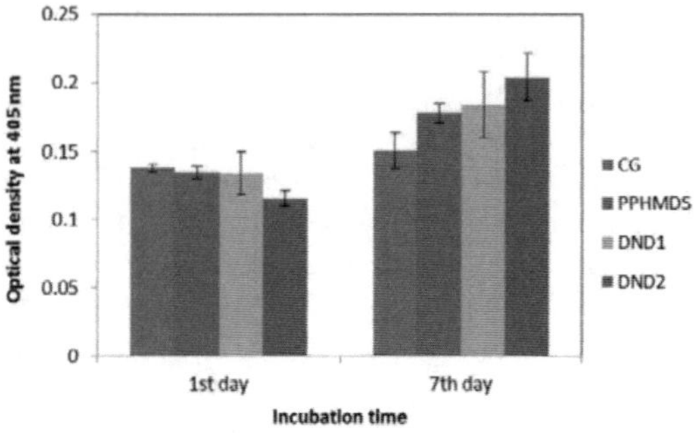

Figure 6. ALP activity of MG-63 cells cultured for 7 days on different surfaces.

4. Conclusion

Generalizing our results we can conclude that the modification of biomaterials with nanodiamond particles can improve the adhesion, growth and function of osteoblast-like cells and therefore these materials can be used to induce osteogenic differentiation even without the addition of soluble osteogenic factors. The method of the modification however results in different surface properties that alter the biological response of cells.

Acknowledgements

This work was supported by Bulgarian National Scientific Research Fund project №: DFNI_B01_18/2012.

References

[1] Yang P, Huang X, Wang C, Dang X and Wang K. 2013 *Exp. Biol. Med.* **238** 1331
[2] Stevens MM. 2008 *Mater. Today*, **11** 18
[3] Catledge S A, Fries MD, Vohra YK, Lacefield WR, Lemons JE, Woodard S and Venugopalan R. 2002 *J. Nanosci. Nanotechnol.* **2** 293
[4] Hu Y C and Zhong J P. 2009 *Chin. Med. J. (Engl).* **122** 2386
[5] Ibnabddjalil M, Loh IH, Chu CC, Blumenthal N, Alexander H and Turner D. 1994 *J. Biomed. Mater. Res.* **28** 289
[6] Mochalin VN, Shenderova O, Ho D and Gogotsi Y. 2012 Nat. Nanotech. **7** 11
[7] Roy RK and Lee KR. 2007 *J. Biomed. Mater. Res* .B. **83** 72
[8] Amaral M, Gomes PS, Lopes MA, Santos JD, Silva RF and Fernandes MH. 2008 *J. Nanomater.*, **2008** 1
[9] Pramatarova L *et al* 2011 In: On Biomimetics, ed L Pramatarova (InTech), chapter 14 pp. 297
[10] Mitev D, Dimitrova R, Spassova m, Minchev C and Stavrev S. 2007 *Diam. Relat. Mater.* **16** 776
[11] Krasteva NA, Toromanov G, Hristova KT, Radeva EI, Pecheva EV, Dimitrova RP, Altankov GP and Pramatarova LD. 2010 *AIP Conference Proc* **1203** 688
[12] Anselme K, Bigerelle M, Noel B, Dufresne E, Judas D, Iost A and Hardouin P. 2000 *J. Biomed. Mater. Res.* **49** 155
[13] Zinger O, Zhao G, Schwartz Z, Simpson J, Wieland M, Landolt D and Boyan B. 2005 *Biomaterials* **26** 1837

[14] Altankov G, Grinell F and Groth T. 1996 *J. Biomed. Mater. Res.* **30** 385

Journal of Physics: Conference Series **558** (2014) 012058 doi:10.1088/1742-6596/558/1/012058

Si nanoparticles embedded in pseudo binaries alloys with Al_2O_3

P Vitanov[1], A Harizanova[1], T Ivanova[1] and A Ulyashin[2]

[1]Central Laboratory of Solar Energy and New Energy Sources, Bulgarian Academy of Sciences, Blvd. Tzarigradsko chaussee 72, Sofia, Bulgaria
[2]SINTEF Materials and Chemistry, Metallurgy Department, Forskningsveien 1, Blindern, NO-0314 Oslo, Norway

E-mail: vitanov@phys.bas.bg

Abstract. The present paper reports the incorporation of Si nanoparticles (NPs) in $(Al_2O_3)_x$ $(B_2O_3)_{1-x}$ films. A new approach for involving Si NPs (20 nm) p^+ doped into sol solution was used. After spin coating deposition and additional annealing at 850°C, $(Al_2O_3)_x$ $(B_2O_3)_{1-x}$ films with Si NPs have been obtained with thickness of 75 and 120 nm. The $(Al_2O_3)_x$ $(B_2O_3)_{1-x}$ films were deposited on p-type and n-type Si wafers. The optical study reveals that Si NPs embedded in $(Al_2O_3)_x(B_2O_3)_{1-x}$ leads to higher absorption in the visible spectral range compared to dielectric layer without nanoparticles. AFM studies show that the incorporated Si nanoparticles into the dielectric layer have probably a uniform distribution on film surface. This conclusion is confirmed by SEM observation. The current - voltage (I-V) characteristics of a structure $(Al_2O_3)_x(B_2O_3)_{1-x}$ layer with Si NPs on p- type Si shows an ohmic type of conductivity. The sheet resistance is 480 Ohm/sq respectively, when the same layer is deposited on quartz substrate. This work presents a new technological approach for integration of Si NPs in dielectric matrix of $(Al_2O_3)_x$ $(B_2O_3)_{1-x}$. The results can find applications for formation of ultra thin emitter for photodiodes and Si solar cells.

1. Introduction

Materials with nano particles (NPs) form a new class of artificial materials with tuneable electronic properties controlled at the nanoscale and composed of close-packed granules varying in size from a few to hundreds of nanometers -often referred to as *nanocrystals*. Their properties discover a new route for potential novel electronic, optical, and optoelectronic applications.

Applications range from light-emitting devices to photovoltaic cells and biosensors. Among traditional methods of preparation of such materials, the most common are thermal evaporation and sputtering techniques. During those processes metallic and insulating components are simultaneously evaporated or sputtered onto a substrate. Diffusion of the metallic component leads to the formation of small metallic grains, usually 3–50 nm in diameter.

Another approach is Si nanocrystallites embedded in dielectric matrices. The Si NPs in oxide matrix exhibit unique electrical, optical, and optoelectronical properties [1-4]. Among different dielectric materials, silicon oxide is the most addressed as a host for Si NPs [4, 5]. During the last

[1] To whom any correspondence should be addressed.

decades, the properties of Si NPs-SiO_2 systems have been widely investigated. Several approaches have been used to form Si-NPs in amorphous and/or crystalline Al_2O_3. Most known methods are Si ion implantation [7, 8] and electron beam evaporation followed by subsequent high-temperature annealing as well as laser ablation [9].

The present paper reports the fabrication of Si NPs in $(Al_2O_3)_x\ (B_2O_3)_{1-x}$ films. A new approach for involving Si nanoparticles (20 nm) p^+ doped into sol solution was used. After spin coating deposition and additional annealing at 850°C, thin $(Al_2O_3)_x\ (B_2O_3)_{1-x}$ films with Si NPs have been obtained.

2. Experimental

The sol solution preparation for depositing $(Al_2O_3)_x(B_2O_3)_{1-x}$ films has been previously reported [10]. The used Si nanoparticles are p^+ doped and 20 nm sized, fabricated by SINTEF.

Suitable receipt for homogenization of Si NPs has been developed. Homogeneous, stable suspensions in sol solution were obtained by mixing the respective amount of nanoparticles with sol solution followed by ultrasonic treatment of the dispersion. Thin films with thickness of 75 and 120 nm were obtained by spin coating method on Si wafers or quartz substrates and substantial additional annealing at 850°C.

For comparison, $(Al_2O_3)_x(B_2O_3)_{1-x}$ films with and without Si NPs have been deposited.

Scanning Electron Microscopy (SEM), High Resolution Transmission Electron microscopy (HRTEM) and Atomic Force microscopy (AFM) were used to determine the morphology of layers. Optical properties were studied by using UV-VIS-NIR Shimadzu 3600 spectrophotometer as for the sol-gel films obtained on quartz substrates.

The electrical properties of the films have been studied by forming MIS structures. The aluminium oxide thin films were deposited on p-type Si wafers, FZ, with 3.6-5.4 Ω.cm resistivity and an orientation <100>. The mercury probe was used as metal contact.

3. Results and discussions

3.1. Optical characterization

Optical measurements of $(Al_2O_3)_x\ (B_2O_3)_{1-x}$ films with and without Si NPs were performed and the results are compared in figure 1.

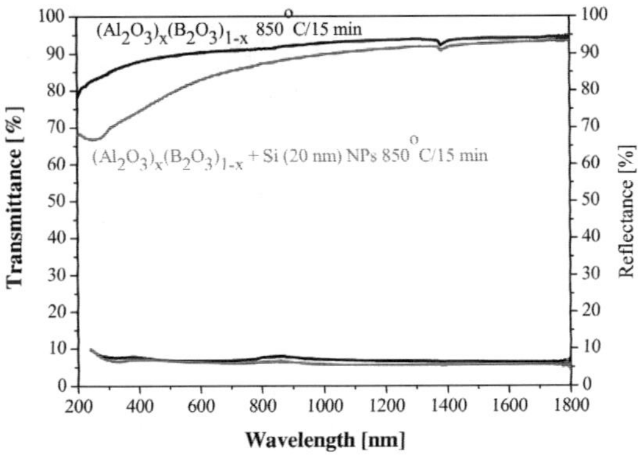

Figure 1. Spectral dependences of transmittance and the reflectance of layers, deposited on quartz substrates.

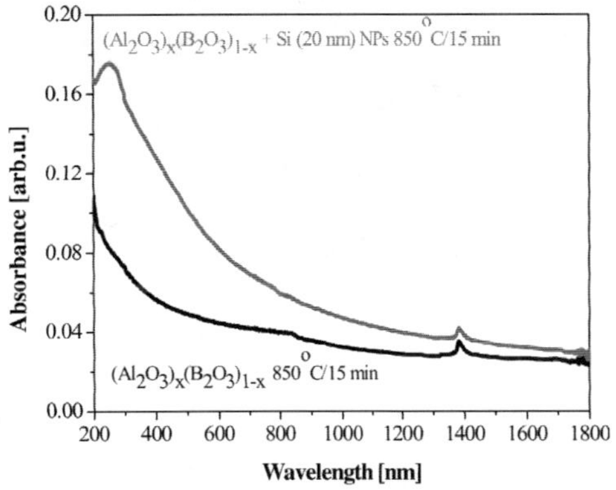

Figure 2. Optical absorbance of the thin layers with and without Si NP on quartz substrates.

The optical study reveals that the incorporation of Si NPs in $(Al_2O_3)_x(B_2O_3)_{1-x}$ leads to higher absorption in the visible spectral range up to 800 nm compared to dielectric layer without nanoparticles (figure 2) In the same time, no significant difference is observed in the mirror reflectance recorded for $(Al_2O_3)_x(B_2O_3)_{1-x}$ layer without and with Si nanoparticles. The reflectance is below 10% in the whole spectral range.

3.2. Morphological studies

The surface morphology and microstructure of the layers were examined by AFM, TEM, SEM techniques.

The surface roughness of $(Al_2O_3)_x(B_2O_3)_{1-x}$ layer with Si nanoparticles (sample 3_3) and $(Al_2O_3)_x(B_2O_3)_{1-x}$ layer without Si nanoparticles (sample 6_3) is determined by AFM and the micrographs are shown in figures 3 and 4.

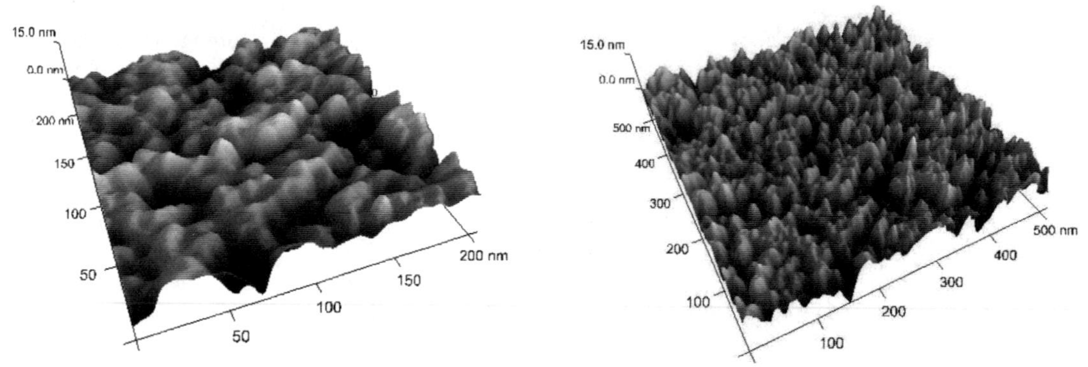

Figure 3. 3-D image of sample 3_3 $(Al_2O_3)_x(B_2O_3)_{1-x}$ layer (120 nm thick) with Si nanoparticles (20 nm).

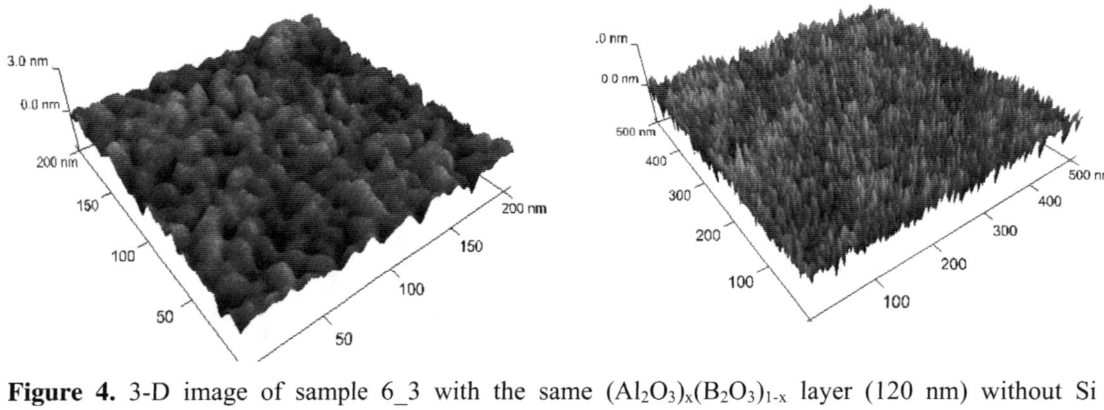

Figure 4. 3-D image of sample 6_3 with the same $(Al_2O_3)_x(B_2O_3)_{1-x}$ layer (120 nm) without Si nanoparticles.

On the surface of sample 3_3 can be clearly observed uniformly distributed granules in contrast to the surface of sample 6_3. RMS roughness for an area of 1 μm is R_q=0.345 nm for $(Al_2O_3)_x(B_2O_3)_{1-x}$ layer (120 nm) without Si NPs. More information for the sizes of the granules can be obtained from figure 5 and figure 6.

The surface of sample 6_3 has the characteristic morphology of $(Al_2O_3)_x(B_2O_3)_{1-x}$ with deviation of 1.5 nm from the middle plane.

The conclusions that can be derived from the AFM studies are that the embedded Si nanoparticles into the dielectric layer have probably a uniform distribution on the film surface.

This conclusion is confirmed by the surface morphology analysis by the means of SEM revealing uniformly distributed grains with sizes of 20 nm as it can be seen from figure 7.

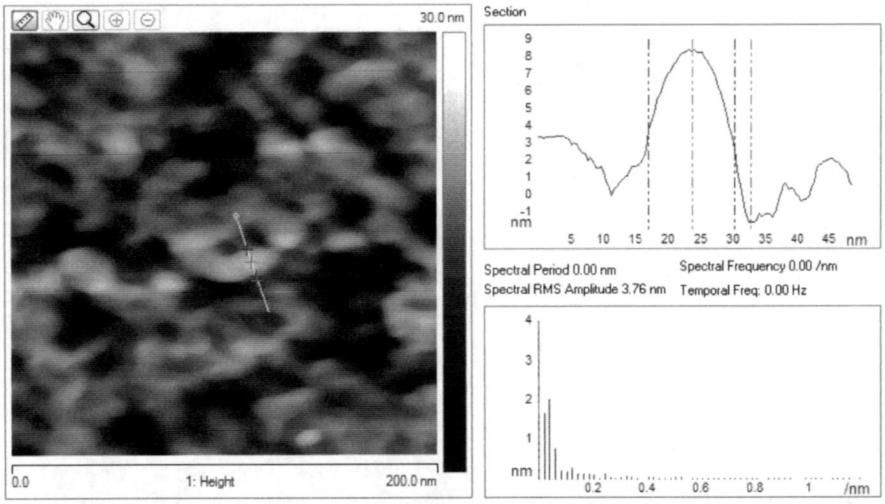

Pair	Horizontal Distance	Vertical Distance	Surface Distance	Angle	Rmax	Rz	Rz Count	Rms	Ra (Freguency C
0	13.236 (nm)	-1.089 (nm)	17.752 (nm)	-4.7...	5.351 ...	0.000 ...	0.000	1.154 ...	3.890 (nm)
1	9.100 (nm)	-9.932 (nm)	14.435 (nm)	-47....	9.489 ...	0.000 ...	0.000	0.650 ...	3.628 (nm)
2	0.000 (nm)	0.000 (nm)	0.000 (nm)	0.00...	0.000 ...	0.000 ...	0.000	0.000 ...	0.000 (nm)

Figure 5. AFM image of sample 3_3 $(Al_2O_3)_x(B_2O_3)_{1-x}$ layer (120 nm thick) with Si nanoparticles (20 nm). The size of the particle is: width 13 nm and height 9 nm.

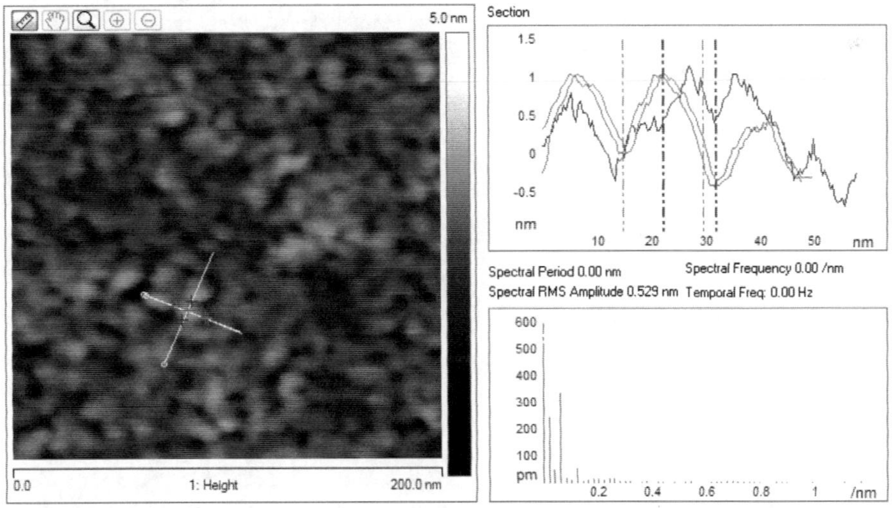

Pair	Horizontal Distance	Vertical Distance	Surface Distance	Angle	Rmax	Rz	Rz Count	Rms	Ra (Freguency C
0	9.786 (nm)	0.029 (nm)	10.097 (nm)	0.16...	0.735 ...	0.000 ...	0.000	0.191 ...	0.417 (nm)
1	14.745 (nm)	-0.109 (nm)	14.986 (nm)	-0.4...	1.134 ...	1.134 ...	2.000	0.319 ...	0.327 (nm)
2	9.690 (nm)	-1.465 (nm)	9.844 (nm)	-8.5...	1.304 ...	0.000 ...	0.000	0.392 ...	0.387 (nm)

Figure 6. AFM image of sample 6_3 $(Al_2O_3)_x(B_2O_3)_{1-x}$ layer without Si NPs.

The results from TEM observation on $(Al_2O_3)_x(B_2O_3)_{1-x}$ layer embedded with Si NPs are presented on figure 8. On the left side, the Si substrate is visible as highly periodic arranged atoms. Inside the dielectric, Si NPs with random crystal orientation are visible. Their size varies between 5 and 8 nm.

Figure 7. SEM micrograph of the dielectric surface embedded with Si NPs.

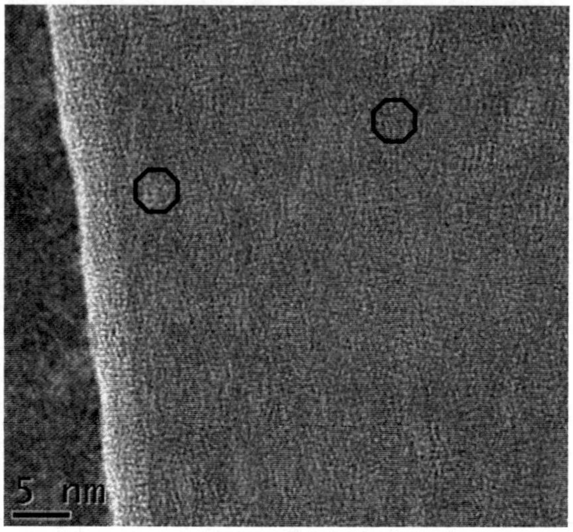

Figure 8. TEM micrograph of the dielectric surface embedded with Si NPs.

3.3. Electrical properties

The sheet resistance of the layer with Si NPs is 480 Ω/□, when the layer is deposited on quartz substrate at the same conditions. The electrical properties of the films have been studied by forming MIS structures. The aluminium oxide thin film with Si NPs was deposited on p-type Si wafers, FZ, with 3.6-5.4 Ω.cm resistivity and an orientation <100>. The mercury probe was used as metal contact.

The electrical measurements of this structure reveal that it has electrical conductivity towards Si substrate (figure 9). The I-V characteristic is similar to unipolar device and shows a symmetric behavior with respect to 0 V.

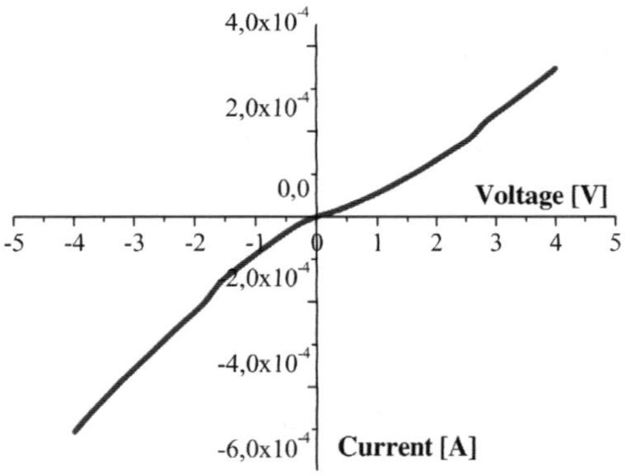

Figure 9. I-V characteristics of $(Al_2O_3)_x(B_2O_3)_{1-x}$ layer (120 nm thick) with Si NPs on p-type Si wafer.

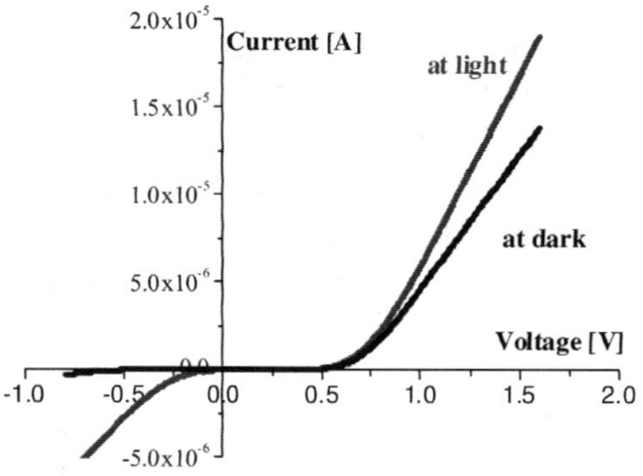

Figure 10. I-V curve of the $(Al_2O_3)_x(B_2O_3)_{1-x}$ layer with Si NPs on n- type Si wafer.

Journal of Physics: Conference Series **558** (2014) 012058 doi:10.1088/1742-6596/558/1/012058

The electric behavior of $(Al_2O_3)_x(B_2O_3)_{1-x}$ layer without Si NPs (sample 6_3) is typical for dielectric. The breakdown voltage of the MIS structure is 19-20 V and the C-V curve shows negative fixed charge.

Figure 10 presents the I-V curve of a structure $(Al_2O_3)_x(B_2O_3)_{1-x}$ layer with Si NPs on n- type Si wafer. The current – voltage characteristic has a typical diode behavior of the p+/n junction. In contrast to the unipolar device (figure 9), this structure shows a clear rectifying diode-like behavior. Under reverse bias the current depends from light illumination. The difference in electrical behaviour between the unipolar device structure and the p/n junction structure proves the observed rectifying characteristic to be the result of the realized p-n junction and not the result of a Schottky contact [11].

4. Conclusions

An approach for integration of p^+ doped Si nanoparticles (20 nm) in $(Al_2O_3)_x(B_2O_3)_{1-x}$ matrix has been developed, using spin coating deposition of sol solution on Si substrates. Uniform distribution of Si nanoparticles has been observed on the film surface by AFM.

The films with embedded Si NPs reveal an electric conductivity towards Si substrate and the sheet resistance is 480 Ω/□, when it is deposited on quartz substrate. A p-n diode structure can been produced on n-type Si wafers using highly doped Si-NPs in combination with a high temperature sintering process.

References

[1] Dutta A, Kimura M, Honda Y, Otobe M, Itoh A and Oda S 1997 *Jpn. J. Appl. Phys.* 36 4038

[2] Choi H, Hwong S W, Kirn G, Shin H C, Kim Y and Kirn E K 1998 *Appl. Phys. Lett.* 73 3129

[3] Inokuma T, Wakayama Y, Muramoto T, Aoki R, Kurata Y and Hasegawa S 1998 *J. Appl. Phys.* 83 2228

[4] Seifarth D H, Gritzschel R, Markwitz A, Matz W, Nitzsche P and Rebohle L 1998 *Thin Solid Films* 330 202.

[5] Khomenkova L, Korsunska N, Yukhimchuk V, Jumaev B, Torchinska T, Vivas Hernandez, A Many A, Goldstein Y, Savir E and Jedrzejewski J 2003 *J. Lumin.* 102/103 705

[6] Chen X Y, Lu Y F, Tang L J, Wu Y H, Cho B J, Xu X J, Dong J R and Song W D 2005 *J. Appl. Phys.* 97 014913

[7] Mikhaylov A, Belov A, Kostyuk A, Zhavoronkov I, Korolev D, Nezhdanov A, Ershov A, Guseinov D, Gracheva T, Malygin N, Demidov E and Tetelbaum D 2012 *Phys Solid State* 54, 368

[8] Yerci S, Serincan U, Dogan I, Tokay S, Genisel M, Aydinli A and Turan R 2006 *J. Appl. Phys.* 100 074301

[9] Núñez-Sánchez S, Serna R, García López J, Petford-Long A, Tanase M, Kabius B 2009 *J. Appl. Phys.* 105 013118

[10] Vitanov P, Harizanova A, Ivanova T and Dimitrova T 2009 *Thin Solid Films* 517 6327

[11] Meseth M, Ziolkowski P, Schierning G, Theissmann R, Petermann N, Wiggers H, Bensona N, and Schmechel R 2012 *Scripta Materialia* 67 265

Chemical vapour deposition growth of graphene layers on metal substrates

Y-C Lai[1], S-C Yu[1], P M Rafailov[2]*, E Vlaikova[2], S Valkov[2], S Petrov[2], J Koprinarova[2], P Terziyska[2], V Marinova[1,3], S H Lin[4], P Yu[1], G C Chi[1], D Dimitrov[2,3], and M M Gospodinov[2]

[1]Department of Photonics, National Chiao Tung University, Hsinchu 30010, Taiwan
[2]Georgi Nadjakov Institute of Solid State Physics, BAS, 72 Tzarigradsko Chaussee blvd., 1784 Sofia, Bulgaria
[3]Institute of Optical Materials and Technologies, BAS, Sofia 1113, Bulgaria
[4]Department of Electrophysics, National Chiao Tung University, Hsinchu 30010, Taiwan

* E-mail: rafailov@issp.bas.bg

Abstract. Graphene layers were grown by chemical vapour deposition (CVD) on Si wafers covered by a SiO_2 substrate layer and a Ni interlayer, and on copper and nickel foil. The obtained graphene layers were characterized by Raman spectroscopy. The films grown on SiO_2/Ni substrate and Ni foil comprise mainly multilayer defect-rich graphene, while those on Cu foil exhibit the spectroscopic fingerprint of relatively defect-free single-layer graphene due to the low carbon solubility in copper and the suitably chosen substrate position in a quasi-closed volume. Optimal growth conditions and the nature of defects in the layers are discussed.

1. Introduction

Graphene is two-dimensional zero band-gap material with charge carriers behaving as Dirac fermions (zero effective mass) [1] with extraordinary mobilities and ballistic transport distances. These properties make it potentially useful for high-frequency electronic devices [2] while its exceptionally low absorbance bears potential for an ideal transparent conductor [3].

Application in integrated devices requires technology for wafer scale deposition with controllable number of graphene layers. On the other hand, implementation as transparent conductor requires reproducible deposition of single- or few-layer graphene. Especially, synthesis of bilayer graphene becomes increasingly desirable as it was reported to develop a tunable band gap upon mechanical stress and electrical gating [4].

Although pristine graphene available through mechanical exfoliation has very low concentration of structural defects, this method is not suitable for industrial application as the thickness, size and location of the obtained graphene are not reproducible. The most perspective, inexpensive and readily accessible approach for deposition of reasonably high-quality graphene turns out to be chemical vapor deposition (CVD) onto transition metal substrates such as Ni or Cu [5].

One fundamental limitation of utilizing Ni as deposition substrate is that single- and few-layered graphene is obtained only over few-micron regions and the film is not homogeneous over the entire

* To whom any correspondence should be addressed.

substrate. The impossibility to control the number of layers is partially due to the fact that the segregation of carbon from the metal carbide upon cooling occurs rapidly within the Ni grains and heterogeneously at the grain boundaries [6]. In addition, the degree of crystallinity and the surface quality of the metal substrate should also play a role. Initially, due to the substantial solubility of carbon in nickel at high temperature a solid solution forms. Subsequent lowering the temperature decreases the solubility, driving the carbon atoms to precipitate at the Ni surface and to arrange in hexagonal lattice for which the (111) orientation of the Ni layer serves as a template. In contrast, the peculiar combination of very low affinity between carbon and copper combined with the ability to form transient soft bonds facilitates formation of nucleation sites on the copper foil surface that form preferentially single-layer graphene through lateral growth. However, for a preferential growth of bilayer graphene, one may have to consider taking advantage of the precipitation process and Ni remains thus interesting for the graphene growth research.

In this paper we examine the CVD growth of graphene on different nickel-containing substrates with the purpose to establish favorable conditions for reproducible growth of single-layer and few-layer graphene. The obtained results are then compared to CVD graphene on copper foil.

2. Experimental

The CVD process was carried out in quartz tubes via thermal decomposition of high-purity methane in argon and hydrogen flow[4]. Initially, the Si wafers were cleaned in diluted HF (1:10) and subsequently their surface was oxygenated to obtain Si/SiO_2 structure. To improve the adhesion of the carbon atoms, the SiO_2 layer was coated with an interlayer of Ni. Besides the Si substrates, small pieces (ca. 1x1 $inch^2$) of Ni and Cu foil 25 um thickness were also used. X-ray diffraction measurements established that while the Ni interlayer sputtered on SiO_2 was mainly amorphous, the foils were polycrystalline with significant amount of (200)- and (111)-oriented grains. For the graphene growth on Si/SiO_2/Ni and Ni foil substrates atmospheric pressure CVD was applied. The reaction times were between 3 and 10 min.

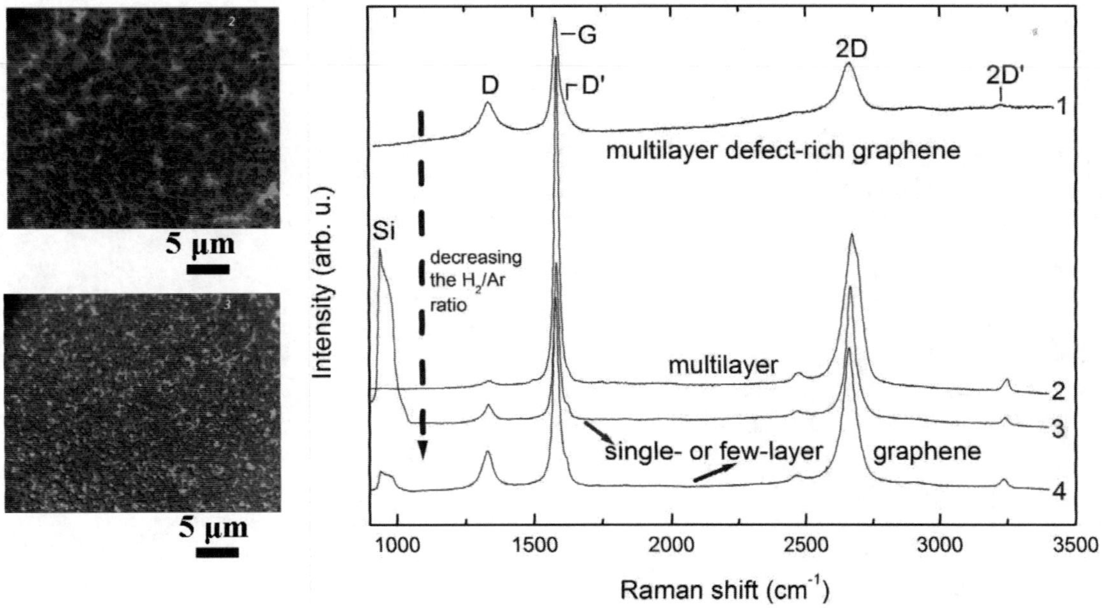

Figure 1. Raman spectra of graphene films on Si/SiO_2/Ni with microscopic images of the corresponding sample regions.

Graphene growth on Cu foil samples was performed by low-pressure CVD. The samples were positioned in a one-zone 3-inch furnace equipped with a 2-inch quartz tube. The CVD was performed in 50 sccm high-purity hydrogen and 5 sccm methane. Prior to this process, the Cu foil was annealed at 1050 °C for 30 min. in hydrogen atmosphere. At a temperature of 1000 °C a methane flow was launched for 300 min. Afterwards, the sample was quenched at 60 °C /min. in hydrogen atmosphere to 300 °C. Cooling to room temperature was done in high-purity argon atmosphere.

The Raman spectra were measured on a Jobin Yvon Labram spectrometer with a CCD detector and entrance slits set to 1.5 cm^{-1} spectral width. The 632.8 nm line of a He/Ne-laser was used for excitation, the absolute accuracy being about 1 cm^{-1}. The laser beam with power of ≈ 1 mW was focused on a spot of about 1 - 2 μm in diameter on the sample surface using microscope optics.

3. Results and Discussion

Figure 1 shows Raman spectra of graphene films grown on various $Si/SiO_2/Ni$ substrates. The first sample (corresponding to spectrum 1 in figure 1) was obtained at a flow ratio H_2:Ar = 5:1. The large linewidth of the 2D band (ca. 65 cm^{-1}) implies a higher number of graphene layers (> 5) [7] and the D band intensity reveals considerable defect density. The graphene quality is greatly improved and the layer number is presumably reduced to two upon decreasing the H_2:Ar flow ratio and for reaction times of 5 – 6 min. judging from spectra 3 and 4 with 2D linewidth of ca. 35 cm^{-1} in figure 1). The unstable 2D position (varying within 5 cm^{-1}) shows non-intentional self-doping [8,9]. The grainy appearance of the films imply only small graphene areas of few μm. This is confirmed by the varying D band intensity and the broadened and upshifted 2D band [10] with frequency surpassing 2655 cm^{-1}. Better quality graphene, though still confined on distinct Ni grains (see the line scan in Fig. 2) is achieved through CVD on Ni foil. This graphene film already contains single-layered grains of 1-2 μm which exhibit intensity ratio I(2D)/I(G) approaching 3 and no D band at all. This is presumably due to the better surface quality of the foil with larger single-crystalline grains, as can be appreciated from Fig. 2. On the other hand, larger areas of about 5 μm with high-quality single-layer graphene are obtained on copper foil positioned in a quasi-closed volume as inferred from the uniform Raman response over a wrinkle-limited grain (see inset) represented by the spectrum in Fig. 3. This spectrum displays an intensity ratio I(2D)/I(G) above 4, the 2D linewidth is about 30 cm^{-1} and there is again no D band which emphasizes the low defect concentration.

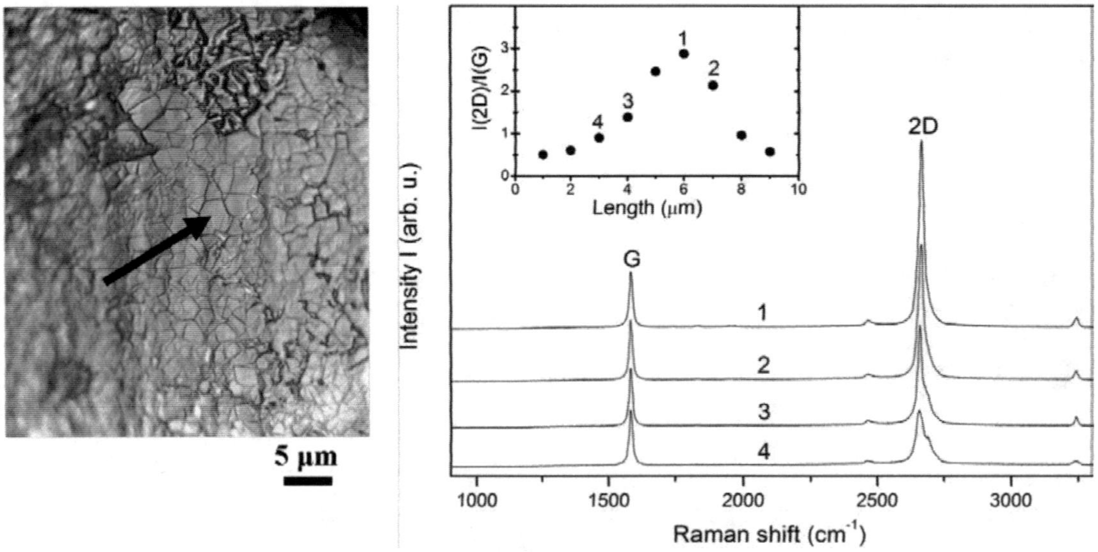

Figure 2. Selected spectra (*right*) from a Raman line-scan in the arrow-marked region of graphene film grown on a Ni foil (*left*). Inset: intensity ratio of 2D and G mode I(2D)/I(G) along the scan line.

We find that depending on the growth conditions (low H_2:Ar flow ratio and reaction times of 5 – 6 min.), films of preferentially double- or few-layer graphene can be grown by CVD on Si/SiO_2/Ni and on Ni foil with occasional occurrence of even single-layered grains on the Ni foil. Raman characterization of the obtained graphene films reveal that their quality depends on the type and the degree of crystallinity of the metal substrate layer. Cu foil facilitates growth of high-quality single-layer graphene by coalescence of nucleation sites without precipitation.

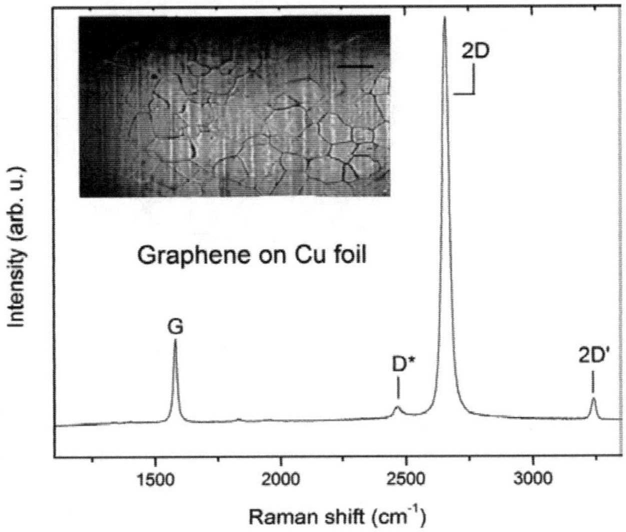

Figure 3. Raman spectrum of graphene on Cu foil with optical microscope image of the corresponding sample spot. The scale bar represents 5 μm.

4. Conclusions

Our results demonstrate that CVD layers of graphene on Ni substrates are strongly non-uniform with single-layered graphene grains of up to 1-2 microns being obtained only occasionally, presumably due to the heterogeneous segregation of carbon atoms at the Ni grain boundaries. Depending on the growth conditions, films of preferentially double- or few-layer graphene can be grown by CVD on Si/SiO_2/Ni and on Ni foil, while Cu foil facilitates growth of single-layer graphene by coalescence of nucleation sites without precipitation. Raman characterization of the graphene films on different Ni substrates reveal that their quality depends on the surface quality and the degree of crystallinity of the substrate layer.

Acknowledgments

The authors acknowledge support from the FP7 Project INERA 316309 with the EU and the Ministry of Science and Technology, Taiwan under the contracts 103-2911-I-009-516, 101-2221-E-009-111-MY3, 103-2221-E-009-079 and AUT Ministry of Education, Taiwan. S. P. acknowledged the Elite intership program NCTU, Taiwan.

References

[1] Neto A H C, Guinea F, Peres N M R, Novoselov K S and Geim A K 2009 *Rev. Mod. Phys.* **81** 109
[2] Lin Y-M, Dimitrakopoulos C, Jenkins K A, Farmer D B, Chiu H-Y and Grill A 2010 *Science* **327** 662
[3] Nair R R, Blake P, Grigorenko A N, Novoselov K S, Booth T J, Stauber T, Peres N M R and Geim A K 2008 *Science* **320** 1308

Journal of Physics: Conference Series **558** (2014) 012059 doi:10.1088/1742-6596/558/1/012059

[4] Zhang Y, Tang Ts-T, Girit C, Hao Zh, Martin M C, Zettl A, Crommie M F, Shen Y R and Wang, F 2009 *Nature* **459** 820
[5] Mattevi C, Kima H and Chhowalla M 2011 *J. Mater. Chem.* **21** 3324
[6] Obraztsov A N, Obraztsova E A, Tyurnina A V and Zolotukhin A A 2007 *Carbon* **45** 2017
[7] Ferrari A C, Meyer J C, Scardaci V, Casiraghi C, Lazzeri M, Mauri F, Piscanec S, Jiang D, Novoselov K S, Roth S and Geim A K 2007 *Phys. Rev. Lett.* **97** 187401
[8] Ferrari A C, 2007 *Solid State Commun.* **143** 47
[9] Poncharal P, Ayari A, Michel T and Sauvajol J-L 2008 *Phys. Rev. B* **78** 113407
[10] Ferrari A C and Basko D M 2013 *Nature Nanotechnology* **8** 235

Studying the influence of nanodiamonds over the elasticity of polymer/nanodiamond composites for biomedical application

T Hikov[1], D Mitev[1], E Radeva[1], A Iglic[2], R Presker[2], M. Daniel[3], J Sepitka[3], N Krasteva[4], M Keremidarska[4] , I Cvetanov[1] and L Pramatarova[1]

[1]*Georgi Nadjakov Institute of Solid State Physics, BAS, 1784 Sofia, Bulgaria*
[2]*University of Ljubljana, 1000 Ljubljana, Slovenia*
[3]*Faculty of Mechanical Engineering, Czech Technical University, 166 07 Prague, Czech Republic*
[4]*Institute of Biophysics and Biomedical engineering, BAS, 1113 Sofia, Bulgaria*

E-mail: detri@abv.bg

Abstract. The combined unique properties offered by organic and inorganic constituents within a single material on a nanoscale level make nanocomposites attractive for the next generation of biocompatible materials. The composite materials of the detonation nanodiamond/polymer type possess spatial organization of components with new structural features and physical properties, as well as complex functions due to the strong synergistic effects between the nanoparticles and the polymer [1]. The plasma polymerization (PP) method was chosen to obtain composites of silicon-based polymers, in which detonation generated nanodiamond (DND) particles were incorporated. The composite layers are homogeneous, chemically resistant, thermally and mechanically stable, thus allowing a large amount of biological components to be loaded onto their surface and to be used in tissue engineering, regenerative medicine, implants, stents, biosensors and other medical and biological devices. Mesenchymal stem cells (MSCs) are the main focus of research in regenerative medicine due to their extraordinary potential to differentiate into different kinds of cells including osteoblasts, which are needed for various bone disease treatments. However, for optimal usage of MSCs knowledge about the factors that influence their initial distribution in the human system, tissue-specific activation and afterwards differentiation into osteoblasts is required. In recent studies it was found that one of these factors is the elasticity of the substrates [2]. The choice of the proper material which specifically guides the differentiation of stem cells even in the absence of growth factors is very important when building modern strategy for bone regeneration. One of the reasons for there not being many studies in this area worldwide is the lack of suitable biomaterials which support these kinds of experiments. The goal of this study is to create substrates suitable for cell culture with a range of mechanical properties (namely elasticity and hardness) using composite layers (PPHMDS-DND) of plasma polymerized (PP) hexamethyldisiloxane (HMDS) and detonation generated nanodiamond (DND). The samples' elastic modulae and hardness were measured by CSM Ultra Nanoindentation Tester.

---

[1]E-mail: detri@abv.bg

Journal of Physics: Conference Series **558** (2014) 012060 doi:10.1088/1742-6596/558/1/012060

1. Introduction

The structure of a given biomaterial is crucial when determining the cell response, and respectively, the variants for its biomedical applications. The combined unique properties offered by organic and inorganic constituents within a single material on a nanoscale level make nanocomposites attractive for the next generation of biocompatible materials. Recently, there has been growing interest in the synthesis of composite functional materials with new physicochemical properties, involving integration of inorganic nanomaterials into a polymer matrix [1,3,4,5]. Plasma polymerized hexamethyldisilozane (PHMDS) has a long history of exploitation in a variety of applications because it is non-toxic, transparent, with a very low surface tension and it neither dissolves nor swells in a cell culture medium [6]. On the other hand, nanodiamond structures are of interest due to the combination of the unique properties inherent to diamond and the specific surface structure of the particles facilitating its fictionalisation [1].

2. Materials

The substrates used were cover glass (CG) discs with a diameter of 15 mm.

The detonationalnanodiamonds (DND), synthesized by a detonation of carbon-containing explosives, are produced with particles of mostly 4 nm in size and can be easily modified by chemical reactions. The diamonds (NSFPA and NASHCl) that we used were cleansed with different acid mixtures (HNO_3, H_2SO_4, $HClO_4$, HF - NSFPA and HNO_3, H_2SO_4, HCl - NASHCl) at temperatures of up to 533 K and pressures of up to $1.1x10^7$ Pa in a hyperbaric chamber.

Poly-hexamethyldisiloxane (PPHMDS), layers were prepared by plasma polymerization technology (PPHMDS) as previously described [7]. The deposition of the composites was carried out in the same plasma polymerization equipment by the following procedure: the mixture of DND powder and HMDS was ultrasonically treated for 15 minutes and then put in the plasma chamber.

In order to prepare composites with different elasticity we have used three different concentration of DND particles: 0,01 mg/ml; 0,05 mg/ml and 0,1 mg/ml. Certain amount of nanodiamond (10mg, 50mg and 100mg) was added to 100ml HMDS before polymerization.

3. Characterization

The nanodiamond powders were characterized by FTIR. The measurements were carried out using a Brucker FTIR spectrometer at ambient temperature in the range of 400 to 4000 cm^{-1}, using OPUS software, with an average of 64 scans and a resolution of 2 cm^{-1}. The assignment of the absorption bands was based on experience with organic compounds and the literature data.

The nanodiamonds' cytotoxicity was studied by estimating cell proliferation using CCK assay. Briefly, MG63 osteoblast-like cells were seeded on sterile cover glasses, placed in 24-well plates and incubated in DMEM (Sigma, D5796) with 10% FBS (Sigma). On the next day the medium was removed and the suspensions of DND particles were added. 24- and 72 hours after the addition of DNDs the cells were washed with PBS. To evaluate cytotoxicity CCK reagent (Sigma, 96992) was used following manufacture instructions. The extinction was measured at λ=450 nm.

The composites were also characterized by:

Nanoindentation – performed on a Hysitron TI 950 TriboIndenter™ nanomechanical test instrument.

Contact angle – water contact angle was measured through the sessile drop shape method under ambient conditions. The static water contact angle was measured on a Kruss contact-angle system (DSA10, Krus, GmbH, Germany) on freshly prepared surfaces. A 20µl deionized water was dropped onto the investigated surface and the water drop was photographed. The shape of the drop was then analysed using a sessile drop fitting model. Contact angles on five different regions on each sample were measured and averaged.

4. Results

On figure 1a and figure 1b we have shown FTTR spectra of two different type of DND particles: NASHCl and NSFPA. We found differences in the intensity of several peaks which indicated different concentrations of chemically active groups. The peak around 1100 cm^{-1} shows sp^2 bonding (non-diamond carbon) which is more pronounced in NASHCL.

Further, we have estimated the cytotoxity of DND particles using an osteoblast-like cell-line: MG63 (fig 2). As can be seen MG-63 cells demonstrated high proliferation ability during the whole 3-days incubation period and compared to the control (cells, incubated without DND particles) the cell growth rates were not decreased meaning absent of cytotoxity effect of DND particles on MG63 cells. We did not observe also any statistically significant differences between different types of DND particles.

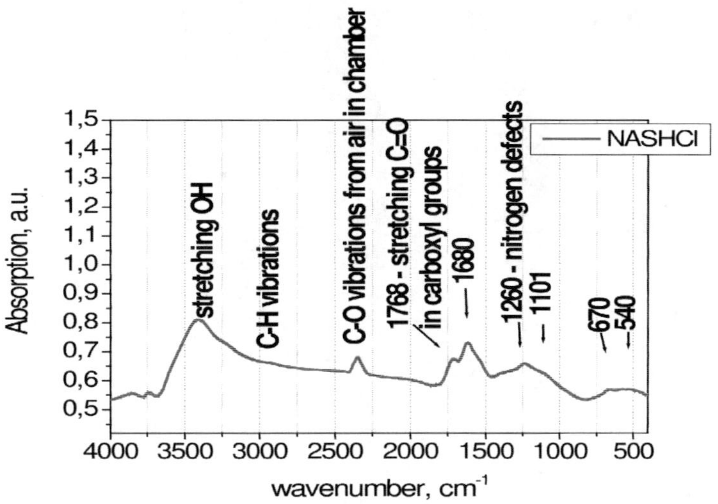

Figure 1a. FTIR spectra of NASHCl DND powder.

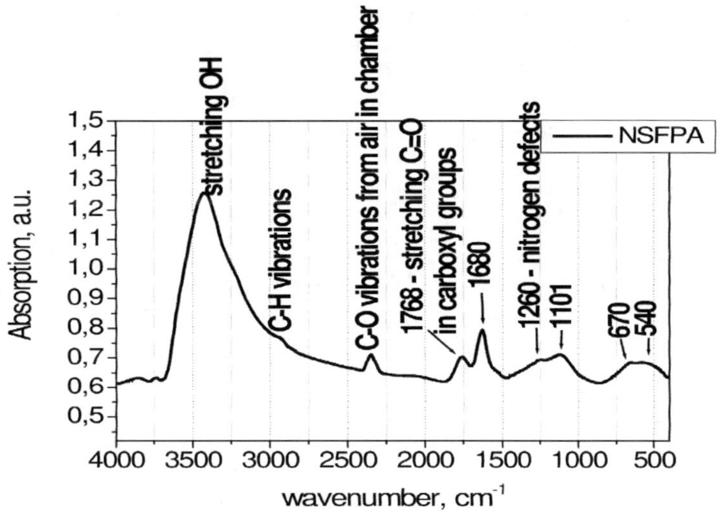

Figure 1b. FTIR spectra of NSFPA DND powder.

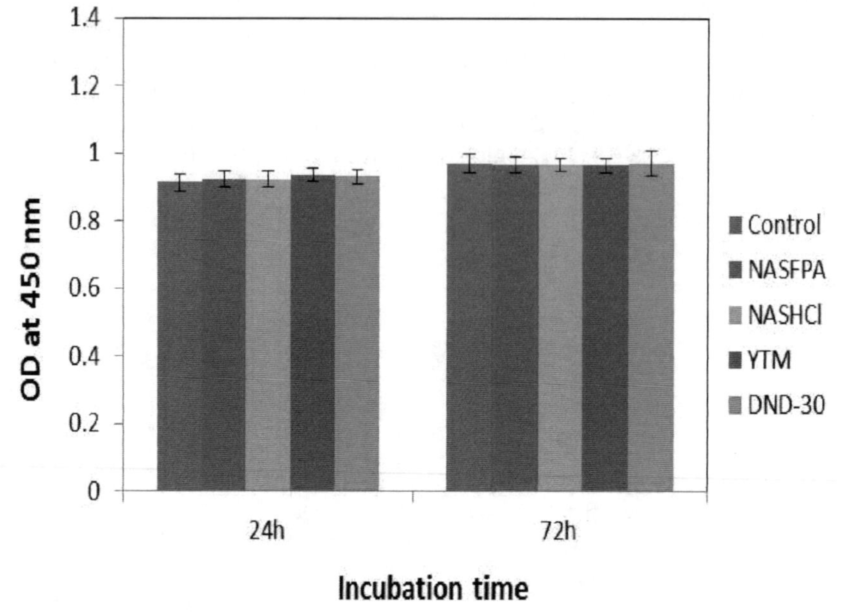

Figure 2. Proliferative ability of MG-63 cells in the presence of DND particles.

Based on the results NASHCl particles were chosen to prepare thin plasma polymerized composite films with different elasticity. NASHCl particles have also the lowest particles size among the studied particles as we have shown in our previous study [8].

The results of the measurements of the hardness and elastic modulus of the composite layers with different nanodiamond concentrations are shown in figure 3.The sample with the highest concentration of DND particles (0,1 mg/ml) shows the highest hardness and elastic modulus (258 MPa and 3.81 GPa, respectively). However, the other two samples with both lower concentrations of DND particles (0,01 and 0,05 mg/ml) demonstrated similar values of hardness and modulus approx. 101 MPa and 0.5385 GPa respectively. These results suggested that the hardness and the elastic modulus can be varied with DND concentration.

In addition, we have measured water contact angles (WCA) of composite layer in order to study if addition of DND in HMDS would change the wettabilty. The results (figure 4) demonstrated that there is no significant difference between WCAs of both samples with lower concentration of DND, but with increasing the concentration up to 0,1 mg /100 ml of the DND, composite layer became more wettable, e.g. hydrophilic.

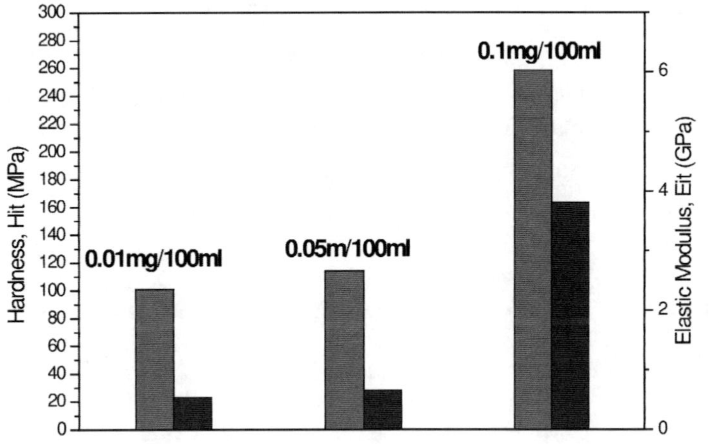

Figure 3. Hardness and elastic modulus of the testes samples (DND/polymer composites with different nanodiamond concentrations).

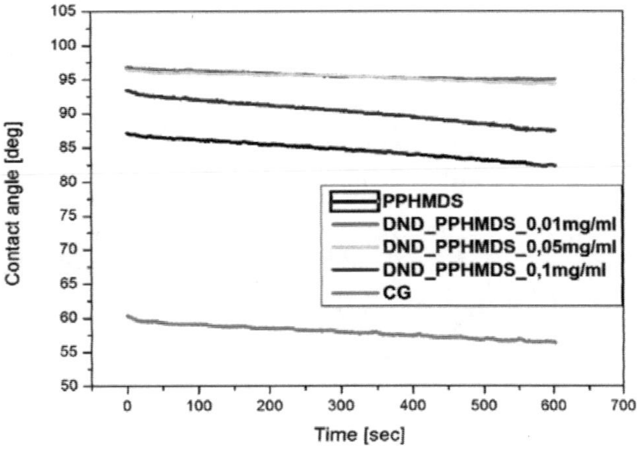

Figure 4. Contact angle of DND/polymer composites with different nanodiamond concentrations.

5. Conclusion

The plasma polymerization of the well known hexamethyldisiloxane monomer has been used for the deposition of nanodiamond (DND) containing composites with excellent adhesion on different substrates and remarkable properties. The composite layers were precisely characterized by different physicochemical methods. The results show that varying nanodiamonds concentration it is possible to influence the hardness, elastic modulus and wettabity of the composites in order to control the cell response. The cell studies showed that the used DND particles for preparation of PPHMDS/DND composites are not cytotoxic and therefore can be used to study the cell response.

Acknowledgments
This work was supported by Bulgarian national Scientific Research Fund project DFNI_B01_18/2012.

References
[1] Pramatarova L, Radeva E, Pecheva E, Hikov T, Krasteva N, Dimitrova R, Mitev D, Montgomery P, Sammons R and Altankov G, 2011, *on Biomimetics*, ed L Pramatarova, (Rijeka:InTech), pp 297
[2] Evans N D, Minelli C, Gentleman E, LaPointe V, Patankar S N, Kallivretaki M, Chen X, Roberts C J, Stevens M M, 2009, *European Cells & Materials,* vol**18**, pp 1
[3] Shenderova O A, Zhirnov V V and Brenner D W, 2002, *Slid State mater Sci*, **27**, pp 227
[4] Dolmatov V Yu, 2007, *Russian Chemical reviews,* **70**, pp 607
[5] Borjanovic V, Lawrence W G, Hens S, jarkic M, Zamboni I, Edson C, Vlasov I, Shenderova P and McGuire G E, 2009, *Nanotechnology*, **19**, pp 1
[6] Min-Hsien, 2009, *Surf Interface Anal,* **41**, pp 11
[7] Radeva E, Tsankov D, Bobev K and Spassov L, *J. Appl. Polym. Sci.*, **50**, pp 165
[8] Keremidarska E, Ganeva A, Mitev D, Hikov T, Presker R, Pramatarova L and Krasteva N, 2014, *Biotechnology & Biotechnological Equipment* (in press)

Silicon-ion-implanted PMMA with nanostructured ultrathin layers for plastic electronics

G B Hadjichristov[1], **Tz E Ivanov, Y G Marinov**
Georgi Nadjakov Institute of Solid State Physics, Bulgarian Academy of Sciences, 72 Tzarigradsko chaussee, Sofia 1784, Bulgaria

E-mail: georgibh@issp.bas.bg

Abstract. Being of interest for plastic electronics, ion-beam produced nanostructure, namely silicon ion (Si^+) implanted polymethyl-methacrylate (PMMA) with ultrathin nanostructured dielectric (NSD) top layer and nanocomposite (NC) buried layer, is examined by electric measurements. In the proposed field-effect organic nanomaterial structure produced within the PMMA network by ion implantation with low energy (50 keV) Si^+ at the fluence of 3.2×10^{16} cm^{-2}, the gate NSD is ion-nanotracks-modified low-conductive surface layer, and the channel NC consists of carbon nanoclusters. In the studied ion-modified PMMA field-effect configuration, the gate NSD and the buried NC are formed as planar layers both with a thickness of about 80 nm. The NC channel of nano-clustered amorphous carbon (that is an organic semiconductor) provides a huge increase in the electrical conduction of the material in the subsurface region, but also modulates the electric field distribution in the drift region. The field effect via the gate NSD is analyzed. The most important performance parameters, such as the charge carrier field-effect mobility and amplification of this particular type of PMMA-based transconductance device with NC *n*-type channel and gate NSD top layer, are determined.

1. Introduction

The electrical properties of the ion-modified polymers are of great importance for their electronic applications [1-6]. In particular, the low-energy (keV range) ion irradiation and ion implantation have shown a significant enhancement of the electrical conductivity of normally insulating plastics, such as poly-(methyl methacrylate) (PMMA), very interesting material for plastic electronics. The ion-beam sensitivity of PMMA makes it possible to considerably modify and engineer the electrical conductivity of this polymer [7, 8] that opens ways to use ion-implanted PMMA in flexible electronics, optoelectronics and in various high-tech and industry areas. Nowadays there is an increasing interest of fabrication of organic thin-film transistors (TFT) utilizing the electrical properties of ultrathin organic layers on the surface of ion-implanted polymers [9, 10]. Previous our investigations [11, 12] indicated that the implantation of bulk PMMA by silicon ions (Si^+) accelerated to a relatively low ion energy of 50 keV and at the fluence of 3.2×10^{16} Si^+/cm^2 does result in the formation of carbonized ion-modified material close to the very surface within this plastic. In the present work, we report on the electrical response of an organic electronic element with simple configuration based on field effect in Si^+-implanted PMMA, observed in a sandwich structure of the type *dielectric/semiconductor/*

[1] To whom any correspondence should be addressed.

Journal of Physics: Conference Series **558** (2014) 012061 doi:10.1088/1742-6596/558/1/012061

insulator created on the surface of PMMA by the same ion implantation. It is of scientific and technology interest to estimate the electrical parameters relevant to the performance of such a device based on ion-implanted polymer. We demonstrated that in the ion-beam produced nanostructure driven by a buried ion-implanted nanocomposite (NC) ultrathin layer (as a channel) and a gate nanostructured dielectric (NSD) ultrathin top layer, the field effect and the device functionality can be efficiently controlled.

2. Experimental

The sample was a plate of size 1 cm × 1 cm from PMMA commercially available (Röhm Plexiglas) with a thickness of 5 mm. It was implanted with Si^+ ions at a relatively low energy of 50 keV and Si^+ ion fluence of 3.2×10^{16} cm^{-2}. The experimental details have been described elsewhere [11].

For electrical measurements, the Si^+-implanted PMMA sample was electroded with silver paste. Electric contacts (1 mm-narrow, 1 cm-long) from silver paste were deposited at two opposite edges of the implanted surface of the PMMA sample, as schematically shown in figure 1. Direct current (DC) measurements were conducted by Keithley 617 Programmable Electrometer controlled by computer. The DC transconductance of Si^+-implanted PMMA was studied upon applying a gate voltage on a third (gate) electrode from copper (width $W = 1$ cm, length $L = 3$ mm) placed on the implanted PMMA surface between the drain and the source side-electrodes (figure 1).

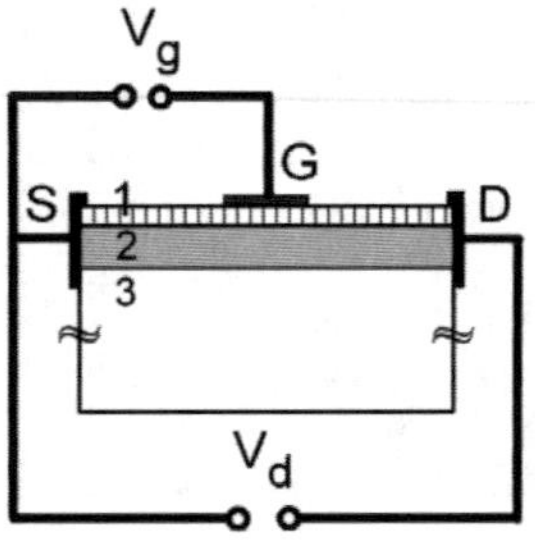

Figure 1. Schematic of the field-effect nanostructured Si^+-implanted PMMA device with source (S), drain (D) and gate (G) electrodes. The spatial structure: (1) ion-nanotracks-modified low-conductive top layer, being a gate nanostructured dielectric (NSD); (2) conductive carbon nano-clustered ion-implanted nanocomposite (NC) buried layer; (3) the bulk of the non-implanted PMMA (the substrate from pristine PMMA).

In addition to the DC measurements, the electrical response of the Si^+-implanted PMMA was also examined under alternating (AC) electric field. In this case, a sinusoidal voltage with amplitude of 1 V was applied and the field frequency was swept in the range from 1 Hz to 30 kHz. The sampling signal from a Stanford Research Systems SR830 DSP lock-in build-in generator was acquired by a computer that controls the frequency sweep. A reference sample of non-implanted PMMA (the same as the one used as a substrate for the ion-implanted sample) was also measured. These measurements consisted in determination of the frequency dependent current flowing through the samples when the electric field was longitudinally applied to the ion-implanted layer.

The measurements were carried out under ambient environment (in air and at room temperature).

3. Results and Discussion

As known, the modification of electrical properties of ion implanted materials is due to chemical and structural changes upon ion implantation [1-6]. In particular, for hydrocarbon polymers like PMMA, these changes include polymer decomposition, loss of hydrogen and hydrocarbons, outgassing of light fragments and densification of remaining material, chain scissions accompanying with formation of double bonds (C=C) where the side groups containing carbonyl groups are broken. As a result, a well defined subsurface layer from a more compacted material with carbonaceous structure having conjugated double bonds is formed [1-6] that actually leads to increased electrical conductivity even at low implantation fluence (of the order of $10^{14} - 10^{15}$ cm^{-2}) and relatively low ion energy (keV range).

In our case, the implantation of PMMA with 50 keV Si^+ ions at the fluence of 3.2×10^{16} cm^{-2} results in formation of nanostructured material within the PMMA substrate. Optical spectroscopy reveals the presence of carbonized material (namely hydrogenated amorphous carbon) with nano-clustered structure (carbon nanoclusters, CNCs) in the ion-implanted region in the Si^+-implanted PMMA [11, 13]. This carbonaceous structure has a three-dimensional network of independent clusters of sp^2 bonded carbon atoms. These rings fairly resemble the disordered graphite-like structures [1]. As a sensitive probe for local atomic bonding, Raman spectroscopy clearly evidences the nanometer-sized CNCs in the ion-implanted layer beneath the surface of Si^+-implanted PMMA, and the polarized micro-Raman spectroscopy allows estimating even the orientation of the CNCs. The average size of the CNCs was estimated to be 3 – 5 nm [11, 13], and the material formed in the ion-implanted layer in Si^+-implanted PMMA can be considered as nanocomposite (NC), i.e. CNCs within PMMA network.

Upon the conditions of the ion implantation we have performed, the 50 keV Si^+ ions were implanted in PMMA in a very small depth (< 100 nm) beneath the surface, forming a buried planar NC layer of a thickness of about 80 nm under the surface [14] (figure 1). Further, by the present Si^+-ion implantation a real doping of PMMA with silicon at a doping level of 1 – 2 at. %) is achieved in this ultrathin layer #2, as expected at the ion energy of 50 keV and the ion fluence higher than 10^{16} ions/cm^2 [15]. Si^+-implanted PMMA becomes a typical organic semiconducting material [14].

The DC electrical response of the examined Si^+-implanted PMMA is reported in figure 2(a). The linear current-voltage plot reveals Ohmic conductance of the formed NC layer that. This experimental fact is consistent with the electron hopping mechanism of conductivity [2-4, 16] (a variable range electron hopping conductivity through high-conducting CNCs separated by material with much lower conductivity, similarly to other examples of PMMA subjected to low-energy ion implantation [7, 9, 17]. From data in figure 2(a), the DC conductivity of the considered NC layer of Si^+-implanted PMMA was calculated to be $\approx 6 \times 10^{-5}$ S/cm, much higher than that ($< 10^{-15}$ S/cm) of the originally insulating PMMA substrate (the pristine PMMA, highly insulating plastic), i.e. an enhancement by more than ten orders of magnitude takes place as a result from ion implantation. Thus, at the implantation fluence of 3.2×10^{16} Si^+/cm^2 Si^+-implanted PMMA attains a DC conductivity typical for highly conductive amorphous phase of carbon and conducting polymers at room temperature [18].

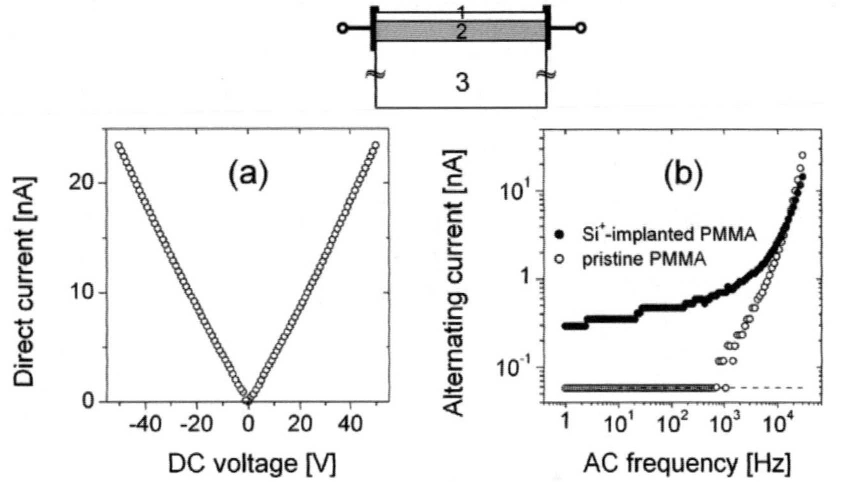

Figure 2. (a) DC current-voltage dependence in positive and negative voltage sweep (a) and AC frequency behavior (b) of Si^+-implanted PMMA. For comparison, data for pristine PMMA is given in (b) (open circles). The dashed line indicates the sensitivity limit of the apparatus. At the top is schematically shown the measured structure with unconnected (floating) gate electrode.

The electrical response of the studied Si^+-implanted PMMA was also characterized by AC measurements (figure 2(b)). They indicated that Si^+ ion implantation of PMMA (in the present case at an ion energy of 50 keV and at the fluence of 3.2×10^{16} Si^+/cm^2) leads to a considerable enhancement of the AC conductivity of the NC layer, most pronounced at electric-field frequency below 1 kHz.

Te spatial structure formed in the host PMMA by considered Si^+-ion implantation consists of the following regions (in the depth) (recall figure 1): (i) ion-modified top layer (#1) with a thickness of about 80 – 100 nm [14]; (ii) a buried ion-implanted layer (#2) with a thickness of about 80 nm [14]; and (iii) the non-intact PMMA insulating bulk region (#3). Due to the ion-beam-produced nano-sized ion tracks along the trajectory of the impact ions (cylindrically-shaped tracks having nanoscale transversal dimension [1]) within the top ultrathin layer #1, the electronic and electrical properties of this ion-modified surface region differ from those of the pristine PMMA substrate (#3). Our complementary measurements of the leakage current through the gate in the examined electronic system (with the same electrode configuration) indicated a presence of an amount of free charge carriers (electrons) in the layer #1, i.e. it acts as a source of electrons. As such, the ion-tracks-modified surface layer #1 should be considered as a low-conductive (i.e. not ideal insulator) gate nanostructured dielectric (NSD), in contrast to the highly conductive ion-implanted NC layer #2 (actually, semiconducting *n*-type owing to the hydrogenated amorphous CNCs [19]). Due to the distribution of the ion-implanted material, both ion-modified ultrathin organic layers (the top NSD layer #1 and the buried NC layer #2) have diffusive interface with no sharp boundary between them.

Upon suitable conditions, the sandwich-like *dielectric/semiconductor/insulator* structure formed within the polymer PMMA allows a field-effect-transistor (FET)-like operation by the buried ultrathin NC layer #2 used as a channel of organic thin film FET. Thus, with an appropriate electrode configuration, both spatial structure and the formed organic nanomaterial in Si^+-implanted PMMA enable the control of the channel current by a potential difference upon applying an external electric field (gate voltage). Our investigation has shown that by the above electrode geometry this complex organic structure exhibits a field-effect transconductance ~ 0.35 nS, and thereby the ion-implanted material considered here is of practical interest for assembling of FET-like electronic elements.

Figure 3(a) reports output electrical characteristics of FET-like device based on the channel structure of the studied Si^+-implanted PMMA with two side electrodes (source and drain) and a top electrode (gate) as shown in figure 1. The results for the dependence of the drain-to-source current I_D versus the drain-to-source voltage V_D are in accordance with the output characteristics that correspond to an *n*-channel with electron accumulation mode. The field effect was relatively weak, as seen also by the transfer characteristics of the studied Si^+-implanted PMMA FET-like configuration (figure 3(b)).

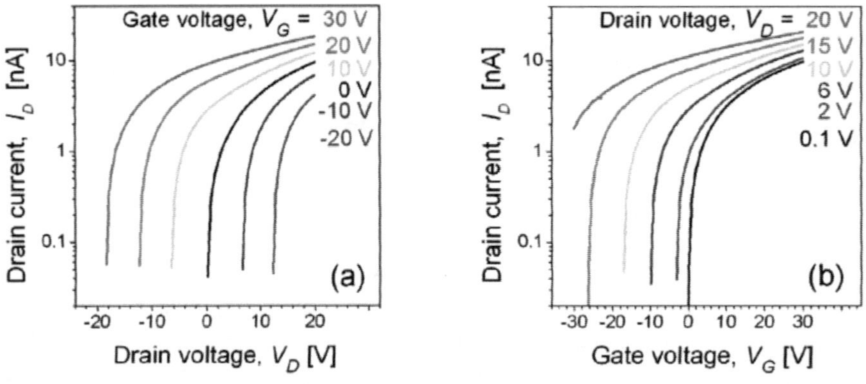

Figure 3. Output characteristics at various gate voltage V_G (a) and transfer characteristics at various drain voltage V_D (b) of the examined Si^+-implanted PMMA transconductance device (schematically drawn in figure 1), all measured at room temperature.

The amplification taken from the transfer curve measured at $V_D = 0.1$ V as the maximum current on/off ratio obtained at fixed gate voltages 20 V and 0 V, was ≈ 300. It should be noted that such a value is much lower as compared to those reported for high-performance organic FETs [20, 21]. Further, the on/off ratio sharply decreases with the increasing drain-to-source voltage V_D (figure 4).

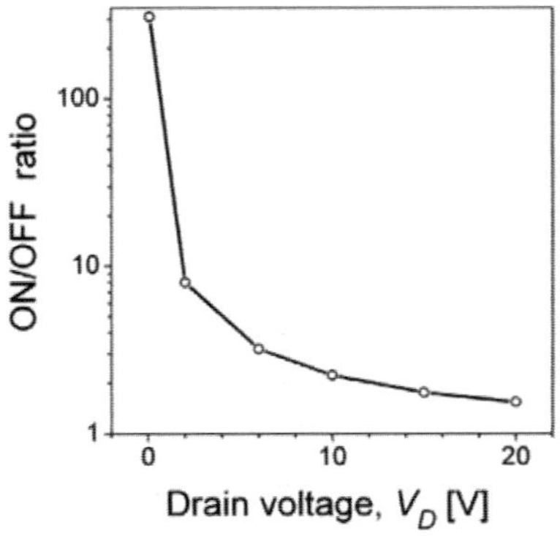

Figure 4. The maximum current on/off ratio obtained at fixed gate voltages 20 V and 0 V from the transfer characteristic curves of the examined Si+-implanted PMMA transconductance device at several values of the drain-to-source voltage V_D.

Using data from DC measurements, one can assess the effective mobility (μ) of the charge carriers relevant to the performance of the organic transconductance device. For the purpose, we have applied a procedure used for electrical characterization of TFTs [22], namely a differential method in a linear regime of transistor operation, i.e. at $V_{GS} >> V_{DS}$. In our modeling, the curve measured at $V_D = 0.1$ V was chosen from the transfer characteristics family given in figure 3(b). We calculated $\mu \approx 0.02$ cm^2 V^{-1} s^{-1} (at room temperature, a result obtained by using the values $W = 1$ cm, $L = 3$ mm and $C_i = 40$ nF/cm^2, where W and L are the width and the length of the gate, respectively, and C_i is the capacitance per unit area of the NSD). Such a value of the charge carrier field-effect mobility in the considered Si^+-implanted PMMA FET-like device is rather low as compared to the values (e.g., 1 – 5 cm^2 V^{-1} s^{-1}) reported for the high-performance thin-film organic FETs [20, 21]. Reasonably, the device figures of merits strongly depend on the production implant process and on the device configuration.

4. Conclusions

Electrical properties of ion-implanted PMMA with ultrathin planar gate nanostructured dielectric (NSD) and nanocomposite (NC) buried layer are characterized, and a field-effect configuration is inspected as operated in such complex nanostructure of PMMA being implanted with low energy (50 keV) Si^+ ions at the fluence of 3.2×10^{16} cm^{-2}. The electrical system of the all-organic transconductive device based on Si^+-implanted PMMA examined here is driven through the formed NC ion-implanted layer (as a channel with a thickness of about 80 nm) of nano-clustered hydrogenated amorphous carbon as n-type organic semiconductor nanomaterial produced in PMMA by the ion implantation. The structure under study exhibits a feasible field effect which has a potential for electronic applications, e.g. in organic FET-like devices based on ion implantation-modified PMMA.

The formed field-effect organic structure and FET-like device on such layered nanomaterial may be of practical interest for plastic electronics and nano-electronics, in particular, for electrical sensing (e.g., charge-based sensing), as well as electrically-controlled bio-medical applications utilizing the high porosity, the nano-sized ion tracks within the NSD in the polymer, the bio-compatibility of PMMA (and ion-modified PMMA) and other useful properties of the ion-irradiated polymer surface and subsurface region.

The values of important electrical parameters, such as the charge carrier field-effect mobility and amplification of the considered type of polymer-based transconductance device appear to be accessible to optimization. The improving of the performance needs careful engineering (by the implant process) of the formed NC ion-implanted layer (channel) that is the most important, as well as the gate NSD. The implantation conditions (the ion energy and fluence, as well as the angle of incidence of the ion beam) determine the carbonaceous nanostructure and its in-depth distribution, and thereby, the in-depth electronic and electrical properties of the formed organic nanomaterial of both NSD and buried

NC layer, as well as the dielectric-semiconductor interface between them and its properties. These conditions are very important for the formation of an effective transconductance structure.

Acknowledgements
INERA EU project Research Potential (FP7-316309-REGPOT-2012-2013-1) "Research and Innovation Capacity Strengthening of ISSP-BAS in Multifunctional Nanostructures" is acknowledged.

References
[1] Fink D 2004 *Fundamentals of Ion Irradiated Polymers* (Berlin: Springer)
[2] Fink D 2004 *Transport Processes in Ion-Irradiated Polymers* (Berlin: Springer)
[3] Popok V N 2012 *Rev. Adv. Mater.* Sci. **30** 1
[4] Sviridov D V, Odzhaev V D and Kozlov I P 1998 Ion-implanted polymers *Electrical and Optical Polymer Systems – Fundamentals, Methods and Applications* eds D L Wise, G E Wnek, D J Trantolo, T M Cooper and J D Gresser (New York: Marcel Dekker) chapter 11 p. 387
[5] Giedd R E, Moss M G, Kaufmann J, Wang Y Q 1998 Electrical applications of ion-implanted polymer films *Electrical and Optical Polymer Systems – Fundamentals, Methods and Applications* eds D L Wise, G E Wnek, D J Trantolo, T M Cooper and J D Gresser (New York: Marcel Dekker) chapter 29 p. 1011
[6] Kondyurin A V and Bilek M 2008 *Ion Beam Treatment of Polymers: Application Aspects from Medicine to Space* (Oxford: Elsevier)
[7] Koval Y, Fistul M V and Müller P 2005 *J. Vac. Sci. Technol.* **A 23** 1375
[8] Gupta R, Kumar V, Goyal P K, Kumar S and Goyal S L 2012 *Adv. Appl. Sci. Res.* **3** 2766
[9] Koval Y, Lazareva I, Müller P, Müller K, Henkel K and Friedrich D 2011 *Phys. Status Solidi (b)* **248** 299
[10] Popok V N, Karpovich I A, Odzhaev V B and Sviridov D V 1999 *Nucl. Instrum. Meth. B* **148** 1106
[11] Hadjichristov G B, Ivanov V and Faulques E 2008 *Appl. Surf. Sci.* **254** 4820
[12] Ivanov V G, Hadjichristov G and Faulques E 2009 *Appl. Spectrosc.* **63** 1022
[13] Ivanov V G and Hadjichristov G B 2011 *J. Raman Spectrosc.* **42** 1340
[14] Hadjichristov G B, Gueorguiev V K, Ivanov T E, Marinov Y G, Ivanov V G and Faulques E 2008 *Org. Electron.* **9** 1051
[15] Moliton A, Moreau C and Moliton J P 1993 *Nucl. Instrum. Meth. Phys. Res. B* **80-81 Part 2** 1028
[16] Hesto P Nature of electronic conduction 1986 *Instabilities in Silicon Devices* eds G Barbottin and A Vapaille Vol. 1 (Amsterdam: North Holland) chapter 5 p. 263
[17] Lazareva I, Koval Y, Alam M, Strömsdörfer S and Müller P 2007 *Appl. Phys. Lett* **90** 262108
[18] Bhattacharyya S and Subramanyam S V 1998 *Electrical and Optical Polymer Systems – Fundamentals, Methods and Applications* eds D L Wise, G E Wnek, D J Trantolo, T M Cooper and J D Gresser (New York: Marcel Dekker) chapter 7 p. 201
[19] Jones D I and Stewart A D 1982 *Philos. Mag. B* **46** 423
[20] Horowitz G Organic Transistors 2006 *Organic Electronics: Materials, Manufacturing, and Applications* ed H Klauk (Weinheim: Wiley-VCH) chapter 1 p. 3
[21] Dimitrakopoulos C D and Malenfant P R L 2002 *Adv. Mater.* **14** 99
[22] Natali D, Fumagallli L and Sampietro M 2007 *J. Appl. Phys.* **101** 014501

Refractive index investigations of nanoparticles dispersed in water

I Bodurov[1], T Yovcheva[1] and S Sainov[2]

[1]Department of Experimental Physics, Plovdiv University "Paisii Hilendarski", 24 Tzar Assen str., 4000 Plovdiv, Bulgaria
[2]"HOLOBUL" Ltd., Acad. Georgi Bonchev str., block 109, 1113 Sofia, Bulgaria

E-mail: bodurov@uni-plovdiv.net

Abstract. The refractive index of nanoparticles dispersed in water is measured, using the total internal reflection method. The critical angle is determined by the disappearance of diffraction orders from a metal grating. The investigated nanoparticles are titanium dioxide (anatase phase), (35 nm diameter), zinc oxide, (<50 nm diameter), zirconium dioxide, (<100 nm diameter). The refractive index is measured with the experimental uncertainty of 1%. The Lorentz-Lorenz, Maxwell Garnett and Bruggeman relations are applied in the nanoparticle's refractive indices calculations.

1. Introduction

More than 120 years ago LeBlanc [1] has proposed a critical angle method for refractive index (RI) measurement of powders. In fact, that is an immersion technique, where critical angle is fixed when particle's RI is equal to the RI of the immersion liquid. In this case the light scattering is minimized. The Abbe refractometer ensures satisfactory measurement precision. However, the LeBlanc's method is time consuming; limited to the short list of high RI immersions; and high RI particles are measured with inflammable and toxic immersion liquids. In 1991 one of the authors [2] has proposed RI measuring method based on the critical angle determination by the disappearance of the diffraction pattern, created by a metal grating, placed in contact with the liquid or solid sample with a thickness varying from 250 nm to several μm. Applying this method, RI of turbid samples, cow butter for example, could be measured [3].The maximum experimental uncertainty is ±0.001 [4].

In the present paper the RI of high RI nanoparticles are measured. The samples are water dispersions of ZnO, TiO_2 and ZrO_2 nanoparticles with filling factors (f) 0.15, 0.012 and 0.02, respectively. The thickness of the liquid layer's is 6 μm. After RI determination of water dispersion's RI, RI of the nanoparticles are calculated using well-known Lorentz-Lorenz [5], Maxwell Garnett [6] and Bruggeman [7] mixing models. He-Ne laser (632.8nm) and semiconductor laser (532nm) are utilized as light sources in the present RI measurements.

The RI data at these two laser wavelengths is used for one oscillator Sellmeier dispersion relation.

2. Theoretical background

Our considerations are based on the most popular mixing theories. According Lorentz-Lorenz [5], in the optical spectral range we have:

$$\frac{n_e^2-1}{n_e^2+2}=f\frac{n_i^2-1}{n_i^2+2}+(1-f)\frac{n_m^2-1}{n_m^2+2}, \quad (1)$$

where n_e is experimentally obtained effective RI; n_i is RI of inclusions; n_m is RI of matrix.

Filling factor f is defined as ratio between the total particles' volume and full volume:

$$f=\frac{V_i}{V_m+V_i}. \quad (1a)$$

Using the effective dielectric function, Maxwell Garnett [6] obtained:

$$n_e^2=n_m^2\frac{(n_i^2+2n_m^2)+2f(n_i^2-n_m^2)}{(n_i^2+2n_m^2)-f(n_i^2-n_m^2)}. \quad (2)$$

In 1935, Bruggeman [7, 8] proposed another mixing relation:

$$f\frac{(n_i^2-n_e^2)}{(n_i^2+2n_e^2)}+(1-f)\frac{(n_m^2-n_e^2)}{(n_m^2+2n_e^2)}=0. \quad (3)$$

Solving the equation (1), (2) and (3), for RI of embedded nanoparticles we have:

$$n_i^{L-L}=\left[\frac{2(n_e^2-n_m^2)+fn_m^2(n_e^2+2)}{(n_m^2-n_e^2)+f(n_e^2+2)}\right]^{\frac{1}{2}}, \quad (4)$$

$$n_i^{MG}=n_m\left[\frac{2(n_e^2-n_m^2)+f(n_e^2+2n_m^2)}{f(n_e^2+2n_m^2)-(n_e^2-n_m^2)}\right]^{\frac{1}{2}}, \quad (5)$$

$$n_i^{Br}=n_e\left[\frac{3fn_m^2-2(n_m^2-n_e^2)}{3fn_e^2+(n_m^2-n_e^2)}\right]^{\frac{1}{2}}, \quad (6)$$

where n_i^{L-L}, n_i^{MG} and n_i^{Br} are the RI of nanoparticles, embedded into liquid or solid matrix and calculated using different mixing rules.

3. Experimental

3.1. Samples

All investigated samples are in the form of nanoparticles dispersed in water. The ZnO nanoparticles with diameter less than 50 nm, product of "Sigma-Aldrich", are 50 wt.% dispersion concentration in water. Titanium dioxide nanoparticles with diameter 35 nm are in anatase form and 5 wt.% dispersion concentration in water. The ZrO_2 nanoparticles with diameter less than 100 nm are also product of "Sigma-Aldrich" and 10 wt.% dispersion concentration in water. The 5 μl of the liquid sample is spread over 800 mm^2 of the grating and its thickness of is calculated to be 6 μm.

3.2. Refractive index measurements

The refractive index of the liquid sample is measured with a two-wavelength laser micro-refractometer at 632.8 nm and 532 nm. A 1 mW "Spectra Physics" He-Ne laser and second harmonic 20 mW semiconductor laser are used in the present investigations as light sources.

The laboratory device is illustrated in figure 1.

As described [9], the critical angle φ_c is determined by the observed diffraction pattern disappearing on the screen (10).

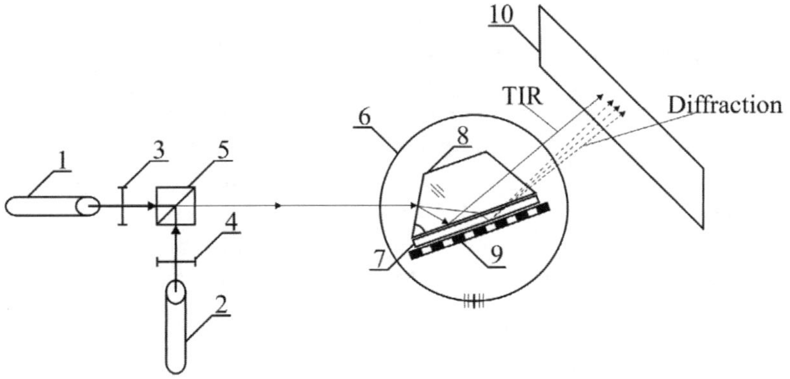

Figure 1. Two-wavelengths laser micro-refractometer: 1 – He-Ne laser (632.8 nm); 2 – 532 nm semiconductor laser; 3 and 4 – choppers; 5 – beam splitter; 6 – rotary stage with vernier; 7 – sample; 8 – glass prism; 9 – metal diffraction grating; 10 – screen.

The refractive index is calculated by the following equation:

$$n = N\sin\left[A \pm \arcsin\frac{\sin\varphi_{cr}}{N}\right]. \quad (7)$$

The sign (+) and (-) corresponds to the clockwise and the counter-clockwise determination of φ_{cr}, respectively. In our case the sign is (-). In equation (7) $A = 65°$ is the reflecting angle of the prism, N is the refractive index of the prism and has the values 1.7347 and 1.7480 for the used wavelength of 632.8 nm and 532 nm, respectively.

The optical properties of the materials are usually presented by their dispersion dependences. The most appropriate dispersion relation in dielectric and non-magnetic materials far from the fundamental absorption band is the Sellmeier relation:

$$n^2 - 1 = \frac{s\lambda^2}{\lambda^2 - \lambda_s^2}. \quad (8)$$

In the one-oscillator Sellmeier's model the dependence of refractive index as a function of wavelength λ is expressed by two constants, s and λ_s, which are called Sellmeier's coefficients.

4. Results and discussion

The obtained experimental results calculated by equations (4), (5) and (6) are listed in table 1. The RI of water n_m is measured by the laser micro-refractometer (see part 3.2) with the experimental uncertainty 0.01%. The nanoparticle concentration uncertainty is 0.05%. The nanoparticles RI uncertainty is $\Delta n = \pm 0.02$ and mainly depended from effective RI measurements.

Table 1. Refractive index of investigated nanoparticles.

Sample	**TiO_2**		**ZrO_2**		**ZnO**	
λ, nm	532	632.8	532	632.8	532	632.8
n^{MG}, $\Delta n = \pm 0.02$	2.52	2.46	2.19	2.14	2.06	2.02
n^{Br}, $\Delta n = \pm 0.02$	2.54	2.49	2.20	2.16	2.04	2.01
n^{L-L}, $\Delta n = \pm 0.02$	3.79	2.81	3.94	2.76	2.37	2.13

Journal of Physics: Conference Series **558** (2014) 012062 doi:10.1088/1742-6596/558/1/012062

Table 1 represents a good correlation, between experimentally obtained RI of nanoparticles and those described in the literature within our experimental uncertainty. The results obtained by Maxwell Garnett and Bruggeman mixing relations match. According to the data published in [10] refractive index of anatase is 2.49 at 589.3 nm wavelength. According to the data published on Internet (www.refractiveindex.info) and references therein the averaged RI of ZnO is $n^{532} = 2.03$ and $n^{633} = 1.99$. For ZrO_2 $n^{532} = 2.17$ and $n^{633} = 2.15$. The experimental results are very close, which has been theoretically proved earlier [9]. The Lorentz-Lorenz mixing rule is not applicable for RI investigations of nanoparticle dispersions. The experimental uncertainty in the present research is not high because scattered light background hinders the exact critical angle determination. Due to the Mie scattering theory [11], the scattered light intensity is very high - proportional to $\left|\left(n_i^2 - n_m^2\right)\right|^2$.

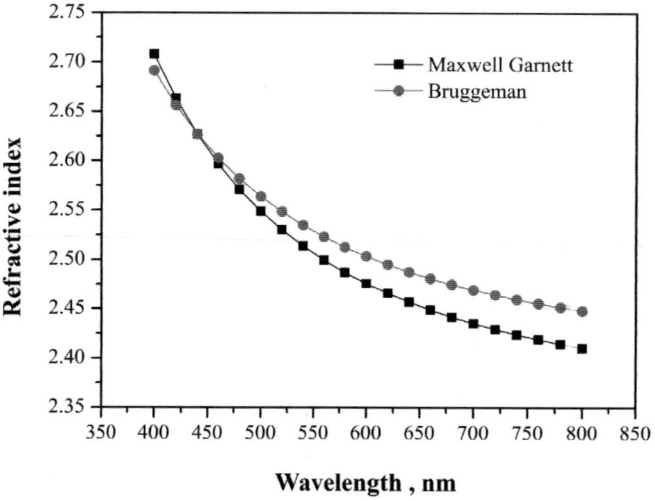

Figure 2. Sellmeier approximation of the dispersion dependences of the refractive index of TiO_2 nanoparticles obtained by Maxwell Garnett and Bruggeman relations.

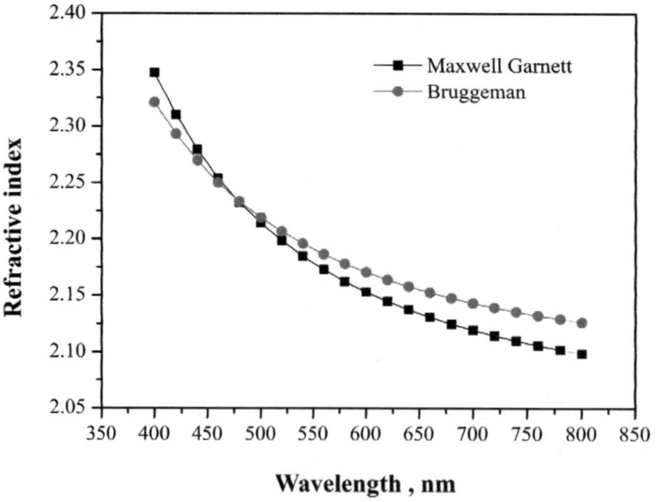

Figure 3. Sellmeier approximation of the dispersion dependences of the refractive index of ZrO_2 nanoparticles obtained by Maxwell Garnett and Bruggeman relations.

In order to attain an experimental uncertainty better than ±0.001 it is necessary to increase signal-to-noise ratio in diffraction pattern detection. It could be achieved by CCD camera detection followed by computer processing for scattered light background suppression. The simultaneous sample's thickness diminishing in the range 0.5 μm – 1.0 μm will also increase signal-to-noise ratio.

The refractive index dispersion dependencies of the investigated nanoparticles are obtained by fitting using equation (8) and listed below.

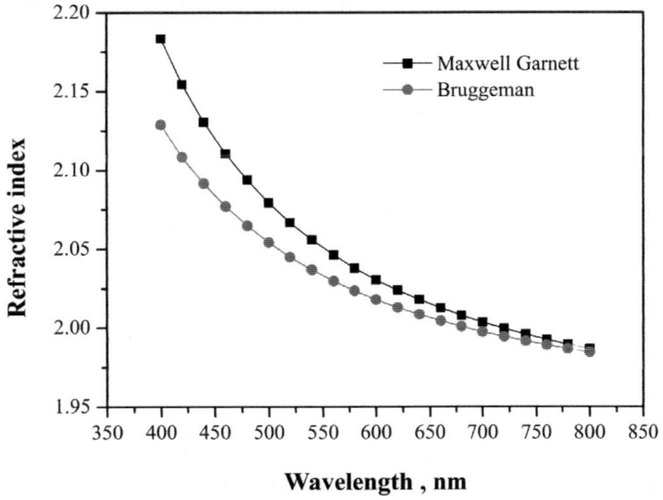

Figure 4. Sellmeier approximation of the dispersion dependences of the refractive index of ZnO nanoparticles obtained by Maxwell Garnett and Bruggeman relations.

5. Conclusion

The refractometric method of the disappearing diffraction pattern is applicable for RI investigation of nanoparticles. In the present work high RI nanoparticles are dispersed in water - i.e. the case with the biggest RI difference and consequently, very high light scattering background.

Acknowledgements

S. Sainov is grateful for the financial support under the contracts BG161PO003-1. 1.005-0085 –C0001 and COST Action MP 1102, contract DCOST 01/7-13.20.2012 with National Science Fund.

References

[1] LeBlanc M 1892 *Z. Phys. Chem.* **10** 433
[2] Sainov S 1991 *Rev. Sci. Instr.***62** 3106
[3] Nikolova K, Panchev I and Sainov S 2007 *Eur. Food Res. Technol.* **225** 139
[4] Nikolova K, Panchev I and Sainov S 2009 *CR Acad. Bulg. des Sci.* **62** 559
[5] Ioffe B 1983 *Refractometric methods in chemistry* (Leningrad: Khimiya) (in Russian)
[6] Maxwell Garnett J C 1904 *Philos Trans. R. Soc.* **203** 385
[7] Bruggeman D A G 1935 *Ann. Phys.* **24** 636
[8] Petroff Yu 1982 *Physics of small particles* (Moscow: Nauka Publishers) (in Russian)
[9] Sainov S, Vlaeva I, Yovcheva T, Dragostinova V and Stavrev S 2010 J. Phys. Conf. Ser. 253 012067
[10] Hanaor D A H and Sorrell Ch C 2011 *J. Mater. Sci.* **46** 855
[11] Mie G 1909 *Ann. Phys.* **25** 377

Functionalization of surfaces by ultrafast laser micro/nano structuring

C Hairaye[1,2], F Mermet[1], T Engel[1,2], P C Montgomery[2] and J Fontaine[1,2]
[1] IREPA LASER, Parc d'Innovation, 67400 Illkirch, France.
[2] Laboratoire des Sciences de l'Ingénieur, de l'Informatique et de l'Imagerie (ICube), UDS-CNRS-INSA-ENGEES, UMR 7357, 300 bd Sébastien Brant, CS 10413, 67412 Illkirch, France.

E-mail: chairaye@unistra.fr

Abstract. In surface functionalization, micro- and nanotexturing with ultrashort pulse lasers is of increasing interest in industry. These two scales of structures can be obtained by a self-organization of the matter after laser irradiation, or by a micro-machining of the sample with successive ablations. Controlling key-parameters such as wavelength, laser fluence, spot overlap, number of repetitions and polarization, allows the generation of self-organized nano-structures called ripples or the creation of square-shaped micro-structures. In the first case, a surface textured with these nano-gratings presents a diffractive aspect or a permanent marking, but changes little the initial wetting behaviour of the surface. In the second case, the micro-structures improve significantly the wettability of the surface, with an increase of almost 20° of the contact angle with a water droplet. So a large range of surface texturing can be performed with ultrashort pulsed laser technology, offering a wide range of applications, from biomedical to aeronautical.

1. Introduction

Functionalizing a surface with micro- or nanostructures is particularly challenging today as the structural size becomes smaller and the variety of materials requiring to be treated becomes wider. Texturing a surface with an ultrashort pulse laser is a fast and simple process that can be used to modify the properties of the irradiated area. Different micro- or nanostructures can be produced endowing certain specifications to the surface, which can be used to change its visual aspect [1, 2] or it's wetting behaviour, for instance to imitate the surface state of the lotus leaf [3, 4].

Ultrashort pulsed laser sources present a growing interest for industry because of the recent increase in their mean power, repetition rate and reliability. The resulting peak power that can be reached gives the possibility of processing almost all materials with high precision and a greatly reduced heat-affected zone [5].

The interaction between ultrashort laser pulses (100's of femtoseconds) and a material surface is a complex process, in which many parameters come into play. Depending on the nature of the material and on the beam characteristics (wavelength, pulse duration, polarization, quality factor M^2), several process parameters such as the laser fluence, the spot overlap and the number of repetitions can be varied to control the interaction [5, 6]. In fact, surface features rely greatly on the process parameters settings.

In this paper we present some initial results of a study aiming at the control of these key-parameters, in order to create different types of micro- and nanostructures. We first performed tests on samples of stainless steel, textured by self-organized nanogratings, known as "ripples". Then, by modifying the laser beam parameters, microstructures were created. In some cases, a colouring aspect change is obtained [2], and in others a modification of the wetting behaviour of the textured surface is observed [6, 7].

2. Experimental process

2.1. Material and laser system

Samples of AISI 316L stainless steel having a thickness of 0.5 mm were prepared with an ultrasonic bath of acetone followed by deionized water. No additional mechanical polishing was performed, leaving relatively significant striations and an initial roughness R_a of 69 nm.

The irradiating laser source used was the ultrafast pulsed fibre laser Tangerine (Amplitude Systèmes, France), with a central wavelength of 1030 nm, 300 femtoseconds of pulse durations and 100 µJ pulse energy at a repetition rate of 200 kHz. A Third Harmonic Generation (THG) module allows operation in the UV range, at 343 nm. Texturing tests were accomplished at IR and UV wavelengths. The spatial profile of the laser beam was nearly Gaussian ($M^2 < 1.2$).

The micro-machining MUSE-3D platform (Micro-Usinage Système Expérimental 3D, Optec, Belgium) includes an attenuator in the optical path to modulate the laser power and a beam expander to easily change the diameter of the laser beam, which is focused with a 100 mm focal length F-theta lens. Samples, placed on an X-Y translation stage, are scanned by the laser beam thanks to a galvanometric head. The laser beam polarization is linear, and it is possible to work with circular polarization by inserting a quarter-wave plate in the optical path.

2.2. Characterization devices

After laser irradiation, the samples' topography was analysed with three different optical methods.

The InfiniteFocus 3D profilometer (Alicona Imaging GmbH, Austria) was used for measuring the size of the microstructures and the roughness. This instrument has a lateral resolution of 0.5 µm and a vertical resolution of up to 10 nm. It works on the focus variation method in white light, combining an optical system with a limited depth of field and vertical scanning.

The Leitz-Linnik microscope is an interference microscopy system developed at ICube, based on white light scanning interferometry and described in [8]. Its Linnik type objective (x50, NA = 0.85) gives a lateral resolution of 0.43 µm in visible light, an axial resolution of 1 to 15 nm depending on the algorithm used and nature of the surface to be measured, with a Basler avA1000-100gc camera (1024x1024 pel).

The Atomic Force Microscope XE-70 (Park Systems, Korea), placed within an acoustic enclosure, works in the non-contact mode. The 2 nm sized tip used as a point probe is a non-contact high frequency. The field size used was 25x25 µm for 256x256 pixels, with a lateral resolution of 0.1 µm.

The wettability of flat and textured samples was determined with the drop shape analyser DSA25 from Krüss. The sessile drop method was employed to measure the values of the contact angles, for 5 µL droplets of deionized water. Since no additional coating was applied on the textured samples, the wetting behaviour of the surface was therefore only due to its topography.

3. Results

Two approaches have been considered to texture the surface of the stainless steel samples: the first consists in a self-organization of the matter by irradiation with IR-light laser, and the second is achieved by micro-machining the metal with successive ablations in UV-light.

3.1. Self-organized micro- and nanostructures in Infra-Red

The phenomena of laser energy absorption and transfer during the laser-matter interaction, just below or at the ablation threshold of the material, are responsible for the self-organization of nano-structures that are known as "ripples" or "Laser Induced Periodic Surface Structures" (LIPPS) [6]. At a

Journal of Physics: Conference Series **558** (2014) 012063 doi:10.1088/1742-6596/558/1/012063

wavelength of 1030 nm, the laser parameters studied to generate ripples are the fluence, the spots overlap, the number of repetitions and the beam polarization (linear or circular).

The results presented of the ripples are obtained for a one-step process with a focused diameter of about 47 μm, for a laser fluence of 0.81 J/cm². The overlap is maintained constant at 75% for both directions of scanning (longitudinal ≡ scanning rate / transversal ≡ hatch distance).

Linear and circular polarizations are used, leading to different types of nano-structures. When the laser beam is linearly polarized, fine ripples are created, as can be observed in the AFM results in figure 1. With the image field size being 13.8x13.8 μm, the lateral resolution is 50 nm. Their preferential orientation is perpendicular to the polarization direction. Analysis of these ripples shows a periodicity of about 1 μm, and a height of 0.5 μm. This analysis also underlines the presence of nanometric nodules superimposed on the ripples with a lateral size of 400-500 nm and with a height of up to a hundred of nanometers.

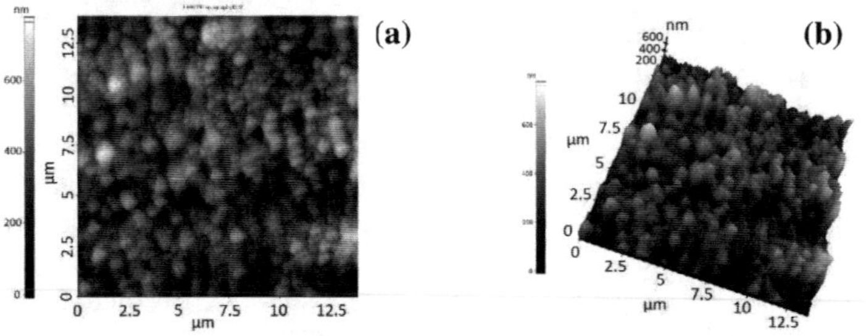

Figure 1. AFM analysis of a textured surface by ripples in linear polarization: (a) 2D-view and (b) 3D-view.

By adding a quarter-wave plate, the linear polarization of the laser beam becomes circular, resulting in the nano-structures self-organizing into a circular shape. The laser parameters are the same as in the previous case, with only the polarization changing. The circular nano-structures are about 1 μm wide and ~600 nm high, measurable by AFM.

To distinguish easily between both types of ripples, those created with linear polarization will be referred to as "linear ripples", and those generated with circular polarization as "circular ripples".

Roughness measurements were performed with the profilometer on the initial and textured surfaces. For the "linear" ripples, the roughness R_a is just a bit higher than that of the untextured surface (resp. 75 nm and 69 nm), and for the "circular" ripples R_a it is equal to 70 nm.

The results in figure 2 show views of the ripples obtained with the linear and circular polarization. The AFM observations show the difference in the topography for each test. In addition, comparisons are made with the results from Scanning Electron Microscopy (SEM), using a FEI Quanta-400 microscope. Comparable nodules observed on linear ripples by AFM in figure 1 are also present on the SEM analyses.

The results in figure 4 show these linear and circular ripples as measured with the optical profilometer and the Leitz-Linnik microscope (figure 4 (a), (b), (c) and (f), (g), (h)).

By varying the laser parameters, fine ripples evolve into coarse ripples [2], which are parallel to the direction of polarization. These kinds of structures are also obtained for a laser fluence of 0.81 J/cm² and a diameter of 47 μm, but with a greater transversal overlap equal to 99%. Measurements of these coarse ripples show a period of 4.3 μm for a mean height of 2.8 μm (see figure 4, pictures (d) and (i)).

3.2. Microstructures obtained by successive ablations in the Ultra-Violet

The use of the UV wavelength of 343 nm for femtosecond pulses has the advantage of higher laser fluences on smaller spot diameters.

This test consists in the micro-machining of parallel lines (vertical and horizontal), in order to create micro-protrusions in squares, as inspired from [7]. The focused laser beam of 31 μm diameter successively ablates the surface, at a low fluence of 0.12 J/cm². It scans the sample with a longitudinal overlap of 95%, and a transverse overlap of 15%. In order to obtain the desired depth of micro-structures, this ablation is repeated 50 times.

Analysis with the InfiniteFocus profilometer and the interferometric microscope gives the following measurements: the depth equal to about 12 µm, the width equal to 16 µm and the distance between the micro-squares of 10 µm (see figure 4, pictures (e) and (j)).

Besides, the InfiniteFocus underlines the different depths for these structures: the vertical lines are deeper than the horizontal ones. There is also a small difference in the widths of the protrusions. These deviations are assumed to be due to the slight ellipticity of the laser beam profile.

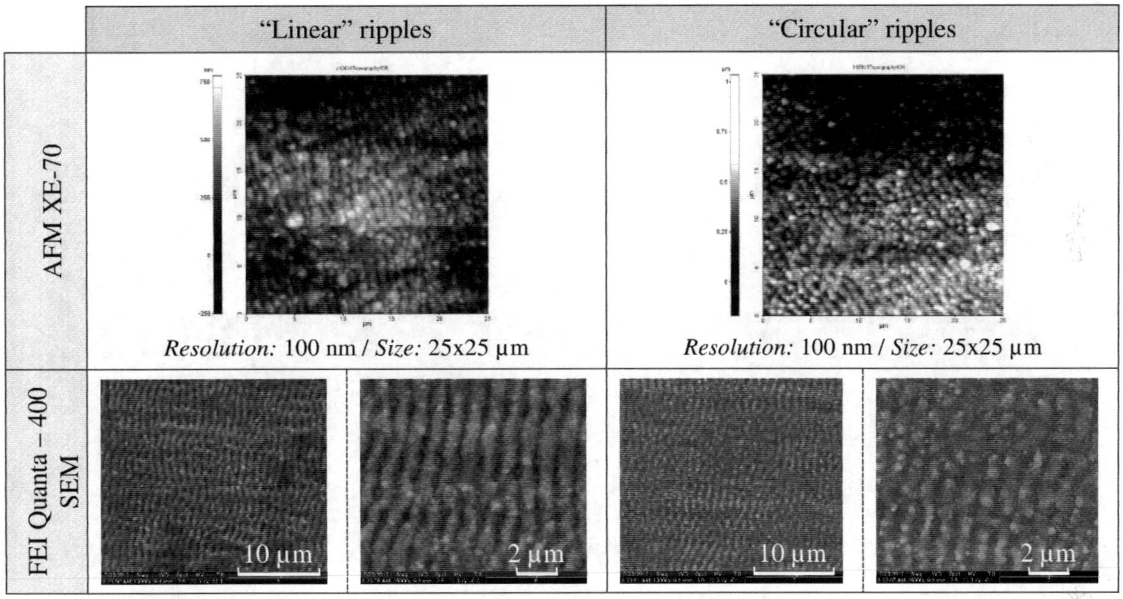

Figure 2. "Linear" and "circular" ripples on stainless steel observed with AFM and SEM.

4. Discussion

Nano-structures such as ripples can be used to change the visual aspect of a surface, by diffracting white light on the created nano-gratings (see figure 3). As for the coarse ripples and the square-shaped micro-structures, the textured surface is still darker and blackened: the marking is thus permanent.

Figure 3. Colouring effect of the different textured surfaces.

The wetting behaviour linked with the roughness of the surface and its topography was then investigated. The results in figure 4 show the evolution of the contact angle of the droplet on the textured and untextured surfaces with the dimensions of the micro- and nano-structures. The contact angle values were obtained for a single series of measurements.

On the untextured sample ($R_a \approx$ 69 nm), the contact angle θ_c is equal to 75°: the surface is hydrophilic. With fine "linear" ripples, the roughness is a little higher ($R_a \approx$ 75 nm) and the contact angle decreases to 55°: the hydrophilic nature is reinforced. Even if "circular" ripples are almost similar to the "linear" ones in terms of dimensions, the R_a of this textured surface is almost equal to the untextured one and the contact angle is thus nearly the same ($\theta_c \approx$ 79°). The roughness therefore enhances the wetting nature of the surface [9].

In the case of the coarse ripples, the droplet is totally absorbed by the micro-structures, making it impossible to measure any contact angle.

Concerning the ablated micro-structures created in UV, an increase of 17° for the contact angle is observed, being slightly higher than 90°. This result shows that it is possible to obtain a hydrophobic behaviour on a metallic surface by laser texturing, without any additional coating.

		Self-organized structures in IR			Ablated structures in UV
	Initial surface	Linear ripples	Circular ripples	Coarse ripples	Micro-squares
InfiniteFocus 3D profilometer	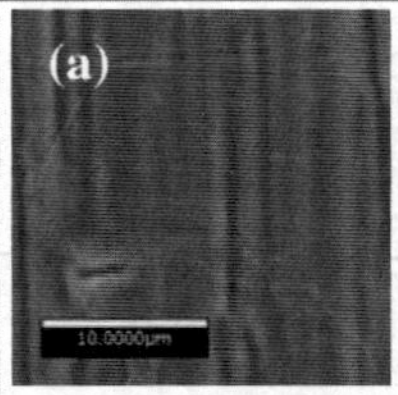	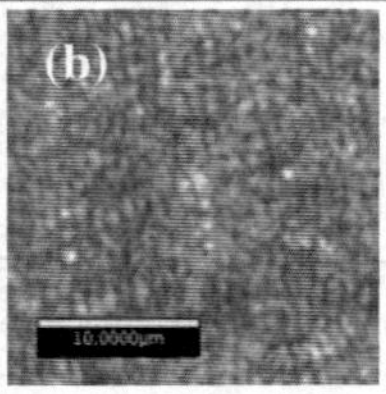	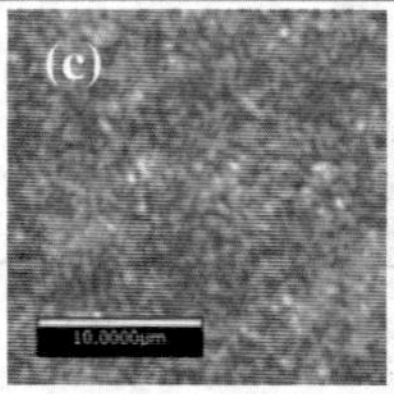	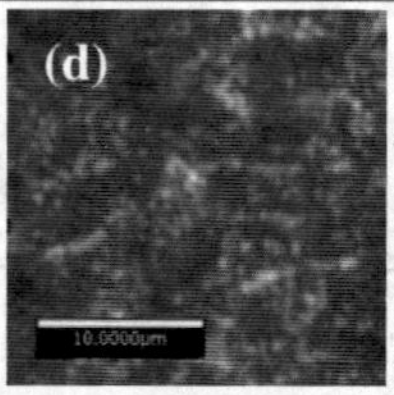	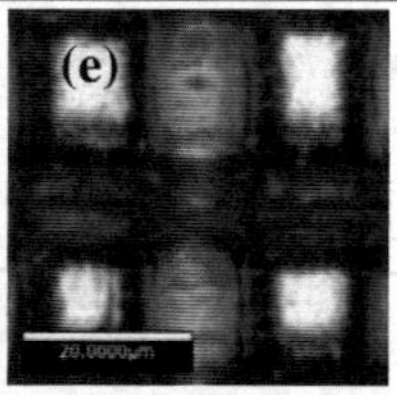
Leitz-Linnik microscope	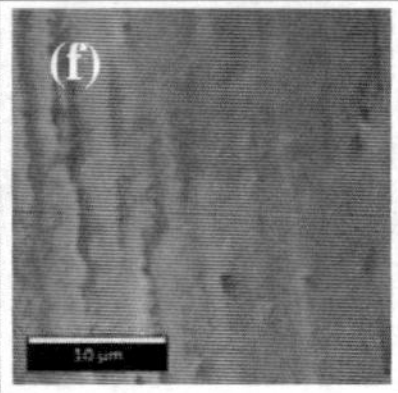		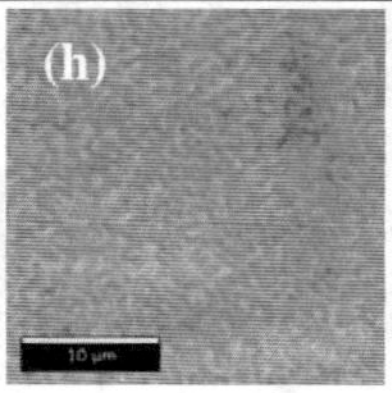	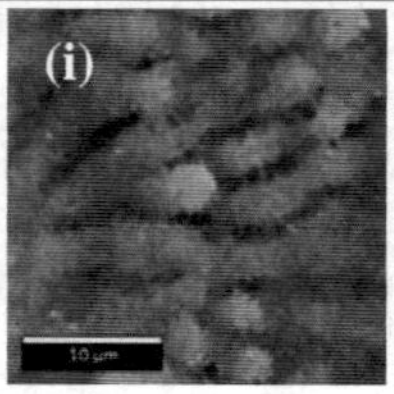	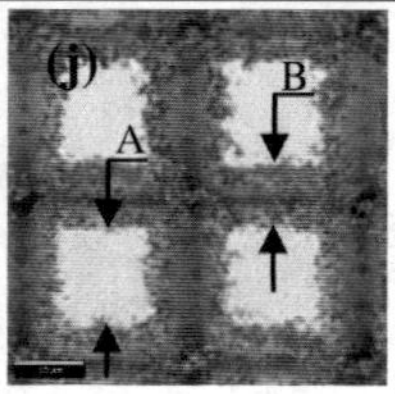
Size & roughness	R_a = 69 nm	Period = 1 μm Height = 0.5 μm R_a = 75 nm	Width = 1 μm Height = 600 nm R_a = 70 nm	Period = 4.3 μm Height = 2.8 μm	D[a] = 12 μm A[b] = 16 μm B[c] = 10 μm
Photographs of 5 μL drops	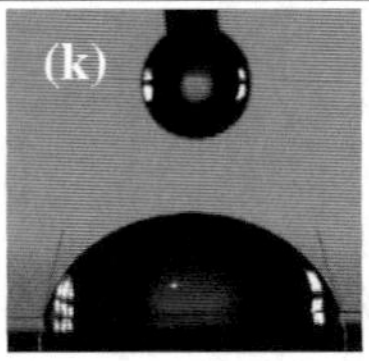	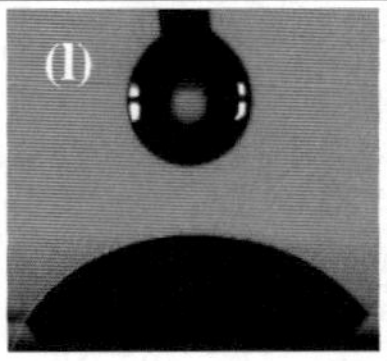	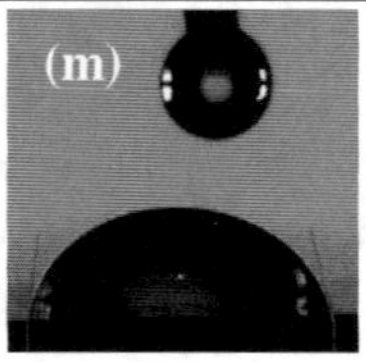	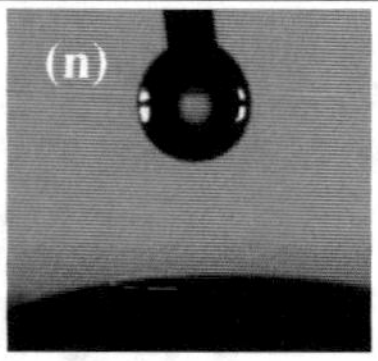	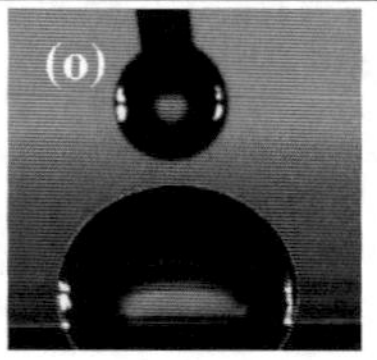
Contact angle	$\theta_c \approx 75°$	$\theta_c \approx 55°$	$\theta_c \approx 79°$	Not measurable	$\theta_c \approx 92°$

[a] Depth of the square-shaped micro-structures.
[b] Width of the square-shaped micro-structures.
[c] Distance between the square-shaped micro-structures

Figure 4. Summary of the different textured surfaces: measurements with the optical profilometer [(a) to (e)] and the Leitz-Linnik microscope [(f) to (j)], their size and roughness, photographs of 5 μL deionized water [(k) to (o)] and the contact angle measurements. (Scale: the black rectangle represents 10 μm for pictures (a) to (d), and for (f) to (j), and 20 μm for picture (e)).

Journal of Physics: Conference Series **558** (2014) 012063 doi:10.1088/1742-6596/558/1/012063

5. Conclusions

In this paper we have demonstrated the potential of the micro- and nanotexturing of a stainless steel sample using irradiation with an ultrashort pulse laser. Two approaches have been considered.

The first consists in a self-organization of the nanostructured ripples in the IR-range (linear, circular and coarse ripples) allowing the creation of a colouring effect by diffraction or a permanent dark marking. In addition, the roughness of the ripples makes it possible to enhance the hydrophilic behaviour of the textured surface.

As for the second approach, surface texturing was performed by successive ablations of parallel lines at a UV-wavelength. Square-shaped micro-structures were created and their presence increased the contact angle of a water droplet to 17° by comparison with the untextured surface: a hydrophobic behaviour was observed, with no additional coating modifying the wetting behaviour.

We can thus see that there is real potential for surface micro- and nano-texturing with ultrashort pulsed laser technology, for the aeronautical, biomedical or luxury goods fields.

Acknowledgements

Thanks are extended to P. Pfeiffer and A. Leong-Hoï (ICube, Illkirch, France), G. Chabrol (ECAM, Schiltigheim, France), and L. Vonna (IS2M, Mulhouse, France) for their permission and help to use their characterization systems for this study.

References

[1] Vorobyev A Y and Guo C 2008 Colorizing metals with femtosecond laser pulses *Appl. Phys. Lett.* **92** 041914

[2] Dusser B, Sagan Z, Soder H, Faure N, Colombier J-P, Jourlin M and Audouard E 2010 Controlled nanostructrures formation by ultra fast laser pulses for color marking *Opt. Express* **18** 2913

[3] Groenendijk M and Meijer J 2005 Microstructuring using femtosecond pulsed laser ablation International Conference on Applications of Lasers and Electro-optics, ICALEO'05 (Miami, USA: LIA) pp M219

[4] Kam D H, Bhattacharya S and Mazumder J 2012 Control of the wetting properties of an AISI 316L stainless steel surface by femtosecond laser-induced surface modification *J. Micromechanics Microengineering* **22** 105019

[5] Nayak B K and Gupta M C 2010 Self-organized micro/nano structures in metal surfaces by ultrafast laser irradiation *Opt. Lasers Eng.* **48** 940

[6] Mellor L, Edwardson S P, Perrie W, Dearden G and Watkins K 2009 Surface plasmon polaritons for micro and nano-texturing of metal surfaces *Proc. ICALEO* vol 102 pp 1329

[7] Römer G-W, Arnaldo del Cerro D, Sipkema R C and Groenendijk M N W 2009 Ultra short pulse laser generated surface textures for anti-ice applications in aviation

[8] Pecheva E, Montgomery P, Montaner D and Pramatarova L 2007 White light scanning interferometry adapted for large-area optical analysis of thick and rough hydroxyapatite layers *Langmuir* **23** 3912

[9] Badre C 2007 *Étude de la réactivité de surface par mesure d'angle de contact: influence de la fonctionnalisation et de la structure. Applications aux films d'oxyde de zinc électrodéposés* Ph.D. thesis Université Pierre et Marie Curie Paris VI Paris France

Journal of Physics: Conference Series **558** (2014) 012064 doi:10.1088/1742-6596/558/1/012064

Photoelectrical characterization of a new low molecular weight compound

V Siderov[1], G H. Dobrikov[2], I Zhivkov[1, 3], G M. Dobrikov[4], Y Georgiev[2], R Yordanov[2], J Honova[1], and M Weiter[1]

[1] Faculty of Chemistry, Centre for Materials Research, Brno University of Technology, Purkynova 118, 612 00 Brno, Czech Republic
[2] Faculty of Electronic Engineering and Technologies, Department of Microelectronics, Technical University of Sofia, Kl. Ochridski Blvd. 8, 1756 Sofia, Bulgaria
[3] Institute of Optical Materials and Technologies "Acad. J. Malinowski", Bulgarian Academy of Sciences, Acad. G. Bonchev Str. bl. 101/109, 1113 Sofia, Bulgaria
[4] Institute of Organic Chemistry with Center of Phytochemistry, Bulgarian Academy of Sciences, Acad. G. Bonchev Str. 9, 1113 Sofia, Bulgaria

Email: zhivkov@fch.vutbr.cz

Abstract. Photoelectrical characterization of a newly synthesized low molecular weight compound was carried out. 1,8-naphtalimide (chemical formula $C_{32}H_{34}N_4O_5S$) was originally synthesized and analyzed by NMR spectroscopy. Thin films were deposited in vacuum on commercially pre-patterned ITO covered glass substrates and the samples were prepared in clean room environment. The films deposited were characterized by SEM. Photoelectrical characteristics of the samples prepared were estimated by dark current-voltage measurement, spectral dependence of the photoconductivity and measurement under exposure with light, produced by solar simulator. Finally electroluminescence measurements were performed. It was found that the samples exhibit diode behaviour. The low values characterizing photovoltaic parameters obtained could be connected with the relative higher series resistance (R_{series}). The predominant influence of R_{series} is assumed as the relative high photoluminescence, measured from solution should be related to a relatively strong charge carrier photogeneration. This result is supported by electroluminescent measurement. Another reason for the low values of the photovoltaic parameters measured could be the non-optimized film thickness leading to a non-optimal light absorption and increased charge carrier recombination. The assumption for the predominant influence of R_{series} is supported by the electroluminescent measurements.

1. Introduction

Organic materials are potential candidates as active thin semiconductor films in the microelectronic devices due to their easy processing and low cost.

Owing to their potential applications in the organic electronics, the development of new organic semiconductor materials has been an ever hot issue in the past decades. Application of pure oligomers and polymers bearing strictly defined substituents and degree of polymerization is crucial for successful preparation of organic electronic devices. However, preparation of such organic materials usually requires expansive multistep synthetic procedures, leading to low overall yield of the final products. The low solubility of many light emitting polymers in organic solvents can be an additional obstacle for their purification.

Because of their strong fluorescence and excellent photostability, 1,8-naphthalimide derivatives have been widely used in light emitting diodes, in most cases as part of oligomers and polymers [1, 2]. The strong photoluminescence measured also shows that under some conditions effective solar cells could be prepared. To the best of our knowledge, the synthesis and application of low-molecular 1,8-naphtalimides for an application in the organic electronics is comparatively less investigated [3, 4, 5].

Thin films of small molecular weight semiconductors are usually prepared by means of a variety of complex techniques including physical or chemical vapour deposition, organic molecular beam epitaxy or solution-based deposition techniques. The performance of small molecular weight organic devices has been shown to be highly sensitive to film morphology and processing conditions. Often, the solution deposited active layers of devices (e.g. spin cast films) exhibit a high portion of microcrystallites and aggregates whereas the vapour deposition techniques provide high quality crystalline films, characterized by improved charge transport properties compared with those of solution deposited films.

This work reports a photoelectrical characterization of vacuum deposited thin films from newly synthesized by simple procedures and cheap commercially available precursors – low-molecular 1,8-naphthalimide derivatives.

2. Experimental

1,8-naphtalimide (chemical formula $C_{32}H_{34}N_4O_5S$) hereafter denoted as MAX-06-02 (figure 1) was synthesized according to a procedure explained elsewhere. The synthesized material was purified by flash column chromatography. Elementary analysis was performed by NMR spectroscopy.

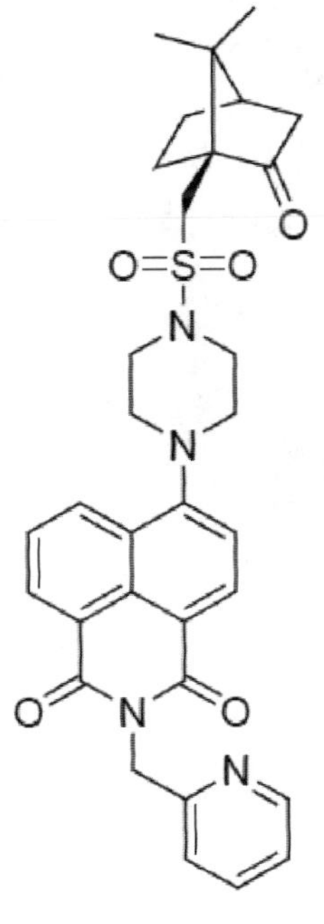

Figure 1. Chemical structure of the active organic semiconductor material MAX-06-02 ($C_{32}H_{34}N_4O_5S$).
Molecular Weight: 586.71.
Elemental Analysis: C, 65.51; H, 5.84; N, 9.55; O, 13.63; S, 5.46.

For fabrication of the samples "Ossila OLED/OPV Pixelated Anode Substrate System" and Ossila encapsulation system were used. It consists of pre-patterned ITO-covered glass substrate with standard active area of 0.06 cm^2, mask for the organic film deposition, cathode deposition mask, encapsulation epoxy, encapsulation glass slides and electrical connection legs with standard 0.1 inch (2.54 mm) pitch. The pre-patterned ITO structure consists of one cathode strip by the long side of the substrate and six fingers which combined with deposition masks forms six active pixels with dimensions 4×1.5 mm and six anodes connected to them. The pre-patterned ITO structures were cleaned by sonication in 10% NaOH water solution, de-ionized water and isopropyl alcohol followed by nitrogen blowing for drying.

The subsequent active semiconductor and electrode films were prepared in the clean room class C, to prevent the dust and particles from disrupting the thin layer. The clean lab is equipped with MB-200B MBRAUN gloveboxes (M. Braun Inertgas-Systeme GmbH, Garching, Germany). Oxygen sensitive materials were stored and manipulated exclusively in the nitrogen gloveboxes, where also most of device fabrication step took place.

The MAX-06-02 thin film was deposited onto Ossila substrates in the vacuum system from thermally heated sources at evaporation temperatures of 180 ºC and deposition rate of 2.5 Å/s through the active area deposition mask. The deposition rate was controlled by quartz crystal microbalance. After deposition of the active composite layer, the samples were taken out in nitrogen atmosphere, the mask was changed and vacuum deposition of aluminum was performed. The aluminum cathode of 150 nm film thickness was thermally evaporated in 1×10^{-6} mbar at a deposition rate of 1 Å/s.

Finally the structures were encapsulated by epoxide resin, which was hardened for 30 minutes under UV lamp irradiation. After that the samples were removed from the glovebox and provided with contacts for electrical measurements.

Figure 2. SEM image of the deposited MAX-06-02 thin films taken at a magnification of 20000.

Then the samples were checked by optical microscopy for film imperfections and a possible overlapping of the layers.

Selected samples were studied by Scanning Electron Microscope (SEM) Philips 515. The SEM morphology characterization of the MAX-06-02 films was carried out on the Ossila substrates without implementing the final encapsulation procedure.

The setup for measurement of dark *I-V* characteristic and the spectral dependencies of the photocurrent consists of Keithley 6517A electrometer including a voltage source. Monochromatic light was produced by LOT-Oriel halogen lamp LSH502 and LOT-Oriel monochromator MSH101. Light power was measured by the S120VC - Standard Si Photodiode Power Sensor and Keithley 485 picoampermeter.

Dark *I-V* characteristics were measured in both directions of the voltage scale. Spectral dependence of the photocurrent was measured at zero applied voltage in the range of 250÷550 nm then the dependence of the photocurrent on the incidental light power (irradiance) was measured. For this measurement the sample was exposed to monochromatic light at a wavelength of 395 nm.

Selective photoconductive samples were measured under exposure of a standard light produced by solar simulator LS0916 LOT Oriel class AAA, which provides defined light source according to standard AM 1.5. Incident light intensity was about 1000 $mWcm^{-2}$ and was controlled by calibrated reference silicon cell RR2000 from ReRa systems (ReRa Solutions BV; Nijmegen, Netherlands). For the studied samples, the exact irradiation of 953 $mWcm^{-2}$ was measured. In this case the current voltage measurements were executed by electrometer Keithley 2601B (Keithley Instruments Inc., Cleveland, OH, U.S.A.).

3. Results

3.1. Characterization

The film surface characterization of the MAX-06-02 films was performed by SEM. The SEM image presented in figure 2 is taken at a magnification of 20000 before applying the encapsulation procedure. The SEM image shows that smooth and uniform MAX-06-02 films without a presence of pinholes were obtained. The predominant part of the surface area is flat which facilitates the deposition of subsequent layers. In general it could be concluded that the smoothness and the uniformity of the films obtained is a good prerequisite for a formation of an interface without voids between the contacted electrode and semiconductor surfaces. As a consequence the ITO|MAX-06-02|Al samples prepared will be suitable for electrical measurements in a "sandwich" type electrode configuration.

3.2. Photoelectrical measurements

I-V characteristics measured in dark on ITO| MAX-06-02|Al structures with active MAX-06-02 films of about 150 nm thickness are presented in figure 3. To estimate the electrical parameters of the samples, the graphs are plotted in a semilogarithmic scale (negative values of the current are multiplied by -1).

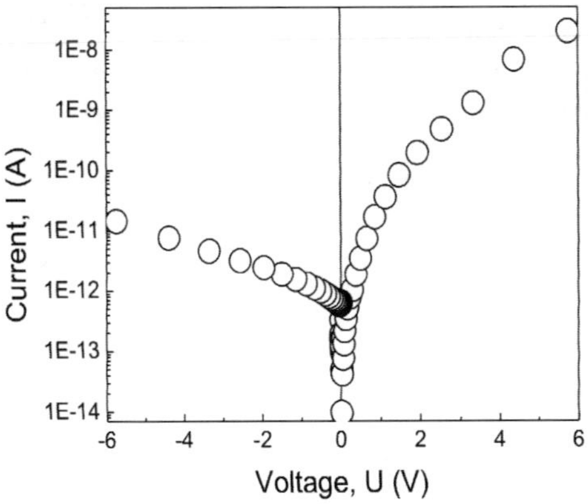

Figure 3. *I-V* characteristics measured in the dark.

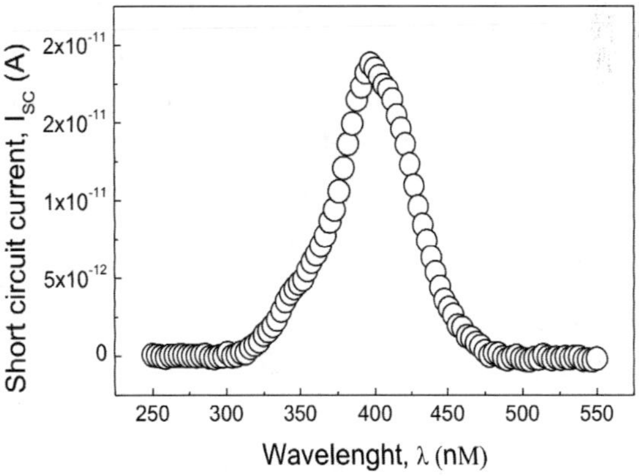

Figure 4. Spectral dependence of the photocurrent at zero applied voltage.

From the dark current measurements, non-linear and unsymmetrical characteristics are obtained. Comparing the curves measured in forward and backward direction on the voltage scale a rectification ratio of more than 3 orders of magnitude was found. This clearly shows that the conductivity is predominantly contact limited and the sample behaves as a diode. It should be mentioned that no

Journal of Physics: Conference Series **558** (2014) 012064 doi:10.1088/1742-6596/558/1/012064

additional intermediate layers serving to reduce the contact barrier were used. The optimization of the stacked multilayer structures could be a subject of a future study.

Spectral dependences of the photocurrent, measured at zero applied voltage between the electrodes from ITO| MAX-06-02 |Al structures with active MAX-06-02 films are presented in figure 4. The spectrum consists of a wide single peak centered at 395 nM. The wide peak shows that a relative broad sensitivity in an extended spectral region could achieve. For this purpose the representative samples were exposed to a white standard light. The data obtained from the solar simulator are summarized in table 1. The measurements show low values of ISC and VMP, and consequently VPP, and η. This result could be related to the relative high value of the series resistance. This effect is more probable as other experiments show a relative high photoluminescence measured from solution.

Table 1. Photovoltaic parameters of the samples prepared.

parameter	dimension	value
Size	[cm^2]	0.06
Irradiance	[W/m^2]	952,9
I_{SC}	[mA]	$2,67\times10^{-6}$
V_{OC}	[mV]	542
I_{MP}	[mA]	$1,86\times10^{-6}$
V_{MP}	[mV]	318
P_{MP}	[mW]	5.93×10^{-7}

The electroluminescence (figure 5) measurements could add more information about the photoelectrical properties of the samples prepared. The *I-V* characteristic measured in an extended area of voltages applied on the samples confirms the diode behaviour of the samples. The first part at the I-V characteristic measured at low voltages is clearly related to a slope less than 1, which a typically connected with a contact limited current. The absence of space charge limited and double injection areas at higher voltages ($I{\sim}U^{2\text{-}4}$) also supports this state. In general the slope of the *I-V* characteristics is gradually changed from less than 1 to 8 instead.

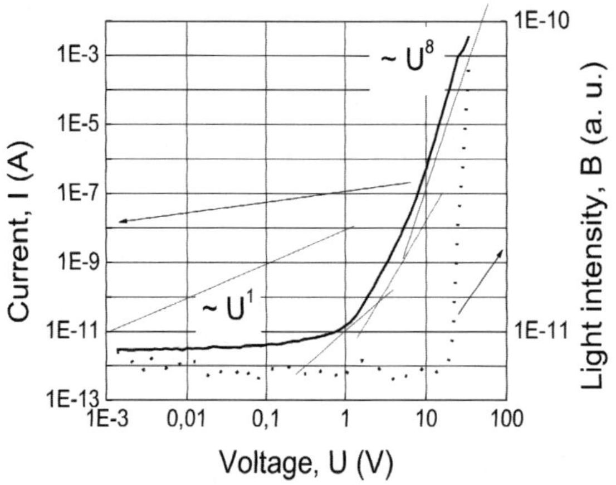

Figure 5. Electroluminescence measurements of the samples prepared.

4. Conclusion

Photoelectrical characterization of a newly synthesized low molecular weight compound was carried out. The samples exhibit clear diode-like behavior. The low values for a characterizing photovoltaic parameters obtained could be connected with the relative higher series resistance. The predominant influence of R_{series} is assumed as the relative high photoluminescence, measured from solution should be related to a relatively strong charge carrier photogeneration. This result is supported by electroluminescent measurement. Another reason for the low values of the photovoltaic parameters measured could be the non-optimized film thickness leading to a non-optimal light absorption and increased charge carrier recombination. The assumption for the predominant influence of R_{series} is supported by the electroluminescent measurements.

Acknowledgements

The work was supported by the Grant Agency of the Czech Republic (project No. 13-29358S) and by the Ministry of Education Youth and Sports of the Czech Republic (project No. LO1211).

References

[1] Yang J-X, Wang X-H 2012 *Adv. Mater. Res.* **557-559** 1031
[2] Liu J, Li Y, Wang Y, Sun H, Lu Zh., Wub H, Peng J, and Huang Y 2012 *Opt. Mater.* **34**, 1535
[3] Ding G, Xu Zh, Zhong G, Jing S, Li F, and Zhu W 2008 Res. *Chem. Intermed.* **34**, 299
[4] Mikroyannidis J. A, Ye S, and Liu Y 2009 *Synth. Met.* **159**, 492
[5] Ulla H, Garudachari B, Satyanarayan M N, Umesh G, and Isloor A M 2014 *Opt. Mater.* **36**, 704